HUMAN BIOLOGY

HUMAN BIOLOGY

Second Edition

Sylvia S. Mader

WCB Wm. C. Brown Publishers

Book Team

Editor *Kevin Kane*
Developmental Editor *Carol Mills*
Visuals/Design Consultant *Marilyn Phelps*
Designer *David C. Lansdon*
Art Editor *Donna Slade*
Production Editor *Diane S. Clemens*
Photo Editor *Carol M. Smith*
Permissions Editor *Mavis M. Oeth*
Visuals Processor *Jodi Wagner*

WCB **Wm. C. Brown Publishers**

President *G. Franklin Lewis*
Vice President, Editor-in-Chief *George Wm. Bergquist*
Vice President, Director of Production *Beverly Kolz*
Vice President, National Sales Manager *Bob McLaughlin*
Director of Marketing *Thomas E. Doran*
Marketing Communications Manager *Edward Bartell*
Marketing Manager *Craig S. Marty*
Executive Editor *Edward G. Jaffe*
Manager of Visuals and Design *Faye M. Schilling*
Production Editorial Manager *Colleen A. Yonda*
Production Editorial Manager *Julie A. Kennedy*
Publishing Services Manager *Karen J. Slaght*

Human Anatomy insert following page 220:
ERNEST W. BECK, medical illustrator
in collaboration with
HARRY MONSEN, Ph.D.
Professor of Anatomy, College of Medicine, University of Illinois
COPYRIGHT © THE C. V. MOSBY COMPANY

For my family

BRIEF CONTENTS

CONTENTS

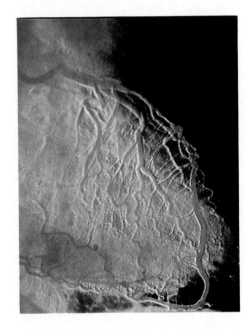

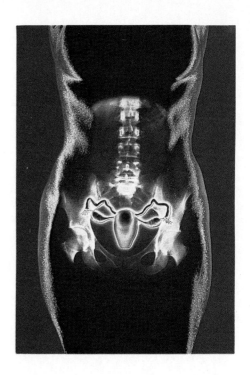

H *uman Biology* is written primarily for one-semester biology courses that emphasize human physiology and the role that humans play in the biosphere. Although this text draws from my other book, *Inquiry into Life,* the material has been reworked to produce a stronger human approach to biological study. All young people should leave college with a firm grasp of how their bodies normally function and how the human population can become more fully integrated into the biosphere. This knowledge can be applied daily and helps assure our continued survival as individuals and as a species.

The application of biological principles to practical human concerns is a relatively new approach for instructional courses in biology. It has gained even wider acceptance because it fulfills a great need. Human beings are frequently called upon to make decisions about their bodies and their environment. Wise decisions require adequate knowledge.

The second edition of *Human Biology* has basically the same style and organization as the previous edition, with minor modification (Parts Four and Five have been reversed). Each chapter presents the topic clearly, simply, and distinctly so that the student will feel capable of achieving an adult level of understanding. Detailed, high-level scientific data and terminology are not included because I believe that true knowledge consists of working concepts rather than technical facility.

Students and instructors alike will find this text stimulating and a pleasure to read and study. Even so, we have not neglected routine matters, and students will appreciate the new end-of-chapter summary and key-term glossary for each chapter. Those who share the current concern about critical thinking will be pleased to know that there are two opportunities for students to develop this ability. The Human Issue boxes ask students to thoughtfully consider and come to a decision regarding complex social issues. Also, end-of-chapter thought questions have been provided. These questions require no more input than is given in the text, but they do ask the student to use this material in a creative manner.

ORGANIZATION OF THE TEXT

An introductory chapter precedes the twenty-two chapters of *Human Biology,* which are grouped in six parts.

Introduction: A Human Perspective

The introductory chapter lays a foundation and rationale for the text as a whole. It discusses the biological characteristics of humans, and in this edition more emphasis has been placed on our cultural heritage, an influence that helps set us apart from our closest relatives, the great apes. In keeping with adopters' wishes, the section on the scientific method has been expanded.

Part One Human Organization

Part One, Chapters 1–3, presents some basic biological principles before reviewing the organization of the human body itself and discussing homeostasis, the general theme for the parts that follow.

In this edition, the chemistry chapter presents the structure and function of nucleic acids in greater detail. Mitotic cell division is no longer covered in the cell chapter; it has been placed in chapter 17 along with meiosis.

Part Two Human Processing and Transporting

Part Two covers those systems of the human body that could be described as vegetative and tells how they contribute to homeostasis.

In this edition, the nutrition portion of chapter 4 has been completely rewritten. This section now reflects more closely the current research in this area and gives appropriate applications for our daily lives. The other chapters have been reorganized and portions rewritten as was necessary to improve their content.

Part Three Human Integration and Coordination

The control of homeostasis by the nervous and hormonal systems is discussed in Part Three. It also includes the musculoskeletal system and sense organs.

In this edition, the portion of chapter 10, Nervous System, that considers drugs has been completely rewritten. It now stresses drugs of abuse and presents detailed information regarding alcohol, marijuana, heroin, and cocaine. Instructors will find this new section a highly relevant addition to their curriculums. The other chapters have been reorganized and portions rewritten as was needed to improve their content.

Part Four Human Reproduction

Part Four deals with human reproduction and development. In the first edition, this part appeared after Human Inheritance. It appears before Inheritance in this edition because of valuable feedback from instructors who felt strongly about the tighter logic of the current sequence. The chapter on sexually transmitted diseases includes some coverage of microbes in general.

The coverage of AIDS has been updated, of course. Also, the development chapter (chap. 16) contains a greatly expanded section on aging, a topic of greater interest now that a large segment of the U.S. population is over age 65.

Part Five Human Inheritance

The transference of traits from one generation to the next and associated topics are considered in Part Five, Chapters 17–19. Genetics is that area of biology that is undergoing the most change and advancing most rapidly. It is fitting that this text should present the most up-to-date information in a student-orientated manner.

There are two completely new chapters in this part: Genes and Medical Genetics, (chap. 18) and DNA and Biotechnology (chap. 19). In the first of these chapters, great emphasis is placed on current knowledge regarding human genetic disorders . In the second, the way modern-day biotechnology is expected to revamp many areas of human concern is shown.

Part Six Human Evolution and Ecology

This part discusses the evolution of humans (chap. 20) before presenting two ecology chapters with highly practical emphases. Chapter 21 contrasts the functioning of natural ecosystems with the human ecosystem, and chapter 22 reviews the history of the world population and concurrent concerns.

Human Biology includes a number of aids that will help students study biology successfully and enjoyably.

Text Introduction

The introduction discusses the characteristics of humans and presents an overview of the book. It outlines the biological principles that will be important to understanding the text and awakens students to the possibility of using these principles to make bioethical decisions.

Part Introductions

An introduction to each part highlights the central ideas of that part and specifically tells the reader how the topics within each part contribute to biological knowledge.

Chapter Concepts

Each chapter begins with a list of concepts stressed in the chapter. This list introduces the reader to the chapter by organizing its content into a few meaningful sentences. The concepts provide a framework for the content of each chapter.

Boldfaced Words

Terms that are pertinent to the topic being discussed appear in boldfaced print. The first time these terms appear, they are defined in context. In any case, however, all boldfaced terms are defined in the glossary. Each entry in the glossary is accompanied by its phonetic pronunciation.

Tables and Illustrations

Numerous tables and illustrations appear in each chapter and are placed near their related textual discussion. The tables clarify complex ideas and summarize sections of the narrative. Once students have achieved an understanding of the subject matter by examining the chapter concepts and the text, these tables can be used as an important review tool. The photographs and drawings have been carefully chosen and designed to help students visualize structures and processes.

Boxed Readings

Two types of boxed readings are included in the text. Readings chosen from popular magazines illustrate the applications of concepts to modern concerns. These spark interest by illustrating that biology is an important part of everyday life. The second type of reading is designed to expand, in an interesting way, on the core information presented in each chapter. Topics pertaining to human concerns are addressed in these readings.

In-Chapter Summaries

Summary statements are placed strategically within the body of each chapter. These give students periodic reinforcement of the information being presented. Such statements are highlighted for easy identification.

Human Issue Boxes

Human Issue Boxes are included in every chapter throughout the text. These general discussion boxes are designed to stimulate student interest and thought, especially about how the chapter topics can be applied to human concerns. Most of the Issue boxes have been thoroughly revised for this edition and many new ones have been added on topics like AIDS, woodburning stoves, euthanasia, and prenatal screening.

Chapter Questions

Objective questions, study questions, and thought questions are at the close of each chapter. The objective questions allow students to quiz themselves with short fill-in-the-blank objective questions. Answers to these questions appear on the same page. The study questions allow students to test their understanding of the information in the chapter. Thought questions require students to use the information presented in this and previous chapters in a creative manner. The thought questions, along with the Human Issue boxes, provide students with an opportunity to think critically.

Chapter Summaries

Chapter summaries offer a concise review of material in each chapter. Students may read them before beginning the chapter to preview the topics of importance, and they may also use them to refresh their memories after they have a firm grasp of the concepts presented in each chapter.

Chapter Glossaries

New to this edition are the chapter-ending, key-term glossaries. Selected key terms are listed with their phonetic pronunciations, definitions, and page references. All boldfaced terms are still listed alphabetically with their pronunciations, definitions, and page references in the text glossary at the end of the book.

Further Readings

For those students who would like more information about a particular topic or who are seeking references for a research paper, each part ends with a listing of articles and books to help them get started. Usually the entries are *Scientific American* articles and specialty books that expand on the topics covered in the chapter.

Cancer Supplement

Because of the current interest and recent discoveries in cancer research, a special section has been set aside for extensive treatment of this disease. This section includes the characteristics, causes, types, detection and treatment, survival rates, and suggestions for possibly preventing the occurrence of cancer. The cancer supplement is inserted between chapters 16 and 17 of the text.

Human Anatomy Acetate Overlays

New to this edition, following page 220, is a full-color, ten-page insert of Human Anatomy Acetate Overlays. This intriguing and clearly-labeled art provides a quick reference for determining specific location of organs, bones and tissues, and the overall organization of the human body.

Appendixes and Glossary

The appendixes contain optional information for student referral. Appendix I is the expanded Periodic Table of the Elements; Appendix II compares the light and electron microscopes and discusses the metric system.

The text glossary defines the terms most necessary for making the study of biology successful. By using this tool, students can review the definitions of the most frequently used terms.

Index

The text also includes an index in the back matter of the book. By consulting the index it is possible to determine on what page or pages various topics are discussed.

ADDITIONAL AIDS

Instructor's Manual/Test Item File

The *Instructor's Manual* is designed by the author and Trudy McKee to assist instructors as they plan and prepare for classes using *Human Biology*. An outline and a general discussion are provided for each chapter; together these give the overall rationale for the chapter. There are many objective test questions and several essay questions on each chapter. A list of suggested films for the various topics and a list of film suppliers are included at the end of the *Instructor's Manual*.

Student Study Guide

To ensure close coordination with the text, the author has written the *Student Study Guide* that accompanies this text. Each text chapter has a corresponding study guide chapter that includes a listing of behavioral objectives, a pretest, study exercises, and a posttest. Answers to study guide questions appear at the end of sections, giving students immediate feedback.

Laboratory Manual

The author has also written the *Laboratory Manual* that accompanies *Human Biology*. With few exceptions, each chapter in the text has an accompanying laboratory exercise in the manual (some chapters have more than one accompanying exercise). In this way, instructors will be better able to emphasize particular portions of the curriculum if they wish. The nineteen laboratory sessions in the manual are designed to further help students appreciate the scientific method and to learn the fundamental concepts of biology and the specific content of each chapter. All exercises have been tested for student interest, preparation time, and feasibility.

Laboratory Resource Guide

More extensive information regarding preparation is found in the *Laboratory Resource Guide*. The guide includes suggested sources for materials and supplies, directions for making up solutions and otherwise setting up the laboratory, expected results for the exercises, and suggested answers to all questions in the *Laboratory Manual*. It is available for free to all adopters of the *Laboratory Manual*.

Transparencies

This edition is accompanied by 80 transparencies in two and four colors. The transparencies feature text illustrations with oversized labels, facilitating their use in large lecture rooms. They are available for free to all adopters.

Lecture Enrichment Kit

The Lecture Enrichment Kit is a series of optional lecture notes to accompany the text's transparencies. For each transparency there is a corresponding lecture unit. Each unit contains a summary of all text material pertinent to the process or phenomenon depicted in the transparency. The summary is followed by three to five "extensions," topics not discussed in the text. Extensions are drawn from popular periodicals of general interest, scientific periodicals, or more advanced texts. They vary in detail and degree of rigor, and are available free to all adopters.

Ancillary Box

An attractive slip-case box for housing the text and its ancillaries is available for free to all adopters of the new edition.

COMPUTERIZED ANCILLARIES

wcb QuizPak

A student computer software program is available with *Human Biology*. **wcb QuizPak,** the interactive self-testing, self-scoring quiz program, will help your students review text material from any chapter by testing themselves on an Apple IIe, IIc, or Macintosh, or an IBM PC. Adopters will receive the QuizPak program, question disks, and an easy-to-follow user's guide. QuizPak may be used at a number of workstations simultaneously and requires only one disk drive.

wcb TestPak

wcb TestPak is a computerized system that enables you to make up customized exams quickly and easily. Test questions can be found in the Test Item File, which is printed in your Instructor's Manual or as a separate packet. For each exam you may select up to 250 questions from the file and either print the test yourself or have **wcb** print it.

ACKNOWLEDGMENTS

The personnel at Wm. C. Brown Publishers are due many thanks for their help in developing and producing this edition of *Human Biology*. Kevin Kane, biology editor, not only oversaw the development of the manuscript, but also made important production decisions. He was assisted by Carol Mills, developmental editor, who helped on a more daily basis. David Lansdon designed the book with skill, and Carol Smith picked just the right photographs. Diane Clemens was the production editor who coordinated the efforts of many.

There are many new illustrations in this edition of *Human Biology*. Kathleen Hagelston provided most of these; the clarity and beauty of her work are well known to all those familiar with my texts. Also, Tom Waldrop did strikingly beautiful anatomical drawings. In addition, Anne Greene, Laurie O'Keefe, Marjorie C. Leggitt, and Precision Graphics provided artwork that is most appealing and competent.

Finally, I wish to express appreciation to my family for their constant support. My children, Karen and Eric, and my sister, Rhetta, were always ready to offer advice and encouragement.

Many instructors have contributed to this revision of *Human Biology,* and I especially want to thank Robin W. Tyser of the University of Wisconsin–LaCrosse, who revised the previous edition's Human Issue boxes and provided additional Human Issue boxes for this edition. Many other instructors commented on the entire or a portion of the text. The author is extremely thankful to each one, for we have all worked diligently to remain true to our calling and to provide a product that will be the most useful to our students. In particular, it seems proper to acknowledge the help of the following individuals:

D. J. Burks *Wilmington College*

Vic Chow *City College of San Francisco*

Sheldon R. Gordon *Oakland University*

Laszlo Hanzely *Northern Illinois University*

Keith Knutson *St. Cloud State University*

Cran Lucas *LSU-Shreveport*

Joyce B. Maxwell *California State University-Northridge*

John McCue *St. Cloud State University*

Sally Sapp Olson *Winthrop College*

Raymond R. White *City College of San Francisco*

Roberta B. Williams *University of Nevada-Las Vegas*

A HUMAN PERSPECTIVE

INTRODUCTORY CONCEPTS

1 Human beings have characteristics in common with all living things.

2 The scientific method is the process by which scientists gather information about the material world.

3 All persons have to be prepared to use scientific information to make value judgments.

Human beings share the environment with all living things.

INTRODUCTION

This book has two primary functions. The first is to explore human anatomy and physiology so that you will know how the body functions (fig. I.1). The second is to take a look at human evolution and ecology in order to understand the place of humans in nature. Both the human body and the environment are self-regulating systems that can be thrown out of kilter by misuse and mismanagement. An appreciation of the delicate balance present in both systems provides the perspective by which future decisions can be made.

Y ou are about to launch on a study of human biology. Before you begin, it is appropriate to discuss the biological characteristics of human beings. The following have been singled out as being especially pertinent.

Human beings are a product of the evolutionary process. Life has a history that began with the evolution of the first cell(s) about 3.5 billion years ago. Thereafter living things became increasingly complex. Figure I.2 only shows the evolutionary history of vertebrate animals. *Human beings are vertebrate animals* because they have a dorsal hollow nerve cord protected by vertebrae. The repeating units of the vertebrae indicate that we are segmented. Segmentation leads to specialization of parts as is well exemplified in humans (fig. I.2).

Today most biologists classify living things into the five kingdoms or groups noted in figure I.3. Because *human beings are related to other living things,* they contain the same type of chemicals, like DNA, the substance of genes, and ATP, the energy currency of cells. It's even possible to do research with bacteria (kingdom Monera) and have the results apply to humans. For example, it is common practice today to test food additives first by applying the chemicals to bacterial culture plates. If the chemical causes mutations in bacteria, it is considered unsafe for human consumption.

Human beings reproduce. Reproduction is that part of the human life cycle that assures continuance of the human species. It also permits evolution to occur. It is possible to trace human ancestry through a series of prehistoric ancestors until modern-day humans finally evolved. Human beings are distinguishable from their closest relatives, the apes, by their highly developed brains, completely upright stance, and the power of creative language.

When human beings reproduce, they pass on their organization to their offspring. The sperm and egg contain chromosomes contributed by each parent, and these chromosomes carry the blueprint of life that is essential to the offspring. While evidence is strong that genes on chromosomes control our physical traits, we are less certain and willing to entertain the belief that genes also control our behavior.

FIGURE I.2 *Human beings, like all living things, have an evolutionary history. Therefore their organ systems are similar to those of other vertebrates. The evolutionary tree of life has many branches and the vertebrate line of descent is just one of many.*

Hagelston/Leggit

FIGURE I.3 *All living things are placed in one of these five kingdoms, the major groupings of classification.*

Name of Kingdom	Representative Organisms	Descriptions
Monera	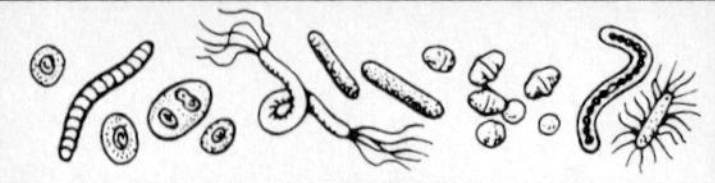	Bacteria and cyanobacteria
Protista	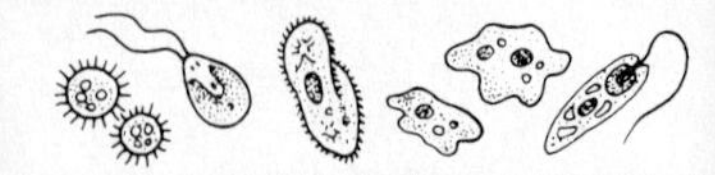	Protozoans, algae of various types
Fungi		Molds and mushrooms
Plants		Mosses, ferns, various trees, and flowering plants
Animals		Sponges, worms, insects, fishes, amphibians, reptiles, birds, and mammals

Human beings are highly organized. The organs depicted in figure I.2 are composed of tissues that contain cells of a particular type. In order to maintain this organization, living things must take chemicals and energy from the environment. Only plants and plantlike organisms are capable of utilizing inorganic chemicals and energy of the sun in the process of making their food. Other organisms, such as humans, must take in preformed food as a source of chemicals and energy.

Human beings need an internal environment that remains fairly constant. The body's organ systems work together to maintain homeostasis in which bodily activities fluctuate minimally. Certain of the systems are directly concerned with maintenance (digestive, circulatory, immune, respiratory, and excretory), while others are concerned with coordination of these systems and interaction with the environment (nervous, sensory, muscular and skeletal, and hormonal). The ability to respond to the environment is a unique ability of living things.

Humans are also a product of a cultural heritage. We are born without knowledge of civilized ways of behavior, and we gradually acquire these by adult instruction and imitation of role models. Unfortunately, it is our cultural inheritance that separates us from nature and makes it difficult for us to see our dependence on the natural world. On the contrary, our industrialized society has an extreme ability to exploit and alter the environment (fig. I.4) for its own purposes. The end result is often a degradation of the environment that is harmful to all living things, including human beings. In recent years human beings have become aware of their destructive influence and are seeking ways to work with nature rather than against nature. The desire to exploit nature is being replaced by a concern to be wise managers of nature.

Like all living things, humans are a product of the evolutionary process. They reproduce, are highly organized, and maintain a dynamic constancy of the internal environment. Unlike other living things, humans are also a product of a cultural inheritance.

HUMAN ISSUE

You are constantly exposed to all kinds of information. Information comes to you from a variety of sources, such as television, newspapers, magazines, the *Farmer's Almanac*, and horoscopes. Even as you read this textbook you are encountering more information! In your opinion, is there anything special and unique about scientific information based on scientific data compared to conclusions that are not based on such data?

FIGURE 1.4 *Natural areas versus one developed by humans. Humans need both types of places.* a. *They often benefit psychologically from visiting natural areas that provide a home for many plants and animals.* b. *Humans carry on most of their activities in developed areas.*

a.

b.

. . . US AND THE APES

Most agree that the anatomical differences between us and the apes stem from our habit of walking upright all the time. Apes tend to knuckle-walk, with hands bent, while on the ground. In keeping with walking on all fours, the ape pelvis is long and narrow and the hands and feet tend to resemble each other. In humans the shorter pelvis is designed to bear the weight of the trunk, and it also better serves as a place of attachment for the muscles of the longer and more muscular legs. The human foot is markedly different from the hand; it supports the body on a broad heel and a thick-skinned ball cushioned by fat. Still, the foot has a springy arch needed for our striding gait.

The upright stance of humans frees the hands so that they are ready to do manipulative work. Although both apes and humans have a thumb that is "opposable" in the sense that it can reach over and touch the fingers, the human thumb is longer and rotates. It is this remarkable thumb that allows us to be such great tool users. Tool use and the human brain are believed to have evolved together, each one promoting an increase in the other:

tool use ⟷ intelligence

Apes use tools but they don't make highly specialized tools for specific purposes. They also don't carry tools about with them for when they might be needed.

The human brain really sets us apart from the apes. If an ape were the size of a human, its brain would still be only ⅓ as large. Not surprising is the fact that the lobes of the brain needed for thinking and language are proportionately larger in humans also. Apes make faces and sounds that serve to communicate with others, but only humans are capable of communicating many different types of ideas. It's possible that human intelligence and language evolved as a means to understand and function within a complex social system.

Because the large brain of humans develops after birth, infants are born in an immature state and are dependent upon their mothers for quite some time. Perhaps this caused the women to "stay home" while the men went out to hunt for food that they later brought home to share with the other members of their group. Again, apes form social groups but an organized system of food sharing is not characteristic of the group.

Learning how to think and use tools within a social group is the mainstay of human culture. Today we humans are enveloped by our culture; most of us live in cities, existing on food and using materials grown or made far away. We no longer have a sense of our evolutionary past nor of our place in nature. We don't realize that even though we can control our environment, to a large extent we are still dependent upon natural ecological systems. Consider, for example, our dependence on the rains to bring us fresh water, on plants to supply us with food and to purify the air, and on the past remains of plants and animals to give us a supply of energy. Only when we fully appreciate our place in nature will it be possible for us to wisely guide our future destiny. If we greatly interfere with the cycles of nature, as when we overdam rivers, cut down forests, or burn fossil fuels to excess, it can only be to our detriment.

Was it the human hand that led to evolution of culture? a. Apes use their hands primarily to climb trees. b. They live in social groups where offspring are cared for. c. Humans use their hands primarily to manipulate tools. d. They live in a society where children learn the culture of that society.

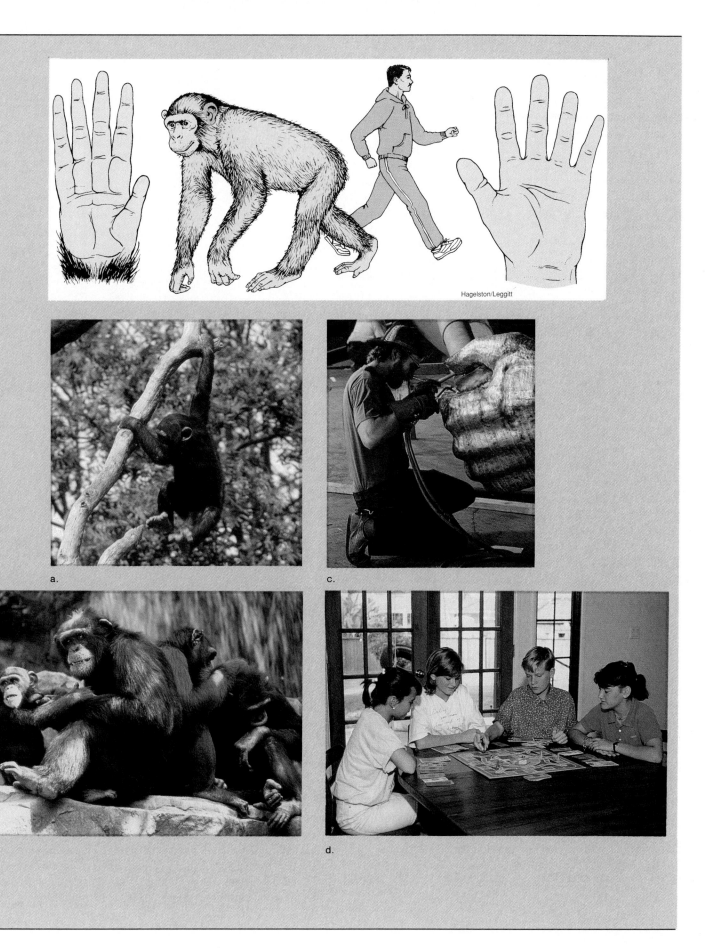

a.

b.

c.

d.

Hagelston/Leggitt

Scientific Method

Science is a process resulting in a body of knowledge that allows us to understand and manipulate the material world. Scientists often employ a methodological approach such as is described in figure I.5. For example, suppose physiologists wanted to determine if sweetener S was a safe additive for foods. First they would study any *previous data,* or objective information, collected on the subject. Then on the basis of this information they might *hypothesize* that sweetener S would be safe at a low concentration but unsafe at a high concentration. Next, the physiologists would design an experiment in order to collect new data. They might decide to feed sweetener S to groups of mice at ever greater concentrations.

Group 1: diet contains no sweetener S
Group 2: 5% of diet contains sweetener S
Group 3: 10% of diet contains sweetener S

Group 11: 50% of diet contains sweetener S

No doubt the physiologists would keep the different groups of mice in separate cages in the laboratory. The laboratory setting is much preferred for scientific experiments because it is here that all aspects of the experiment can be controlled. For example, all the groups of mice would be kept under the same environmental conditions and fed the same diet except for the amounts of sweetener S. Notice that one group is fed no sweetener S at all. This is the *control group,* the sample that goes through all the steps of the experiment except the one being tested. Use of a control group, in which no effect is expected, gives greater validity to the results of the experiment.

The hypothetical results of the sweetener S study are given in figure I.6. On the basis of this new data, the physiologists might *conclude* that mice fed an even greater amount of sweetener S above 20% are more likely to develop bladder cancer at an ever increasing rate. Scientists often prefer such mathematical data because it is objective and easily evaluated.

Any experiment must be repeatable. Other scientists using the same design and carrying out the experiment under the same conditions are expected to get the same results. If they do not, the original experiment is called into question.

In order to make the experimental design and results available to the scientific community, scientists often publish their results in scientific journals, where they may be read and studied by all those interested. It's also possible that scientists might go on to design and do other experiments with sweetener S.

FIGURE I.5 *The scientific method often includes these steps.*

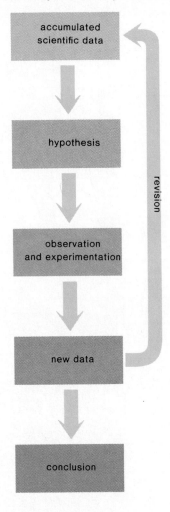

The ultimate goal of science is to understand the natural world in terms of concepts, interpretations that take into account the results of many experiments and observations. These concepts are stated as theories. In a movie a detective might claim to have a "theory" about the crime, or you might say that you have a "theory" about the win-loss record of your favorite baseball team. But in science the word *theory* is applied to those hypotheses that have been supported by a large number of observations and are considered valid by an overwhelming majority of scientists.

Social Responsibility

There are many ways in which science has improved our lives. The most obvious examples are in the field of medicine. The discovery of antibiotics such as penicillin and the vaccines for polio, measles, and mumps have increased our life spans by decades. Cell biology research is helping us understand the mechanisms that cause cancer. Genetic research has produced new strains of agricultural plants that have eased the burden of feeding our burgeoning world population.

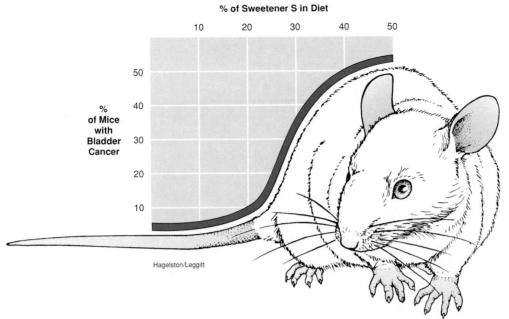

FIGURE I.6 *Hypothetical results of sweetener S study.*

% of Sweetener S in Diet

10   20   30   40   50

%
of Mice
with
Bladder
Cancer

50

40

30

20

10

Hagelston/Leggitt

Science has also produced conclusions that we find disturbing and has fostered technologies that have proven to be ecologically disastrous if not controlled properly. Too often we blame science for this and think that scientists are duty bound to pursue only those avenues of research that will not conflict with our system of values and/or result in environmental degradation, particularly when applied in an irresponsible manner. Yet science, by its very nature, is impartial and simply attempts to study natural phenomena. Science does not make ethical or moral decisions. Instead, all men and women have a responsibility to decide how best to use scientific knowledge so that it benefits the human species and all living things. Therefore, this text includes suggestions for the discussion of various human issues so that you will have an opportunity to think about the application of scientific, especially biological, information to everyday human concerns.

Scientists make use of the scientific method to discover information about the material world. It is the task of all persons to use this information as they make value judgments about their own lives and about the environment.

HUMAN ISSUE

Scientists frequently hold differing opinions about the same social issue. For example, there are some scientists who believe that the world is not overpopulated and population control is not necessary, while there are other scientists who believe that the world is overpopulated and we should do all we can to control population growth. How would you decide which group of scientists is correct? Are value judgments made by scientists any more valid than those made by nonscientists? Who should be involved in making social value judgments—only scientists, only nonscientists, or both?

SUMMARY

Human beings, just like other organisms, are a product of the evolutionary process. They are members of the animal kingdom and are most closely related to vertebrates and specifically to the apes. Like other living things, human beings reproduce, are highly organized, and maintain a fairly constant internal environment. Unlike other living things, they are also the product of a cultural inheritance.

The scientific method often includes studying previous data, making hypotheses, doing experiments, and coming to a conclusion based on the results (data) from these experiments. It is the responsibility of all to make ethical and moral decisions about how best to make use of the results of scientific investigations.

PART ONE

HUMAN ORGANIZATION

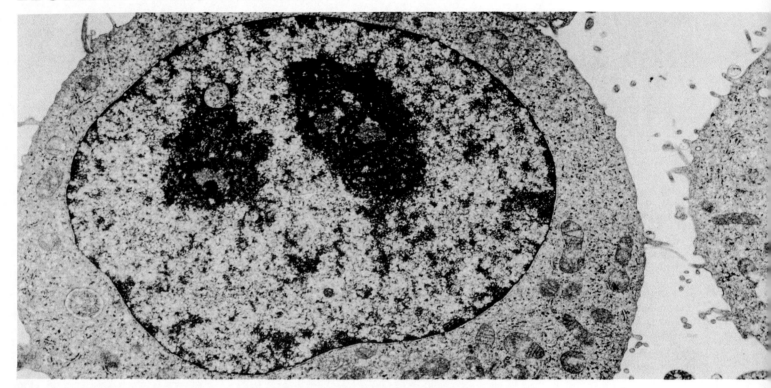

The human body is composed of cells, the smallest units of life. An understanding of cell structure, physiology, and biochemistry serves as a foundation to understanding how the human body functions.

Principles of inorganic and organic chemistry are discussed before a study of human cell structure is undertaken. The human cell is bounded by a membrane and contains organelles, many of which are also membranous. It is membrane that regulates the entrance and exit of molecules and determines how cellular organelles carry out their functions.

The many cells of the body are specialized into tissues that are found within the organs of the various systems of the body. All body systems help maintain a dynamic constancy of the internal environment so that the proper physical conditions exist for each cell.

CHAPTER ONE

CHEMISTRY OF LIFE

CHAPTER CONCEPTS

1 Humans are composed of inorganic and organic compounds.

2 Atoms, the smallest units of matter, react with one another to form molecules.

3 Some important inorganic compounds in humans are water, acids, bases, and salts.

4 Some important organic compounds in living organisms are proteins, carbohydrates, fats, and nucleic acids, each of which is composed of smaller molecules joined together.

Diagrams of the structure of molecules and an atom.

FIGURE 1.1 *In nature, everything is connected to everything else. For example, the gas carbon dioxide we breathe out is taken up by plants that produce food and oxygen, substances we and other living things need.*

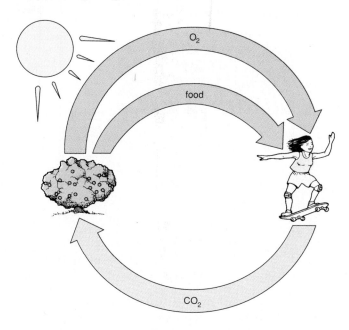

FIGURE 1.2 *Representation of an atom. The nucleus contains protons and neutrons; the shells contain electrons. The first shell is complete with two electrons, and every shell thereafter may contain as many as eight electrons.*

p = protons
n = neutrons
⬤ = electrons

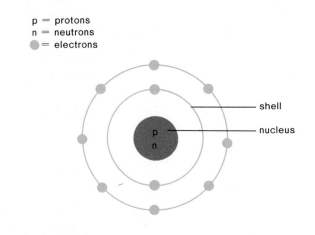

shell

nucleus

INTRODUCTION

The human body breathes out the gas carbon dioxide with every breath. You might think there is not much use for this molecule, but plants make use of it. When plants photosynthesize they use the energy of the sun to convert carbon dioxide from the air and water from the soil into energy-packed molecules we can use as food. In the process they give off oxygen, the gas we breathe in (fig. 1.1). No wonder there are bumper stickers that read Have You Thanked a Green Plant Today?

INORGANIC CHEMISTRY

Although inorganic chemistry pertains to nonliving matter, inorganic chemicals are important constituents of all living things. Also, some knowledge of inorganic chemistry is necessary for considering the unique molecules of life. For example, all types of molecules are composed of atoms.

Atoms

An **atom** is the smallest unit of matter, nondivisible by chemical means. While it is possible to split an atom by physical means, an atom is the smallest unit to enter into chemical reactions. For our purposes, it is satisfactory to think of an atom as having a central *nucleus*, where subatomic particles called **protons** and **neutrons** are located, and *shells,* where **electrons** orbit about the nucleus (fig. 1.2). Two important features of protons, neutrons, and electrons are their weight and charge, which are indicated in table 1.1.

TABLE 1.1 *SUBATOMIC PARTICLES*

NAME	CHARGE	WEIGHT
Electron	One negative unit	Almost no weight
Proton	One positive unit	One atomic unit
Neutron	No charge	One atomic unit

From Mader, Sylvia S., *Inquiry Into Life,* 4th ed. © 1976, 1979, 1982, 1985 Wm. C. Brown Publishers, Dubuque, Iowa. All Rights Reserved. Reprinted by permission.

The Periodic Table of the Elements (Appendix A) shows all the elements that are presently known. An **element** is any substance that contains just one type of atom. Figure 1.3 gives a simplified table highlighting the elements that are most common to living things. Each type of atom has an *atomic number;* for example, carbon is number 6 and nitrogen is number 7. Notice, too, that in the table each specific atom has a *symbol;* for example, C = carbon and N = nitrogen. *The atomic number equals the number of protons.* Also, each type of atom has an *atomic weight,* or mass. Carbon has an atomic weight or a mass of 12, and nitrogen has an atomic weight of 14. *The atomic weight equals the number of protons plus the number of neutrons.*

Figure 1.4 diagrams a specific *electrically neutral* atom; in which the number of protons (+) is equal to the number of electrons (−). The first shell of an atom can contain up to two electrons; thereafter each shell of those atoms in the simplified table (fig. 1.3) can contain up to eight electrons.

FIGURE 1.3 Periodic Table of the
Elements (simplified). See appendix A
for the complete table. Each element has
an atomic number, atomic symbol, and
atomic weight. The elements in dark
color are the most common, and those in
light color are also common in living
things.

I	II	III	IV	V	VI	VII	VIII
1 H hydrogen 1							2 He helium 4
3 Li lithium 7	4 Be beryllium 9	5 B boron 11	6 C carbon 12	7 N nitrogen 14	8 O oxygen 16	9 F fluorine 19	10 Ne neon 20
11 Na sodium 23	12 Mg magnesium 24	13 Al aluminum 27	14 Si silicon 28	15 P phosphorus 31	16 S sulfur 32	17 Cl chlorine 35	18 Ar argon 40
19 K potassium 39	20 Ca calcium 40						

atomic number
atomic symbol
atomic weight

FIGURE 1.4 Carbon atom. The diagram of the atom shows
that the number of protons (the atomic number) equals the
number of electrons when the atom is electrically neutral.
Carbon may also be written in the manner shown below the
diagram. The subscript is the atomic number, and the
superscript is the weight.

p = protons
n = neutrons
● = electrons

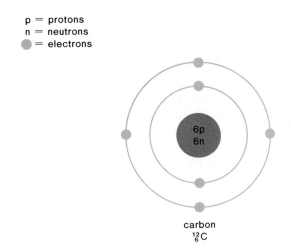

carbon
$^{12}_{6}$C

Although the table gives only one weight for each type
of atom, actually, the atoms of an element can vary as to
weight. Such atoms, called **isotopes,** have the same number
of protons but differ as to the number of neutrons. For example, there are isotopes of carbon, designated as ^{12}C, ^{13}C,
and ^{14}C, that have atomic weights of 12, 13, and 14 because
of the varying number of neutrons in the nucleus. Certain
isotopes called *radioactive isotopes* are unstable and emit radiation, which may be detected by a Geiger counter. Among

those isotopes of carbon listed, carbon-14 is radioactive. Radioactive isotopes are widely used in biological research because it is possible to trace their presence in chemical
substances and tissues.

All matter is composed of atoms, which are arranged in the
Periodic Table of the Elements according to increasing
weight. The weight of an atom is dependent on the number
of protons and neutrons in the nucleus, but its chemical properties are dependent on the number and arrangement of electrons in the shells.

Reactions between Atoms

Atoms react with one another to form *molecules*. In one type
of reaction, there is a transfer of an electron(s) from one atom
to another in order to form a molecule. Such atoms are thereafter called **ions,** and the reaction is called an *ionic reaction.*
For example, figure 1.5 depicts a reaction between sodium
(Na) and chlorine (Cl) in which chlorine takes an electron
from sodium. Now the sodium ion (Na$^+$) carries a positive
charge, and the chlorine ion (Cl$^-$) carries a negative charge.
Notice that a negative charge indicates that the ion has more
electrons ($-$) than protons ($+$) and a positive charge indicates that the ion has more protons ($+$) than electrons ($-$).
Oppositely charged ions are attracted to one another, and this
attraction is called an **ionic bond.**

In another type of reaction, atoms form a molecule by
sharing electrons. The bond that forms between these atoms
is called a **covalent bond.** For example, when oxygen reacts
with two hydrogen atoms, water (H_2O) is formed (fig. 1.6).
Sometimes the atoms in a covalently bonded molecule
share electrons evenly, but in water the electrons spend more
time encircling the larger oxygen than the smaller hydro-

FIGURE 1.5 *Formation of the salt sodium chloride. During this ionic reaction, an electron is transferred from the sodium atom to the chlorine atom. Each resulting ion carries a charge as shown. Most people use the term* salt *to refer only to sodium chloride, but chemists use the term to refer to similar combinations of positive and negative ions.*

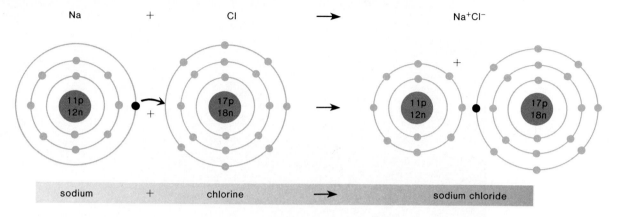

Na + Cl → Na⁺Cl⁻

sodium + chlorine → sodium chloride

FIGURE 1.6 *Formation of water. Following a covalent reaction, oxygen is sharing electrons with two hydrogen atoms.*

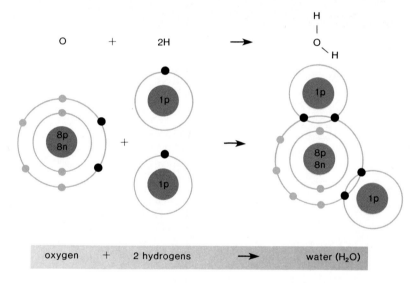

O + 2H →

oxygen + 2 hydrogens → water (H_2O)

gens. Therefore, there is a slight positive charge on the hydrogen atoms and a slight negative charge on the oxygen atom. For this reason water is called a *polar molecule,* and hydrogen bonding occurs between water molecules (fig. 1.7). A **hydrogen bond** occurs whenever a partially positive hydrogen is attracted to a partially negative atom. The hydrogen bond is represented by a dotted line in figure 1.7 because it is a weak bond that is easily broken.

Atoms react with one another to form molecules. In one type of reaction, positively and negatively charged ions are formed when an electron(s) is (are) transferred from one atom to another. In another type of reaction, the atoms share electrons in a molecule where the atoms are covalently bonded to one another.

FIGURE 1.7 *Water molecules are polar; each hydrogen carries a partial positive charge and each oxygen carries a partial negative charge. The polarity of the water molecules brings about hydrogen bonding between the molecules in the manner shown. The dotted lines represent hydrogen bonds. (δ = partial)*

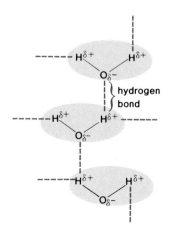

FIGURE 1.8 *The weather is more moderate near the sea because the ocean cools down slowly in the winter and absorbs a lot of heat in the summer.*

Life is dependent on the various characteristics of water.

Characteristics of Water

Hydrogen bonds are relatively weak, but they still cause water molecules to cling together. Without hydrogen bonding between molecules, water would boil at −80° C and freeze at −100° C, making life impossible. Instead, *water is a liquid at body temperature. It tends to cool down slowly and to absorb a great deal of heat before it vaporizes.* Aside from giving the seashore a more moderate climate (fig. 1.8), water helps keep the body temperature within normal limits. This is particularly noticeable when sweating occurs. When humans sweat, body heat is used to vaporize water.

Because water molecules cling together, water fills tubular vessels and is an excellent transport medium for distributing substances and heat throughout the body. Being a polar molecule, *water acts as a solvent and dissolves various chemical substances,* particularly other polar molecules. This property of water greatly facilitates chemical reactions in cells.

Dissociation

Polarity also causes water molecules to tend to *dissociate,* or split up, in this manner:

$$H - O - H \rightarrow H^+ + OH^-$$

The hydrogen ion (H^+) has lost an electron; the hydroxide ion (OH^-) has gained the electron. Because very few molecules actually dissociate, few hydrogen ions and hydroxide ions result (fig. 1.9).

Acids and Bases

Acids dissociate in water and release hydrogen ions. For example, an important inorganic acid is hydrochloric acid (HCl), which dissociates in this manner:

$$HCl \rightarrow H^+ + Cl^-$$

Dissociation is almost complete, and this acid is called a strong acid. If HCl is added to a beaker of water (fig. 1.10), the number of hydrogen ions increases.

Bases dissociate in water and release hydroxide ions (OH^-). For example, an important inorganic base is sodium hydroxide (NaOH), which dissociates in this manner:

$$NaOH \rightarrow Na^+ + OH^-$$

Dissociation is complete, and sodium hydroxide is called a strong base. If NaOH is added to a beaker of water (fig. 1.11), the number of hydroxide ions increases.

FIGURE 1.9 *In water there are always a few water molecules that have dissociated. Dissociation produces an equal number of hydrogen and hydroxide ions.*

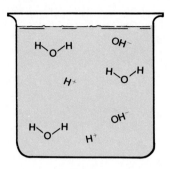

FIGURE 1.10 *HCl is an acid that releases hydrogen ions as it dissociates in water. Notice that the addition of HCl to this beaker has caused it to have more hydrogen ions than hydroxide ions.*

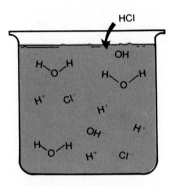

FIGURE 1.11 *NaOH is a base that releases hydroxide ions as it dissociates. Notice that the addition of NaOH to this beaker has caused it to have more hydroxide ions than hydrogen ions.*

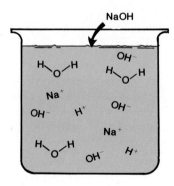

pH

The **pH** scale ranges from 0–14. Any pH value below 7 is acidic, with ever increasing acidity toward the lower numbers. Any pH value above 7 is basic (or alkaline), with ever increasing basicity toward the higher numbers. A pH of exactly 7 is neutral. Water has an equal number of H^+ and OH^- ions, and thus one of each is formed when water dissociates. The fraction of water molecules that dissociate is 10^{-7} (or 0.0000001), which is the source of the pH value for neutral

FIGURE 1.12 *The pH scale. The proportionate amount of hydrogen ions (H^+) to hydroxide ions (OH^-) is indicated by the diagonal line. Any pH above 7 is basic, while any pH below 7 is acidic.*

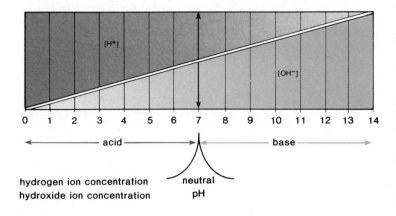

solutions. The pH scale was devised to simplify discussion of the hydrogen ion concentration $[H^+]$, without using cumbersome numbers. For example,

a. 1×10^{-6} $[H^+]$ = pH 6   Each lower pH unit has ten
b. 1×10^{-7} $[H^+]$ = pH 7   times the amount of H^+ as
c. 1×10^{-8} $[H^+]$ = pH 8   the next higher unit.

Which of the preceding items (*a–c*) indicates a higher concentration of hydrogen ions and therefore refers to an acid? The numbers with the smaller negative exponents indicate a greater quantity of hydrogen ions than those with a larger negative exponent. Therefore, (*a*) refers to an acid. Bases add hydroxide ions to solutions and increase the OH^- ion concentration of water. Basic solutions, then, have fewer H^+ ions than OH^- ions. In the preceding list, (*c*) refers to a base because it indicates a lesser concentration of H^+ than OH^- ions compared with that of water. Figure 1.12 gives the complete pH scale with proper notations.

All living things need to maintain the hydrogen ion concentration, or pH, at a constant level. For example, the pH of the blood is held constant at about 7.4, or we become ill. The presence of buffers helps keep the pH constant. A **buffer** is a chemical or a combination of chemicals that can take up excess hydrogen ions or excess hydroxide ions. When an acid is added to a buffered solution, a buffer takes up excess hydrogen ions, and when a base is added to a buffered solution, a buffer takes up excess hydroxide ions. Therefore, the pH changes minimally whenever a solution is buffered.

Acids have a pH that is less than 7, and bases have a pH that is greater than 7. The presence of buffers helps keep the pH of body fluids constant at about neutral, or pH 7, because a buffer can absorb both hydrogen and hydroxide ions.

ACID RAIN AND COLON CANCER ARE LINKED

Acid rain could be responsible for elevated death rates from cancer of the colon and other organs in the northeastern United States according to two California researchers. They say carbon dioxide gas, a major component of acid rain, absorbs ultraviolet light that normally would fuel the body's production of vitamin D. Without vitamin D, people cannot absorb enough calcium to protect tissues from becoming cancerous, said Frank Garland, chief of occupational medicine at the Navy Health Research Center near San Deigo.

Scientists have long suspected that acid rain contributes to lung cancer, but Garland said it also could be responsible for deaths from cancer of the colon, which is dramatically higher in the Northeast than in other areas of the nation. "If we know sulfur dioxide levels, we can predict with 60% accuracy the rate of colon cancer in that area," Garland said. Sulfur dioxide, released mainly from plants burning coal and oil, also could cause cancer of the breast and ovaries, he said.

"That is a very imaginative and exciting hypothesis," said Dr. Richard Rivlin of the Memorial Sloan-Kettering Cancer Center in New York. Lack of vitamin D could be one of several dietary factors responsible for colon cancer, Rivlin said, and further research is needed on air pollution's role in blocking production of the vitamin. Dr. Kurt Isselvacher, director of Massachusetts General Hospital's cancer center, called the theory "surprising" but "worthy of examination and review." And Senator John Kerry, a leader in pushing for acid rain controls, said the new research would "add to the impetus" for a strong cleanup program, "although this is a tragic way to have to add to the impetus."

. . . Garland said he first suspected something was wrong when, 12 years ago, he saw the National Cancer Institute's earliest maps showing cancer deaths across the country. Red dots indicated the greatest numbers of colon cancer deaths, he explained, and almost every single red spot was in the northeastern quadrant of the United States. Diet has long been suspected as a cause of colon cancer, but nothing seemed to distinguish diets in this part of the country, said Garland, who conducted the research with his brother Cedric, a senior cancer researcher at the University of California in San Diego.

The pair noticed, however, that the map on skin cancers was "exactly opposite" from the one on colon cancers. Sunlight causes skin cancer, and the pair

HUMAN ISSUE

A number of organic and inorganic compounds are used as drugs for medical and nonmedical purposes. As you know, the federal government maintains a close watch over the production and marketing of drugs. Some believe that taking virtually any drug should be a matter of personal choice; the government should merely be required to inform us of the biological effects of the drugs and then we can make our own choice. Others believe that it should be illegal to produce and sell certain types of drugs to the public. Which approach do you think is better?

TABLE 1.2 *INORGANIC VERSUS ORGANIC CHEMISTRY*

INORGANIC COMPOUNDS	ORGANIC COMPOUNDS
Usually contain metals and nonmetals	Always contain carbon and hydrogen
Usually ionic bonding	Always covalent bonding
Always contain a small number of atoms	May be quite large with many atoms
Often associated with nonliving elements	Associated with living organisms

From Mader, Sylvia S., *Inquiry Into Life,* 4th ed. © 1976, 1979, 1982, 1985 Wm. C. Brown Publishers, Dubuque, Iowa. All Rights Reserved. Reprinted by permission.

ORGANIC CHEMISTRY

Table 1.2 contrasts inorganic compounds with organic compounds. (A *compound* is a substance that contains two or more different atoms.) Both types of compounds are necessary to the proper functioning of the human body.

Unit Molecules

The chemistry of carbon accounts for the formation of the very large number of organic compounds we associate with living organisms. Carbon shares electrons with as many as four other atoms. Many times, carbon atoms share electrons

Pollutants in the air such as sulfur dioxide contribute to the acidity of rain and may also be detrimental to our health.

suspected that certain rays of the sun were blocked in the Northeast, but not in other regions with comparable climates, increasing colon cancer cases and decreasing the number of skin cancer cases. Further studies showed that sulfur dioxide absorbs a limited spectrum of ultraviolet light—precisely the same waves that fuel production of vitamin D in the skin. While ozone and other pollutants also block sunlight, he said, they do not affect the rays that produce vitamin D.

. . . The research, if it holds up, could have major implications for the heated debate over acid rain, which so far has focused mainly on damage to lakes, trees, and buildings. But Garland said his work was not designed with political motivations. "Sulfur dioxide did not become acid rain to us," he said, "until someone mentioned, 'That's the same stuff that makes acid rain. Do you realize the implications of that?' "

By Larry Tye. *The Boston Globe* February 10, 1988. Reprinted courtesy of *The Boston Globe.*

with each other to form rings or chains of carbon atoms. These act as skeletons for the unit molecules found in the life molecules—carbohydrates, fats, and nucleic acids. Thus the properties of carbon are essential to life as we know it.

Synthesis and Hydrolysis

Figure 1.13 diagrammatically illustrates that the life molecules are synthesized or made when small unit molecules join together. A bond that joins two unit molecules together is created after the removal of H^+ from one molecule and OH^- from the next molecule. As water forms, dehydration **synthesis** occurs.

Life molecules are often **polymers,** or chains of unit molecules joined together. They can be broken down in a manner opposite to synthesis: the addition of water leads to the disruption of the bonds linking the unit molecules together. During this process, called **hydrolysis,** one molecule takes on H^+ and the next takes on OH^-.

FIGURE 1.13 Synthesis and hydrolysis of an organic polymer. When the unit molecules, called monomers, join together to form the polymer (synthesis), water is released; when the polymer is broken down (hydrolysis), water is added.

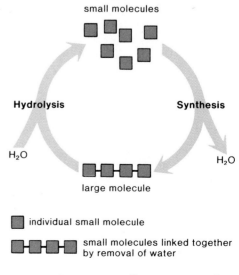

small molecules

Hydrolysis   Synthesis

H_2O H_2O

large molecule

■ individual small molecule

■■■■ small molecules linked together by removal of water

Proteins

Functions

Proteins are large, complex *macromolecules* that sometimes have mainly a structural function. For example, in humans, keratin is a protein that makes up hair and nails, and collagen is a protein found in all types of connective tissue, including ligaments, cartilage, bone, and tendons. The muscles (fig. 1.14) contain proteins (actin and myosin) that account for their ability to contract.

Some proteins function as **enzymes,** necessary contributors to the chemical workings of the cell and therefore of the body. Enzymes are organic catalysts that speed up chemical reactions. They work so quickly that a reaction that might normally take several hours or days takes only a fraction of a second when an enzyme is present.

Structure

The unit molecules found in proteins are called **amino acids.** The name amino acid refers to the fact that the molecule has two functional groups, an *amino group* and an *acid group*.

$$H-N-C-C \begin{matrix} O \\ O-H \end{matrix}$$

R = remainder of molecule

Amino acids differ from one another by their *R groups,* which vary from being a single hydrogen atom to a complicated ring. Because there are about twenty different common amino acids found in the proteins of living things, there are also about twenty different types of R groups.

The bond that joins two amino acids together is called a **peptide bond.** As you can see in figure 1.15, when synthesis occurs, the acid group of one amino acid reacts with the amino group of another amino acid and water is given off. A dipeptide contains only two amino acids, but when ten or twenty amino acids have joined together, the resulting chain is called a **polypeptide.** A very long polypeptide of approximately seventy-five amino acids is called a protein.

Levels of Structure   Proteins are said to have at least three levels of structure: primary, secondary, and tertiary. The *primary structure* is simply the sequence, or order, of the different amino acids. Any number of the twenty different amino acids may be joined in various sequences, and each type of protein has its own particular sequence. The resulting chain is like a necklace comprising up to twenty different types of beads that reoccur and are linked in a set way. The *secondary structure* is the usual orientation of the amino acid chain. One common arrangement of the chain is the alpha helix, a right-handed spiral held in place by hydrogen bonding between members of the various peptide bonds. The *tertiary structure* of a protein refers to its final three-dimensional shape. In a structural protein like collagen, the helical chains lie parallel to one another, but in enzymes the helix bends and twists in different ways. The final shape of a protein is maintained by various types of bonding between the R groups. Covalent, ionic, and hydrogen bonding are all seen. Figure 1.16 illustrates the main features of protein chemistry.

Some proteins have more than one type of polypeptide chain, each with its own primary, secondary, and tertiary structures. Within the protein, these separate chains are arranged to give a fourth level of structure termed the *quaternary structure.* Hemoglobin is a complex protein having a quaternary structure.

Amino acids are the unit molecule for peptides and proteins. Proteins have both structural and metabolic[1] functions in the human body.

Carbohydrates

Carbohydrates are characterized by the presence of H — C — OH groupings in which the ratio of hydrogen atoms to oxygen atoms is approximately 2:1. Since this ratio is the same as the ratio in water, the meaning of this compound's name—hydrates of carbon—is very appropriate. If the number of carbon atoms in the compound is low (from about three to seven), the carbohydrate is a simple sugar, or monosaccharide. Larger carbohydrates are created by joining together monosaccharides in the manner described in figures 1.13 and 1.17.

[1]Metabolism is all the chemical reactions that occur in a cell.

FIGURE 1.15 *Formation of a dipeptide. Notice that as the peptide bond forms, water is given off. In other words, the water molecule on the right-hand side of the equation is derived from components removed from the amino acids on the left-hand side.*

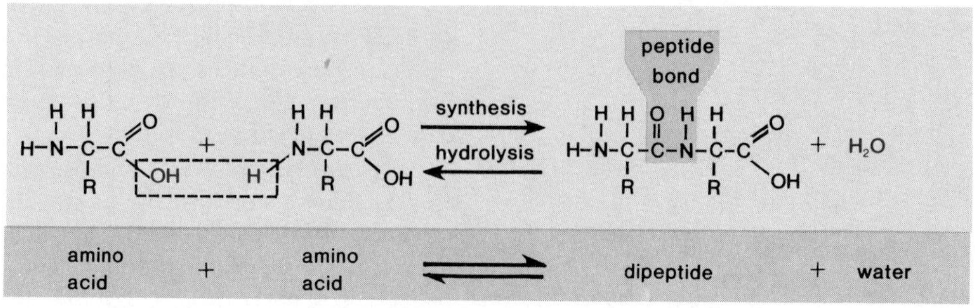

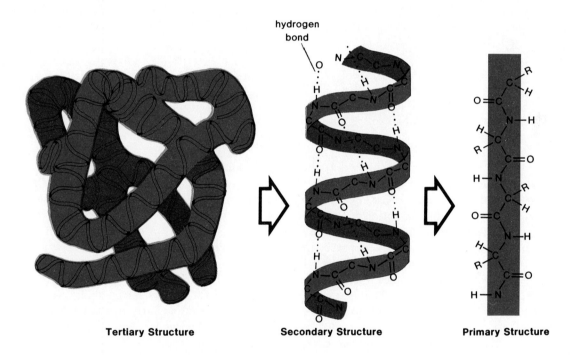

FIGURE 1.16 *Proteins have at least three levels of structure. The tertiary structure is often a twisting and turning of the helix that takes place because of bonding between the R groups; the secondary structure is often a helix; and the primary structure is the order of the amino acids.*

Tertiary Structure　　**Secondary Structure**　　**Primary Structure**

FIGURE 1.17 *Each glucose molecule has the structure shown. When two glucose molecules combine, the disaccharide maltose is formed. During synthesis, a bond forms between the two glucose molecules as a molecule of water is formed. During hydrolysis, the components of water are added as the bond is broken.*

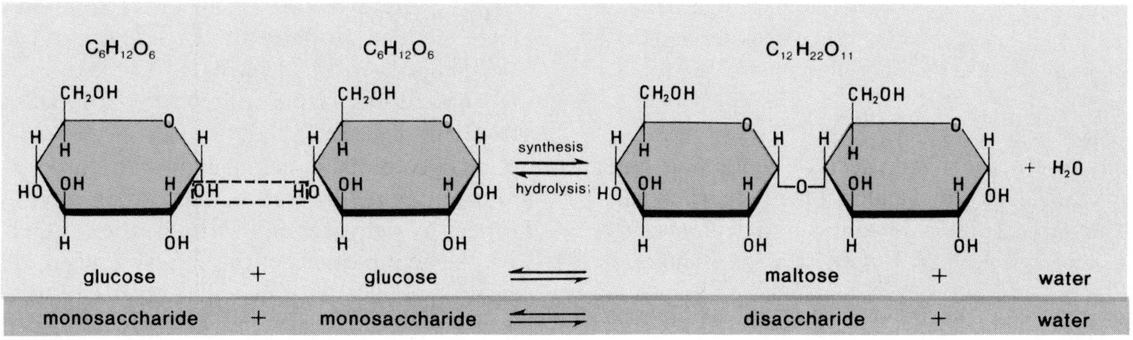

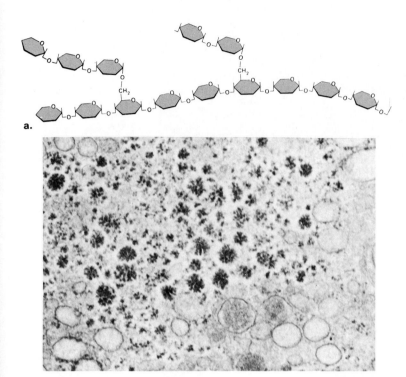

a.

b.

Monosaccharides and Disaccharides

As their name implies, **monosaccharides** are simple sugars having only one unit. These compounds are often designated by the number of carbons they contain; for example, pentose sugars have five carbons, and hexose sugars have six carbons. **Glucose** is a six-carbon sugar, with the structural formula shown in figure 1.17. Although there are other monosaccharides with the molecular formula $C_6H_{12}O_6$, in this text we will use the molecular formula $C_6H_{12}O_6$ to mean glucose, since glucose is the most common six-carbon monosaccharide found in cells. Cells use glucose as an immediate energy source.

The term **disaccharide** tells us that there are two monosaccharide units joined together in the compound. When two glucose molecules join together, maltose is formed (fig. 1.17). You may also be interested to know that when glucose and another monosaccharide, fructose, are joined together, the disaccharide called sucrose is formed. *Sucrose* is derived from plants and is commonly used at the table to sweeten foods.

Polysaccharides

A **polysaccharide** is a carbohydrate that contains a large number of monosaccharide molecules. There are three polysaccharides that are common in animals and plants: glycogen, starch, and cellulose. All of these are polymers, or

chains, of glucose, just as a necklace might be made up of only one type of bead. Even though all three polysaccharides contain only glucose, they are distinguishable from one another.

Glycogen, a molecule having many side branches (fig. 1.18), is the storage form of glucose in humans. After eating, the liver stores glucose as glycogen; in between eating, the liver releases glucose so that the blood concentration of glucose is always 0.1%.

The polymers *starch* and *cellulose* are found in plants (p. 23). Plants store glucose as starch, a polymer similar in structure to glycogen except that it has few side branches. Starch is an important source of glucose energy in our diet because it can be hydrolyzed to glucose by digestive enzymes. In cellulose, a common structural compound in plants, the glucose units are joined by a slightly different type of linkage compared to that of glycogen and starch. For this reason we are unable to digest cellulose, and it passes through our digestive tract as roughage. Recently it has been suggested that the presence of roughage in the diet is necessary to good health and prevention of colon cancer.

In this text, $C_6H_{12}O_6$ stands for glucose, the unit molecule in glycogen and starch, two important energy sources for humans.

Lipids

Many **lipids** are nonpolar and therefore are insoluble in water. This is true of fats, the most familiar lipids, such as lard, butter, and oil, which are used in cooking or at the table. In the body, fats serve as long-term energy sources. Adipose tissue is composed of cells that contain many molecules of neutral fat.

Neutral Fats (Triglycerides)

A neutral (nonpolar) fat contains two types of unit molecules: **glycerol** and **fatty acids.** Each fatty acid has a long chain of carbon atoms, with hydrogens attached, ending in an acid group (fig. 1.19). Fatty acids are either *saturated* or *unsaturated*. Saturated fatty acids have no double bonds between the carbon atoms. The carbon chain is saturated, so to speak, with all the hydrogens that can be held. Unsaturated fatty acids have double bonds in the carbon chain wherever the number of hydrogens is less than two per carbon atom. Unsaturated fatty acids are most often found in vegetable oils and account for the liquid nature of these oils. Vegetable oils are hydrogenated to make margarine. Polyunsaturated margarine still contains a large number of unsaturated, or double, bonds.

Glycerol is a compound with three $H—C—OH$ groups attached by way of the carbon atoms. When fat is formed, by dehydration synthesis, the $—OH$ groups react with the acid portions of three fatty acids so that three molecules of water are formed. The reverse of this reaction represents hydrolysis of the fat molecule into its separate components (fig. 1.20).

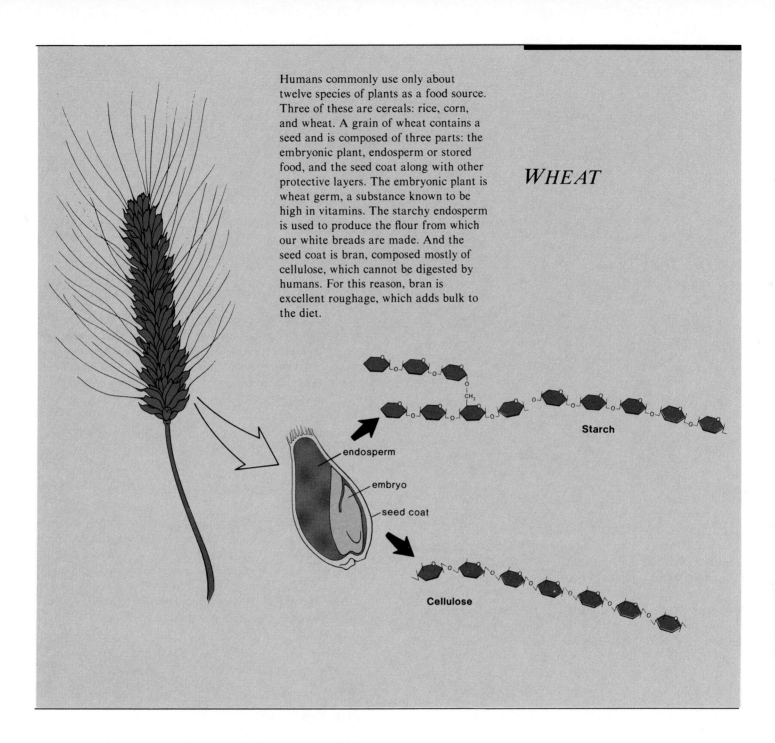

Humans commonly use only about twelve species of plants as a food source. Three of these are cereals: rice, corn, and wheat. A grain of wheat contains a seed and is composed of three parts: the embryonic plant, endosperm or stored food, and the seed coat along with other protective layers. The embryonic plant is wheat germ, a substance known to be high in vitamins. The starchy endosperm is used to produce the flour from which our white breads are made. And the seed coat is bran, composed mostly of cellulose, which cannot be digested by humans. For this reason, bran is excellent roughage, which adds bulk to the diet.

WHEAT

endosperm

embryo

seed coat

Starch

Cellulose

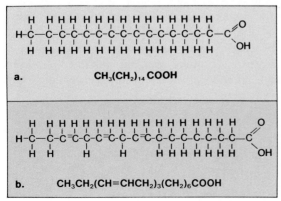

a. $CH_3(CH_2)_{14}COOH$

b. $CH_3CH_2(CH=CHCH_2)_3(CH_2)_6COOH$

FIGURE 1.19 *Fatty acids. a. Saturated fatty acids have no double bonds because each carbon is bonded to the maximum of two hydrogen atoms. b. Unsaturated fatty acids have double bonds because each carbon is bonded to less than two hydrogens.*

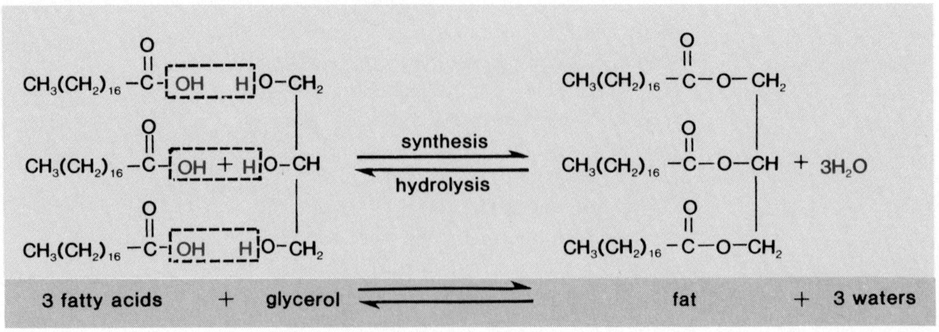

Cells use fat molecules, composed of glycerol and three fatty acids, for long-term energy storage.

Soaps

A soap is a salt formed by a fatty acid and an inorganic base; for example,

$$NaOH + RCOOH \rightarrow RCOO^-Na^+$$

sodium   fatty   soap
hydroxide acid

Although fats do not mix with water because they are nonpolar, a soap, being polar, will mix with water. When soaps are added to oils, then oils too will mix with water. Figure 1.21 shows how a soap positions itself about an oil droplet so that the polar ends project outward. Now the droplet will be soluble in water. This process of causing an oil to disperse in water is called **emulsification,** and it is said that an emulsion has been formed. Emulsification occurs when dirty clothes are washed with soaps and detergents. Also, prior to the digestion of fatty foods, fats are emulsified by bile. Usually a person who has had the gallbladder removed has trouble digesting fatty foods because the gallbladder stores bile for use at the proper time during the digestive process.

Phospholipids

Essentially, **phospholipids** are constructed like neutral fats, except for the third fatty acid, which is replaced by a phosphate group or a grouping that contains both phosphate and nitrogen:

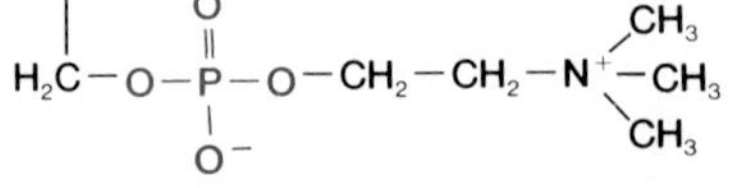

These molecules are not electrically neutral as are the fats because the phosphate group is polar. Notice, then, that phospholipids have both a nonpolar (uncharged) region and a polar (charged) region. Thus, phospholipids are soluble in water. This latter property makes them very useful compounds in the body, as we will see in the next chapter.

Unlike neutral fats, soaps and phospholipids have polar ends that are attracted to polar water molecules.

Cholesterol and Steroids

Cholesterol and **steroids** have similar structures. They are constructed of four fused rings of carbon atoms to which is usually attached a carbon chain of varying length (fig. 1.22). Today there is a great deal of interest in cholesterol because a high blood cholesterol level is associated with development of coronary heart disease as discussed on page 114. Steroids, however, are very necessary compounds in the body; for example, the sex hormones are steroids.

Steroids are metabolically important in the human body and are very much implicated in our state of health.

Nucleic Acids

Nucleic acids are huge, macromolecular compounds with very specific functions in cells; for example, the genes are composed of a nucleic acid called **DNA** (deoxyribonucleic acid). DNA has the ability to replicate, or make a copy of itself. It also controls protein synthesis. Another important nucleic acid, **RNA** (ribonucleic acid), works in conjunction with DNA to bring about protein synthesis.

FIGURE 1.21 *Fat molecules, being nonpolar, will not disperse in water. An emulsifier contains molecules that have a polar end and nonpolar end. When an emulsifier is added to a beaker containing a layer of nondispersed oil, the polar ends are attracted to the water, and the nonpolar ends arrange themselves as shown. This causes droplets of oil to disperse in water.*

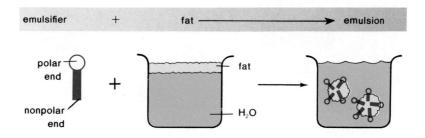

FIGURE 1.22 *Like cholesterol (a), steroid molecules have four adjacent rings, but their effects on the body largely depend on the type of chain attached at the location indicated. The chain in (b) is found in aldosterone, which is involved in the regulation of sodium and potassium blood content. The chain in (c) is found in testosterone, the male sex hormone.*

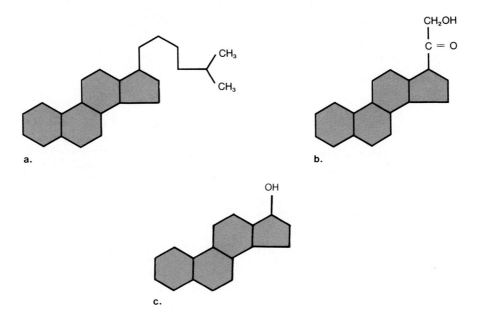

Both DNA and RNA are polymers of nucleotides and therefore are chains of nucleotides joined together. Just like the other synthetic reactions we have studied in this section, when these units are joined together to form nucleic acids, water molecules are removed.

Both DNA and RNA are composed of nucleotides. DNA makes up the genes and along with RNA controls protein synthesis.

Nucleotides

Every **nucleotide** is a molecular complex of three types of unit molecules: phosphoric acid (phosphate), a pentose (five-carbon) sugar, and a nitrogen base. In DNA the sugar is deoxyribose and in RNA the sugar is ribose, and this difference accounts for their respective names. There are four different types of nucleotides in DNA and RNA. Figure 1.23 shows the two different types of nucleotides that are present in DNA. The base can be one of the **purines,** adenine or guanine, which have a double ring, or one of the **pyrimidines,** thymine or cytosine, which have a single ring. These structures are called bases because they have basic characteristics that raise the pH of a solution. RNA differs from DNA in that the base uracil is used in place of the base thymine.

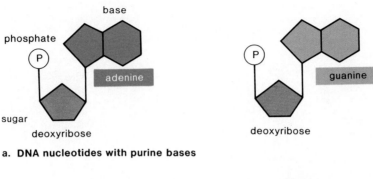

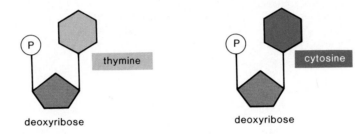

a. DNA nucleotides with purine bases

b. DNA nucleotides with pyrimidine bases

Strands

When nucleotides join together, they form a linear molecule called a strand in which the so-called backbone is made up of phosphate-sugar-phosphate-sugar, with the bases projecting to one side of the backbone. RNA is single stranded (fig. 1.24), but DNA is double stranded. The two strands of DNA twist about one another in the form of a **double helix** (fig. 1.25a and b). The two strands are held together by hydrogen bonds between purine and pyrimidine bases. Thymine (T) is always paired with adenine (A), and guanine (G) is always paired with cytosine (C). This is called complementary base pairing. If we unwind the DNA helix, it resembles a ladder (fig. 1.25c). The sides of the ladder are made entirely of phosphate and sugar molecules, and the rungs of the ladder are made only of the *complementary paired bases*. The bases can be in any order, but A is always paired with T, and G is always paired with C, and vice versa. Therefore, no matter what the order or the quantity of any particular base pair, the number of purine bases always equals the number of pyrimidine bases.

DNA has a structure like a twisted ladder: sugar-phosphate backbones make up the sides of the ladder, and hydrogen-bonded bases make up the rungs of the ladder. The base A is always paired with the base T, and the base C is always paired with the base G. RNA differs from DNA in several respects (table 1.3).

FIGURE 1.24 *RNA is a single-stranded polymer. Nucleic acid polymers contain a chain of nucleotides. Each strand has a backbone made of sugar and phosphate molecules. The bases project to the side.*

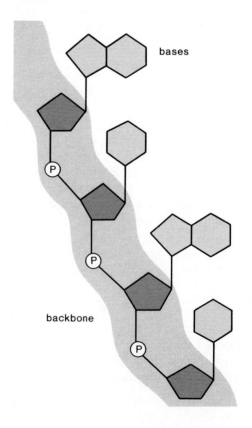

FIGURE 1.25 Overview of DNA structure. a. Double helix. b. Complementary base pairing. c. Ladder configuration. Notice that the uprights are composed of sugar and phosphate molecules and the rungs are complementary paired bases.

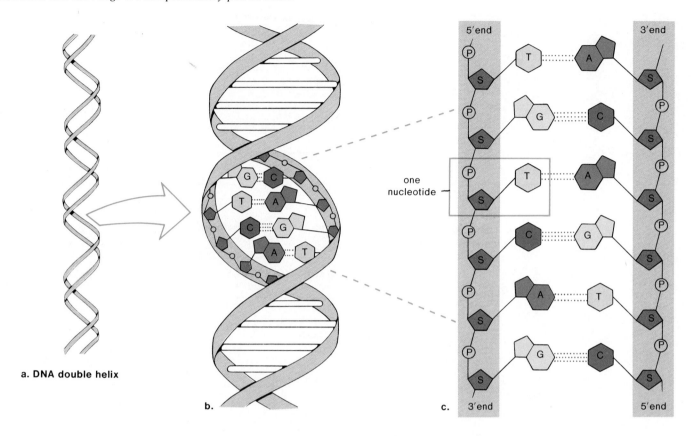

a. DNA double helix

b.

one nucleotide

c.

TABLE 1.3 DNA STRUCTURE COMPARED TO RNA STRUCTURE

	DNA	RNA
Sugar	Deoxyribose	Ribose
Bases	Adenine, guanine, thymine, cytosine	Adenine, guanine, uracil, cytosine
Strands	Double stranded with base pairing	Single stranded
Helix	Yes	No

ATP

ATP, adenosine triphosphate (fig. 1.26), is a very special type of nucleotide. It is composed of the base adenine and the sugar ribose (together called adenosine) and three phosphate groups. The wavy lines in the formula for ATP indicate high-energy phosphate bonds. When these bonds are broken, an unusually large amount of energy is released. Because of this property, ATP is the energy currency of cells; when cells "need" something, they "spend" ATP.

ATP is used in body cells for synthetic reactions, active transport, nervous conduction, and muscle contraction. When energy is required for these processes, the end phosphate group is removed from ATP, breaking down the molecule to ADP (adenosine diphosphate) and Ⓟ (phosphate) (fig. 1.26).

ATP Cycle

The reaction shown in figure 1.26 occurs in both directions; not only is ATP broken down, it is also built up when ADP joins with Ⓟ. Since ATP breakdown is constantly occurring, there is always a ready supply of ADP and Ⓟ to rebuild ATP again.

Figure 1.27 illustrates the ATP cycle in a diagrammatic way. Notice that when ATP is broken down, energy is released, and when it is built up, energy is required. We shall see that aerobic cellular respiration, a metabolic pathway that takes place largely within mitochondria, produces the energy needed for ATP buildup.

ATP is the energy molecule of cells because it contains high-energy phosphate bonds.

FIGURE 1.26 *ATP, the energy molecule in cells, has two high-energy phosphate bonds (indicated in the figure by wavy lines). When cells require energy, the last phosphate bond is broken and a phosphate molecule is released.*

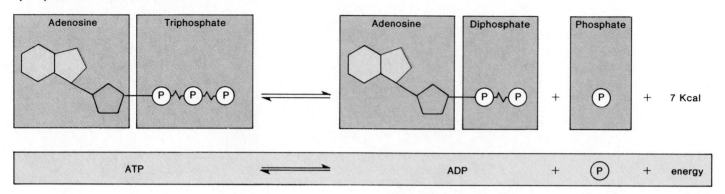

FIGURE 1.27 *The ATP cycle. When ADP joins with a P group, energy is required; but when ATP breaks down to ADP and a P group, energy is given off.*

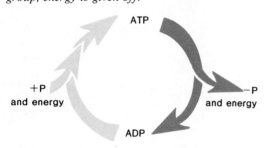

TABLE 1.4 *ORGANIC COMPOUNDS OF LIFE*

MACROMOLECULES	UNIT MOLECULE	USUAL ATOMS
Protein	Amino acid	C, H, O, N, S
Carbohydrate, e.g., starch	Glucose	C, H, O
Lipid	Glycerol and fatty acids	C, H, O
Nucleic acid	Nucleotide	C, H, O, N, P

SUMMARY

All matter is made up of atoms, each having a weight that is dependent on the number of protons and neutrons in the nucleus and chemical properties that are dependent on the number of electrons in the outermost shell. Atoms react with one another in order to form molecules. In ionic reactions, one atom gives electrons to another and in covalent reactions atoms share electrons.

Water, acids, and bases are important inorganic compounds. Water has a neutral pH; acids decrease and bases increase the pH of water. The organic molecules of interest are proteins, carbohydrates, lipids, and nucleic acids, each of which has (a) particular unit molecule(s) (table 1.4). Dehydration synthesis joins unit molecules together

and hydrolytic degradation releases them. All enzymes are proteins; carbohydrates serve as immediate energy sources; and fats are a long-term energy source for the individual. Nucleic acids are of two types, DNA and RNA. DNA is the genetic material, and both of these have functions related to protein synthesis, which will be discussed in chapter 19.

OBJECTIVE QUESTIONS

1. _____ are the smallest units into which matter can be chemically broken.
2. Isotopes differ by the number of _____ in the nucleus.
3. The two primary types of reactions and bonds are _____ and _____ .
4. A type of weak bond, called _____ bonding, exists between water molecules.

5. Acidic solutions contain more _____ ions than basic solutions, but they have a _____ pH.
6. The primary structure of a protein is the sequence of _____ _____ ; the secondary structure is very often an alpha _____ ; the tertiary structure is the final _____ of the protein.
7. All _____ are proteins and function to speed up chemical reactions.

8. Glycogen is a polymer of _____ , molecules that serve to give the body immediate _____ .
9. A neutral fat hydrolyzes to give one _____ molecule and three _____ molecules.
10. The genes are composed of _____ , a nucleic acid made up of _____ joined together.

Answers to Objective Questions

10. DNA, nucleotides
8. glucose, energy 9. glycerol, fatty acid
acids, helix (or spiral), shape 7. enzymes
4. hydrogen 5. hydrogen, lower 6. amino
1. Atoms 2. neutrons 3. ionic, covalent

STUDY QUESTIONS

1. Describe the composition of an atom, and give the weight and charge of an atom's components. (p. 13)
2. Give an example of an ionic reaction, and define the term *ion*. (p. 14)
3. Give an example of a covalent reaction, and define the term *covalent bond*. (p. 14)
4. On the pH scale, which numbers indicate a basic solution? an acidic solution? (p. 17)
5. What are buffers, and why are they important to life? (p. 17)
6. Name four general differences between inorganic and organic compounds. (p. 18)
7. Explain synthesis by dehydration and breakdown by hydrolysis of organic compounds. (p. 19)
8. Describe the primary, secondary, and tertiary structure of proteins. What functions do proteins serve in the body? (p. 20)
9. Name some monosaccharides, disaccharides, and polysaccharides, and state appropriate functions. What is the most common unit molecule for these? (p. 21)
10. What type molecules react to form a neutral fat? Explain the difference between a saturated and an unsaturated fatty acid. (p. 22)
11. Name several types of lipids, and state their functions. (pp. 22–23)
12. What are the two types of nucleic acids in cells and what is their function? What is the unit molecule of a nucleic acid? (pp. 24–25) Name four differences between DNA and RNA. (p. 27)

THOUGHT QUESTIONS

1. The human body is composed of organic molecules but also contains inorganic molecules. Argue for the importance of inorganic molecules in the body.
2. Compare a glycogen molecule to a protein molecule. How are the differences you note related to the function of these molecules in the body?
3. A physiologist removed a liver from a dog and placed it in a dish with water. The next morning he found sugar in the water. Does this prove that the liver is alive? Why or why not?

KEY TERMS

acid (as'id) a solution in which pH is less than 7; a substance that contributes or liberates hydrogen ions (protons) in a solution. *16*

amino acid (ah-me'no ās'id) a unit of protein that takes its name from the fact that it contains an amino group (NH_2) and an acid group (COOH). *20*

atom (at'om) smallest unit of matter nondivisible by chemical means. *13*

ATP adenosine triphosphate; a compound containing adenine, ribose, and three phosphates, two of which are high-energy phosphates. It is the "common currency" of energy for most cellular processes. *27*

base (bās) a solution in which pH is more than 7; a substance that contributes or liberates hydroxide ions in a solution; alkaline; opposite of acidic. Also, a term commonly applied to one of the components of a nucleotide. *16*

buffer (buf'er) a substance or compound that prevents large changes in the pH of a solution. *17*

DNA deoxyribonucleic acid; a nucleic acid, the genetic material that replicates and directs protein synthesis in cells. *24*

electron (e-lek'tron) a subatomic particle that has almost no weight and carries a negative charge; travels in an orbital, called a shell, about the nucleus of an atom. *13*

emulsification (e-mul"si-fi-ka'shun) the act of dispersing one liquid in another. *24*

enzyme (en'zīm) a protein catalyst that speeds up a specific reaction or a specific type of reaction. *20*

hydrogen bond (hi'dro-jen bond) a weak attraction between a hydrogen atom carrying a partial positive charge and another atom carrying a partial negative charge. *15*

ion (i'on) an atom or group of atoms carrying a positive or negative charge. *14*

isotopes (i'so-tōps) atoms with the same number of protons and electrons but differing in the number of neutrons and therefore in weight. *14*

lipid (lip'id) a group of organic compounds that are insoluble in water; notably fats, oils, and steroids. *22*

neutron (nu'tron) a subatomic particle that has a weight of one atomic mass unit, carries no charge, and is found in the nucleus of an atom. *13*

nucleic acid (nu-kle'ik as'id) a large organic molecule made up of nucleotides joined together; for example, DNA and RNA. *24*

peptide bond (pep'tīd bond) the bond that joins two amino acids. *20*

pH a measure of the hydrogen ion concentration; any pH below 7 is acid and any pH above 7 is basic. *17*

polysaccharide (pol"e-sak'ah-rīd) a macromolecule composed of many units of sugar. *22*

protein (pro'te-in) a macromolecule composed of one or several long polypeptides; polypeptides contain many amino acids joined together. *20*

proton (pro'ton) a subatomic particle found in the nucleus of an atom that has a weight of one atomic mass unit and carries a positive charge; a hydrogen ion. *13*

RNA ribonucleic acid; a nucleic acid that assists DNA in the production of proteins within the cell. *24*

CHAPTER TWO

CELL STRUCTURE AND FUNCTION

CHAPTER CONCEPTS

1 The fundamental unit of life is the cell, which is highly organized and contains organelles that carry out specific functions.

2 The cytoplasm contains metabolic pathways, each a series of reactions controlled by enzymes.

3 The cell membrane regulates the entrance and exit of molecules to and from the cell.

4 The nucleus, a centrally located organelle, controls the metabolic functioning and structural characteristics of the cell.

5 Endoplasmic reticulum, the Golgi apparatus, and lysosomes are all membranous tubules or vesicles concerned with the entrance, production, digestion, excretion, or transportation of molecules.

6 Mitochondria are organelles concerned with the conversion of glucose energy into ATP molecules.

7 Centrioles, cilia, and flagella all contain microtubules.

A light micrograph of stained human nerve cells in a culture.

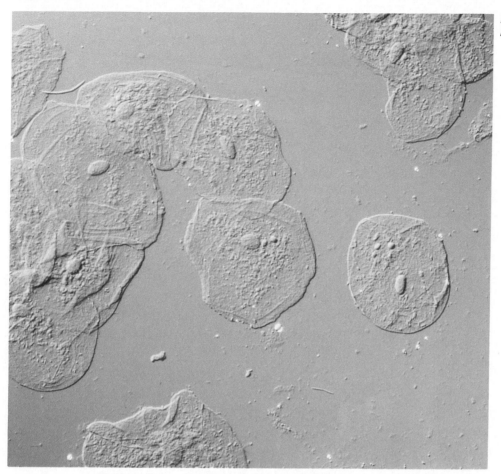

FIGURE 2.1 *Epithelial cells from a human cheek.*

INTRODUCTION

We are multicellular animals. The **cell** (fig. 2.1) is the fundamental unit of our bodies, and it is at the cellular level that we must understand health and disease. Because cells are microscopic it is sometimes hard to imagine that it is not the intestine or heart that is causing the difficulty—it is the cells that make up the intestine or heart. Nowhere is this more evident than when we study the cause of cancer of the uterus, the lungs, the colon, and so forth. In all cases, cancer is characterized by uncontrolled growth of cells due to irregularities of cell structure and function.

HUMAN ISSUE

Most scientists, on the basis of data, believe that a chemical evolution produced the first cell(s), and thereafter all organisms evolved from this (these) cell(s). Many lay people, on the basis of faith, believe that God created all living things. Advocates of divine creationism argue that biology texts should include creationism because students have the right to be presented with an alternative viewpoint to evolution. The vast majority of scientists argue that no acceptable scientific data supporting creationism exists and that only ideas generated by scientific investigation should be presented in biology texts. Should biology texts be required to include the theory of divine creation? If you were a school board member and this issue arose in your school district, what position would you take?

The cells of your body perform specific functions, and therefore their structure varies greatly. Even so, because all cells have the same basic organization, it is possible for us to begin our study of cell structure by examining a generalized cell. Our knowledge of the generalized animal cell depicted in figure 2.2 was obtained by using the light microscope and the electron microscope. The *light microscope,* which utilizes light to view the object, does not show much detail, but the *electron microscope,* which uses electrons to view the object, allows cell biologists to discern cell structure in great detail. Table 2.1 contrasts these two types of microscopes, which are also discussed in Appendix B.

FIGURE 2.2 *Animal cell. This drawing of a generalized cell is based on electron micrographs.*

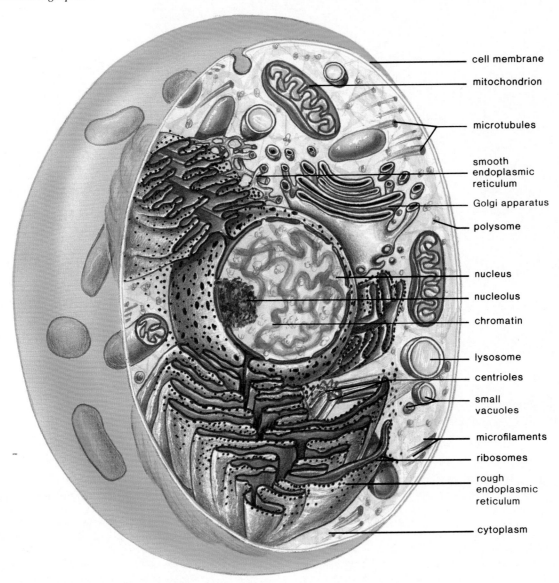

cell membrane
mitochondrion
microtubules
smooth endoplasmic reticulum
Golgi apparatus
polysome
nucleus
nucleolus
chromatin
lysosome
centrioles
small vacuoles
microfilaments
ribosomes
rough endoplasmic reticulum
cytoplasm

TABLE 2.1 A COMPARISON OF A LIGHT AND AN ELECTRON MICROSCOPE

LIGHT	ELECTRON
1. Glass lenses	1. Electromagnetic lenses
2. Illumination by light	2. Illumination due to beam of electrons
3. Resolution[a] $\simeq$ 0.1 μm	3. Resolution $\simeq$ 1 nm
4. Magnifies to 2,000$\times$	4. Magnifies to 100,000$\times$
5. Cost: up to thousands	5. Cost: up to hundreds of thousands
6. Specimen may be living or dead	6. Specimen must be dead

[a]Resolution is the ability to distinguish two points as being separate.
See Appendix C for metric units of measurement.

Electron micrographs, photographs obtained by use of the electron microscope, have helped biologists develop an understanding of cell structure.

Figure 2.2 shows that an animal cell is surrounded by an outer membrane, or **cell membrane,** within which is found the **cytoplasm,** the substance of the cell outside the nucleus. Within the cytoplasm there are various **organelles,** small bodies with specific structures and functions. Table 2.2 summarizes the organelles that will be studied.

TABLE 2.2 ORGANELLES (SIMPLIFIED)

NAME	STRUCTURE	FUNCTION
Cell membrane	Bilayer of phospholipid and globular proteins	Passage of molecules into and out of cell
Nucleus	Nuclear envelope surrounds chromatin, nucleolus, and nucleoplasm	Control of cell
Nucleolus	Concentrated area of RNA in the nucleus	Ribosome formation
Chromatin (chromosomes)	Composed of DNA and protein	Contains hereditary information
Endoplasmic reticulum	Folds of membrane forming flattened channels and tubular canals	Transport by means of vesicles
Rough	Studded with ribosomes	Protein synthesis
Smooth	Having no ribosomes	Lipid and carbohydrate synthesis
Ribosome	RNA and protein in two subunits	Protein synthesis
Golgi apparatus	Stack of membranous saccules	Packaging and secretion
Vacuole and vesicle	Membranous sacs	Containers of material
Lysosome	Membranous container of hydrolytic enzymes	Intracellular digestion
Mitochondrion	Inner membrane (cristae) within outer membrane	Cellular respiration
Cytoskeleton	—	Cell shape and subcellular movement
Microfilament	Actin and myosin proteins	Same
Microtubule	Tubulin protein	Same
Cilium and flagellum	9 + 2 pattern of microtubules	Locomotion
Centriole	9 + 0 pattern of microtubules	Organization of microtubules; associated with cell division

MEMBRANE

Membrane not only surrounds the cell, it also makes up many of the organelles.

Membrane Structure

The *fluid mosaic model* of membrane structure tells us that protein molecules have a changing pattern (form a mosaic) within a bilayer of phospholipid molecules that are fluid, having a consistency of light oil. Notice the manner in which the phospholipid molecules arrange themselves in figure 2.3. Their structure, discussed in chapter 1, causes each molecule to have a polar head and nonpolar tails. Within the phospholipid bilayer, the tails face inward and the heads face outward, where they are likely to encounter a watery environment. The protein molecules may reside above or below the phospholipids, extend from top to bottom of the membrane, or simply penetrate a short distance. On the outer surface some of the protein molecules form *receptor sites* for receiving chemicals, such as hormones, that influence the metabolism of the cells. On the inner surface the proteins tend to be enzymes, and those that penetrate the cell membrane are involved in transport as discussed later.

Short chains of sugars are attached to the outer surface of some protein and lipid molecules. There is evidence that these carbohydrate chains allow cells to recognize one another and/or a cell as belonging to a particular individual. If this is the case, their presence may explain in part why a patient's system sometimes rejects an organ transplant.

Membrane Function

The cell membrane forms a boundary between the outside of the cell and the inside of the cell. It allows only certain molecules to enter and exit the cytoplasm freely; therefore, the cell membrane is said to be **selectively permeable.** Small molecules that are lipid soluble, such as oxygen and carbon dioxide, can pass through the membrane easily. Certain other small molecules, like water, are not lipid soluble and still penetrate the membrane quickly because they pass through pores, passageways formed by protein molecules.

The cell membrane, composed of phospholipid and protein molecules, is selectively permeable and regulates the entrance and exit of molecules from the cell.

Diffusion and Osmosis

Some small molecules pass through the membrane by **diffusion** (table 2.3), the movement of molecules from the area of greater concentration to the area of lesser concentration until they are equally distributed. To illustrate diffusion, imagine opening a perfume bottle in the corner of a room. The smell of the perfume soon penetrates the room because the molecules that make up the perfume have drifted to all parts of the room. Another example is putting a tablet of dye into water. The water eventually takes on the color of the dye as the tablet dissolves.

FIGURE 2.3 *Fluid mosaic model of a cell membrane. Protein molecules are embedded in and project to either side of a double layer of phospholipid molecules. Note that some of these are receptors that specifically bind with certain molecules acting as chemical messengers, thus influencing the activity of the cell. The right side of the illustration shows the effect of fracturing a frozen membrane.*

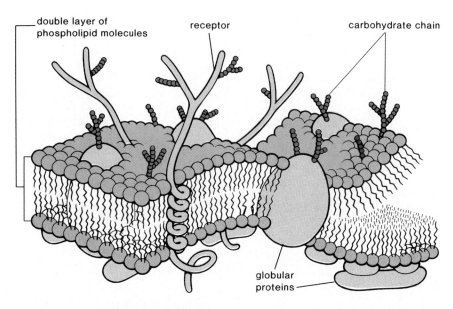

TABLE 2.3 PASSAGE OF MOLECULES INTO AND OUT OF CELLS

NAME	DIRECTION	REQUIREMENTS	EXAMPLES
Diffusion	Toward lesser concentration	————	Lipid-soluble molecules Water Gases (oxygen and carbon dioxide)
Transport			
Facilitated	Toward lesser concentration	Carrier	Sugars and amino acids
Active	Toward greater concentration	Carrier plus energy	Sugars, amino acids, and ions
Exocytosis	Toward greater concentration	Vacuole release	Secretion of substances
Endocytosis	Toward greater concentration	Vacuole formation	Phagocytosis of substances

Osmosis is the diffusion of water across a cell membrane. It occurs whenever there is an unequal concentration of water on either side of a selectively permeable membrane. For example, figure 2.4 represents a thistle tube covered with a selectively permeable membrane. The tube contains a protein suspension, and the beaker contains distilled water. Because of the presence of the protein (solute), there is a lesser concentration of water (solvent) inside the tube than there is outside the tube. Since the protein cannot cross the membrane, there will be a net movement of water to the inside of the tube. Once water enters, a "back pressure" builds up that prevents any further net gain of water. This is called **osmotic pressure,** the amount of force that must be exerted to stop osmosis from continuing. Osmotic pressure is potential pressure, and water moves across the membrane toward the region of greater osmotic pressure.

Since cytoplasm contains proteins and salts and is surrounded by a selectively permeable membrane, a cell exerts osmotic pressure when it is placed in a **hypotonic** solution, which contains a greater concentration of water (lesser concentration of solute) than does the cell. Under these circumstances the cell will swell or even burst (fig. 2.5). When a cell is placed in a **hypertonic** solution, which contains a lesser concentration of water (greater concentration of solute) than does cytoplasm, the cell loses water. In an **isotonic** solution, the osmotic pressure is similar on both sides of the membrane, and there is no net movement of water.

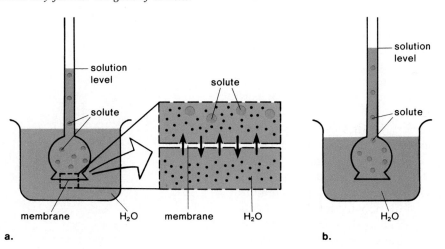

FIGURE 2.4   *Osmosis demonstration.* a. *A thistle tube, covered at the broad end by a membrane, contains a solute (large circles) in addition to a solvent (small circles). The beaker contains only solvent. The solute is unable to pass through the membrane, but the solvent passes through in both directions.* b. *There is a net movement of solvent toward the inside of the thistle tube. This causes the solution to rise in the thistle tube until a back pressure develops that prevents any further net gain of solvent.*

Carriers

Most solutes (ions and molecules other than water) do not simply diffuse across a cell membrane; rather, they are transported by means of protein molecules within the membrane. These proteins are called **carriers.** During **facilitated transport,** a molecule (e.g., an amino acid or glucose) is transported across the cell membrane from the side of higher concentration to the side of lower concentration. The cell does not need to expend energy for this type of transport because the molecule is moving in the normal direction.

During **active transport** a molecule is moving contrary to the normal direction; that is, from lower to higher concentration. Iodine collects in the cells of the thyroid gland; sugar is completely absorbed from the gut by cells that line the digestive tract; and sodium (Na^+) is sometimes almost completely withdrawn from urine by cells lining the kidney tubules. Active transport requires a protein carrier and the use of cellular energy obtained from the breakdown of ATP (p. 28). When ATP is broken down, energy is released and in this case the energy is used by a carrier to carry out active transport.

Certain small molecules, like water and gases, diffuse across a cell membrane. The diffusion of water, termed osmosis, can cause a cell to swell or dehydrate depending on the environmental medium. Other molecules must be transported by means of protein carriers found in the membrane.

FIGURE 2.5   *Effect of tonicity on an animal cell. In a hypotonic solution, there is a net movement of water (arrow) to the inside of the cell, and the cell swells to bursting. In a hypertonic solution, there is a net movement of water (arrow) to the outside of the cell, and the cell shrinks. In an isotonic solution, there is no net movement of water and the appearance of a red blood cell remains the same.*

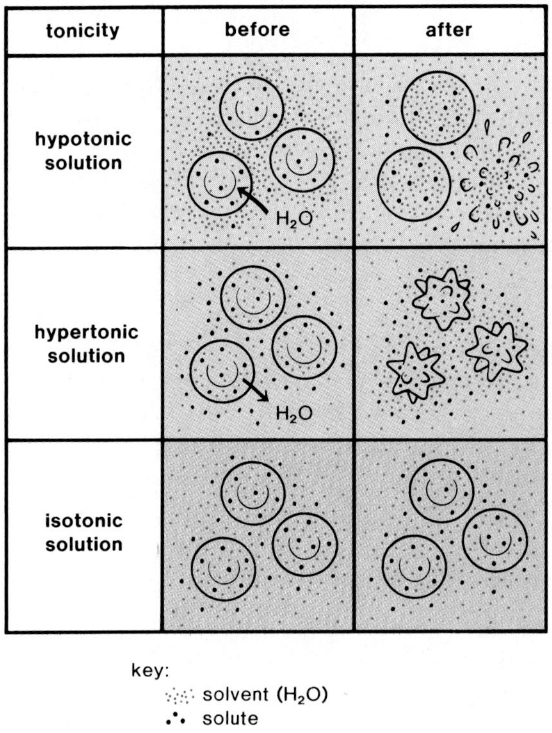

tonicity	before	after
hypotonic solution	H_2O	
hypertonic solution	H_2O	
isotonic solution		

key:
- solvent (H_2O)
- solute

CHAPTER TWO   CELL STRUCTURE AND FUNCTION   **35**

CELLULAR ORGANELLES

As mentioned previously, the cell contains a number of organelles, small bodies with specific structure and functions. These help the cell carry out its many activities.

Nucleus

The **nucleus,** the largest organelle found within the cell, is enclosed by a double-layered **nuclear envelope** that is actually continuous with the endoplasmic reticulum discussed following. As illustrated in figure 2.6, there are pores, or openings, in this membrane through which large molecules pass from the *nucleoplasm,* the fluid portion of the nucleus, to the cytoplasm or vice versa.

The nucleus is of primary importance in the cell because it is the control center that oversees the metabolic functioning of the cell and ultimately determines the cell's characteristics. Within the nucleus there are masses of threads called **chromatin,** so called because they take up stains and become colored. Chromatin is indistinct in the nondividing cell, but it condenses to rodlike structures called **chromosomes** at the time of cell division. Chemical analysis shows that chromatin, and thus chromosomes, contain the chemical DNA (deoxyribonucleic acid) along with certain proteins and some RNA (ribonucleic acid). This is not surprising because we already know that chromosomes contain the genes and that the genes are composed of DNA. DNA, with the help of the type of RNA found in the chromosomes, controls protein (enzyme) synthesis within the cytoplasm, and it is this function that allows DNA to control the cell.

Nucleoli

One or more **nucleoli** are present in the nucleus. These dark-staining bodies are actually specialized parts of chromatin in which another type of RNA called ribosomal RNA (rRNA) is produced. Ribosomal RNA joins with proteins before migrating to the cytoplasm where it becomes part of the ribosomes, organelles to be discussed in the following sections.

The nucleus contains chromatin, which condenses into chromosomes just prior to cell division. Chromosomes contain DNA that, with the help of RNA, directs protein synthesis in the cytoplasm. Another type of RNA, rRNA, is made within the nucleolus before migrating to the cytoplasm, where it is incorporated into ribosomes.

Membranous Canals and Vacuoles

Endoplasmic reticulum, the Golgi apparatus, vacuoles, and lysosomes (fig. 2.2) are structurally and functionally related membranous structures. Ribosomes are not composed of membrane but are included in this category because they are often intimately associated with the endoplasmic reticulum.

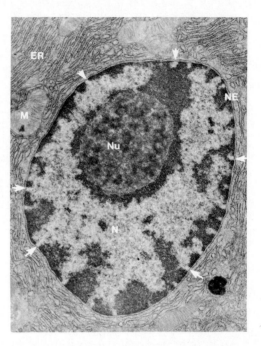

FIGURE 2.6 *An electron micrograph of a nucleus (N) with a clearly defined nucleolus (NU) and irregular patches of chromatin scattered throughout the nucleoplasm. The nuclear envelope (NE) contains pores indicated by the arrows. This nucleus is surrounded by endoplasmic reticulum (ER), and its size may be compared to the mitochondrion (M) that appears to the left.*

Endoplasmic Reticulum

The **endoplasmic reticulum (ER)** forms a membranous system of tubular canals that begins at the nuclear envelope and branches throughout the cytoplasm. Small granules, called ribosomes, are attached to some portions of the endoplasmic reticulum. If they are present, the reticulum is called **rough endoplasmic reticulum;** if they are not present, it is called **smooth endoplasmic reticulum.** Apparently, smooth endoplasmic reticulum contains, within its membrane, enzymes that synthesize lipids. Therefore, smooth endoplasmic reticulum is abundant in cells that produce steroid hormones. Also, it is known that the administration of drugs increases the amount of smooth endoplasmic reticulum in the liver. It would seem then that the reticulum has enzymes that detoxify drugs. It is quite possible that these are detoxified within structures called *peroxisomes,* membrane-bound vacuoles often attached to smooth ER that contain enzymes capable of carrying out oxidation of various substances, including alcohol.

Ribosomes **Ribosomes** look like small, dense granules in low-power electron micrographs (fig. 2.7a), but a higher resolution shows that each one contains two subunits (fig. 2.7c). As their name implies, ribosomes contain RNA, but they also contain proteins. The larger of the two subunits contains at least thirty different proteins, and the smaller unit contains at least twenty different proteins.

FIGURE 2.7 *Rough endoplasmic reticulum. a. Electron micrograph showing that in some cells the structure has this appearance. b. A three-dimensional drawing gives a better idea of the structure's actual shape. c. This three-dimensional model of a ribosome is based on high-power electron microscopy studies that indicate that a ribosome is composed of a small and a large subunit.*

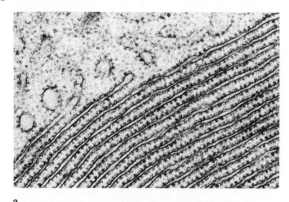

a.

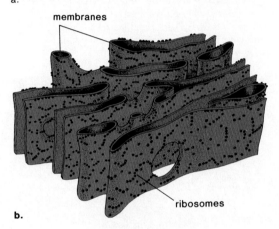

membranes

ribosomes

b.

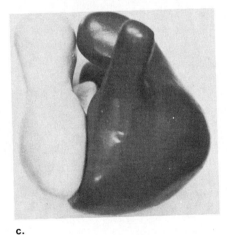

c.

Ribosomal RNA produced in the nucleolus joins with proteins before migrating to the cytoplasm. Once the ribosomes are fully assembled within the cytoplasm, they function in the process of protein synthesis. Synthesis, as discussed in chapter 1, refers to the joining together of small organic molecules to make larger ones. In this case, amino acids are joined together to make a protein.

Ribosomes are very often attached to endoplasmic reticulum (fig. 2.7), but are also found unattached within the cytoplasm. In the cytoplasm, several ribosomes, each of which is producing the same type protein, are arranged in a functional group called a *polysome*. Most likely these proteins are for use inside the cell. Protein that is destined for export outside the cell is prepared at the rough ER and temporarily stored in the channels of the reticulum. Small portions of the endoplasmic reticulum then break away to form membrane-enclosed *vesicles* (small *vacuoles*) that migrate to the Golgi apparatus, where the product is received and repackaged for export.

The endoplasmic reticulum is a membranous system of tubular canals that may be smooth or rough. Smooth ER functions in lipid metabolism and detoxification of drugs. Rough ER functions in protein synthesis.

Golgi Apparatus

The **Golgi apparatus** (fig. 2.8) is named for Camillo Golgi, who first discovered its presence in cells. It is composed of a stack of about a half-dozen or more saccules that look like flattened vacuoles. One side of the stack faces the nucleus, and the other side faces the cell membrane. Vesicles are seen especially at the rims of the saccules, but vesicles also occur along the length of either face of the apparatus.

These observations, along with biochemical evidence, suggest that the Golgi apparatus receives protein-filled vesicles from the smooth ER at its inner face. After the proteins are processed, they are packaged into vesicles that often move to the cell membrane, where their contents are released. This process, called **exocytosis,** is the manner in which secretion occurs. During secretion a cell product leaves a cell.

The Golgi apparatus processes and packages those proteins that will be secreted at the cell membrane. It also produces lysosomes.

Lysosomes

Lysosomes are a special type of vesicle formed by the Golgi apparatus. All lysosomes are concerned with intracellular digestion and contain powerful enzymes called *hydrolytic enzymes* (p. 74). Lysosomes sometimes fuse with phagocytic vesicles that form at the cell membrane (fig. 2.8) in order to bring substances into the cell. This type of vesicle formation is sometimes called **endocytosis** (table 2.3). After fusion and digestion, the breakdown products enter the cytoplasm, and the nondigested residue is expelled from the cell (fig. 2.8).

Occasionally, a person is unable to manufacture an enzyme normally found within the lysosome. In these cases, the lysosome fills to capacity as the substrate for that enzyme accumulates. The cells may become so filled with lysosomes of this type that it brings about the death of the individual.

FIGURE 2.8 *Golgi apparatus function. The Golgi apparatus receives vesicles from the endoplasmic reticulum and thereafter forms at least two types of vesicles, lysosomes and secretory vesicles. Lysosomes contain hydrolytic enzymes that can break down large molecules. Sometimes lysosomes join with vesicles, bringing large molecules into the cell. Thereafter, the molecules are digested. The secretory vesicles formed at the Golgi apparatus also discharge their contents at the cell membrane.*

FIGURE 2.9 *Epididymal epithelium for lysosomes. These organelles vary widely, but all function as a cellular digestive system in assemblying multiple acid hydrolases.*

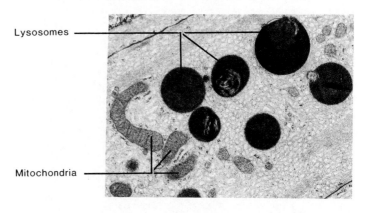

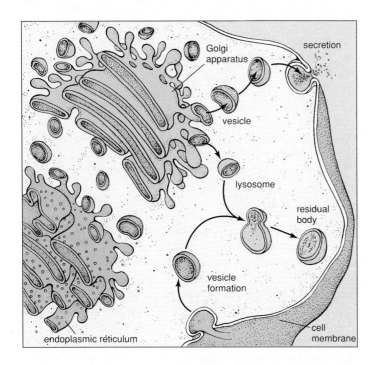

Lysosomes also carry out *autodigestion,* or the disposal of worn-out or damaged cell components, such as mitochondria (fig. 2.9), which have a short life span in the cell. This is an essential part of the normal process of cytoplasmic maintenance and turnover. By turnover it is meant that the cell is constantly breaking down and remaking its parts.

Lysosomes contain hydrolytic enzymes that function in digestion of substances taken in at the cell membrane and also of the cell parts themselves.

Mitochondria

A **mitochondrion** (fig. 2.10) is bounded by both an outer and an inner membrane. The inner membrane forms shelflike projections called *cristae* where enzymes are located in a set order. There are also enzymes in the *matrix,* a background substance. These enzymes participate in **aerobic cellular respiration** (fig. 2.11), a series of reactions that break down glucose into carbon dioxide and water. As chemical bond energy is released, ATP molecules are built up. Mitochondria are often referred to as the powerhouses of the cell because just as a powerhouse burns fuel to produce electricity, the mitochondria "burn" glucose products to produce ATP molecules. Notice in fig. 2.11 that aerobic cellular respiration, a process discussed in more detail on page 42, does require oxygen.

FIGURE 2.10 *Mitochondrion structure. a. Scanning electron micrograph of a mitochondrion surrounded by rough endoplasmic reticulum. Note the shelflike cristae formed by the inner membrane. b. A diagrammatic drawing shows outer and inner structure more clearly.*

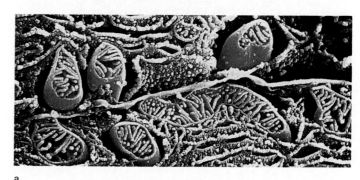

a.

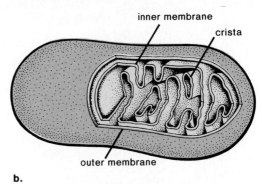

b.

FIGURE 2.11 *Overall equation for aerobic cellular respiration. ATP buildup is indicated by a curved arrow above the reaction arrow, which shows that as glucose is oxidized, ATP is produced.*

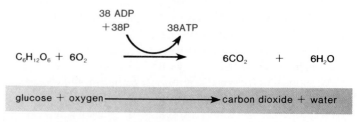

$$C_6H_{12}O_6 + 6O_2 \xrightarrow[\quad]{38 \text{ ADP} + 38P \quad 38ATP} 6CO_2 + 6H_2O$$

glucose + oxygen ⟶ carbon dioxide + water

FIGURE 2.12 *Cytoskeleton of cell. Notice that the various organelles are suspended in a cytoplasm that includes microtubules and microfilaments. a. Electron micrograph. b. Drawing showing placement of microtubules and microfilaments in cell. c. Detailed structure of a microtubule. d. Detailed structure of a microfilament.*

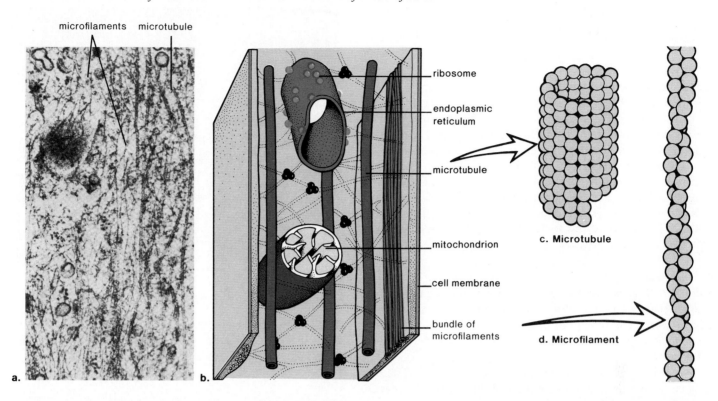

Mitochondria are the sites of cellular respiration, a process that provides ATP energy molecules to the cell.

FIGURE 2.13 *A scanning electron micrograph of an individual cell in a tissue culture. Notice the fingerlike projections on the "ruffle," which marks the leading edge of the cell. Microfilaments and microtubules are most likely present in this ruffle.*

Other Organelles

Several types of filamentous protein structures form a **cytoskeleton** (fig. 2.12) that helps maintain the cell's shape and either anchors the organelles or allows them to move as appropriate. The cytoskeleton includes microfilaments and microtubules. **Microfilaments** are long, extremely thin fibers that usually occur in bundles or other groupings. Microfilaments have been isolated from a number of cells. When analyzed chemically, their composition is similar to that of actin or myosin, the two proteins responsible for muscle contraction.

Microtubules are shaped like thin cylinders and are several times larger than microfilaments. Each cylinder contains thirteen rows of tubulin, a globular protein, arranged in a helical fashion. Aside from existing independently in the cytoplasm, microtubules are also found in certain organelles, such as cilia, flagella, and centrioles.

Remarkably, both microfilaments and microtubules assemble and disassemble within the cell. When they are assembled, the protein molecules are bonded together, and when they are disassembled, the protein molecules are not attached to one another. When microfilaments and microtubules are assembled, the cell has a particular shape, and when they disassemble, the cell can change shape (fig. 2.13).

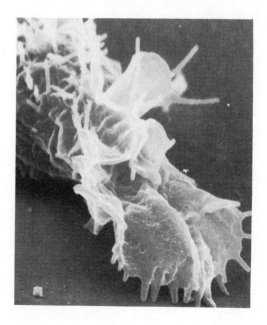

The cytoskeleton contains microfilaments and microtubules. Microfilaments, thin actin or myosin strands, and microtubules, thirteen rows of tubulin protein molecules arranged to form a hollow cylinder, maintain the shape of the cell and also direct the movement of cell parts.

Centrioles

Centrioles are short cylinders with a 9 + 0 pattern of microtubule triplets (fig. 2.15c). There is always one pair lying at right angles to one another near the nucleus (fig. 2.2). Before a cell divides, the centrioles duplicate, and the members of each pair are also at right angles to one another.

Centrioles give rise to basal bodies that direct the formation of cilia and flagella. Centrioles may also be involved in other cellular processes that use microtubules, such as the movement of material throughout the cell or the appearance and disappearance of the spindle apparatus (p. 369). Their exact role in these processes is uncertain, however.

Cilia and Flagella

Cilia and **flagella** are hairlike projections of cells that can move either in an undulating fashion, like a whip, or stiffly, like an oar. Cells that have these organelles are capable of movement. For example, sperm cells, carrying genetic material to the egg, move by means of flagella (fig. 2.14). The cells that line our upper respiratory tract are ciliated. These cilia sweep debris trapped within mucus back up the throat, and this action helps keep the lungs clean.

Cilia are much shorter than flagella, but even so they both are constructed similarly (fig. 2.15a). They are membrane-bound cylinders enclosing a matrix area. In the matrix are nine microtubule doublets arranged in a circle around two central microtubules (fig. 2.15b). This is called the 9 + 2 pattern of microtubules. Each doublet also has pairs of arms projecting toward a neighboring doublet and spokes extending toward the central pair of microtubules. Recent evidence indicates that cilia and flagella move when the microtubule doublets slide along one another. The clawlike arms and spokes seem to be involved in causing this sliding action, which requires ATP energy.

Each cilium and flagellum has a basal body (fig. 2.15c) lying in the cytoplasm at its base. **Basal bodies,** which are short cylinders with a circular arrangement of nine microtubule triplets called the 9 + 0 pattern, are believed to organize the structure of cilia and flagella.

Centrioles have a 9 + 0 pattern of microtubules and give rise to basal bodies that organize the 9 + 2 pattern of microtubules in cilia and flagella. Centrioles may be connected in some way to the origination of microtubules and to the spindle fibers that are seen during cell division.

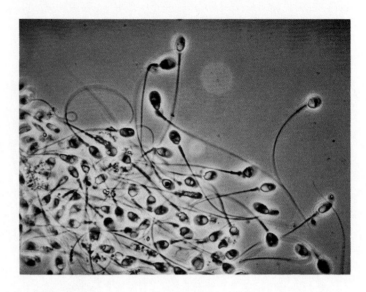

FIGURE 2.14 *Sperm cells use long, whiplike flagella to move about. Cilia and flagella have the structure depicted in figure 2.15.*

CELLULAR METABOLISM

Cellular **metabolism** includes all the chemical reactions that occur in a cell. Quite often these reactions are organized into metabolic pathways:

$$A \xrightarrow{1} B \xrightarrow{2} C \xrightarrow{3} D \xrightarrow{4} E \xrightarrow{5} F \xrightarrow{6} G$$

The letters, except *A* and *G,* are *products* of the previous reaction and the *reactants* for the next reaction. *A* represents the beginning reactant(s), and *G* represents the end product(s). The numbers in the pathway refer to different enzymes. *Every reaction in a cell requires a specific enzyme.* In effect, no reaction occurs in a cell unless its enzyme is present. For example, if enzyme number 2 in the diagram is missing, the pathway cannot function; it will stop at *B*. Since enzymes are so necessary in cells, their mechanism of action has been studied extensively.

Metabolic pathways contain many enzymes that are arranged to perform their reactions in a sequential order.

Enzymes

When an enzyme speeds up a reaction, the reactant(s) that participates in the reaction is called the enzyme's *substrate(s)*. Enzymes are often named for their substrate(s) (table 2.4). Enzymes have a specific region, called an **active site,** where the substrates are brought together so that they can react. An enzyme's specificity is caused by the shape of the active site, where the enzyme and its substrate(s) fit together in a specific way, much as the pieces of a jigsaw puzzle

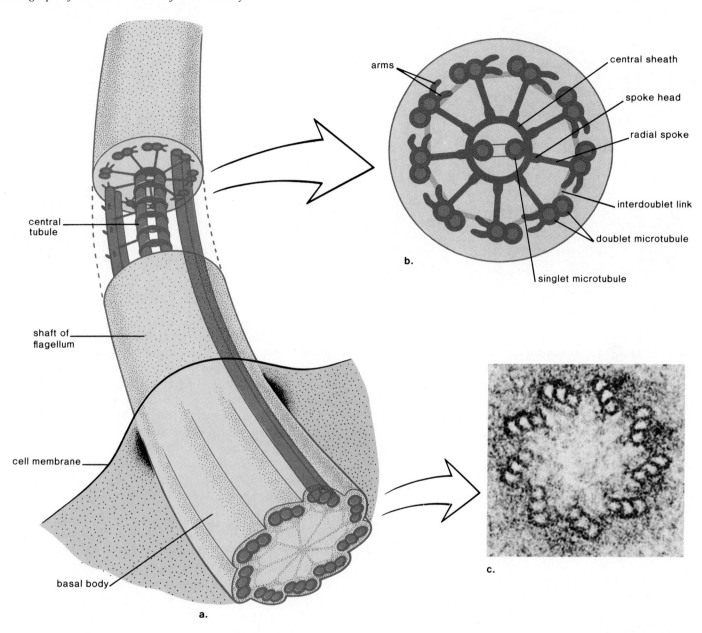

FIGURE 2.15 *Anatomy of cilia and flagella.* a. *Drawing showing that the 9 + 2 pattern of microtubules within a cilium or flagellum is derived (in some unknown way) from the 9 + 0 pattern in a basal body.* b. *Cross-sectional drawing of the 9 + 2 pattern showing the exact arrangement of microtubules. Notice the clawlike arms of the outer doublets and the spokes that connect them to the central pair.* c. *Electron micrograph of the cross section of a basal body.*

fit together (fig. 2.16). After one reaction has been completed, the product or products are released, and the enzyme is ready to catalyze another reaction. What we have said can be summarized in the following manner:

$$E + S \rightarrow ES \rightarrow E + P$$

(where E = enzyme, S = substrate, ES = enzyme substrate complex, and P = product).

TABLE 2.4 *ENZYMES NAMED FOR THEIR SUBSTRATES*

SUBSTRATE	ENZYME
Lipid	Lipase
Urea	Urease
Maltose	Maltase
Ribonucleic acid	Ribonuclease
Lactose	Lactase

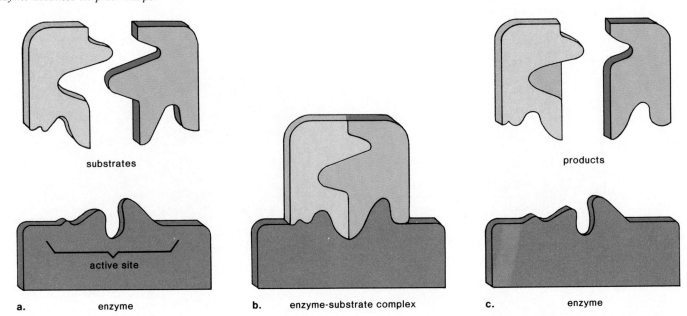

substrates

products

active site

a. enzyme

b. enzyme-substrate complex

c. enzyme

Coenzymes

Many enzymes have **coenzymes,** nonprotein portions that assist the enzyme and may even accept or contribute atoms to the reaction. It is of interest that vitamins are often components of coenzymes. The vitamin niacin is a part of the coenzyme NAD that removes hydrogen atoms from one substrate and passes them on to another molecule. Therefore, NAD functions as a *dehydrogenase,* an agent of **oxidation** (removal of hydrogen atoms) and **reduction** (addition of hydrogen atoms).

Like enzymes, coenzymes are reusable.

Enzymes are specific because they have an active site that accommodates their substrates. Enzymes often have nonprotein helpers called coenzymes. NAD is the usual coenzyme of oxidation and reduction within a cell.

Cellular Respiration

Cellular respiration is an important part of cellular metabolism because it provides ATP energy to cells. Cellular respiration includes *aerobic* (requires oxygen) respiration and fermentation, an *anaerobic* (does not require oxygen) process.

Aerobic Cellular Respiration

The overall equation for aerobic cellular respiration is given in figure 2.11. The oxidation of glucose does not actually occur in one step, however. The entire process requires three subpathways (*glycolysis,* the *Krebs cycle,* and the *respiratory chain*) and the *transition reaction,* as illustrated in figure 2.17.

In the figure, each arrow represents a different enzyme, and the letters represent the product of the previous reaction and the substrate for the next. Notice how each pathway resembles a conveyor belt in which a beginning substrate continuously enters at the start and, after a series of reactions, end products leave at the termination of the belt. It is important to realize, too, that all three pathways are going on at the same time. They can be compared to the inner workings of a watch, in which all parts are synchronized.

It is possible to relate the reactants and products of the overall reaction (fig. 2.11) to the subpathways in figure 2.17:

1. Glucose, $C_6H_{12}O_6$, is to be associated with **glycolysis,** the breakdown of glucose to two molecules of pyruvic acid (PYR). Thus far enough energy has been released to allow the buildup of two ATP molecules.
2. Carbon dioxide, CO_2, is to be associated with the transition reaction and the Krebs cycle. During the transition reaction, PYR is oxidized to active acetate (AA). AA enters the **Krebs cycle,** a cyclical series of oxidation reactions that give off CO_2 and produce one ATP molecule. Notice that since glycolysis ends with two molecules of PYR, the transition reaction and the Krebs cycle occur twice per glucose molecule. Altogether, then, the Krebs cycle accounts for two ATP molecules per glucose molecule.
3. Oxygen, O_2, and water, H_2O, are to be associated with the respiratory chain. The **respiratory chain** begins with $NADH_2$, the coenzyme that carries most of the hydrogens to the chain. The respiratory chain is a

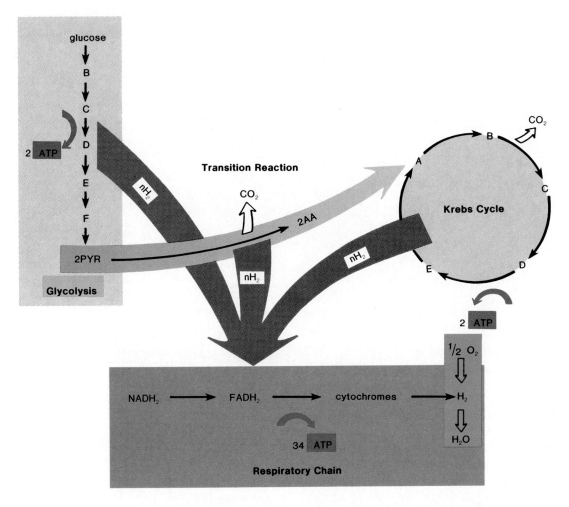

FIGURE 2.17 *Aerobic cellular respiration contains three subpathways: glycolysis, the Krebs cycle, and the respiratory chain. As the reactions occur, a number of hydrogen atoms (nH_2) and carbon dioxide molecules are removed from the various substrates. Oxygen acts as the final acceptor for the hydrogen atoms.*

series of molecules that pass hydrogen and/or electrons from one to the other until they are finally received by oxygen, which is then reduced to water. As every two electrons pass down the chain, enough energy is released for the buildup of approximately three ATP.

4. ATP is to be associated with glycolysis, the Krebs cycle, and the respiratory chain. Most ATP, however, is produced by the respiratory chain. It is usually said that the chain produces a total of thirty-four ATP molecules per glucose molecule.

Table 2.5 summarizes our discussion of aerobic cellular respiration. This table assumes that aerobic cellular respiration produces thirty-eight ATP per glucose molecule. In actuality, the number can vary depending on the current conditions.

TABLE 2.5 *OVERVIEW OF AEROBIC CELLULAR RESPIRATION*

NAME OF PATHWAY	EVENTS	ATP (PER GLUCOSE MOLECULE)
Glycolysis	Removal of $-H_2$ from substrates	2
Transition reaction	Removal of $-H_2$ from substrates Releases CO_2	
Krebs cycle	Removal of $-H_2$ from substrates Releases CO_2	2 (after two turns)
Respiratory chain	Accepts $-H_2$ from other pathways and passes them on to O_2, producing H_2O	34
Total		38

Wine, beer, and whiskey production all require yeast fermentation. To produce wine, grape juice is allowed to ferment. After the grapes are picked, they are crushed in order that the juice may be collected. In the old days, wine makers simply relied on spontaneous fermentation by yeasts that were on the grape skins, but now many add specially selected cultures of yeast. Also, it is now common practice to maintain the temperature at about 20° C for white wines and 28° C for red wines. Fermentation ceases after most of the sugar has been converted to alcohol. Various methods are used to clarify the wine; that is, to remove any suspended materials. Also many fine wines improve when they are allowed to "age" during barrel or bottle storage.

Fermentation of corn (above), *grapes* (near right), *and barley* (far right) *produces whiskey, wine, and beer respectively. a. © Carolina Biological Supply Company.*

Brewing beer is more complicated than wine production. Usually grains of barley are first *malted;* that is, allowed to germinate for a short time so that amylase enzymes are produced that will break down the starch content of the grain. After the germinated grains have been crushed and mixed with water, the *malt wort* is separated from the spent grains and traditionally boiled with hops (an herb derived from the hop plant) to give flavor to the beer. Now, however, the *hop wort* is seeded with a strain of yeast that converts the sugars in the wort to alcohol and carbon dioxide. At the end of fermentation, the yeast is

---

Although glycolysis occurs outside the mitochondria, the transition reaction, Krebs cycle, and respiratory chain occur within the mitochondria. Evidence suggests that the enzymes of the Krebs cycle are located in the matrix, while the enzymes for the respiratory chain are located along the cristae. These enzymes are arranged in an assembly-line fashion on these membranous shelves. The membrane is divided into functional units, and a very small area of each crista contains one complete set of enzymes. The inner membrane lends itself to this type of arrangement, and so we see that structure aids function.

Comment It is interesting to think about how our bodies provide the reactants for aerobic cellular respiration and how they dispose of the products. The air we breathe contains oxygen, and the food we eat contains glucose. These enter the bloodstream, which carries them to the body's cells, where they diffuse into each and every cell. In mitochondria, glucose products are broken down to carbon dioxide and water as ATP is produced. All three of these leave the mitochondria. The ATP is utilized inside the cell for energy-requiring processes. Carbon dioxide diffuses out of the cell into the bloodstream. The bloodstream takes the carbon dioxide to the lungs, where it is exhaled. The water molecules produced, called metabolic water, only become important if by chance this is the organism's only supply of water. In these cases, it can help prevent dehydration of the organism.

Aerobic cellular respiration requires glycolysis, which takes place in the cytoplasm; the Krebs cycle, which is located in the matrix of the mitochondria; and the respiratory chain, which is located on the cristae of the mitochondria.

Fermentation

Fermentation is an anaerobic process. When oxygen is not available to cells, the respiratory chain soon becomes inoperative because oxygen is not present to accept hydrogen atoms. In this case, most cells have a safety valve so that some ATP can still be produced. The glycolytic pathway will run as long as it is supplied with "free" NAD; that is, NAD that can pick up hydrogen atoms. Normally, $NADH_2$ passes hydrogen to the respiratory chain and thereby becomes "free" of hydrogen atoms. However, if the chain is not working due to lack of oxygen, $NADH_2$ passes its hydrogen atoms to PYR as shown in the following reaction:

The Krebs cycle and respiratory chain do not function as part of fermentation, but when oxygen is available again, lactic acid can be converted back to pyruvic acid and metabolism can proceed as usual.

separated from the beer, which is then allowed to mature for an appropriate period. After filtration and pasteurization, the beer is packaged.

The production of whiskey (from grains), brandy (from grapes), and rum (from molasses) differs from wine and beer production chiefly in that the alcohol is removed from the fermented substance by distillation. Most often in the United States, corn or rye is used in the production of whiskey. These grains are ground up and mashed to release their starch content. Then amylase enzymes are added to convert the starch to fermentable sugars. Now yeast is added so that fermentation can occur. Following fermentation, the alcohol is concentrated by distillation. A warm temperature causes the alcohol to become gaseous and rise in a column where it condenses to a liquid before entering a collecting vessel. The alcohol content of the collecting vessel is much higher following this distillation process. The distillate is usually stored, quite often in an oak barrel, to improve the aroma and taste of the final product.

Fermentation is an impractical process for two reasons. First, it produces only two ATP per glucose molecule, and second, it results in lactic acid buildup. Lactic acid is toxic to cells and causes muscles to cramp and fatigue. If fermentation continues for any length of time, death will follow.

It is of interest to know that fermentation takes its name from yeast fermentation. Yeast fermentation produces alcohol and carbon dioxide (instead of lactic acid). When yeast is used to leaven bread, it is the carbon dioxide that produces the desired effect. When yeast is used to produce alcoholic beverages, a process discussed in the reading above, it is the alcohol that humans make use of.

Fermentation is an anaerobic process, a process that does not require oxygen, but produces very little ATP per glucose molecule and results in lactic acid buildup.

SUMMARY

The human cell is surrounded by a cell membrane, which regulates the entrance and exit of molecules. Some molecules, such as water and gases, diffuse through the membrane. The direction in which water diffuses is dependent on the tonicity of the cell.

Table 2.2 lists the animal cell organelles we have studied in the chapter. The nucleus is a large organelle of primary importance because it controls the rest of the cell. Within the nucleus lies the chromatin, which condenses to become chromosomes during cell division.

Proteins are made at the rough endoplasmic reticulum before being packaged at the Golgi apparatus. Golgi-derived lysosomes fuse with incoming vacuoles to digest any material enclosed within, and lysosomes also carry out autodigestion of old parts of cells.

Mitochondria are the powerhouses of the cell. During the process of cellular respiration mitochondria convert carbohydrate energy to ATP energy.

Microfilaments and microtubules are found within a cytoskeleton that maintains the shape and permits movement of cell parts. Centrioles are associated with the spindle apparatus during cell division, and they also produce basal bodies that give rise to cilia and flagella.

Cellular metabolism uses pathways in which there are a series of reactions that proceed in an orderly step-by-step manner. Each of these reactions requires a specific enzyme. Sometimes enzymes require coenzymes, nonprotein portions that participate in the reaction. NAD is a coenzyme.

Aerobic cellular respiration (the breakdown of glucose to carbon dioxide and water) requires three subpathways: glycolysis, the Krebs cycle, and the respiratory chain. If oxygen is not available in cells, the respiratory chain is inoperative and fermentation (an anerobic process) occurs. Fermentation makes use of glycolysis only, plus one more reaction in which PYR is reduced to lactic acid.

OBJECTIVE QUESTIONS

For questions 1–5, match the organelles in the key to their functions.

Key: a. mitochondria
    b. nucleus
    c. Golgi apparatus
    d. rough ER
    e. centrioles

1. packaging and secretion _____
2. cell division _____
3. powerhouses of the cell _____
4. protein synthesis _____
5. control center for cell _____

6. Microfilaments and microtubules are a part of the _____ , the framework of the cell that provides its shape and regulates movement of organelles.
7. Water will enter a cell when it is placed in a _____ solution.
8. Substrates react at the _____ _____ , located on the surface of their enzyme.
9. During aerobic cellular respiration, most of the ATP molecules are produced at the _____

_____ , a series of carriers located on the _____ of mitochondria.
10. Fermentation produces only _____ ATP compared to the maximum possible number of _____ ATP produced by aerobic cellular respiration.

Answers

STUDY QUESTIONS

1. Describe the structure and biochemical makeup of membrane. (p. 33)
2. What are the three mechanisms by which substances enter and exit cells? Define isotonic, hypertonic, and hypotonic solutions. (p. 34)
3. Describe the nucleus and its contents, including the terms DNA and RNA in your description. (p. 36)
4. Describe the structure and function of endoplasmic reticulum. Include the terms rough and smooth ER and ribosomes in your description. (p. 36)
5. Describe the structure and function of the Golgi apparatus. Mention vacuoles and lysosomes in your description. (p. 37)
6. Describe the structure of mitochondria, and relate this structure to the pathways of cellular metabolism. (pp. 38, 44)
7. Describe the composition of the cytoskeleton. (p. 39)
8. Describe the structure and function of centrioles, cilia, and flagella. (p. 40)
9. Discuss and draw a diagram for a metabolic pathway. Discuss and give a reaction to describe the lock-and-key theory of enzymatic action. Define coenzyme. (pp. 40–41)
10. Name and describe the events within the three subpathways that make up cellular respiration. Why is fermentation wasteful and potentially harmful to the human body? (pp. 43, 45)

THOUGHT QUESTIONS

1. What is the advantage of organelles in complex cells?
2. Present evidence that the cell is dynamic rather than static as it appears to be in drawings and micrographs.
3. Show that a human being is indeed multicellular by discussing the relation of homeostasis to cells, to growth and repair of the body, and to human reproduction.

KEY TERMS

active site (ak′tiv sīt) the region on the surface of an enzyme where the substrate binds and where the reaction occurs. *40*

cell (sel) the structural and functional unit of an organism; the smallest structure capable of performing all the functions necessary for life. *31*

centriole (sen′tre-ōl) a short, cylindrical organelle in animal cells that contains microtubules in a 9 + 0 pattern and is associated with the formation of the spindle apparatus during cell division. *40*

chromosomes (kro′mo-sōmz) rod-shaped bodies in the nucleus, particularly during cell division, that contain the hereditary units or genes. *36*

coenzyme (ko-en′zīm) a nonprotein molecule that aids the action of an enzyme, to which it is loosely bound. *42*

cytoplasm (si′to-plazm″) the ground substance of cells located between the nucleus and the cell membrane. *32*

cytoskeleton (si″to-skel′ĕ-ton) filamentous protein structures found throughout the cytoplasm that help maintain the shape of the cell. *39*

endoplasmic reticulum (en-do-plaz′mic rĕ-tik′u-lum) a complex system of tubules, vesicles, and sacs in cells; sometimes having attached ribosomes. *36*

fermentation (fer″men-ta′shun) anaerobic breakdown of carbohydrates that results in organic end products such as alcohol and lactic acid. *44*

glycolysis (gli-kol′ĭ-sis) the metabolic pathway that converts sugars to simpler compounds and ends with pyruvate. *42*

Golgi apparatus (gol'ge ap''ah-ra'tus) an organelle that consists of concentrically folded membranes and functions in the packaging and secretion of cellular products. *37*

Krebs cycle (krebz si'kl) a series of reactions found within the matrix of mitochondria that give off carbon dioxide; also called the citric acid cycle because the reactions begin and end with citric acid. *42*

lysosome (li'so-sōm) an organelle in which digestion takes place due to the action of hydrolytic enzymes. *37*

metabolism (mĕ-tab'o-lizm) all of the chemical changes that occur within a cell. *40*

microfilament (mi''kro-fil'ah-ment) an extremely thin fiber found within the cytoplasm that is involved in the maintenance of cell shape and movement of cell contents. *39*

microtubule (mi''kro-tu'būl) an organelle composed of thirteen rows of globular proteins; found in multiple units in several other organelles, such as the centriole, cilia, and flagella. *39*

mitochondrion (mi''to-kon'dre-on) an organelle in which aerobic respiration produces the energy molecùle ATP. *38*

nucleolus (nu-kle'o-lus) an organelle found inside the nucleus; composed largely of RNA for ribosome formation. *36*

nucleus (nu'kle-us) a large organelle containing the chromosomes and acting as a control center for the cell. *36*

organelles (or''gan-elz') specialized structures within cells, such as the nucleus, mitochondria, and endoplasmic reticulum. *32*

respiratory chain (re-spi'rah-to''re chān) a series of carriers within the inner mitochondrial membrane that pass electrons one to the other from a higher energy level to a lower energy level; the energy released is used to build ATP; also called the electron transport system; the cytochrome system. *42*

ribosomes (ri'bo-somz) minute particles, found attached to endoplasmic reticulum or loose in the cytoplasm, that are the site of protein synthesis. *36*

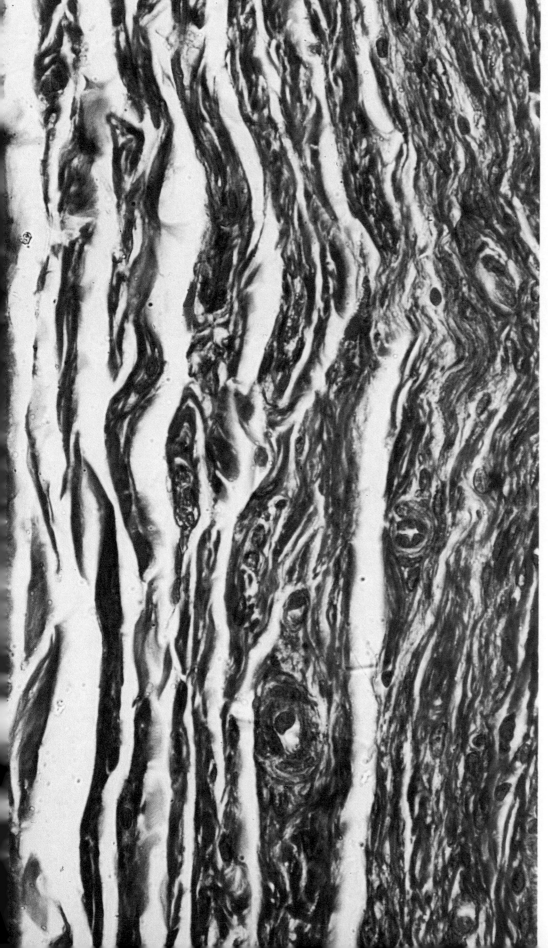

CHAPTER THREE

BODY ORGANIZATION

CHAPTER CONCEPTS

1 The human body has increasingly complex levels of organization from cells to tissues to organ systems.

2 Human tissues can be categorized into four major types: epithelial, connective, muscle, and nervous tissues.

3 Each organ usually contains several types of tissues. For example, although skin is primarily composed of epithelial and connective tissue, it also contains muscle and nerve fibers.

4 All of the organ systems contribute to maintaining a relatively constant internal environment.

Light micrograph of an Azan stain of connective tissue of the urinary bladder.

FIGURE 3.1 Which of
these is the human embryo?
It's hard to tell because all
vertebrates (animals with
backbones) resemble each
other during early
development. Only later
does the human form (a)
become especially evident.
The other photo (b) is a pig
embryo. (a. © Carolina
Biological Supply
Company.)

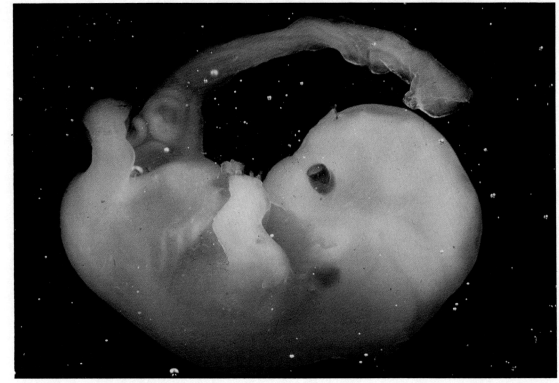

a.

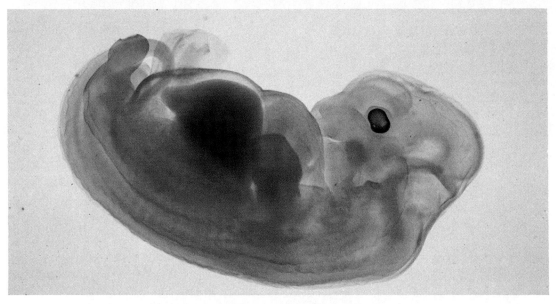

b.

INTRODUCTION

A human being begins life when the sperm fertilizes the egg. This single cell divides to produce many cells that become specialized and organized into tissues. At first, a human embryo resembles those of closely related species (fig. 3.1), but by the end of the second month of development human organization has been achieved and the embryo, soon to be called the fetus, is recognizably human.

The human body has structural and functional levels of organization. Cells of the same type are joined together to form a **tissue.** Different tissues are found in an organ, and various types of organs are arranged into an organ system. Finally, the organ systems make up the organism (fig. 3.2).

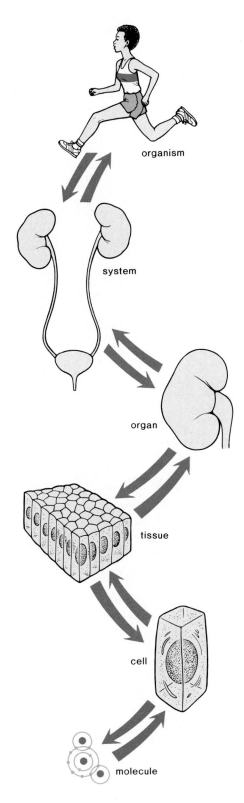

FIGURE 3.2 *Levels of organization in the human body. Cells are composed of molecules; tissues are made up of cells; organs are composed of tissues, and the organism contains organ systems.*

organism

system

organ

tissue

cell

molecule

HUMAN ISSUE

If human organization is not readily apparent for some time after fertilization, when does human life begin? The answer to this question affects the right to an abortion, fetal research, and various legal considerations. Some would argue that as long as the potential for human life is present (even in the single cell) then human life has begun. Others argue that human life doesn't really begin until the fetus can exist on its own outside the womb. In your view, is the answer to the question of when life begins a matter of scientific fact or a matter of opinion?

TISSUES

The tissues of the human body can be categorized into four major types: *epithelial tissue,* which covers body surfaces and lines body cavities; *connective tissue,* which binds and supports body parts; *muscular tissue,* which causes parts to move; and *nervous tissue,* which responds to stimuli and transmits impulses from one body part to another (fig. 3.3).

Epithelial Tissue

Epithelial tissue forms a continuous layer, or sheet, over the entire body surface and most of the body's inner cavities. On the external surface, it protects the body from drying out, injury, and bacterial invasion. On internal surfaces, this tissue may be specialized for other functions in addition to protection; for example, it secretes mucus along the digestive tract; it sweeps up impurities from the lungs by means of hairlike extensions called cilia; and it efficiently absorbs molecules from kidney tubules because of fine cellular extensions called microvilli.

There are three types of epithelial tissue. **Squamous epithelium** (fig. 3.4a) is composed of flat cells and is found lining the lungs and blood vessels. **Cuboidal epithelium** (fig. 3.4b) contains cube-shaped cells and is found lining the kidney tubules. In **columnar epithelium** (fig. 3.4c), the cells resemble pillars or columns, and nuclei are usually located near the bottom of each cell. This epithelium is found lining the digestive tract. Each type of epithelium may have microvilli or cilia as appropriate for its particular function. For example, the oviducts are lined by ciliated cells whose beat propels the egg toward the uterus or womb.

Each of these types of epithelium may be stratified. **Stratified** means to exist as layers piled one over the other. The nose, mouth, anal canal, and vagina are all lined by stratified squamous epithelium. As we shall see, the outer layer of skin is also stratified squamous epithelium, but here the cells have been reinforced by keratin, a protein that strengthens cells. Some epithelial cells are **pseudostratified** and appear to be layered; because each cell touches the baseline, however, true layers do not exist. The lining of the windpipe, or trachea, is called *pseudostratified ciliated columnar epithelium* (fig. 3.5).

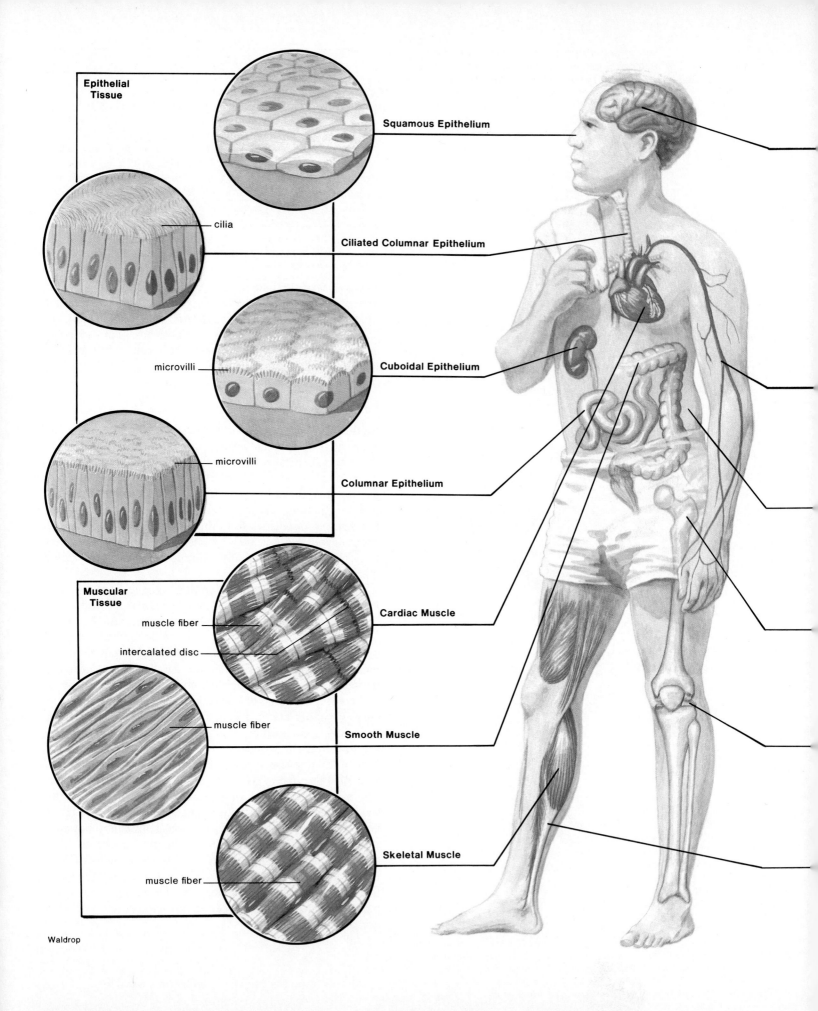

Epithelial Tissue

Squamous Epithelium

Ciliated Columnar Epithelium

cilia

Cuboidal Epithelium

microvilli

Columnar Epithelium

microvilli

Muscular Tissue

Cardiac Muscle

muscle fiber

intercalated disc

Smooth Muscle

muscle fiber

Skeletal Muscle

muscle fiber

Waldrop

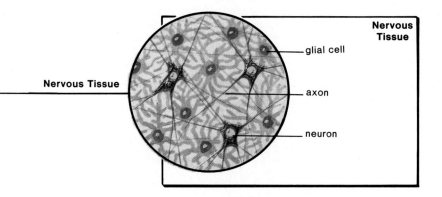

Nervous Tissue

Nervous Tissue

- glial cell
- axon
- neuron

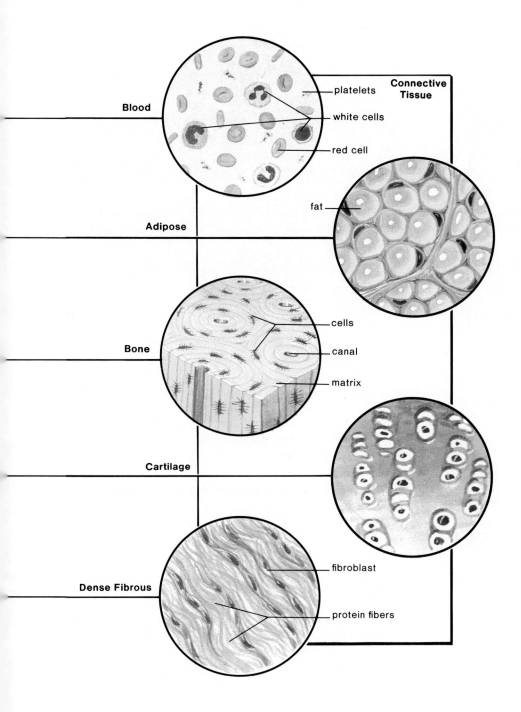

Connective Tissue

Blood
- platelets
- white cells
- red cell

Adipose
- fat

Bone
- cells
- canal
- matrix

Cartilage

Dense Fibrous
- fibroblast
- protein fibers

FIGURE 3.3 *The major tissues in the human body. Observe that nervous tissue contains specialized cells called neurons. Connective tissue includes blood, adipose tissue, bone, cartilage, and dense fibrous tissue. Muscular tissue is of three types: skeletal, smooth, and cardiac. Epithelial tissue includes columnar, cuboidal, ciliated columnar, and squamous epithelium.*

FIGURE 3.4 *Simple epithelial tissue. a. Simple squamous consists of a single layer of thin cells. b. Simple cuboidal is composed of cells that look like cubes. c. Simple columnar cells resemble columns because they are elongated. (The arrow points to a goblet cell.)*

FIGURE 3.5 *a. Pseudostratified ciliated columnar epithelium from the lining of the windpipe. When you cough, material trapped in the mucus secreted by goblet cells is moved upward to the throat, where it can be swallowed. b. Photomicrograph of pseudostratified ciliated columnar epithelium.*

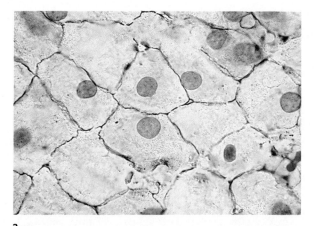

a.

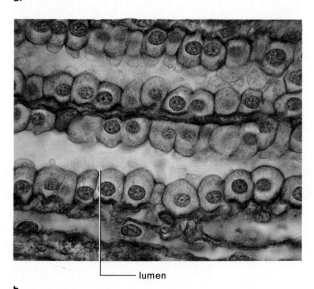

lumen

b.

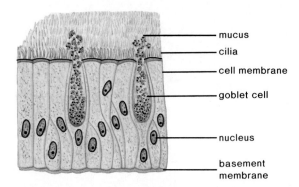

mucus
cilia
cell membrane
goblet cell
nucleus
basement membrane

a.

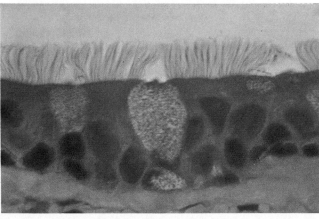

b.

Epithelial cells sometimes secrete a product, in which case they are described as glandular. A gland can be a single cell, as in the case of the mucus-secreting goblet cells found within the columnar epithelium lining the digestive tract (fig. 3.4c), or a gland can contain numerous cells. Glands that secrete their products into ducts are called **exocrine glands,** and those that secrete products directly into the bloodstream are called **endocrine glands.**

Epithelial tissue is classified according to the shape of the cell. There can be a single layer or many layers of cells, and the layer lining a cavity can be ciliated and/or secretory (table 3.1).

Junctions between Cells

Epithelial cells are tightly packed and joined to one another in one of three ways: spot desmosomes, tight junctions, or gap junctions (fig. 3.6). In a *spot desmosome* internal cytoplasmic plaques, firmly attached to the cytoskeleton within each cell, are joined by intercellular filaments. The cells are more closely joined in a *tight junction* because adjacent cell

TABLE 3.1 *EPITHELIAL TISSUES*

TYPE	FUNCTION	LOCATION
Simple squamous	Filtration, diffusion, osmosis	Air sacs of lungs, walls of capillaries, linings of blood vessels
Simple cuboidal	Secretion, absorption	Surface of ovaries; linings of kidney tubules
Simple columnar	Protection, secretion, absorption	Linings of uterus; tubes of the digestive tract
Pseudostratified columnar	Protection, secretion, movement of mucus and sex cells	Linings of respiratory passages; various tubes of the reproductive systems
Stratified squamous	Protection	Outer layer of skin; linings of mouth cavity, vagina, and anal canal

From John W. Hole, Jr., *Human Anatomy and Physiology*, 4th ed. © 1987 Wm. C. Brown Publishers, Dubuque, Iowa. All Rights Reserved. Reprinted by permission.

FIGURE 3.6 *Epithelial cells are held tightly together by (a) spot desmosomes and (b) tight junctions. c. Gap junctions allow materials to pass from cell to cell.*

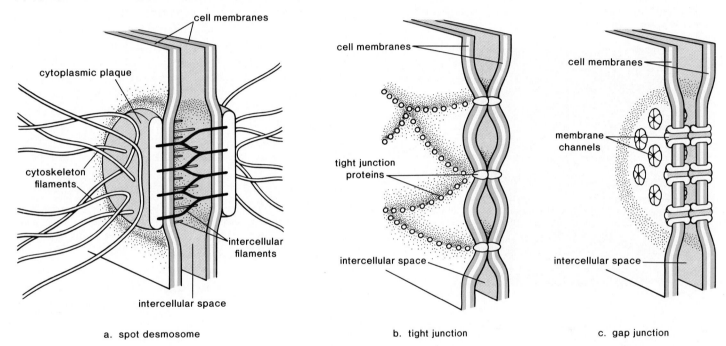

a. spot desmosome   b. tight junction   c. gap junction

membrane proteins actually attach to each other, producing a zipperlike fastening. A *gap junction* is formed when two identical cell membrane channels join. This does lend strength, but it also allows substances to pass between the two cells.

As mentioned earlier, some epithelial tissues have many layers. In any case, the bottom layer is often joined to underlying connective tissue by a so-called *basement membrane*. We now know that the basement membrane is glycoprotein reinforced by fibers supplied by the connective tissue.

Connective Tissue

Connective tissues (table 3.2) bind structures together, provide support and protection, fill spaces, store fat, and form blood cells. As a rule, connective tissue cells are widely separated by a noncellular **matrix.** The matrix may have fibers of two types. White fibers contain collagen, a substance that gives them flexibility and strength. Yellow fibers contain elastin, a substance that is not as strong as collagen but is more elastic.

Loose Connective Tissue

Loose connective tissue binds structures together (fig. 3.7). The cells of this tissue, which are mainly fibroblasts, are located some distance from one another and are separated by a jellylike intercellular material that contains many white collagen and yellow elastic fibers. The collagen fibers occur in bundles and are strong and flexible. The elastic fibers form networks that are highly elastic—when stretched they return

TABLE 3.2 CONNECTIVE TISSUES

TYPE	FUNCTION	LOCATION
Loose connective	Binds organs together	Beneath the skin, beneath most epithelial layers
Adipose	Insulation; storage of fat	Beneath the skin; around the kidneys
Fibrous connective	Binds organs together	Tendons; ligaments
Hyaline cartilage	Support; protection	Ends of bones; nose; rings in walls of respiratory passages
Elastic cartilage	Support; protection	External ear; part of the larynx
Fibrocartilage	Support; protection	Between bony parts of backbone and knee
Bone	Support; protection	Bones of skeleton

FIGURE 3.7 *Loose connective tissue has plenty of space between components. This type of tissue is found surrounding and between the organs.*

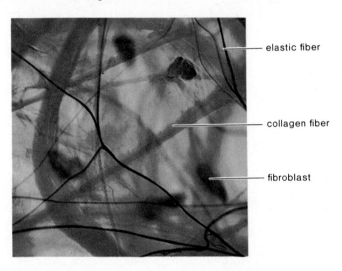

elastic fiber

collagen fiber

fibroblast

FIGURE 3.8 *Adipose tissue cells look like white "ghosts" because they are filled with fat. The nucleus of one cell is indicated by the arrow.*

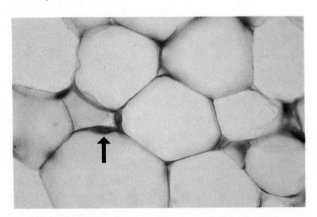

to their original length. As discussed previously, loose connective tissue commonly lies beneath epithelium. In certain instances, epithelium and its underlying connective tissue form the body membranes (p. 61). In addition, adipose tissue (fig. 3.8) is a type of loose connective tissue in which the fibroblasts enlarge and store fat and the intercellular matrix is reduced.

Fibrous Connective Tissue
Fibrous connective tissue contains large numbers of collagenous fibers that are closely packed together. This type of tissue has more specific functions than does loose connective tissue. For example, fibrous connective tissue is found in **tendons,** which connect muscles to bones, and **ligaments,** which connect bones to other bones at joints. Tendons and ligaments take a long time to heal following an injury because their blood supply is relatively poor.

Loose and fibrous connective tissues, which bind body parts together, differ according to the type and abundance of fibers in the matrix.

Cartilage
In **cartilage** the cells lie in small chambers called **lacunae,** separated by a matrix that is solid yet flexible. Unfortunately, because this tissue lacks a direct blood supply, it heals very slowly. There are three types of cartilage, depending on the type of fiber in the matrix.

Hyaline cartilage (fig. 3.9), the most common type, contains only very fine collagenous fibers. The matrix has a clear, milk-glass appearance. This type of cartilage is found in the nose, at the ends of the long bones and ribs, and in the supporting rings of the windpipe. The fetal skeleton is also made of this type of cartilage. Later, the cartilaginous skeleton is replaced by bone.

Elastic cartilage has more elastic fibers than hyaline cartilage. For this reason, it is more flexible and is found, for example, in the framework of the outer ear.

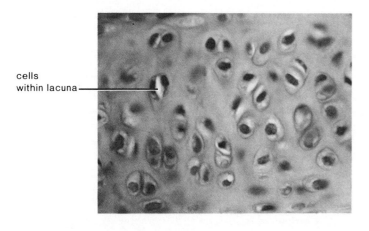

cells within lacuna

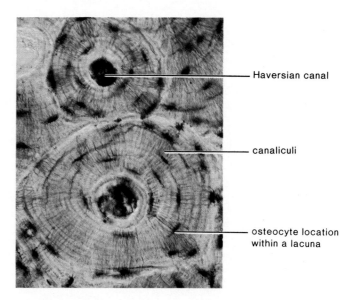

Haversian canal

canaliculi

osteocyte location within a lacuna

Fibrocartilage has a matrix containing strong collagenous fibers. It is found in structures that withstand tension and pressure, such as the pads between the vertebrae in the backbone and the wedges in the knee joint.

HUMAN ISSUE

Human beings put a great deal of emphasis on safety—safety crossing the street, in the workplace, in a car or plane, for example. Yet, humans will play football, take to the slopes, and do all manner of recreational activities that might injure the body. Every year thousands of operations are done to remove damaged cartilage from the knee. Are such injuries justifiable?

Bone

Bone (fig. 3.10) is the most rigid of the connective tissues. It consists of an extremely hard matrix of calcium salts deposited around protein fibers. The minerals give it rigidity, and the protein fibers provide elasticity and strength, much as steel rods do in reinforced concrete.

The outer portion of a long bone contains compact bone. In **compact bone,** bone cells (osteocytes) are located in lacunae that are arranged in concentric circles around tiny tubes called Haversian canals. There are nerve fibers and blood vessels in these canals. The latter bring the nutrients that allow bone to renew itself. The nutrients can reach all of the cells because there are minute canals (canaliculi) containing thin processes of the osteocytes that connect them with one another and with the Haversian canals.

The inner portion of a long bone contains spongy bone, which has an entirely different structure. Spongy bone contains numerous bony bars and plates separated by irregular spaces. Although lighter than compact bone, **spongy bone** is still designed for strength. Just as braces are used for support in buildings, the solid portions of spongy bone follow lines of stress.

Cartilage and bone are support tissues. Cartilage is more flexible than bone because the matrix is rich in protein, not calcium salts like that of bone.

Blood

Blood (fig. 3.11) is a connective tissue in which the cells are separated by a liquid called **plasma,** the contents of which are listed in table 3.3. Blood cells are of two types: **red** (erythrocytes), which carry oxygen, and **white** (leukocytes), which aid in fighting infection. Also present are **platelets,** which are important to the initiation of blood clotting. Platelets are not complete cells; rather, they are fragments of giant cells found in the bone marrow.

Blood is a connective tissue in which the matrix is plasma.

Muscular Tissue

The cells that make up **muscular tissue** contain **actin** and **myosin** microfilaments. The interaction of these microfilaments accounts for the movements we associate with animals. There are three types of vertebrate muscles: *skeletal, smooth,* and *cardiac* (table 3.4).

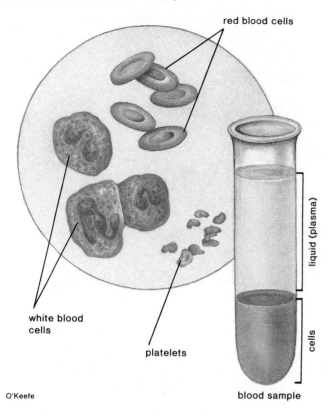

FIGURE 3.11 *Blood is classified as connective tissue. Plasma, the liquid portion of blood, contains several types of cells (red blood cells, white blood cells, and platelets).*

red blood cells

liquid (plasma)

cells

white blood cells

platelets

O'Keefe

blood sample

Skeletal muscle is attached to the bones of the skeleton and functions to move body parts. It is under our *voluntary control* and has the fastest contraction of all the muscle types. The cylindrical cells of this muscle have characteristic light and dark bands perpendicular to the length of the cell or fiber. These bands give the muscle a **striated** appearance (fig. 3.12a).

Smooth muscle is so named because it lacks striations. The spindle-shaped cells that make up smooth muscle are not under voluntary control and are said to be *involuntary*. Smooth muscle, which is found in the viscera (intestine, stomach, and so on) and blood vessels, contracts more slowly than skeletal muscle but can remain contracted for a longer time. The cells tend to form layers in which the thick middle portion of one cell is opposite the thin ends of adjacent cells. Consequently, the nuclei form an irregular pattern in the tissue (fig. 3.3).

Cardiac muscle seems to combine features of both smooth and skeletal muscle. It has *striations* like those of skeletal muscle, but the contraction of the heart is involuntary for the most part. Heart muscle cells also differ from skeletal muscle cells in that they are branched and seemingly fused, one with the other, so that the heart appears to be composed of one large, interconnecting mass of muscle cells (fig. 3.12b). Actually, however, cardiac muscle cells are separate and individual.

All muscle tissue contains actin and myosin microfilaments; these form a striated pattern in skeletal and cardiac muscle but not in smooth muscle.

Nervous Tissue

The brain and nerve cord (also called the spinal cord) contain conducting cells called neurons. A **neuron** (fig. 3.13) is a specialized cell that has three parts: (1) *dendrites* that conduct impulses (send a message) to the cell body; (2) the *cell body* that contains most of the cytoplasm and the nucleus of the neuron; and (3) the *axon* that conducts impulses away from the cell body.

When axons and dendrites are long, they are called *nerve fibers*. Outside the brain and spinal cord, nerve fibers are bound together by connective tissue to form **nerves.** Nerves

---

TABLE 3.3 BLOOD PLASMA

Water	92% of plasma
Inorganic ions (salts)	Na^+, Ca^{++}, K^+, Mg^{++}; Cl^-, HCO_3^-, HPO_4^-, SO_4^{--}
Gases	O_2 and CO_2
Plasma proteins	Albumin, globulin, fibrinogen
Organic nutrients	Glucose, fats, phospholipids, amino acids, etc.
Nitrogenous waste products	Urea, ammonia, uric acid
Regulatory substances	Hormones, enzymes

TABLE 3.4 MUSCULAR TISSUES

TYPE	FIBER APPEARANCE	LOCATIONS	CONTROL
Skeletal	Striated	Attached to skeleton	Voluntary
Smooth	Spindle shaped	Internal organs	Involuntary
Cardiac	Striated and branched	Heart	Involuntary

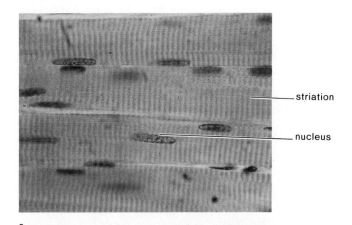

striation

nucleus

a.

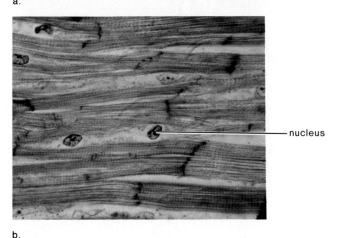

nucleus

b.

FIGURE 3.13 *Conduction of the nerve impulse is dependent on neurons, each of which has the three parts indicated. A dendrite takes nerve impulses to the cell body, and an axon takes them away from the cell body.*

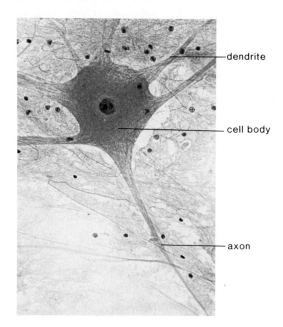

dendrite

cell body

axon

conduct impulses from sense organs to the spinal cord and brain, where the phenomenon called sensation occurs. They also conduct nerve impulses away from the spinal cord and brain to the muscles, causing them to contract.

In addition to neurons, nervous tissue contains **glial cells.** These cells maintain the tissue by giving support and protection to neurons. They also provide nutrients to neurons and help keep the tissue free of debris.

Nerve cells, called neurons, have fibers (processes) called axons and dendrites. Outside the brain and cord these fibers are found in nerves.

ORGANS AND ORGAN SYSTEMS

Some organs in the body are composed largely of one type of cell. For example, muscles are made up of muscle cells, and glands are most often made up of epithelial cells. Many organs, however, are a composite of different types of tissue. We will consider skin as an example.

Skin

Skin (fig. 3.14) covers the body, protecting it from loss of water and deterring invasion by microorganisms. It also helps regulate body temperature and contains sense organs for touch, pressure, temperature, and pain. Skin has an outer *epidermal* and inner *dermal* layer. Beneath the dermis there is a *subcutaneous layer* that binds the skin to underlying organs.

A burn can affect one or all of these layers. A first-degree burn, which affects only the epidermal layer, is characterized by redness, pain, and swelling. As in a sunburn, the skin usually peels in a few days. A second-degree burn, which affects both the epidermis and dermis, usually causes blistering. A third-degree burn is most serious because it leaves underlying parts with no protection at all. When a person is burned over a large portion of the body, it is sometimes difficult to find enough skin to make autografting (graft from skin remaining) possible. Under these circumstances, physicians can now make use of artificial skin, consisting of two layers. The inner layer is a lattice made from shark cartilage and collagen fibers from cowhide. The outer layer is a rubberlike silicone plastic. After the artificial skin is sewn in place, the lattice is slowly digested away and replaced by the patient's own cells. Now the silicone layer can be safely removed.

FIGURE 3.14 *Skin anatomy. Skin is composed of two layers: the epidermis and the dermis. The subcutaneous layer is not strictly a part of skin. Most cells in the epidermal layer are no longer living. Skin cancer brought on by UV radiation from the sun starts in the lower epidermal cells because they are mitotically active. The dermis contains the various structures depicted.*

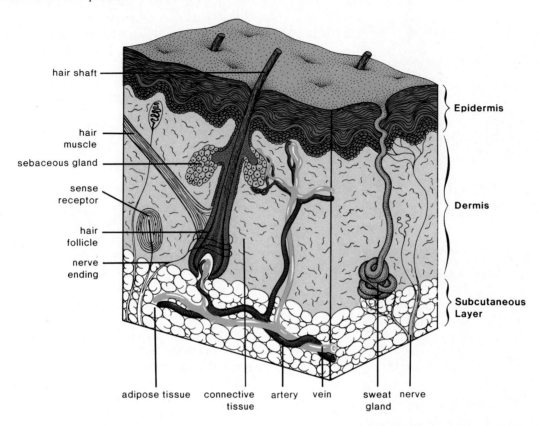

Epidermis

The **epidermis** is made up of stratified squamous epithelial cells, which are continually produced by a germinal layer that lies next to the dermis. Here, constant cell division of *basal cells* produces many new cells that gradually rise to the surface when outer layers of cells are worn off. Specialized cells in this layer, called melanocytes, produce **melanin,** the pigment responsible for skin color. A suntan is caused by the appearance of melanin granules in the outer surface of the skin. Unfortunately, ultraviolet (UV) radiation from the sun can cause dividing skin cells to become cancerous. In recent years a great increase of skin cancers due to sunbathing has prompted physicians to strongly recommend that everyone stay out of the sun—or at least, use sunscreen lotions to protect the skin. The most dangerous type of skin cancer is malignant melanoma, recognized by dark brown or black patches that look like moles. The more common types of skin cancer are basal-cell cancer and squamous-cell cancer, which often appear as rough scaly patches.

Cells gradually become flattened and hardened as they are pushed toward the surface of the skin. Hardening is caused by keratinization, cellular production of a fibrous, waterproof protein called **keratin.** Over much of the body, keratinization is normally minimal and patches of overly keratinized cells, called spots of keratosis, can be early indications of basal cell cancer or squamous cell cancer. In contrast, both the palm of the hand and the sole of the foot normally have a particularly thick outer layer of dead, keratinized cells arranged in spiral and concentric patterns. These patterns, unique to each person, are known as fingerprints and footprints. Hair is present whenever the skin is less keratinized, although nails and hair are themselves composed of tightly packed keratinized cells.

Dermis

The **dermis** is largely dense connective tissue. Sunlight contributes to the aging process because it causes this tissue to lose its elasticity, producing "wrinkles" in the skin.

From above, the epidermis regularly dips down into the dermis, serving as a location for sweat glands and hair follicles with their associated sebaceous glands. The sweat glands help regulate body temperature. They become active and release fluid onto the skin when the body temperature rises. The evaporation of this fluid cools the body. The sebaceous glands lubricate the adjoining hairs and skin. At the time of puberty, they are also responsible for acne.

Blood vessels and nerve endings are also present in the dermis. Some of the latter control a small muscle attached to each hair follicle. When this muscle contracts, the hair stands on end, causing what is commonly known as "goose bumps." Encapsulated (membrane-surrounded) nerve endings are found in the microscopic *sense organs* for touch, pressure, and temperature. Stimulation of free nerve endings produces pain. Regulation of the size of the arteries in skin helps maintain a constant internal body temperature. When the arteries increase in size, more blood is brought to the surface of the skin to cool it off. At the same time, sweat glands become active and the person perspires.

Subcutaneous Layer

The **subcutaneous layer** of skin is a layer of loose connective tissue containing adipose cells. A well-developed subcutaneous layer gives a rounded appearance to the body. Excessive development accompanies obesity.

The skin, composed of the epidermis, dermis, and subcutaneous layers, is an organ that performs various functions, including protection, sense reception, and temperature regulation.

Organ Systems

In this text, we are going to study the organ systems depicted in figure 3.15. Each of these systems has a specific location within the body. The central nervous system is located in a *dorsal* (towards the back) cavity, where the brain is protected by the skull, and the spinal cord, which gives off spinal nerves, is protected by the vertebrae (fig. 3.16). The repeating units of vertebrae and spinal nerves show that humans are segmented animals, meaning that body parts reoccur at regular intervals.

The other internal organs are found within a *ventral* (front) body cavity (fig. 3.16). This cavity is divided by a muscular diaphragm that assists breathing. The heart, a pump for the closed circulatory system, and the lungs are located in the upper *thoracic* (chest) cavity. The major portion of the digestive system and the kidneys are located in the *abdominal* cavity; much of the reproductive system, the urinary bladder, and the terminal portion of the large intestine are located in the *pelvic* cavity.

The musculoskeletal system is another internal organ system. The skeleton provides the surface area for attachment of striated muscles, which are well developed and powerful. The musculoskeletal system makes up most of the body weight and is specialized for locomotion.

Body Membranes

The term *membrane* on the organ level generally refers to a thin lining or covering that is composed of a layer of epithelial tissue overlying a layer of loose connective tissue. For example, mucous membrane lines the organs of the respiratory and digestive systems. This type of membrane, as its name implies, secretes mucus. Serous membrane lines enclosed cavities and covers the organs that lie within these cavities, such as the heart, lungs, and kidneys. This type of membrane secretes a watery lubricating fluid.

There are numerous cavities within the body. These cavities are usually lined with membrane. Some even call the skin cutaneous membrane.

Homeostasis

Homeostasis means that the internal environment remains relatively constant regardless of the conditions in the external environment. In humans, for example:

1. Blood glucose concentration remains at about 0.1%.
2. The pH of the blood is always near 7.4.
3. Blood pressure in the brachial artery remains near 120/80.
4. Blood temperature remains around 37° C (98.6° F).

The ability of the body to keep the internal environment within a certain range allows humans to live in a variety of habitats, such as the arctic regions, deserts, or the tropics.

FIGURE 3.15 *Organ systems of the human body.*

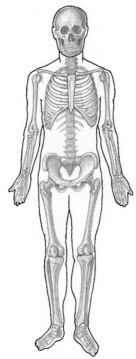

Skeletal system function: Internal
support and flexible framework for body
movement; production of blood cells

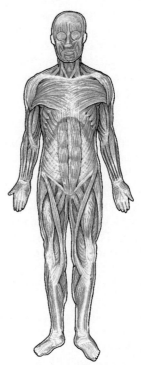

Muscular system function: Body
movement; production of body heat

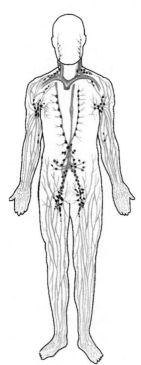

Lymphatic system function: Body
immunity; absorption of fats; drainage of
tissue fluid

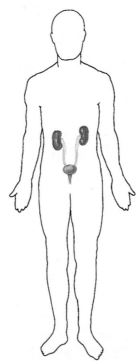

Urinary system function: Filtration of
blood; maintenance of volume and
chemical composition of the blood

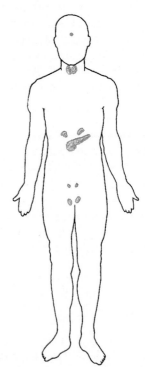

Endocrine system function: Secretion of
hormones for chemical regulation

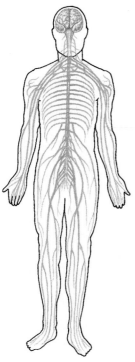

Nervous system function: Regulation of all body activities; learning and memory

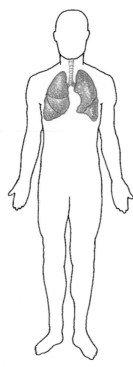

Respiratory system function: Gaseous exchange between external environment and blood

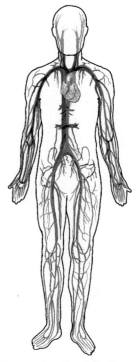

Circulatory system function: Transport of life-sustaining materials to body cells; removal of metabolic wastes from cells

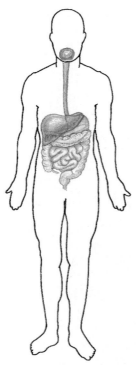

Digestive system function: Breakdown and absorption of food materials

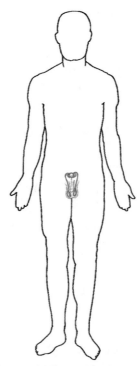

Male reproductive system function: Production of male sex cells (sperm); transfer of sperm to reproductive system of female

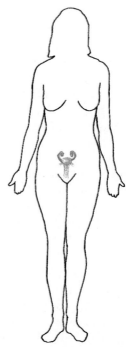

Female reproductive system function: Production of female sex cells (ova); receptacle of sperm from male; site for fertilization of ovum, implantation, and development of embryo and fetus; delivery of fetus

FIGURE 3.16 *Organs are located in cavities of the human body. Notice that we are obviously vertebrates, having a backbone composed of vertebrae.*

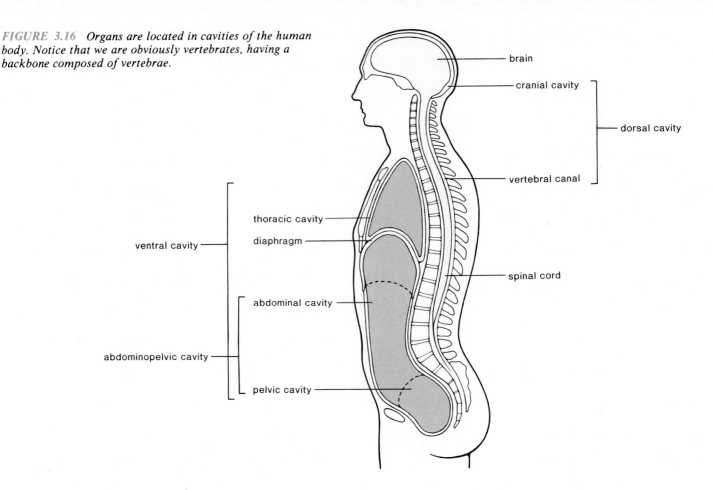

The internal environment includes a tissue fluid that bathes all of the tissues of the body. The composition of tissue fluid must remain constant if cells are to remain alive and healthy. Tissue fluid is created when water (H_2O), oxygen (O_2), and nutrient molecules leave a capillary (the smallest of the blood vessels). Tissue fluid is purified when water, carbon dioxide (CO_2), and other waste molecules enter a capillary from the fluid (fig. 3.17). Tissue fluid remains constant only as long as blood composition remains constant. Although we are accustomed to using the word *environment* to mean the external environment of the body, it is important to realize that it is the internal environment of tissues that is ultimately responsible for our health and well-being.

The internal environment of the body consists of blood and tissue fluid, which bathes the cells.

Most systems of the body contribute to maintenance of a constant internal environment. The digestive system takes in and digests food, providing nutrient molecules that enter the blood and replace the nutrients that are constantly being used up by the body cells. The respiratory system adds oxygen to the blood and removes carbon dioxide. The amount of

FIGURE 3.17 *The internal environment of the body is the blood and tissue fluid. Tissue cells are surrounded by tissue fluid, which is continually refreshed because nutrient molecules constantly exit from and waste molecules continually enter the bloodstream as shown.*
From John W. Hole, Jr., Human Anatomy and Physiology, *4th ed. Copyright © 1987 Wm. C. Brown Publishers, Dubuque, Iowa. All Rights Reserved. Reprinted by permission.*

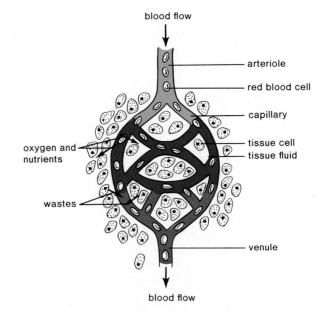

Are tanning machines safe? Most dermatologists feel they are not. Tanning booths are almost certainly capable of causing skin to age, degenerate, or develop cancer.

On the whole, people who use tanning machines probably get away with it, but they are taking a chance. Tanning machines expose their patrons to intense ultraviolet light, which is capable of producing the acute and chronic side effects of exposure to sunlight. Tanning machines use mainly UVA light, which is somewhat less potent in producing biological changes than the slightly shorter UVB waves, but sufficient doses of UVA can be deleterious. Also, extra UVA may enhance the carcinogenic potential of exposure to natural sunlight. People taking medications that induce photosensitivity may have severe reactions to the light from tanning machines. Certain diseases, notably lupus erythematosus, which are exacerbated by sunlight, are also made worse by tanning machines.

The light in these booths is intense so as to achieve in a few minutes the skin reaction that would otherwise take a longer period of baking in natural sunlight. People who tan in booths may do so without any clothing; thus they expose areas of skin that lack protective pigmentation built up from long-term exposure to sunlight. UV light has the potential to injure the retina; people who patronize tanning parlors should always wear protective goggles.

The American Academy of Dermatology has formed a Task Force on Photobiology. Its chairman, Dr. Leonard C. Harber, is also chairman of the Department of Dermatology at Columbia University College of Physicians and Surgeons. He says, "We want health warnings to appear in tanning parlors as they do on cigarette packs."

Excerpted from "That Oh, So Nice, Tan" from the March 1988 issue of *The Harvard Medical School Health Letter* © 1988 President and Fellows of Harvard College.

THAT OH, SO NICE, TAN

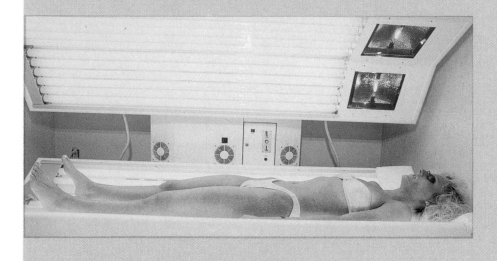

It is possible tanning booths can help cause skin to age, degenerate, or develop cancer.

oxygen taken in and carbon dioxide given off can be increased to meet bodily needs. The chief regulators of composition, however, are the liver and the kidneys. They monitor the chemical composition of plasma (table 3.3) and alter it as required. Immediately after glucose enters the blood, it can be removed by the liver for storage as glycogen. Later glycogen can be broken down to replace the glucose used by the body cells; in this way, the glucose composition of the blood remains constant. (The hormone insulin, secreted by the pancreas, regulates glycogen storage.) The liver also removes toxic chemicals, such as ingested alcohol and drugs, and ammonia given off by the cells. These are converted to waste molecules that can be excreted by the kidneys. The kidneys are also under hormonal control as they excrete wastes and salts, substances that can affect the pH level of the blood.

FIGURE 3.18 *Diagram illustrating the principle of feedback control. A receptor (sense organ) responds to a stimulus, such as high or low temperature, and notifies a regulator center that directs an adaptive response, such as sweating. Once normalcy, such as a normal temperature, is achieved, the receptor is no longer stimulated.*

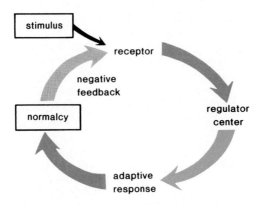

FIGURE 3.19 *Temperature control. When the body temperature rises, the blood vessels dilate and the sweat glands become active. When the body temperature lowers, the blood vessels constrict and shivering may occur. In between these extremes the receptor is not stimulated and thus body temperature fluctuates above and below normal.*

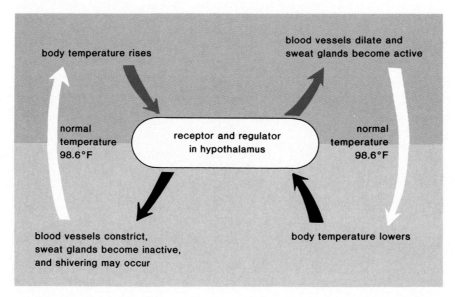

All of the systems of the body contribute to homeostasis; that is, maintenance of the relative constancy of the internal environment.

Although homeostasis is, to a degree, controlled by hormones, it is ultimately controlled by the *nervous system.* The brain contains centers that regulate such factors as temperature and blood pressure. Maintaining proper temperature and blood pressure levels requires sensors that detect unacceptable levels and signal a control center. If a correction is required, the center then directs an adaptive response (fig. 3.18). Once normalcy is obtained, the receptor no longer stimulates the center. This is called control by **negative feedback** because the control center shuts down until it is stimulated to be active once again. This type of homeostatic regulation results in fluctuation between two levels, as illustrated for temperature control in figure 3.19. We may also note that feedback control is a self-regulatory mechanism.

Control of normalcy by negative feedback is a self-regulatory mechanism that results in slight fluctuations within narrow limits.

SUMMARY

Human tissues are categorized into four groups. Epithelial tissues cover the body and line its cavities. Connective tissues often bind body parts together. Contraction of muscle tissues permit movement of the body and its parts. Nerve impulses conducted by neurons within nervous tissue help bring about coordination of body parts.

Tissues are joined together to form organs, each one having a specific function. Organs are grouped into organ systems. In humans, the brain and spinal cord are located in a dorsal cavity and the internal organs are located in a ventral cavity that contains the thoracic, abdominal, and pelvic cavities.

All organ systems contribute to the constancy of the internal environment. The nervous and hormonal systems regulate the other systems. Both of these are controlled by a feedback mechanism that results in fluctuation above and below the desired level.

OBJECTIVE QUESTIONS

1. Most organs contain several different types of _____ .
2. Kidney tubules are lined by cube-shaped cells called _____ epithelium.
3. Pseudostratified ciliated columnar epithelium contains cells that appear to be _____ , have projections called _____ , and are _____ in shape.
4. Both cartilage and blood are classified as _____ tissue.
5. Cardiac muscle is _____ but involuntary.
6. Nerve cells are called _____ .
7. Skin has three layers: epidermis, _____ , and subcutaneous.
8. Outer skin cells are filled with _____ , a waterproof protein that strengthens them.
9. Mucous membrane contains _____ tissue overlying _____ tissue.
10. Homeostasis is maintenance of the relative _____ of the internal environment; that is, the blood and _____ fluid.

Answers

1. tissues 2. cuboidal 3. layered (stratified), cilia, columnar (elongated) 4. connective 5. striated 6. neurons 7. dermis 8. keratin 9. epithelial, loose connective 10. constancy, tissue

STUDY QUESTIONS

1. State in order the levels of organization of the human body. (p. 51)
2. Name the four major groups of tissues. (pp. 52–58)
3. What are the functions of epithelial tissue? Name the different kinds, and give a location for each. (p. 51)
4. What are the functions of connective tissue? Name the different kinds, and give a location for each. (p. 55)
5. What are the functions of muscular tissue? Name the different kinds, and give a location for each. (p. 57)
6. Nervous tissue contains what types of cells? Which organs in the body are made up of nervous tissue? (p. 58)
7. Describe the structure of skin, and state at least two functions of this organ. (p. 59)
8. In general terms, describe the location of the human organ systems. (p. 65)
9. Distinguish between cell membrane and body membrane. (p. 62)
10. What is homeostasis? What is its importance, and how is it achieved in the human body? (p. 62)

THOUGHT QUESTIONS

1. Some texts add the integumentary system to those pictured in figure 3.15. Show that this is proper by indicating which structures should be included in an integumentary system aside from the skin.
2. Classification schemes are arbitrary in nature. Make up a new classification system for the tissues described in this chapter.
3. Some lower animals lack body cavities altogether. Of what use are body cavities to more complex animals?

KEY TERMS

bone (bōn)   connective tissue having a hard matrix of calcium salts deposited around protein fibers.   *57*

cartilage (kar'tĭ-lij)   a connective tissue in which the cells lie within lacunae embedded in a flexible matrix.   *56*

compact bone (kom'pakt bōn)   hard bone consisting of Haversian systems cemented together.   *57*

connective tissue (kŏ-nek'tiv tish'u)   a type of tissue, characterized by cells separated by a matrix, that often contains fibers.   *55*

dermis (der'mis)   the thick skin layer that lies beneath the epidermis.   *61*

epidermis (ep''ĭ-der'mis)   the outer skin layer composed of stratified squamous epithelium.   *60*

epithelial tissue (ep''ĭ-the'le-al tish'u)   a type of tissue that lines cavities and covers the external surface of the body.   *51*

homeostasis (ho''me-o-sta'sis)   the constancy of conditions, particularly the internal environment of birds and mammals: constant temperature, blood pressure, pH, and other body conditions.   *61*

hyaline cartilage (hi'ah-lĭn kar'tĭ-lij)   cartilage composed of very fine collagenous fibers and a matrix of a clear, milk-glass appearance.   *56*

lacuna (lah-ku'nah)   a small pit or hollow cavity, as in bone or cartilage, where a cell or cells are located.   *56*

ligament (lig'ah-ment)   dense connective tissue that joins bone to bone.   *56*

muscular tissue (mus'ky-lar tish'u)   a type of tissue that contains cells capable of contracting: skeletal muscles are attached to skeleton; smooth muscle is found within walls of internal organs; and cardiac muscle comprises the heart.   *57*

negative feedback (neg'ah-tiv fēd'bak)   a mechanism that is activated by a surplus imbalance and acts to correct it by stopping the process that brought about the surplus.   *66*

neuron (nu'ron)   nerve cell that characteristically has three parts: dendrite, cell body, axon.   *58*

pseudostratified (su''-do strat'e-fĭd)   the appearance of layering in some epithelial cells when actually each cell touches a baseline and true layers do not exist.   *51*

spongy bone (spun'je bōn)   porous bone found at the ends of long bones.   *57*

stratified (strat'ĭ-fĭd)   layered as in stratified epithelium, which contains several layers of cells.   *51*

striated (stri'āt-ed)   having bands; cardiac and skeletal muscle are striated with bands of light and dark.   *58*

subcutaneous layer (sub''ku-ta'ne-us)   a tissue layer found in vertebrate skin that lies just beneath the dermis and tends to contain adipose tissue.   *61*

tendon (ten'don)   dense connective tissue that joins muscle to bone.   *56*

FURTHER READINGS FOR PART I

Allen, R. D. February 1987. The microtubule as an intracellular engine. *Scientific American.*

Berns, M. W. 1983. *Cells.* 3d ed. New York: Holt, Rinehart & Winston.

Bretscher, M. S. October 1985. The molecules of the cell membrane. *Scientific American.*

Caplan, A. I. October 1984. Cartilage. *Scientific American.*

Dautry-Varsat, A., and H. F. Lodish. May 1984. How receptors bring proteins and particles into cells. *Scientific American.*

Dustin, P. August 1980. Microtubules. *Scientific American.*

Hayflick, L. January 1980. The cell biology of human aging. *Scientific American.*

Hole, J. W. 1987. *Human anatomy and physiology.* 4th ed. Dubuque, Iowa: Wm. C. Brown Publishers.

Kessel, R. G., and R. H. Kardon. 1979. *Tissues and organs: a text-atlas of scanning electron microscopy.* San Francisco: W. H. Freeman.

Lack, J. A. August 1981. Ribosome. *Scientific American.*

Porter, K. R., and J. B. Tucker. March 1981. The ground substance of the living cell. *Scientific American.*

Sloboda, R. D. 1980. The role of microtubules in cell structure and cell division. *American Scientist* 68(3):290.

Swanson, C. P., and P. L. Webster. 1985. *The cell.* 5th ed. Englewood Cliffs, New Jersey: Prentice-Hall.

Weber, K., and M. Osborn. October 1985. The molecules of the cell matrix. *Scientific American.*

Weinberg, R. A. October 1985. The molecules of life. *Scientific American.*

PART TWO

HUMAN PROCESSING AND TRANSPORTING

All of the systems of the body help maintain homeostasis, the relative constancy of the internal environment. The internal environment is the fluid that surrounds the cells of the tissues. The heart pumps the blood and sends it to the tissues, where exchange occurs with tissue fluid. At the same time, blood is continually being renewed by the digestive, respiratory, and excretory systems. Nutrients enter the blood at the small intestine, external gas exchange occurs in the lungs, and waste products are excreted at the kidneys. The immune system prevents microorganisms from taking over the body and interfering with its proper functioning.

CHAPTER FOUR

DIGESTION AND NUTRITION

CHAPTER CONCEPTS

1 Small molecules, such as amino acids, glucose, and fatty acids, that can cross cell membranes are the products of digestion that nourish the body.

2 Regions of the digestive tract are specialized to carry on specific functions; for example, the mouth is specialized to receive and chew food, and the small intestine is specialized to absorb the products of digestion.

3 Digestion requires enzymes, which are specific to their particular functions.

4 Proper nutrition requires that the energy needs of the body be met and that the diet be balanced so that all vitamins and essential amino acids are included.

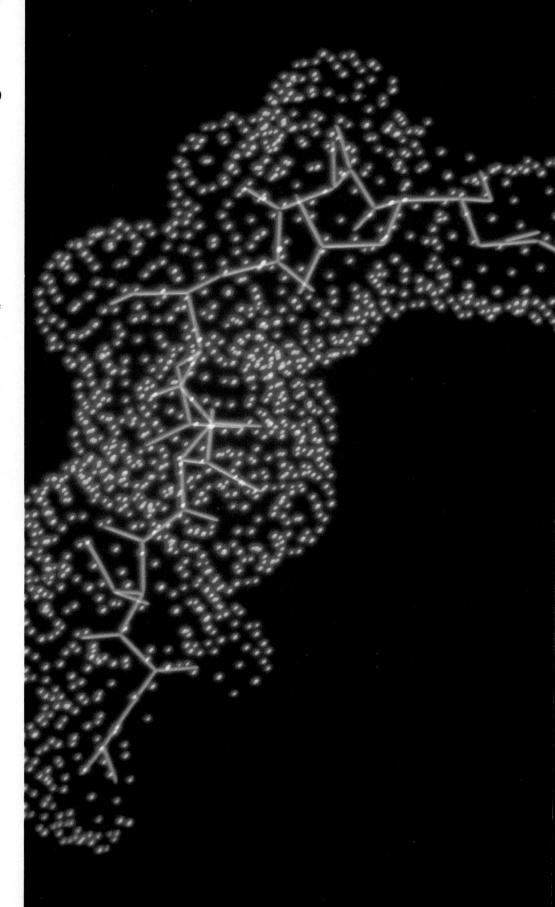

Molecular structure of Carboxy Proteinase Enzyme.

FIGURE 4.1 *Life is dependent on a flow of energy from the prey to the predator. The herbivore feeds on plant material, and the carnivore feeds on the herbivore (or other carnivores).*

INTRODUCTION

In the world of living things there are the *herbivores* that feed on plant material, the *carnivores* that feed on the herbivores (fig. 4.1), and the *omnivores* that feed on both plants and animals. Human beings are omnivores as any anatomist would be able to decide by an examination of our digestive tracts. For example, human dentition is nonspecific—humans don't have large flattened molars like that of a herbivore nor the sharp pointed canine teeth of the carnivore. The digestive system is elaborate but not as elaborate as that of a herbivore, which needs special chambers for bacterial action to break down the cellulose of plant cell walls. On the other hand, the human digestive system is more extensive than that of a carnivore, whose only type of food does not require much digestive action. Altogether then, you would be able to tell that humans are generalists when it comes to the food we eat.

TABLE 4.1 *PATH OF FOOD*

ORGAN	SPECIAL FEATURE(S)	FUNCTION(S)
Mouth	Teeth	Chewing of food; digestion of starch to maltose
Esophagus		Passageway
Stomach	Gastric glands	Digestion of protein to peptides
Small intestine	Intestinal glands Villi	Digestion of all foods and absorption of unit molecules
Large intestine		Absorption of water
Anus		Defecation

Digestion takes place within a tube, often called the gut, which begins with the mouth and ends with the anus (table 4.1 and fig. 4.2). Digestion of food in humans is an extracellular process. Digestive enzymes are secreted into the gut by glands that reside in the lining or lie nearby. Food is never found within these *accessory glands,* only within the gut itself.

FIGURE 4.2 *The human digestive system. In the box, the liver is drawn smaller than normal size and moved back to show the gallbladder and to expose the stomach and duodenum.*

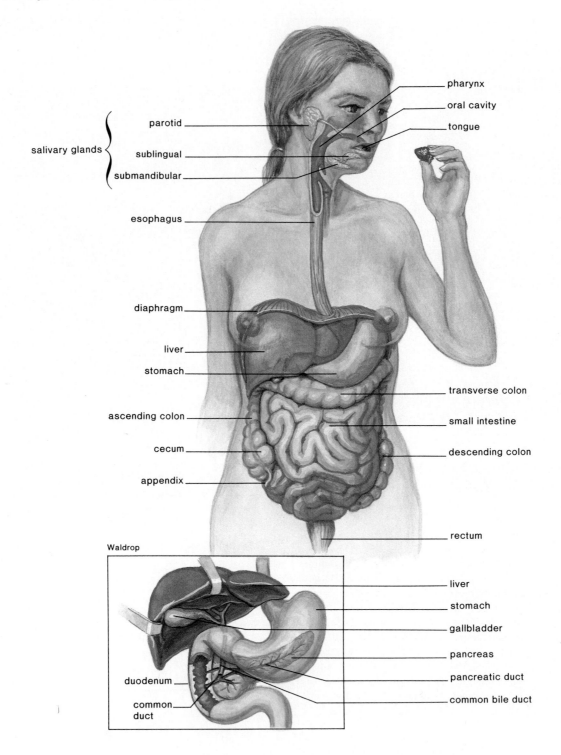

Digestion reduces food to small, soluble molecules that can cross cell membranes and be absorbed by the gut lining. Too often we are inclined to think that since we eat meat (protein), potatoes (carbohydrate), and butter (fat), these are the substances that nourish our bodies. Instead, it is the amino acids from the protein, the sugars from the carbohydrate, and the glycerol and fatty acids from the fat that actually enter the blood and are transported about the body to nourish our cells. Any component of food, such as cellulose, that is incapable of being digested to small molecules leaves the gut as waste material.

Digestion of food requires a cooperative effort between different parts of the body. We shall see that both the production of hormones and the nervous system bring about the coordination of events needed to achieve cooperation of body parts.

DIGESTIVE SYSTEM

Mouth

The *mouth* receives the food in humans. Most people enjoy eating because of the combined sensations of smelling and tasting food. The olfactory receptors, located in the nose, are responsible for smelling; tasting is, of course, a function of the taste buds, located on the tongue. (See chapter 12 for a description of these sense organs.)

Normally, adults have thirty-two teeth (fig. 4.3), which chew the food into pieces convenient to swallow. One-half of each jaw has teeth of four different types: two chisel-shaped incisors for biting; one pointed canine for tearing; two fairly flat premolars for grinding; and three molars, more flattened, for crushing. The last molar, or wisdom tooth, may fail to erupt, or if it does, it is sometimes crooked and useless.

Each *tooth* (fig. 4.4) has a layer of enamel, an extremely hard outer covering of calcium compounds; dentin, a thick layer of bonelike material; and an inner pulp, which contains the nerves and blood vessels. Tooth decay, or *caries,* commonly called cavities, occurs when the bacteria within the mouth metabolize sugar and give off acids that corrode the teeth. Two measures may prevent tooth decay: eating a limited amount of sweets and daily brushing and flossing of teeth. It has also been found that fluoride treatments can make the enamel stronger and more resistant to decay. Gum disease is more apt to occur as you age. Inflammation of the gums (gingivitis) may spread to the periodontal membrane (fig. 4.4) that lines the tooth socket. Then the individual has **periodontitis,** characterized by a loss of bone and loosening of the teeth, with the possibility that these teeth may have to be pulled. Stimulation of the gums in a manner advised by dentists has been found helpful in controlling this condition. Laboratory tests are now available to determine what type of bacteria may be causing gum disease so that antibiotic therapy can be initiated if it seems advisable.

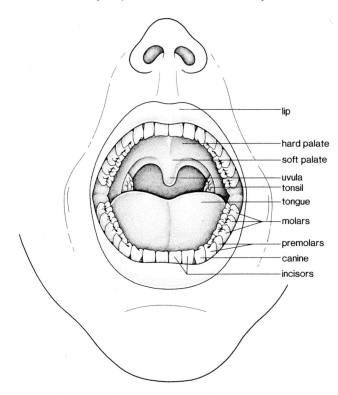

FIGURE 4.3 Diagram of the mouth showing the adult teeth. The sizes and shapes of teeth correlate with their functions.

lip
hard palate
soft palate
uvula
tonsil
tongue
molars
premolars
canine
incisors

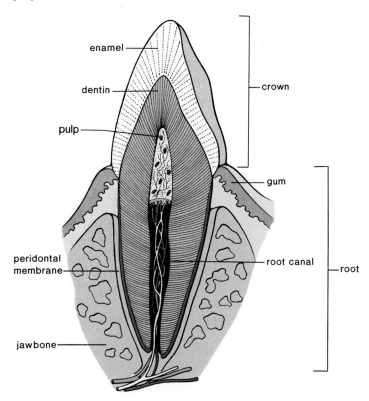

FIGURE 4.4 Longitudinal section of a canine tooth. The tooth pulp contains nerves and blood vessels.

enamel
dentin
pulp
crown
gum
peridontal membrane
root canal
root
jawbone

The roof of the mouth separates the air passages from the mouth cavity. The roof has two parts: an anterior **hard palate** and a posterior **soft palate** (fig. 4.3). The hard palate contains several bones, but the soft palate is merely muscular. The soft palate ends in the *uvula,* a suspended process often mistaken by the lay person for the tonsils, but as figure 4.3 shows, the tonsils lie to the sides of the throat.

There are three pairs of **salivary glands,** which send their juices by way of ducts to the mouth. The parotid glands lie at the sides of the face immediately below and in front of the ears. They become swollen when a person has the mumps, a viral infection most often seen in children. Each parotid gland has a duct that opens on the inner surface of the cheek just at the location of the second upper molar. The sublingual glands lie beneath the tongue, and the submandibular glands lie beneath the lower jaw. The ducts from these glands open into the mouth under the tongue. You can locate all these openings if you use your tongue to feel for small flaps on the inside of your cheek and under your tongue.

Saliva, secreted by the salivary glands, has a neutral pH. It contains mostly water, mucus, and the digestive enzyme **salivary amylase.** This enzyme acts on starch. Like all of the digestive enzymes, salivary amylase is a **hydrolytic enzyme.** This means that its substrate is broken down upon the addition of water:

$$starch + H_2O \xrightarrow{\text{salivary amylase}} maltose$$

In this equation, salivary amylase is written above the arrow to indicate that it is neither a reactant nor a product in the reaction. It merely speeds up the reaction in which its substrate, starch, is digested to many molecules of maltose. Maltose is not one of the small molecules that can be absorbed by the gut lining. Additional digestive action is usually required to convert maltose to glucose. This occurs farther along the digestive tract.

No other digestive process occurs in the mouth. The tongue takes the chewed food and forms it into a mass called a *bolus* in preparation for swallowing.

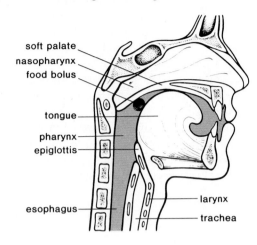

FIGURE 4.5 *When food is swallowed, the soft palate covers the nasopharyngeal openings, and the epiglottis covers the glottis so that the food bolus must pass down the esophagus. Therefore you do not breathe during swallowing.*

soft palate
nasopharynx
food bolus
tongue
pharynx
epiglottis
esophagus
larynx
trachea

In the mouth, food is chewed and acted upon by salivary amylase before it is swallowed.

Pharynx

Swallowing (fig. 4.5) occurs in the **pharynx,** a region between the mouth and the esophagus, which is a long muscular tube that leads to the stomach. Swallowing is a *reflex action,* which means the action is usually performed automatically and does not require conscious thought. During swallowing, food normally enters the esophagus because the air passages are blocked. Unfortunately, we have all had the unpleasant experience of having food "go down the wrong way." The wrong way may be either into the nose or into the windpipe (trachea). If it is the latter, coughing usually forces the food up out of the trachea into the pharynx again. Food usually goes into the esophagus because the openings to the nose, called the *nasopharyngeal openings,* are covered when the soft palate moves back. The opening to the larynx (voice box) at the top of the trachea, called the **glottis,** is covered when the trachea moves up under a flap of tissue, called the **epiglottis.** This is easy to observe in the up-and-down movement of the Adam's apple, a part of the larynx, when a person eats. Breathing does not occur when you swallow. Why not?

The air passage and food passage cross in the pharynx. When you swallow, the air passage is usually blocked off and food must enter the esophagus.

Esophagus

After swallowing occurs, the **esophagus** conducts the bolus through the thoracic cavity. The wall of the esophagus is representative of the gut in general (fig. 4.6). A *mucous membrane* lines the lumen (space within the tube); this is followed

FIGURE 4.6 *The gut wall usually has four layers: a mucous membrane next to the lumen; a submucosa; a smooth muscle layer; and a serous layer next to the abdominal cavity. The serous layer becomes a part of a body membrane within the abdominal cavity.*

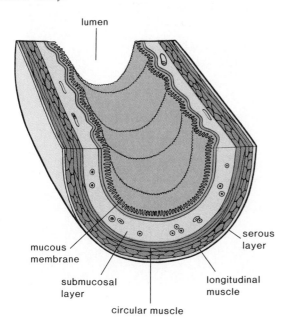

lumen

mucous membrane

submucosal layer

circular muscle

serous layer

longitudinal muscle

FIGURE 4.7 *Peristalsis in the digestive tract. Rhythmic waves of muscle contraction move material along the digestive tract. The three drawings show how a peristaltic wave moves through a single section of gut over time.*

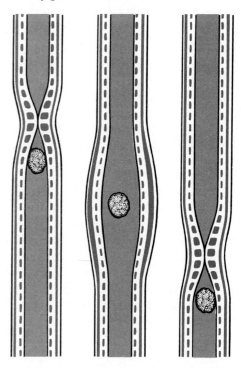

FIGURE 4.8 *Photomicrograph of the mucosa of the stomach. Gastric glands produce gastric juice rich in pepsin, an enzyme that digests protein to peptides.*

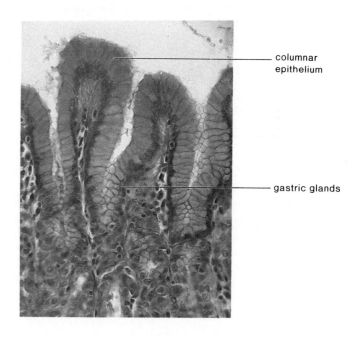

columnar epithelium

gastric glands

by a *submucosal layer* of connective tissue that contains nerves and blood vessels, a *smooth muscle layer* having both a longitudinal and circular muscle, and finally a *serous membrane layer.*

A rhythmical contraction of the esophageal wall, called **peristalsis** (fig. 4.7), pushes the food along. Occasionally peristalsis begins even though there is no food in the esophagus. This produces the sensation of a lump in the throat.

The entrance of the esophagus into the stomach is marked by the presence of a constrictor. When food is swallowed, the constrictor relaxes, allowing the *bolus* to pass into the stomach. Normally, this constrictor prevents foods from moving up out of the stomach, but when vomiting occurs, a reverse peristaltic wave causes the constrictor to relax, and the contents of the stomach are propelled upward through the esophagus.

Stomach

The *stomach* is a thick-walled, J-shaped organ that lies on the left side of the body beneath the diaphragm. It is an enlarged portion of the gut, which can stretch to hold about a half-gallon of liquids and solids. The walls contain three layers of muscle instead of two layers, and contraction of these muscles causes the stomach to churn and mix its contents. Hunger pangs are felt when an empty stomach churns.

The mucosal lining of the stomach (fig. 4.8) contains millions of microscopic digestive glands called **gastric glands** (the term *gastric* always refers to the stomach). The gastric glands produce a gastric juice. Gastric juice contains pepsinogen and hydrochloric acid (HCl). When *pepsinogen* is exposed to hydrochloric acid within the stomach, it becomes the digestive enzyme pepsin. **Pepsin** is a hydrolytic enzyme that prefers an acid pH in order to digest protein to peptides:

$$\text{protein} + H_2O \xrightarrow{\text{pepsin}} \text{peptides}$$

Peptides vary in length but always consist of a number of amino acids joined together. Peptides are too large to be absorbed by the gut lining. However, they are later broken down to amino acids in another part of the digestive tract.

The presence of hydrochloric acid causes the contents of the stomach to have an acid pH of about 2–3. This low pH is beneficial in that it usually kills most bacteria present in food. Normally, the wall is protected by a thick layer of mucus, but if by chance HCl does penetrate this mucus, pepsin starts to digest the stomach lining, and an ulcer results. An ulcer (fig. 4.9) is an open sore in the wall caused by the gradual disintegration of tissues. It is believed that the most frequent cause of an ulcer is oversecretion of gastric juice due to too much nervous stimulation. Persons under stress tend to have a greater incidence of ulcers. However, there is now evidence that a bacterial (*Campylobacter pyloridis*) infection may impair the ability of cells to produce mucus, thus leading to an ulcer.

HUMAN ISSUE

Ulcers, we are told, are more apt to occur in individuals who have a limited ability to cope with stress. Although formerly the medical profession was inclined to consider only physiological symptoms, more and more doctors have also begun to take into account the psychological state of their patients. Is it the duty of the physicians to consider the "whole" patient, or is it sufficient to treat only the obvious symptoms of disease? Some would argue that there are plenty of psychologists and psychiatrists to deal with psychological symptoms, and it is too time consuming for physicians to consider these. Others feel that the practice of medicine is incomplete if all of a patient's symptoms are not considered. What do you think?

Normally, the stomach empties in about two to six hours. By this time, the bolus of food has become a semi-liquid food mass called *acid chyme.* Acid chyme leaves the stomach and enters the small intestine by way of a sphincter. **Sphincters** are muscles that encircle tubes and act as valves; tubes close when sphincters contract, and they open when sphincters relax. In this case the sphincter repeatedly opens and closes, allowing chyme to enter the small intestine in small squirts only. This assures that digestion in the small intestine will proceed at a slow and thorough rate.

Gastric glands produce HCl and a precursor of pepsin, an enzyme that breaks down protein.

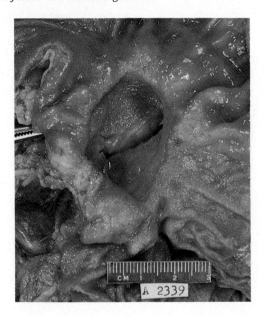

FIGURE 4.9 *Peptic ulcer in the stomach. Normally the digestive tract produces enough mucus to protect itself from digestive juices. An ulcer begins when this protection fails and digestive juices do reach the gut wall.*

Small Intestine

Digestion

The small intestine gets its name from its small diameter (compared to that of the large intestine). But perhaps it should be called the long intestine because it averages about 6.0 m (20 ft.) in length compared to the large intestine, which is about 1.5 m (5 ft.) long. The first 25 cm (10 in.) of the small intestine are called the **duodenum.** Duodenal ulcers sometimes occur because the acid and pepsin within the acid chyme from the stomach may corrode and digest the internal wall in this region.

Liver and Pancreas   Two very important accessory glands, the **liver** and the **pancreas,** send secretions to the duodenum (fig. 4.2). The liver produces **bile,** which is stored in the **gallbladder** and sent by way of the bile duct to the duodenum. Bile looks green because it contains pigments that are products of hemoglobin breakdown. This green color is familiar to anyone who has observed the color changes of a bruise. Within a bruise, the hemoglobin is breaking down into the same type of pigments found in bile. However, bile also contains bile salts, which are emulsifying agents that break up fat into fat droplets:

$$\text{fat} \xrightarrow{\text{bile salts}} \text{fat droplets}$$

Following emulsification (fig. 1.21), the resulting fat droplets are ready for chemical digestion. Almost all fat digestion occurs in the small intestine, as discussed in the following paragraphs.

The study of the control of digestive gland secretion began in the late 1800s. At that time, Ivan Pavlov showed that dogs would salivate at the ringing of a bell because they had learned to associate the sound of the bell with being fed. Pavlov's experiments demonstrated that even the thought of food can bring about the secretion of digestive juices. Certainly if food is present in the mouth, stomach, and small intestine, digestive secretion occurs. This is attributable to a simple reflex occurrence. The presence of food sets off nerve impulses that travel to the brain. Thereafter, the brain stimulates the digestive glands to secrete.

In this century, investigators have discovered that specific control of digestive secretions is achieved by hormones. A *hormone* is a substance that is produced by one set of cells but affects a different set of cells, called the target cells. Hormones are transported by the bloodstream. For example, when a person has eaten a meal particularly rich in protein, the hormone **gastrin,** produced by the lower part of the stomach, enters the bloodstream (see figure) and soon reaches the upper part of the stomach, where it causes the gastric glands to secrete more pepsinogen.

Experimental evidence has also shown that the duodenal wall produces hormones, the most important of which are **secretin** and **CCK** (cholecystokinin). Acid, especially HCl present in the acid chyme, stimulates the release of secretin, while partially digested protein and fat stimulate the release of CCK. These hormones enter the bloodstream (see figure) and signal the pancreas and the gallbladder to send secretions to the duodenum.

Still another hormone has only recently been discovered. **GIP** (gastric inhibitory peptide) produced by the small intestine apparently works in opposition to gastrin because it inhibits gastric acid secretion. This is not surprising because very often the body has hormones that have opposite effects.

CONTROL OF DIGESTIVE GLAND SECRETION

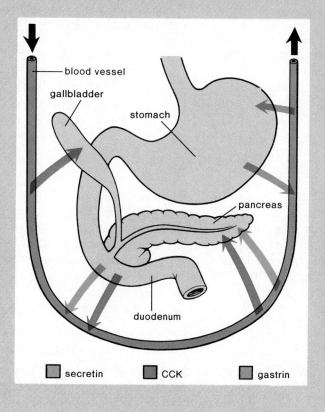

secretin   CCK   gastrin

Hormonal control of digestive gland secretions. Especially after eating a protein-rich meal, gastrin produced by the lower part of the stomach enters the bloodstream and thereafter stimulates the upper part of the stomach to produce more digestive juices. Acid chyme from the stomach causes the duodenum to secrete secretin and CCK. Secretin and CCK stimulate the pancreas, and CCK alone stimulates the gallbladder to release bile.

The pancreas sends *pancreatic juice* into the duodenum by way of the pancreatic duct (fig. 4.2). You may be more familiar with the pancreas as the source of the hormone insulin, but some other pancreatic cells produce a juice that contains digestive enzymes and *sodium bicarbonate* ($NaHCO_3$). The latter makes pancreatic juice basic, which is also called alkaline (pH 8.5). The alkaline pancreatic juice neutralizes the acid of the acid chyme and causes the pH of the small intestine to be basic. The enzymes found in the small intestine prefer a basic pH and only function to their optimum at this pH.

Pancreatic juice contains hydrolytic enzymes that act on every major component of food. There is a pancreatic enzyme called **pancreatic amylase** that digests starch:

$$\text{starch} + H_2O \xrightarrow{\text{pancreatic amylase}} \text{maltose}$$

Trypsin is an example of a pancreatic enzyme that digests protein:

$$\text{protein} + H_2O \xrightarrow{\text{trypsin}} \text{peptides}$$

Trypsin is secreted as trypsinogen and changes to trypsin in the gut.

Lipase digests the fat droplets:

$$\text{fat droplets} + H_2O \xrightarrow{\text{lipase}} \text{glycerol} + \text{fatty acids}$$

The digestion of fat is now complete and the molecules of glycerol and fatty acids are small enough to be absorbed by the lining of the small intestine.

In the small intestine, fat is emulsified by bile salts to fat droplets before being acted upon by pancreatic lipase. Protein is digested by pancreatic trypsin, and starch is digested by pancreatic amylase.

Intestinal Enzymes

The cells that line the wall of the small intestine (fig. 4.10b) produce the intestinal enzymes. The enzymes in this juice complete the digestion of protein and carbohydrate. Instead of being deposited in the lumen, these enzymes remain attached to the microvilli with their active sites exposed.

Peptides, which result from the first step in protein digestion, are digested to amino acids:

$$\text{peptides} + H_2O \xrightarrow{\text{peptidases}} \text{amino acids}$$

Maltose, which results from the first step in starch digestion, is digested to glucose:

$$\text{maltose} + H_2O \xrightarrow{\text{maltase}} \text{glucose}$$

Other disaccharides, each of which has its own enzyme, are digested in the small intestine. The absence of any one of these enzymes may cause illness. For example, although infants produce lactase, many adults, including as many as 75% of American blacks, cannot digest lactose, the sugar found in milk, because they lack the enzyme. Drinking untreated milk often gives these individuals severe diarrhea. In most areas it is possible to purchase milk made lactose-free by the addition of lactase, an enzyme that converts lactose to its components, glucose and galactose.

Absorption

The small intestine is specialized for absorption. First, it is quite long with convoluted walls. Secondly, the absorptive surface is increased by the presence of fingerlike projections called **villi,** and the villi themselves have tiny microvilli (fig. 4.10). The huge number of villi that cover the entire surface of the small intestine give it a soft, velvety appearance. Each villus (fig. 4.10b) has an outer layer of columnar cells and contains blood vessels and a small lymph vessel called a **lacteal.** The lymphatic system is an adjunct to the circulatory system and returns fluid to the veins.

Absorption takes place across the wall of each villus, continuing until all small molecules have been absorbed. Thus absorption is an active process involving active transport of molecules across cell membranes and requiring an expenditure of cellular energy (p. 35). Sugars and amino acids cross the columnar cells to enter the blood, but glycerol and fatty acids enter the lacteals.[1]

The wall of the small intestine is lined by villi covered by microvilli. Cells of the villi produce enzymes that hydrolyze disaccharides, peptides, and other substances. These enter the blood vessels, and reformed fats enter the lacteals of the villi.

Liver

Blood vessels from the villi merge to form the **hepatic portal vein,** which leads to the liver (fig. 4.11). The liver acts in some ways as the gatekeeper to the blood; it removes poisonous substances from the blood and works to keep the contents of the blood constant. In particular, we may note that the glucose level of the blood is always about 0.1% even though we eat intermittently. Any excess glucose that is present in the hepatic portal vein is removed and stored by the liver as glycogen:

$$\text{glucose} \longrightarrow \text{glycogen} + H_2O$$

Between eating periods, glycogen is broken down to glucose, which enters the hepatic vein, and in this way the glucose content of the blood remains constant.

If by chance the supply of glycogen or glucose runs short, the liver will convert amino acids to glucose molecules:

$$\text{amino acids} \longrightarrow \text{glucose} + \text{amino groups}$$

[1]For the fate of these molecules see also page 83.

FIGURE 4.10 *Anatomy of intestinal lining.* a. *The products of digestion are absorbed by villi, fingerlike projections of the intestinal wall.* b. *Each villus contains blood vessels and a lacteal.* c. *The scanning electron micrograph shows that the villi themselves are covered with microvilli (Mv).* d. *A transmission electron micrograph shows that the microvilli contain microfilaments (Mf). These allow limited motion of the microvilli.*

(c and d) From: Tissues and Organs: A Text-Atlas of Scanning Electron Microscopy *by R. G. Kessel and R. Kardon. W. H. Freeman and Company,* © 1979.

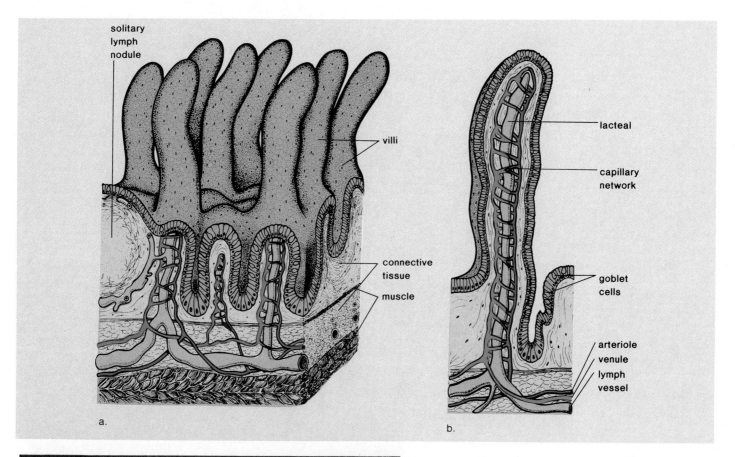

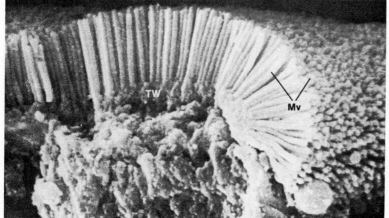

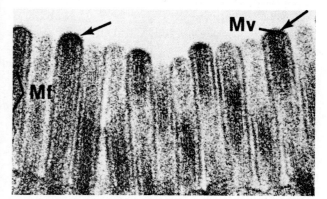

FIGURE 4.11 *The hepatic portal vein takes the products of digestion from the digestive system to the liver where they are processed before entering the circulatory system proper.*

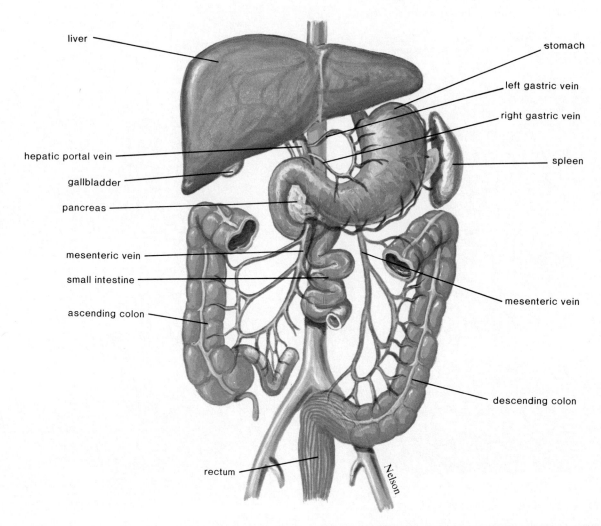

You will recall that amino acids contain nitrogen in the form of amino groups, whereas glucose contains only carbon, oxygen, and hydrogen. Therefore, before amino acids can be converted to glucose molecules, **deamination,** or the removal of amino groups from the amino acids, must take place. By an involved metabolic pathway, the liver converts these amino groups to urea:

$$H_2N - \overset{\overset{\textstyle O}{\|}}{C} - NH_2$$

Urea is the common nitrogen waste product of humans. After its formation in the liver, it is transported to the kidneys for excretion.

The liver also makes blood proteins from amino acids. These proteins are not used as food for cells; rather, they serve important functions within the blood itself.

Altogether in the past several pages we have mentioned the following functions of the liver:

1. Conversion of hemoglobin breakdown products to bile pigments (bilirubin and biliverdin)
2. Production of bile, which is stored in the gallbladder before entering the small intestine, where it emulsifies fats
3. Storage of glucose as glycogen after eating and breakdown of glycogen to glucose between eating to maintain a constant glucose concentration of the blood
4. Production of urea from the breakdown of amino acids
5. Production of the blood proteins
6. Detoxification of the blood by removing and metabolizing poisonous substances

---

Blood from the small intestine enters the hepatic portal vein, which goes to the liver, a vital organ that has numerous important functions as listed.

Liver Disorders

Jaundice   When a person is jaundiced there is a yellowish tint to the skin due to an abnormally large amount of bilirubin (hemoglobin breakdown pigment) in the blood. In one type of jaundice, called *hemolytic jaundice,* red blood cells are broken down in such quantity that the liver cannot excrete the bilirubin fast enough, and an extra amount spills over into the bloodstream. In *obstructive jaundice,* there is an obstruction of the bile duct or damage to the liver cells, and this causes an increased amount of bilirubin to enter the bloodstream.

Obstructive jaundice often occurs when crystals of cholesterol precipitate out of bile and form **gallstones,** which on occasion also contain calcium carbonate. The stones may be so numerous that passage of bile along the bile duct is blocked, and the gallbladder must be removed due to severe pain (fig. 4.12). In the meantime, the bile leaves the liver by way of the blood, and a jaundiced appearance results.

Jaundice is also frequently caused by liver damage due to *viral hepatitis,* a term that includes two separate but similar diseases. *Type A* virus causes *infectious hepatitis,* transmitted by unsanitary food or water. In recent years, for example, persons have been known to acquire the disease after eating shellfish from polluted waters. *Type B* virus causes *serum hepatitis,* commonly spread by means of blood transfusions, kidney dialysis, and injection with inadequately sterilized needles. However, there is evidence that both types of hepatitis can also be sexually transmitted. To recover from hepatitis, a long recuperation period is commonly required, during which time the patient is in a very weakened condition. To prevent the possibility of passing on the disease, a person who has had serum hepatitis cannot give blood.

Cirrhosis   Cirrhosis is a chronic disease of the liver in which the organ first becomes fatty and then liver tissue is replaced by inactive fibrous scar tissue. This condition is common among alcoholics, in which case it is most likely caused by the need for the liver to detoxify alcohol and to break it down. When the liver detoxifies alcohol, it readies it for excretion by the kidneys. When the liver and other type cells break down alcohol it is converted to active acetate, and these molecules can be synthesized to fatty acids. Smooth endoplasmic reticulum increases dramatically in the liver of alcoholics. This may be the first step toward cirrhosis.

The liver is a very critical organ, and any malfunctioning is a matter of considerable concern.

Large Intestine

The large intestine includes the colon and rectum. The **colon** has three parts: the ascending colon goes up the right side of the body to the level of the liver; the transverse colon crosses the abdominal cavity just below the liver and stomach; and the descending colon passes down the left side of the body to

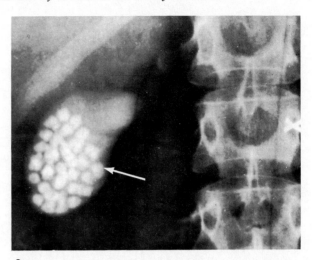

FIGURE 4.12   a. *An X ray of a gallbladder filled with gallstones.* b. *After removal, this gallbladder was cut open to show its contents—numerous gallstones. The dime was added later merely to indicate the size of the stones.*

a.

b.

the rectum. The **rectum** is the last approximately 20 cm (7.5 in.) of the intestinal tract. The opening of the rectum to the exterior is called the **anus.** Water is the primary substance absorbed by the cells that line the colon. If too little water is absorbed, diarrhea results, and if too much water is absorbed, constipation occurs. These conditions are discussed in the following section.

Indigestible remains finally enter the rectum. As the rectum fills, it distends until it is sufficiently stimulated to give rise to a nervous reflex called the defecation reflex (fig. 4.13). In addition to indigestible remains, **feces** also contain certain excretory substances such as bile pigments and heavy metals and large quantities of the usually noninfectious bacterium *E. coli.*

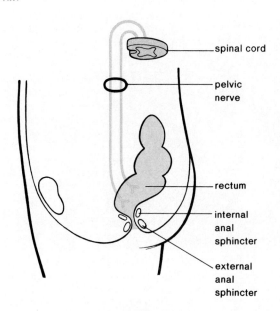

- spinal cord
- pelvic nerve
- rectum
- internal anal sphincter
- external anal sphincter

The large intestine normally contains a large population of bacteria that live off any substances that were not digested earlier. When they break down this material, they give off odorous molecules that cause the characteristic odor of feces. Some of the vitamins, amino acids, and other growth factors produced by these bacteria spill out into the gut and are absorbed by the gut lining. In this way *E. coli* performs a service for us.

Water is considered unsafe for swimming when the *E. coli* count reaches a certain level. This is not because *E. coli* normally causes disease, but because a high count is an indication of the amount of fecal material that has entered the water. The more fecal material present, the greater the possibility that pathogenic, or disease-causing, organisms are also present.

Diarrhea and Constipation

Two common, everyday complaints associated with the large intestine are diarrhea and constipation.

The major causes of *diarrhea* are infection of the lower tract and nervous stimulation. In the case of infection, such as food poisoning caused by eating contaminated food, the intestinal wall becomes irritated and peristalsis increases. Lack of absorption of water is a protective measure, and the diarrhea that results serves to rid the body of the infectious organisms. In nervous diarrhea, the nervous system stimulates the intestinal wall and diarrhea results. Loss of water due to diarrhea may lead to dehydration, a serious condition in which the body tissues lose their normal water content.

When a person is *constipated,* the feces are dry and hard. One cause of this condition is that socialized persons have learned to inhibit defecation to the point that the normal reflexes are often ignored. Two components of the diet can help to prevent constipation: water and roughage. A proper intake of water prevents the drying out of the feces. Dietary inclusion of roughage (fiber), or nondigestible plant substances, provides the bulk needed for elimination and may even help protect against colon cancer. It's believed that the less time that feces are in contact with the membrane of the colon, the better.

The frequent use of laxatives is certainly discouraged, but if it should be necessary to take a laxative, a bulk laxative is the most natural because, like roughage, it produces a soft mass of cellulose in the colon. Lubricants like mineral oil make the colon slippery, and saline laxatives like milk of magnesia act osmotically; they prevent water from exiting or even cause water to enter the colon depending on the dosage. Some laxatives are irritants; they increase peristalsis to the degree that the contents of the colon are expelled.

Chronic constipation is associated with the development of *hemorrhoids,* a condition that is discussed on page 117.

The large intestine does not produce digestive enzymes; it does absorb water. In diarrhea, too little water has been absorbed; in constipation, too much water has been absorbed.

Appendectomy and Colostomy

Two surgical procedures associated with the large intestine are appendectomy and colostomy.

Appendectomy The small intestine joins the large intestine in such a way that there is a blind end to one side of the large intestine (fig. 4.14). This blind sac, or cecum, has a

FIGURE 4.14 The anatomical relationship between the small intestine and the colon is shown. The cecum is the blind end of the ascending colon. The appendix is attached to the cecum.

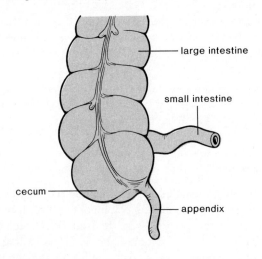

- large intestine
- small intestine
- cecum
- appendix

small projection about the size of the little finger, called the **appendix.** In humans, the appendix is vestigial, meaning that the organ is underdeveloped (in other animals, it is developed and serves as a location for bacterial digestion of cellulose). Unfortunately, the appendix can become infected. When this happens, the individual has *appendicitis,* a very painful condition in which the fluid content of the appendix may rise to the point that it bursts. The appendix should be removed before this occurs because it may lead to a generalized infection of the serous membrane of the abdominal cavity.

Colostomy   The colon is subject to the development of *polyps,* small growths that generally appear on epithelial tissue, such as the epithelial tissue that lines the digestive tract. Whether polyps are benign or cancerous, they can be individually removed along with a portion of the colon if necessary. Should it be necessary to remove the last portion of the rectum and the anal canal, then the intestine is sometimes attached to the abdominal wall in a procedure termed colostomy, and the digestive remains are collected in a plastic bag fastened around the opening. Recently, the use of metal staples has permitted surgeons to join the colon to a piece of rectum that formerly was considered too short.

Two serious conditions associated with the large intestine require surgical removal of its parts. An infection of the appendix (appendicitis) requires removal of the appendix, and cancer of the colon requires the removal of cancerous tissue.

NUTRITION

The body is quite complex, containing many different types of organic molecules and a smaller number of various types of inorganic ions and compounds. To maintain this structure we must continually take in food that contains the nutrients we need. Nutrition involves an interaction between food and the living organism, and a **nutrient** is a substance in food that is used by the body for the maintenance of health. In order to be sure that the forty-six known essential nutrients (table 4.2) are included in the diet, it is recommended that we eat a balanced diet. A *balanced diet* is obtained by eating a variety of foods from the four food groups pictured in figure 4.15.

In order to get a daily supply of the forty-six known essential nutrients it is necessary to have a balanced diet.

Food largely consists of proteins, carbohydrates, and fats. Therefore we will begin by considering these substances.

Proteins

Foods rich in protein include meat, fish, poultry, dairy products, legumes, nuts, and cereals. Following digestion of protein, amino acids enter the blood and are transported to the tissues. Most are incorporated into structural proteins found in muscles, skin, hair, and nails. Others are used to synthesize such proteins as hemoglobin, plasma proteins, enzymes, and hormones.

TABLE 4.2   THE ESSENTIAL FORTY-SIX NUTRIENTS

MAJOR CATEGORY	ESSENTIAL NUTRIENT[a]	
Carbohydrate	Glucose	
Lipid or fat	Linoleic acid	
Protein	Amino acids:	lysine, methionine, phenylalanine, threonine, tryptophan, leucine, valine, isoleucine, histidine
Vitamins	Fat-soluble:	A[b] (retinol) D (cholecalciferol, ergocalciferol) E (tocopherol) K (menadione)
	Water-soluble:	C (ascorbic acid), B (thiamin[c] [vitamin B_1], riboflavin [vitamin B_2], niacin[c] [nicotinamide, nicotinic acid], vitamin B_6 [pyridoxine, pyridoxal, pyridoxamine], pantothenic acid, biotin, folacin[b] [folic acid, folate], cobalamin [cyanocobalamin])
Minerals	Macrominerals:	calcium, phosphorus, sodium[d], potassium, magnesium, chlorine, sulfur
	Microminerals:	iron[b], zinc[b], iodine[c], selenium, copper, manganese, cobalt, chromium, molybdenum, vanadium, tin, nickel, silicon

[a]Experts disagree on the exact number of essential nutrients. This is because our knowledge of the role of some nutrients in the human body is less than perfect. This list is agreed on by most experts to be the *known* essential nutrients.
[b]Intake may be too low among some people in the United States.
[c]Intake was low in the recent past in the United States. The problem has been corrected.
[d]Intake may be too high among some people in the United States.

From Vincent Hegarty, *Decisions in Nutrition.* Copyright © C. V. Mosby Company, St. Louis, Missouri.

FIGURE 4.15 *In order to have a balanced diet, it is recommended that you eat a variety of foods from each food group every day.*

TABLE 4.3 *ANALYSIS OF MEAL FOR 19-YEAR-OLD MALE AND FEMALE*

FOOD	AMOUNT	ENERGY (KCALORIES)	PROTEIN (g)	CALCIUM (mg)	VITAMIN C (mg)	VITAMIN A (IU)
Roast beef (lean & fat)	3 oz.	165	25	11	—	10
Baked potato	1 medium	145	4	14	31	Trace
Broccoli	½ cup	20	2.5	68	70	3,880
Roll	1	155	5	24	Trace	Trace
Margarine	1 pat	35	Trace	1	0	170
Milk, low-fat (1%)	1 cup	100	8	300	2	500
Apple	1 medium	80	Trace	10	6	120
Totals		700	44.5	428	109	4,680
RDA 19-year-old woman			44	800	60	4,000
% RDA			101%	53%	182%	117%
RDA 19-year-old-man			56	800	60	5,000
% RDA			80%	52%	182%	94%

From Vincent Hegarty, *Decisions in Nutrition.* Copyright © C. V. Mosby Company, St. Louis, Missouri.

Protein formation requires twenty different types of amino acids. Of these, nine are required in the diet because the body is unable to produce them. These are termed the **essential amino acids.** The body produces the other amino acids by simply transforming one type into another type. Some protein sources, such as meat, are *complete* in the sense that they provide all the different types of amino acids. Vegetables and grains do supply us with amino acids, but each one alone is an *incomplete* source because at least one of the essential amino acids is absent. However, it is possible to combine foods in order to acquire all of the essential amino acids. For example, the combination of cereal with milk or beans with rice will provide all the essential amino acids.

A complete source of protein is absolutely necessary to ensure a sufficient supply of the essential amino acids.

Even though we put a lot of stress in this country on protein intake, it does not take very much to be sure that the daily requirement has been met. For example, for the meal described in table 4.3, the woman has met her recommended *dietary (daily) allowance (RDA)* for protein with a single serving of meat. In the United States, the RDAs are determined by the National Research Council, a part of the National Academy of Sciences. The RDA for the "reference woman" (120 lb. or 55 kg) is 44 g of protein a day. For the "reference man" (154 lb. or 70 kg) it is 56 g of protein a day.

Unfortunately, in this country the manner in which we meet our daily requirement for protein is not always the healthiest way. The foods that are richest in protein are apt to be the ones that are also richest in fat (fig. 4.16). In the sample meal (table 4.3) the roast beef serving provides 25 g of protein and 165 **Kcalories.**[2] There are 4 Kcalories in each gram of protein (table 4.4); therefore, 65 of the Kcalories in the roast beef serving were from fat.

[2]The amount of heat required to raise 1 g of water 1° C.

FIGURE 4.16 *Protein-rich foods are frequently high in fat.*

FIGURE 4.17 *To meet our energy needs, dieticians recommend complex carbohydrates like those shown here rather than simple carbohydrates like candy and ice cream. The latter are more likely to cause weight gain than the recommended complex ones displayed.*

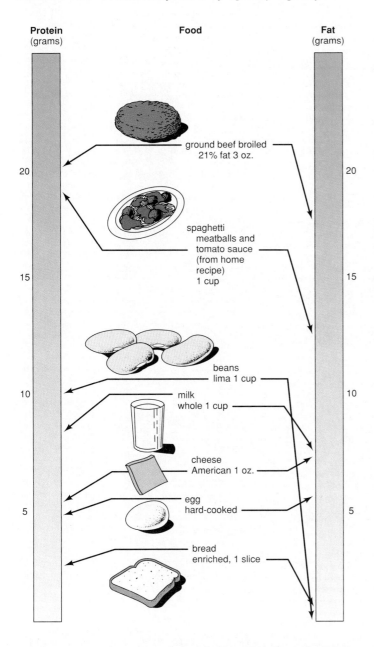

Protein (grams) **Food** **Fat** (grams)

ground beef broiled 21% fat 3 oz.

spaghetti meatballs and tomato sauce (from home recipe) 1 cup

beans lima 1 cup

milk whole 1 cup

cheese American 1 oz.

egg hard-cooked

bread enriched, 1 slice

Foods richest in protein also tend to be richest in fat.

While it is very important to receive the RDA for protein, consuming more can actually be detrimental. Calcium loss in the urine has been noted when dietary protein intake is over twice the RDA. Everything considered, then, it is probably a good idea to depend on protein from plant origins (eg., whole grain cereals, dark breads, legumes) to a greater extent than is the custom in this country.

Carbohydrates

The quickest, most readily available source of energy for the body is carbohydrates, which can be complex as in breads and cereals or simple as in candy, ice cream, and soft drinks (fig. 4.17). As mentioned previously, starches are digested to glucose, which is stored by the liver in the form of glycogen. Between eating, the blood glucose is maintained at about 0.1% by the breakdown of glycogen or by the conversion of amino acids to glucose. If necessary, these amino acids are taken from the muscles, even from heart muscle. To avoid this situation, it is suggested that the diet contain at least 100 g of carbohydrate daily.

Carbohydrate is needed in the diet to maintain the blood glucose level.

TABLE 4.4 *CALORIC ENERGY RELEASE*

	KCALORIES/GRAM
Carbohydrate	4.1
Fat	9.3
Protein	4.1

HOW DOES YOUR DIET RATE FOR VARIETY?

Check the blank that best describes your eating habits.

How often do you eat

	Seldom or never	1 or 2 times a week	3 to 4 times a week	Almost daily
1. at least six servings of bread, cereals, rice, crackers, pasta, or other foods made from grains (a serving of one slice of bread or a half-cup cereal, rice, etc.) per day?	_____	_____	_____	_____
2. foods made from whole grains?	_____	_____	_____	_____
3. three different kinds of vegetables per day?	_____	_____	_____	_____
4. cooked dry beans or peas?	_____	_____	_____	_____
5. a dark-green leafy vegetable, such as spinach or broccoli?	_____	_____	_____	_____
6. two kinds of fruit or fruit juice per day?	_____	_____	_____	_____
7. two servings (three if teenager, pregnant, or breastfeeding) of milk, cheese, or yogurt per day?	_____	_____	_____	_____
8. two servings of lean meat, poultry, fish, or alternates, such as eggs, dry beans, or nuts per day?	_____	_____	_____	_____

Source: United States Department of Agriculture, *Human Nutrition Information Service Home and Garden Bulletin No. 232–1*, April 1986.
Note: See page 98 for answers.

Actually the dietary guidelines produced jointly by the U.S. Department of Argiculture and the Department of Health and Human Services recommend that we increase the proportion of carbohydrate Kcalories in the diet (table 4.5) from the present level of 46% to 58%. Further, it is assumed that these carbohydrates will be of the complex type. The simple carbohydrates (eg., sugars) are labeled "empty calories" by some because they contribute to energy needs and weight gain and are not a part of foods that supply other nutritional requirements. Table 4.6 gives suggestions on how to cut down on your consumption of dietary sugars.

In contrast to simple sugars, complex carbohydrates are likely to be accompanied by a wide range of other nutrients and fiber. Fiber is the nondigestible portion of food from a

TABLE 4.5 *COMPARISON OF TYPICAL AMERICAN DIET OF THE 1980S AND RECOMMENDED DIET*

	TYPICAL AMERICAN DIET(%)[a]	DIETARY GUIDELINES(%)
Protein	12	12
Carbohydrate	46	58
Fat	42	30

[a]of total Kcalories

From Vincent Hegarty, *Decisions in Nutrition.* Copyright © C. V. Mosby Company, St. Louis, Missouri.

TABLE 4.6 *REDUCING DIETARY SUGAR*

To reduce dietary sugar, the following suggestions are recommended.
1. Eat fewer sweets, such as candy, soft drinks, ice cream, and pastry.
2. Eat fruit that is fresh or canned fruit without heavy syrup.
3. Use less sugar—white, brown, or raw—and less honey and syrups.
4. Avoid sweetened breakfast cereals.
5. Eat less jelly.
6. Drink pure fruit juices, not imitations.
7. When cooking, use spices like cinnamon to flavor foods instead of using sugar.
8. Do not put sugar in tea or coffee.

plant source such as bran, whole wheat cereals and breads, fruits, and vegetables. Fiber does not occur in meat or other animal products, and it is refined out of sugar and white flour. Fiber is not digested so it provides bulk that aids the elimination process.

Some studies have suggested that "soluble fibers" found in fruit, oat bran, and beans lower serum cholesterol. It's reasoned that cholesterol binds to this type fiber and is eliminated with the feces. Other studies have suggested that "insoluble fibers" found in wheat bran, whole wheat, beans, fruits, and vegetables reduce the risk of colon cancer. This type fiber has a laxative effect, and it is reasoned that its presence reduces the amount of time that potential mutagens in the feces have to act on the colon.

While the diet should be adequate in the amount of fiber, a high-fiber diet may be detrimental. Some evidence suggests that the absorption of iron, zinc, and calcium may be impaired by a high-fiber diet.

Complex carbohydrates, along with fiber, are considered to be beneficial to health.

Carbohydrates provide most of the dietary Kcalories. There are only 4 Kcalories per gram of carbohydrate, but since carbohydrates are the bulk of the diet, they provide the most calories in the end (table 4.4).

Fats

Fats are present not only in butter, margarine, and oils, but also in foods high in protein (fig. 4.16). Fats from an animal origin tend to have saturated fatty acids, and those from plants tend to have unsaturated fatty acids. After being absorbed, the products of fat digestion are transported by the lymph and blood to the tissues. The liver can alter ingested fats to suit the body's needs, except it is unable to produce the fatty acid linoleic acid. Since this is required for phospholipid production, it is considered to be an essential fatty acid.

Fats have the highest caloric content, but they should not be avoided entirely because of the essential fatty acid linoleic acid.

While we need to be sure to ingest some fat in order to satisfy our need for linoleic acid, the dietary guidelines mentioned earlier (table 4.5) suggest that we reduce the amount of fat from the typical 40% of total Kcaloric intake to 30%. Dietary fat has been implicated in cancer of the colon, pancreas, prostate, and breast (fig. 4.18). Many animal studies have shown that a high-fat diet stimulates the development of mammary tumors while a low-fat diet does not. Similarly it has been found that women who have a high-fat diet are more likely to develop breast cancer. Mammary tumors in animals and breast cancer in women produce abundant amounts of prostaglandins (p. 287), and fatty acids are precursors for these substances. Surprisingly, it's been discovered that reduced amounts of linoleic acid in the diet help prevent the occurrence of breast cancer. Linoleic acid is found in corn, safflower, sunflower, and other common plant oils, but it is not abundant in olive oil or in fatty fishes and marine animals.

There is very strong evidence that women who have a diet high in fat are more apt to develop breast cancer.

As a nation we have increased our consumption of fat from plant sources and have decreased our consumption from animal sources, such as red meat and butter (fig. 4.19). Most likely this is due in part to recent information concerning the link between saturated fatty acids and cholesterol with hypertension and heart attack (p. 114).

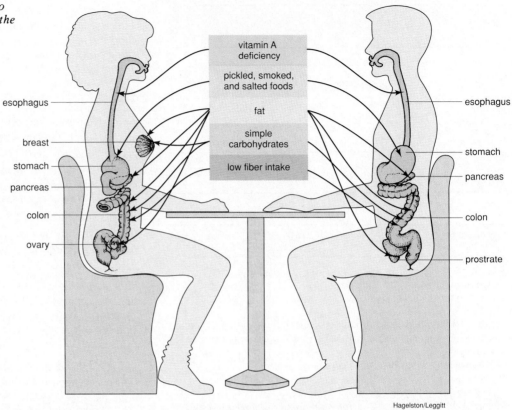

FIGURE 4.18 *Evidence is growing to suggest dietary factors that influence the development of cancer.*

vitamin A deficiency

pickled, smoked, and salted foods

fat

simple carbohydrates

low fiber intake

esophagus

breast

stomach

pancreas

colon

ovary

esophagus

stomach

pancreas

colon

prostrate

Hagelston/Leggitt

FIGURE 4.19 *Fat content in the diet.* a. *Americans acquire 43% of fat intake from animal fats like butter and plant oils like cooking oils; 36% comes from meat, poultry, and fish; and lesser amounts come from the sources shown.* b. *The amount of fat acquired from vegetable sources is now larger than it was in the early 1900s.*

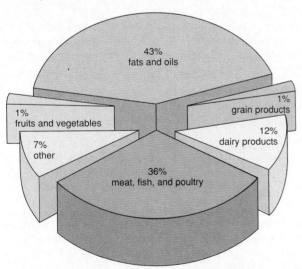

43% fats and oils

1% grain products

1% fruits and vegetables

12% dairy products

7% other

36% meat, fish, and poultry

a. **Sources of Fat in the U.S. Diet, 1980**

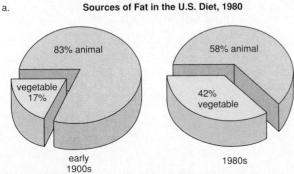

83% animal

vegetable 17%

58% animal

42% vegetable

early 1900s

1980s

b. **Types of Fat in the Diet**

TABLE 4.7   REDUCING DIETARY FAT

To reduce dietary fat, the following suggestions are recommended.
1. Choose lean meat, poultry, fish, and dry beans and peas as a protein source.
2. Trim fat off meat before cooking, and remove skin from poultry before cooking.
3. Cook meat or poultry on a rack so that fat will drain off.
4. Broil, boil, or bake, rather than fry.
5. Limit your intake of butter, cream, hydrogenated oils, shortenings, and coconut oil.
6. Use herbs and spices to season vegetables instead of butter, margarine, or sauces. Use lemon juice instead of salad dressing.
7. Drink skim milk instead of whole milk, and use skim milk in baking.
8. Eat eggs and such organ meats as liver in moderation because they are high in cholesterol. Substitute egg whites for whole eggs; for example, use two egg whites instead of one whole egg when baking. Use one egg yolk per person when scrambling eggs.

Fats are the highest energy foods (table 4.4); they contain about 9 Kcalories per gram. Raw potatoes have about 0.9 Kcalories per gram, but when they are cooked in fat, the number of Kcalories jumps up to 6 Kcalories per gram. Another problem for those trying to limit their Kcaloric intake is that fats are not always highly visible; butter melts on toast or potatoes and salad dressings coat lettuce leaves. Table 4.7 gives suggestions for cutting down on the amount of fat in the diet.

a.

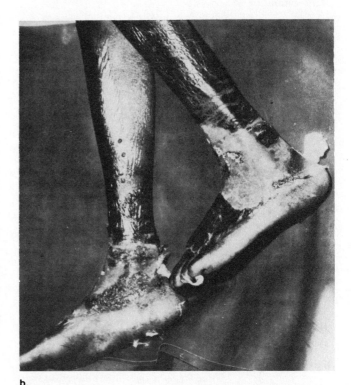

b.

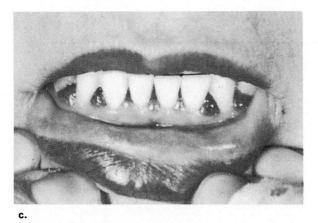

c.

d.

Vitamins and Minerals

Vitamins

Vitamins are organic compounds (other than carbohydrates, fats, and proteins) that the body is unable to produce but needs for metabolic purposes. Therefore, vitamins must be present in the diet; if they are lacking, various symptoms develop (fig. 4.20). There are many substances that are advertised as being vitamins, but in reality there are only thirteen vitamins (table 4.8). Table 4.8 also gives the RDA vitamin values for the "reference female" and the "reference male."

In general, carrots, squash, turnip greens, and collards are good sources of vitamin A. Citrus fruits and other fresh fruits and vegetables are natural sources of vitamin C. Sunshine and irradiated milk are primary sources of vitamin D, and whole grains are a good source of B vitamins.

As table 4.3 shows, it is not difficult to acquire the RDAs for vitamins if your diet is balanced (fig. 4.15) because each vitamin is needed in small amounts only. Many vitamins are portions of coenzymes, or enzyme helpers. For example, niacin is part of the coenzyme NAD, and riboflavin is part of another dehydrogenase, FAD. Coenzymes are needed in only small amounts because each one can be used over and over again.

TABLE 4.8 *VITAMINS, THEIR FUNCTIONS AND SOURCES*

NAME	BODY FUNCTIONS	MAJOR FOOD SOURCE	RDA	
Fat-soluble vitamins			**Female**[a]	**Male**
Vitamin A	Vision; health of skin; growth of teeth, nails, hair, bones, and glands	Dairy products, liver, carotene in deep green and orange vegetables	800 mg	1000 mg
Vitamin D	Bones and teeth	Dairy products, egg yolk	7.5 mg	7.5 mg
Vitamin E	Antioxidant—prevents cell damage	Oils, nuts, seeds	8 mg	10 mg
Vitamin K	Blood clotting	Green leafy vegetables, meats		
Water-soluble vitamins				
Vitamin C	Antioxidant; collagen formation; health of teeth and gums	Citrus fruits, green pepper, broccoli, cantaloupe	60 mg	60 mg
Thiamin (vitamin B_1)	Nerve function; aids in energy metabolism	Whole grains and cereals, meats (especially pork)	1.1 mg	1.5 mg
Riboflavin (vitamin B_2)	Aids in energy metabolism; protects against skin and eye disorders	Whole or enriched grains, milk, eggs, cheese, meats, green vegetables	1.3 mg	1.7 mg
Niacin	Aids in energy metabolism	Lean meats, fish, whole grains, cheese, peanuts, vegetables, eggs	14 mg	19 mg
Vitamin B_6	Aids in amino acid–protein metabolism; nerve function	Liver, fish, nuts, meats, potatoes, some vegetables	2.0 mg	2.2 mg
Pantothenic acid	Aids in energy metabolism; nerve function	Most foods of plant and animal origin	—	—
Biotin	Aids in energy metabolism	Most foods of plant and animal origin	—	—
Folacin	Synthesis of DNA, RNA (genetic material); prevents a type of anemia; red blood cell formation	Liver, green leafy vegetables, peanuts	400 mg	400 mg
Vitamin B_{12}	Red blood cell formation; synthesis of DNA and RNA; component of sheath around nerves	Foods of animal origin, micro-organisms (fermented foods)	3.0 mg	3.0 mg

[a] females 19–22, 120 lbs., 5'4''
males 19–22, 154 lbs., 5'10''
From Vincent Hegarty, *Decisions in Nutrition*. Copyright © C. V. Mosby Company, St. Louis, Missouri.

The National Academy of Sciences suggests that we eat more fruits and vegetables in order to acquire a good supply of vitamins C and A because these two vitamins may help guard against the development of cancer. Nevertheless they discourage the intake of excess vitamins by way of pills because this practice can possibly lead to illness. For example, excess vitamin C can cause kidney stones, and this excess is converted to oxalic acid, a molecule that is toxic to the body. Vitamin A taken in excess over long periods can cause loss of hair, bone and joint pains, and loss of appetite. Excess vitamin D can cause an overload of calcium in the blood, which in children leads to loss of appetite and retarded growth. Megavitamin therapy should always be supervised by a physician.

A properly balanced diet will include all the vitamins and minerals needed by most individuals to maintain health.

Minerals

In addition to vitamins, various **minerals** are also required by the body (table 4.9). Minerals are divided into macrominerals, which are recommended in amounts more than 100 mg per day, and microminerals (trace elements), which are recommended in amounts less than 20 mg per day. The macrominerals sodium, magnesium, phosphorus, chlorine, potassium, and calcium serve as constituents of cells and body fluids and as structural components of tissues. For example, calcium is needed for the construction of bones and teeth and also for nerve conduction and muscle contraction.

The microminerals seem to have very specific functions. For example, iron is needed for the production of hemoglobin, and iodine is used in the production of thyroxin, a hormone produced by the thyroid gland. As research continues, more and more elements have been added to the list of those considered essential. During the past three decades, molybdenum, selenium, chromium, nickel, vanadium, silicon, and even arsenic have been found to be essential to good health in very small amounts.

TABLE 4.9 *MINERALS, THEIR FUNCTIONS AND SOURCES*

NAME	BODY FUNCTIONS	MAJOR FOOD SOURCE
Macrominerals		
Calcium (Ca)	Strong bones and teeth; nerve conduction and muscle contraction	Milk, milk products, leafy green vegetables
Phosphorus (P)	Strong bones and teeth; metabolic compounds and reactions	Meats, poultry, fish, cheese, nuts, whole grain cereals, milk, legumes
Potassium (K)	Nerve conduction and muscle contraction; metabolic reactions	Avocados, dried apricots, meats, nuts, potatoes, bananas
Sulfur (S)	Found in certain amino acids and other metabolites	Meats, milk, eggs, legumes
Sodium (Na)	Water balance; nerve conduction and muscle contraction; regulation of pH	Table salt, cured ham, sauerkraut, cheese, graham crackers
Chlorine (Cl)	Closely associated with sodium	Same as for sodium
Magnesium (Mg)	Metabolic reactions; production of ATP	Milk, dairy products, legumes, nuts, leafy green vegetables
Microminerals		
Iron (Fe)	Part of hemoglobin and certain enzymes	Liver, lean meats, dried apricots, raisins, enriched whole grain cereals, legumes, molasses
Manganese (Mn)	Fatty acid synthesis; formation of urea; functioning of nervous system	Nuts, legumes, whole grain cereals, leafy green vegetables, fruits
Copper (Cu)	Synthesis of hemoglobin; development of bone; formation of myelin	Liver, oysters, crabmeat, nuts, whole grain cereals, legumes
Iodine (I)	Found in thyroid hormones	Iodized table salt
Cobalt (Co)	Synthesis of enzymes	Liver, lean meats, poultry, fish, milk
Zinc (Zn)	Found in several enzymes necessary for normal wound healing and healthy skin	Seafoods, meats, cereals, legumes, nuts, vegetables

[a]females 19–22, 120 lbs. 5'4"
males 19–22, 154 lbs, 5'10"

Occasionally it has been found that individuals do not receive enough iron (in women), calcium, magnesium, or zinc in their diets. Adult females need more iron in the diet than males (RDA of 18 mg compared to 10 mg) because they lose hemoglobin each month during menstruation. Stress can bring on a deficiency of magnesium, and an inexpensive diet might lack zinc. However, a varied and complete diet will supply the RDAs for minerals.

Calcium   There has been much interest of late in calcium supplements (fig. 4.21) to counteract the possibility of developing osteoporosis, a degenerative bone disease that afflicts an estimated ¼ of older men and ½ of older women in the United States. These individuals have porous bones that tend to break easily because their bones lack sufficient calcium. In 1984 a National Institutes of Health conference on osteoporosis advised postmenopausal women to increase their intake of calcium to 1,500 mg and all others to 1,000 mg (compared with the RDA of 800 mg).

FIGURE 4.21   *There are now many over-the-counter calcium supplements available.*

However, recent studies have shown that calcium supplements cannot prevent osteoporosis after menopause even when the dosage is 3,000 mg a day. The reason is that the body becomes less able to take in calcium after about the age of 35. The best effective defense against osteoporosis in older women is now believed to be estrogen replacement and exercise. In one survey, women ages 35 to 65 who took a fifty-minute aerobics class three times a week lost only 2.5% of the density in their forearm bones, compared with 9.5% for women who did not exercise.

HUMAN ISSUE

In 1984 the National Institutes of Health conference on osteoporosis recommended that postmenopausal women increase their intake of calcium. For a period of time it was suggested that drinking more milk, taking calcium tablets, and even taking Tums tablets would lead to stronger bones. These recommendations were made prior to additional studies that showed increasing dietary calcium apparently has little effect on osteoporosis. Should scientists and scientific bodies wait to make recommendations until they are sure beyond a reasonable doubt as to their veracity? Or are they duty bound to make recommendations based on incomplete information if these recommendations could possibly benefit the health of millions of people?

Young women can guard against the possiblity of developing osteoporosis when they grow older by forming strong, dense bones before menopause occurs. Eighteen-year-old women are apt to get only 679 mg of calcium a day when the RDA is 800 mg. They should consume more calcium-rich foods such as milk and dairy products. Taking calcium supplements may not be as effective. A cup of milk supplies 270 mg of calcium while a 500 mg tablet of calcium carbonate provides only 200 mg. The rest is just not taken up by the body; it is not in a form that is *bioavailable*. However, an excess of bioavailable calcium can lead to kidney stones.

Dietary calcium and exercise, plus estrogen therapy if needed, are the best safeguards against osteoporosis.

Sodium The recommended amount of sodium intake per day is 400–3,300 mg, and the average American takes in 4,000–4,700 mg. In recent years this imbalance has caused concern because high sodium intake has been linked to hypertension in some people. About ⅓ of the sodium we consume occurs naturally in foods; another ⅓ is added during commercial processing; and we add the last ⅓ either during home cooking or at the table in the form of table salt.

Clearly it is possible for us to cut down on the amount of sodium in the diet. Table 4.10 gives recommendations for doing so.

TABLE 4.10 *REDUCING DIETARY SODIUM*

To reduce dietary sodium, the following suggestions are recommended.
1. Use spices to flavor foods instead of salt.
2. Add little or no salt to foods at the table, and add only small amounts of salt when you cook.
3. Eat unsalted crackers, pretzels, potato chips, nuts, and popcorn.
4. Avoid frankfurters, ham, bacon, luncheon meat, smoked salmon, sardines, and anchovies.
5. Avoid processed cheese and canned or dehydrated soups.
6. Avoid prepared catsup and horseradish.
7. Avoid brine-soaked foods such as pickles and olives.

Excess sodium in the diet can lead to hypertension; therefore, excess sodium intake should be guarded against.

DIETING

When persons go on a diet, they usually begin by determining how many Kcalories they intend to consume each day.

Daily Energy Requirements

Humans need energy primarily for basal metabolism and also for physical activities. *Basal metabolism* involves breathing, circulating blood, and maintaining body temperature and muscle tone. The basal metabolic rate (*BMR*) is best measured fourteen hours after the last meal with the subject lying down at complete physical and mental rest. The BMR is usually lower for women than for men and, in general, is affected by size, shape, weight, age, activity of the endocrine glands, and so forth. The BMR for the "reference female" of 120 pounds is 1,320, and for the "reference male" weighing 154 pounds, the BMR is 1,848 Kcalories.

The energy used for various voluntary activities can be determined by consulting tables such as table 4.11 that lists the number of Kcalories per hour that are required for all sorts of activities. From such tables, it is possible to estimate the number of Kcalories that will be needed each day for voluntary activities.

The average daily Kcalorie requirement for our "reference female" and "reference male" is expected to be as shown in the following.

	Female	Male
Sedentery	1,700	2,500
Moderately active	2,100	2,900
Very active	2,500	3,300

TABLE 4.11 CALORIC EXPENDITURES FOR SOME PHYSICAL ACTIVITIES

ACTIVITY	KCALORIES per minute	KCALORIES per hour
Sleeping	1.2	71
Watching television	1.2	72
Eating	1.4	85
Classwork	1.7	101
Writing	1.9	111
Driving	3.0	180
Walking 2 m.p.h.	3.6	213
4½ m.p.h.	6.7	401
going upstairs	17.5	1052
Ping pong	3.9	235
Baseball—infield-outfield	4.7	284
pitching	6.0	362
Bicycling—slow	5.0	300
strenuous	10.8	650
Swimming—leisurely	5.0	300
rapidly	10.8	650
Tennis—doubles	5.8	350
singles	7.5	450
Bowling(nonstop)	6.8	405
Badminton & volleyball		
moderate	5.8	345
vigorous	9.8	590
Fast dancing	7.5	450
Basketball—moderate	7.1	426
vigorous	8.5	512
Football	8.3	500
Skiing—downhill	9.8	585
cross country 5 m.p.h.	11.8	710
Jogging	10.0	600
Wrestling, judo, karate	12.9	775

From Jordan, H. A., *Finding Your Weigh to Slimming*, 1983. No. 141. Carolina Biology Reader Series. Carolina Biological Supply Company, Burlington, North Carolina.

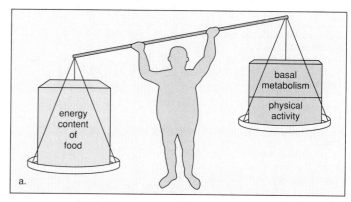

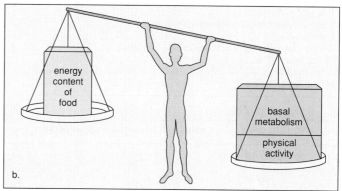

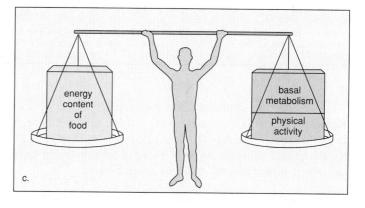

FIGURE 4.22 Diagram illustrating the relationship between caloric intake and weight gain or loss. In each instance, energy needs are divided between basal metabolism (basal metabolic rate) and physical activity. a. Energy content of food is greater than energy needs of body—weight gain occurs. b. Energy content of food is less than energy needs of body—weight loss occurs. c. Energy content of food equals energy needs of body—no weight change occurs.

Figure 4.22 indicates that a loss of weight will occur if the caloric intake is reduced while the same level of activity is maintained. The net Kcalorie restriction needed for losing 1 pound of body weight is 3,500. So, if you want to lose a half-pound in one week, you have to reduce the Kcaloric intake by 250 calories a day. If you are a sedentary male weighing 154 pounds, this would mean you are striving for a total intake of 2,250 Kcalories per day.

Reducing the caloric intake and/or increasing the amount of exercise will eventually result in weight loss.

The recommended body weight for men and women can be determined in the following way.[3]

[3]Add or subtract 10 pounds depending on frame.

Women: 100 pounds (45.4 kg) for first 5 feet (1.5 m) of height; 5 pounds per inch (2.5 cm) over 5 feet

Men: 140 pounds for first 5 feet of height; 5 pounds per inch over 5 feet

Once you have determined how much you want to weigh and how quickly you wish to obtain this weight, then you will know approximately how long it will take you to achieve your goal.

Breads
1 slice of bread or any of the following:
¾ cup ready-to-eat cereal
⅓ cup corn
1 small potato
(1 bread = 15 g carbohydrate, 20 g protein, and 70 Kcal)
Milk
1 cup skim milk or   (2% milk—add 5 g fat)
1 cup skim-milk yogurt, plain   (Whole milk—add 8 g fat)
1 cup buttermilk
½ cup evaporated skim milk or milk dessert
(1 milk = 12 g carbohydrate, 8 g protein, and 80 Kcal)
Vegetables
½ cup greens
½ cup carrots
½ cup beets
(1 vegetable = 5 g carbohydrate, 1 g protein, and 25 Kcal)
Fruits
½ small banana or
1 small apple
½ cup orange juice or ½ grapefruit
(1 fruit = 10 g carbohydrate and 40 Kcal)
Meats (lean)
1 oz. lean meat or
1 oz. chicken meat without the skin
1 oz. any fish
¼ cup canned tuna or 1 oz. low-fat cheese
(1 oz. low-fat meat = 7 g protein, 3 g fat, and 55 Kcal)

Meats (medium fat)
1 oz. pork loin
1 egg
¼ cup creamed cottage cheese
(1 medium-fat meat = 7 g protein, 5½ g fat and about 80 Kcal)
Meats (high fat)
1 oz. high-fat meat is like
1 oz. country-style ham
1 oz. cheddar cheese
a small hot dog (frankfurter)
(1 high-fat meat = 7 g protein, 8 g fat, and 100 Kcal)
Peanut butter
Peanut butter is like a meat in terms of its protein content but is
    very high in fat. It is estimated as (2 Tbs. peanut butter = 7 g
    protein, 15½ g fat, and about 170 Kcal)
Fats
1 tsp. butter or margarine
1 tsp. any oil
1 Tbs. salad dressing
1 strip crisp bacon
5 small olives
10 whole Virginia peanuts
(1 fat = 5 g fat and 45 Kcal)
Legumes (beans & peas)
Legumes are like meats because they are rich in protein and iron,
    but are lower in fat than meat. They contain much starch.
    They can be treated as (½ cup legumes = 15 g carbohydrate,
    9 g protein, 3 g fat, and 125 Kcal)

Miscellaneous Foods

	Grams			Kcal		Grams			Kcal
	Protein	Fat	Carbohydrate			Protein	Fat	Carbohydrate	
Ice Cream (1 cup)	5	14	32	274	Beer (1 can)	1	0	14	60
Cake (1 piece)	3	1	32	149	Soft drink	0	0	37	148
Doughnuts (1)	3	11	22	199	Soup (1 cup)	2	3	15	95
Pie	3	15	51	351	Coffee and tea	0	0	0	0
Caramel candy (1 oz.)	1	3	22	119					

*For caloric values not included in this report, consult Whitney and Hamilton, 1984. *Understanding nutrition.* 3d ed. New York: West Publishing Co.
Reprinted by permission from B. L. Frye and R. L. Neill, "A Laboratory Exercise in Nutrition" in *American Biology Teacher,* September 1987, p. 372.

In order to determine whether you are meeting your Kcalorie goal, it is necessary to keep track of the Kcalories in the food you are eating. Table 4.12 offers a shortcut method for doing this. If you wish to be more precise, you will have to consult larger and more complicated tables.

Fad Diets versus Behavior Modification

Nutritionists tell us that the best way to lose weight is to modify our behavior and to lose the weight slowly. Behavior modification requires that we examine our eating behaviors (fig. 4.23), identify the situations that cause us to snack unnecessarily, and work at changing these. For example, if you are used to having cake and ice cream for dessert, substitute fruit instead. If you pass an ice cream shop every day and tend to stop in, then change your route and don't go by this shop.

Very important is the addition of physical activity to your daily routine. Aside from helping to firm up the body after weight loss, it also keeps you in the mood to keep it off, and it burns off calories.

The reading on page 96 lists and discusses some of the types of fad diets that are popular. Several are dangerous to your health and don't work because the weight loss is simply regained once the diet ceases. Our bodies are not adapted for rapid weight loss. First, if calories are severely and suddenly restricted, the body turns first to burning protein rather than fat. This protein is removed from the muscles, even the heart muscle. Second, the body apparently has a "set point" for its usual amount of fat. If the amount of stored fat falls below this point, the fat cells are believed to signal the brain by the release of a chemical substance. In response, the metabolic rate is lowered so that fewer calories are needed to stay at the same weight. This set point hypothesis also explains why people tend to immediately regain any weight that has been lost through dieting. There is some evidence that exercise and the avoidance of fatty foods lowers the set point for body fat.

The overall conclusion, then, is that in order to keep unwanted pounds off, a long-term program is needed. This program should include regular exercise and a balanced diet that avoids fatty foods.

Fad diets can be dangerous to your health. A balanced diet containing a reduced number of calories, along with an exercise program, offers long-term weight control.

Eating Disorders

Authorities recognize three primary eating disorders: obesity, bulimia, and anorexia nervosa. Although they exist in a continuum as far as body weight is concerned, there is much overlap between them.

Obesity is defined as a body weight of more than 120% of the ideal weight. It is most likely caused by a combination of factors, including endocrine, metabolic, and social factors. The social factors include the eating habits of other family members. Obese individuals need to consult a physician if they wish to bring their body weight down to normal and keep it there permanently.

Obesity has many complex causes that possibly can be detected by a physician.

Bulimia may coexist with either obesity or anorexia nervosa. People, usually young women, who are afflicted have the habit of eating to excess and then purging themselves by some artificial means, such as vomiting or laxatives. These individuals are usually depressed, but whether the depression causes or is caused by the bulimia cannot be determined. While individual psychological help does not seem to be effective, there is some indication that group therapy does help. The possibility of a hormonal disorder has not been ruled out, however.

Anorexia nervosa is diagnosed when an individual is extremely thin but still claims to "feel fat" and continues to diet. It's possible that these individuals have an incorrect body image that makes them think they are fat. It's also possible they have various psychological problems, including a desire to suppress their sexuality. Menstruation ceases in very thin women.

Both bulimia and anorexia nervosa are serious disorders that require the assistance of competent medical personnel.

"CRASH" DIETS

Many people end up willing to try any gimmick that is supposed to lead to permanent weight loss. The diets that have been promoted in recent years range from the merely ineffective to the outright dangerous, with a scattering of reasonably sound ones. Most weight-loss aids, schemes, and plans fall into one of the following categories.

Pills. Some of the widely marketed over-the-counter drugs sold for dieting are supposed to suppress the appetite; others claim to "burn" fat. Appetite suppressants may work at first, but weight-control researchers say that to lead to permanent weight loss, a pill would have to be taken for life, like blood-pressure pills. Doctors say most diet pills are useless, and some—in particular, the amphetamines and similar prescription drugs—are addictive. So-called starch-blocker pills, which bind to starch and make it undigestible, are now illegal in the United States but are still sold on the black market. They cause a number of adverse side effects, including abdominal cramps.

Low-carbohydrate diets. Severely cutting back on carbohydrates upsets the body's chemical balance in such a way that fluids are depleted. While this gives the illusion of weight loss, fat is not lost, and the water weight will eventually be regained. Besides, carbohydrates are the body's prime source of energy. Despite common opinion, starches are not fattening—rather, fat is fattening.

Fasting. Some years ago, a popular diet had people forgoing food entirely and living on liquid protein drinks and vitamins. A few people on this regime died, probably because their bodies were forced to digest so much muscle that their heart muscles failed.

Liquid diets. While some liquid diets provide sufficient protein and vitamins, they may restrict the dieter to 400 calories a day or less. Because the body cannot burn fat quickly enough to compensate for so few calories, however, muscle is also digested. Most doctors do not recommend cutting calories to fewer than 1,200 a day on any diet.

Single-category diets. These programs call for a diet restricted entirely to one kind of food, such as fruits or vegetables or rice alone. However, no single category of food provides enough nutrients to maintain healthy body tissue. Some dieters in recent years had a dramatic revelation of the inadequacy of such diets—their hair and fingernails fell out!

High-carbohydrate, low-fat diets. Most of the sound diets fall into this category. High-carbohydrate, high-fiber foods are a good source of energy and nutrients, and most are low in fat. Combined with exercise, this kind of diet promotes gradual loss of body fat. Still, the number of calories allotted per day must be sufficient to prevent the body from consuming muscle tissue. A few high-carbohydrate diet plans on the market cut calories too severely.

Reprinted from the 1988 *Medical and Health Annual,* copyright 1987, with the permission of Encyclopaedia Britannica, Inc., Chicago, Illinois.

Dieting in particular causes people to purchase all sorts of specially prepared foods. Nutritionists advise that this is not necessary—simply eat a variety of wholesome foods in small quantities and exercise. Dieting should be considered a life-long project rather than a short-term one.

SUMMARY

In the mouth, food is chewed and starch is acted upon by salivary amylase. After swallowing, peristaltic action moves the food along the esophagus to the stomach. Here pepsin, in the presence of HCl, acts on protein. In contrast, the small intestine has an alkaline environment. Here fat is emulsified by bile salts to fat droplets before being acted upon by pancreatic lipase. Protein is digested by pancreatic trypsin, and starch is digested by pancreatic amylase. The cells that line the intestinal wall secrete intestinal enzymes to finish the digestion of proteins and carbohydrates. Only nondigestible material passes from the small intestine to the large intestine. The large intestine absorbs water from this material. It also contains a large population of bacteria that can use the material as food. In the process, the bacteria produce vitamins that can be absorbed and used by our bodies.

The walls of the small intestine have fingerlike projections called villi within which are blood capillaries and a lymphatic lacteal. Amino acids and glucose enter the blood; glycerol and fatty acids enter the lymph. The blood from the small intestine moves into the hepatic portal vein, which goes to the liver, an organ that monitors and contributes to blood composition.

A balanced diet is required for good health. Food should provide us with all necessary vitamins, amino acids, fatty acids, and an adequate amount of energy. If the caloric value of food consumed is greater than that needed for bodily functions and activity, weight gain will occur.

OBJECTIVE QUESTIONS

1. In the mouth, salivary _____ digests starch to _____ .
2. When swallowing, the _____ covers the opening to the larynx.
3. The _____ takes food to the stomach where _____ is primarily digested.
4. The gallbaldder stores _____ , a substance that _____ fat.
5. The pancreas sends digestive juices to the _____ , the first part of the small intestine.
6. Pancreatic juice contains _____ for digesting protein, _____ _____ for digesting starch, and _____ for digesting fat.
7. Whereas pepsin prefers a _____ _____ pH, the enzymes found in pancreatic juice prefer a _____ _____ pH.
8. The products of digestion are absorbed into the cells of the _____ , fingerlike projections of the intestinal wall.
9. After eating, the liver stores glucose as _____ .
10. The diet should include a complete protein source, one that includes all the _____ _____ .

Answers

1. amylase, maltose 2. epiglottis 3. esophagus, protein 4. bile, emulsifies 5. duodenum 6. trypsin, pancreatic amylase, lipase 7. strongly acidic, slightly basic 8. villi 9. glycogen 10. essential amino acids

STUDY QUESTIONS

1. List the parts of the digestive tract, anatomically describe them, and state the contribution of each to the digestive process. (pp. 71–72)
2. List the accessory glands and describe the part that they play in the digestion of food. (p. 76)
3. Discuss the absorption of the products of digestion into the circulatory system. (p. 78)
4. List six functions of the liver. How does the liver maintain a constant glucose level in the blood? (p. 80)
5. What is jaundice? cirrhosis of the liver? (p. 81)
6. What is the common intestinal bacterium? What do these bacteria do for us? (p. 82)
7. What are gastrin, secretin, and CCK? Where are they produced? What are their functions? (p. 77)
8. Discuss the digestion of starch, protein, and fat, listing all the steps that occur to bring about digestion of each of these. (pp. 74–78)
9. What factors determine how many calories should be ingested? (pp. 92–93)
10. Give reasons why carbohydrates, fats, proteins, vitamins, and minerals are all necessary to good nutrition. (pp. 83–92)

THOUGHT QUESTIONS

1. Why would you expect hormones rather than the nervous system to control the release of digestive enzymes?
2. Discuss this sentence: Because humans have a "tube within a tube" body plan (1) the lumen of the gut is outside the body and (2) there is specialization of parts.
3. Reexamine figure 4.2 and give examples to show that structure suits the function.

KEY TERMS

amylase (am′i-lās)   a starch-digesting enzyme secreted by the salivary glands (salivary amylase) and the pancrease (pancreatic amylase). *74, 78*

CCK   cholecystokinin; hormone produced by the duodenum that stimulates release of bile from the gallbladder. *77*

colon (ko′lon)   the large intestine of vertebrates. *81*

epiglottis (ep″i-glot′is)   a structure that covers the glottis during the process of swallowing. *74*

esophagus (ē-sof′ah-gus)   a tube that transports food from the mouth to the stomach. *74*

gastric gland (gas′trik gland)   gland within the stomach wall that secretes gastric juices. *75*

gastrin (gas′trin)   a hormone secreted by stomach cells to regulate the release of pepsin by the stomach wall. *77*

glottis (glot′is)   slitlike opening between the vocal cords. *74*

hard palate (hard pal′at)   anterior portion of the roof of the mouth that contains several bones. *74*

hydrolytic enzyme (hi-dro-lit′ik en′zīm)   an enzyme that catalyzes a reaction in which the substrate is broken down with the addition of water. *74*

lipase (li′-pās)   a fat-digesting enzyme secreted by the pancreas. *78*

pepsin (pep′sin)   a protein-digesting enzyme secreted by the gastric glands. *76*

peristalsis (per′i-stal′sis)   a rhythmical contraction that serves to move the contents along in tubular organs such as the digestive tract. *75*

pharynx (far′ingks)   a common passageway (throat) for both food intake and air movement. *74*

salivary gland (sal′i-ver-e gland)   a gland associated with the mouth that secretes saliva. *74*

secretin (se-kre′tin)   hormone secreted by the small intestine that stimulates the release of pancreatic juice. *77*

soft palate (soft pal′at)   entirely muscular posterior portion of the roof of the mouth. *74*

sphincter (sfingk′ter)   a muscle that surrounds a tube and closes or opens the tube by contracting and relaxing. *76*

trypsin (trip′sin)   a protein-digesting enzyme secreted by the pancreas. *78*

villi (vil′i)   fingerlike projections that line the small intestine and function in absorption. *78*

vitamin (vi′tah-min)   essential requirement in the diet, needed in small amounts, that is often a part of coenzymes. *89*

ANSWER BOX—HOW DOES YOUR DIET RATE FOR VARIETY?

Compare your answers to the best answer listed below.

1. **ALMOST DAILY.** Many people believe that eating breads and cereals will make you fat. That's not true for most of us. Extra calories often come from the fat and/or sugar you MAY eat with them. Both whole grain and enriched breads and cereals provide starch and essential nutrients.

2. **ALMOST DAILY.** Whole grain breads and cereals contain vitamins, minerals, and dietary fiber that are low in the diets of many Americans. Select whole grain cereals and bakery products—those with a whole grain listed first on the ingredient label, or make your own and use whole wheat flour.

3. **ALMOST DAILY.** Vegetables vary in the amounts of vitamins and minerals they contain, so it's important to include several kinds every day.

4. **3 to 4 TIMES A WEEK.** Dry beans and peas fit into two food groups because of the nutrients they provide. They can be used as an alternate to meat, poultry, and fish. They are also an excellent vegetable choice.

5. **3 to 4 TIMES A WEEK.** Popeye gulped down spinach to build his superior strength. Although this effect of spinach was exaggerated, spinach and other dark green leafy vegetables are excellent sources of some nutrients that are low in many diets.

6. **ALMOST DAILY.** Fruits are nature's sweets. They taste good and are good for you. Choose several different kinds each day.

7. **ALMOST DAILY.** Adults as well as children need the calcium and other nutrients found in milk, cheese, and yogurt.

8. **ALMOST DAILY.** Most Americans include some meat, poultry, or fish in their diets regularly. Dry beans and peas, peanuts (including peanut butter), nuts and seeds, and eggs can be used as alternates.

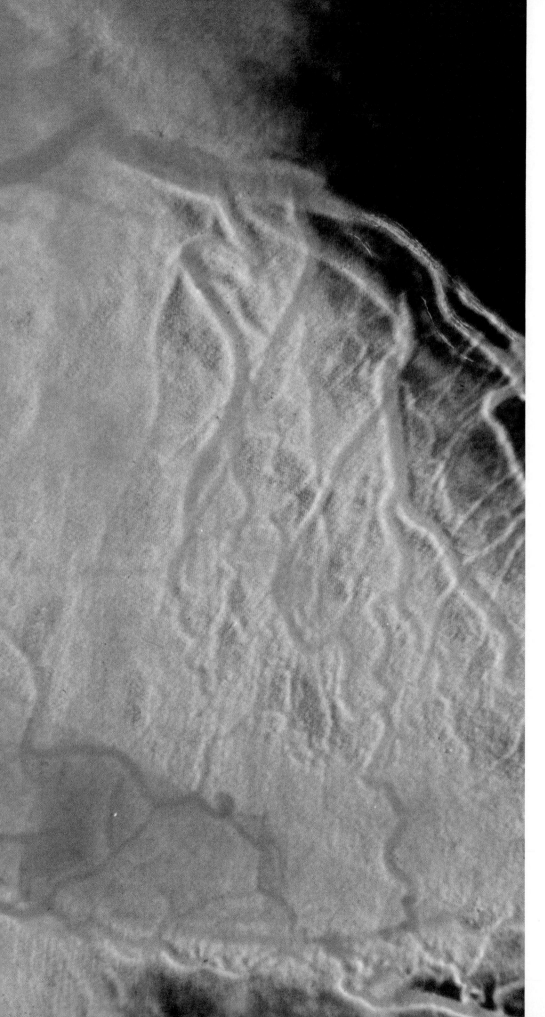

CHAPTER FIVE

CIRCULATION

CHAPTER CONCEPTS

1 In human beings, the blood, kept in motion by the pumping of the heart, circulates through a series of vessels.

2 The heart is actually a double pump: the right side pumps blood to the lungs and the left side pumps blood to the rest of the body.

3 The lymph vessels form a one-way lymphatic system that transports lymph from the tissues to certain cardiovascular veins.

4 Although the circulatory system is very efficient, it is still subject to various degenerative illnesses that in large measure can be prevented by good health habits.

An arteriograph (angiograph) of the heart showing evidence of multiple stenosis (narrowing) of a coronary artery on the top right side of the heart.

Life is dependent on the proper functioning of the coronary blood vessels (fig. 5.1), little tubes that serve the needs of cardiac muscle. A common circulatory problem today is blockage of the coronary arteries so that they are unable to function properly. Although coronary heart disease develops slowly over the years, a heart attack may come on quite suddenly. Evidence is growing that coronary heart disease (CHD) may be preventable in part, but there is no quick fix. A lifetime of devotion to these little vessels is required, and good health habits are a necessity. A now-famous study for over 38 years of 6,000 residents in the city of Framingham, Massachusetts, has helped investigators determine that cigarette smoking, elevated blood cholesterol, and the presence of hypertension all predispose one to CHD. Other important factors are lack of exercise, obesity, stress, and a family history of coronary artery disease.

Single-celled organisms don't have need of a circulatory system. Their watery environment brings them their food and removes their wastes. But most of our 60–100 trillion cells are far removed from the external environment and need to be serviced. It is the circulatory system that brings them their daily supply of nutrients, such as amino acids and glucose, and takes away their wastes, such as carbon dioxide and ammonia. At the center of the system is the heart (fig. 5.2), which keeps the blood moving along its predetermined circular path. Circulation of the blood is so important that if the heart discontinues beating for only a few minutes, death will result.

HUMAN ISSUE

People hold a number of different views about the pros and cons of a healthy life-style. We are told that eating and exercising correctly and avoiding smoking and drinking can add years to our lives. Yet, many people choose to smoke and drink and have not committed themselves to an exercise program. Is it wrong for people to live this way? Should living longer be the ultimate goal of adopting a healthier life-style?

CARDIOVASCULAR SYSTEM

Blood Vessels

The blood vessels are arranged so that they continually carry blood from the heart to the tissues and then return it from the tissues to the heart. Blood vessels (fig. 5.3) are of three types: the **arteries** (and **arterioles**) carry blood away from the heart; the **capillaries** exchange material with the tissues; and the **veins** (and **venules**) return blood to the heart.

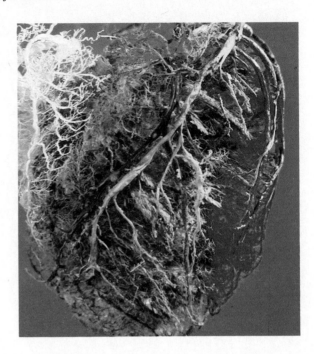

FIGURE 5.1 *A cast of the coronary blood vessels and their major branches.*

Arteries and Arterioles

Arteries have thick walls (fig. 5.4) because in addition to an inner endothelial layer and an outer connective tissue layer they have a thick middle layer of elastic and muscle fibers. The elastic fibers enable an artery to expand to accommodate the sudden increase in blood volume that results after each heartbeat. Arterial walls are so thick that the walls themselves are supplied with blood vessels. The *arterioles* are small arteries just visible to the naked eye. The middle layer of these vessels has some elastic tissue but is composed mostly of smooth muscle having fibers that encircle the arteriole. The contraction of the smooth muscle cells is under involuntary control by the autonomic nervous system (p. 207). If the muscle fibers contract, the bore of the arteriole gets smaller; if the fibers relax, the bore of the arteriole enlarges. Whether arterioles are constricted or dilated affects blood pressure. The greater the number of vessels dilated, the lower the blood pressure.

Capillaries

Arterioles branch into small vessels called capillaries. Each one is an extremely narrow, microscopic tube with a wall composed of only one layer of endothelial cells (fig. 5.4). *Capillary beds* (a network of many capillaries) are present in all regions of the body; consequently, a cut to any body tissue draws blood. The capillaries are the most important part of a closed circulatory system because an exchange of nutrient and waste molecules takes place across their thin walls. Oxygen and glucose diffuse out of a capillary into the tissue fluid that surrounds cells, and carbon dioxide and ammonia diffuse into the capillary (fig. 6.9). Since it is the capillaries that serve the needs of the cells, the heart and other

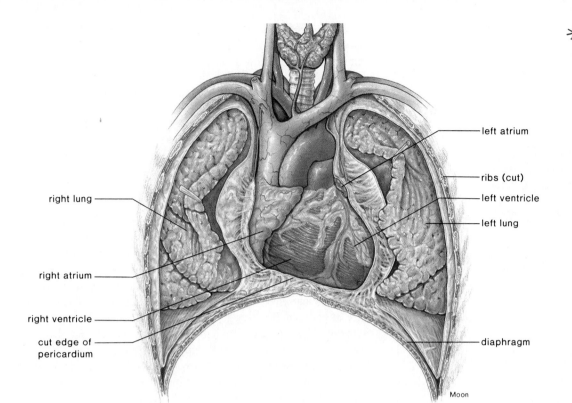

left atrium

ribs (cut)

left ventricle

left lung

diaphragm

right lung

right atrium

right ventricle

cut edge of pericardium

Moon

vessels of the circulatory system can be thought of as a means by which blood is conducted to and from the capillaries.

Not all capillary beds (fig. 5.5) are open or in use at the same time. After eating, the capillary beds of the digestive tract are usually open; during muscular exercise, the capillary beds of the skeletal muscles are open. Most capillary beds have a shunt that allows blood to move directly from the arteriole to the venule when the capillary bed is closed. There are sphincter muscles that encircle the entrance to each capillary. These are constricted, preventing blood from entering the capillaries, when the bed is closed and relaxed when the bed is open. As would be expected, the larger the number of capillary beds open, the lower the blood pressure.

Veins and Venules

Veins and venules take blood from the capillary beds to the heart (fig. 5.3). First, the *venules* drain the blood from the capillaries and then join together to form a vein. The wall of a venule is much thinner than that of an arteriole or artery because the middle layer of muscle and elastic fibers is poorly developed (fig. 5.4). Within some veins, especially in the major veins of the arms and legs, there are **valves** (fig. 5.17) that allow blood to flow only toward the heart when they are open and prevent the backward flow of blood when they are closed.

At any given time, more than half of the total blood volume is found in the veins and venules. If a loss of blood occurs, for example, due to hemorrhaging, nervous stimulation causes the veins to constrict, providing more blood to the rest of the body. In this way, the veins act as a blood reservoir.

FIGURE 5.3 *Diagram illustrating the path of blood. Blood leaving the heart moves from an artery to arterioles to capillaries to venules and then returns to the heart by way of a vein. Thus arteries are vessels that take blood away from the heart, and veins are vessels that return blood to the heart.*

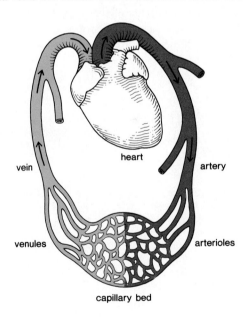

vein

heart

artery

venules

arterioles

capillary bed

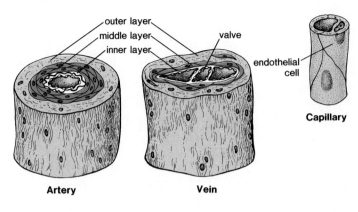

Arteries and arterioles carry blood away from the heart, veins and venules carry blood to the heart, and capillaries join arterioles to venules.

Heart

The **heart** is a cone-shaped, muscular organ, about the size of a fist (fig. 5.6). It is located between the lungs, directly behind the sternum, and is tilted so that the apex is directed to the left. The major portion of the heart is called the **myocardium,** which consists largely of cardiac muscle tissue. The muscle fibers within the myocardium are branched and joined

to one another so tightly that, prior to studies with the electron microscope, it was thought that they formed one continuous muscle; now it is known that there are individual fibers. The inner surface of the heart is lined with endothelial tissue called *endocardium,* which resembles squamous epithelium. The outside of the heart is covered with an epithelial and fibrous tissue called *pericardium,* which forms a sac called the pericardial sac, within which the heart is located. Normally, this sac contains a small quantity of liquid to lubricate the heart.

Internally, the heart has a right and a left side, separated by the **septum** (fig. 5.7). The heart has four chambers: two upper, thin-walled **atria** (singular, atrium), sometimes called auricles, and two lower, thick-walled **ventricles.** The atria are much smaller than the strong, muscular ventricles.

The heart also has valves that direct the flow of blood and prevent a backflow. The valves that lie between the atria and ventricles are called the **atrioventricular valves.** The valves are supported by strong fibrous strings called chordae tendineae. These cords, which are attached to muscular projections of the ventricular walls, support the valves and prevent them from inverting. The atrioventricular valve on the right side is called the *tricuspid valve* because it has three leaflets, or flaps; and the valve on the left side is called the *bicuspid,* or *mitral, valve* because it has two flaps. There are also **semilunar valves,** which resemble half-moons, between the ventricles and their attached vessels.

Humans have a four-chambered heart (two atria and two ventricles), in which the right side is separated from the left by a septum.

FIGURE 5.5 *Anatomy of a capillary bed. Capillary beds form a matrix of vessels that lie between an arteriole and a venule. a. Sphincter muscles are found at the junctions between an arteriole and capillaries. When these are contracted, the 0capillary bed is closed. Blood moves from the arteriole to the venule by way of a shunt. b. When a capillary bed is open, blood moves freely in the matrix of vessels making up the bed. If all capillary beds were open at the same time, an individual would suffer very severe low blood pressure.*

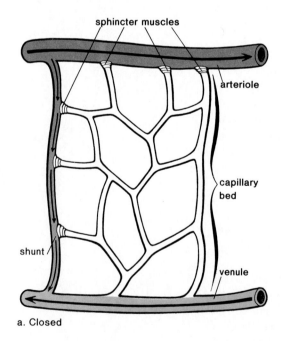

a. Closed

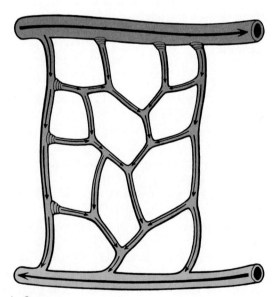

b. Open

FIGURE 5.6 a. *External heart anatomy. The coronary arteries bring oxygen and nutrients to the heart muscle. Should they fail to do so, the individual suffers a heart attack.* b. *An arteriogram of the left coronary artery. Injection of a dye into the coronary arteries makes them visible on X-ray film.*

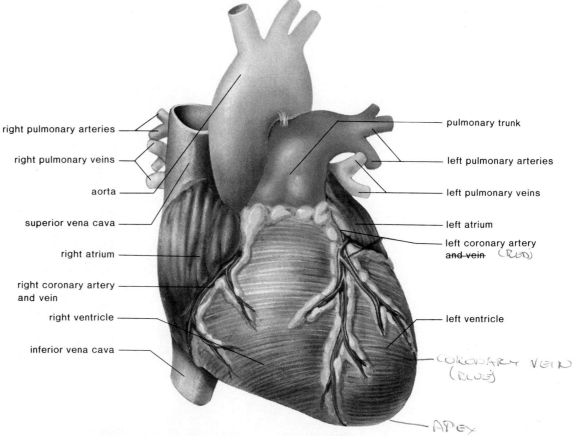

right pulmonary arteries

right pulmonary veins

aorta

superior vena cava

right atrium

right coronary artery
and vein

right ventricle

inferior vena cava

pulmonary trunk

left pulmonary arteries

left pulmonary veins

left atrium

left coronary artery
~~and vein~~ (RLED)

left ventricle

CORONARY VEIN
(RLUE)

APEX

a.

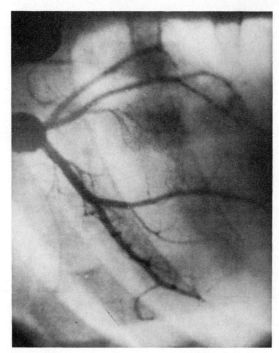

b.

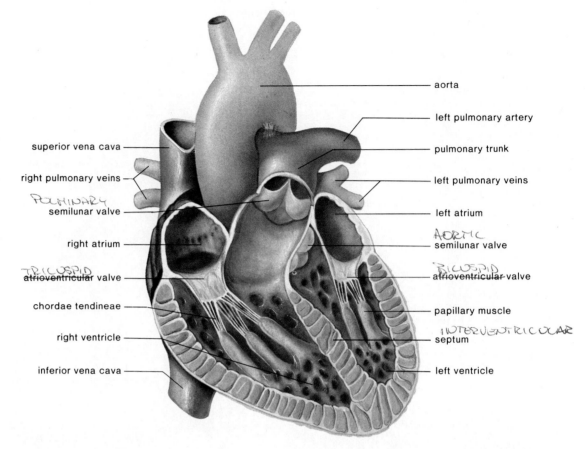

aorta

left pulmonary artery

pulmonary trunk

superior vena cava

right pulmonary veins

POLMINARY
semilunar valve

left pulmonary veins

left atrium

AORTIC
semilunar valve

right atrium

TRICUSPID
atrioventricular valve

BICUSPID
atrioventricular valve

chordae tendineae

papillary muscle

INTERVENTRICULAR
septum

right ventricle

inferior vena cava

left ventricle

Double Pump

The right side of the heart sends blood through the lungs, and the left side sends blood throughout the body (fig. 5.8). Therefore, there are actually two circular paths (circuits) of the blood: (1) from the heart to the lungs and back to the heart and (2) from the heart to the body and back to the heart. The right side of the heart is a pump for the first of these circuits, and the left side of the heart is a pump for the second. Thus the heart is a double pump. Since the left ventricle has the harder job—it pumps blood to all the body—its walls are much thicker than those of the right ventricle.

Path of Blood in the Heart   It is possible to trace the path of blood through the heart in the following manner (figs. 5.7 and 5.8). Blood low in oxygen and high in carbon dioxide enters the right atrium from the **superior** and **inferior venae cavae,** the largest veins in the body. Contraction of the right atrium forces the blood through the tricuspid valve to the right ventricle. The right ventricle pumps it through a semi-lunar valve, called the *pulmonary semilunar valve,* which allows blood to enter the pulmonary trunk. The pulmonary trunk divides into the **pulmonary arteries,** which take blood to the lungs. From the lungs, blood high in oxygen and low in carbon dioxide enters the left atrium from the **pulmonary**

FIGURE 5.8 *Diagram of pulmonary and systemic systems. The blue-colored vessels carry deoxygenated blood, while the red-colored vessels carry oxygenated blood. Notice that the blood cannot move from the right side of the heart to the left side without passing through the lungs.*

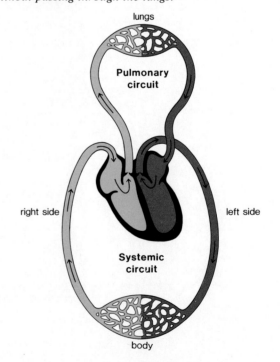

lungs

Pulmonary circuit

right side

left side

Systemic circuit

body

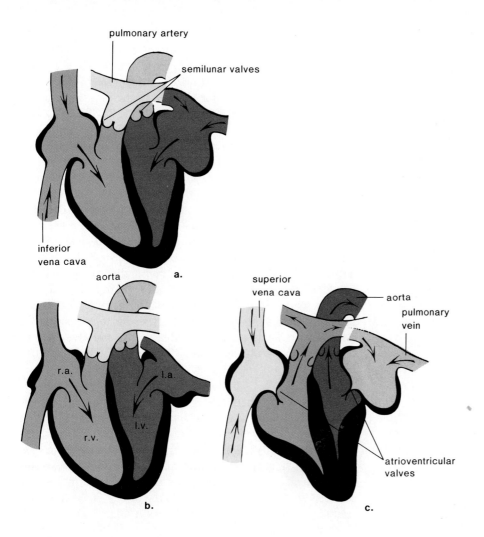

pulmonary artery

semilunar valves

inferior
vena cava

a.

aorta

r.a.　l.a.

r.v.　l.v.

b.

superior
vena cava

aorta
pulmonary
vein

atrioventricular
valves

c.

FIGURE 5.9 *Stages in the cardiac cycle*. a. *When the heart is relaxed, both atria and ventricles are filling with blood.* b. *When the atria contract, the ventricles are relaxed and filling with blood.* c. *When the ventricles contract, the atrioventricular valves are closed, the semilunar valves are open, and blood is pumped into the pulmonary artery and aorta.*

veins. Contraction of the left atrium forces blood through the bicuspid valve into the left ventricle. The left ventricle pumps it through a semilunar valve, called the *aortic semilunar valve,* into the **aorta,** the largest artery in the body. The aorta sends blood to all body tissues. Notice that oxygen-poor blood never mixes with oxygen-rich blood and that blood must pass through the lungs before entering the left side of the heart.

The right side of the heart pumps blood to the lungs, and the left side pumps blood to the tissues.

Heartbeat

Cardiac Cycle

From this description of the path of blood through the heart, it might seem that the right and left sides of the heart beat independently of one another, but actually they contract together. First, the two atria contract simultaneously; then the two ventricles contract at the same time. The word **systole** refers to contraction of heart muscle, and the word **diastole** refers to relaxation of heart muscle; thus, atrial systole is followed by ventricular systole. The heart contracts, or beats,

about seventy times a minute, and each heartbeat lasts about 0.85 second. Each heartbeat, or *cardiac cycle* (fig. 5.9), consists of the following elements:

Time	Atria	Ventricles
0.15 sec.	Systole	Diastole
0.30 sec.	Diastole	Systole
0.40 sec.	Diastole	Diastole

This shows that while the atria contract, the ventricles relax, and vice versa, and that all chambers rest at the same time for 0.40 second. The short systole of the atria is appropriate since the atria send blood only into the ventricles. It is the muscular ventricles that actually pump blood out into the circulatory system proper. When the word *systole* is used alone, it usually refers to the left ventricular systole.

The heartbeat is divided into two phases. First the atria contract, and then the ventricles contract. When the atria are in systole the ventricles are in diastole, and vice versa.

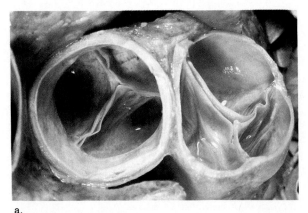

a.

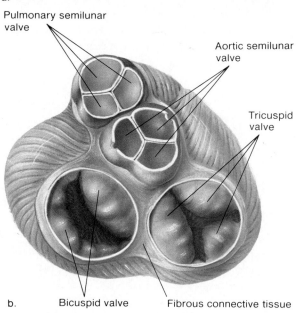

Pulmonary semilunar valve

Aortic semilunar valve

Tricuspid valve

b.   Bicuspid valve   Fibrous connective tissue

Heart Sounds

When the heart beats, the familiar lub-DUPP sound may be heard as the valves of the heart close. The lub is caused by vibrations of the heart when the atrioventricular valves close, and the DUPP is heard when vibrations occur due to the closing of the semilunar valves (fig. 5.10). Heart murmurs, or a slight slush sound after the lub, are often due to ineffective valves that allow blood to pass back into the atria after the atrioventricular valves have closed. Rheumatic fever resulting from a strep infection is one cause of a faulty valve, particularly the mitral valve. If operative procedures are unable to open and/or restructure the valve, it may be replaced by an artificial valve.

The heart sounds are due to the closing of the heart valves.

Cardiac Conduction System

The heart will beat independently of any nervous stimulation. In fact, it is possible to remove the heart of a small animal, such as a frog, and watch it undergo contraction in a petri dish. The reason for this lies in the fact that there is a unique type of tissue called nodal tissue, having both muscular and nervous characteristics, located in two regions of the heart. The first of these, the **SA (sinoatrial) node,** is found in the upper dorsal wall of the right atrium; the other, the **AV (atrioventricular) node,** is found in the base of the right atrium very near the septum (fig. 5.11). The SA node, or the pacemaker, initiates the heartbeat and automatically sends out an excitation impulse every 0.85 second to cause the atria to contract. When the impulse reaches the AV node, it signals the ventricles to contract by way of specialized fibers called Purkinje fibers. The SA node is called the **pacemaker** because it usually keeps the heartbeat regular. If the SA node fails to work properly, the heart will still beat, but irregularly. To correct this condition, it is possible to implant in the body an artificial pacemaker that automatically gives an electric shock to the heart every 0.85 second. This causes the heart to beat regularly again.

Electrocardiogram (EKG)   With the contraction of any muscle, including the myocardium, ionic changes occur that can be detected by electrical recording devices. Therefore, it is possible to study the heartbeat by recording voltage changes that occur when the heart contracts. (Voltage, which in this case is measured in millivolts, is the difference in polarity between two electrodes attached to the body.) The record that results is called an **electrocardiogram** (fig. 5.11b), which clearly shows an atrial phase and a ventricular phase. The first wave in the electrocardiogram, called the P wave, represents the excitation and contraction of the atria. The second wave, or the QRS wave, occurs during ventricular excitation and contraction. The third, or T, wave is caused by the recovery of the ventricles. An examination of the electrocardiogram indicates whether the heartbeat has a normal or an irregular pattern.

The conduction system of the heart includes the SA node, the AV node, and the Purkinje fibers. With an EKG, it is possible to determine if the conduction system, and therefore the beat of the heart, is regular.

Nervous Control of the Heartbeat

The rate of the heartbeat is also under nervous control. There is a cardiac center in the medulla oblongata (p. 209) of the brain, which can alter the beat of the heart by way of the *autonomic nervous system* (p. 207). This system is made up of two divisions: the *parasympathetic,* which promotes those functions we tend to associate with normal activities; and the *sympathetic system,* which brings about those responses we associate with times of stress. For example, the parasym-

FIGURE 5.11 *Control of the heart cycle.* a. *The SA node sends out a stimulus that causes the atria to contract. When this stimulus reaches the AV node it signals the ventricles to contract by way of the Purkinje fibers.* b. *A normal EKG indicates that the heart is functioning properly. The P wave indicates that the atria have contracted; the QRS wave indicates that the ventricles have contracted; and the T wave indicates that the ventricles are recovering from contraction.*

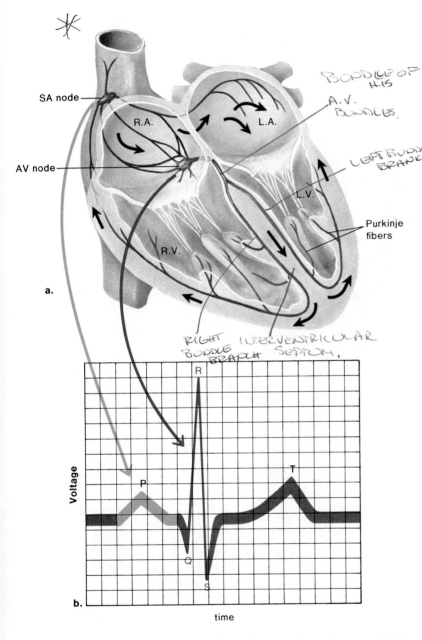

Vascular System

The vascular system, which is represented in figure 5.12, can be divided into two circuits: the **pulmonary circuit,** which circulates blood through the lungs, and the **systemic circuit,** which serves the needs of the body's tissues.

Pulmonary Circuit

The path of blood through the lungs can be traced as follows. Blood from all regions of the body first collects in the right atrium and then passes into the right ventricle, which pumps it into the pulmonary trunk. The pulmonary trunk divides into the **pulmonary arteries,** which divide up into the arterioles of the lungs. The arterioles take blood to the pulmonary capillaries, where carbon dioxide and oxygen are exchanged. The blood then enters the pulmonary venules that lead to the **pulmonary veins** and through them back to the left atrium. Since the blood in the pulmonary arteries is low in oxygen and the blood in the pulmonary veins is high in oxygen, it is not correct to say that all arteries carry blood high in oxygen and all veins carry blood low in oxygen. It is just the reverse in the pulmonary system.

The pulmonary arteries take deoxygenated blood to the lungs, and the pulmonary veins return oxygenated blood to the heart.

Systemic Circuit

The systemic circuit includes all of the other arteries and veins shown in figure 5.12. The largest artery in the systemic circuit is the **aorta,** and the largest veins are the **superior** (anterior) and **inferior** (posterior) **venae cavae.** The superior vena cava collects blood from the head, chest, and arms, and the inferior vena cava collects blood from the lower body regions. Both enter the right atrium. The aorta and the venae cavae serve as the major pathways for blood in the systemic system.

The path of systemic blood to any organ in the body begins in the left ventricle, which pumps blood into the aorta. Branches from the aorta go to the major body regions and organs. For example, the path of blood to the kidneys may be traced as follows: left ventricle—aorta—renal artery—renal arterioles, capillaries, venules—renal vein—inferior vena cava—right atrium. To trace the path of blood to any organ in the body, you need only mention the aorta, the proper branch of the aorta, the organ, and the returning vein to the vena cava. In most instances the artery and vein that serve the same organ are given the same name (fig. 5.12). In the systemic circuit, unlike the pulmonary system, arteries contain oxygenated blood and appear a bright red, but veins contain deoxygenated blood and appear a purplish color.

The systemic circuit takes blood from the left ventricle of the heart to the right atrium of the heart. It serves the body proper.

---

pathetic system causes the heartbeat to slow down, and the sympathetic system increases the heartbeat. Various factors, such as the relative need for oxygen or the blood pressure level, determine which of these systems becomes activated.

The heart rate is regulated largely by the autonomic nervous system.

FIGURE 5.12 *Blood vessels in the pulmonary and systemic circulatory circuits. The blue-colored vessels carry deoxygenated blood, and the red-colored vessels carry oxygenated blood; the arrows indicate the flow of blood. Compare this figure to figure 5.13 for a more realistic view of the placement of the blood vessels.*

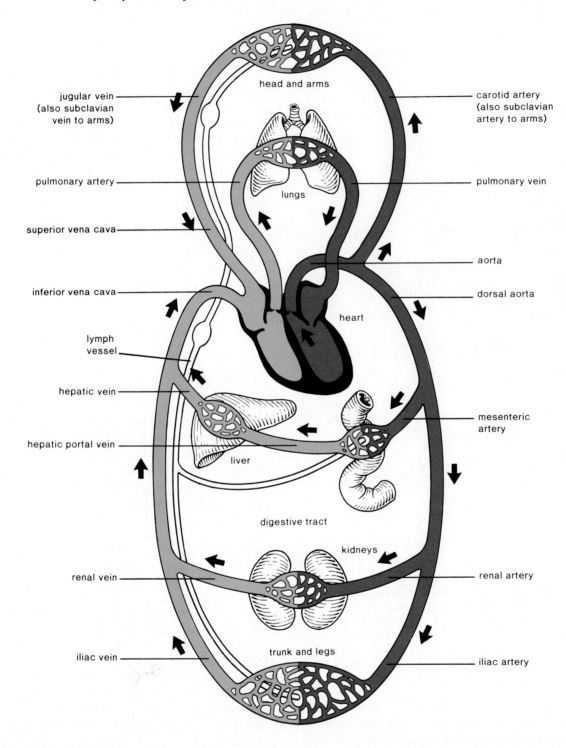

The **coronary arteries** (fig. 5.6), which are a part of the systemic circuit, are extremely important arteries because they serve the heart muscle itself. (The heart is not nourished by the blood in its chambers.) The coronary arteries arise from the aorta just beyond the aortic semilunar valve. They lie on the exterior surface of the heart, where they branch off in various directions into arterioles. The coronary capillary beds join to form venules. The venules converge into the coronary vein, which empties into the right atrium. Although the coronary arteries receive blood under high pressure, they have a very small diameter and may become blocked as discussed on page 114.

FIGURE 5.13 *Human circulatory system. A more realistic representation of the major blood vessels in the body shows that arteries and veins go to all parts of the body. The superior and inferior venae cavae take their names from their relationship to which organ?*

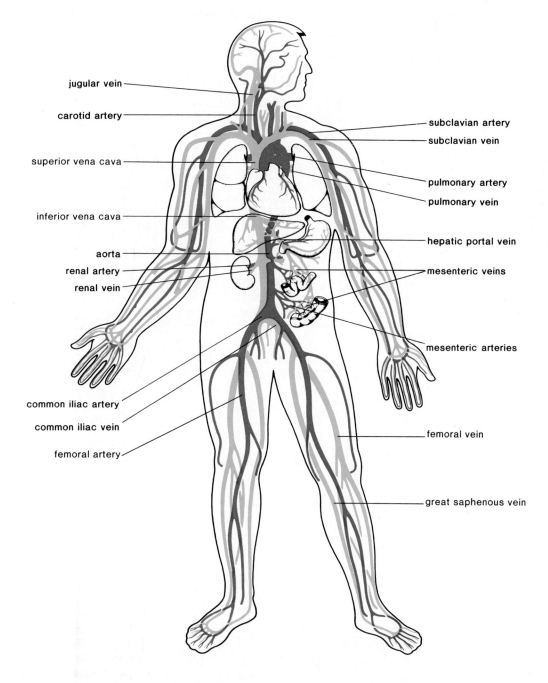

The body has an important portal system, the **hepatic portal system** (fig. 4.11), that takes blood from the intestine to the liver. A portal system is one that begins and ends in capillaries; in this case, the first set of capillaries occurs at the villi of the small intestine, and the second occurs in the liver. Blood passes from the capillaries of the villi into venules that join to form the **hepatic portal vein,** a vessel that enters the liver. Emerging from the liver, the hepatic vein enters the vena cava.

Two systemic circulations of special interest consist of the coronary blood vessels that serve the heart and the hepatic portal system that takes blood to the liver.

While figure 5.12 is helpful in tracing the path of the blood, it must be remembered that all parts of the body receive both arteries and veins, as illustrated in figure 5.13.

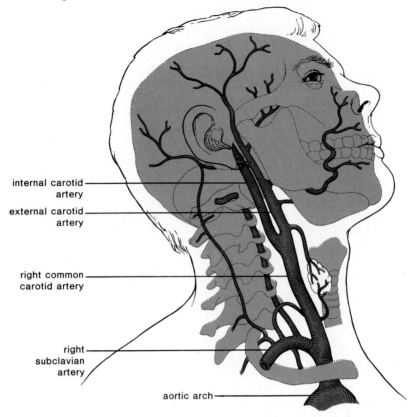

internal carotid artery

external carotid artery

right common carotid artery

right subclavian artery

aortic arch

FEATURES OF THE CIRCULATORY SYSTEM

When the left ventricle contracts, the blood is sent out into the aorta under pressure.

Pulse

The surge of blood entering the arteries causes their elastic walls to swell, but then they almost immediately recoil. This alternate expanding and recoiling of an arterial wall can be felt as a **pulse** in any artery that runs close to the surface. It is customary to feel the pulse by placing several fingers on the radial artery, which lies near the outer border of the palm side of the wrist. The carotid artery is another good location to feel the pulse (fig. 5.14). Normally the pulse rate indicates the rate of the heartbeat because the arterial walls pulse whenever the left ventricle contracts.

The pulse rate indicates the heartbeat rate.

Blood Pressure

Blood pressure is the pressure of the blood against the wall of a blood vessel.

Measurement of Blood Pressure

To measure blood pressure (fig. 5.15), a sphygmomanometer is used. This consists of a hollow cuff connected by tubing to a compressible bulb and to a pressure gauge. The cuff is placed about the upper arm over the brachial artery and inflated with air. Eventually the brachial artery becomes squeezed shut, and, when a stethoscope is placed just beneath the cuff, no sounds can be heard. Air is slowly released, and the cuff is deflated until a sharp sound can be heard through the stethoscope. The examiner glances at the manometer, or pressure gauge, and notes the pressure at this point. This is the value to be assigned to **systolic blood pressure,** the highest arterial pressure, reached during ejection of blood from the heart. The systolic pressure has overcome the pressure exerted by the cuff and has caused the blood to flow in the artery. The cuff is further deflated while the examiner continues to listen. The tapping sounds become louder as the pressure is lowered. Finally, the sounds become abruptly dull and muffled just before there are no sounds at all. Now the

FIGURE 5.15 *Determination of blood pressure using a sphygmomanometer. The technician inflates the cuff with air and then as he or she gradually reduces the pressure, he or she listens by means of a stethoscope for the sounds that indicate the blood is moving past the cuff. A pressure gauge on the cuff is used to tell the systolic and diastolic blood pressure.*

FIGURE 5.16 *Diagram illustrating how velocity and blood pressure are related to the total cross-sectional area of blood vessels. Capillaries have the greatest cross-sectional area and the least pressure and velocity. Skeletal muscle contraction, not blood pressure, accounts for the velocity of blood in the veins.*

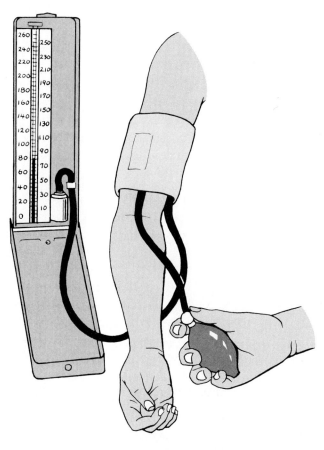

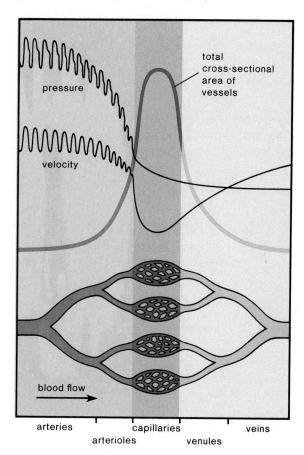

examiner again notes the pressure. This is **diastolic blood pressure,** the lowest arterial pressure. Diastolic pressure occurs while the heart ventricles are relaxing. Normal resting blood pressure for a young adult is said to be 120 mm of mercury (Hg) over 80 mm, or simply 120/80; the higher number is the systolic pressure and the lower number is the diastolic pressure.

Blood Pressure throughout the Body
Actually, 120/80 is the expected blood pressure in the brachial artery of the arm; blood pressure decreases with distance from the left ventricle. Blood pressure is, therefore, higher in the arteries than in the arterioles. Further, there is a sharp drop in blood pressure when the arterioles reach the capillaries. The decrease may be correlated with the increase in the total cross-sectional area of the vessels as blood moves through arteries, arterioles, and then into capillaries. There are more arterioles than arteries, and many more capillaries than arterioles (fig. 5.16).

Blood pressure steadily decreases from the aorta to the veins.

Velocity of Blood Flow

The velocity of blood varies in different parts of the circulatory system (fig. 5.16). Blood pressure accounts for the velocity of the blood flow in the arterial system and, therefore, as blood pressure decreases due to the increased cross-sectional area of the arterial system, so does velocity. The blood moves more slowly through the capillaries than it does through the aorta. This is important because the slow progress allows time for the exchange of molecules between the blood and the tissues.

FIGURE 5.17 *Skeletal muscle contraction moves blood in veins.* a. *Muscle contraction exerts pressure against vein and blood moves past valve.* b. *Once blood has passed the valve, it cannot slip back.*

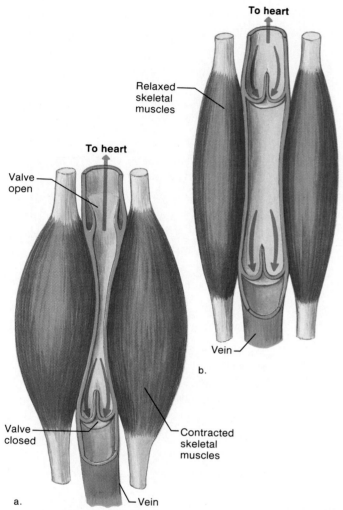

Blood pressure cannot account for the movement of blood through the venules and veins since they lie on the other side of the capillaries. Instead, movement of the blood through the venous system is due to skeletal muscle contraction. When the skeletal muscles contract, they press against the weak walls of the veins. This causes the blood to move past a *valve* (fig. 5.17). Once past the valve, the blood will not fall back. The importance of muscle contraction in moving blood in the venous system can be demonstrated by forcing a person to stand rigidly still for a number of hours. Frequently, fainting will occur because the blood collects in the limbs, robbing the brain of oxygen. In this case, fainting is beneficial because the resulting horizontal position aids in getting blood to the head. Blood flow gradually increases in the venous

system (fig. 5.16) due to a progressive reduction in the cross-sectional area as small venules join to form veins. The two venae cavae together have a cross-sectional area of only about double that of the aorta. The blood pressure is lowered in the chest cavity whenever the chest expands during inspiration. This also aids the flow of venous blood into the chest area because blood flows in the direction of reduced pressure.

Blood pressure accounts for the flow of blood in the arteries and arterioles; skeletal muscle contraction accounts for the flow of blood in the venules and veins.

CIRCULATORY DISORDERS

During the past thirty years, the number of deaths due to cardiovascular disease has declined more than 30%. Even so, more than 50% of all deaths in the United States are still attributable to cardiovascular disease. The number of deaths due to hypertension, stroke, and heart attack is greater than the number due to cancer and accidents combined.

Cardiovascular disease is the number one killer in the United States.

Hypertension

It is estimated that about 20% of all Americans suffer from *hypertension,* high blood pressure indicated by a blood pressure reading of over 160/95. Even though hypertension is easily detected by blood pressure readings, it is believed that at least one-third of people in whom hypertension has been detected were unaware they had this condition, which can lead to failure of the cardiovascular system.

The reasons for the development of hypertension are various. One possible scenario (fig. 5.18) has been described. Blood pressure normally rises with excitement or alarm due to the involvement of the sympathetic nervous system (p. 207), which causes arterioles to constrict and the heart to beat faster. When arterioles are constricted, reduced blood flow to the kidneys causes them to release renin, a molecule that brings about retention of sodium in the blood. This excess sodium in turn leads to water retention and high blood pressure. The same effect can be brought about directly by an excess intake of salt (sodium chloride) in the diet.

Medical treatment for hypertension is based on this sequence of events (fig. 5.18). Sympathetic-blocking agents act at arrow #1 and prevent action of the sympathetic nervous system. Vasodilators act at arrow #2 and prevent the arteries from constricting. Diuretics act at arrow #3 and cause the kidneys to excrete excess salts and fluids.

Diet, stress, and kidney involvement are implicated in the development of hypertension in some persons.

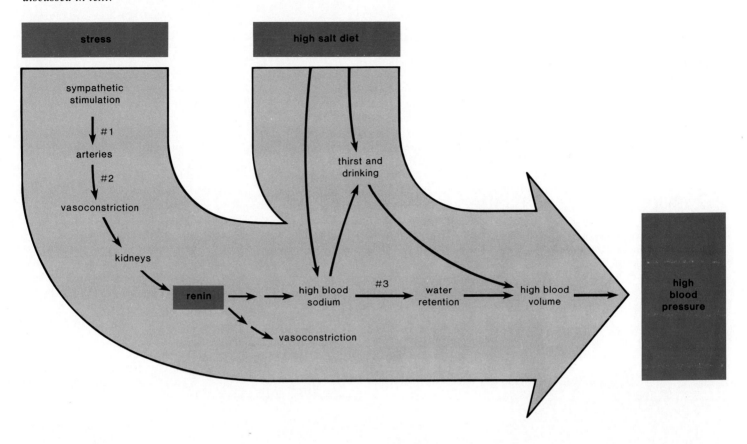

FIGURE 5.18 *A scheme that explains the development of high blood pressure due to either stress or a high salt diet. The numbers (#1–3) indicate the action site of hypertensive drugs, as discussed in text.*

Atherosclerosis

Hypertension is also seen in individuals who have *athero-sclerosis,* an accumulation of soft masses of fatty materials, particularly cholesterol, beneath the inner linings of arteries. Such deposits are called *plaque,* and as they develop they tend to protrude into the vessel and interfere with the flow of blood. Atherosclerosis begins in early adulthood, develops progressively through middle age, but symptoms may not appear until an individual is 50 or older. To prevent its onset and development, a diet low in saturated fats and cholesterol is recommended by the American Heart Association and other organizations, as discussed in the reading on page 114.

The occurrence of plaque can cause a clot to form on the irregular arterial wall. As long as the clot remains stationary, it is called a *thrombus,* but when and if it dislodges and moves along with the blood, it is called an *embolus.* If *thromboembolism* is not treated, complications can arise as mentioned in the following section.

Development of atherosclerosis, which is associated with a high level of blood cholesterol, can lead to thromboembolism.

Stroke and Heart Attack

Both strokes and heart attacks are associated with hypertension and atherosclerosis. A *stroke* occurs when a portion of the brain dies due to a lack of oxygen. A stroke, characterized by paralysis or death, often results when a small arteriole bursts or becomes blocked by an embolus.

A *heart attack* occurs when a portion of the heart dies because of a lack of oxygen. Due to atherosclerosis, the coronary artery may be partially blocked. The individual may then suffer from *angina pectoris,* characterized by a radiating pain in the left arm. When a coronary artery becomes completely blocked, perhaps because of thromboembolism, a heart attack occurs.

Stroke and heart attack are associated with both hypertension and atherosclerosis.

CORONARY HEART DISEASE

a. *Plaques (yellow) in the coronary artery of a heart patient. b. Cross section of a plaque shows its composition and indicates how it bulges out into the lumen of an artery, obstructing blood flow.*

Even though the United States has seen a decline in deaths due to cardiovascular disease, more than 800,000 heart attacks occurred during 1987 and of these only 300,000 survived. It seems desirable, then, to identify those factors that predispose one to developing coronary heart disease.

A physician can make two routine office measurements that will indicate if a person is at greater risk of a heart attack than average. He or she can measure the resting blood pressure, and if it is 160/95 or higher, there is reason for concern, and antihypertensive drugs can be prescribed (p. 112). A physician can also measure the blood cholesterol level. The National Heart, Lung, and Blood Institute, which believes that all Americans over the age of 20 should be tested at least every five years, has recently set cholesterol-level standards for adults. If the cholesterol level is

200–239 mg/100 ml the person should take precautionary measures and be retested annually. If the level is greater than 240 mg/100 ml then more refined cholesterol testing is needed immediately.

Cholesterol is a substance found within the cell membrane of all cells. It is converted to hormones by certain glands. We now know that cholesterol is ferried in the bloodstream by either low-density lipoproteins (LDL) or high-density lipoproteins (HDL). LDL takes cholesterol to the tissues, and HDL transports cholesterol out of the tissues. The cholesterol-LDL molecular combination is atherogenic—it leads to atherosclerosis. When these molecules adhere to arterial walls a series of events occur that eventually result in an accumulated mass known as plaque (see the figure). This plaque can eventually

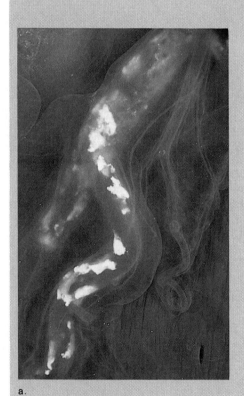

a.

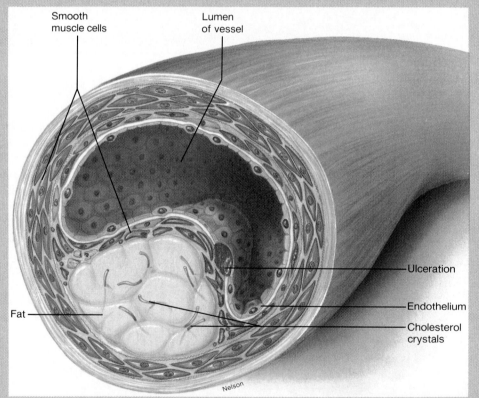

Smooth muscle cells

Lumen of vessel

Ulceration

Endothelium

Fat

Cholesterol crystals

Nelson

b.

grow so large that it hinders or even stops blood flow through the artery. A plaque also can break open and release blood clots.

The NHLBI recommends that a person with a cholesterol level of over 240 mg/100 ml should be further tested to determine the level of LDL-cholesterol. Those with an LDL level of 160 mg/100 ml and above are considered at high risk, and those between 130 and 159 mg/100 ml also require treatment. First and foremost, treatment consists of adopting a diet that is low in saturated fat and cholesterol (see table). Although the prescribed diet does not lower blood cholesterol in all persons, it is expected to do so for most individuals.

If diet alone does not bring down the cholesterol level, drugs can be prescribed. The drugs cholestyramine and colestipol act in the intestines to remove bile, which is derived in part from cholesterol, and a new drug, lovastatin, inhibits the production of cholesterol in the liver. These drugs do reduce the blood level of cholesterol, but the long-term side effects are not completely known and may be serious.

Aside from the presence of hypertension and an elevated blood cholesterol level, certain behaviors are associated with the development of coronary heart disease. These are smoking cigarettes and excessive weight gain. Smoking cigarettes has been found to be a major contributor to the development of coronary heart disease, and filter cigarettes carry as great a risk as the nonfiltered brands. Excessive weight is accompanied by both increased blood pressure and atherosclerosis. Since it is very difficult for obese individuals to lose weight, it is recommended that weight control begin before it gets out of hand. No doubt the prescribed low-cholesterol diet (see table) would be helpful.

Investigators have identified two behaviors that may help reduce the risk of heart attack and stroke. Exercise seems to be critical. Sedentary individuals have a risk of coronary heart disease that is about double that of those who are very active. One investigator, for example, recommends that his patients walk for one hour, three times a week. Reduction of stress is also desirable. The same investigator recommends meditating and doing yoga-like stretching and breathing exercises everyday to reduce stress.

Genetics also seems to play a role in the development of coronary heart disease—the disease tends to run in families. Although nothing can be done about this factor, the other factors are apt to be controlled if the individual believes it is worth the effort.

TABLE A *DIETARY REDUCTION OF BLOOD CHOLESTEROL LEVELS*

EAT LESS	EAT MORE
Fatty meats	Fish, poultry, lean cuts of meat
Organ meats (liver, kidney, brain, pancreas), shrimp	
Sausage, bacon, processed meats	Fruits, vegetables, cereals, starches
Whole milk	Skim or low-fat milk
Butter, hard margarine	Soft margarine
High-fat cheese (bleu, cheddar)	Low-fat cheese
Ice cream and cream	Yogurt
Egg yolks	Egg whites
Foods fried in animal fats	Vegetable fats for frying and salad dressing
Commercial baked goods	

Reprinted by permission from the 1986 *Medical and Health Annual*, copyright 1985, Encyclopaedia Britannica, Inc., Chicago, Illinois.

Surgical Treatment

Surgical treatments are now available for blocked coronary arteries.

Thrombolytic Therapy

Recently, two new technologies have been developed that do away with the obstruction without the need for a major operation. In both of these, a plastic tube is threaded into an artery of an arm or leg and guided through a major blood vessel toward the heart. In one procedure, when the tube reaches the clot, a balloon attached to the end of the tube inflates and breaks up the clot. In the other procedure, a drug called streptokinase is injected to dissolve the clot. Then the artery opens and the blood begins to flow again. The latter procedure is done even while a person is suffering a heart attack. If the clot is removed within six hours after a heart attack begins, there is usually no damage to the heart muscle. Streptokinase and the new drug tPA (tissue plasminogen activator) can also be given intravenously. tPA, produced by recombinant DNA technology (p. 429), has the advantage of activating the body's own enzyme—plasmin—to dissolve the clot.

There are procedures to clear blocked arteries that do not require major surgery.

Coronary Bypass Operations

As many as 100,000 persons a year have *coronary bypass surgery*. During this operation, surgeons take segments from another blood vessel, often a large vein in the leg, and stitch one end to the aorta and the other end to a coronary artery past the point of obstruction (fig. 5.19). Between 75% and 90% of those who have had bypass surgery say their angina pain has been relieved.

Once the heart is exposed, some physicians have used lasers to open up clogged coronary vessels. Presently the technique is used in conjunction with coronary bypass operations, but eventually it may be possible to use lasers without the need to open the thoracic cavity.

Donor Heart Transplants and Artificial Heart Implants

Persons with weakened hearts may eventually suffer from *congestive heart failure,* meaning that the heart is no longer able to pump blood adequately. These individuals are candidates for a donor heart transplant or even implantation of an artificial heart. The difficulties with a donor heart transplant are, first, one of availability and, second, the tendency of the body to reject foreign organs. It would be helpful to find ways to repair the heart instead of replacing it. For example, in one recent procedure, a surgeon took a back muscle and wrapped it around a heart weakened by the removal of myocardial tissue. Later he implanted an artificial pacemaker that caused the muscle to contract regularly and help pump the blood.

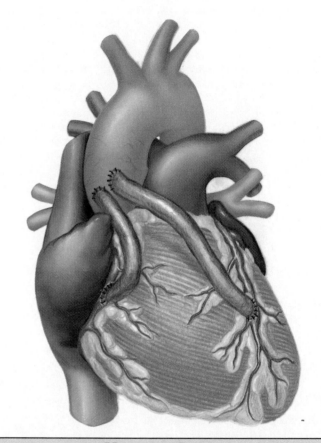

FIGURE 5.19 *Coronary bypass operation. During this operation, the surgeon grafts segments of another blood vessel between the aorta and the coronary vessels, bypassing areas of blockage. Patients who are ill enough to require such surgery often receive two or three bypasses in a single operation.*

HUMAN ISSUE

In June of 1986, the unwed parents of an infant known as "Baby Jesse" went on television to plead a case for their child. Baby Jesse had been turned down as a candidate for a heart transplant because hospital officials had doubted that his parents could properly care for him. While they were on television a heart was donated by a watching couple whose child was brain-dead at birth. Amidst the rejoicing came nagging questions about exactly how the recipients of donor hearts should be chosen. First, should various criteria—state of health, ability to pay, age, etc.—be used in choosing the recipient? Second, if such criteria are used, should independent citizens or only medical officials be involved in judging the ability of potential recipients to meet these criteria?

On December 2, 1982, Barney Clark became the first person to receive an artificial heart. The heart's two polyurethane ventricles were attached to Clark's own atria and blood vessels by way of Dacron fittings. Two long tubes were stretched between the artificial heart and an external machine that periodically sent bursts of air into the ventricles, forcing the blood out into the aorta and pulmonary trunk.

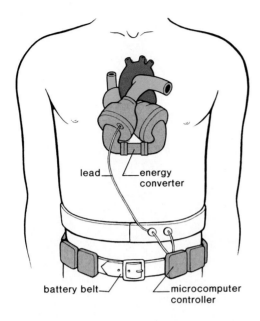

lead—— ——energy converter

battery belt—— ——microcomputer controller

Clark's body was permanently attached to an external machine, but it is hoped that eventually an artificial heart can be powered by batteries so that the patient will be completely mobile (fig. 5.20).

Coronary bypass, donor heart transplants, and artificial heart implantation all require major surgery.

Varicose Veins and Phlebitis

Varicose veins are abnormal and irregular dilations in superficial (near the surface) veins, particularly those in the lower legs. Varicose veins in the rectum, however, are commonly called piles, or more properly, *hemorrhoids.* Varicose veins develop when the valves of the veins become weak and ineffective due to a backward pressure of the blood. The problem can be aggravated when venous blood flow is obstructed by crossing the legs or by sitting in a chair so that its edge presses against the back of the knees.

Phlebitis, or inflammation of a vein, is a more serious condition, particularly when a deep vein is involved. Blood in the inflamed vessel may clot, in which case thromboembolism occurs. An embolus that originates in a systemic vein may eventually come to rest in a pulmonary arteriole, blocking circulation through the lungs. This condition, termed *pulmonary embolism,* can result in death.

Veins have weak walls, and this occasionally leads to medical disorders.

LYMPHATIC SYSTEM

The **lymphatic system** is closely associated with the cardiovascular system because it consists of vessels that take up excess tissue fluid and transport it to the bloodstream. **Tissue fluid** is the fluid that surrounds cells.

The lymphatic system is a one-way system that starts in the tissues and empties into the cardiovascular system.

Lymphatic Vessels

Lymph vessels consist of *lymph capillaries* and *lymph veins.* The latter have a construction similar to cardiovascular veins, including the presence of valves (fig. 5.17).

The lymphatic system (fig. 5.21) is a one-way system rather than a circulatory system. The system begins with lymph capillaries that lie near blood capillaries and take up fluid that has diffused from the capillaries and has not been reabsorbed by them (fig. 6.10). Once tissue fluid enters the lymph vessels, it is called **lymph.** Lymph also contains *lymphocytes,* a type of white blood cell (p. 133). Some lymphocytes produce antibodies, proteins that are capable of combining with foreign proteins called antigens. At times, the foreign proteins are associated with disease-causing bacteria and viruses, and therefore the lymphatic system helps fight this source of infection.

Lymph is collected in vessels that join to form two main trunks: the right lymphatic duct, which drains the upper right portion of the body, and the thoracic duct, which drains the rest of the body. The former empties into the right subclavian vein, and the latter into the left subclavian vein.

The *lacteals* are blind ends of lymph vessels found in the villi of the small intestine. As previously mentioned, the products of fat digestion enter the lacteals. These products eventually enter the cardiovascular system when the lymph ducts join the subclavian veins.

The importance of the lymphatic system is illustrated when edema occurs. **Edema** is localized swelling due to excess tissue fluid not collected by this system. In one condition, which occurs in the tropics, the lymph vessels are blocked by a filarial worm infection. Blockage caused by these worms, which are spread by mosquitoes, can cause the tissues to swell to the point that they take on an elephantlike appearance (fig. 5.22).

The lymphatic system has three main functions: (1) to transport excess tissue fluid back to the cardiovascular system; (2) to absorb fat from the intestine and to transport it to the blood; (3) to fight infection by cleansing lymph and producing lymphocytes.

FIGURE 5.21 *Lymphatic system. The lymphatic system drains excess fluid from the tissues and returns it to the cardiovascular system. The thoracic duct is one of the major lymph vessels. The enlargement shows the lymph vessels, called lacteals, which are present in the intestinal villi.*

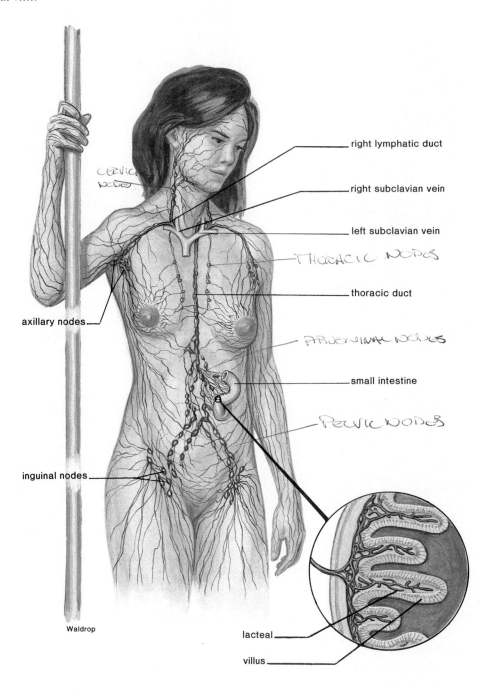

right lymphatic duct

right subclavian vein

left subclavian vein

THORACIC NODES

thoracic duct

ABDOMINAL NODES

small intestine

PELVIC NODES

CERVICAL nodes

axillary nodes

inguinal nodes

Waldrop

lacteal

villus

FIGURE 5.22 *Elephantiasis. An infection with a filarial worm has caused this individual to experience extreme swelling in regions where the worms have blocked the lymph vessels.*

Lymphatic Organs

At certain strategic points along medium-sized lymph vessels, there occur small ovoid, or round, structures called **lymph nodes** (fig. 5.21). Lymph nodes produce and are packed full of lymphocytes.

Lymph nodes also filter lymph of any damaged cells and debris, helping to purify the blood. When a local infection is present, such as a sore throat, the lymph nodes in that region swell and become painful. Lymph nodes may be removed in cancer operations because stray cancer cells are sometimes dispersed by the lymphatic system.

There are two other lymphatic organs having a function similar to that of a lymph node. The **spleen,** the larger of the two, is located in the abdominal cavity below the stomach (fig. 4.11). Not only does the spleen contain white blood cells, it also stores blood, contracting when the blood pressure drops. The bilobed **thymus** gland is located in the upper thoracic cavity and becomes progressively smaller with age (fig. 7.4). The thymus has an important function in the production and maturation of some lymphocytes, and its decrease in size may be important in the aging process.

SUMMARY

Movement of blood in the circulatory system is dependent on the beat of the heart. During the cardiac cycle, the SA node (pacemaker) initiates the beat and causes the atria to contract. The AV node picks up the stimulus and initiates contraction of the ventricles. The heart sounds, lub-DUPP, are due to the closing of the atrioventricular valves followed by the closing of the semilunar valves.

The circulatory system is divided into the pulmonary and systemic circuits.

In the pulmonary circuit the pulmonary artery takes blood from the right ventricle to the lungs and the pulmonary veins return it to the left atrium. To trace the path of blood in the systemic circuit, start with the aorta from the left ventricle. Follow its path until it branches to an artery going to a specific organ. It may be assumed that the artery will divide into arterioles and capillaries and that the capillaries will lead to venules. The vein that takes blood to the vena cava most likely has the same name as the artery.

Lymph vessels, or veins, are constructed similarly to cardiovascular veins and contain valves to keep lymph moving from the tissues to the veins. The lymphatic system is a one-way system, taking excess tissue fluid to the subclavian veins. The lymphatic organs, such as lymph nodes and the thymus, are involved in the infection-fighting capacity of the body.

OBJECTIVE QUESTIONS

1. Arteries are blood vessels that take blood _____ from the heart.
2. When the left ventricle contracts blood enters the _____ .
3. The right side of the heart pumps blood to the _____ .
4. The _____ node is known as the pacemaker.
5. The blood vessels that serve the heart are the _____ arteries and veins.
6. The pressure of blood against the walls of a vessel is termed _____ .
7. Blood moves in arteries due to _____ and in veins due to _____ .
8. Reducing the amount of _____ in the diet reduces the chances of plaque buildup in arteries.
9. Varicose veins develop when _____ become weak and ineffective.
10. There are only two types of lymph vessels, the lymph _____ and the lymph _____ .

Answers to Objective Questions

STUDY QUESTIONS

1. What types of blood vessels are there? Discuss their structure and function. (pp. 100–101)
2. Trace the path of blood in the pulmonary circuit as it travels from and returns to the heart. (p. 104)
3. Describe the cardiac cycle (using the terms systole and diastole) and explain the heart sounds. (pp. 105–166)
4. Describe an EKG and tell how its components are related to the cardiac cycle. (p. 106)
5. Trace the path of blood from the mesenteric arteries to the aorta, indicating which of the vessels are in the systemic circuit and which are in the pulmonary circuit. (p. 107)
6. What is blood pressure, and why is the average normal arterial blood pressure said to be 120/80? (pp. 110–111)
7. In which type of vessel is blood pressure highest? lowest? Velocity is lowest in which type vessel and why is it lowest? Why is this beneficial? What factors assist venous return of the blood? (pp. 111–112)
8. What is atherosclerosis? (p. 113) Name two illnesses associated with hypertension and thromboembolism. (p. 113)
9. What is a lymph vessel? Give three functions of the lymphatic system and tell how these functions are carried out. (pp. 117–119)

THOUGHT QUESTIONS

1. The ancients did not know that the blood circulates. Instead they thought that the veins and arteries were two different, unrelated systems of tubes. What evidence can you provide to prove that the blood does indeed circulate?
2. Fishes have only a two-chambered heart (one atrium and one ventricle); amphibians have a three-chambered heart (two atria and one ventricle); humans have a four-chambered heart. Why is this complexity of value in the human heart?
3. Argue both for and against the suggestion that the lymphatic system should be considered a part of the blood circulatory system.

KEY TERMS

aorta (a-or'tah) major systemic artery that receives blood from the left ventricle. *105*

arteries (ar'ter-ēz) vessels that take blood away from the heart; characteristically possessing thick elastic walls. *100*

arterioles (ar-te're-ōlz) vessels that take blood from arteries to capillaries. *100*

atria (a'tre-ah) chambers; particularly the upper receiving chambers of the heart that lie above the ventricles. (*sing.* atrium) *102*

AV node (a-ve nōd) a small region of neuromuscular tissue that transmits impulses received from the SA node to the ventricular walls. *106*

capillaries (kap'ĭ-lar''ēz) microscopic vessels connecting arterioles to venules having thin walls through which molecules either exit or enter the blood. *100*

coronary arteries (kor'ŏ-na-re ar'ter-ēz) arteries that supply blood to the wall of the heart. *108*

diastole (di-as'to-le) relaxation of the heart chambers. *105*

lymph (limf) fluid having the same composition as tissue fluid and carried in lymph vessels. *117*

lymphatic system (lim-fat'ik sis'tem) vascular system that takes up excess tissue fluid and transports it to the bloodstream. *117*

pulmonary circuit (pul'mo-ner''e ser'kit) that part of the circulatory system that takes deoxygenated blood to and oxygenated blood away from the lungs. *107*

SA node (es a nōd) small region of neuromuscular tissue that initiates the heartbeat; also called the pacemaker. *106*

systemic circuit (sis-tem'ik ser'kit) that part of the circulatory system that serves body parts other than the gas-exchanging surfaces in the lungs. *107*

systole (sis'to-le) contraction of the heart chambers. *105*

tissue fluid (tish'u floo'id) fluid found about tissue cells containing molecules that enter from and exit to the capillaries. *117*

valves (valvz) openings that open and close, insuring one-way flow; common to the systemic veins, the lymphatic veins, and the heart. *101*

veins (vānz) vessels that take blood to the heart; characteristically having nonelastic walls. *100*

venae cavae (ve'nah ka'vah) large systemic veins that return blood to the right atrium of the heart. *107*

ventricles (ven'trĭ-k'lz) cavities in an organ, such as the lower pumping chambers of the heart. *102*

venules (ven'ūlz) vessels that take blood from capillaries to veins. *100*

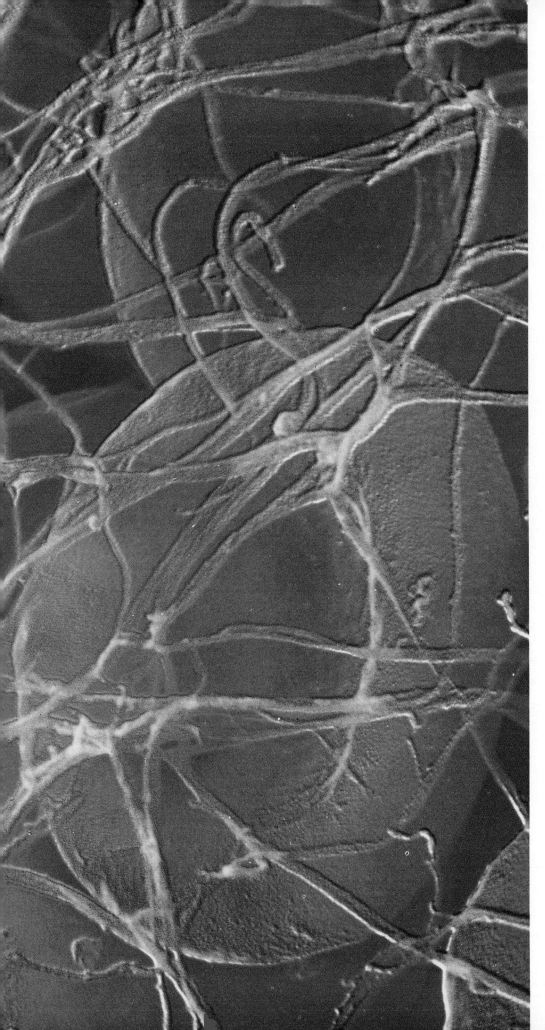

CHAPTER SIX

BLOOD

CHAPTER CONCEPTS

1 Blood, which is composed of cells and a fluid containing many inorganic and organic molecules, has three primary functions: transport, clotting, and fighting infection.

2 Exchange of molecules between blood and tissue fluid takes place across capillary walls.

3 Blood is typed according to the antigens present on the red blood cells.

4 All of the functions of blood may be correlated to the ability of the body to maintain a constant internal environment.

Fibrin threads entrap red blood cells within a blood clot.

a.

b.

INTRODUCTION

In persons with sickle cell anemia, the red blood cells aren't biconcave discs (fig. 6.1a) like normal red blood cells; they are elongated. In fact, they are sickle shaped (fig. 6.1b). The defect is caused by an abnormal hemoglobin that piles up inside the cells. Because the sickle-shaped cells can't pass along narrow capillary passageways like disc-shaped cells, they clog the vessels and break down. No wonder, then, that persons with sickle cell anemia suffer from poor circulation, anemia, and poor resistance to infection. Internal hemorrhaging leads to further complications like jaundice, episodic pain of the abdomen and joints, and damage to internal organs. The importance of a normal structure and function of blood components is dramatically illustrated by sickle cell anemia.

I f blood is transferred from a person's vein to a test tube and prevented from clotting, it separates into two layers (fig. 6.2). The lower layer consists of red blood cells (**erythrocytes**), white blood cells (**leukocytes**), and blood platelets (**thrombocytes**). Collectively, these are called the **formed elements** (fig. 6.3) and they take up about 45% of the volume of whole blood. The upper layer, called **plasma,** contains a variety of inorganic and organic substances dissolved or suspended in water. Plasma accounts for about 55% of the volume of whole blood.

Table 6.1 lists the components of blood, which we will discuss in terms of three functions: transport, clotting, and infection fighting. All of these can be related to the blood's primary function of maintaining a constant internal environment, or *homeostasis.*

FIGURE 6.2 *Volume relationship of plasma and formed elements (cells) in blood. Red blood cells are by far the most prevalent blood cell, which accounts for the color of blood.*

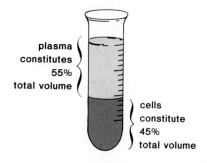

plasma constitutes 55% total volume

cells constitute 45% total volume

Blood is a liquid tissue; the liquid portion is termed plasma, and the solid portion consists of the formed elements.

TRANSPORT

The transport function of the blood helps maintain the constancy of tissue fluid. The blood transports oxygen from the lungs and nutrients from the intestine to the capillaries, where they enter tissue fluid. Here, it also takes up carbon dioxide and nitrogen waste (i.e., ammonia) given off by the cells and transports them away. Carbon dioxide exits the blood at the lungs, and ammonia exits at the liver, where it is converted to urea, a substance that later travels by way of the bloodstream to the kidneys and is excreted. Figure 6.4 diagrams the major transport functions of blood, indicating the manner in which these functions help keep the internal environment relatively constant.

Homeostasis is only possible because blood brings nutrients to the cells and removes their wastes.

FIGURE 6.3 *The formed elements in blood are the cells and platelets that are fragments of cells. a. Diagram of formed elements. Erythrocytes are red blood cells. Leukocytes (both granular and agranular) are white blood cells. Thrombocytes are platelets. b. A representation of formed elements as they would appear within a blood vessel.*

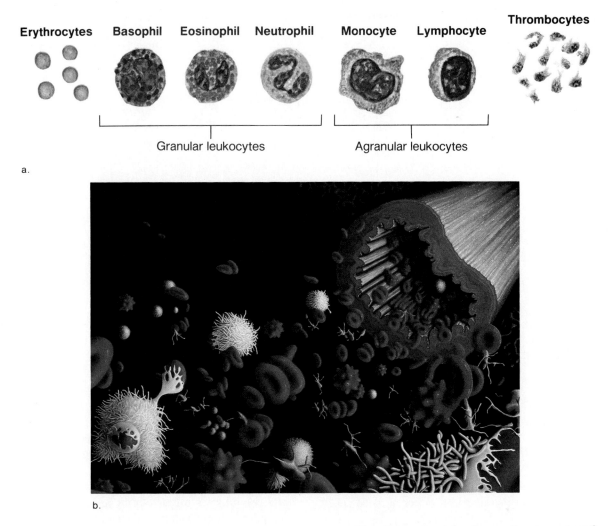

Erythrocytes   Basophil   Eosinophil   Neutrophil   Monocyte   Lymphocyte   **Thrombocytes**

Granular leukocytes   Agranular leukocytes

a.

b.

Plasma

Plasma is the liquid portion of blood (fig. 6.2). Small organic molecules such as glucose and urea simply dissolve in plasma, but large organic molecules combine with proteins for transport.

Plasma Proteins

Plasma proteins make up 8–9% of plasma. These molecules assist in transporting large organic molecules in the blood. For example, the molecule bilirubin, a breakdown product of hemoglobin, is transported by **albumin,** and the **alpha** and **beta globulins**[1] transport hormones and fat-soluble vitamins.

The plasma proteins have other transport functions as well. Blood needs a certain volume in order to exert a pressure, the force that moves blood in the arteries. Plasma proteins, together with salts, create an osmotic pressure that draws water from the tissues into the blood. You will recall that water moves across cell membranes from the area of greater concentration to the area of lesser concentration of water. Since proteins are too large to pass through or across a capillary wall, the fluid within the capillaries is always the area of lesser concentration of water, and water will therefore tend to pass into capillaries. This function of the plasma proteins is associated particularly with albumin, the smallest and most plentiful of the plasma proteins (table 6.2).

[1]When globulins undergo electrophoresis (are put in an electrical field), they separate into three major components called alpha globulin, beta globulin, and gamma globulin. Almost all circulating antibodies are found in the gamma globulin fraction and, as a result, this term is used for circulating antibodies.

TABLE 6.1 COMPONENTS OF BLOOD

BLOOD	FUNCTION	SOURCE
Formed elements		
Red blood cells	Transport oxygen	Bone marrow
Platelets	Clotting	Bone marrow
White blood cells	Fight infection	Bone marrow and lymphoid tissue
Plasma[a]		
Water	Maintain blood volume and transport molecules	Absorbed from intestine
Plasma proteins	Maintain blood osmotic pressure and pH	
Albumin	Transport	Liver
Fibrinogen	Clotting	Liver
Globulins		
Alpha and beta	Transport	Liver
Gamma	Fight infection	Lymphocytes
Gases		
Oxygen	Cellular respiration	Lungs
Carbon dioxide	End product of metabolism	Tissues
Nutrients		
Fats, glucose, amino acids, etc.	Food for cells	Absorbed from intestinal villi
Salts	Maintain blood osmotic pressure and pH; aid metabolism	Absorbed from intestinal villi
Wastes		
Urea and ammonia	End products of metabolism	Tissues
Hormones, vitamins, etc.	Aid metabolism	Varied

[a]Plasma is 90–92% water, 7–8% plasma proteins, not quite 1% salts; all other components are present in even smaller amounts.

Certain of the plasma proteins (table 6.2) we will be discussing have a type of function that is not duplicated by any other. For example, *fibrinogen* is necessary to blood clotting, and the *gamma globulins* are antibodies that help fight infection.

Plasma proteins assist in the transport function of blood. They serve as carriers for some molecules, and they help maintain the volume of the blood.

Red Blood Cells (Erythrocytes)

The red blood cells are also involved in the transport function of blood. They contain the respiratory pigment **hemoglobin,** which carries oxygen. Since hemoglobin is a red pigment, the cells appear red, and their color also makes the blood red. There are between 4 and 6 million red blood cells (fig. 6.5) per mm³ of whole blood, and each of these cells contains about 200 million hemoglobin molecules. If this much hemoglobin were suspended within the plasma rather than enclosed within the cells, the blood would be so thick the heart would have difficulty pumping it.

Each hemoglobin molecule (fig. 6.6) contains four polypeptide chains that make up the protein **globin,** and each chain is joined to **heme,** a complex iron-containing structure. It is the iron that forms a loose association with oxygen and in this way carries it in the blood.

Humans are active warm-blooded animals; the brain and the muscles often require much oxygen within a short period of time. Plasma carries only about 0.3 ml of oxygen per 100 ml, but whole blood carries 20 ml of oxygen per 100 ml. This shows that the presence of hemoglobin increases the carrying capacity of blood sixty times. Although the iron portion of hemoglobin carries oxygen, the equation for oxygenation of hemoglobin is usually written as

$$Hb + O_2 \underset{tissues}{\overset{lungs}{\rightleftharpoons}} HbO_2$$

The hemoglobin on the right, which is combined with oxygen, is called *oxyhemoglobin*. Oxyhemoglobin forms in the lungs and is a bright red color. The hemoglobin on the left, which has given up oxygen to tissue fluid, is called *reduced hemoglobin* and is a dark purple color.

Hemoglobin is an excellent carrier for oxygen because it forms a loose association with oxygen in the cool, neutral conditions of the lungs. It then readily gives up oxygen under the warm and more acidic conditions of the tissues.

FIGURE 6.4 *Diagram illustrating the transport function of blood. Oxygen (O₂) is transported from the lungs to the tissues, and carbon dioxide (CO₂) is transported from the tissues to the lungs. Ammonia (NH₃) is transported from the tissues to the liver, where it is converted to urea, a molecule excreted by the kidneys. Glucose (C₆H₁₂O₆) is absorbed by the gut and may be temporarily stored in the liver as glycogen before it is transported to the tissues.*

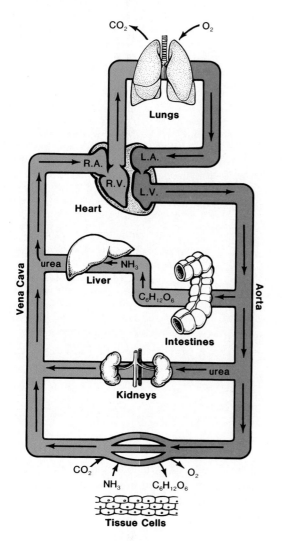

FIGURE 6.5 *a. Photomicrograph of red blood cells contained within a blood vessel. When viewing arterioles in the living animal, it is possible to see the red blood cells scurrying along. b. Electron micrograph of red blood cells contained within each of two adjacent capillaries. Capillaries are just large enough for the small red blood cells to squeeze through.*

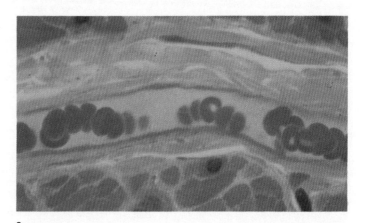

FIGURE 6.6 *The hemoglobin molecule is a protein that contains four polypeptide chains, two of which are alpha (α) and two of which are beta chains (β). The plane in the center of each chain represents an iron-containing heme group. When hemoglobin carries oxygen, it combines with the iron.*

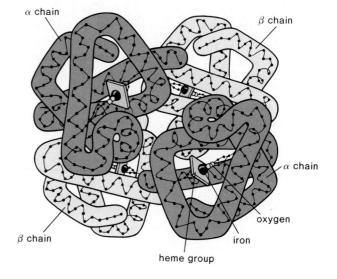

TABLE 6.2 *PLASMA PROTEINS*

NAME	LOCATION	FUNCTION
Albumin	Plasma	Maintain osmotic pressure and transport bilirubin
Globulins		
Alpha and beta	Plasma	Transport hormones and fat-soluble vitamins
Gamma	Plasma and lymph	Antibodies to fight infection
Fibrinogen	Plasma	Blood clotting

ARTIFICIAL BLOOD

There are risks to having a blood transfusion, such as the possibility of an immune reaction and acquiring an infection. A cross-matching test between the donor's blood and the recipient's blood usually detects if the recipient has antibodies in the plasma that will react against antigens on the membrane of the donor's red blood cells and vice versa. Blood is also screened for the presence of two types of viruses that are especially troublesome. They are the hepatitis B virus and the AIDS virus. Blood donors are questioned carefully and their blood is tested for the presence of these viruses. Despite the care that is taken to avoid immune reactions and the transference of disease, it would be advantageous to develop an artificial blood that would have neither of these risks.

Investigation is proceeding in three directions. Enrico Bucci, a blood-substitute specialist at the University of Maryland School of Medicine, is working with modifying hemoglobin itself. He links several hemoglobin molecules together so that they don't become "lost" from the circulatory system and uses chemicals to modify the complexes' oxygen affinity so that they are more likely to give up the oxygen when needed. Many other problems remain, however. The hemoglobin is taken from whole blood, and it alone may contain infectious material. Modified hemoglobin also seems to cause a generalized constriction of the body's blood vessels, making oxygenation of tissues more difficult.

Anthony Hunt and colleagues at the University of California at San Francisco are working with artificial red blood cells called neohemocytes (NHCs). To make the cells, purified human hemoglobin, taken from outdated donor blood, is encapsulated in a lipid bilayer membrane. The artificial cells are much smaller than normal human red blood cells, and they don't contain as much hemoglobin. However, when tested in rats, the animals did survive until they were sacrificed for gross toxicity studies. The investigators believe that the tests are successful enough to warrant further study. They want to improve the stability and vascular retention time of the cells since they are removed from the bloodstream and broken down at a faster rate than normal cells.

Carbon monoxide, present in automobile exhaust, combines with hemoglobin more readily than does oxygen, and it stays combined for several hours, regardless of the environmental conditions. Accidental death or suicide from carbon monoxide poisoning occurs because the hemoglobin of the blood is not available for oxygen transport. This transport function of blood is so important that life can be temporarily sustained by giving a patient a hemoglobin substitute transfusion when whole blood is not available or cannot be given. The reading on these pages discusses the possible benefits of this "artificial blood."

Oxygen is transported to the tissues in combination with hemoglobin, a pigment found in red blood cells.

Red blood cells that are not engaged in carrying oxygen assist in transporting carbon dioxide. First, reduced hemoglobin will combine with carbon dioxide to form *carbaminohemoglobin:*

$$Hb + CO_2 \underset{lungs}{\overset{tissues}{\rightleftharpoons}} HbCO_2$$

However, such a combination with hemoglobin actually represents only a small portion of the carbon dioxide in the blood. Most of the carbon dioxide is transported as the *bicarbonate ion* HCO_3^-. This ion is formed after carbon dioxide has combined with water. Carbon dioxide combined with water forms carbonic acid. This dissociates (breaks down) to a hydrogen ion and a bicarbonate ion:

$$CO_2 + H_2O \underset{lungs}{\overset{tissues}{\rightleftharpoons}} H_2CO_3 \underset{lungs}{\overset{tissues}{\rightleftharpoons}} H^+ + HCO_3^-$$

George Groveman is director of new-products marketing at Alpha Therapeutic, a subsidiary of a Japanese firm that is based in Los Angeles. His firm is working on a third possibility, a substance called perfluorocarbon oil (PFC) emulsion that can be transfused and will carry oxygen much like hemoglobin does. This substance has served as a blood substitute for humans in emergency situations where only the oxygen-carrying function of blood was required. FDA approval has been sought by a Japanese corporation to market PFC under the trade name of Fluosol-DA, but thus far the FDA has denied permission on the grounds that the clinical trials were not successful enough. There is some hope, though, that PFC will be approved for localized use. For example, it may be helpful to administer Fluosol-DA directly to the heart when a person is suffering a heart attack and/or undergoing thrombolytic therapy (p. 116).

Although researchers have been working for twenty years to produce a blood substitute for general use, the prospects are still in the future. Anthony Hunt says, "Physiological systems always turn out to be more complicated than we thought."

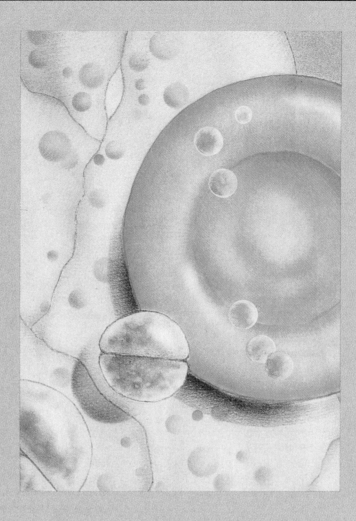

View inside a capillary, in which blood has been replaced with a 25% suspension of hemoglobin-containing synthetic neohemocytes (NHCs). A normal red blood cell is shown for scale. Unlike the living red blood cell, neohemocytes are nearly spherical and may have one, two, or three chambers.

There is an enzyme within red blood cells, called *carbonic anhydrase,* that speeds up this reaction. The released hydrogen ions, which could drastically change the pH, are absorbed by the globin portions of hemoglobin, and the bicarbonate ions diffuse out of the red blood cells to be carried in the plasma. Reduced hemoglobin, which combines with a hydrogen ion, may be symbolized as HHb. This combination plays a vital role in maintaining the pH of the blood.

Once systemic venous blood has reached the lungs, the reaction just described takes place in the reverse: the bicarbonate ion joins with a hydrogen ion to form carbonic acid, and this splits into carbon dioxide and water. The carbon dioxide diffuses out of the blood into the lungs for expiration. Now hemoglobin is ready again to transport oxygen. Table 6.3 summarizes the structure and function of hemoglobin.

TABLE 6.3 *HEMOGLOBIN*

HEME	*GLOBIN*
Nonprotein	Protein
Contains iron	
Carries oxygen	Carries carbon dioxide; acts as a buffer; absorbs H^+
Becomes bile pigments	May be reused

Hemoglobin also participates in the transport of carbon dioxide in the blood.

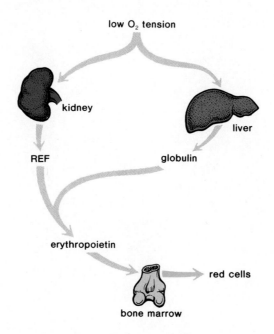

HUMAN ISSUE

The ancients believed that the blood carried vital spirits and the heart was the center of the emotions. In other words, the cardiovascular system was seen as possessing a life-force and contributing to our human nature. Today there is an artificial heart, and there may soon be artificial blood. Other structures in the body can also be successfully replaced by artificial structures. This raises an interesting question: How many or what structures would have to be replaced by artificial parts before we considered the individual a robot rather than a human?

Life Cycle of Red Blood Cells

Red blood cells are continuously manufactured in the red bone marrow of the skull, ribs, vertebrae, and ends of the long bones (fig. 11.8). The number produced increases whenever arterial blood carries a reduced amount of oxygen, as will happen when an individual first takes up residence at a high altitude. It is now known that if the kidneys receive less oxygen than usual, they produce a substance called *renal erythropoietic factor*, REF. REF joins with liver globulin to give a hormone that stimulates the *red bone marrow* to produce more red blood cells (fig. 6.7).

Before they are released from the bone marrow into the blood, red blood cells pass through several developmental stages during which time they lose the nucleus and acquire hemoglobin (fig. 6.8). Possibly because they lack a nucleus,

red blood cells only live about 120 days. They are destroyed chiefly in the *liver and spleen,* where they are engulfed by large phagocytic cells. When red blood cells are broken down, the hemoglobin is released. The iron is recovered and returned to the red bone marrow for reuse. The heme portion of the molecule undergoes chemical degradation and is excreted by the liver in the bile as bile pigments. These bile pigments are primarily responsible for the color of feces.

Anemia

When there is an insufficient number of red blood cells or the cells do not have enough hemoglobin, the individual suffers from **anemia**[2] and has a tired, run-down feeling. In iron deficiency anemia, the hemoglobin count is low. It may be that the diet does not contain enough iron. Certain foods, such as raisins and liver, are rich in iron, and the inclusion of these in the diet can help prevent this type of anemia.

In another type of anemia, called pernicious anemia, the digestive tract is unable to absorb enough vitamin B_{12}. This vitamin is essential to the proper formation of red blood cells; without it, immature red blood cells tend to accumulate in the bone marrow in large quantities. A special diet and administration of vitamin B_{12} by injection is an effective treatment for pernicious anemia.

Illness (anemia) results when the blood has too few red blood cells and/or not enough hemoglobin.

---

[2]Sickle cell anemia and Cooley anemia are discussed in chapter 18.

FIGURE 6.8 *Maturation of red blood cells. Red blood cells are made in red bone marrow, where stem cells continuously divide. During the maturation process, a red blood cell loses its nucleus and gets much smaller.*

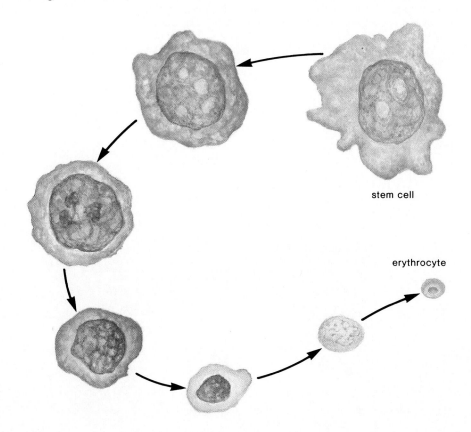

stem cell

erythrocyte

Capillary Exchange Within the Tissues

When blood arrives at the tissues, an exchange takes place that maintains the constancy of tissue fluid.

Arterial Side

When arterial blood enters the tissue capillaries (fig. 6.9), it is bright red because the red blood cells are carrying oxygen. It is also rich in nutrients that are dissolved in the plasma. At this end of the capillary, blood pressure (40 mm Hg) is higher than the osmotic pressure of the blood (25 mm Hg). The blood pressure, you will recall, is created by the pumping of the heart. The osmotic pressure is caused by the presence of salts and also, in particular, by the plasma proteins that are too large to pass through the wall of the capillary. Since the blood pressure is higher than the osmotic pressure, fluid, together with oxygen and nutrients (glucose and amino acids), will leave the capillary. This is a *filtration* process because large substances, such as red blood cells and plasma proteins, remain, but small substances, such as water and nutrient molecules, leave the capillaries. Tissue fluid, created by this process, consists of all of the components of plasma except the proteins.

Midsection

Along the length of the capillary, molecules will follow their concentration gradient as diffusion occurs. Diffusion, you will recall, is the movement of molecules from an area of greater concentration to an area of lesser concentration. The area of greater concentration for nutrients is always the blood because after these molecules have passed into the tissue fluid, they are taken up and metabolized by the tissue cells. The cells use glucose and oxygen in the process of cellular respiration, and they use amino acids for protein synthesis. Following cellular respiration, the cells give off carbon dioxide and water. Whenever the cells break down amino acids, they remove the amino group, which is released as ammonia. Carbon dioxide and ammonia, being waste products of metabolism, leave the cell by diffusion. Since tissue fluid is always the area of greater concentration for these waste materials, they diffuse into the capillary.

Oxygen and nutrient molecules (e.g., glucose and amino acids) exit a capillary near the arterial end; waste molecules (e.g., carbon dioxide and ammonia) enter a capillary near the venous end.

FIGURE 6.9 *Diagram of a capillary illustrating the exchanges that take place and the forces that aid the process. At the arterial end of a capillary, the blood pressure is higher than the osmotic pressure so water, oxygen, and glucose tend to leave the bloodstream. At the venous end of a capillary, the osmotic pressure is higher than the blood pressure so water, ammonia, and carbon dioxide tend to enter the bloodstream. Notice that the red blood cells and plasma proteins are too large to exit a capillary.*

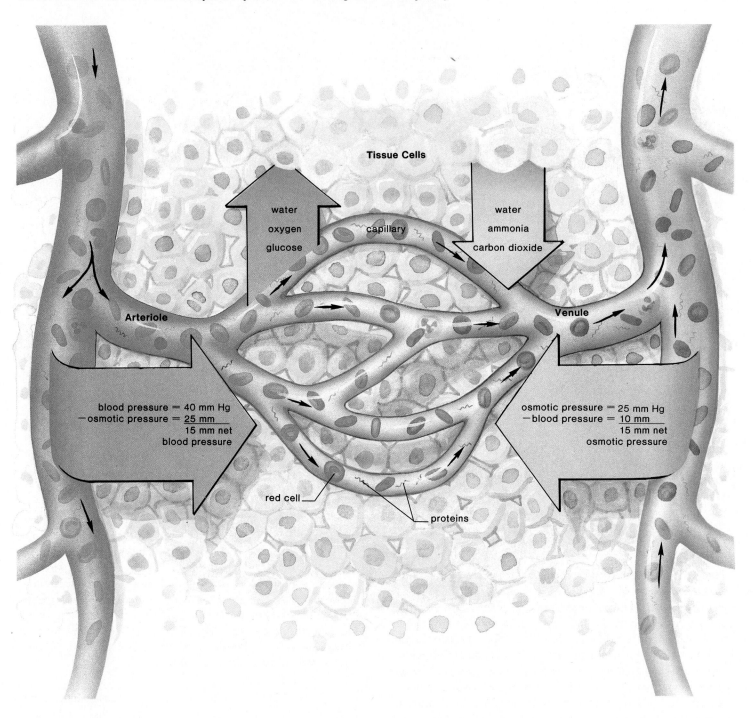

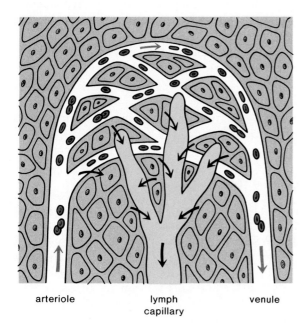

arteriole   lymph   venule
capillary

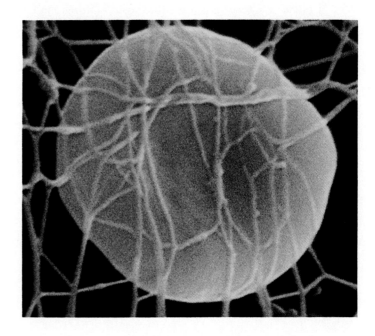

Venous Side

At the venous end of the capillary, blood pressure is much reduced (10 mm Hg), as can be verified by reviewing figure 5.16 in the previous chapter. However, there is no reduction in osmotic pressure (25 mm Hg), which tends to pull fluid into the capillary. As water enters a capillary, it brings with it additional amounts of waste molecules (carbon dioxide and ammonia). The blood that leaves the capillaries is deep purple in color because the red blood cells contain reduced hemoglobin. Carbon dioxide becomes the bicarbonate ion, and this, along with ammonia, is dissolved in the plasma.

This method of retrieving fluid by means of osmotic pressure is not completely effective. There is always some fluid that is not picked up at the venous end. This excess tissue fluid enters the lymph vessels (fig. 6.10). Lymph is tissue fluid contained within lymph vessels. Lymph is returned to the systemic venous blood when the major lymph vessels enter the subclavian veins (p. 117).

Lymphatic capillaries lie in close proximity to cardiovascular capillaries, where they collect excess tissue fluid.

BLOOD CLOTTING

When an injury occurs to a blood vessel, **clotting,** or coagulation, of the blood takes place. This is obviously a protective mechanism to prevent excessive blood loss. As such, blood clotting is another mechanism by which blood components maintain homeostasis.

Portions of the blood that have been identified as necessary for clotting are (1) platelets, (2) prothrombin, a globulin protein, and (3) fibrinogen. **Platelets,** or thrombocytes, (fig. 6.3) result from the fragmentation of certain large cells, called megakaryocytes, in the red bone marrow. They are produced at a rate of 200 billion a day, and the bloodstream possesses more than a trillion. **Fibrinogen** and **prothrombin** are manufactured and deposited in the blood by the liver. Vitamin K is necessary for the production of prothrombin, and if by chance this vitamin is missing from the diet, hemorrhagic disorders develop.

The steps necessary for blood clotting are quite complex but may be summarized in this simplified manner:

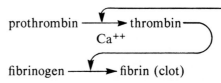

platelets and injured tissue cells → prothrombin activator

prothrombin → thrombin

Ca^{++}

fibrinogen → fibrin (clot)

When a blood vessel is damaged, platelets clump at the site of the puncture and partially seal the leak. They and the injured tissues release an enzyme, called prothrombin activator, that converts prothrombin to thrombin. **Thrombin,** in turn, acts as an enzyme that severs two short amino acid chains from each fibrinogen molecule. These activated fragments then join end to end, forming long threads of **fibrin.** Fibrin threads wind around the platelet plug in the damaged area of the blood vessel and provide the framework for the clot. Red blood cells are also trapped within the fibrin threads (fig. 6.11), and their presence makes a clot appear red.

A blood clot consists of red blood cells entangled within fibrin threads.

If blood is placed in a test tube, blood clotting can be prevented by adding citrate or any other substance that combines with calcium. This is because calcium ions (Ca++) are required for the blood-clotting reactions to occur. In contrast, since blood clotting is an enzymatic process, clotting takes place at a faster rate if blood is warmed than if it is kept cool. If the blood is allowed to clot, a yellowish fluid comes to lie above the clotted material (fig. 6.12). This fluid is called **serum,** and it contains all the components of plasma except fibrinogen. Since we have now used a number of different terms to refer to portions of the blood, table 6.4 reviews these terms for you.

A fibrin clot is only a temporary way to repair the blood vessel. An enzyme called plasmin destroys the fibrin network and restores the fluidity of plasma. This is a protective measure because a blood clot can act as a thrombus or an embolus that interferes with circulation. In fact, the new drug tPA (p. 000) is an enzyme that converts a precursor to plasmin.

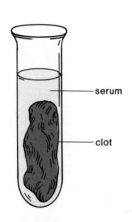

FIGURE 6.12 *When blood clots, serum is squeezed out as a solid plug is formed. In a blood vessel this plug helps prevent further blood loss.*

serum

clot

INFECTION FIGHTING

The body defends itself against parasites, such as bacteria and viruses, in several ways. The so-called first line of defense is the outer covering (skin and mucous membranes), which resists invasion by parasites. The second line of defense is dependent on two components of blood: white blood cells and gamma globulins. Infection fighting is the third of the three ways in which the blood components contribute to homeostasis.

White blood cells fight infection. They attack bacteria and viruses that have invaded the body.

White Blood Cells (Leukocytes)

White blood cells may be distinguished from red blood cells in that they are usually larger, have a nucleus, and without staining, would appear to be white in color. With staining, white blood cells characteristically appear a bluish shade in color. White blood cells are less numerous than red blood cells, with only 7,000–8,000 cells per mm[3].

Table 6.5 lists the different types of white blood cells and figure 6.13 shows two in detail. On the basis of structure, it is possible to divide white blood cells into the granular leukocytes and the agranular leukocytes (fig. 6.3). The **granular leukocytes** have granules in the cytoplasm and a many-lobed nucleus joined by nuclear threads; therefore, they are called polymorphonuclear. Granular leukocytes, like red blood cells and platelets, are formed in the red bone marrow. The **agranular leukocytes** do not have granules and have either a circular or indented nucleus. They are produced in lymphoid tissue found in the spleen, lymph nodes, and tonsils.[3] *Leukemia* is a form of cancer characterized by an uncontrolled production of leukocytes in either the red bone marrow or in

[3]Stem cells in the bone marrow produce specialized cells. Certain ones seed the thymus, spleen, and lymph nodes and produce lymphocytes there.

TABLE 6.4 *BODY FLUIDS*

NAME	COMPOSITION
Blood	Formed elements and plasma
Plasma	Liquid portion of blood
Serum	Plasma minus fibrinogen
Tissue fluid	Plasma minus proteins
Lymph	Tissue fluid within lymph vessels

TABLE 6.5 *WHITE BLOOD CELLS (LEUKOCYTES)*

	GRANULAR LEUKOCYTES (POLYMORPHONUCLEAR)	
	Size	Granules stain
Neutrophils	9–12µm	Lavender
Eosinophils	9–12µm	Red
Basophils	9–12µm	Deep blue
	AGRANULAR LEUKOCYTES	
	Size	Type of nucleus
Lymphocytes	8–10µm	Large
Monocytes	12–20µm	Indented

lymphoid tissue. In both types of leukemia, the white blood cells are numerous but nonfunctional; therefore, the victim has a lowered resistance to infections.

White blood cells are divided into the granular leukocytes and the agranular leukocytes.

Infection fighting by white blood cells is primarily dependent on the *neutrophils,* which comprise 60–70% of all leukocytes, and the *lymphocytes,* which make up 25–30% of the leukocytes. Neutrophils are phagocytic; they destroy bacteria and viruses by traveling to the site of invasion and engulfing the foe. Like neutrophils, monocytes and eosinophils are phagocytic. Lymphocytes secrete gamma globulins, called immunoglobulins, or antibodies, that combine with foreign substances to deactivate them. Neutrophils and lymphocytes may be contrasted in the following manner.

Neutrophils	*Lymphocytes*
Granules in cytoplasm	No granules in cytoplasm
Polymorphonuclear	Mononuclear
Produced in bone marrow	Produced in lymphoid tissue
Phagocytic	Make antibodies

The neutrophils, which are amoeboid and engulf invaders, may be contrasted to lymphocytes, which produce antibodies that combine with antigens.

Antibodies

Parasites and their toxins cause lymphocytes to produce antibodies. Each lymphocyte produces one type of antibody that is specific for one type of antigen. **Antigens** are most often proteins and polysaccharides located, for example, in the outer covering of a parasite or present in its toxin. **Antibodies** combine with their antigens (fig. 6.14) in such a way that the antigens are rendered harmless. Sometimes the antibodies cause precipitation of the antigen, agglutination (clumping) of the antigen, or simply prepare it for phagocytosis. In any case, it is well to keep in mind that the antigen is the foreigner and the antibody is the substance prepared by the body. The antigen-antibody reaction may be symbolized as follows:

antigen + antibody → inactive complex
(foreign substance) (globulin protein)

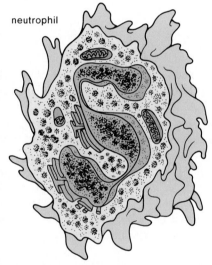

neutrophil

FIGURE 6.13 Two leukocytes of interest. a. *Scanning electron micrograph and an artist's representation of a neutrophil.* b. *Scanning electron micrograph and an artist's representation of a lymphocyte.*

a.

lymphocyte

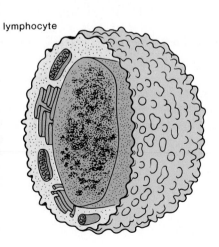

b.

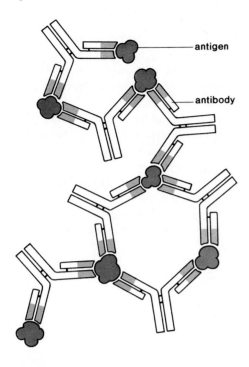

antigen

antibody

The antigen-antibody reaction is a lock-and-key reaction in which the two molecules fit together as do a lock and key. This seems surprising at first because it has been shown that all antibodies have the same overall shape. Even so, each type of antibody has a variable region, a unique sequence of amino acids that results in a receptor site that is capable of combining with one type of antigen. In other words, this particular sequence of amino acids shapes a site where the antibody fits the antigen.

Immunity An individual is actively immune when the body is capable of producing antibodies that can react to a specific disease-causing antigen. The blood in these individuals contains lymphocytes that are capable of reacting to these antigens so that antibodies can be produced. Exposure to the antigen, either naturally or by way of a vaccine, can cause active immunity to develop. Chapter 7 is about immunity and explores the topic in detail.

Lymphocytes are responsible for immunity, the presence of antibodies that specifically combine with antigens, foreign substances that are a part of or are produced by bacteria and viruses.

Diseases

Often illnesses cause an increase in a particular type of white blood cell. For this reason, a differential white blood cell count, involving the microscopic examination of a blood sample and the counting of each type of white blood cell, may be done as part of the diagnostic procedure. For example, the characteristic finding in the viral disease **mononucleosis** is a great number of lymphocytes that are larger than mature lymphocytes and stain more darkly. This condition takes its name from the fact that lymphocytes are mononuclear.

Inflammatory Reaction

Whenever the skin is broken due to a minor injury, a series of events occur that are known as the **inflammatory reaction** because there is swelling and reddening at the site of the injury. Figure 6.15 illustrates the participants in the inflammatory reaction. One participant, the mast cells, are derived from basophils, a type of white blood cell that takes up residence in the tissues.

When an injury occurs, a capillary and several tissue cells are apt to rupture and release certain precursors that lead to the presence of **bradykinin,** a chemical that (1) initiates nerve impulses resulting in the sensation of pain, (2) stimulates mast cells to release **histamine,** another chemical that together with bradykinin (3) causes the capillary to become enlarged and more permeable. The enlarged capillary causes the skin to redden, and its increased permeability allows proteins and fluids to escape so that swelling results.

Any breakage of the skin will allow bacteria and viruses to enter the body. In figure 6.15, a lymphocyte is releasing antibodies that attack the bacteria (or viruses), preparing them for **phagocytosis.** When a neutrophil or monocyte phagocytizes a foreign substance, an intracellular vacuole is formed. The engulfed material is destroyed or neutralized by hydrolytic enzymes when the vacuole combines with a lysosome. (The granules of a neutrophil are in fact lysosomes.)

Once monocytes have arrived on the scene, they swell to five to ten times their original size and become **macrophages,** large phagocytic cells that are able to devour a hundred invaders and still survive. Some tissues, particularly connective tissues, have resident macrophages that routinely act as scavengers, devouring old blood cells, bits of dead tissue, and other debris. Such macrophages are also capable of bringing about an explosive increase in the number of leukocytes by liberating a substance that passes by way of the blood to the bone marrow, where it stimulates the production and release of white blood cells, usually neutrophils.

As the infection is being overcome, some neutrophils die. These, along with dead tissue, cells, bacteria, and living white blood cells, form **pus,** a thick yellowish fluid. The presence of pus indicates that the body is trying to overcome the infection.

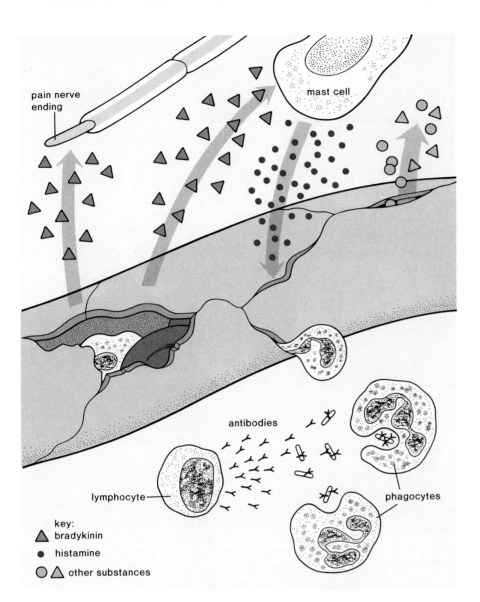

pain nerve ending

mast cell

antibodies

lymphocyte

phagocytes

key:
△ bradykinin
● histamine
◯ △ other substances

The inflammatory reaction is a "call to arms"—it marshals phagocytic white blood cells and antibody-producing lymphocytes to a site of invasion by bacteria and viruses.

BLOOD TYPING

ABO Grouping

Two antigens that may be present on the red blood cells have been designated as A and B. As table 6.6 shows, an individual may have one of these antigens (i.e., type A or type B), or both (type AB), or neither (type O). Thus, blood type is dependent on which antigens are present on the red blood cells. As you can see, type O blood is most common in the United States.

TABLE 6.6 *BLOOD GROUPS*

TYPE	ANTIGEN	ANTIBODY	% U.S.[a] BLACK	% U.S.[a] CAUCASIAN
A	A	b	27	41
B	B	a	21	10
AB	A,B	none	4	4
O	none	a,b	48	45

[a]Blood type frequency for other races is not available.

In the ABO blood grouping system there are four types of blood: A, B, AB, and O. Type O blood has no antigens on the red blood cells; the other types of blood are designated by the antigens present on the red blood cells.

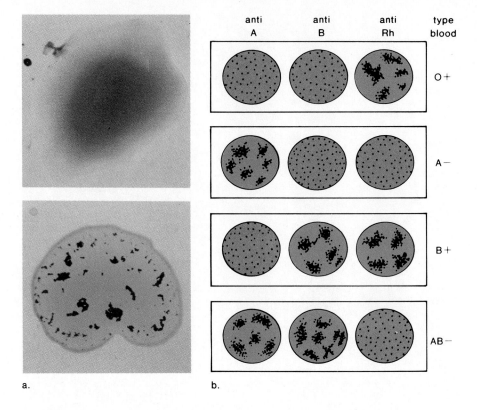

a.   b.

Within the plasma of an individual, there are antibodies to the antigens that are *not* present on that individual's red blood cells. Thus, for example, type A blood has antibody b. Type AB blood has no antibodies because both antigens are on the red blood cells. This is reasonable because if the same antigen and antibody are present, **agglutination,** or clumping, of red blood cells will occur. Agglutination of red blood cells can cause the blood to stop circulating, and death follows.

For a recipient to receive blood from a donor, the recipient's plasma must not have an antibody that would cause the donor's cells to agglutinate. For this reason, it is important to determine each person's blood type. Figure 6.16 demonstrates a way to use the antibodies derived from plasma to determine blood type. If clumping occurs after a sample of blood is exposed to a particular antibody, the person has that type of blood.

HUMAN ISSUE

When people undergo operations they often require a blood transfusion. Some people today are storing blood that they themselves might use later. That's appropriate if they expect to undergo an operation sometime soon, but there is not enough space for blood banks to store blood for each individual indefinitely. Therefore, it seems that we do need a blood bank for everyone's use. Do you think that all should be required to give blood to a blood bank or is it all right to expect to be given blood even when none has been contributed?

Rh System

Another important antigen in matching blood types is the **Rh factor.** Persons with this particular antigen on the red blood cells are Rh positive; those without it are Rh negative. Rh negative individuals do not normally make antibodies to the Rh factor, but they will make them when exposed to the Rh factor. It is possible to extract these antibodies and use them for blood type testing. When Rh positive blood is mixed with Rh antibodies, agglutination occurs.

The designation of blood type usually also includes whether the person has the Rh factor (Rh⁺) or does not have the Rh factor (Rh⁻) on the red blood cells.

The Rh factor is particularly important during pregnancy (fig. 6.17). If the mother is Rh negative and the father is Rh positive, the child may be Rh positive. The Rh positive red blood cells begin leaking across into the mother's circulatory system as placental tissues normally break down before and at birth. This causes the mother to produce Rh antibodies. If the mother becomes pregnant with another Rh positive baby, Rh antibodies (but not antibodies a and b discussed earlier) may cross the placenta and cause destruction of the child's red blood cells. This is called erythroblastosis of the newborn, or **fetal erythroblastosis.**

This problem has been solved by giving Rh negative women an Rh immune globulin injection called RhoGAM just after the birth of any Rh positive child. This injection

FIGURE 6.17 *Diagram describing the development of fetal erythroblastosis.*
a. Baby's red blood cells carry Rh antigen. b. Some of these cells escape into mother's system. c. Mother begins manufacturing Rh antibodies. d. During a subsequent pregnancy, mother's Rh antibodies cross placenta to destroy the new baby's red blood cells.

● Rh antigen   🡒 Rh antibody

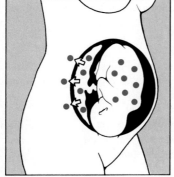

a. during pregnancy   b. before delivery   c. months and years later   d. subsequent pregnancy

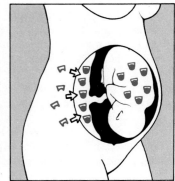

contains Rh antibodies that attack the child's red blood cells before these cells can stimulate the mother to produce her own antibodies.

The possibility of fetal erythroblastosis exists when the mother is Rh negative and the father is Rh positive.

SUMMARY

Nutrients and wastes are transported in plasma, but oxygen is combined with hemoglobin within red blood cells. The end result of transport is capillary exchange with tissue fluid, regulated by blood pressure and osmotic pressure. Blood clotting requires a series of enzymatic reactions involving platelets, prothrombin, and fibrinogen. In the final reaction fibrinogen becomes fibrin threads entrapping red blood cells.

White blood cells and gamma globulin proteins are required in the process of fighting infections. The two most prevalent white blood cells are the phagocytic neutrophils and the antibody-producing lymphocytes. The first of these is involved in the inflammatory reaction and the second is involved in immunity development.

Blood transfusions require that the types of blood be compatible. Of consideration are the antigens (A and/or B) on the red blood cells and antibodies (a and/or b) in the plasma. Another important antigen is the Rh antigen, particularly because an Rh negative mother may possess antibodies that will attack the red blood cells of an Rh positive fetus.

OBJECTIVE QUESTIONS

1. The liquid part of blood is called
   _____ .
2. Red blood cells carry _____ ,
   and white blood cells _____
   _____ .
3. Hemoglobin that is carrying oxygen
   is called _____ .
4. Human red blood cells lack a _____
   and only live about _____ days.
5. When a blood clot occurs, fibrinogen
   has been converted to _____
   threads.

6. The most common granular leukocyte
   is the _____ , a phagocytic
   white blood cell.
7. Lymphocytes are made in _____
   tissue and produce _____ that
   react with antigens.
8. At a capillary, _____ , _____
   _____ , and _____ leave
   the arterial end, and _____
   _____ and _____ enter
   the venous end.

9. AB blood has the antigens _____
   and _____ on the red blood
   cells and _____ antibodies in
   the plasma.
10. Fetal erythroblastosis can occur when
    the mother is _____
    and the father is _____ .

Answers

1. plasma 2. oxygen, fight infection
3. oxyhemoglobin 4. nucleus, 120 5. fibrin
6. neutrophil 7. lymphoid, antibodies
8. oxygen, amino acids, glucose; carbon dioxide, ammonia 9. A, B, no 10. Rh⁻, Rh⁺

1. Define blood, plasma, tissue fluid, lymph, and serum. (pp. 122, 132)
2. Name three functions of blood and tell how they are related to the maintenance of homeostasis. (p. 122)
3. State the major components of plasma. Name the plasma proteins and tell their common function as well as their specific functions. (p. 123)
4. Give the equation for the oxygenation of reduced hemoglobin. Where does this reaction occur? Where does the reverse reaction occur? (p. 124)
5. Give an equation that indicates how CO_2 is commonly carried in the blood. Indicate the direction of the reaction in the tissues and in the lungs. In what ways does hemoglobin aid the process of transporting CO_2? (p. 126)
6. Compare the structure and function of heme to globin. (p. 127) Discuss the life cycle of red blood cells. (p. 128)
7. What forces operate to facilitate exchange of molecules across the capillary wall? (pp. 129–131)
8. Name the steps that take place when blood clots. Which substances are present in the blood at all times and which appear during the clotting process? (p. 131)
9. Name and discuss two ways that blood fights infection. Associate each of these with a particular type of white blood cell. (pp. 132–133)
10. Describe the inflammatory reaction, and give a role for each type cell and chemical that participates in the reaction. (pp. 134–135)
11. What are the four ABO blood types in humans? For each, state the antigen(s) on the red blood cells and the antibody(ies) in the plasma. (p. 135)
12. Problems can arise during childbearing if the mother is which Rh type and the father is which Rh type? Explain why this is so. (p. 136)

THOUGHT QUESTIONS

1. Multicellular animals have exchange areas with the external environment in order to maintain the internal environment. Explain with reference to figure 6.4.
2. Blood coagulation is a series of enzymatic reactions that lead from one to the other. Of what value is such a complicated system?
3. In what way is the inflammatory reaction a lifesaving event?

KEY TERMS

agglutination (ah-gloo″tǐ-na′shun) clumping of cells, particularly in reference to red blood cells involved in an antigen-antibody reaction. *136*

agranular leukocytes (ah-gran′u-lar lu′ko-sītz) white blood cells that do not contain distinctive granules. *132*

antibody (an′tǐ-bod″e) a protein produced in response to the presence of some foreign substance in the blood or tissues. *133*

antigen (an′tǐ-jen) a foreign substance, usually a protein, that stimulates the immune system to produce antibodies. *133*

clotting (klot′ing) process of blood coagulation, usually when injury occurs. *131*

fetal erythroblastosis (ě-rith″ro-blas-to′sis) destruction of Rh⁺ fetal red blood cells due to antibodies produced by a mother who is Rh⁻. *136*

fibrinogen (fi-brin′o-jen) plasma protein that is converted into fibrin threads during blood clotting. *131*

formed element (form′d el′ě-ment) a constituent of blood that is either cellular (red blood cells and white blood cells) or at least cellular in origin (platelets). *122*

granular leukocytes (gran′u-lar lu′ ko-sītz) white blood cells that contain distinctive granules. *132*

inflammatory reaction (in-flam′ah-to″re re-ak′shun) a tissue response to injury that is characterized by dilation of blood vessels and an accumulation of fluid in the affected region. *134*

macrophage (mak′ro-fāj) an enlarged monocyte that ingests foreign material and cellular debris. *134*

phagocytosis (fag″o-si-to′sis) the taking in of bacteria and/or debris by engulfing. *134*

platelet (plāt′let) a formed element that is necessary to blood clotting. *131*

prothrombin (pro-throm′bin) plasma protein that is converted to thrombin during the process of blood clotting. *131*

pus (pus) thick yellowish fluid composed of dead phagocytes, dead tissue, dead bacteria, and living white blood cells. *134*

serum (se′rum) light-yellow liquid left after clotting of the blood. *132*

thrombin (throm′bin) an enzyme that converts fibrinogen to fibrin threads during blood clotting. *131*

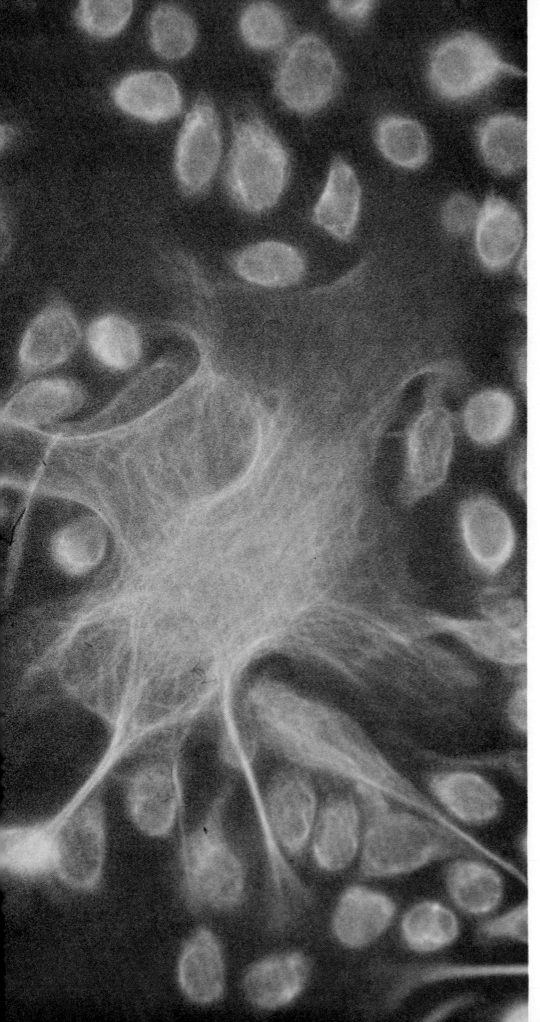

CHAPTER SEVEN

IMMUNITY

CHAPTER CONCEPTS

1 The body has general mechanisms to protect itself from disease. If these fail, the immune system comes into play.

2 There are two types of immune responses. One type of lymphocyte, the B cell, is responsible for antibody-mediated immunity, and another type, the T cell, is responsible for cell-mediated immunity.

3 Active immunity, the ability of the body to produce specific antibodies, can be promoted by immunization.

4 Passive immunity, acquired when antibodies are received from an outside source, is now possible by using monoclonal antibodies produced in the laboratory.

5 While the immune system preserves our existence, it is also responsible for certain undesirable effects, such as tissue rejection, allergies, and autoimmune diseases.

Fluorescent micrograph of a macrophage in motion (400×).

INTRODUCTION

Most of us don't spend much time thinking about the immune system, and that's because it usually functions pretty well. Not so for David (fig. 7.1) who spent twelve years in a germ-free plastic bubble. David was born with SCID (*severe combined immunodeficiency disease*) and couldn't produce the cells that are needed to protect one's self from disease organisms. A treated bone marrow transplant was given by his sister, but unfortunately, it was infected with EB (*Epstein-Barr*) viruses that caused him to develop a cancer called immunoblastic sarcoma. The virus hadn't caused this difficulty in his sister because she already had a functioning immune system when she was exposed. A well-functioning immune system permits us to live in a world that is filled with all sorts of disease-causing agents.

The body is prepared to protect itself from foreign substances and cells, including infectious microbes. The first line of defense is immediately available because it consists of mechanisms that are nonspecific. The second line of defense takes a little longer to come into play because it is highly specific and contains mechanisms that are tailored to a particular threat.

GENERAL DEFENSE

The environment contains many organisms that are able to invade and infect the body. The general defense mechanisms are useful against all of them.

Barriers to Entry

We have already mentioned that the skin and the mucous membrane lining of the respiratory and digestive tracts serve as mechanical barriers to possible entry by bacteria and viruses. Also, the secretions of the sebaceous glands in the skin

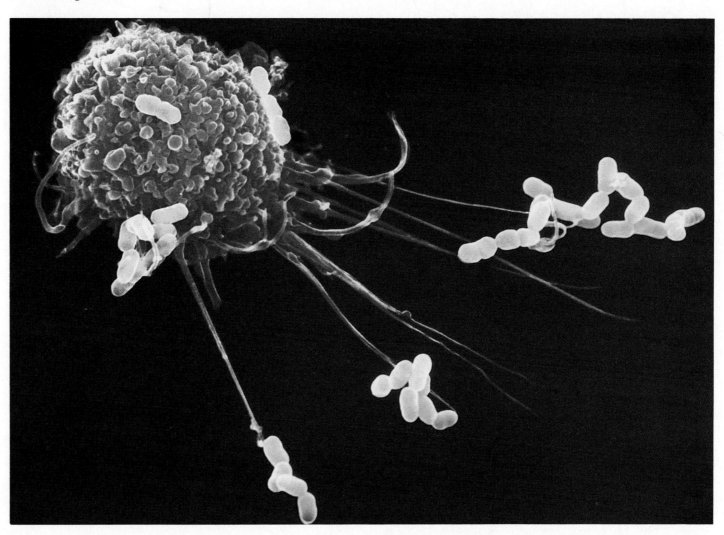

FIGURE 7.2 *Macrophages are the body's scavengers. They engulf microbes and debris in the body's fluids and tissues. Here the microphage is colored red and the bacteria are green.*

contain chemicals that weaken or kill bacteria. The respiratory tract is lined by cilia that sweep mucus and any trapped particles up into the throat, where they may be swallowed. The stomach has an acid pH that inhibits the growth of many types of bacteria. The intestine and organs such as the vagina contain a mix of bacteria that normally reside there; these prevent potential pathogens from also taking up residence.

Phagocytic White Blood Cells

If microbes do gain entry into the body, as described under inflammatory reaction on page 134, other nonspecific forces come into play. For example, neutrophils (fig. 6.13) and *monocyte*-derived macrophages (fig. 7.2) are phagocytic white blood cells that can engulf bacteria and viruses upon contact. *Macrophages* engaged in fighting infection give off a chemical that stimulates general defense and also causes fever. Fever may also be protective because some microbes cannot tolerate a rise in normal body temperature.

Protective Chemicals

Also present in the blood and tissues are chemicals that may prevent the multiplication of bacteria and viruses.

Complement

Complement is a series of proteins produced by the liver that is present in the plasma. When complement is activated, a cascade of reactions occur. Every protein molecule in the series activates many others in a predetermined sequence. In the end, certain of the proteins form pores in bacterial cell walls. This allows fluids and salts to enter until the cell bursts (fig. 7.3).

FIGURE 7.3 *Complement is a number of proteins always present in the plasma. When activated, some of these form pores in bacterial membranes, allowing fluids and salts to enter until the cell eventually bursts.*

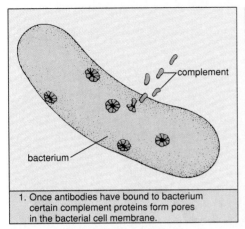 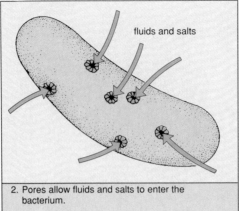 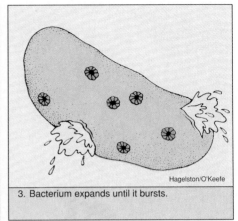

Hagelston/O'Keefe

1. Once antibodies have bound to bacterium certain complement proteins form pores in the bacterial cell membrane.

2. Pores allow fluids and salts to enter the bacterium.

3. Bacterium expands until it bursts.

If complement is unable to destroy the microbes directly it still coats them and releases chemicals that attract phagocytes to the scene. Although complement is a general defense mechanism, it also plays a role in specific defense as we shall see later.

Interferon

If a virus should escape phagocytosis and enter a tissue cell, then the infected cell will produce and secrete interferon. **Interferon** binds to receptors on noninfected cells, and this causes them to prepare for possible attack by producing substances that will interfere with viral replication.

A cell under the influence of interferon will be protected against any type virus. However, the interferon itself is specific to the species; only human interferon can be used in humans. It used to be quite a problem to collect enough interferon for clinical and research purposes, but now interferon is made by recombinant DNA technology (p. 429). Thus far, however, it has been possible to find few illnesses that benefit greatly from interferon therapy.

The first line of defense against disease is nonspecific. It consists of barriers to entry, phagocytic white blood cells, and protective chemicals.

SPECIFIC DEFENSE

Sometimes we are threatened by an illness that cannot be successfully counteracted by the general defense mechanisms. In such cases, the immune system becomes activated to provide a specific defense. The immune system (fig. 7.4) consists of about a trillion agranular leukocytes (lymphocytes and monocytes); the spleen, thymus, and lymph nodes,

where lymphocytes and monocytes are produced; and the lymph vessels, where these cells are primarily found. Following maturation, agranular leukocytes are also found in the bloodstream.

The cells of the immune system respond to **antigens,** usually a protein (or polysaccharide) they recognize as not normally found within the body. Since antigens can be part of a bacterial cell wall or viral coat, the immune system protects us from disease. However, an antigen can also be part of a foreign cell or cancerous cells. These, too, are destroyed by the immune system. Ordinarily the immune system does not attack the body's own cells; therefore, it is said that the immune system is able to distinguish self from nonself.

Quite often immunity lasts for some time. For example, once we have recovered from measles we usually cannot be infected a second time.

The immune system includes the lymphoid organs and the agranular leukocytes, which respond to the presence of antigens.

Immunity is dependent on two types of lymphocytes, the B cells and the T cells.

B Cells versus T Cells

The different types of blood cells are derived from bone marrow stem cells, which divide to give a specific precursor for each cell type. Certain of these precursor cells leave the bone marrow and seed the lymphoid organs, where they divide and produce functioning monocytes and lymphocytes. There are two main types of lymphocytes that are distinguishable by their maturation process. The precursor cells that will become **T lymphocytes** have passed through the *thymus* (fig. 7.4); those that will become **B lymphocytes** have not passed through the thymus and therefore are derived more directly from *bone marrow* stem cells.

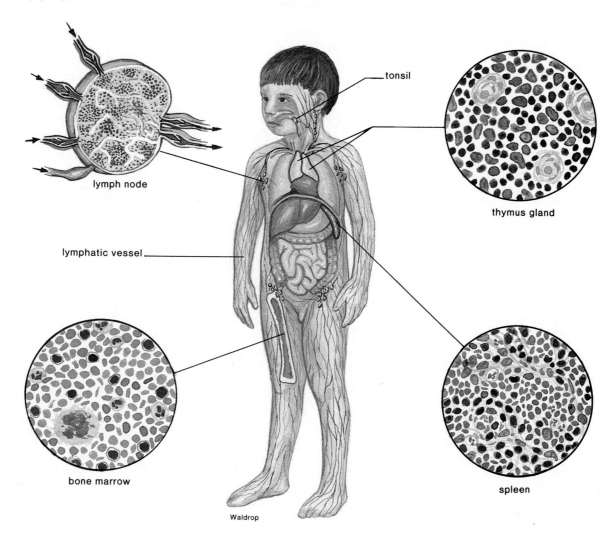

FIGURE 7.4 *The immune system includes all those organs in which lymphocytes are produced and found, such as lymph nodes, thymus, and spleen. The thymus is important to the maturation of T cells, a particular type of lymphocyte. The lymphocytes and monocytes within lymph nodes help purify lymph, and those within the spleen help purify the blood.*

T cells and B cells have different functions. B cells produce **antibodies,** proteins that are capable of combining with and inactivating antigens. These antibodies are secreted into the blood and lymph. In contrast, T cells do not produce antibodies. Instead, the T lymphocytes themselves directly attack cells that bear antigens they recognize.

Lymphocytes are capable of recognizing an antigen because they bear *receptor* molecules on their surfaces. The shape of the receptor on any particular lymphocyte is complementary to a pattern found on one specific antigen. It is often said that the receptor and antigen fit together as do a lock and key. It is estimated that during our lifetimes, we encounter a million different antigens, and we need the same number of different lymphocytes to protect ourselves against those antigens. It is remarkable to think that so much diversification occurs during the maturation process that in the end there is a different type of lymphocyte for each possible

antigen. Despite the diversity, none of these lymphocytes is supposed to attack the body's own cells. It's believed that if by chance a lymphocyte is made that could respond to the body's own proteins, it is normally suppressed and develops no further.

There are two types of lymphocytes. B cells produce and secrete antibodies that combine with an antigen. T cells directly attack antigen-bearing cells.

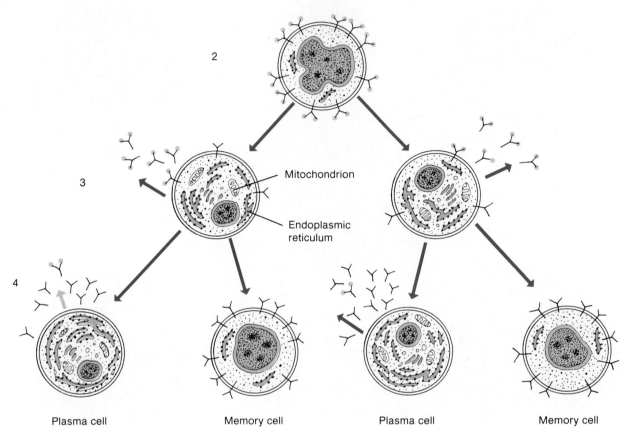
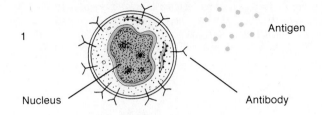

FIGURE 7.5 *Clonal selection theory as it applies to B cells. The presence of an antigen causes the appropriate B cell to clone, producing by the fifth day many mature plasma cells. These mature cells actively secrete antibodies and memory cells that retain the ability to secrete these antibodies at a future time.*

B Lymphocytes

The receptor present on a B lymphocyte is termed a membrane-bound antibody because it is shaped exactly like an antibody. When a B lymphocyte encounters a bacterial cell or toxin bearing an antigen that combines with its receptor, the B cell is stimulated to divide and produce **plasma cells** that actively secrete antibodies (fig. 7.5). All of the plasma cells derived from one parent lymphocyte are called a clone, and they all produce the same type of antibody. Notice that the particular antigen determines which lymphocyte will be stimulated to produce antibodies. This is called the *clonal selection theory* because the antigen has selected which B cells will produce a clone of plasma cells. (T cells also undergo clonal selection as discussed following.)

Once antibody production is high enough, the antigen disappears from the system, and the development of plasma cells ceases. Some members of the clone do not participate in the current antibody production; instead they remain in the bloodstream as **memory B cells,** capable for some time of producing the antibody specific to a particular antigen. As long as this is the case, the individual is said to be *actively immune* because a certain number of antibodies are always present and also because memory cells can produce more plasma cells if the same antigen invades the system again.

Defense by B cells is called *antibody-mediated immunity* because B cells produce antibodies (table 7.1). Sometimes it is also called humoral immunity because these antibodies are present in the bloodstream.

B cells are responsible for antibody-mediated immunity.

TABLE 7.1 SOME PROPERTIES OF B CELLS AND T CELLS

PROPERTY	B CELLS	T CELLS
Antigen recognition	Direct recognition	Must be presented by macrophages
Early response	Produce plasma cells	Enlarge, multiply, and secrete lymphokines
Later response	Antibody production	Helper cells stimulate other cells; killer cells attack antigen-bearing cells
Final response	Memory cells	Memory cells; suppressor cells shut down immune response

FIGURE 7.6 Antibody structure. a. Computer model of an antibody. b. Schematic drawings. (1) An antibody contains two heavy (long) amino acid chains and two light (short) amino acid chains. Part of each chain contributes to a variable region where a particular antigen is capable of binding with the antibody in a lock-and-key manner. The rest of each chain is a constant regions. (2) Quite often the antigen-antibody reaction produces complexes of antigen combined with antibodies.

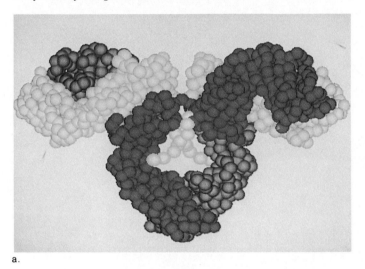

a.

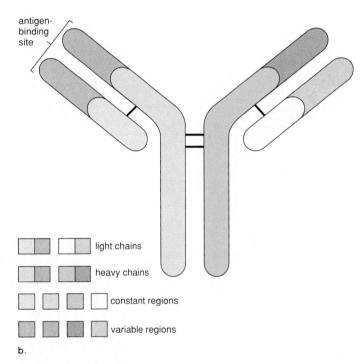

b.

light chains

heavy chains

constant regions

variable regions

antigen-binding site

Antibodies

An antibody is a Y-shaped molecule having two long, "heavy" chains and two short, "light" chains of amino acids. At the end of each arm of every antibody is a tiny segment called the *variable region* that binds to an antigen in a lock-and-key manner. In contrast to the variable regions, the other segments of the chains are called *constant regions* (fig. 7.6).

The constant regions are not the same among all antibodies. Instead, antibodies can be classified according to which constant regions are present in the heavy chains. All antibodies that have the same basic kinds of constant regions in their heavy chains are placed in a particular class. Table 7.2 lists the different classes of antibodies and their specific functions. Most antibodies belong to class IgG (immunoglobulin G).

TABLE 7.2 ANTIBODIES (THE IMMUNOGLOBULINS)

ANTIBODY CLASS	EXAMPLES OF FUNCTIONS
IgG	Main form of antibodies in circulation; production increased after immunization
IgM	Present on lymphocyte surface prior to immunization; secreted during primary response
IgA	Main antibody type in external secretions, such as saliva and mother's milk
IgD	Present on lymphocyte surface prior to immunization; other functions unknown
IgE	Protect against parasitic infections; responsible for allergic symptoms in immediate hypersensitivity reactions

After Fox, Stuart I., *Human Physiology*, 2d ed. © 1987 Wm. C. Brown Publishers, Dubuque, Iowa. All Rights Reserved. Reprinted by permission.

TREATMENT FOR AIDS

AIDS viruses. The viruses are covered by a membrane that is derived from the host cell membrane. The protein gp-120 extends beyond the membrane.

AIDS (acquired immune deficiency syndrome) is characterized by the destruction of the immune system, which leads to various infectious diseases and a virulent type of skin cancer (Kaposi's sarcoma). Many victims also show early brain impairment with ever-increasing neuromuscular and psychological consequences. Since the number of deaths of AIDS reported thus far in the United States is at least 50,000 and victims rarely live longer than four years, there is great interest in developing techniques to treat those who are infected and to prevent the spread of infection to others.

Investigators have now identified several viruses that can cause AIDS.

These are termed human immunodeficiency viruses and are numbered as HIV-1 and HIV-2, for example. The HIV viruses (see figure) primarily attack helper T lymphocytes because the viruses have an outer coat protein (gp-120) that combines with a receptor (CD4) on these lymphocytes. Monocytes, which have this receptor in fewer number, are therefore less susceptible to viral infection. They act as a reservoir for the virus and distribute it about the body, including the lungs and brain. Neurons lack the receptor and therefore are rarely infected so research is focusing on the possibility that the viruses cause a chemical imbalance leading to the psychiatric disorders seen in AIDS patients.

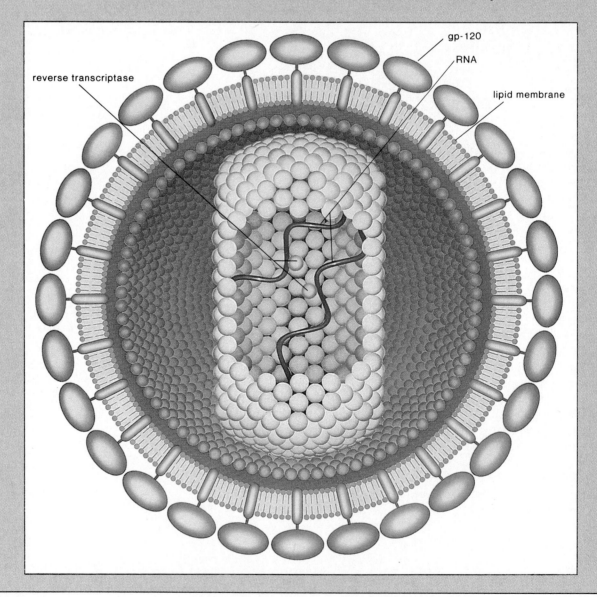

The AIDS viruses are retroviruses (p. 318) that have the capability of incorporating a DNA copy of their genome into the host cell DNA. When this occurs the virus is latent and does not multiply, and the individual does not display any symptoms of having AIDS. Eventually, about 1% of infected people (fig. 15.5) come down with AIDS. There is great interest in determining what triggers activation of the viruses so that they multiply and burst forth from the T cell, destroying it. It's possible that activation of the T lymphocyte by an antigen also activates the AIDS viruses. Certainly, AIDS is more often seen in individuals that have another type of sexually transmitted disease in addition to AIDS.

Since antibodies to the viruses are found in the bloodstream, it would seem that at first the body is able to mount an offensive to the invader. Eventually, however, the body succumbs. Various explanations have been given. 1. Infected lymphocytes are known to fuse with others some of which are not infected. If this is the way the viruses usually spread, then they need never return to the bloodstream where antibodies are located. 2. There is evidence that the AIDS viruses alter the MHC proteins that appear on the surfaces of infected cells. This would make T cells incapable of recognizing and destroying infected cells. 3. The AIDS viruses mutate so rapidly that antibody production lags behind and never catches up enough to be useful. Then, too, the viruses invade the very cells that are sent to destroy them, so that antibody production is never adequate.

Development of a treatment for AIDS is taking several avenues. There are drugs that will interfere with the viruses' normal reproductive cycle. The new drug AZT (azidothymidine) interferes with the ability of the viruses to incorporate their genomes into host DNA, and others are being developed that prevent viral replication. Most of these, like AZT, are abnormal nucleotides that take the place of normal nucleotides, preventing DNA

continued

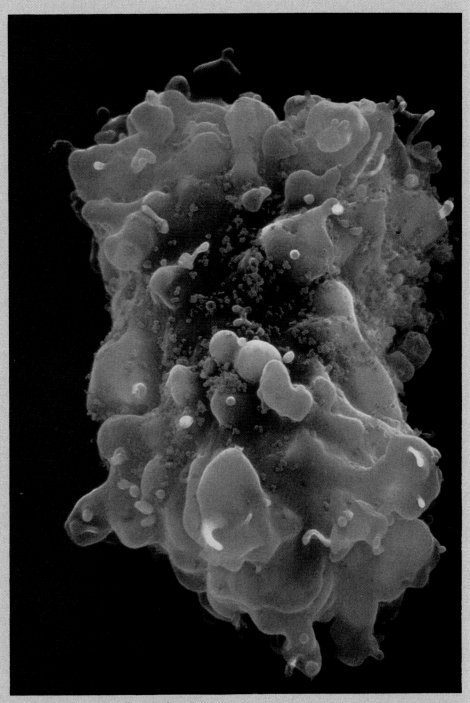

T cell (pink) attacked by AIDS virus (blue).

TREATMENT FOR AIDS
continued

replication. Clinical trials are underway and the results will be forthcoming. One concern is that the drugs will also interfere with the production of lymphocytes themselves. In any case, it will be necessary to reconstitute the normal white blood cell count in AIDS victims, and it is hoped that the growth factors mentioned on page 156 will be helpful in this regard.

Two experimental vaccines have been developed against the AIDS viruses and they are undergoing clinical trials. Both vaccines utilize gp-120, the outer viral coat protein labeled in the figure, as the antigen. One uses gp-120 directly. For the other, the gene coding for gp-120 is incorporated into the vaccinia virus, a much manipulated virus that used to be the smallpox vaccine. The hope is that after this vaccine enters the body, the gene will use the machinery of invaded cells to make gp-120 proteins, which will then appear on the cell surface and be recognized by the body's immune system.

A novel treatment based on the viruses' strong attraction to the CD4 receptor has been proposed, but thus far has only been tried in vitro (within laboratory glassware). It's possible to make the CD4 protein by recombinant DNA technology. When copies of this protein are added to samples of the AIDS viruses and helper T lymphocytes, the manufactured CD4 proteins attach to the viruses, preventing them from combining with the lymphocytes. Testing of this approach in animals and humans is planned.

While there is much to be learned about AIDS, we do at least have knowledge of its means of transmission. We know that AIDS can be transmitted by blood-to-blood contact. This puts intravenous drug abusers at great risk. Infected semen (and possibly saliva) also passes the disease to a sexual partner when it comes in contact with the partner's bloodstream. This can happen if a woman is menstruating, if the partner has an open abrasion, or if the sexual act itself damages the mucosal lining.

There is little danger of contracting AIDS by casual encounters between individuals, such as shaking hands, touching the same utensils, or breathing the same air. However, the public is advised to avoid promiscuity because it may cause them to unknowingly have sex with a carrier of AIDS. It is estimated that there may be as many as two million persons in the United States who are infected with AIDS, but who display no symptoms. It's believed that the use of condoms is a protection against contracting AIDS. Even so, individuals can best protect themselves by practicing monogamy (always the same sexual partner) with someone who is free of the disease.

TABLE 7.3 REACTIONS OF ANTIBODIES

TYPES OF ANTIBODIES	REACTIONS
Antitoxins	Neutralize toxins of infective agents
Agglutinins	Cause clumping of certain infective agents
Opsonins	Make certain infective agents more susceptible to work of phagocytes
Lysins	Dissolve certain infective agents
Precipitins	Bring about a precipitation of extracts of infective agents

From *Living, Health, Behavior and Environment*, 5th edition by Hein, Fred V., Farnsworth, Dana L. and Charles E. Richardson. Copyright © 1970 by Scott Foresman and Company.

Antigen-Antibody Complexes   The antigen-antibody reaction can take several forms, as indicated in table 7.3. Quite often the antigen-antibody reaction produces complexes of antigens combined with antibodies (fig. 6.14). Such an antigen-antibody complex, sometimes called the **immune complex**, marks the antigen for destruction by other forces. For example, the complex may be engulfed by neutrophils or macrophages or it may activate a portion of blood serum called complement. As mentioned earlier, complement refers to a series of different proteins that stimulate phagocytic cells. This series of proteins also brings about lysis of bacterial cells—several of the proteins form a channel that allows water and salts to enter the cell until it bursts (fig. 7.3).

An antibody combines with its antigen in a lock-and-key manner. The antibody-antigen reaction can lead to complexes that contain several antibodies and antigens.

T Lymphocytes

As the T cells pass through the thymus, they differentiate into types that have receptors for foreign antigens. Although these receptors combine with antigens in a lock-and-key manner, they are shaped differently from those on B cells. Ordinarily none of the receptors will combine with the body's own proteins. It is said that the T cells have learned to be *tolerant* of (do not pay attention to) the body's own proteins.

Recognition

Unlike B cells, T cells are unable to recognize an antigen unless it is presented to them by a (monocyte-derived) macrophage (fig. 7.7). When a *macrophage* engulfs a bacteria or virus, it enzymatically breaks it down and displays only a protein fragment on the cell membrane, together with an

FIGURE 7.7

FIGURE 7.7 *T cell activation and proliferation. a. Photomicrograph shows four T cells on surface of macrophage. b. Macrophage has ingested an antigen-bearing bacterium so that a portion, along with the MHC complex, can be presented to T cell. This activates the T cell, which produces lymphokines. The T cell now clones to produce many cells capable of responding to this particular antigen. Activation and proliferation are enhanced by the presence of lymphokines produced by both macrophages and T cells.*

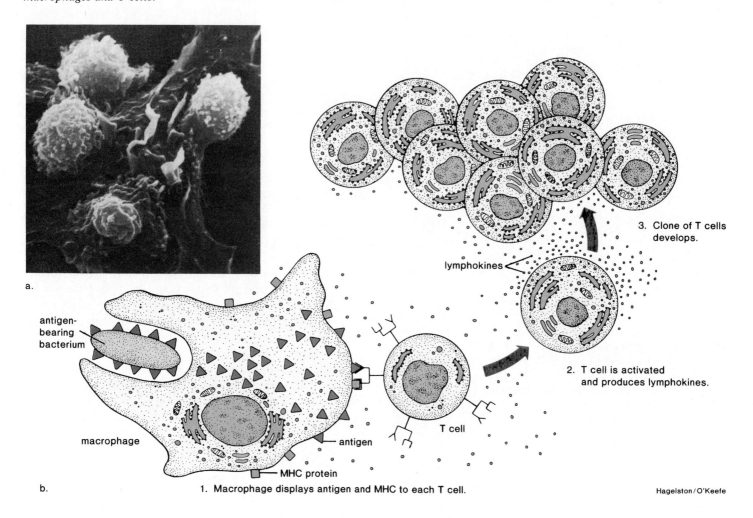

a.

3. Clone of T cells develops.

lymphokines

2. T cell is activated and produces lymphokines.

antigen-bearing bacterium

macrophage

antigen

T cell

MHC protein

b.

1. Macrophage displays antigen and MHC to each T cell.

Hagelston/O'Keefe

MHC (major histocompatibility complex) **protein.** The importance of these proteins was first recognized when it was discovered that they contribute to the specificity of tissues and make it difficult to transplant a tissue from one person to another. In other words, the donor and the recipient must be histo- (tissue) compatible (the same or nearly so) for a transplant to be successful without the administration of immunosuppressive drugs.

Once a T cell recognizes an antigen, it proliferates, giving rise to a clone of cells that contains the types discussed next. Therefore, it can be seen that T cells also undergo clonal selection.

In order for a T cell to recognize an antigen, the antigen must be presented along with an MHC protein to the T cell by a macrophage.

Types of T Lymphocytes

There are several different types of T cells that look alike but can be distinguished by their function.

Helper T (T_H) cells are believed to orchestrate the immune response. Once they recognize an antigen, they enlarge and secrete **lymphokines,** including interferons and interleukins. Lymphokines are stimulatory chemicals for all types of immune cells, including B cells (fig. 7.8). Because the AIDS (acquired immune deficiency syndrome) viruses attack helper T cells, they inactivate the immune system. This is discussed in the reading on page 146.

Killer T (T_K) cells are the type of T cell that attack and destroy cells that bear a foreign antigen (fig. 7.9). *Cytotoxic T cells* mainly do away with virus-infected cells. The so-called *natural killer cells* are thought to restrict their activities to tumor cells and perhaps also to cells infected by agents other

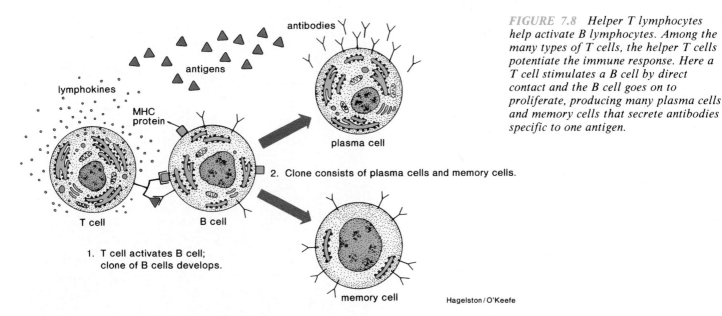

antibodies

antigens

lymphokines

MHC protein

T cell

B cell

plasma cell

2. Clone consists of plasma cells and memory cells.

1. T cell activates B cell; clone of B cells develops.

memory cell

Hagelston / O'Keefe

FIGURE 7.8 *Helper T lymphocytes help activate B lymphocytes. Among the many types of T cells, the helper T cells potentiate the immune response. Here a T cell stimulates a B cell by direct contact and the B cell goes on to proliferate, producing many plasma cells and memory cells that secrete antibodies specific to one antigen.*

than viruses. Regardless of this distinction, they both function similarly. They have storage granules that contain a chemical called perforin because it perforates cell membranes. Perforin joins with others of its kind to form a pore in the membrane that admits water and salts. Now the cell swells and eventually bursts.

Suppressor T (T_S) cells increase in number more slowly than the other two types of T cells just discussed. Once there is sufficient number, however, the immune response ceases. Following suppression, a population of **memory T (T_M) cells** also persists, perhaps for life.

T_K cells are responsible for cell-mediated immunity. T_H cells promote the immune response, and T_S cells suppress the immune response.

IMMUNOLOGICAL SIDE EFFECTS AND ILLNESSES

The immune system protects us from disease because it can tell self from nonself. Sometimes, however, the immune system is underprotective, as when an individual develops cancer, or is overprotective, as when an individual has allergies.

Allergies

Allergies are caused by an overactive immune system that forms antibodies to substances that are not usually recognized as being foreign substances. Unfortunately, allergies are usually accompanied by coldlike symptoms or, even at times, severe systemic reactions such as a sudden low blood pressure, termed shock.

Among the five varieties (table 7.2) of antibodies—immunoglobulin A (IgA), IgD, IgE, IgG, and IgM—it is IgE that causes allergies. IgE antibodies are found in the bloodstream, but they, unlike the other types of antibodies, also reside in the membrane of mast cells found in the tissues. Some investigators contend that mast cells are basophils that have left the bloodstream and taken up residence in the tissues. In any case, when the *allergen,* an antigen that provokes an allergic reaction, attaches to the IgE antibodies on mast cells, these cells release histamine and other substances (fig. 7.10) that cause secretion of mucus and constriction of the airways, resulting in the characteristic wheezing and labored breathing of someone with asthma. On occasion, basophils and other white blood cells release these chemicals into the bloodstream. The increased capillary permeability that results from this can lead to fluid loss and shock.

Allergy shots sometimes prevent the occurrence of allergic symptoms. Injections of the allergen cause the body to build up high quantities of IgG antibodies, and these combine with allergens received from the environment before they have a chance to reach the IgE antibodies located in the membrane of mast cells.

Allergic symptoms are caused by the release of histamine and other substances from mast cells.

Tissue Rejection

Certain organs, such as skin, the heart, and the kidneys, could easily be transplanted from one person to another if the body did not attempt to reject them. It is obvious that the transplanted organ is foreign to the individual, and for this reason the immune system reacts to it. At first T cells appear on the scene, and later antibodies bring about disintegration of the foreign tissue.

FIGURE 7.9 *Cell-mediated immunity.* a. *Scanning electron micrograph showing killer T cells attacking cancer cell.* b. *Cancer cell is now destroyed.* c. *During the killing process, granules in a T cell fuse with the cell membrane and release units of the protein perforin. These combine to form pores in the target cell membrane. Thereafter fluid and salts enter so that the target cell eventually bursts.*

a.

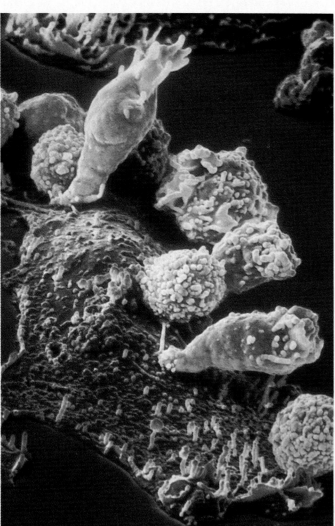

b.

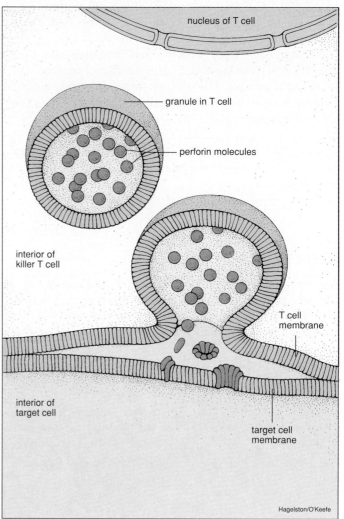

c.

nucleus of T cell

granule in T cell

perforin molecules

interior of
killer T cell

T cell
membrane

interior of
target cell

target cell
membrane

Hagelston/O'Keefe

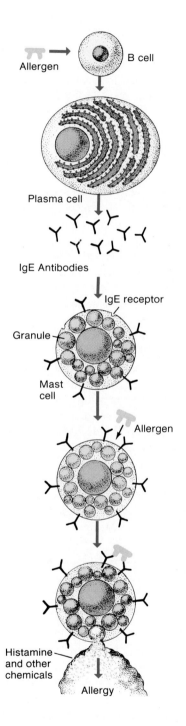

FIGURE 7.10 *Allergic response. An allergen causes a plasma cell to secrete IgE antibodies, some of which bind to mast cells. Then when the individual is exposed again, the allergen attaches to these IgE antibodies and the mast cell releases histamine and other chemicals that cause the allergic response.*

Allergen

B cell

Plasma cell

IgE Antibodies

IgE receptor

Granule

Mast cell

Allergen

Histamine and other chemicals

Allergy

Organ rejection can be controlled in two ways: careful selection of the organ to be transplanted and the administration of *immunosuppressive drugs*. In the first instance, it is best if the organ is made up of cells having the same type of MHC proteins (p. 149) as those on the cells of the prospective recipient. It is these that the T cell recognizes as foreign to the recipient. In regard to immunosuppression, there is now available a drug called Cyclosporine, which suppresses cell-mediated immunity only as long as it is administered. This means that after the body has adjusted to the new organ, the drug can be withdrawn and immunity will return. This is desirable because a person is more apt to catch an infectious disease when immunity is suppressed.

When an organ is rejected, the immune system is attacking cells that bear different MHC proteins from those of the individual.

Autoimmune Diseases

Certain human illnesses are believed to be due to the production of antibodies that act against an individual's own tissues. In myasthenia gravis, autoantibodies attack the neuromuscular junctions so that the muscles do not obey nervous stimuli. Muscular weakness results. In MS (multiple sclerosis), antibodies attack the myelin sheath of nerve fibers, causing various neuromuscular disorders. A person with SLE (systemic lupus erythematosus) forms various antibodies to different constituents of the body, including the DNA of the cell nucleus. The disease sometimes results in death, usually due to kidney damage. In rheumatoid arthritis it is the joints that are affected. When an autoimmune disease occurs, a viral infection of tissues has often set off an immune reaction to the body's own tissues in an attempt to attack the virus. There is evidence to suggest diabetes is also the result of this sequence of events.

Autoimmune diseases seem to be preceded by a viral infection that fools the immune system into attacking the body's own tissues.

Aging

The thymus (fig. 7.4) is large in relation to the rest of the body during fetal development and childhood; however, it stops growing by puberty and then begins to atrophy. This has led some researchers to suggest that aging may be associated with a general decline in the immune system. Certainly some diseases, such as cancer, are more prevalent with age.

The thymus gland becomes smaller with age, and perhaps this occurrence contributes to the aging process.

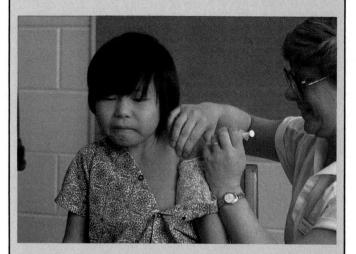

FIGURE 7.11 *Suggested immunization schedule for infants and young children. Children who are not immunized are subject to childhood diseases that can cause serious health consequences.*

AGE	SHOTS
2 months old	1 DTP[a] immunization 1 polio immunization
4 months old	2 DTP immunizations 2 polio immunizations
6 months old	3 DTP immunizations 2 polio immunizations[b]
15 months old	3 DTP immunizations 2 polio immunizations[b] 1 measles immunization[c] 1 rubella immunization[c,d] 1 mumps immunization[c]
18 months and older	4 DTP immunizations 3 polio immunizations[b]
4 to 6 years, before starting to school	A DTP booster (the 5th immunization) A polio booster (the 4th immunization)[b]
Thereafter	Tetanus-diphtheria (Td) booster should be given every 10 years or following a dirty wound if a booster has not been given in the preceding 5 years.

[a]D = diphtheria, T = tetanus, P = pertussis (whooping cough)
[b]Some doctors give one additional dose when the child is 6 months old.
[c]Only one shot needed. Some doctors combine these vaccines in a single injection.
[d]rubella (German measles)

DHRS Immunization Program, State of Florida.

IMMUNOTHERAPY

The immune system can be manipulated to help people avoid or recover from diseases. Some of these techniques have been utilized for quite a long time, and some are relatively new.

Active Immunity

Active immunity, which provides long-lasting protection against a disease-causing organism, develops after an individual becomes infected with a virus or bacterium. In many instances today, however, it is not necessary to suffer an illness to become immune because it is possible to be medically immunized against a disease. One possible recommended immunization schedule for children is given with figure 7.11. Immunization requires the use of **vaccines,** which are traditionally bacteria and viruses that have been treated so that they are no longer virulent (able to cause disease). New methods of producing vaccines are being developed. For example, it is possible to use the recombinant DNA technique to mass-produce a protein that can be used as a vaccine (p. 429).

After a vaccine is injected, it is possible to determine the amount of antibody present in the bloodstream—this is called the *antibody titer.* After the first injection, a primary response occurs. There is a period of several days during which no antibodies are present; then there is a slow rise in the titer, which is followed by a gradual decline (fig. 7.12). After a second injection, a secondary response occurs. The titer rises rapidly to a level much greater than before. A second injection is often called the "*booster shot*" since it boosts the antibody titer to a high level. The antibody titer is now high enough to prevent disease symptoms even if the individual is exposed to the disease. For some time thereafter, the individual is immune to that particular disease.

Primary and secondary responses occur whenever an individual is exposed twice to the same disease-causing agent. The difference in the responses may be related to the number of plasma and memory cells. Upon the second exposure, these cells are already present, and antibodies can be rapidly produced.

Vaccines can be used to make people actively immune.

Passive Immunity

Passive immunity occurs when an individual is given antibodies to combat a disease. Since these antibodies are not produced by the individual's B cells, passive immunity is short-lived. For example, newborn infants possess passive immunity because antibodies have crossed the placenta from the mother's blood. These antibodies soon disappear, however, so that within a few months infants become more susceptible to infections. Breast feeding (fig. 7.13) prolongs the passive immunity an infant receives from the mother because there are antibodies in the mother's milk.

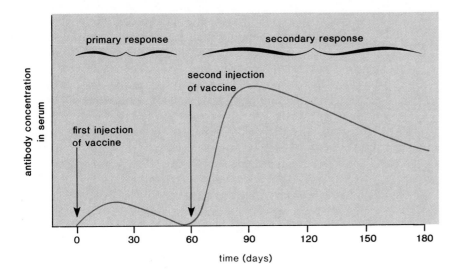

FIGURE 7.12 *Immunization responses. The primary response after the first injection of a vaccine is minimal, but the secondary response after the second injection shows a dramatic rise in the amount of antibody present in serum.*

Even though passive immunity is not lasting, it is sometimes used to prevent illness in a patient who has been unexpectedly exposed to an infectious disease. Usually, the person receives an injection of a serum containing antibodies. This may have been taken from donors who have recovered from the illness. In other instances, horses have been immunized, and serum has been taken from them to provide the needed antibodies. Horses have been used to produce antibodies against diphtheria, botulism, and tetanus. Occasionally a patient who receives these antibodies becomes ill because the serum contains proteins that the individual's immune system recognizes as foreign. This is called serum sickness.

Passive immunity is short-lived because the antibodies are administered to and not made by the individual.

Monoclonal Antibodies

There is much hope that the technique of producing *mono-* (one) *clonal* (clone) antibodies will one day make plentiful amounts of pure and specific antibodies available. One method of producing monoclonal antibodies is depicted in figure 7.14. Lymphocytes are removed from the body and exposed in vitro (in laboratory glassware) to a particular antigen. Then they are fused with a cancer cell because cancerous cells, unlike normal cells, will divide an unlimited number of times. The fused cells are called **hybridomas** (hybrid- refers to their being a fusion of two different cells, and -oma stands for the fact that one of the cells is a cancer [carcinoma] cell).

The antibodies produced by hybridoma cells are called **monoclonal antibodies** because all of them are the same type and all are produced by cells derived from the same parent cell. At present monoclonal antibodies are being used to allow quick and certain diagnosis of various conditions. For example, a certain hormone is present in the urine when a woman is pregnant (p. 343). A monoclonal antibody can be used to detect the hormone and indicate that the woman is

FIGURE 7.13 *Example of passive immunity. Breast feeding is believed to provide a newborn with antibodies during the period of time when the child is not yet producing antibodies.*

FIGURE 7.14 *One possible method for producing human monoclonal antibodies. a. Blood sample is taken from a patient. b. Inactive lymphocytes from a sample are exposed to an antigen. c. Activated lymphocytes are fused with cancer cells. d. Resulting hybridomas divide repeatedly, giving many cells that produce monoclonal antibodies.*

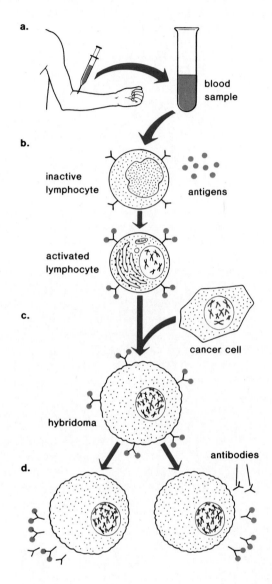

a.

blood sample

b.

inactive lymphocyte   antigens

activated lymphocyte

c.

cancer cell

hybridoma

antibodies

d.

Lymphokines   Both interferon and various interleukins have been used as immunotherapeutical drugs, particularly to potentiate the ability of the individual's own T cells (and possibly B cells) to fight cancer.

Interferon is a substance produced by leukocytes, fibroblasts, and probably most cells in response to a viral infection. When it is produced by T cells, it is called a lymphokine. Interferon is still being investigated as a possible cancer drug, but thus far it has proven to be effective only in certain patients, and the exact reasons cannot as yet be discerned. For example, interferon has been found to be effective in up to 90% of patients with a type of leukemia known as hairy-cell leukemia (because of the hairy appearance of the malignant cells).

When and if cancer cells carry an altered protein on their cell surface, by all rights they should be attacked and destroyed by T_K cells. Whenever cancer develops, it's possible that the T_K cells have not been activated. In that case, the use of lymphokines might awaken the immune system and lead to the destruction of the cancer. In one technique being investigated, researchers first withdraw T cells from the patient and activate the cells by culturing them in the presence of an interleukin. The cells are then reinjected into the patient, who is given doses of interleukin to maintain the killer activity of the T cells.

Growth Factors   Several growth factors that raise the blood cell count of particular white blood cells have been discovered. The best known of these is *GM-CSF* (granulocyte-macrophage colony-stimulating factor), which enhances production of macrophages, neutrophils, eosinophils, and possibly red blood cells. This growth factor has been given to patients with AIDS, and it did raise their white blood cell count. Presently, the growth factor is being produced by recombinant DNA technology and more extensive clinical trials are expected soon.

Lymphokine therapy and white blood cell growth factors show some promise of potentiating the individual's own immune system.

---

pregnant. Soon monoclonal antibodies may also be available to detect and fight certain types of cancer within the individual. It may be possible to have these antibodies carry chemicals that will help destroy the cancerous cells.

Monoclonal antibodies are produced in pure batches—they are specific against just one antigen.

Lymphokines and Growth Factors
Lymphokines and blood cell growth factors are being investigated as possible adjunct therapy for cancer and AIDS.

SUMMARY

The body is prepared to defend itself in both a generalized and a specific manner. Barriers to entry, phagocytic white blood cells, and protective chemicals will react to any threat, and therefore are factors in the generalized response. The immune system is capable of specific defense. It consists of lymphocytes and associated structures. T lymphocytes responsible for cell-mediated immunity have passed through the thymus, and B lymphocytes responsible for antibody-mediated immunity have not.

Although B cells can directly recognize an antigen and thereafter produce antibody-secreting plasma cells, T cells must have the antigen, along with an MHC protein, presented by a macrophage. There are three types of T cells: T_K cells kill cells on contact; T_H cells stimulate B cells and other T cells; and T_S cells suppress the immune response. Following an immune response, memory T and B cells remain in the body, providing long-lasting immunity.

Immunity has certain unwanted side effects. Allergies are due to an overactive immune system that forms antibodies to substances not normally recognized as foreign. T_K cells attack transplanted organs, although immunosuppressive drugs are available. Autoimmune illnesses occur when antibodies form against the body's own cells.

Immunity can be fostered by immunotherapy. Vaccines are available to promote active immunity, and monoclonal antibodies are sometimes available to provide an individual with short-term passive immunity. Lymphokines, notably interferon, the interleukins, and white blood cell growth factors, are used to promote the body's ability to recover from cancer.

OBJECTIVE QUESTIONS

1. T lymphocytes have passed through the _____ .
2. A stimulated B cell produces antibody-secreting _____ cells and _____ cells that are ready to produce the same type antibody at a later time.
3. B cells are responsible for _____ -mediated immunity.
4. In order for a T cell to recognize an antigen, it must be presented by a _____ along with an MHC protein.
5. T cells produce _____ that are stimulatory chemicals for all types of immune cells.
6. Killer T cells are responsible for _____ -mediated immunity.
7. Allergic reactions are associated with the release of _____ from mast cells.
8. The body recognizes foreign cells because they bear different _____ proteins than the body's cells.
9. Immunization with _____ brings about active immunity.
10. Hybridomas produce _____ antibodies.

Answers

1. thymus 2. plasma, memory 3. antibody 4. macrophage 5. lymphokines 6. cell 7. histamine 8. MHC 9. vaccines 10. monoclonal

STUDY QUESTIONS

1. List the organs of the immune system and give the function of each. (p. 142)
2. What is the clonal selection theory? (p. 144)
3. B cells are said to be responsible for what type immunity? (p. 144)
4. Explain the process that allows a T cell to recognize an antigen. (p. 149)
5. List the three types of T cells and state their functions. (p. 149)
6. What are lymphokines and how are they used in immunotherapy? (p. 150)
7. Killer T cells are said to be responsible for what type immunity? (p. 150)
8. Discuss allergies, tissue rejection, autoimmune diseases, and aging as they relate to the immune system. (p. 151)
9. Relate active immunity to the presence of plasma cells and memory cells. (p. 154)
10. How is active immunity achieved? passive immunity achieved? (p. 154)

THOUGHT QUESTIONS

1. There are more cells in the immune system than any other system. What does this tell us about the environment of organisms?
2. The immune system has various ways of attacking foreign antigens. Does this seem to be a duplication of effort or a necessity?
3. What would be anti-antibodies and why would you expect B cells to produce such a protein? What could be the function of anti-antibodies?

B lymphocytes (lim′fo-sīts) lymphocytes that react against foreign substances in the body by producing and secreting antibodies. *142*

complement (kom′plĕ-ment) a group of proteins in plasma that produce a variety of effects once an antigen-antibody reaction has occurred. *141*

helper T cells (hel′per te selz) T lymphocytes that stimulate certain other T and B lymphocytes to perform their respective functions. *150*

hybridomas (hi″brid-o′mahz) fused lymphocytes and cancer cells used in the manufacture of monoclonal antibodies. *155*

immune complex (ĭmūn′ kom′pleks) the product of an antigen-antibody reaction. *149*

interferon (in″ter-fēr′on) a protein formed by a cell infected with a virus that can increase the resistance of other cells to the virus. *142*

killer T cells (killer te selz) T lymphocytes that attack cells bearing foreign bodies. *150*

lymphokines (lim′fo-kīnz) chemicals secreted by T cells that have the ability to affect the characteristics of lymphocytes and monocytes. *150*

memory B cells (mem′o-re be selz) cells derived from B cells that are ever present within the body and that produce a specific antibody and account for the development of active immunity. *144*

MHC protein (em āch se pro′te-in) major histocompatibility protein; a surface molecule that serves as a genetic marker. *149*

monoclonal antibodies (mon″-o-klōn′al an′tĭ-bod″ēz) antibodies of one type that are produced by cells that are derived from a lymphocyte that has fused with a cancer cell. *155*

plasma cells (plaz′mah selz) cells derived from B cell lymphocytes that are specialized to mass-produce antibodies. *144*

suppressor T cells (su-pres′ or te selz) T lymphocytes that suppress certain other T and B lymphocytes from continuing to divide and perform their respective functions. *151*

T lymphocytes (lim′fo-sīts) lymphocytes that interact directly with antigen-bearing cells and are responsible for cell-mediated immunity. *142*

vaccine (vak′sēn) antigens prepared in such a way that they can promote active immunity without causing disease. *154*

CHAPTER EIGHT

RESPIRATION

CHAPTER CONCEPTS

1 The respiratory tract of humans is designed in such a way that air is filtered, warmed, and saturated with water before gas exchange takes place across a very extensive moist surface.

2 Breathing is a required process for human life because it brings to the blood the oxygen needed by the cells for cellular respiration and it rids the body of carbon dioxide, a by-product of cellular respiration.

3 The respiratory pigment hemoglobin has chemical and physical characteristics that promote its combination with oxygen in the lungs and its release of oxygen in the tissues. It also aids in the transport of carbon dioxide from the tissues to the lungs, largely by its ability to buffer.

4 The respiratory tract is especially subject to disease because it serves as an entrance for infectious agents. Polluted air contributes to the cause of two major lung disorders—emphysema and cancer.

Human lung tissue and red blood cells.

INTRODUCTION

Fish take their oxygen from the water. Why must we take our oxygen with us when we descend into the depths (fig. 8.1)? For one thing we are warm blooded and therefore require more oxygen intake per unit time than do fish. Water contains 5–7 ppt (parts of oxygen per thousands parts of water), but air contains 210 ppt. If our oxygen demands were lower, could we breathe water? The answer seems to be yes, particularly if the water is cold. Scientists have studied the mammalian diving reflex, which allows air-breathing whales and seals to remain submerged in cold water. When frigid water splashes over the forehead and nose, nerves signal the brain to divert oxygen-rich blood from the limbs to the heart and brain. Water entering the lungs further cools the body, slowing down the metabolic rate, thereby reducing the brain's oxygen requirements. It was not viewed as astonishing, then, when physicians were able to revive Jimmy Tontlewicz, a lad who accidently plunged into the icy waters of Lake Michigan and was not found for twenty minutes.

Breathing is more eminently necessary than eating. Although it is possible to stop eating altogether for several days, it is not possible to remain alive for longer than several minutes without breathing. Breathing supplies the body with the oxygen needed for cellular respiration, as indicated in the following equation.[1]

$$38\ ADP + 38\ \textcircled{P} \longrightarrow 38\ ATP$$
$$C_6H_{12}O_6 + 6\ O_2 \longrightarrow 6\ H_2O + 6\ CO_2$$

The equation indicates that the body requires oxygen to convert the energy within glucose to phosphate-bond energy. Therefore, the more energy expended, the greater the need for oxygen. The minimum amount of oxygen a person consumes at complete rest, without eating previously, is related to the BMR (p. 92). The average young adult male utilizes about 250 ml of oxygen per minute in a basal, or restful, state. Exercise and the digestion of food raise the need for oxygen. The average amount of oxygen needed with mild exercise is 500 ml of oxygen per minute. The equation for cellular respiration also indicates that cells produce carbon dioxide. This metabolic end product must be eliminated from the body by the breathing process.

Breathing is necessary to supply the body with oxygen so that ATP can be formed by cellular respiration.

---

[1] The body also requires oxygen for the respiration of fats and amino acids in addition to glucose.

FIGURE 8.1 *When individuals scuba dive, they carry a tank that supplies them with air. One of the major problems with deep dives is due to the increased pressure under water. Nitrogen dissolves in the blood and then begins to bubble out and collect in the joints when the diver rises to the surface. To prevent this so-called decompression sickness, it is necessary to decompress slowly so that the nitrogen is expired by the respiratory system.*

Altogether, respiration may be used to refer to the complete process of getting oxygen to body cells for cellular respiration and to the reverse process of ridding the body of the carbon dioxide given off by cells (fig. 8.2). Respiration may be said to include the following:

1. **breathing**—entrance and exit of air into and out of the lungs
2. **external respiration**—exchange of gases ($O_2 + CO_2$) between air and blood
3. **internal respiration**—exchange of gases between blood and tissue fluid
4. **cellular respiration**—production of ATP in cells

In this chapter, we are studying the first three portions of the respiratory process. Cellular respiration is discussed in chapter 2.

FIGURE 8.2 *Respiration is divided into four components: breathing brings gas into the lungs; external respiration is the exchange of gases in the lungs; internal respiration is the exchange of gases in the tissues; and cellular respiration is the production of ATP in cells—an oxygen-requiring process.*

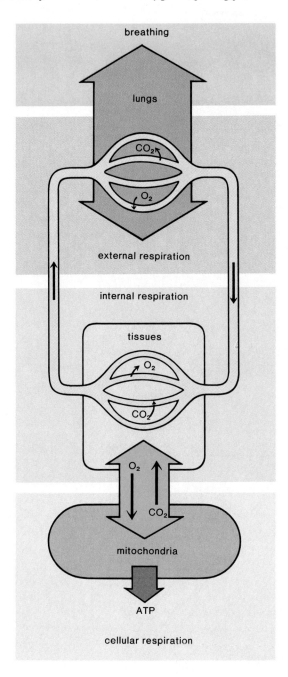

TABLE 8.1 PATH OF AIR

STRUCTURE	FUNCTION
Nasal cavities	Filter, warm, and moisten air
Nasopharynx	Passage of air from nose to throat
Pharynx (throat)	Connection to surrounding regions
Glottis	Passage of air
Larynx (voice box)	Sound production
Trachea (windpipe)	Passage of air to thoracic cavity
Bronchi	Passage of air to each lung
Bronchioles	Passage of air to each alveolus
Alveoli	Air sacs for gas exchange

Passage of Air

During inspiration and expiration, air is conducted toward or away from the lungs by a series of cavities, tubes, and openings, which are listed in order in table 8.1 and illustrated in figure 8.3.

As air moves in along the air passages, it is filtered, warmed, and moistened. The filtering process is accomplished by coarse hairs and cilia in the region of the nostrils and by cilia alone in the rest of the nose and windpipe. In the nose, the hairs and cilia act as a screening device. In the trachea, cilia beat upward, carrying mucus, dust, and occasional bits of food that "went down the wrong way" into the pharynx where the accumulation may be swallowed or expectorated. The inspired air is warmed by heat given off by the blood vessels lying close to the surface of the lining of the air passages, and it is moistened by the wet surface of these passages.

On the other hand, as air moves out during expiration, it becomes progressively cooler and loses its moisture. As the gas cools, it deposits its moisture on the lining of the windpipe and nose, and the nose may even drip as a result of this condensation. But the air still retains so much moisture that upon expiration on a cold day it condenses and forms a small cloud.

Air is warmed, filtered, and moistened as it moves from the nose toward the lungs.

Each portion of the air passage also has its own unique structure and function, as described in the sections that follow.

BREATHING

The normal breathing rate is about fourteen to twenty times per minute. Breathing consists of taking air in, **inspiration,** and forcing air out, **expiration.** Expired air contains less oxygen and more carbon dioxide than inspired air, indicating that the body takes in oxygen and gives off carbon dioxide.

FIGURE 8.3 *Diagram of human respiratory tract, with an enlargement of alveoli, surrounded by a capillary network. Each lung contains hundreds of thousands of alveoli where gas exchange occurs.*

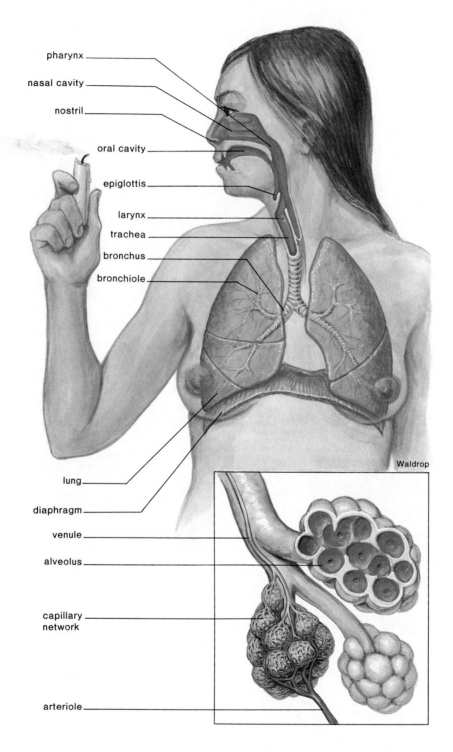

pharynx

nasal cavity

nostril

oral cavity

epiglottis

larynx

trachea

bronchus

bronchiole

Waldrop

lung

diaphragm

venule

alveolus

capillary network

arteriole

Nose

The **nose** contains two nasal cavities, narrow canals with convoluted lateral walls that are separated from one another by a septum. Up in the narrow recesses of the nasal cavities are special ciliated cells (fig. 12.5) that act as odor receptors. Nerves lead from these cells and go to the brain, where the impulses are interpreted as smell.

The nasal cavities have a number of openings. The tears produced by tear (lacrimal) glands drain into the nasal cavities by way of tear ducts. For this reason, crying produces a runny nose. The nasal cavities open into the cranial sinuses, air-filled spaces in the skull, and empty into the nasopharynx, a chamber just beyond the soft palate. The *eustachian tubes* lead into the nasopharynx from the middle ears (fig. 12.17).

Pharynx

The air and food channels cross in the **pharynx** so that the trachea (windpipe) lies in front of the esophagus, which normally opens only during the process of swallowing food. Just below the pharynx lies the larynx, or voice box (fig. 8.3).

The nasal cavities contain the sense receptors for smell and open into the pharynx. The pharynx is the back of the throat where the air passage and food passage cross.

Larynx

The **larynx** may be imagined as a triangular box with its apex, the Adam's apple, located at the front of the neck. At the top of the larynx is a variable-sized opening called the **glottis.** When food is being swallowed, the glottis is covered by a flap of tissue called the **epiglottis** so that no food passes into the larynx. If, by chance, food or some other substance does gain entrance to the larynx, reflex coughing usually occurs to expel the substance. If this reflex is not sufficient, it may be necessary to resort to the *Heimlich maneuver* (fig. 8.4).

At the edges of the glottis, embedded in mucous membrane, are elastic ligaments called the **vocal cords** (fig. 8.5). These cords, which stretch from the back to the front of the larynx just at the sides of the glottis, vibrate when air is expelled past them through the glottis. Vibration of the vocal cords produces sound. The high or low pitch of the voice depends upon the length, thickness, and degree of elasticity of the vocal cords and the tension at which they are held. The loudness, or intensity, of the voice depends upon the amplitude of the vibrations, or the degree to which vocal cords vibrate.

At the time of puberty, the growth of the larynx and the vocal cords is much more rapid and accentuated in the male than in the female, causing the male to have a more prominent Adam's apple and a deeper voice. The voice "breaks" in the young male due to his inability to control the longer vocal cords.

FIGURE 8.4 *Heimlich maneuver. More than eight Americans choke to death each day on food lodged in the trachea. A simple process termed the abdominal thrust (Heimlich) maneuver can save the life of a person who is choking. The abdominal thrust maneuver is performed as follows: If the victim is standing or sitting, (1) Stand behind the victim or the victim's chair, and wrap your arms around his or her waist; (2) Grasp your fist with your other hand, and place the fist against the victim's abdomen, slightly above the navel and below the rib cage; (3) Press your fist into the victim's abdomen with a quick upward thrust; (4) Repeat several times if necessary. If the victim is lying down, (1) Position the victim on his or her back; (2) Face the victim, and kneel on his or her hips; (3) With one of your hands on top of the other, place the heel of your bottom hand on the abdomen, slightly above the navel and below the rib cage; (4) Press into the victim's abdomen with a quick upward thrust; (5) Repeat several times if necessary. If you are alone and choking, use anything that applies force just below your diaphragm. Press into a table or a sink, or use your own fist.*

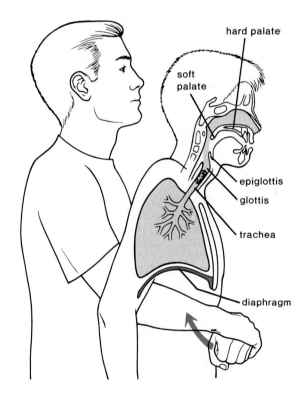

hard palate

soft palate

epiglottis

glottis

trachea

diaphragm

The larynx is the voice box because it contains the vocal cords at the sides of the glottis, an opening sometimes covered by the epiglottis.

Trachea

The larynx is continuous with the **trachea,** which is held open by cartilaginous rings. Ciliated mucous membrane (fig. 3.5) lines the trachea, and normally these cilia keep the windpipe free of debris. Smoking is known to destroy the cilia, and consequently the soot in cigarette smoke collects in the lungs. (Smoking will be discussed more fully at the end of this chapter.)

FIGURE 8.5 *Vocal cords lie at the edge of the glottis. When air is expelled from the larynx, the cords vibrate, producing the voice.* a. *Closed position.* b. *Open position.*

FIGURE 8.6 *Site of tracheotomy to allow air to enter the trachea beneath the location of an obstruction.*

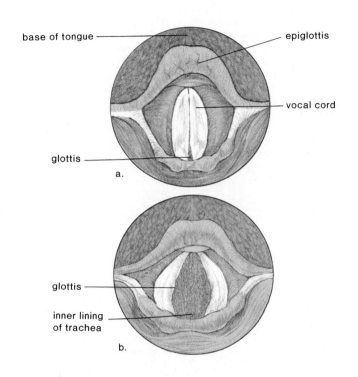

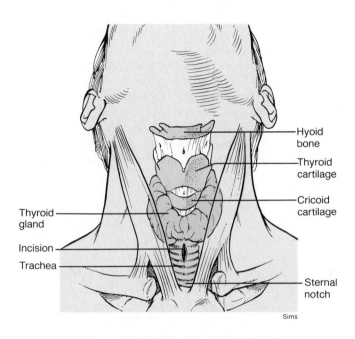

FIGURE 8.7 *Cast of lungs showing a large number of airways. The bronchi branch into the bronchioles that branch and rebranch until they terminate in the alveoli. Each lobe of the lung is in a different color in this cast.*

If the trachea is blocked because of illness or accidental swallowing of a foreign object, it is possible to insert a tube by way of an incision made in the trachea. This tube acts as an artificial air intake and exhaust duct. The operation is called a *tracheotomy* (fig. 8.6).

Bronchi

The trachea divides into two **bronchi** that enter the right and left lungs and branch into a great number of smaller passages called the **bronchioles.** The two bronchi resemble the trachea in structure, but as the bronchial tubes divide and subdivide, their walls become thinner and the small rings of cartilage do not occur (fig. 8.7). Each bronchiole terminates in an elongated space that is enclosed by a multitude of air pockets, or sacs, called **alveoli** (fig. 8.3), which make up the lungs.

Lungs

Within the lungs each alveolar sac is only one layer of squamous epithelium surrounded by blood capillaries. Gas exchange occurs between the air in the alveoli and the blood in the capillaries (fig. 8.3).

A film of lipoprotein that lines the alveoli of mammalian lungs lowers the surface tension and prevents them from closing up. Some newborn babies, especially premature infants, lack this film, resulting in lung collapse. This condition, called infant respiratory distress syndrome, often results in death.

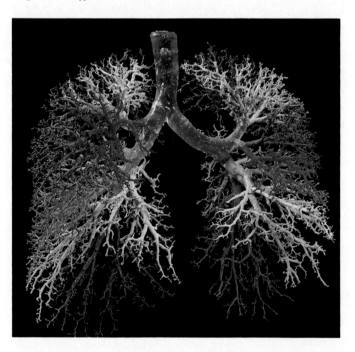

There are approximately 300 million alveoli per lung. Their total surface area is 50–70 m², about forty times the surface area of the skin. Because of their many air spaces, the lungs are very light; normally, a piece of lung tissue dropped in a glass of water will float.

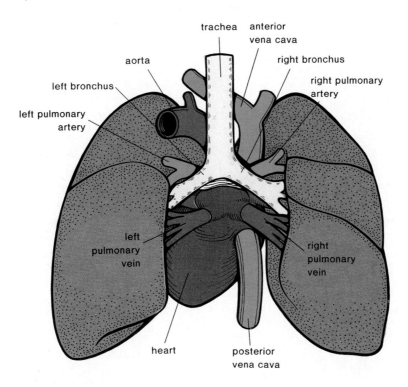

Externally, the lungs are cone-shaped organs that lie on both sides of the heart in the *thoracic cavity*. The branches of the pulmonary artery accompany the bronchial tubes and form a mass of capillaries around the alveoli. The four pulmonary veins collect blood from these capillaries and empty into the left atrium of the heart. Figure 8.8 shows the relationship of the pulmonary vessels to the trachea and bronchial tubes.

Air moves from the trachea and the two bronchi, held open by cartilaginous rings, into the lungs. The lungs are composed of air sacs, called alveoli.

Mechanism of Breathing

In order to understand **ventilation,** the manner in which air is drawn into and expelled out of the lungs, it is necessary to remember first that when you are breathing there is a continuous column of air from the pharynx to the alveoli of the lungs; that is, the air passages are open.

Secondly, we may note that the lungs lie within the sealed-off chest (thoracic) cavity. The **ribs,** which are hinged to the vertebral column at the back and to the *sternum* (breastbone) at the front, along with the muscles that lie between them, make up the top and sides of the chest cavity. The **diaphragm,** a dome-shaped horizontal muscle, forms the floor of the chest cavity. The lungs themselves are enclosed by the **pleural membranes** (fig. 8.9b), one of which adheres closely to the walls of the chest and diaphragm, while the other is fused to the lungs. The two pleural layers lie very close to one another, being separated only by a thin film of fluid. Normally, the intrapleural pressure is less than atmospheric pressure. The importance of this reduced pressure is demonstrated when by design or accident air enters the intrapleural space: the lung is no longer able to expand to allow inspiration, and instead it actually collapses.

The lungs are completely enclosed and by way of the pleural membranes adhere to the chest cavity walls.

Inspiration

It can be shown that carbon dioxide and hydrogen ions are the primary stimuli that cause us to breathe. When the concentration of CO_2 and subsequently H^+ reaches a certain level in the blood, the *breathing center* in the **medulla oblongata**—the stem portion of the brain—is stimulated. This center is not affected by low oxygen levels, but there are chemoreceptors in the *carotid bodies* (located in the carotid arteries) and in the *aortic bodies* (located in the aorta) that do respond to low blood oxygen in addition to carbon dioxide and hydrogen ion concentration.

FIGURE 8.9 *Inspiration versus expiration.* a. *When the rib cage lifts up and outward and the diaphragm lowers, the lungs expand so that (b) air is drawn in. This sequence of events is only possible because the pressure within the intrapleural space, containing a thin film of water, is less than atmospheric pressure.* c. *When the rib cage lowers and the diaphragm rises (d), the lungs recoil so that air is forced out.*

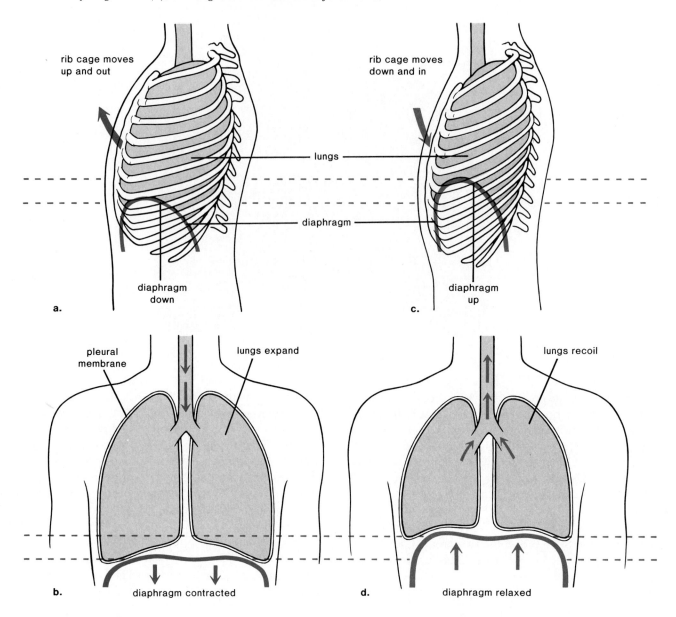

When the breathing center is stimulated, a nerve impulse goes out by way of nerves to the diaphragm and rib cage (fig. 8.10). In its relaxed state, the *diaphragm* is dome shaped, but upon stimulation it contracts and lowers. When the rib muscles contract, the *rib cage* moves upward and outward. Both of these contractions serve to increase the size of the chest cavity. As the chest cavity increases in size, the lungs expand. When the lungs expand, air pressure within the enlarged alveoli lowers and is immediately rebalanced by air rushing in through the nose or mouth.

Inspiration (fig. 8.9a and b) is the active phase of breathing. It is during this time that the diaphragm contracts, the rib muscles contract, and the lungs are pulled open, with the result that air comes rushing in. Note that the air comes in because the lungs have already opened up; the air does not force the lungs open. This is why it is sometimes said that *humans breathe by negative pressure.* It is the creation of a partial vacuum that sucks air into the lungs.

Stimulated by nerve impulses, the rib cage lifts up and out and the diaphragm lowers to expand the chest cavity and lungs, allowing inspiration to occur.

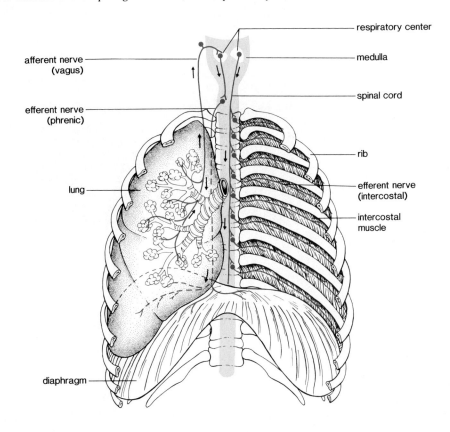

afferent nerve
(vagus)

efferent nerve
(phrenic)

lung

diaphragm

respiratory center

medulla

spinal cord

rib

efferent nerve
(intercostal)

intercostal
muscle

Expiration

When the lungs are expanded, the stretching of the alveoli stimulates special receptors in the alveolar walls, and these receptors initiate nerve impulses from the inflated lungs to the breathing center. When the impulses arrive at the medulla oblongata, the center is inhibited and stops sending signals to the diaphragm and the rib cage. The *diaphragm* relaxes and resumes its dome shape (fig. 8.9c and d). The abdominal organs press up against the diaphragm. The *rib cage* moves down and inward. The elastic lungs recoil and air is pushed outward. Table 8.2 summarizes the events causing inspiration and expiration. It is clear that while inspiration is an active phase of breathing, normally expiration is passive since the breathing muscles automatically relax following contraction. But it is possible in deeper and more rapid breathing for both phases to be active because there is another set of rib muscles whose contraction can forcibly cause the chest to move downward and inward. Also, when the abdominal wall muscles are contracted, there is an increase in pressure that helps expel air.

When nervous stimulation ceases, the rib cage lowers and the diaphragm rises, allowing the lungs to recoil and expiration to occur.

TABLE 8.2 BREATHING PROCESS

INSPIRATION	EXPIRATION
Medulla sends stimulatory message to diaphragm and rib muscles.	Stretch receptors in lungs send inhibitory message to medulla.
Diaphragm contracts and flattens.	Diaphragm relaxes and resumes dome position.
Rib cage moves up and out.	Rib cage moves down and in.
Lungs expand.	Lungs recoil.
Negative pressure develops in lungs.	Positive pressure develops in lungs.
Air is pulled in.	Air is forced out.

Consequences   There are certain physiological consequences that result from the manner in which breathing is regulated. First of all, we may note that breathing is initiated and continues due to the presence of carbon dioxide in the blood. Therefore, when necessary, it is better to give a person oxygen gas containing carbon dioxide rather than pure oxygen alone. The mixture of gases stimulates the resumption of breathing, whereas pure oxygen does not.

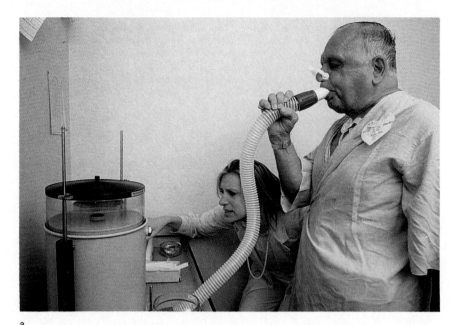

FIGURE 8.11 *Vital capacity. a. This individual is using a spirometer that measures the amount of air that can be maximally inhaled and exhaled. When he inspires, a pen moves up, and when he expires, a pen moves down. b. The resulting pattern, such as the one shown here, is called a spirograph.*

a.

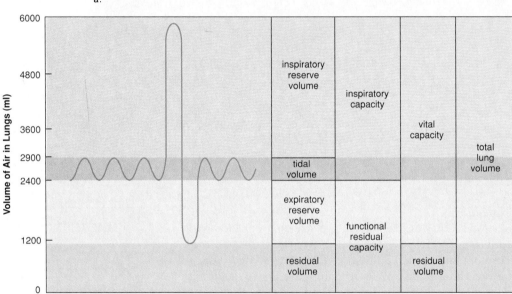

b.

The breathing center may stimulate deeper and more rapid breathing or this may be done voluntarily. It is possible for a person to deliberately breathe more rapidly or more slowly or to hold the breath for a short time. However, it is impossible to commit suicide by holding your breath; eventually, carbon dioxide buildup in the blood forces the resumption of breathing.

Lung Capacities

When we breathe, the amount of air moved in and out with each breath is called the **tidal volume.** Normally the tidal volume is about 500 ml, but we can increase the amount inhaled and exhaled by deep breathing. The total volume of air that can be moved in and out during a single breath is called the **vital capacity** (fig. 8.11). First, we can increase inspiration by as much as 3,100 ml of air. This is called the *inspiratory reserve volume.* Similarly, we can increase expiration by contracting the chest muscles. This is called the *expiratory reserve volume* and measures approximately 1,400 ml of air. Vital capacity is the sum of tidal, inspiratory-reserve, and expiratory-reserve volumes.

It can be noted in figure 8.11 that even after very deep breathing, some air (about 1,000 ml) remains in the lungs; this is called the **residual volume.** This air is no longer useful for gas exchange purposes. In some lung diseases, such as emphysema (p. 173) and asthma (p. 151), the residual volume builds up because the individual has difficulty clearing the lungs. This means that the lungs tend to be filled with useless air and, as you can see from examining figure 8.11, the vital capacity is reduced.

FIGURE 8.12

Distribution of air in lungs. The air between (a) and (b) does not immediately reach the alveoli; therefore, this is called dead space. The air below (c) represents the amount of residual air that has not left the lungs. Only the air between b and c brings with it additional oxygen for respiration.

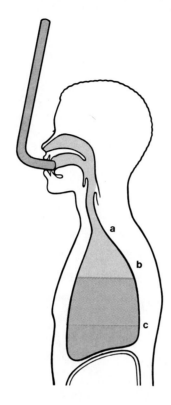

External And Internal Respiration

External Respiration

The term *external respiration* refers to the exchange of gases between the air in the alveoli and the blood within the pulmonary capillaries (fig. 8.3). The wall of an alveolus consists of a thin, single layer of cells, and the wall of a blood capillary also consists of such a layer. Since neither of the walls offers resistance to the passage of gases, *diffusion* is believed to govern the exchange of oxygen and carbon dioxide between alveolar air and the blood. Active cellular absorption and secretion do not appear to play a role. Rather, the direction in which the gases move is determined by the concentration gradients between blood and inspired air.

Atmospheric air contains little carbon dioxide, but blood flowing into the lung capillaries is almost saturated with this gas. Therefore, *carbon dioxide diffuses out of the blood into the alveoli.* The pressure pattern is the reverse for oxygen. Blood coming into the pulmonary capillaries is oxygen poor, and alveolar air is oxygen rich; therefore, *oxygen diffuses into the capillary.* Breathing at high altitudes is less effective than at low altitudes because the air pressure is lower, making the concentration of oxygen (and other gases) lower than normal; therefore less oxygen diffuses into the blood. Breathing problems do not occur in airplanes because the cabin is pressurized to maintain an appropriate pressure. Emergency oxygen is available in case the pressure should, for one reason or another, be reduced.

As blood enters the pulmonary capillaries (fig. 8.13), most of the carbon dioxide is being carried as the bicarbonate ion, HCO_3^-. As the little free carbon dioxide remaining begins to diffuse out, the following reaction is driven to the right:

$$H^+ + HCO_3^- \rightarrow H_2CO_3 \rightarrow H_2O + CO_2$$

The enzyme carbonic anhydrase (p. 127), present in red blood cells, speeds up the reaction. As the reaction proceeds, the respiratory pigment **hemoglobin** (p. 124) gives up the hydrogen ions it has been carrying; HHb becomes Hb.

Now hemoglobin more readily takes up oxygen and becomes oxyhemoglobin.

$$Hb + O_2 \rightarrow HbO_2$$

It is a remarkable fact that at the partial pressure[2] of oxygen in the lungs (P_{O_2} = about 100 mm Hg) hemoglobin is about 95% saturated (fig. 8.14). Hemoglobin takes up oxygen in increasing amounts as the P_{O_2} increases and likewise gives it up as P_{O_2} decreases. The curve begins to level off at about 90 mm Hg. This means that hemoglobin easily retains oxygen in the lungs but tends to release it in the tissues. This effect is potentialed by the fact that hemoglobin

[2]Air exerts pressure, and the amount of pressure each gas exerts in air is called its partial pressure, symbolized by a capital P.

Dead Space

Figure 8.12 shows that some of the inspired air never reaches the lungs; instead it fills the conducting airways. These passages are not used for gas exchange and therefore are said to contain *dead space.* To ventilate the lungs, then, it is better to breathe more slowly and deeply because it insures that a greater percentage of the tidal volume will reach the lungs.

If we breathe through a tube, we increase the amount of dead space and increase the amount of air that never reaches the lungs. Any device that increases the amount of dead space beyond maximal inhaling capacity spells death to the individual because the air inhaled would never reach the alveoli.

The manner in which we breathe has physiological consequences.

HUMAN ISSUE

Many Americans are fighting high home heating costs by switching to wood-burning stoves and furnaces. This has the added benefit of reducing our country's dependence on foreign sources of oil. But "there is no free lunch," as that old saying goes. Like car exhaust fumes, smoke from burning wood contains hydrocarbons and other cancer-causing chemicals. Approximately 80% of the hydrocarbons found in recent wintertime air samples in a New Mexico city were generated by burning wood. Also, nearly 60% of the cancer-causing potential of this city's winter air was attributable to wood smoke. These statistics raise an interesting dilemma. After weighing the economic and medical pro's and con's, do you think people should be encouraged or discouraged from using wood-burning stoves and furnaces?

FIGURE 8.13 Diagram illustrating external and internal respiration. During external respiration in the lungs, CO_2 leaves the blood and O_2 enters the blood. During internal respiration in the tissues, O_2 leaves the blood and CO_2 enters the blood. Steps necessary for gas exchange are shown for the lungs (above) and for the tissue (below).

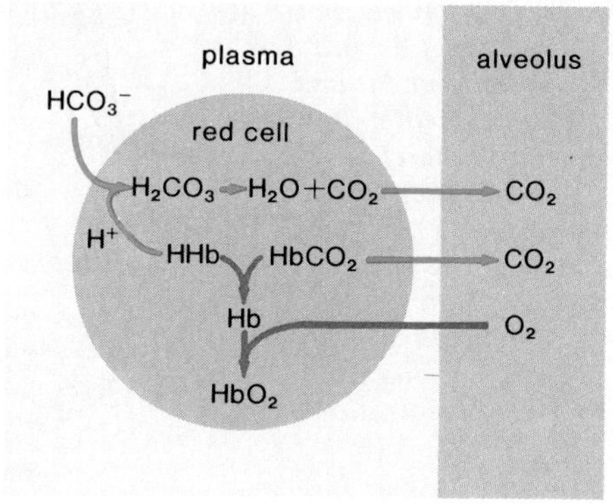

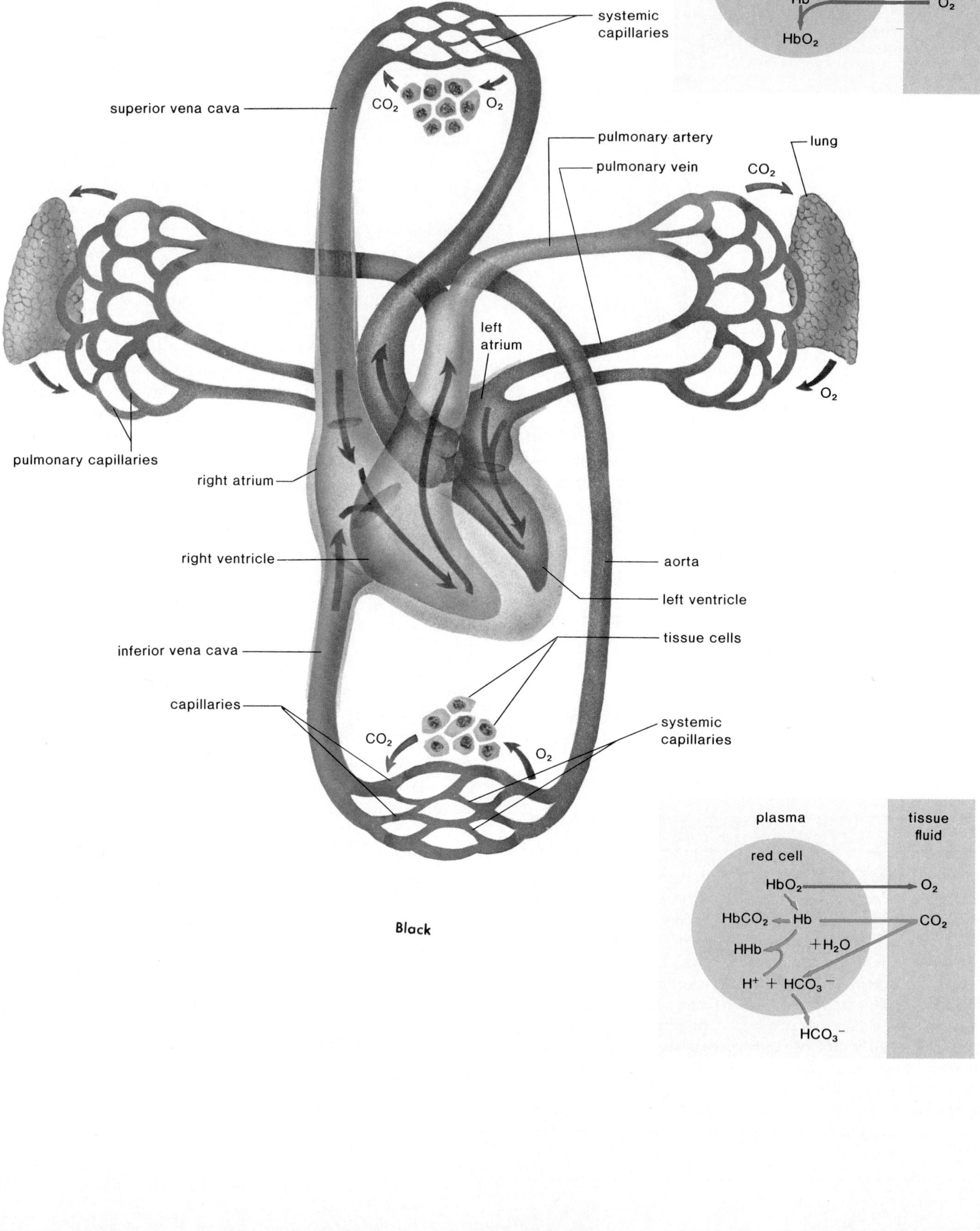

FIGURE 8.14 *Percent saturation of hemoglobin with oxygen varies with the partial pressure of oxygen. Hemoglobin will be more saturated in the lungs where the temperature and acidity are lower than in the tissues where the temperature and acidity are higher for the following reasons.* a. *This data shows that it takes a lower partial pressure of oxygen to saturate hemoglobin when the temperature is 10° C than when the temperature is 43° C.* b. *This data shows that it takes a lower partial pressure of oxygen to saturate hemoglobin when the acidity is low than when the acidity is high.*

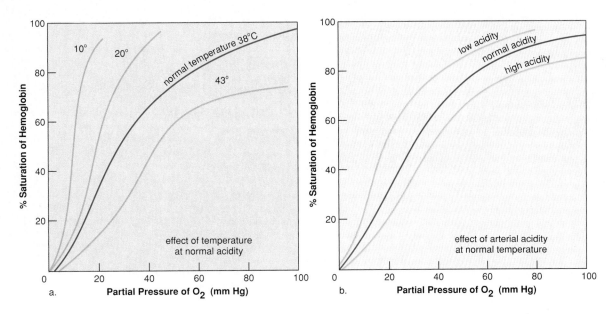

takes up oxygen more readily in the cool temperature (fig. 8.14a) and neutral pH (fig. 8.14b) of the lungs. On the other hand, it will give up oxygen more readily at the warmer temperature and more acidic pH of the tissues.[3]

External respiration, the exchange of oxygen for carbon dioxide between air within the alveoli and the blood in pulmonary capillaries, is dependent on the process of diffusion.

Internal Respiration

As blood enters the systemic capillaries (fig. 8.13), oxygen leaves hemoglobin and diffuses out into tissue fluid:

$$HbO_2 \rightarrow Hb + O_2$$

Diffusion of oxygen out of the blood and into the tissues occurs because the oxygen concentration in tissue fluid is low due to the fact that the cells are continuously using it up in cellular respiration. On the other hand, carbon dioxide concentration is high because carbon dioxide is continuously being produced by the cells. Therefore, *carbon dioxide will diffuse into the blood.*

Carbon dioxide enters the red blood cells, where a small amount is carried by the formation of carbaminohemoglobin (p. 126). But most carbon dioxide combines with water to

form carbonic acid, which dissociates to $H^+ + HCO_3^-$. The enzyme carbonic anhydrase, present in red blood cells, speeds up the reaction:

$$CO_2 + H_2O \rightarrow H_2CO_3 \rightarrow H^+ + HCO_3^-$$

The globin portion of hemoglobin combines with the excess hydrogen ions produced by the reaction, and Hb becomes HHb. In this way, the pH of the blood remains fairly constant. The bicarbonate ion, HCO_3^-, diffuses out of the red blood cells to be carried in the plasma.

Internal respiration, the exchange of oxygen for carbon dioxide between blood in the tissue capillaries and tissue fluid, is dependent on the process of diffusion.

RESPIRATION AND HEALTH

We have seen that the full length of the respiratory tract is lined with a warm, wet mucosal lining that is constantly exposed to environmental air. The quality of this air, determined by the pollutants and germs it contains, can affect the health of the individual.

Germs frequently spread from one individual to another by way of the respiratory tract. Droplets from one single sneeze may be loaded with billions of bacteria or viruses. The mucous membranes are protected by the production of mucus

[3]

	pH	Temperature
Lungs	7.40	37° C (98.6° F)
Body	7.38	38° C (100.4° F)

FIGURE 8.15 a. Millions of viruses can be carried in the droplets of moisture in our breath visible on a cold day.
b. Computer graphics allows us to see the structure of a cold virus. These germs are spread through the air when a person with a cold sneezes or coughs.

a.

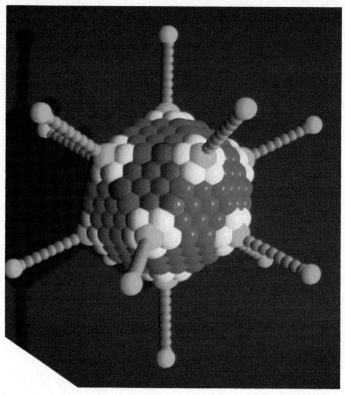

b.

and by the constant beating of the cilia; but if the number of infective agents is large and/or the resistance of the individual is reduced, an upper respiratory infection may result.

Upper Respiratory Infections

A URI (upper respiratory infection) is one that affects only the nasal cavities, throat, trachea, bronchi, and associated organs.[4] Vaccines for these infections are not in wide use, and if they are viral in nature, antibiotics are not helpful. Since viruses take over the machinery of the cell when they reproduce, it is difficult to develop drugs that will affect the virus without affecting the cell itself.

Common Cold

A cold is a viral infection that usually begins with a scratchy sore throat, followed by a watery mucous discharge from the nasal cavities. There is rarely a fever, and symptoms are usually mild, requiring little or no medication. Although colds have a short duration, the immunity they provide is also brief. Since there are estimated to be over 150 cold-causing viruses (fig. 8.15) it is very difficult to gain immunity to all of them. A nasal spray containing interferon has shown some promise in helping prevent colds, but the cost remains prohibitive for most people.

[4]Allergies, including asthma, are discussed on page 151.

Influenza

"Flu" is also a viral infection, but although it begins as an upper respiratory infection, it spreads to other parts of the body, causing aches and pains in the joints. There is usually a fever and the illness lasts for a longer length of time than a cold. Immunity is possible, but usually only the vaccine developed for the particular virus prevalent that season is successful in protecting the individual during a current flu epidemic. Since flu viruses constantly mutate, there can be no buildup in immunity, and a new viral illness rapidly spreads from person to person and from place to place. Pandemics, in which a newly mutated flu virus spreads about the world, have occurred on occasion.

Bronchitis

Viral infections can spread from the nasal cavities to the sinuses (sinusitis), to the middle ears (otitis media), to the larynx (laryngitis), and to the bronchi (bronchitis). Acute bronchitis is usually caused by a secondary bacterial infection of the bronchi, resulting in a heavy mucoid discharge with much coughing. Acute bronchitis usually responds to antibiotic therapy.

Chronic bronchitis is not necessarily due to infection. It is often caused by a constant irritation of the lining of the bronchi, which as a result undergoes degenerative changes with the loss of cilia, thus preventing the normal cleansing action. There is frequent coughing, and the individual is more susceptible to upper respiratory infections. Chronic bronchitis most often affects cigarette smokers.

Strep Throat

Strep throat is a very severe throat infection caused by the bacterium *Streptococcus*. Swallowing is very difficult, and there is a fever. Unlike a viral infection, strep throat should be treated with antibiotics. If not treated, it may lead to complications such as rheumatic fever, from which the heart valves may be permanently affected.

Upper respiratory infections due to a viral infection are not treatable by antibiotics, but bacterial infections can be treated by antibiotics.

Lung Disorders

Pneumonia and tuberculosis, two infections of the lungs, formerly caused a large percentage of deaths in the United States. Now they are controlled by antibiotics. Two other illnesses discussed in the following paragraphs, emphysema and lung cancer, are not due to infections; in most instances, they are due to cigarette smoking.

Pneumonia

Most forms of pneumonia are caused by bacteria or viruses that infect the lungs. The demise of AIDS patients is usually due to a particularly rare form of pneumonia caused by a protozoan, *Pneumocystis carinii*. Sometimes pneumonia is localized in specific lobes of the lungs, and these become inoperative as they fill with mucus and pus. The more lobes involved, the more serious the infection.

Tuberculosis

Tuberculosis is caused by the tubercle bacillus. It is possible to tell if a person has ever been exposed to tuberculosis using a skin test in which a highly diluted extract of the bacilli is injected into the skin of the patient. A person who has never been in contact with the bacillus will show no reaction, but one who has developed immunity to the organism will show an area of inflammation that peaks in about forty-eight hours. If these bacilli do invade the lung tissue, the cells build a protective capsule about the foreigners to isolate them from the rest of the body. This tiny capsule is called a *tubercle*. If the resistance of the body is high, the imprisoned organisms may die, but if the resistance is low, the organisms may eventually be liberated. If a chest X ray detects the presence of tubercles, the individual is put on appropriate drug therapy to ensure the localization of the disease and the eventual destruction of any live bacterial organisms.

Emphysema

Emphysema refers to the destruction of lung tissue, with accompanying ballooning or inflation of the lungs due to trapped air. The trouble stems from the destruction and collapse of the bronchioles. When this occurs, the alveoli are cut off from renewed oxygen supply and the air within them is trapped. The trapped air very often causes rupturing of the alveolar

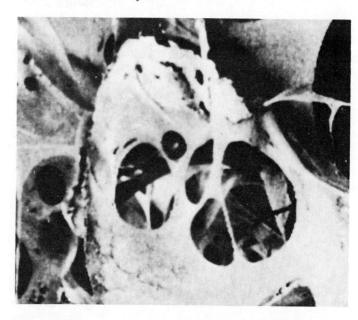

FIGURE 8.16 *Scanning electron micrograph of the lungs of a person with emphysema. There are large cavities in the lungs due to the breakdown of alveoli.*

walls (fig. 8.16), together with a fibrous thickening of the walls of the small blood vessels in the vicinity (see the section on emphysema in the reading on p. 174).

Chronic bronchitis and emphysema, two conditions most often caused by smoking, together are called chronic obstructive pulmonary disease (COPD).

Pulmonary Fibrosis

Inhaling particles such as silica (sand), coal dust, and asbestos (fig. 8.17) can lead to pulmonary fibrosis, in which fibrous connective tissue builds up in the lungs. Breathing capacity can be seriously impaired, and the development of cancer is not rare. Since asbestos has been so widely used as a fireproofing and insulating agent, unwarranted exposure has occurred.

Lung Cancer

Lung cancer used to be much more prevalent in men than women, but recently lung cancer has surpassed breast cancer as a cause of death in women. This can be linked to an increase in the number of women who smoke. Autopsies on smokers have revealed the progressive steps by which the most common form of lung cancer develops (see the section on lung cancer in the reading on p. 174).

A cancerous tumor may grow until the bronchus is blocked, cutting off the supply of air to that lung. The lung then collapses, and the secretions trapped in the lung spaces become infected, with the result pneumonia or the formation of a lung abscess. The only treatment that offers a possibility of cure before secondary growths have had time to form is complete removal of the lung. This operation is called *pneumonectomy*.

RISKS OF SMOKING VERSUS BENEFITS OF QUITTING

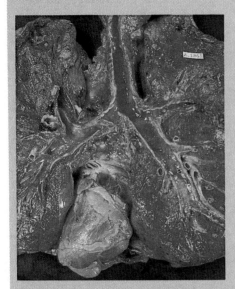

a.

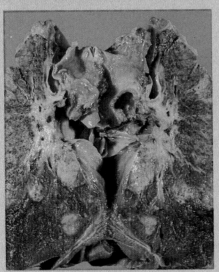

b.

a. *Normal lung with heart in place. Notice the healthy red color.* b. *Lungs of a heavy smoker. Notice how black the lungs are except where cancerous tumors have formed.*

Based on available statistics, the American Cancer Society informs us that smoking carries a high risk. Among the risks of smoking are the following:

Shortened life expectancy A twenty-five-year-old who smokes two packs of cigarettes a day has a life expectancy 8.3 years shorter than a nonsmoker. The greater the number of packs smoked, the shorter the life expectancy.

Lung cancer . . . The first event appears to be a thickening of the cells that line the bronchi. Then there is a loss of cilia so that it is impossible to prevent dust and dirt from settling in the lungs. Following this, cells with atypical nuclei appear in the thickened lining. A disordered collection of cells with atypical nuclei may be considered to be cancer in situ (at one location). The final step occurs when some cells break loose and penetrate the other tissues, a process called metastasis. This is true cancer (*b*).

Cancer of the larynx, mouth, esophagus, bladder, and pancreas The chances of developing these cancers are from 2 to 17 times higher in cigarette smokers than in nonsmokers.

Emphysema Cigarette smokers have 4 to 25 times greater risk of developing emphysema. Damage is seen in the lungs of even young smokers. Smoking causes the lining of the bronchioles to thicken. . . . If a large part of the lungs is involved, the lungs are permanently inflated and the chest balloons out due to this trapped air. The victim is breathless and has a cough. Since the surface area for gas exchange is reduced, not enough oxygen reaches the heart and brain. The heart works furiously to force more blood through the lungs, which may lead to a heart condition. Lack of oxygen for the brain may make the person feel depressed, sluggish, and irritable.

Coronary heart disease Cigarette smoking is the major factor in 120,000 additional U.S. deaths from coronary heart disease each year.

Reproductive effects Smoking mothers have more stillbirths and low-birthweight babies who are more vulnerable to disease and death. Children of smoking mothers are smaller and underdeveloped physically and socially even seven years after birth.

In the same manner, the American Cancer Society informs smokers of the benefits of quitting. These benefits include the following:

Risk of premature death is reduced Do not smoke for 10 to 15 years, and the risk of death due to any one of the cancers mentioned approaches that of the nonsmoker.

Health of respiratory system improves The cough and excess sputum disappear during the first few weeks after quitting. As long as cancer has not yet developed, all the ill effects mentioned can reverse themselves and the lungs can become healthy again. In patients with emphysema, the rate of alveoli destruction is reduced and lung function may improve.

Coronary heart disease risk sharply decreases After only one year the risk factor is greatly reduced, and after 10 years an exsmoker's risk is the same as that of those who never smoked.

The increased risk of having stillborn children and underdeveloped children disappears Even for women who do not stop smoking until the fourth month of pregnancy, such risks to infants is decreased.

People who smoke must ask themselves if the benefits of quitting outweigh the risks of smoking.

American Cancer Society

FIGURE 8.17 *Asbestos fibers.* a. *Polarized light photograph shows the long, thin, flexible, and strong fibers that make up asbestos.* b. *A scanning electron micrograph of macrophages reveals an "asbestos body" in the lung tissue of a person exposed to asbestos for some time. Asbestos bodies are fibers that have been coated with iron, plasma proteins, and other materials.*

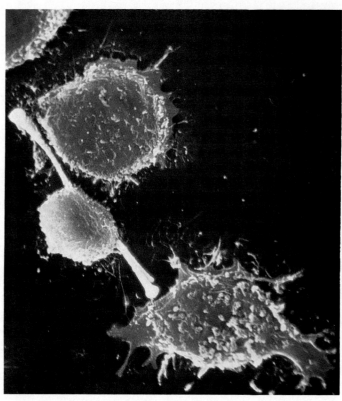
b.

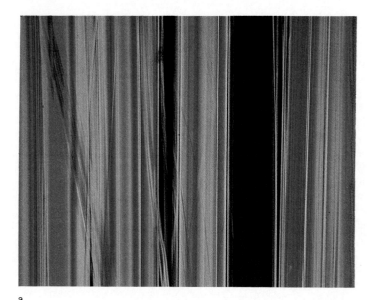
a.

Statistical studies have shown that smoking cigarettes may be associated with other illnesses, including other types of cancer, as discussed in the reading on page 174. If a person does stop smoking and if the body tissues are not already cancerous, they return to normal.

The incidence of lung cancer is much higher in individuals who smoke compared to those who do not smoke.

HUMAN ISSUE

On the basis that passive smoking is detrimental to their health, nonsmokers have been slowly able to restrict those areas where smokers are permitted to smoke. A new law in New York prohibits smoking altogether in enclosed public places such as taxis, stores, and rest rooms. Smoking in federal buildings across the country has recently been restricted to specific, confined areas designated for smokers. Then, too, some employers are now refusing to hire smokers because they raise the insurance premiums too high, and statistics show that they take too many sick days. Smokers feel that their rights are being denied. Should a majority be able to prevent others from indulging in whatever habits they prefer? Do you think that smokers are being unfairly discriminated against?

SUMMARY

Air enters and exits the lungs by way of the respiratory tract (table 8.1). Inspiration begins when the breathing center in the medulla oblongata sends excitatory nerve impulses to the diaphragm and rib cage. As they contract, the diaphragm lowers and the rib cage moves up and out; the lungs expand, creating a partial vacuum that causes air to rush in. Nerves within the expanded lungs then send inhibitory impulses to the breathing center. As the diaphragm relaxes, it resumes its dome shape, and as the rib cage retracts, air is pushed out of the lungs during expiration.

External respiration occurs when carbon dioxide leaves the blood and oxygen enters the blood at the alveoli. Oxygen is transported to the tissues in combination with hemoglobin. Internal respiration occurs when oxygen leaves the blood and carbon dioxide enters the blood at the tissues. Carbon dioxide is carried to the lungs in the form of the bicarbonate ion.

There are a number of illnesses associated with the respiratory tract. In addition to colds and flu, pneumonia and tuberculosis are serious lung infections. Two illnesses that have been attributed to breathing polluted air are emphysema and lung cancer.

OBJECTIVE QUESTIONS

1. In tracing the path of air, the _____ immediately follows the pharynx.
2. The lungs contain air sacs called _____ .
3. The breathing rate is primarily regulated by the amount of _____ and _____ in the blood.
4. Air enters the lungs after they have _____ .

5. Carbon dioxide is carried in the blood as the _____ ion.
6. The H^+ ions given off when carbonic acid dissociates are carried by _____ .
7. Gas exchange is dependent on the physical process of _____ .
8. Reduced hemoglobin becomes oxyhemoglobin in the _____ .

9. The most likely cause of emphysema and chronic bronchitis is _____ .
10. Most cases of lung cancer actually begin in the _____ .

Answers to Objective Questions

1. larynx 2. alveoli 3. CO_2, H^+ 4. expanded 5. HCO_3^- 6. hemoglobin 7. diffusion 8. lungs 9. smoking cigarettes 10. bronchi

STUDY QUESTIONS

1. What are the four parts of respiration? In which of these is oxygen actually used up and carbon dioxide produced? (p. 160)
2. List the parts of the respiratory tract. (p. 161) What are the special functions of the nasal cavities, larynx, and alveoli? (pp. 163–164)
3. What are the steps in inspiration and expiration? How is breathing controlled? (pp. 165–167)
4. Why can't we breathe through a very long tube? (p. 167)
5. What two equations are needed to explain external respiration? (p. 167)
6. How is hemoglobin remarkably suited to its job? (p. 167)
7. What physical process is believed to explain gas exchange? (p. 171)
8. What two equations are needed to explain internal respiration? (p. 171)
9. Name some infections of the respiratory tract. (p. 172)
10. What are emphysema and pulmonary fibrosis, and how do they affect one's health? (p. 173)
11. By what steps is cancer believed to develop in the person who smokes? (p. 174)

THOUGHT QUESTIONS

1. If the human respiratory system were a flow-through system, it would have an opening both for intake and outtake. How would this affect various features of the respiratory system?
2. Fishes live in the water and expose their gills to the external environment. We live on land and have our lungs deep within the body. Explain why this is appropriate.
3. Inspiration is dependent on a difference between air pressure in the lungs and external air pressures. Make up a table to show the effect of higher and lower air pressures, for both the lungs and external air, on inspiration. Which of these applies to breathing at higher altitudes?

KEY TERMS

alveoli (al-ve'o-lī) saclike structures that are the air sacs of a lung. *164*

bronchi (brong'ki) the two major divisions of the trachea leading to the lungs. *164*

bronchiole (brong'ke-ōl) the smaller air passages in the lungs of mammals. *164*

diaphragm (di'ah-fram) a sheet of muscle that separates the chest cavity from the abdominal cavity. *165*

expiration (eks''pi-ra'shun) process of expelling air from the lungs; exhalation. *161*

external respiration (eks-ter'nal res''pi-ra'shun) exchange between blood and alveoli of carbon dioxide and oxygen. *160*

hemoglobin (he''mo-glo'bin) a red iron-containing pigment in blood that combines with and transports oxygen. *169*

inspiration (in''spi-ra'shun) the act of breathing in. *161*

internal respiration (in-ter'nal res''pi-ra'shun) exchange between blood and tissue fluid of oxygen and carbon dioxide. *160*

ribs (ribz) bones hinged to the vertebral column and sternum that, with muscle, define the top and sides of the chest cavity. *165*

trachea (tra'ke-ah) a tube that is supported by C-shaped cartilagenous rings that lies between the larynx and the bronchi; the windpipe. *163*

ventilation (ven''ti-la'shun) breathing; the process of moving air into and out of the lungs. *165*

vocal cords (vo'kal kordz) folds of tissue within the larynx that create vocal sounds when they vibrate. *163*

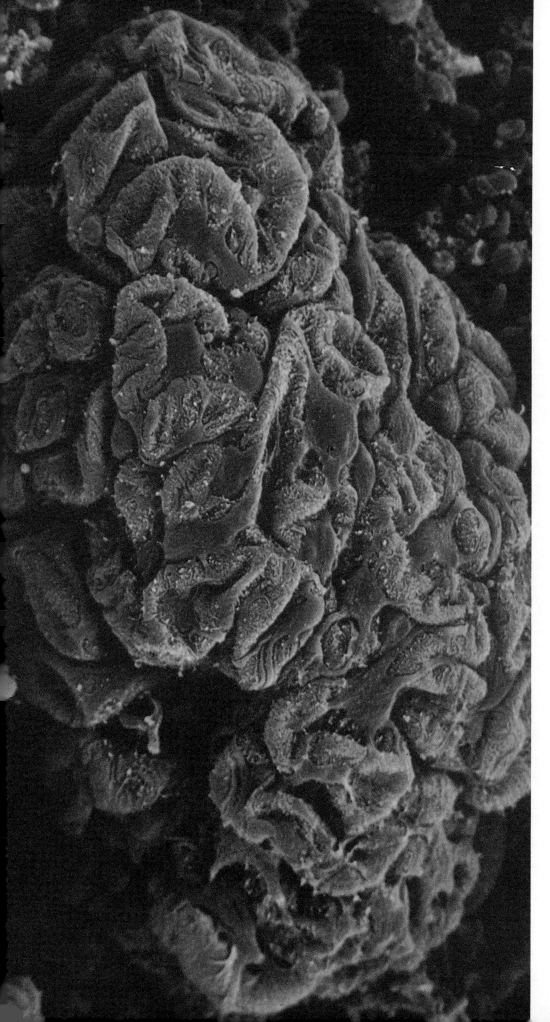

CHAPTER NINE

EXCRETION

CHAPTER CONCEPTS

1 Excretion rids the body of unwanted substances, particularly end products of metabolism.

2 Several organs assist in the excretion process, but the kidneys, which are a part of the urinary system, are the primary organs of excretion.

3 The formation of urine by the more than one million nephrons present in each kidney serves not only to rid the body of nitrogenous wastes but also to regulate the water content, the salt levels, and the pH of the blood.

4 The kidneys, the malfunction of which brings illness and may cause death, are important organs of homeostasis.

A scanning electron micrograph glomerulus of a human kidney.

INTRODUCTION

Everyone knows that if you cross a desert without a supply of fresh water, death is likely. Dehydration occurs because the body loses water and salts through sweating and urination. But what if you were lost at sea (fig. 9.1)? Can you drink salt water? I'm afraid not. Humans have no special way to rid the body of excess salt, and the kidneys would have to excrete more liquid than was consumed in order to wash out the excess salt. If there is no fresh water available, don't drink at all and don't eat salty fish either!

The organs of **excretion** (fig. 9.2) aid in maintaining the constancy of the internal environment by removing unwanted substances from the blood. The composition of blood serving the body cells remains constant due to both the continual addition of substances needed by cells and the continual removal of substances not needed by cells. In previous chapters, we have discussed how the digestive tract and lungs add nutrients and oxygen to the blood. In this chapter we will show how the organs of excretion remove substances from the blood that fall predominantly into two categories: (1) end products of metabolism, and (2) water and other substances, including ions, that may be present in excess.

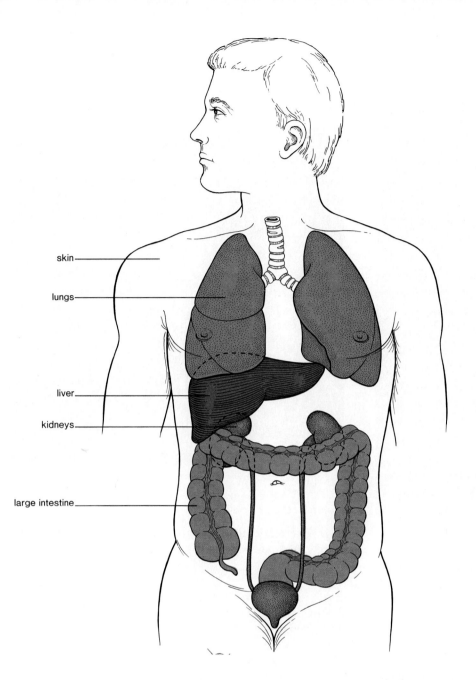

skin

lungs

liver

kidneys

large intestine

FIGURE 9.2 *The organs of excretion include not only the kidneys but also the skin, lungs, liver, and large intestine. The lungs excrete carbon dioxide; the liver excretes hemoglobin breakdown products; and the intestine excretes certain heavy metals. Excretion, ridding the body of metabolic wastes, should not be confused with defecation, ridding the body of nondigestible remains.*

EXCRETORY SUBSTANCES

Table 9.1 shows that the majority of end products of metabolism excreted by humans are related to nitrogen metabolism since amino acids, nucleotides, and creatine all contain nitrogen. There are also other metabolic end products that need to be excreted.

Nitrogenous End Products

Ammonia (NH_3) arises from the **deamination,** or removal, of amino groups from amino acids. Ammonia is extremely toxic to the body, and only animals living in water, who continually flush out their bodies with water, excrete ammonia (fig. 9.3). In our bodies, ammonia is converted to urea by the liver.

TABLE 9.1 *SOME METABOLIC END PRODUCTS*

NAME	END PRODUCT OF	PRIMARILY EXCRETED BY
Nitrogenous wastes		
Ammonia	Amino acid metabolism	Kidneys
Urea	Ammonia metabolism	Kidneys
Uric acid	Nucleotide metabolism	Kidneys
Creatinine	Creatine phosphate metabolism	Kidneys
Other		
Bile pigments	Hemoglobin metabolism	Liver
Carbon dioxide	Cellular respiration	Lungs

FIGURE 9.3 *Fishes can excrete ammonia, a toxic substance, as their nitrogenous waste because they live in an aquatic environment.*

FIGURE 9.4 *Birds excrete uric acid, a solid material, as their nitrogenous waste. It is mixed with fecal material in a common repository for the urinary, digestive, and reproductive systems. Seabirds congregate in such numbers that their droppings build up to give a nitrogen-rich substance called guano. At one time guano was harvested as a natural fertilizer.*

Urea is produced in the liver by a complicated series of reactions called the urea cycle. In this cycle, carrier molecules take up carbon dioxide and two molecules of ammonia to finally release urea:

$$H_2N - \overset{\overset{\textstyle O}{\|}}{C} - NH_2$$

Many terrestrial animals that need to conserve water excrete **uric acid** (fig. 9.4) as their general nitrogenous end product. In humans, uric acid only occurs when nucleotides are metabolically broken down; if uric acid is present in excess, it will precipitate out of the plasma. Crystals of uric acid sometimes collect in the joints, producing a painful ailment called gout.

Creatinine is an end product of muscle metabolism. It results when creatine phosphate, a molecule that serves as a reservoir of high-energy phosphate, breaks down.

Other Excretory Substances

Other excretory substances are bile pigments, carbon dioxide, ions (salts), and water.

Bile Pigments
Bile pigments are derived from the heme portion of hemoglobin and are incorporated into bile within the liver (fig. 9.5). The liver produces bile, which is stored in the gallbladder before passing into the small intestine. If for any reason the bile duct is blocked, bile spills out into the blood, producing a discoloration of the skin called jaundice (p. 81).

Carbon Dioxide
The lungs are the major organs of *carbon dioxide* excretion, although the kidneys are also important. The kidneys excrete bicarbonate ions, the form in which carbon dioxide is carried in the blood.

Ions
Ions (salts) are excreted, not because they are end products of metabolism, but because their proper concentration in the blood is so important to the pH, osmotic pressure, and ion balance of the blood. The balance of potassium (K^+) and sodium (Na^+) is important to nerve conduction. The level of

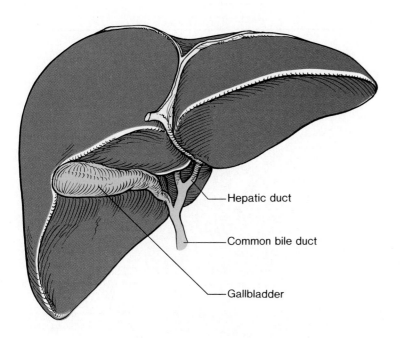

Hepatic duct

Common bile duct

Gallbladder

calcium (Ca^{++}) in the blood affects muscle contraction; iron (Fe^{++}) takes part in hemoglobin metabolism; and magnesium (Mg^{++}) helps many enzymes function properly.

Water

Water is an end product of metabolism; it is also taken into the body when food and liquids are consumed. The amount of fluid in the blood helps determine blood pressure. Treatment of hypertension sometimes includes the administration of a diuretic drug that increases the excretion of sodium and water by the kidneys.

Urea, salts, and water are the primary constituents of human urine. Carbon dioxide is excreted as a gas in the lungs and as the bicarbonate ion in the kidneys.

ORGANS OF EXCRETION

The kidneys are the primary excretory organs, but there are other organs that also function in excretion, such as those described in the discussion that follows.

Skin

The sweat glands in the skin (fig. 3.14) excrete perspiration, a solution of water, salt, and some urea. The sweat glands have a coiled tubule portion in the dermis and a narrow,

straight duct that exits from the epidermis. Although perspiration is an excretion, we perspire not so much to rid the body of waste as to cool the body. The body cools because heat is lost as perspiration evaporates. Sweating keeps the body temperature within normal range during muscular exercise or when the outside temperature rises. In times of renal failure, more urea than usual may be excreted by the sweat glands, to the extent that a so-called urea frost is observed on the skin.

Liver

The liver excretes bile pigments, which are incorporated into bile, before it enters the common bile duct, which passes into the small intestine (fig. 9.5). The yellow pigment found in urine, called urochrome, is also derived from the breakdown of heme, but this pigment is deposited in the blood and therefore is excreted by the kidneys.

Lungs

The process of expiration (breathing out) not only removes carbon dioxide from the body, it also results in the loss of water. The air we exhale contains moisture, as demonstrated by blowing onto a cool mirror.

Intestine

Certain salts, such as those of iron and calcium, are excreted directly into the cavity of the intestine by the epithelial cells lining it. These salts leave the body in the feces.

At this point, it might be helpful to remember that the term *defecation,* and not *excretion,* is used to refer to the elimination of feces from the body. Substances that are excreted are those that are waste products of metabolism. Undigested food and bacteria, which make up feces, have never been a part of the functioning of the body, but salts that are passed into the gut are excretory substances because they were once metabolites in the body.

Kidneys

The kidneys excrete urine, which contains a combination of the end products of metabolism (table 9.2). The kidneys are a part of the urinary system.

There are various organs that excrete metabolic wastes but only the kidneys consistently rid the body of urea.

TABLE 9.2 COMPOSITION OF URINE	
WATER	95%
SOLIDS	5%
Organic wastes (per 1,500 ml of urine)	
Urea	30 g
Creatinine	1–2 g
Ammonia	1–2 g
Uric acid	1 g
Ions (salts)	25 g

Positive	*Negative*
Sodium	Chlorides
Potassium	Sulfates
Magnesium	Phosphates
Calcium	

TABLE 9.3 URINARY SYSTEM	
ORGAN	*FUNCTION*
Kidneys	Produce urine
Ureters	Transport urine
Bladder	Store urine
Urethra	Eliminate urine

FIGURE 9.6 *The urinary system. Urine is only found within the kidneys, ureters, bladder, and urethra.*

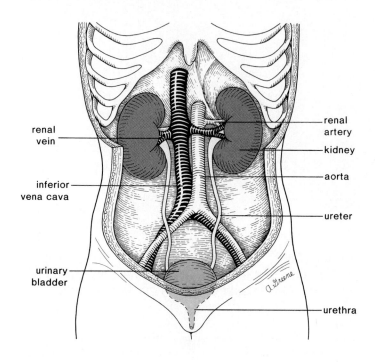

URINARY SYSTEM

The urinary system includes the structures illustrated in figure 9.6 and listed in table 9.3. The organs are listed in order according to the path of urine. The **kidneys** are bean-shaped, reddish-brown organs, each about the size of a fist. They are located one on either side of the vertebral column just below the diaphragm, where they receive some protection from the lower rib cage.

The **ureters** are muscular tubes that convey the urine toward the bladder by peristaltic contractions. Urine enters the bladder at the rate of one to five jets per minute.

The **urinary bladder,** which can hold up to 600 ml of urine, is a hollow muscular organ that gradually expands as urine enters. In the male, the bladder lies ventral to the rectum, seminal vesicles, and vas deferens. In the female, it is ventral to the uterus and upper vagina.

The **urethra,** which extends from the urinary bladder to an external opening, differs in length in females and males. In females, the urethra lies ventral to the vagina and is only about 2.5 cm long. The short length of the female urethra invites bacterial invasion and explains why females are more prone to urethral infections. In males, the urethra averages 15 cm when the penis is relaxed. In males, as the urethra leaves the bladder, it is encircled by the prostate gland (fig. 9.7). In older men, enlargement of the prostate gland may prevent urination, a condition that can usually be corrected.

Notice that there is no connection between the genital (reproductive) and urinary systems in females, but there is a connection in males. When urinating, the urethra in the male carries urine, and during sexual orgasm the urethra transports semen. This double function does not alter the path of urine, and it is important to realize that urine is found only in those structures listed in table 9.3.

Urination

When the bladder (fig. 9.7) fills with urine, stretch receptors send nerve impulses to the spinal cord; nerve impulses leaving the cord then cause the bladder to contract and the sphincters to relax so that urination may take place. In older children and adults, it is possible for the brain to control this reflex, delaying urination until a suitable time.

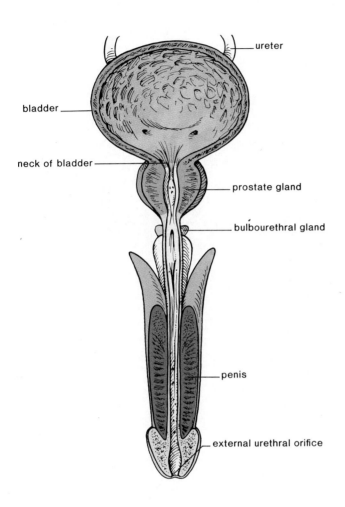

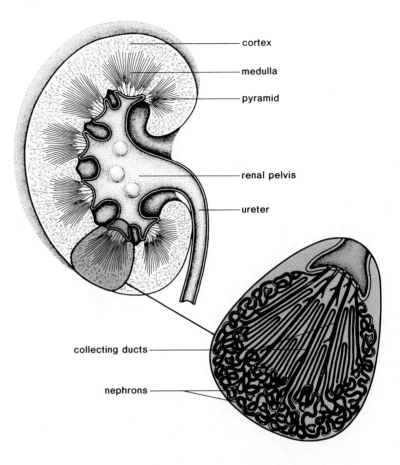

Kidneys

When a kidney is sliced longitudinally, it is shown to be composed of three macroscopic parts (fig. 9.8): (1) an outer granulated layer called the **cortex,** which dips down in between (2) a radically striated, or lined, layer called the **medulla,** and (3) an inner space, or cavity, called the **renal pelvis,** where the urine collects before entering the ureters. Occasionally **kidney stones** form in the pelvis. They are formed from fairly insoluble substances, such as uric acid and calcium salts, that precipitate out of the urine instead of remaining in solution. Previously, either the stones passed naturally or they were surgically removed. Now new methods of treatment are being developed. Sometimes the stones can be crushed by sound waves generated within a special device called a lithotripter. Laser light has also been found to be successful in shattering the stones.

Microscopically, the kidney is composed of over one million **nephrons,** sometimes called renal tubules. Each nephron is made up of several parts (fig. 9.9a). The blind end

of the tubule is pushed in on itself to form a cuplike structure called **Bowman's capsule,** within which there is a capillary tuft called the *glomerulus.* Next, there is the **proximal** (meaning near the Bowman's capsule) **convoluted tubule,** which makes a U-turn to form the portion of the tubule called the **loop of Henle.** This leads to the **distal** (far from Bowman's capsule) **convoluted tubule,** which enters a **collecting duct.** Figure 9.9b indicates the position of a single nephron within the kidney. Bowman's capsules and convoluted tubules lie within the cortex and account for the granular appearance of the cortex. Loops of Henle and collecting ducts lie within the triangular-shaped *pyramids* of the medulla. Since these are longitudinal structures, they account for the striped appearance of the pyramids and medulla (fig. 9.8).

Only the urinary system, consisting of the kidneys, bladder, ureters, and urethra, ever hold urine.

FIGURE 9.9 a. *Diagram of nephron (kidney tubule) gross anatomy. You can trace the path of blood about the nephron by following the arrows. Note that the dotted line indicates which portions of the nephron are in the cortex and which portions are in the medulla of the kidney.* b. *Each kidney receives a renal artery that divides into arterioles within the kidney. Venules leaving the kidney join to form the renal vein. This drawing shows how one nephron is placed in the kidney so that some parts are in the cortex and other parts are in the medulla.*

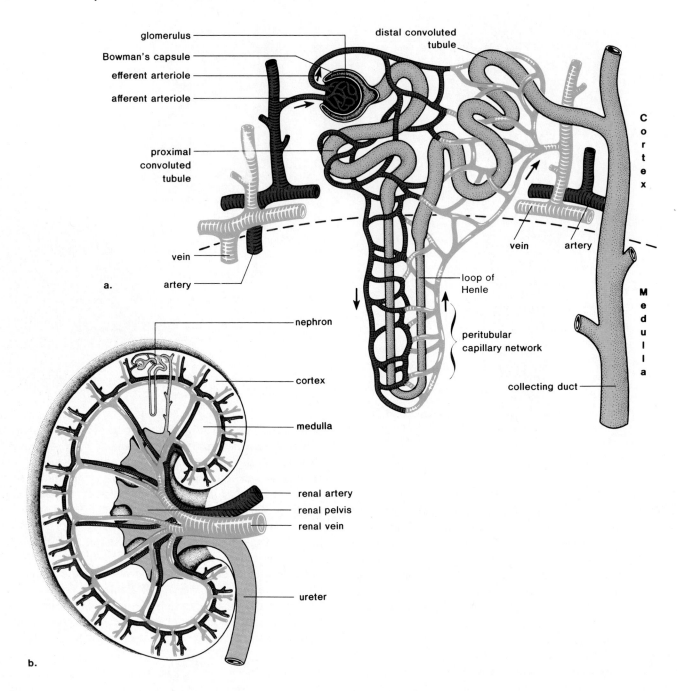

URINE FORMATION

Each nephron has its own blood supply, including two capillary regions: the **glomerulus** is a capillary tuft inside Bowman's capsule, and the **peritubular capillary network** surrounds the rest of the nephron (table 9.4). Urine formation requires the movement of molecules between these capillaries and the nephron (fig. 9.14).

The pattern of blood flow about the nephron is critical to urine formation.

Pressure Filtration

Whole blood, of course, enters the afferent arteriole and the glomerulus (fig. 9.10). Under the influence of glomerular blood pressure, which is usually about 60 mm Hg, small molecules move from the glomerulus to the inside of Bowman's capsule across the thin walls of each. The process is a **pressure filtration** because large molecules and formed elements are unable to pass through. In effect, then, blood that enters the glomerulus is divided into two portions: the filterable components and the nonfilterable components.

Filterable Blood Components	*Nonfilterable Blood Components*
Water	Formed elements (blood cells and platelets)
Nitrogenous wastes	
Nutrients	Proteins
Ions (salts)	

The filterable components form a glomerular **filtrate** that contains small dissolved molecules in approximately the same concentration as plasma. The filtrate stays inside of Bowman's capsule, and the nonfilterable components leave the glomerulus by way of the efferent arteriole.

A consideration of the preceding filterable substances leads us to conclude that if the composition of urine were the same as that of glomerular filtrate, the body would continually lose nutrients, water, and salts. Obviously, death from dehydration, starvation, and low blood pressure would quickly follow. Thus, we can assume that the composition of the filtrate must be altered as this fluid passes through the remainder of the tubule.

During pressure filtration, water, nutrient molecules, and waste molecules move from the glomerulus to the inside of Bowman's capsule. The substances that are filtered are called the glomerular filtrate.

Selective Reabsorption

Both passive and active reabsorption of molecules from the tubule to the blood occur as the filtrate moves along the proximal convoluted tubule.

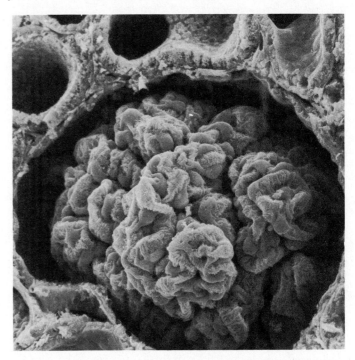

FIGURE 9.10 Scanning electron micrograph of a glomerulus (the outer layer of Bowman's capsule has been removed). Renal failure is due to glomeruli that no longer function properly.

Passive reabsorption involves particularly the movement of water molecules from the area of greater concentration in the filtrate to the area of lesser concentration in the blood. Two factors aid the process. The nonfilterable proteins remain in the blood, where they exert an osmotic pressure that pulls water back into the bloodstream. Following the active reabsorption of sodium (Na^+), which is discussed later, chlorine (Cl^-) follows passively because, being a negative ion, it is attracted to the positive charge of sodium.

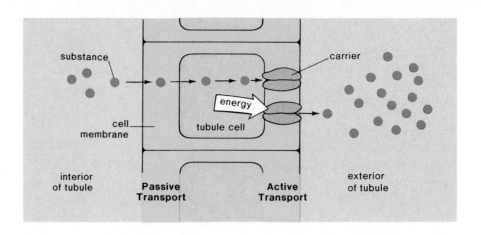

FIGURE 9.11 *Nutrient molecules and sodium are actively reabsorbed from a kidney tubule in the manner illustrated. These molecules move passively into the tubule cell but then are actively transported out of the tubule cell into the blood. Active transport requires the participation of carrier molecules.*

Active reabsorption involves active transport, and this accounts for the large energy needs of the kidney. Figure 9.11 shows a tubule cell that is engaged in active transport. A molecule, such as glucose, diffuses passively into the tubule cell but is then actively transported from the cell into the blood, requiring the use of a carrier molecule. Reabsorption by active transport is selective since only molecules recognized by carrier molecules move across the membrane.

The cells that line the proximal convoluted tubule are anatomically adapted for absorption (fig. 9.12). These cells have numerous microvilli, about 1 μm in length, that increase the surface area for reabsorption. In addition, the cells contain numerous mitochondria, which produce the energy necessary for active transport. **Selective reabsorption** occurs until the threshold level of a particular substance is obtained. Thereafter the substance will appear in the urine. For example, the *threshold level* for glucose is about 180 mg glucose per 100 ml of blood. After this amount is reabsorbed, any excess present in the filtrate will appear in the urine. In diabetes mellitus (sugar diabetes, p. 285), the filtrate contains excess glucose because the liver fails to store glucose as glycogen.

In contrast to the high threshold level of glucose, urea has a very low threshold level that is quickly reached so that nearly all urea remains in the urine.

We have seen that the filtrate that enters the proximal convoluted tubule is divided into two portions: the components that are reabsorbed and the components that are not reabsorbed.

Reabsorbed Filtrate Components	Nonreabsorbed Filtrate Components
Most water	Some water
Nutrients	Nitrogenous wastes
Required ions (salts; e.g., Na$^+$, Cl$^-$)	Excess ions (salts)

FIGURE 9.12 *Electron micrograph of cells that line the lumen (inside) of a proximal convoluted tubule, where selective reabsorption takes place. The cells have a brush border composed of microvilli that greatly increase the surface area exposed to the lumen. Each cell has many mitochondria that supply the energy needed for active transport. A red blood cell (lower right) is seen in the peritubular capillary. (Mv = microvilli; M = mitochondria)*

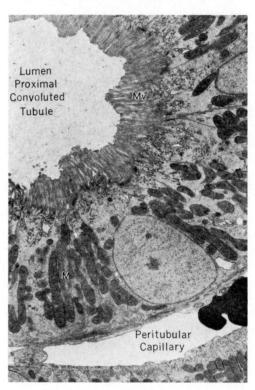

The substances that are not reabsorbed become the tubular fluid that enters the loop of Henle.

During selective reabsorption, nutrient and salt molecules are actively reabsorbed from the proximal convoluted tubule into the peritubular capillary, and water follows passively.

FIGURE 9.13 *The presence of a loop of Henle allows the nephron to concentrate the urine. Salt (Na^+Cl^-) diffuses and is extruded by the ascending limb into the medulla; also, the collecting duct is believed to extrude urea into the tissues of the medulla. This produces a hypertonic environment that draws water out of the descending limb and the collecting duct. This water is returned to the circulatory system.*

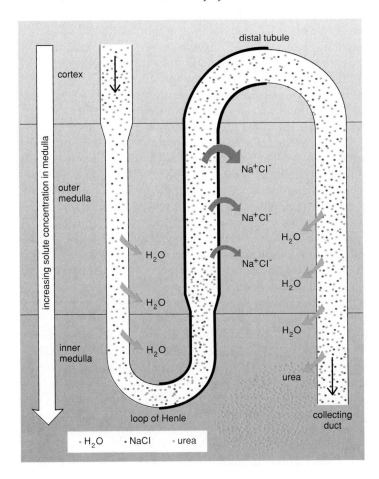

Loop of Henle

The presence of a loop of Henle allows humans to excrete a concentrated urine (i.e., a larger amount of metabolic wastes per volume than that of blood plasma). The loop of Henle, which lies in the medulla (fig. 9.8), is made up of a *descending* (going down) and an *ascending* (going up) limb.

Salt (Na^+ Cl^-) passively diffuses out of the lower portion of the ascending limb, but the upper thick portion actively transports salt out into the tissue of the outer medulla (fig. 9.13). Less and less salt will be available for transport as fluid moves up the thick portion of the ascending limb. This brings about the presence of an osmotic gradient within the tissues of the medulla: the concentration of salt is greater toward the inner medulla. It should be noted that water cannot passively diffuse out of the ascending limb because it is impermeable to water.

If you examine figure 9.13 carefully, you can see that the innermost portion of the inner medulla has the highest concentration of solutes. This cannot be due to salt because active transport of salt does not start until the thick portion of the ascending limb. Some urea is believed to leave the lower portion of the collecting duct, and this molecule contributes to the high osmolarity of this region of the medulla. This urea reenters the loop of Henle and in this way reaches the collecting duct once again.

Because of the solute concentration gradient present in the medulla, water leaves the descending limb of the loop of Henle all along its length (figs. 9.13 and 9.14).

Unique features of the loop of Henle allow humans to excrete a concentrated urine.

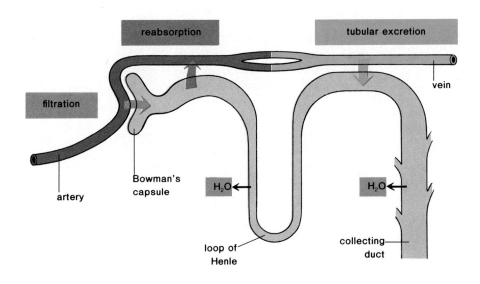

FIGURE 9.14 *Diagram of nephron showing steps in urine formation: filtration, reabsorption, and tubular excretion. Note also that water enters the tissues at the loop of Henle and collecting duct.*

Tubular Excretion

The distal convoluted tubule continues the work of the proximal convoluted tubule in that sodium and water are both reabsorbed. As before, sodium is actively reabsorbed into the blood capillary, and thereafter water follows passively. In this region of the tubule also, substances may be added to the urine by a process called tubular excretion, or augmentation. The cells that line this portion of the tubule have numerous mitochondria. **Tubular excretion** (fig. 9.14) is an active process just like selective reabsorption, but the molecules are moving in the opposite direction. Histamine and penicillin are actively excreted as are ammonia and hydrogen ions (p. 189).

During tubular excretion, certain substances, penicillin and histamine, for example, are actively secreted from the peritubular capillary into the fluid of the tubule.

Collecting Duct

The fluid that enters the collecting duct is isotonic to the cells of the cortex. This means that to this point the net effect of the reabsorption of water and sodium has been to produce a fluid in which the proportion of water to sodium is the same as in most tissues. However, the medulla is hypertonic to the contents of the collecting duct, due to the extrusion of salt by the ascending limb of the loop of Henle (and possibly also due to the presence of urea). Therefore, water diffuses out of the collecting duct into the medulla, and the urine within the collecting duct becomes concentrated (hypertonic to blood plasma). *Urine* (table 9.2) now passes from the collecting duct to the pelvis of the kidney. Table 9.5 summarizes urine formation.

Water diffuses from the collecting duct, and the urine becomes increasingly concentrated as it nears the pelvis of the kidney.

TABLE 9.5 NEPHRON

NAME OF PART	LOCATION IN KIDNEY	FUNCTION
Bowman's capsule	Cortex	Filtrate formation
Proximal convoluted tubule	Cortex	Selective reabsorption
Loop of Henle	Medulla	Extrusion of sodium and reabsorption of water
Distal convoluted tubule	Cortex	Tubular excretion
Collecting duct	Medulla	Reabsorption of water

REGULATORY FUNCTIONS OF THE KIDNEYS

Blood Volume

Maintenance of blood volume and ion balance is under the control of hormones. **ADH (antidiuretic hormone)** is a hormone secreted by the posterior pituitary that primarily maintains blood volume. ADH increases the permeability of the collecting duct so that more water can be reabsorbed. In order to understand the function of this hormone, consider its name. Diuresis means increased excretion of urine, and antidiuresis means suppression of urinary excretion. When ADH is present, more water is reabsorbed, and a decreased amount of urine results. This hormone is secreted according to whether blood volume needs to be increased or decreased. When water is reabsorbed at the collecting duct, blood volume increases, and when water is not reabsorbed, blood volume decreases. In practical terms (table 9.6), if an individual does not drink much water on a certain day, the *posterior lobe of the pituitary* releases ADH; more water is reabsorbed; blood volume is maintained at a normal level; and, consequently, there is less urine. On the other hand, if an individual drinks a large amount of water and does not perspire much, the posterior lobe of the pituitary does not release ADH; more water is excreted; blood volume is maintained at a normal level; and a greater amount of urine is formed.

Drinking alcohol causes diuresis because it inhibits the secretion of ADH. The dehydration that follows is believed to contribute to the symptoms of a "hangover." Drugs called diuretics are often prescribed for high blood pressure. The drugs cause increased urinary excretion and thus reduce blood volume and blood pressure. Concomitantly, any *edema* (p. 117) present is also reduced.

Aldosterone secreted by the adrenal cortex is a hormone that primarily maintains Na^+ (sodium) and K^+ (potassium) balance. It causes the distal convoluted tubule to reabsorb Na^+ and excrete K^+. The increase of Na^+ in the blood causes water to be reabsorbed, leading to an increase in blood volume and blood pressure.

Blood pressure is constantly monitored by the afferent arteriole cells within the juxtaglomerular apparatus. The juxtaglomerular apparatus (fig. 9.15) occurs at a region of contact between the afferent arteriole and the distal convoluted tubule. The afferent arteriole cells in the region secrete

TABLE 9.6 ANTIDIURETIC HORMONE

Increase in ADH	Increased reabsorption of water	Less urine
Decrease in ADH	Decreased reabsorption of water	More urine

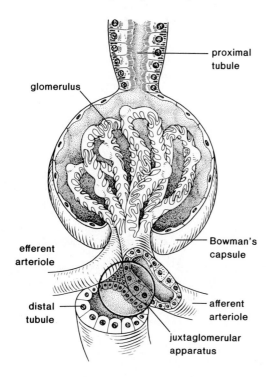

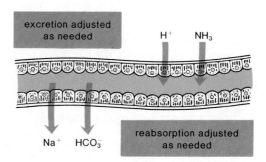

If the blood is alkaline, fewer hydrogen ions are excreted and fewer sodium and bicarbonate ions are reabsorbed. The reabsorption and/or excretion of ions (salts) by the kidneys illustrates their homeostatic ability to maintain not only the pH of the blood but also the osmolarity of the blood. Osmolarity increases as salts are reabsorbed.

Ion Balance

These examples also show that the kidneys regulate the ion balance in the blood by controlling the excretion and reabsorption of various ions. Sodium (Na^+) is an important ion in plasma that must be regulated, but the kidneys also excrete or reabsorb other ions, such as HCO_3^-, K^+, and Mg^{++}, as needed.

The kidneys contribute to homeostasis by excreting urea. They also regulate the volume of the blood, the pH of the blood, and maintain the ion balance of the blood, three very important functions.

PROBLEMS WITH KIDNEY FUNCTION

Because of the great importance of the kidneys to the maintenance of body fluid homeostasis, renal failure is a life-threatening event. There are many types of illnesses that bring on progressive renal disease and renal failure.

Infections of the urinary tract themselves are a fairly common occurrence, particularly in women because the female urethra is considerably shorter than that of the male. If the infection is localized in the urethra, it is called *urethritis*. If it invades the bladder, it is called *cystitis*. And finally, if the kidneys are affected, it is called *nephritis*. Glomerular damage sometimes leads to blockage of the glomeruli so that no fluid moves into the tubules, or it can cause

renin when the blood pressure is insufficient to promote efficient filtration in the glomerulus. Renin is an enzyme that converts a large plasma protein, angiotensinogen, into angiotensin. Angiotensin stimulates the adrenal cortex to release aldosterone. A possible mechanism has also been suggested for suppressing renin secretion. Perhaps the distal tubule cells in the juxtaglomerular apparatus are sensitive to sodium concentration in urine, and when there is a high sodium concentration in the tubule, they inhibit the afferent arteriole cells from secreting renin. If so, this mechanism may be faulty in certain individuals. The renin-angiotensin-aldosterone system seems to be always active in some patients with hypertension.

Adjustment of pH

The kidneys aid in maintaining a constant pH of the blood, and the whole nephron takes part in this process. Figure 9.16 indicates that the excretion of hydrogen ions and ammonia ($H^+ + NH_3$), together with the reabsorption of bicarbonate ions and sodium, is adjusted in order to keep the pH within normal limits. If the blood is acidic, hydrogen ions are excreted in combination with ammonia, and sodium bicarbonate is reabsorbed. This will restore alkalinity because the sodium bicarbonate can react with water to give hydroxide ions, which are balanced by the sodium ions:

$$Na^+HCO_3^- + HOH \rightarrow H_2CO_3 + Na^+OH^-$$

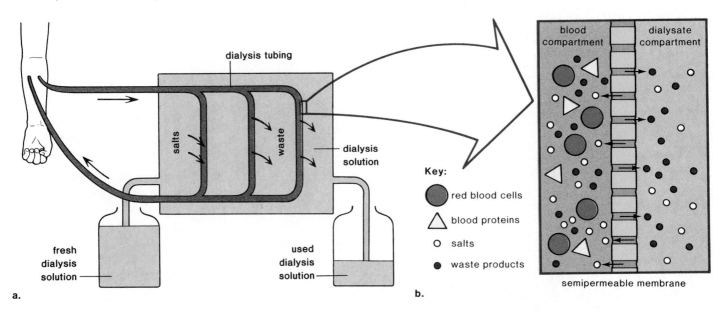

FIGURE 9.17 *Diagram of an artificial kidney.* a. *As the patient's blood circulates through dialysis tubing, it is exposed to a solution.* b. *Wastes exit from the blood into the solution because of a preestablished concentration gradient. In this way the blood is not only cleansed, but the pH can also be adjusted.*

the glomeruli to become more permeable than usual. This is detected when a **urinalysis** is done. If the glomeruli are too permeable, albumin, white blood cells, or even red blood cells may appear in the urine. Trace amounts of protein in the urine are not a matter of concern, however.

When glomerular damage is so extensive that more than two-thirds of the nephrons are incapacitated, waste substances accumulate in the blood. This condition is called **uremia** because urea begins to accumulate in the blood. Although the presence of nitrogenous wastes can cause serious damage, the retention of water and salts is of even more concern. The latter causes edema, fluid accumulation in the body tissues. Imbalance in the ionic composition of body fluids can even lead to loss of consciousness and heart failure.

Kidney Replacements

Kidney Transplant

Patients with renal failure can sometimes undergo kidney transplant operations, during which they receive a functioning kidney from a donor. As with all organ transplants, there is the possibility of organ rejection so that receiving a kidney from a close relative has the highest chance of success. The current one-year survival rate is 97% if the kidney is received from a relative and 90% if it is received from a nonrelative. Recently, investigators have discovered that the drug Cyclosporine is most helpful in preventing organ rejection and, at the same time, allowing the patient to fight infections. Others are hopeful that monoclonal antibodies that react against killer T cells will be available soon.

HUMAN ISSUE

Imagine a situation in which (1) a patient is not doing well on dialysis and doctors believe that only a kidney transplant will allow an improvement in health and (2) there is in the same hospital another patient who is being kept alive only by machines. Under what circumstances—if any—would it be proper to disconnect the machines so that a kidney of the patient (2) could be donated to this other patient (1)?

Dialysis

If a satisfactory donor cannot be found for a kidney transplant, which is frequently the case, the patient may undergo dialysis treatments, utilizing either a kidney machine or CAPD (continuous ambulatory peritoneal [abdominal] dialysis). Dialysis is defined as the diffusion of dissolved molecules through a semipermeable membrane. These molecules will, of course, move across a membrane from the area of greater concentration to one of lesser concentration.

In the case of the kidney machine (fig. 9.17), the patient's blood is passed through a semipermeable membranous tube that is in contact with a balanced salt (dialysis) solution. Substances more concentrated in the blood diffuse into the dialysis solution, also called the dialysate. Conversely, substances more concentrated in the dialysate diffuse into the blood. Accordingly, the artificial kidney can be utilized either to extract substances from the blood, including waste products or toxic chemicals and drugs, or to add substances to the blood, for example, bicarbonate ions if the blood is acidic. In the course of a six-hour dialysis, from

50 to 250 g of urea can be removed from a patient, which greatly exceeds the urea clearance of normal kidneys. Therefore, a patient need undergo treatment only about twice a week.

In the case of CAPD, a fresh amount of dialysate is introduced directly into the abdominal cavity from a bag attached to a permanently implanted plastic tube. Waste and water molecules pass into the fluid from the surrounding organs before the fluid is collected four or eight hours later.

Kidney transplants and dialysis are available procedures for persons who have suffered renal failure.

SUMMARY

The end products of metabolism are, for the most part, nitrogenous wastes, such as ammonia, urea, uric acid, and creatinine, all of which are excreted primarily by the urinary system. The urinary system contains the kidneys, whose macroscopic anatomy is dependent on the presence of nephrons. Urine formation requires three steps: during pressure filtration, small components of plasma pass into Bowman's capsule from the glomerulus due to blood pressure; during selective reabsorption, nutrients and sodium are actively reabsorbed from the proximal convoluted tubule back into the blood; during tubular excretion, a few types of substances are actively secreted into the distal convoluted tubule from the blood.

Water is reabsorbed along the length of the tubule but especially from the loop of Henle and collecting duct. ADH, a hormone produced by the posterior pituitary, controls the reabsorption of water directly, and aldosterone, from the adrenal cortex, controls it indirectly by affecting Na^+ reabsorption. The whole tubule participates in maintaining the pH of the blood by regulating the pH of urine. In practice, hydrogen ions are excreted, and sodium bicarbonate is reabsorbed to maintain the pH.

Various types of problems can lead to kidney failure. In such cases, the person can either receive a kidney from a donor or undergo dialysis treatments by means of the kidney machine or CAPD.

OBJECTIVE QUESTIONS

1. The primary nitrogenous end product of humans is _____ .
2. The large intestine is an organ of excretion because it rids the body of _____ .
3. Urine leaves the bladder in the _____ .
4. The capillary tuft inside Bowman's capsule is called the _____ .
5. _____ is a substance that is found in the filtrate, is reabsorbed, and still is in urine.
6. _____ is a substance that is found in the filtrate, is minimally reabsorbed, and is concentrated in the urine.
7. Tubular excretion takes place at the _____ , a portion of the nephron.
8. Reabsorption of water from the collecting duct is regulated by the hormone _____ .
9. In addition to excreting nitrogenous wastes, the kidneys adjust the _____ and _____ of blood.
10. Persons who have nonfunctioning kidneys are often on _____ machines.

Answers

1. urea 2. certain salts, e.g., calcium and iron 3. urethra 4. glomerulus 5. Water 6. urea 7. distal convoluted tubule 8. ADH 9. volume, pH 10. dialysis

STUDY QUESTIONS

1. Name four nitrogenous end products and explain how each is formed in the body. (p. 179)
2. Name several excretory organs and the substances they excrete. (p. 181)
3. What is the composition of urine? (p. 182)
4. Give the path of urine. (p. 182)
5. Name the parts of the kidney tubule, or nephron. (p. 184)
6. Trace the path of blood about the tubule. (p. 185)
7. Describe how urine is made by telling what happens at each part of the tubule. (p. 187)
8. Explain these terms: pressure filtration, selective reabsorption, and tubular excretion. (pp. 185, 185, 188)
9. How does the nephron regulate the pH of the blood? (p. 189)
10. Explain how the artificial kidney machine and CAPD work. (p. 190)

1. Would animals that live in fresh water or on land have well-developed loops of Henle? Why?

2. Refer to figure 3.2 and state where you would place the kidney tubule in the diagram. To which structure along the digestive tract might you compare a kidney tubule and why?

3. Why is the term filtration used both in regard to a capillary in the tissues and to the glomerulus in the kidney tubule? What differences are there between these two processes?

KEY TERMS

antidiuretic hormone (an″ti-di″u-ret′ik hōr′mōn) ADH; sometimes called vasopressin, a hormone secreted by the posterior pituitary that controls the rate at which water is reabsorbed by the kidneys. *188*

Bowman's capsule (bo′manz kap′sūl) a double-walled cup that surrounds the glomerulus at the beginning of the kidney tubule. *183*

collecting duct (kŏ-lekt′ing dukt) a tube that receives urine from several distal convoluted tubules. *183*

distal convoluted tubule (dis′tal kon′vo-lūt-ed tu′būl) highly coiled region of a nephron that is distant from Bowman's capsule. *183*

excretion (eks-kre′shun) removal of metabolic wastes. *178*

filtrate (fil′trāt) the filtered portion of blood that is contained within Bowman's capsule. *185*

glomerulus (glo-mer′u-lus) a cluster; for example, the cluster of capillaries surrounded by Bowman's capsule in a kidney tubule. *183*

kidneys (kid′nēz) organs in the urinary system that form, concentrate, and excrete urine. *182*

nephron (nef′ron) the anatomical and functional unit of the vertebrate kidney; kidney tubule. *183*

peritubular capillary (per″i-tu′bu-lar kap′i-lar″e) capillary that surrounds a nephron and functions in reabsorption during urine formation. *185*

pressure filtration (presh′ur fil-tra′shun) the movement of small molecules from the glomerulus into Bowman's capsule due to the action of blood pressure. *185*

proximal convoluted tubule (prok′si-mal kon′vo-lūt-ed tu′būl) highly coiled region of a nephron near Bowman's capsule. *183*

renal pelvis (pel′vis) a hollow chamber in the kidney that lies inside the medulla and receives freshly prepared urine from the collecting ducts. *183*

selective reabsorption (sĕ-lek′tiv re″ab-sorp′shun) the movement of nutrient molecules, as opposed to waste molecules, from the contents of the kidney tubule into the blood at the proximal convoluted tubule. *186*

tubular excretion (tu′bu-lar eks-kre′shun) the movement of certain molecules from the blood into the distal convoluted tubule so that they are added to urine. *188*

urea (u-re′ah) primary nitrogenous waste of mammals derived from amino acid breakdown. *180*

uric acid (u′rik as′id) waste product of nucleotide breakdown. *180*

ureters (u-re′terz) tubes that take urine from the kidneys to the bladder. *182*

urethra (u-re′thrah) tube that takes urine from the bladder to outside. *182*

urinary bladder (u′ri-ner″e blad′der) an organ in which urine is stored before being discharged by way of the urethra. *182*

FURTHER READINGS FOR PART II

Ada, G. L., and Sir Gustav Nossal, August 1987. The clonal-selection theory. *Scientific American.*

Buisseret, P. D. August 1982. Allergy. *Scientific American.*

Cohen, I. R. April 1988 The self, the world and autoimmunity. *Scientific American.*

Doolittle, R. F. December 1981. Fibrinogen and fibrin. *Scientific American.*

Gallo, R. C. January 1987. The AIDS virus. *Scientific American.*

Guyton, A. C. 1979. *Physiology of the human body.* 5th ed. Philadelphia: W. B. Saunders.

Hole, J. W. 1984. *Human anatomy and physiology.* 3d ed. Dubuque, IA: Wm. C. Brown Publishers.

Human nutrition: Readings from Scientific American. 1978. San Francisco: W. H. Freeman.

Jarvik, R. K. January 1981. The total artificial heart. *Scientific American.*

Kessel, R. G., and R. H. Kardon. 1979. *Tissues and organs: A text-atlas of scanning electron microscopy.* San Francisco: W. H. Freeman.

Moog, F. November 1981. The lining of the small intestine. *Scientific American.*

Perutz, M. F. December 1978. Hemoglobin structure and respiratory transport. *Scientific American.*

Rose, N. R. February 1981. Autoimmune diseases. *Scientific American.*

Schmidt-Nielson, K. May 1981. Countercurrent systems in animals. *Scientific American.*

Scientific American October 1988. Entire issue devoted to AIDS.

Scrimshaw, N. S., and V. R. Young. September 1976. The requirements of human nutrition. *Scientific American.*

Tonegawa, S. October 1985. The molecules of the immune system. *Scientific American.*

Vander, A. J. 1980. *Human physiology: The mechanisms of body function.* 4th ed. New York: McGraw-Hill.

Wurtman, R. M. and J. J. Wortman, January 1989. Carbohydrates and depression. *Scientific American.*

Young, J. D. and Z. A. Cohn. January 1988. How killer cells kill. *Scientific American.*

Zucker, M. B. June 1980. The functioning of blood platelets. *Scientific American.*

PART THREE

HUMAN INTEGRATION AND COORDINATION

T he nervous and hormonal systems help maintain a relatively constant internal environment, homeostasis, by coordinating the functions of the body's other systems. The nervous system acts quickly but provides short-lived regulation, and the hormonal system acts more slowly but provides a more sustained regulation of body parts.

Organisms must also be able to react with the external environment in order to survive. The sense receptors are the organs that inform the organism about the outside environment. This information is then processed by the nervous system, and the individual responds through the muscular system. These physiological events allow humans to maintain an external environment that is compatible with maintenance of an internal environment.

NERVOUS SYSTEM

CHAPTER CONCEPTS

1 The nervous system is made up of neurons that are specialized to carry nerve impulses. A nerve impulse is an electrochemical change.

2 Transmission between neurons is accomplished by means of chemicals called neurotransmitter substances. Mood-altering drugs affect the transmission of these neurotransmitters.

3 The nervous system consists of the central and peripheral nervous systems. The two systems are joined when a reflex occurs.

4 The central nervous system, made up of the spinal cord and brain, is highly organized. Consciousness is a function only of the cerebrum, which is most highly developed in humans.

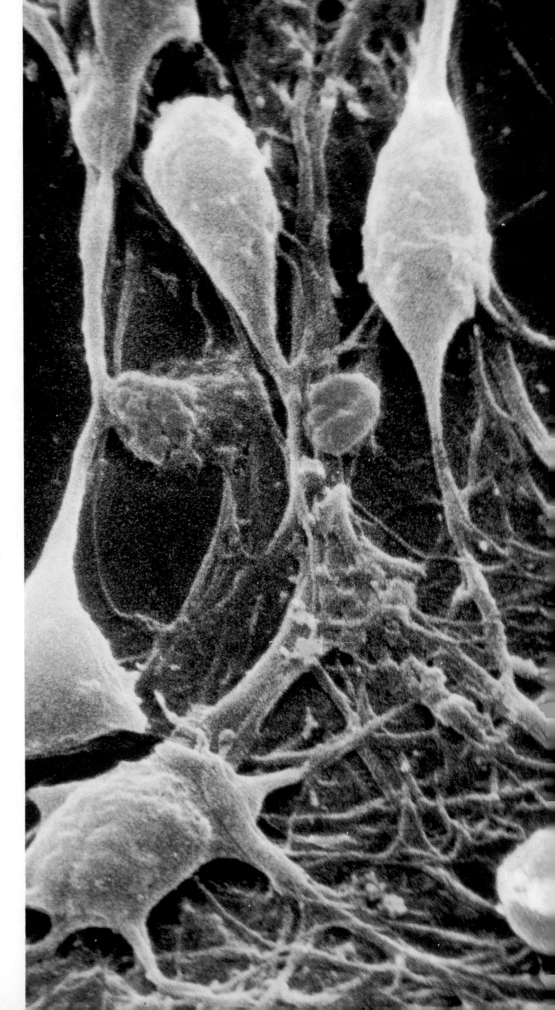

Scanning electron micrograph of neurons from cerebral cortex.

FIGURE 10.8 *Synapse anatomy.* a. *Synapses occur where axon synaptic endings of one neuron lie near the dendrites and cell bodies of the next neuron.* b. *A diagram of a synapse based on electron microscopy studies shows that there is a space called the synaptic cleft between the two neurons.* c. *Transmission across a synapse occurs when a neurotransmitter substance released by vesicles at the presynaptic membrane diffuses across the synaptic cleft to the postsynaptic membrane, where it changes the polarization of the membrane.*

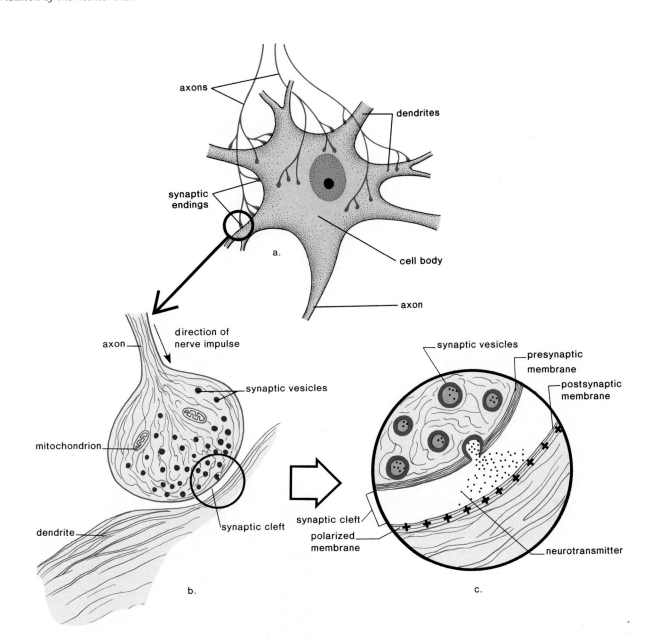

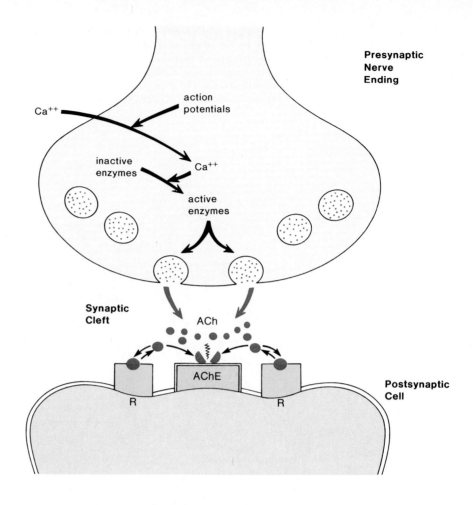

FIGURE 10.9 Transmission of nerve impulse across the synapse. Inflow of calcium (Ca^{++}) leads to enzymatic change in axomembrane permeability so that ACh (acetylcholine) is released by exocytosis and diffuses across the synaptic cleft. After ACh binds to receptors (R), nerve impulses begin and ACh is broken down by AChE (acetylcholinesterase). Further stimulation of the postsynaptic membrane now ceases.

A neuron is on the receiving end of many synapses (fig. 10.8a). Whether a neuron fires or not depends on the summary effect of all the excitatory and/or inhibitory neurotransmitters received. If the amount of excitatory neurotransmitters received is sufficient to overcome the amount of inhibitory neurotransmitters received, the neuron fires. If the amount of excitatory neurotransmitters received is not sufficient, only local excitation occurs. It can be seen, then, that synapses are regions of **integration** where a "summing up" occurs. The presence of synapses allows the nervous system to fine tune its response to the environment.

One-Way Propagation

Transmission across a synapse is *one-way* (fig. 10.10) because only the ends of axons have synaptic vesicles that are able to release neurotransmitter substances to affect the potential of the next neuron. Also, neurons obey the *all-or-none law,* meaning that a neuron either fires maximally or it does not fire at all. A nerve does not obey the all-or-none law because a nerve contains many fibers (fig. 10.11), any number of which may be carrying nerve impulses. Thus a nerve may have degrees of performance.

FIGURE 10.10 Transmission across a synapse can be compared to a relay race. Just as the baton is handed from runner to runner so the nerve impulse is transmitted from neuron to neuron.

FIGURE 10.11 *Diagram of a cross section of a nerve, with one axon extended to show that each fiber is enclosed by myelin. Because nerves contain so many fibers it has been difficult to successfully rejoin them after they are severed in an accident. Scientists have now found that if they hold well-cut pieces together, and then surround them by a solution that resembles cytoplasm, the nerve will repair itself and be functional.*

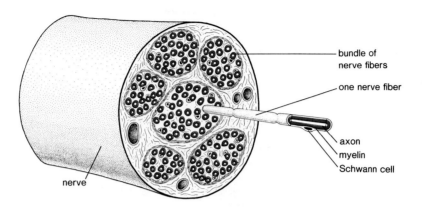

Transmission of a nerve impulse across a synapse is dependent on transmitter substances that change the permeability of the postsynaptic membrane.

PERIPHERAL NERVOUS SYSTEM

Nerves

The PNS (fig. 10.12) consists of nerves that contain only long dendrites and/or long axons. This is so because neuron cell bodies are found only in the brain, spinal cord, and ganglia. **Ganglia** are collections of cell bodies within the PNS.

There are three types of nerves (table 10.2). **Sensory nerves** contain only the long dendrites of sensory neurons; **motor nerves** contain only the long axons of motor neurons; **mixed nerves,** however, contain both the long dendrites of sensory neurons and the long axons of motor neurons. Each nerve fiber within a nerve is surrounded by myelin (fig. 10.11), and therefore nerves have a white, shiny, glistening appearance.

Cranial Nerves

Humans have twelve pairs of **cranial nerves** attached to the brain (fig. 10.13). Some of these are sensory, some are motor, and others are mixed. Notice that although the brain is a part of the CNS, the cranial nerves are a part of the PNS. All cranial nerves, except the vagus, are concerned with the head, neck, and facial regions of the body, but the vagus nerve has many branches to serve the internal organs.

Spinal Nerves

Each **spinal nerve** emerges from the cord (fig. 10.14) by two short branches, or roots, which lie within the vertebral column. The *dorsal root* can be identified by the presence of an enlargement called the **dorsal root ganglion.** This ganglion contains the cell bodies of the sensory neurons whose dendrites conduct impulses toward the cord. The ventral root of each spinal nerve contains the axons of motor neurons that conduct impulses away from the cord. These two roots join just before the spinal nerve leaves the vertebral column. Therefore, all spinal nerves are mixed nerves that contain many sensory dendrites and motor axons.

Human beings have thirty-one pairs of spinal nerves, evidence that humans are segmented animals, especially since the spinal nerves serve the particular region of the body in which they are located.

Cranial nerves take impulses to and/or from the brain. Spinal nerves take impulses to and from the spinal cord.

Somatic Nervous System

The **somatic nervous system** includes all of those nerves that serve the musculoskeletal system and the exterior sense organs, including those in the skin. Exterior sense organs are **receptors** that receive environmental stimuli and then initiate nerve impulses. Muscle fibers are **effectors** that bring about a reaction to the stimulus. Muscle effectors are studied in chapter 11, and receptors are studied in chapter 12.

FIGURE 10.12 *Nerves of the peripheral nervous system consist of the cranial and spinal nerves. The spinal nerves are divided into the cervical (neck region), thoracic (chest region), lumbar (back region), and sacral (pelvic region). Some of the individual nerves are labeled. Each group serves the region of the body mentioned.*

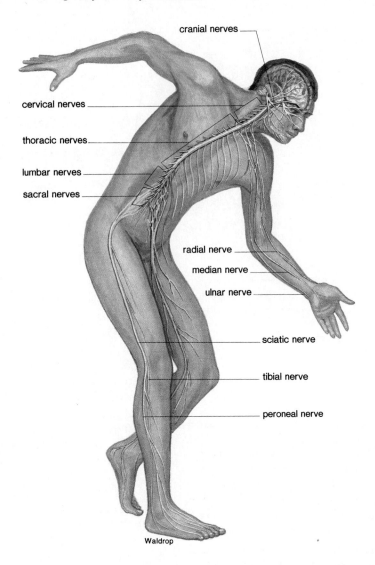

cranial nerves

cervical nerves

thoracic nerves

lumbar nerves

sacral nerves

radial nerve

median nerve

ulnar nerve

sciatic nerve

tibial nerve

peroneal nerve

Waldrop

TABLE 10.2 NERVES

TYPE OF NERVE	CONSISTS OF	FUNCTION
Sensory nerves	Long dendrites only of sensory neurons	Carry message from receptors to CNS
Motor nerves	Long axons only of motor neurons	Carry message from CNS to effectors
Mixed nerves	Both long dendrites of sensory neurons and long axons of motor neurons	Carry message in dendrites to CNS and away from CNS in axons

Note: Compare this table to table 10.1.

FIGURE 10.13 *Underside of the brain showing the origins of the cranial nerves. Many cranial nerves are either sensory nerves that receive impulses from the sense organs or motor nerves that control muscles of the face and neck.*

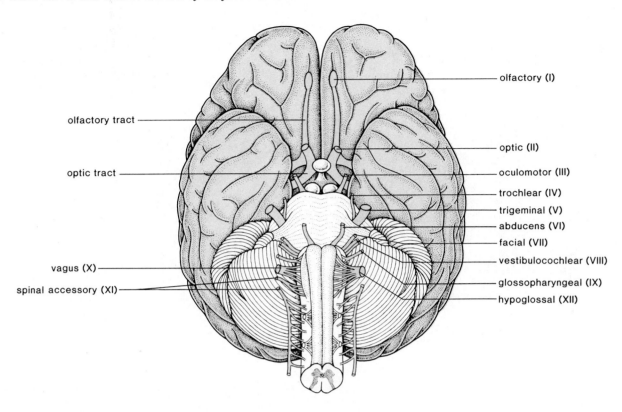

olfactory (I)

olfactory tract

optic (II)

optic tract

oculomotor (III)

trochlear (IV)

trigeminal (V)

abducens (VI)

facial (VII)

vestibulocochlear (VIII)

vagus (X)

glossopharyngeal (IX)

spinal accessory (XI)

hypoglossal (XII)

FIGURE 10.14 *Cross section of spinal cord showing the manner in which spinal nerves leave the cord.*

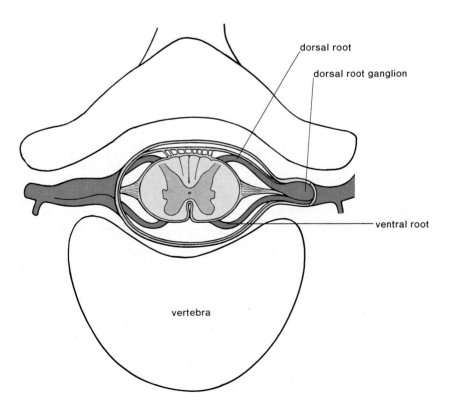

dorsal root

dorsal root ganglion

ventral root

vertebra

FIGURE 10.15 *Diagram of a reflex arc, the functional unit of the nervous system. Trace the path of a reflex by following the black arrows. Name the three types of neurons that are required for a simple reflex, such as the rapid response to touching a hot object with the hand.*

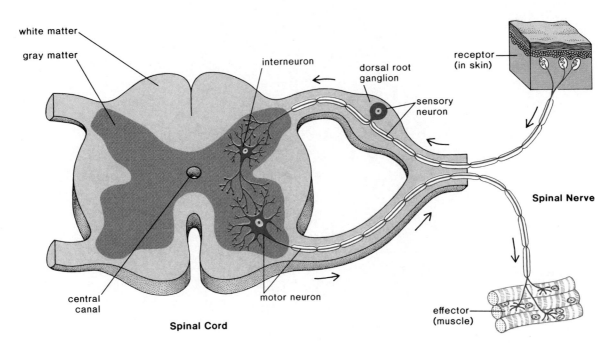

Reflex Arc

Reflexes are automatic, involuntary responses to changes occurring inside or outside the body. In the somatic nervous system, outside stimuli often initiate a reflex action. Some reflexes, such as blinking the eye, involve the brain, but others, such as withdrawing the hand from a hot object, do not necessarily involve the brain. Figure 10.15 illustrates the path of the second type of reflex action. Whenever a person touches a very hot object, a receptor in the skin generates nerve impulses that move along the dendrite of a sensory neuron toward the cell body and CNS. The cell body of a sensory neuron is located in the dorsal root ganglion, just outside the cord. From the cell body, the impulses travel along the axon of the sensory neuron and enter the cord; there they may pass to many interneurons, one of which lies completely within the gray matter and connects with a motor neuron. The short dendrites and cell body of the motor neuron are in the ventral region (horn) of the gray matter, and the axon leaves the cord by way of the ventral root. The nerve impulses travel along the axon to muscle fibers, which then contract so that the hand is withdrawn from the hot object. (See table 10.3 for a listing of these events.) Various other reactions usually accompany a reflex response; the person may look in the direction of the object, jump back, and utter appropriate exclamations. This whole series of responses is explained by the fact that the sensory neuron stimulates several interneurons, which take impulses to all parts of the central nervous system, including the cerebrum, which, in turn, makes the person conscious of the stimulus and his or her reaction to it.

TABLE 10.3 *PATH OF A SIMPLE REFLEX*

1. Receptor (formulates message)[a]	Generates nerve impulses[b]
2. Sensory neuron (takes message to central nervous system)	Impulses move along dendrite (spinal nerve) and proceed to cell body (dorsal root ganglia) and then go from cell body to axon (gray matter of cord)
3. Interneuron (passes message to motor neuron)	Impulses picked up by dendrites and passed through cell body to axon (completely within gray matter)
4. Motor neuron (takes message away from central nervous system)	Impulses travel through short dendrites and cell body (gray matter of cord) to axon (spinal nerve)
5. Effector (receives message)	Receives nerve impulses and reacts: glands secrete and muscles contract

[a]Phrases within parentheses in the left column state overall function.
[b]Phrases within parentheses in the right column indicate location of structure.

The reflex arc is a main functional unit of the nervous system. It allows us to react to internal and external stimuli.

FIGURE 10.16 *Location of ganglia in the sympathetic and parasympathetic nervous systems.* a. *Usually, in the sympathetic nervous system, each ganglion lies close to the spinal cord (CNS), and therefore the preganglionic fiber is short and the postganglionic fiber is long.* b. *In the parasympathetic nervous system, each ganglion lies close to the organ being innervated, and therefore the preganglionic fiber is long and the postganglionic fiber is short.*

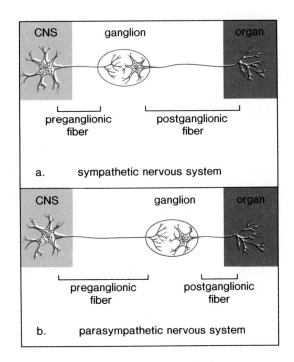

Autonomic Nervous System

The **autonomic nervous system,** a part of the PNS, is made up of motor neurons that control the internal organs automatically and usually without the need for conscious intervention. There are two divisions of the autonomic nervous system: the sympathetic and parasympathetic systems. Both of these (1) function automatically and usually subconsciously in an involuntary manner; (2) innervate all internal organs; and (3) utilize two motor neurons and one ganglion for each impulse. The first neuron has a cell body within the central nervous system and a **preganglionic axon.** The second neuron has a cell body within the ganglion and a **postganglionic axon.**

The autonomic nervous system controls the functioning of internal organs without need of conscious control.

Sympathetic System

The preganglionic fibers of the **sympathetic nervous system** arise from the middle, or thoracic-lumbar, portion of the cord. Usually these fibers almost immediately terminate in ganglia that lie near the cord. Therefore, normally, in this system the preganglionic fiber is short, but the postganglionic fiber that makes contact with the organs is long (fig. 10.16a).

The sympathetic nervous system is especially important during emergency situations and is associated with "fight or flight." For example, it inhibits the digestive tract, but dilates the pupil, accelerates the heartbeat, and increases the breathing rate. It is not surprising, then, that the neurotransmitter released by the postganglionic axon is *noradrenalin,* a chemical close in structure to adrenalin, a well-known heart stimulant.

The sympathetic nervous system brings about those responses we associate with "fight or flight."

Parasympathetic System

The vagus nerve and fibers that arise from the bottom portion of the cord form the **parasympathetic nervous system.** Therefore this system is often called the craniosacral portion of the autonomic nervous system. In the parasympathetic nervous system, the preganglionic fiber is long and the postganglionic fiber is short because the ganglia lie near or within the organ (fig. 10.16b). The parasympathetic system promotes all of those internal responses we associate with a relaxed state; for example, it causes the pupil of the eye to contract, promotes the digestion of food, and retards the heartbeat. The neurotransmitter utilized by the parasympathetic system is *acetylcholine.*

FIGURE 10.17 *Structure and function of the autonomic nervous system. The sympathetic fibers arise from the thoracic and lumbar portion of the cord; the parasympathetic fibers arise from the brain and sacral portion of the cord. Each system innervates the same organs but has contrary effects. For example, the sympathetic system speeds up and the parasympathetic system slows down the beat of the heart.*

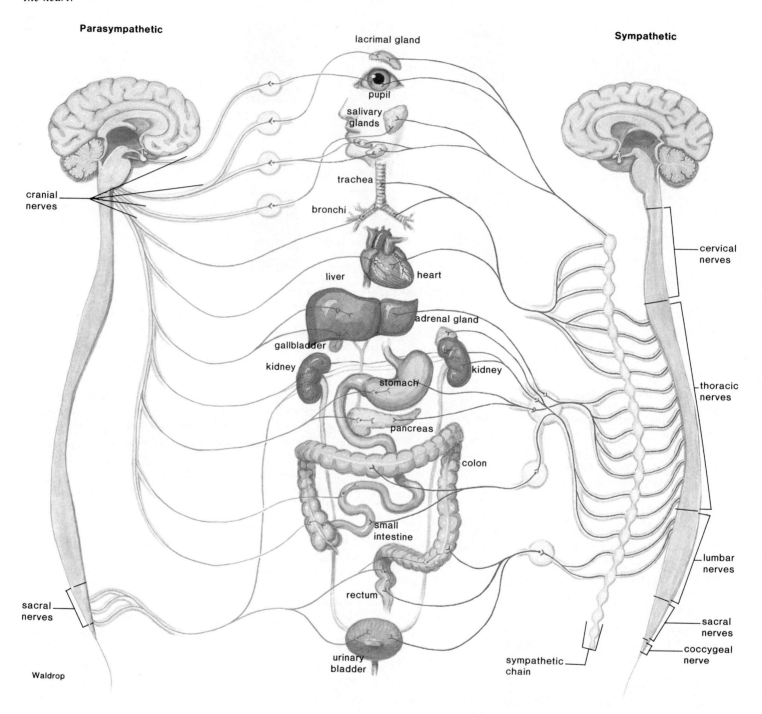

TABLE 10.4 *SYMPATHETIC VERSUS PARASYMPATHETIC SYSTEM*

SYMPATHETIC	PARASYMPATHETIC
Fight or flight	Normal activity
Noradrenalin is neurotransmitter	Acetylcholine is neurotransmitter
Postganglionic fiber is longer than preganglionic fiber	Preganglionic fiber is longer than postganglionic fiber
Preganglionic fiber arises from middle portion of cord	Preganglionic fiber arises from brain and lower portion of cord

Figure 10.17 contrasts the sympathetic and parasympathetic systems, and table 10.4 lists all the differences we have noted between these two systems.

The parasympathetic nervous system brings about those responses we associate with normally restful activities.

CENTRAL NERVOUS SYSTEM

The CNS consists of the spinal cord and the brain. As figure 10.18 illustrates, the CNS is protected by bone: the brain is enclosed within the skull and the spinal cord is surrounded by vertebrae. Also, both the brain and spinal cord are wrapped in three protective membranes known as **meninges;** spinal meningitis is a well-known infection of these coverings. The outer meningeal layer is called the *dura mater,* a tough, white fibrous connective tissue; the middle layer is the *arachnoid membrane,* a weblike covering; and the inner layer is the very thin *pia mater.* The middle and inner layer are separated by the *subarachnoid space,* which is filled with **cerebrospinal fluid.** A small amount of this fluid is sometimes withdrawn for laboratory testing when a spinal tap (i.e., lumbar puncture) is done. Cerebrospinal fluid is also contained within the **central canal** of the spinal cord and the **ventricles** of the brain. The latter are interconnecting spaces that produce and serve as a reservoir for cerebrospinal fluid.

Spinal Cord

The spinal cord (fig. 10.14) contains (1) the central canal filled with cerebrospinal fluid, (2) *gray matter* containing cell bodies and short fibers, and (3) *white matter* containing long fibers of interneurons that run together in bundles called **tracts.** These tracts connect the cord to the brain stem.

The dorsal and ventral sides of the spinal cord are specialized to handle sensory and motor information, respectively. Sensory information from the spinal nerves enters the spinal cord through the dorsal roots, and motor information from the spinal cord is sent to spinal nerves through the ventral roots. In the gray matter, dorsal cells function primarily in receiving sensory information, and ventral cells function primarily in receiving motor information. Within the white matter of the spinal cord, ascending tracts of dorsal axons take information to the brain, and descending tracts in the ventral part of the cord primarily carry information down from the brain. Because the tracts cross over, the left side of the brain controls the right side of the body and the right side of the brain controls the left side of the body.

The central nervous system lies in the midline of the body and consists of the brain and spinal cord, where sensory information is received and motor control is initiated.

Brain

The largest and most prominent portion of the human brain (fig. 10.18) is the cerebrum. Consciousness resides only in the cerebrum; the rest of the brain functions below the level of consciousness.

The Unconscious Brain

The **medulla oblongata** lies closest to the spinal cord and contains centers for regulating heartbeat, breathing, and vasoconstriction (blood pressure), and also reflex centers for vomiting, coughing, sneezing, hiccoughing, and swallowing.

The **hypothalamus** is concerned with homeostasis and contains centers for regulating hunger, sleep, thirst, body temperature, water balance, and blood pressure. The hypothalamus controls the pituitary gland and thereby serves as a link between the nervous and endocrine systems.

The medulla oblongata and the hypothalamus are both concerned with control of the internal organs.

The **midbrain** and **pons** contain tracts that connect the cerebrum with other parts of the brain. In addition, the pons functions with the medulla oblongata to regulate breathing rate, and the midbrain has reflex centers concerned with head movements in response to visual and auditory stimuli.

The **thalamus** is the last portion of the brain before the cerebrum that is involved with sensory input. It serves as a central relay station for sensory impulses traveling upward from other parts of the spinal cord and brain to the cerebrum. It receives all sensory impulses (except those associated with the sense of smell) and channels them to appropriate regions of the cortex for interpretation.

FIGURE 10.18 *Anatomy of the human brain. The cerebrum, the highest and largest part of the human brain, is responsible for consciousness. The medulla, the last part of the brain before the spinal cord, controls various internal organs. The enlargement below shows the anatomy of the meninges.*

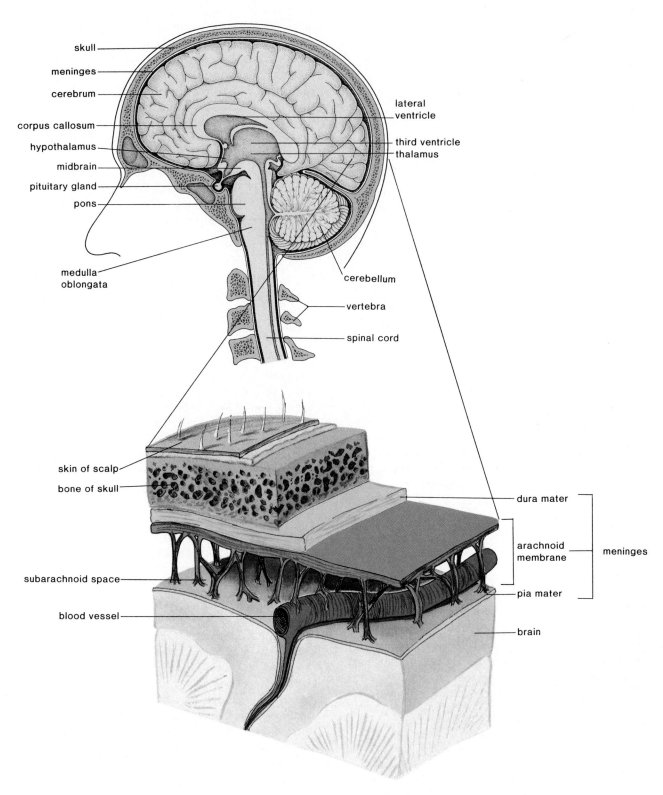

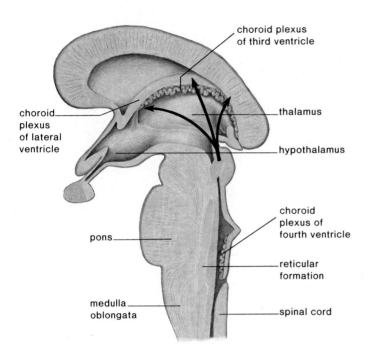

The thalamus has connections to various parts of the brain by way of the diffuse thalamic projection system. This is an extension of the reticular formation (fig. 10.19), a complex network of cell bodies and fibers that extends from the medulla to the thalamus. Together they form the **ARAS (ascending reticular activating system)**, which is believed to sort out incoming stimuli, passing on only those that require immediate attention. For this reason, the thalamus is sometimes called the gatekeeper to the cerebrum because it alerts the cerebrum to only certain sensory input. In this way, it may allow you to concentrate on your homework while the television is on.

The thalamus receives sensory impulses from all parts of the body and channels them to the cerebrum.

The **cerebellum,** a bilobed structure that resembles a butterfly, is the second largest portion of the brain. It functions in muscle coordination (fig. 10.20), integrating impulses received from higher centers to ensure that all the skeletal muscles work together to produce smooth and graceful motions. The cerebellum is also responsible for maintaining normal muscle tone and transmitting impulses that maintain posture. It receives information from the inner ear indicating the position of the body and thereafter sends impulses to those muscles whose contraction maintains or restores balance.

The cerebellum controls balance and complex muscular movements.

The Conscious Brain

The **cerebrum,** which is the only area of the brain responsible for consciousness, is the largest portion of the brain in humans. The outer layer of the cerebrum, called the *cortex,* is gray in color and contains cell bodies and short fibers. The cerebrum is divided into halves known as the right and left **cerebral hemispheres.** Each half contains four types of lobes: **frontal, parietal, temporal,** and **occipital** (fig. 10.21). The cerebrum can be mapped according to the particular functions of each of the lobes (table 10.5). Association areas are believed to be areas for the intellect, artistic and creative ability,

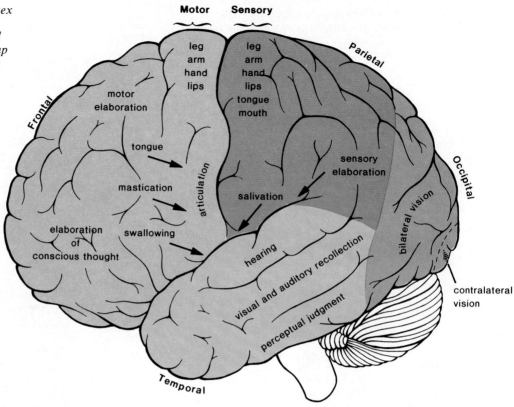

FIGURE 10.21 The convoluted cortex
of the cerebrum is divided into four
lobes: frontal, temporal, parietal, and
occipital. Further, it is possible to map
the cerebrum because each particular
area has a particular function.

TABLE 10.5 FUNCTIONS OF THE CEREBRAL LOBES

LOBE	FUNCTIONS
Frontal lobes	Motor areas control movements of voluntary skeletal muscles Association areas carry on higher intellectual processes, such as those required for concentration, planning, complex problem solving, and judging the consequences of behavior
Parietal lobes	Sensory areas are responsible for the sensations of temperature, touch, pressure, and pain from the skin Association areas function in the understanding of speech and in using words to express thoughts and feelings
Temporal lobes	Sensory areas are responsible for hearing and smelling Association areas are used in the interpretation of sensory experiences and in the memory of visual scenes, music, and other complex sensory patterns
Occipital lobes	Sensory areas are responsible for vision Association areas function in combining visual images with other sensory experiences

learning, and memory. Sensory areas receive nerve impulses from the sense organs and produce what we call sensations. The particular sensation produced is the prerogative of the area of the brain that is stimulated since the nerve impulse itself always has the same nature (as described previously). Motor areas of the cerebrum initiate nerve impulses that control muscle fibers. A momentary lack of oxygen during birth can damage the motor areas of the cerebral cortex so that the individual develops the symptoms of cerebral palsy, a condition characterized by spastic weakness of the arms and legs.

Consciousness is the province of the cerebrum, the most highly developed portion of the brain. It is responsible for higher mental processes, including the interpretation of sensory input and the initiation of voluntary muscular movements.

There has been a great deal of testing to determine whether the right and left halves of the cerebrum serve different functions. These studies have tended to suggest that the left half of the brain is the verbal (word) half and the right half of the brain is the visual (spatial relation) and artistic half. However, other results indicate that such a strict dichotomy does not always exist between the two halves. In any case, the two cerebral hemispheres normally share information because they are connected by a horizontal tract called the **corpus callosum.**

Severing the corpus callosum can control severe epileptic seizures but also results in the person having two brains, each with its own memories and thoughts. Today, the use of the laser permits more precise treatment without these side effects. *Epilepsy* is caused by a disturbance of the normal communication between the ARAS and the cortex. In a *grand mal* seizure, the cerebrum becomes extremely excited. There is a reverberation of signals within the ARAS and cerebrum that continues until neurons become fatigued and the signals cease. In the meantime, the individual loses consciousness, even while convulsions are occurring. Following an attack, the brain is so fatigued the person must sleep for a while.

EEG   The electrical activity of the brain can be recorded in the form of an **electroencephalogram (EEG).** Electrodes are taped to different parts of the scalp, and an instrument called the electroencephalograph records the so-called brain waves (fig. 10.22).

When the subject is awake, two types of waves are usual: *alpha waves,* with a frequency of about 6–13 per second and a potential of about 45 μv, predominate when the eyes are closed, and *beta waves,* with higher frequencies but lower voltage, appear when the eyes are open.

During an eight-hour sleep there are usually five times when the brain waves become slower and larger than alpha waves. During each of these times, there are irregular flurries as the eyes move back and forth rapidly. When subjects are awakened during the latter, called **REM** (rapid eye movement) **sleep,** they always report that they were dreaming. The significance of REM sleep is still being debated, but some studies indicate that REM sleep is needed for memory to occur.

The EEG is a good diagnostic tool; for example, an irregular pattern can signify epilepsy or a brain tumor. A flat EEG signifies lack of electrical activity of the brain, or brain death, and thus, it may be used to determine the precise time of death.

HUMAN ISSUE

Modern medical technology is able to prolong the life of terminally ill patients, and this raises thorny ethical issues. For example, doctors in California removed the feeding tubes from a hopelessly brain-damaged patient who subsequently died. Later the doctors were charged with murder, even though they had acted at the family's request. Presently, a number of states and the District of Columbia recognize "living wills" that request hospitals not take extraordinary measures to prolong the life of the person who made up the will. Do you think this is appropriate, or are there better solutions to the "right-to-die" problem? If you do support the use of such wills, should each of us be required to decide whether or not we want our lives to be prolonged?

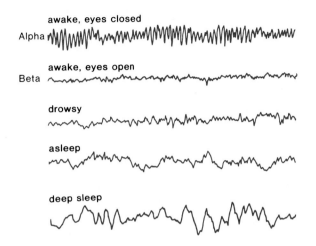

FIGURE 10.22   *Encephalograms are recordings of the electrical activity of the brain. The alpha waves, which appear when the subject is awake with eyes closed, are the most common. Second most common are the beta waves, recorded when the subject is awake with eyes open. Sleep has various stages, as indicated.*

Limbic System

The **limbic system** involves portions of both the unconscious and the conscious brain. It sits above the third ventricle just beneath the cortex (fig. 10.23) and contains neural pathways that connect portions of the frontal lobes, temporal lobes, thalamus, and hypothalamus. Several masses of gray matter that lie deep within each hemisphere of the cerebrum, termed the *basal nuclei,* are also a part of the limbic system.

Stimulation of different areas of the limbic system causes the subject to experience rage, pain, pleasure, or sorrow. By causing pleasant or unpleasant feelings about experiences, the limbic system apparently guides the individual into behavior that is likely to increase the chance of survival.

Learning and Memory

Learning requires memory, but just what permits memory to occur is not definitely known. Investigators have been working with invertebrates, such as slugs and snails, because their nervous systems are very simple and yet they can be conditioned to perform a particular behavior. In order to study this simple type of learning, it has been possible to insert electrodes into individual cells and to alter or record the electrochemical responses of these cells. This type research has recently shown that learning is accompanied by an increase in the number of synapses, while forgetting involves a decrease in the number of synapses. In other words, the nerve-circuit patterns are constantly changing as learning, remembering, and forgetting occur. Within the individual neuron, learning involves a change in gene regulation, nerve protein synthesis, and an increased ability to secrete transmitter substances.

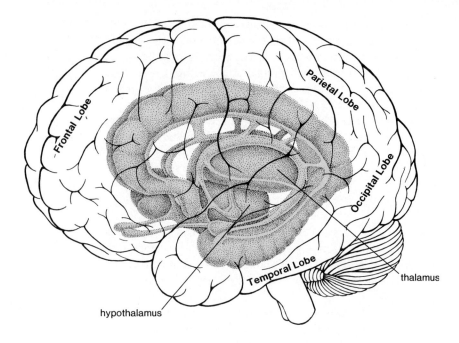

FIGURE 10.23 *The limbic system (in color), which includes portions of the cerebrum, thalamus, and hypothalamus, is concerned mainly with emotion and memory.*

At the other end of the spectrum, some investigators have been studying learning and memory in monkeys. This work has led to the conclusion that the limbic system (fig. 10.23) is absolutely essential to both short-term and long-term memory. An example of short-term memory in humans is the ability to recall a telephone number long enough to dial the number; an example of long-term memory is the ability to recall the events of the day. It's believed that at first impulses move within the limbic circuit but eventually the basal nuclei transmit the neurotransmitter ACh to the sensory areas where memories are stored. The involvement of the limbic system certainly explains why emotionally charged events result in our most vivid memories. The fact that the limbic system communicates with the sensory areas for touch, smell, vision, etc., is the reason any particular sensory stimulus can awaken a complex memory.

The limbic system is particularly involved in the emotions and in memory and learning.

Neurotransmitters in the Brain

As discussed previously, page 200, neurotransmitters take nerve impulses across synapses. Proper functioning of the brain seems to depend on the proper balance of excitatory and inhibitory synaptic transmitters. The *excitatory transmitters* include acetylcholine (ACh), noradrenalin (NA), serotonin, and dopamine. The *inhibitory transmitters* include gamma aminobutyric acid (GABA) and glycine.

Both inhibitory and excitatory neurotransmitters are active in the CNS.

In addition, the enkephalins and endorphins are two neurotransmitters of interest because they are involved in preventing the transmission and perception of pain. When certain cells in the brain, termed SG cells, release the neurotransmitter called substance P, pain is felt. If enkephalins and endorphins occupy receptors on SG cells, substance P is not released and pain is not felt.

Disorders   It has been discovered that several neurological illnesses, such as *Parkinson disease* and *Huntington chorea,* are due to an imbalance in neurotransmitters. Parkinson disease is a condition characterized by a wide-eyed, unblinking expression, an involuntary tremor of the fingers and thumbs, muscular rigidity, and a shuffling gait. All these symptoms are due to dopamine deficiencies. Huntington chorea is characterized by a progressive deterioration of the individual's nervous system that eventually leads to constant thrashing and writhing movements and finally to insanity and death. The problem is believed to be GABA malfunction. Most recently it has been discovered that Alzheimer disease, a severe form of senility with marked memory loss found in 5–10% of all people over age 65, is due to deterioration in cells of the basal nuclei (p. 213) that use ACh as a transmitter. It seems that a gene located on the number-21 chromosome, which is normally active only during development and which directs the production of a protein associated with neuron death, has been inexplicably turned on in these patients.

Treatment of individuals with brain disorders has formerly been directed toward restoring the proper balance of neurotransmitter substances. However, tissue that produces the missing neurotransmitter has been implanted in several patients, and researchers are hopeful that this technique will provide long-lasting cures.

Neurological illnesses are sometimes associated with the deficiency of a particular neurotransmitter in the brain.

DRUG ABUSE FACTS

DRUG ABUSE

A wide variety of drugs can be used to alter the mood and/or emotional state, but our discussion will center on the four most commonly abused drugs: alcohol, marijuana, cocaine, and heroin. *Drug abuse* is evident when a person takes a drug at a dose level and under circumstances that will increase the potential for a harmful effect. The extent of the drug abuse problem can be appreciated from reading the statistics given on this page.

Individuals who are drug abusers are apt to display a *physical dependence* on the drug (formerly called an addiction to the drug). Dependence has developed when the person (1) spends much time thinking about the drug or arranging to get it; (2) often takes more of the drug than was intended; (3) is *tolerant* to the drug—that is, must increase the amount of the drug to get the same effect; (4) has *withdrawal symptoms* when he or she stops taking the drug; and (5) has a repeated desire to cut down on use.

Drug Action

Drugs that affect the nervous system have two general effects: (1) they affect the ARAS (p. 211) and limbic system and (2) they either promote or decrease the action of a particular neurotransmitter. There are a number of different ways drugs can influence the transmission of neurotransmitters, some of which are described in figure 10.24. It is clear,

FIGURE 10.24 *Drug action at synapses.* a. *Drug stimulates release of neurotransmitter.* b. *Drug blocks release of neurotransmitter.* c. *Drug combines with neurotransmitter preventing its breakdown or reuptake.* d. *Drug mimics neurotransmitter.* e. *Drug blocks receptor so that neurotransmitter cannot be received.*

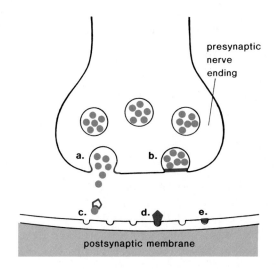

TABLE 10.6   DRUG ACTION

DRUG ACTION	NEUROTRANSMITTER	RESULT
Blocks	Excitatory	Depression
Enhances	Excitatory	Stimulation
Blocks	Inhibitory	Stimulation
Enhances	Inhibitory	Depression

From Mader, Sylvia S., *Inquiry Into Life*, 4th ed. © 1976, 1979, 1982, 1985 Wm. C. Brown Publishers, Dubuque, Iowa. All Rights Reserved. Reprinted by permission.

FIGURE 10.25   Those who abuse drugs, including alcohol, are more likely to be involved in automobile accidents.
Unfortunately, others, in addition to the abuser, often suffer the consequences.

as outlined in table 10.6, that stimulants can either enhance the action of an excitatory transmitter or block the action of an inhibitory transmitter. Also, depressants can either enhance the action of an inhibitory transmitter or block the action of an excitatory transmitter.

Drug abuse often results in physical dependence on the drug because the drug interferes with normal neurotransmitter function in the brain.

Alcohol

The type of alcohol usually consumed is ethanol, the production of which is discussed on page 45. While it is possible to drink alcohol in moderation, the drug is often abused. Alcohol use becomes "abuse" or an illness when alcohol ingestion impairs an individual's social relationships, health, job efficiency, or ability to avoid legal difficulties (fig. 10.25). Table 10.7 lists some of the questions that are used to identify the alcohol-dependent person.

Alcohol effects on the brain are biphasic: after consuming several drinks, blood alcohol concentration rises rapidly and the drinker reports feeling "high" and happy (euphoric). After 90 minutes, and lasting until some 330–400 minutes after consumption, the drinker feels depressed and unhappy (dysphoric) (fig. 10.26). On the other hand, if the drinker continues to drink in order to maintain a high blood level of the alcohol, he or she will experience ever-increasing loss of control. Coma and death are even possible if a substantial amount of alcohol (1¼ pints of whiskey) is consumed within an hour.

Mode of Action and Associated Illnesses
Recent research indicates that alcohol acts on the GABA receptor and potentiates GABA's ability to increase chloride ion uptake. Most likely, it does this by disordering membrane lipids.

Cirrhosis of the Liver   The liver contains the enzyme alcohol dehydrogenase, which begins the breakdown of alcohol to acetic acid, a molecule that can be respired to produce energy. The calories provided by alcohol, however, are termed "empty" because they contribute to energy needs and weight gain without supplying any other nutritional requirements. Worse still, the molecules (glucose and fatty acids) that the liver ordinarily uses as an energy source are converted to fats and the liver cells become engorged with fat droplets. After a few years of being overtaxed, the liver cells begin to die, causing an inflammatory condition known as alcoholic hepatitis. Finally, scar tissue appears in the liver, and it is no longer able to perform its vital functions. This condition is called cirrhosis of the liver, a frequent cause of death among drinkers. Also quite detrimental are the brain impairment and generalized deterioration of other vital organs that are seen in heavy drinkers.

It should be stressed that the early signs of deterioration can be reversed if the habit of drinking to excess is given up.

Alcohol is the most abused drug in the United States. Its abuse often results in well-recognized illnesses and early death.

TABLE 10.7 *SOME QUESTIONS TO IDENTIFY THE ALCOHOL-DEPENDENT PERSON*

1. Do you occasionally drink heavily after a disappointment, a quarrel, or when the boss gives you a bad time?
2. When you are having trouble or feel under pressure, do you always drink more heavily than usual?
3. Have you noticed that you are able to handle more liquor than you did when you were first drinking?
4. Did you ever wake up the "morning after" and discover that you could not remember part of the evening before, even though your friends tell you that you did not "pass out"?
5. When drinking with other people, do you try to have a few extra drinks when others will not know it?
6. Are there certain occasions when you feel uncomfortable if alcohol is not available?
7. Have you recently noticed that when you begin drinking you are in more of a hurry to get the first drink than you used to be?
8. Do you sometimes feel a little guilty about your drinking?
9. Are you secretly irritated when your family or friends discuss your drinking?
10. Have you recently noticed an increase in the frequency of your memory "blackouts"?
11. When you are sober, do you often regret things you have done or said while drinking?
12. Have you often failed to keep the promises you have made to yourself about controlling or cutting down on your drinking?
13. Do more people seem to be treating you unfairly without good reason?
14. Do you eat very little or irregularly when you are drinking?
15. Do you get terribly frightened after you have been drinking heavily?

National Council on Alcoholism.

FIGURE 10.26 *The effects of alcohol have been thoroughly studied. When blood alcohol concentration is high, the user often feels europhic (happy), but as blood alcohol concentration declines, the user feels dysphoric (unhappy). In most states, a person is considered to be legally drunk when the blood alcohol content is 0.1%. This usually requires imbibing three mixed drinks within one and one-half hours.*

FIGURE 10.27 Cannabis sativa, *the plant that contains marijuana, is often smoked in the same manner as tobacco.*

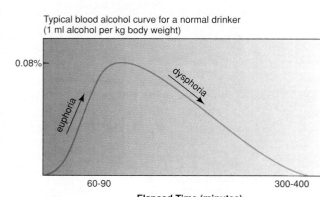

Typical blood alcohol curve for a normal drinker
(1 ml alcohol per kg body weight)

Marijuana

The dried flowering tops, leaves (fig. 10.27), and stems of the Indian hemp plant *Cannabis sativa* contain and are covered by a resin that is rich in THC (tetrahydrocannabinol). The names *cannabis* and *marijuana* can apply to the plant or to THC.

The effects of cannabis differ depending upon the strength and amount consumed, the expertise of the user, and the setting in which it is taken. Usually, the user reports experiencing a mild euphoria along with alterations in vision and judgment that result in distortions of space and time. Also involved may be the inability to concentrate and, to speak coherently and motor incoordination.

Intermittent use of low-potency cannabis is not generally associated with obvious symptoms of toxicity, but heavy use may produce chronic intoxication. Intoxication is recognized by the presence of hallucinations, anxiety, depression, rapid flow of ideas, body image distortions, paranoid reactions, and similar psychotic symptoms. The terms *cannabis psychosis* and *cannabis delirium* refer to such reactions.

Mode of Action

Marijuana is classified as an hallucinogen. It's possible that, like LSD (lysergic acid diethylamide), it has an effect on the action of serotonin.

The use of marijuana does not seem to produce physical dependence but there may be a psychological dependence on the euphoric and sedative effects. Craving or difficulty in stopping may also occur as a part of regular heavy use.

Marijuana has been called a *gateway* drug because adolescents who have used marijuana tend to also try other drugs. For example, in a study of 100 cocaine abusers, 60% had smoked marijuana for more than ten years.

Associated Illnesses

Usually marijuana is smoked in a cigarette form called a joint. Since this allows toxic substances, including carcinogens, to enter the lungs, chronic respiratory disease and lung cancer are considered dangers to long-term heavy use. Some researchers claim that marijuana use leads to long-term brain impairment. Others report that males and females suffer reproductive dysfunctions. *Fetal cannabis syndrome* that resembles fetal alcohol syndrome (p. 345) has been reported.

Some psychologists are very concerned about the use of the drug among adolescents because it can be used as a means to avoid coming to grips with the many personal problems that often have to be worked through at this age.

Although marijuana does not produce physical dependence, it does produce psychological dependence.

Cocaine

Cocaine, currently the second most-popular illegal drug, is an alkaloid derived from the shrub *Erythroxylum cocoa*. *Cocaine* is sold in powder form and as *"crack,"* a more potent extract (fig. 10.28). Users often use the word *rush* to describe the feeling of euphoria that follows intake of the drug. Snorting (inhaling) produces this effect in a few minutes; injection, within 30 seconds; and smoking in less than 10 seconds. Persons dependent upon the drug are, therefore, most

FIGURE 10.28 *Cocaine use.* a. *Crack, the ready-to-smoke form of cocaine that is a more potent, more expensive, and more deadly form than the powder.* b. *Users often smoke crack in a glass water pipe. The high produced consists of a "rush" lasting a few seconds followed by a few minutes of euphoria. Continuous use makes the user extremely dependent on the drug.*

a.

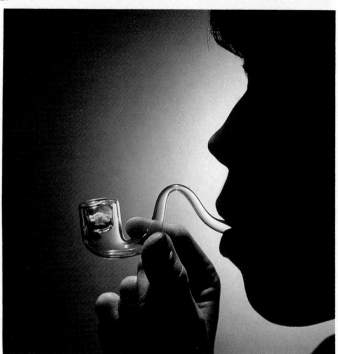

b.

likely to use the last method of intake. The rush only lasts a few seconds and is then replaced by a state of arousal that lasts from 5 to 30 minutes. Then the user begins to feel restless, irritable, and depressed. To do away with these symptoms the user is apt to take more of the drug, repeating the cycle over and over again until there is no more drug left. A binge of this sort can go on for days after which the individual suffers a *"crash."* During the binge period, the user is hyperactive and has little desire for food or sleep, but has an increased sex drive. During the crash period, the user is fatigued, depressed, irritable, has memory and concentration problems, and displays no interest in sex. Indeed, men are often impotent. Other drugs, such as marijuana, alcohol, or heroin, are often taken to ease the symptoms of the crash.

Mode of Action

Cocaine affects the concentration of dopamine in synapses of the brain. After release, the neurotransmitter—in this case, dopamine—is ordinarily withdrawn into the presynaptic cell for recycling and reuse. Cocaine prevents the reuptake of dopamine, and this causes an excess of dopamine in the synaptic cleft so that the user experiences the sensation of a rush. The adrenalin-like effects of dopamine account for the state of arousal that lasts for some minutes after the rush experience.

With continued cocaine use, the body begins to make less dopamine as a compensation for a seemingly excess supply. The user, therefore, now experiences *tolerance* (always needing more of the drug for the same effect), *withdrawal* (symptoms described previously when a drug is not taken), and an intense *craving* for cocaine. These are indications that the person is highly dependent upon the drug or, in other words, that cocaine is extremely addictive.

Associated Illnesses

Overdosing on cocaine is a real possibility. The number of deaths from cocaine and of emergency room admissions for drug reactions involving cocaine has increased greatly. High doses can cause seizures and cardiac and respiratory arrest.

Individuals that snort the drug can suffer damage to the nasal tissues and even perforation of the septum between the nostrils. Whether long-term cocaine abuse causes brain damage in the addict is not yet known but is under investigation. It is known that babies born to addicts suffer withdrawal symptoms and may suffer neurological and developmental problems.

Heroin

Heroin is derived from morphine, an alkaloid of *opium.* Heroin is usually injected. After intravenous injection, the onset of action is noticeable within 1 minute and reaches its peak in 3 to 6 minutes. There is a feeling of euphoria along with relief of pain. Side effects can include nausea, vomiting, dysphoria, and respiratory and circulatory depression leading to death.

Mode of Action

Heroin binds to receptors meant for the body's own opioids, the endorphins and enkephalins. As mentioned previously, the opioids are believed to alleviate pain by preventing the release of a neurotransmitter termed substance P from certain sensory neurons in the region of the spinal cord. When substance P is released, pain is felt, and when substance P is not released, pain is not felt. Evidence also indicates that there are opioid receptors in neurons that travel from the spinal cord to the limbic system and that stimulation of these can cause a feeling of pleasure. This explains why opium and heroin not only kill pain but also produce a feeling of tranquility.

Individuals who inject heroin become physically dependent on the drug. With time, the body begins to produce less of the endorphins and enkephalins; now *tolerance* develops so that the user needs to take more of the drug just to prevent *withdrawal* symptoms. The euphoria originally experienced upon injection is no longer felt.

The withdrawal symptoms include perspiration, dilation of pupils, tremors, restlessness, abdominal cramps, gooseflesh, defecation, vomiting, and increases in systolic pressure and respiratory rate. Those who are excessively dependent may experience convulsions, respiratory failure, and death. Infants born to women who are physically dependent also experience these withdrawal symptoms.

Designer Drugs

Designer drugs are analogues; that is, they are chemical compounds of controlled substances slightly altered in molecular structure. One such drug is MPPP (1-methyl-4-phenylprionoxypiperidine), an analogue of the narcotic fentanyl. Even small doses of the drug are very toxic; MPPP has already caused many deaths on the West Coast.

Cocaine and heroin produce a very strong physical dependence, and an overdose of these drugs can cause death.

HUMAN ISSUE

Much crime and untold human misery in this country are drug related. As you know, there is increasing public sentiment about how to turn the tide away from drug abuse. There are at least three strategies for doing this: drug prevention programs directed toward school-age children, drug rehabilitation programs, and minimizing the flow of drugs into our country. Of these three strategies, which do you think should be emphasized?

SUMMARY

The cell bodies of nerve cells are found in the CNS and ganglia. Axons and dendrites make up nerves. The nerve impulse is a change in permeability of the membrane so that sodium ions move to the inside of a neuron and potassium ions move to the outside. The nerve impulse is transmitted across the synapse by neurotransmitter substances.

During a spinal reflex, a sensory neuron transmits nerve impulses from a receptor to an interneuron, which in turn transmits impulses to a motor neuron, which conducts them to an effector. Reflexes are automatic and some do not require involvement of the brain.

Long fibers of sensory and/or motor neurons make up cranial and spinal nerves of the somatic and autonomic divisions of the PNS. While the somatic division controls skeletal muscle, the autonomic division controls smooth muscle and internal organs.

The CNS consists of the spinal cord and brain. Only the cerebrum is responsible for consciousness; the other portions of the brain have their own specific functions. The cerebrum can be mapped, and each lobe seems to also have particular functions. Neurological drugs, although quite varied, have been found to affect the ARAS and limbic system by either promoting or preventing the action of neurotransmitters.

OBJECTIVE QUESTIONS

1. An _____ carries nerve impulses away from the cell body.
2. During the upswing of the action potential, _____ ions are moving to the _____ of the nerve fiber.
3. The space between the axon of one neuron and the dendrite of another is called the _____ .

4. ACh is broken down by the enzyme _____ after it has altered the permeability of the postsynaptic membrane.
5. Motor nerves innervate _____ .
6. The vagus nerve is _____ _____ nerve that controls _____ .
7. In a reflex arc only the neuron called the _____ is completely within the CNS.
8. The brain and spinal cord are covered by protective layers called _____ .

9. The _____ is that part of the brain that allows us to be conscious.
10. The _____ is the part of the brain responsible for coordination of body movements.

Answers

10. cerebellum
7. interneuron 8. meninges 9. cerebrum
4. AChE 5. muscles 6. cranial,
motor, or parasympathetic; internal organs
1. axon 2. sodium, inside 3. synaptic cleft

STUDY QUESTIONS

1. What are the two main divisions of the nervous system? Explain why these names are appropriate. (p. 195)
2. What are the three types of neurons? How are they similar, and how are they different? (p. 197)
3. What does the term *resting potential* mean, and how is it brought about? (p. 198) Describe the two parts of an action potential and the change that may be associated with each part. (p. 198)
4. What is the sodium/potassium pump, and when is it active? (p. 198)

5. What is a neurotransmitter substance; where is it stored; how does it function; and how is it destroyed? (p. 200) Name two well-known neurotransmitters. (p. 200)
6. What are the three types of nerves, and how are they anatomically different? functionally different? Distinguish between cranial and spinal nerves. (p. 203)
7. Trace the path of a reflex action after discussing the structure and function of the spinal cord and spinal nerve. (p. 206)

8. What is the autonomic nervous system, and what are its two major divisions? (p. 207) Give several similarities and differences between these divisions. (p. 209)
9. Name the major parts of the brain, and give a function for each. (pp. 209–211)
10. Describe the EEG, and discuss its importance. (p. 213)
11. Describe the physiological effects and mode of action of alcohol, marijuana, cocaine, and heroin. (pp. 216–219)

THOUGHT QUESTIONS

1. In what way is a brain neuron even more specialized than a skeletal muscle fiber (cell)?

2. Skeletal muscle fibers are innervated only by somatic motor neurons. Speculate why two different motor neurons (one from the sympathetic system and the other from the parasympathetic system) are used for internal organs and glands.

3. The limbic system is sometimes called the animal brain. Considering the function of the limbic system, why might this term be appropriate? Can you give an explanation for why we have a "primitive" brain beneath the cerebral cortex?

HUMAN ANATOMY

FULL-COLOR PLATES WITH SIX IN TRANSPARENT

"TRANS-VISION"® SHOWING STRUCTURES OF THE HUMAN TORSO

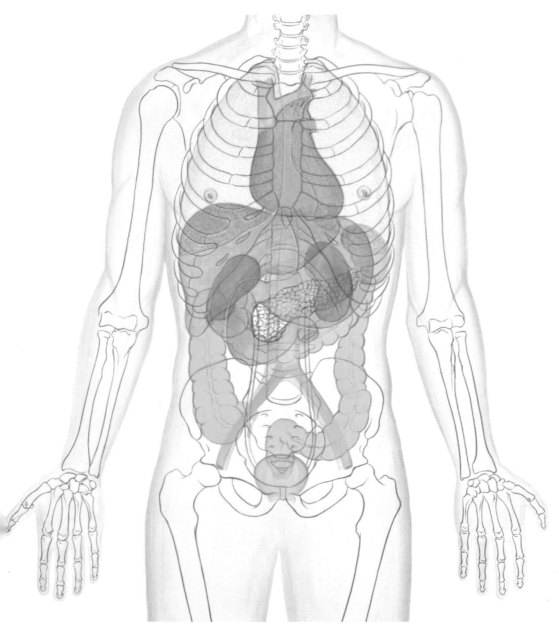

Plate I

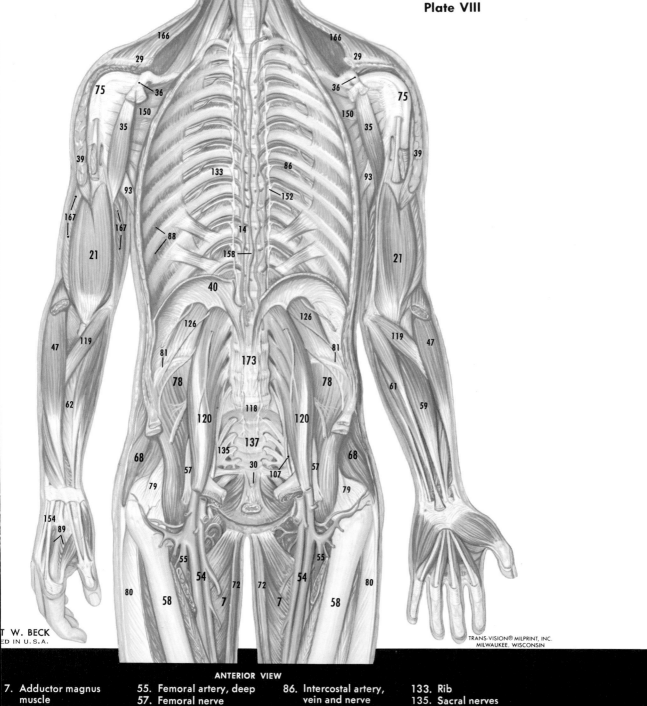

Plate VIII

ANTERIOR VIEW

7. Adductor magnus muscle
14. Azygos veins
21. Brachialis muscle
29. Clavicle
30. Coccyx
35. Coracobrachialis muscle
36. Coracoid process of the scapula
39. Deltoid muscle
40. Diaphragm
47. Extensor carpi radialis longus muscle
54. Femoral artery and vein
55. Femoral artery, deep
57. Femoral nerve
58. Femur
59. Flexor carpi radialis muscle
61. Flexor digitorum profundus muscle
62. Flexor digitorum superficialis muscle
68. Gluteus medius muscle
75. Humerus
78. Iliacus muscle
79. Iliofemoral ligament
80. Iliotibial tract
81. Ilium
86. Intercostal artery, vein and nerve
88. Intercostal muscle, internal
89. Interosseous muscles, dorsal
93. Latissimus dorsi muscle
107. Obturator nerve
118. Promontory
119. Pronator teres muscle
120. Psoas muscles (major and minor)
126. Quadratus lumborum muscle
133. Rib
135. Sacral nerves
137. Sacrum
150. Subscapularis muscle
152. Sympathetic (autonomic) nerve chain
154. Tendons of extensor muscles of hand
158. Thoracic duct
166. Trapezius muscle
167. Triceps brachii muscle
173. Vertebral column

Plate IX

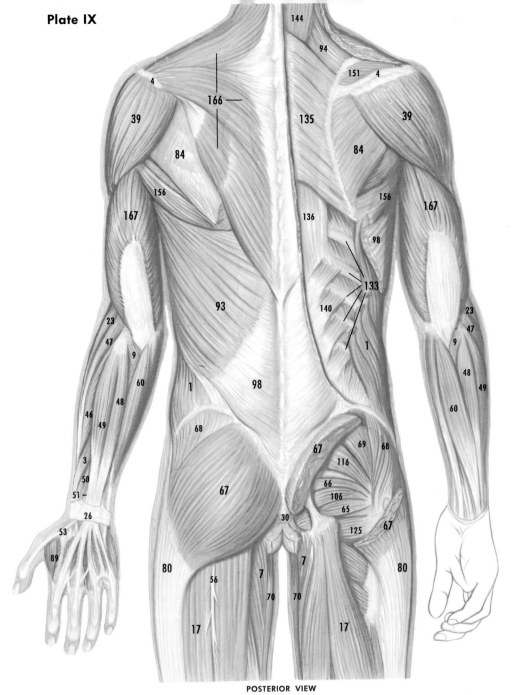

POSTERIOR VIEW

1. Abdominal oblique muscle, external
3. Abductor pollicis longus muscle
4. Acromion process of the scapula
7. Adductor magnus muscle
9. Anconeus muscle
17. Biceps femoris muscle
23. Brachioradialis muscle
26. Carpal ligament, dorsal
30. Coccyx
39. Deltoid muscle
46. Extensor carpi radialis brevis muscle

47. Extensor carpi radialis longus muscle
48. Extensor carpi ulnaris muscle
49. Extensor digitorum communis muscle
50. Extensor pollicis brevis muscle
51. Extensor pollicis longus muscle
56. Femoral cutaneous nerve, posterior
60. Flexor carpi ulnaris muscle
65. Gemellus inferior muscle

66. Gemellus superior muscle
67. Gluteus maximus muscle
68. Gluteus medius muscle
69. Gluteus minimus muscle
70. Gracilis muscle
80. Iliotibial tract
84. Infraspinatus muscle
89. Interosseous muscle, dorsal
93. Latissimus dorsi muscle
94. Levator scapulae muscle
98. Lumbodorsal fascia

106. Obturator internus muscle
116. Piriformis muscle
125. Quadratus femoris muscle
133. Ribs (VII-XII)
135. Rhomboideus muscle
136. Erector spinae muscle
140. Serratus posterior inferior muscle
144. Splenius capitis muscle
151. Supraspinatus muscle
156. Teres major muscle
166. Trapezius muscle
167. Triceps brachii muscle

axon (ak′son) process of a neuron that conducts nerve impulses away from the cell body. *195*

cell body (sel bod′e) portion of a nerve cell that includes a cytoplasmic mass and a nucleus and from which the nerve fibers extend. *195*

central nervous system (sen′tral ner′vus sis′tem) CNS; the brain and spinal cord in vertebrate animals. *195*

cerebral hemisphere (ser′ĕ-bral hem′ĭ-sfēr) one of the large paired structures that together constitute the cerebrum of the brain. *211*

dendrite (den′drīt) process of a neuron, typically branched, that conducts nerve impulses toward the cell body. *195*

effector (ē-fek′tor) a structure, such as the muscles and glands, that allows an organism to respond to environmental stimuli. *203*

ganglion (gang′gle-on) a collection of neuron cell bodies outside the central nervous system. *203*

innervate (in′er-vāt) to activate an organ, muscle, or gland by motor neuron stimulation. *195*

interneuron (in″ter-nu′ron) a neuron that is found within the central nervous system and that takes nerve impulses from one portion of the system to another. *197*

motor neuron (mo′tor nu′ron) a neuron that takes nerve impulses from the central nervous system to the effectors. *195*

myelin (mi′ĕ-lin) the fatty cell membranes that cover long neuron fibers and give them a white glistening appearance. *197*

nerve impulse (nerv im′puls) an electrochemical change due to increased membrane permeability that is propagated along a neuron from the dendrite to the axon following excitation. *198*

neurotransmitter substance (nu″ro-trans′mit′er sub′stans) a chemical made at the ends of axons that is responsible for transmission across a synapse. *200*

parasympathetic nervous system (par″ah-sim″pah-thet′ik ner′vus sis′tem) that part of the autonomic nervous system that usually promotes those activities associated with a normal state. *207*

peripheral nervous system (pĕrif′er-al ner′vus sis′tem) PNS; nerves and ganglia that lie outside the central nervous system. *195*

receptor (re-sep′tor) a sense organ specialized to receive information from the environment. *203*

sensory neuron (sen′so-re nu′ron) a neuron that takes nerve impulses to the central nervous system; afferent neuron. *195*

somatic nervous system (so-mat′ik ner′vus sis′tem) that portion of the PNS containing motor neurons that control skeletal muscles. *203*

sympathetic nervous system (sim′pah-thet′ik ner′vus sis′tem) that part of the autonomic nervous system that usually causes effects associated with emergency situations. *207*

synapse (sin′aps) the region between two nerve cells where the nerve impulse is transmitted from one to the other, usually from axon to dendrite. *200*

CHAPTER ELEVEN

MUSCULO-SKELETAL SYSTEM

CHAPTER CONCEPTS

1 The body's skeleton is divided into the axial and appendicular skeletons.

2 Macroscopically, skeletal muscles work in antagonistic pairs and exhibit certain physiological characteristics related to the fact that muscles are composed of muscle fibers.

3 Microscopically, muscle fiber contraction is dependent on actin and myosin filaments and a ready supply of calcium (Ca^{++}) and ATP.

4 Nerve fibers cause muscle cells to contract, and the sequence of events leading up to contraction is complex.

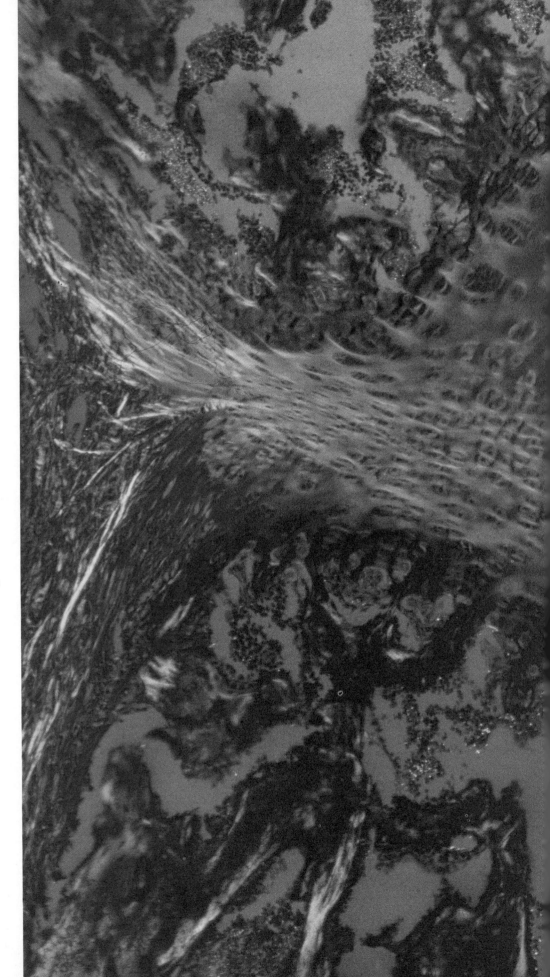

Polarized light micrograph of developing fibrocartilage (100×).

FIGURE 11.1 *Vasily Alexeyev, a Russian weight lifter, strains to lift over 500 pounds.*

INTRODUCTION

Exercise causes a muscle to grow larger—there is an increase in the number and size of the protein filaments within each muscle cell as well as in the amount of myoglobin and the number of mitochondria. Strong muscles also contain more energy sources in the form of glycogen and fat, and they resist fatigue.

Apparently, however, you are born with a certain proportion of slow-twitching versus fast-twitching muscle cells (fibers), and exercise cannot drastically alter the proportion. No wonder, then,

that athletes have their favorite sport at which they perform best. The slow-twitching fibers are generously supplied with mitochondria and capillaries. They are the "dark meat"; they produce a steady pulling power so necessary to swimming, running, and skiing. The fast-twitching fibers aren't as well supplied with blood vessels nor do they have as many mitochondria. They are the "white meat"; they supply bursts of energy for sprinting, weight lifting (fig. 11.1), or swinging a golf club.

T he skeleton, notably the large heavy bones of the legs, supports the body against the pull of gravity. The skeleton also protects soft body parts. For example, the skull forms a protective encasement for the brain, as does the rib cage for the heart and lungs. Flat bones, such as those of the skull, ribs, and breastbone, produce red blood cells in both adults and children. All bones are storage areas for inorganic calcium and phosphorus salts. Bones also provide sites for muscle attachment. The long bones, particularly those of the legs and arms, permit flexible body movement.

The skeleton not only permits flexible movement, it also supports and protects the body, produces red blood cells, and serves as a storehouse for certain inorganic salts.

SKELETON

The skeleton may be divided into two parts: the axial skeleton and the appendicular skeleton (fig. 11.2). The **axial skeleton** lies in the midline of the body and consists of the skull, vertebral column, sternum, and ribs. The **appendicular skeleton** includes the bones of the appendages (arms and legs) plus the girdles, which connect the appendages to the axial skeleton.

Skull

The skull is formed by the cranium and the facial bones.

Cranium

The cranium protects the brain and is composed of eight bones fitted tightly together in adults. In newborns, certain bones are not completely formed and instead are joined by membranous regions called *fontanels,* all of which usually close by the age of 16 months. The bones of the cranium contain the **sinuses,** air spaces lined by mucous membrane, which reduce the weight of the skull and give a resonant sound to the voice. Two sinuses called the mastoid sinuses drain into the middle ear. *Mastoiditis,* a condition that can lead to deafness, is an inflammation of these sinuses.

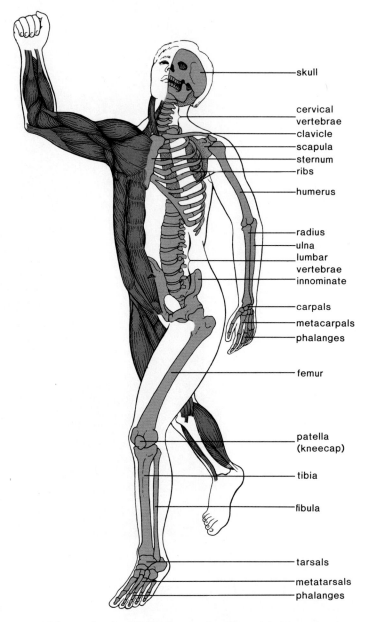

FIGURE 11.2 *Major bones (and muscles) of the human body. The axial skeleton, composed of the skull, vertebral column, sternum, and ribs, lies in the body midline; the rest of the bones belong to the appendicular skeleton.*

The major bones of the cranium have the same names as the lobes of the brain: frontal, parietal, temporal, and occipital. On the top of the cranium (fig. 11.3a), the **frontal bone** forms the forehead, the **parietal bones** extend to the sides, and the **occipital bone** curves to form the base of the skull. Here there is a large opening, the **foramen magnum** (fig. 11.3b), through which the nerve cord passes and becomes the brain stem. Below the much larger parietal bones, each **temporal bone** has an opening that leads to the middle ear. The **sphenoid bone** not only completes the sides of the skull, it also contributes to the floors and walls of the eye sockets. Likewise, the **ethmoid bone,** which lies in front of the sphenoid, is a part of the orbital wall and, in addition, is a component of the nasal septum.

FIGURE 11.3 *Skull. a. Lateral view.*
b. *Inferior view.* c. *Facial view.*

parietal bone

frontal bone

sphenoid bone

ethmoid bone

temporal bone

lacrimal bone

occipital bone

nasal bone

zygomatic bone

external auditory canal

mastoid process

maxilla

mandible

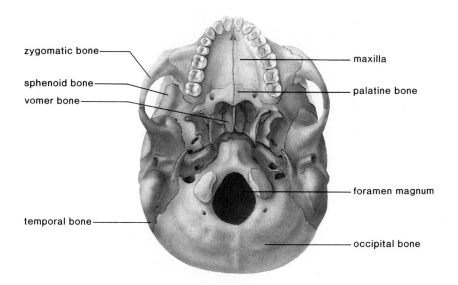

zygomatic bone

maxilla

sphenoid bone

palatine bone

vomer bone

temporal bone

foramen magnum

occipital bone

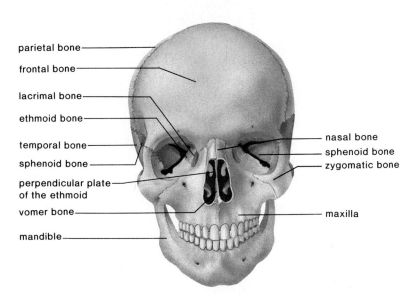

parietal bone

frontal bone

lacrimal bone

ethmoid bone

nasal bone

temporal bone

sphenoid bone

sphenoid bone

zygomatic bone

perpendicular plate
of the ethmoid

vomer bone

maxilla

mandible

The cranium contains eight bones: the frontal, two parietal, the occipital, two temporal, the sphenoid, and the ethmoid.

Facial Bones

The **mandible,** or lower jaw, is the only movable portion of the skull (fig. 11.3c), and its action permits us to chew our food. Tooth sockets are located on this bone and on the **maxillae** (maxillary bones), the upper jaw that also forms the anterior portion of the hard palate. The **palatine bones** make up the posterior portion of the hard palate and the floor of the nasal cavity. The **zygomatic bones** give us our cheekbone prominences, and the nasal bones form the bridge of the nose. Each thin, scalelike **lacrimal bone** lies between an ethmoid bone and a maxillary bone, and the thin, flat **vomer** joins with the perpendicular plate of the ethmoid to form the nasal septum.

The facial bones include the mandible, two maxillae, two palatine, two zygomatic, two lacrimal, two nasal, and the vomer.

Vertebral Column

The **vertebral column** extends from the skull to the pelvis. Normally, the vertebral column has four curvatures that provide more resiliency and strength than a straight column could. The various vertebrae are named according to their location in the body (fig. 11.4). A typical vertebra is shown in figure 11.5. The pedicles and the lamina form a *vertebral arch* about the vertebral foramen (*pl.* foramina). When the vertebrae join, these foramina form a canal through which the nerve cord passes. The *spinous processes* of the vertebrate can be felt as bony projections along the midline of the back.

There are *discs* between the vertebrae that act as a kind of padding. They prevent the vertebrae from grinding against one another and absorb shock caused by movements such as running, jumping, and even walking. Unfortunately, these discs become weakened with age and can slip or even rupture. This causes pain when the damaged disc presses up against the spinal cord and/or spinal nerves. The body may heal itself, or else the disc can be removed surgically. If the latter occurs, the vertebrae can be fused together, but this will limit the flexibility of the body. The presence of the discs allows motion between the vertebrae so that we can bend forward, backward, and from side to side.

The vertebral column, directly or indirectly, serves as an anchor for all the other bones of the skeleton (fig. 11.2). All of the twelve pairs of **ribs** connect directly to the thoracic vertebrae in the back, and all but two pairs connect either directly or indirectly via shafts of cartilage to the **sternum** (breastbone) in the front. The lower two pairs of ribs are called "floating ribs" because they do not attach to the sternum.

FIGURE 11.4 *Curvatures of the spine. The vertebrae are named for their location in the body. The spinal nerves leave the spine at the intervertebral foramina.*

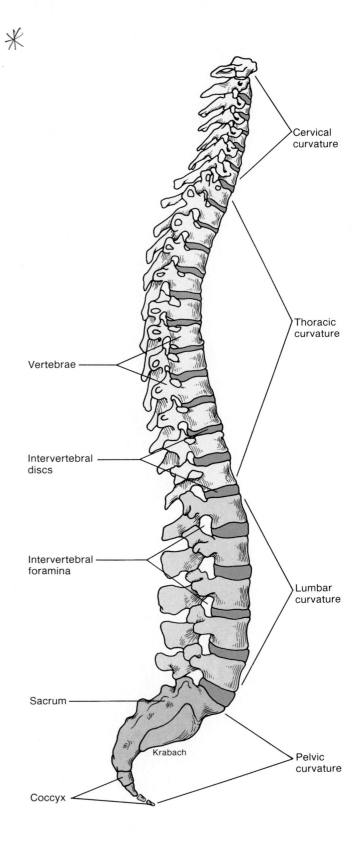

FIGURE 11.5 *Parts of a typical vertebra.*

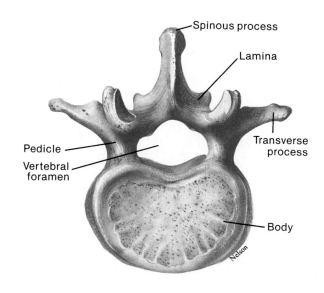

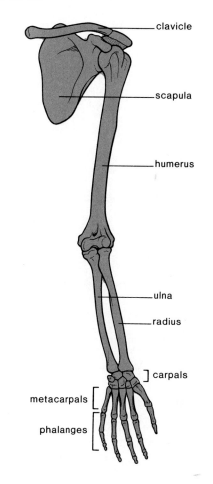

The vertebral column contains the vertebrae and serves as the backbone for the body.

Appendicular Skeleton

The appendicular skeleton consists of the bones within the pectoral and pelvic girdle and the attached appendages. The pectoral (shoulder) girdle and appendages (arms and hands) are specialized for flexibility, but the pelvic girdle (hipbones) and appendages (legs and feet) are specialized for strength.

Pectoral Girdle

The components of the **pectoral girdle** (fig. 11.6) are loosely linked together by ligaments. Each **clavicle** (collarbone) connects with the sternum in front and the **scapula** (shoulder blade) behind, but the scapula is freely movable and held in place only by muscles. This allows it to follow freely the movements of the arm. The single long bone in the upper arm (fig. 11.6), the **humerus,** has a smoothly rounded head that fits into a socket of the scapula. The socket, however, is very shallow and much smaller than the head. Although this means that the arm can move in almost any direction, there is little stability. Therefore, this is the joint that is most apt to dislocate. The opposite end of the humerus meets the two bones of the lower arm, the **ulna** and the **radius,** at the elbow. (The prominent bone in the elbow is the topmost part of the ulna.) When the arm is held so that the palm is turned frontward, the radius and ulna are about parallel to one another. When the arm is turned so that the palm is next to the body, the radius crosses in front of the ulna, a feature that contributes to the easy twisting motion of the lower arm.

The many bones of the hand increase its flexibility. The wrist has eight **carpal** bones, which look like small pebbles. From these, five **metacarpal** bones fan out to form a framework for the palm. The metacarpal bone that leads to the thumb is placed in such a way that the thumb can reach out and touch the other digits. (**Digits** is a term that refers to either fingers or toes.) Beyond the metacarpals are the **phalanges,** the bones of the fingers and the thumb. The phalanges of the hand are long, slender, and lightweight.

Pelvic Girdle

The **pelvic girdle,** or pelvis, (fig. 11.7) consists of two heavy, large **innominate** bones (hipbones). The innominate bones are anchored to the sacrum, and together these bones form a hollow cavity that is wider in females than males. The weight of the body is transmitted through the pelvis to the legs and then onto the ground. The largest bone in the body is the **femur,** or thighbone. Although the femur is a strong bone, it is doubtful that the femurs of a fairy-tale giant could support the increase in weight. If a giant were ten times taller than an ordinary human being, he would also be about ten times

FIGURE 11.7 *The bones of the left portion of the pelvic girdle, left leg, and left foot. The femur is our strongest bone and withstands a pressure of 1,200 pounds per cubic inch when we walk.*

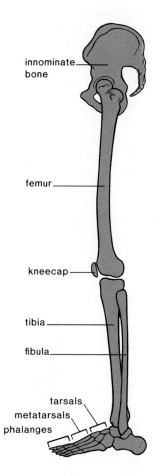

innominate bone

femur

kneecap

tibia

fibula

tarsals
metatarsals
phalanges

TABLE 11.1 *BONES OF THE APPENDICULAR SKELETON*

PART	BONES
Pectoral girdle	Clavicle
	Scapula
Arm	Humerus
	Ulna
	Radius
Hand	Carpals
	Metacarpals
	Phalanges
Pelvic girdle	Innominate
Leg	Femur
	Tibia
	Fibula
Foot	Tarsals
	Metatarsals
	Phalanges

The appendicular skeleton contains the bones that are listed in table 11.1.

Long Bones

A long bone, such as the femur, illustrates principles of bone anatomy. When the bone is split open, as in figure 11.8, the longitudinal section shows that it is not solid, but has a cavity bounded at the sides by compact bone and at the ends by spongy bone. Beyond the spongy bone there is a thin shell of compact bone and finally a layer of cartilage.

Compact bone, as discussed previously (p. 57), contains bone cells in tiny chambers called lacunae, arranged in concentric circles around Haversian canals. The Haversian canals contain blood vessels and nerves. The lacunae are separated by a matrix that contains protein fibers of collagen and mineral deposits, primarily of calcium and phosphorus salts.

Spongy bone contains numerous bony bars and plates separated by irregular spaces. Although lighter than compact bone, spongy bone is still designed for strength. Just as braces are used for support in buildings, the solid portions of spongy bone follow lines of stress. The spaces in spongy bone are often filled with **red marrow,** a specialized tissue that produces red blood cells. The cavity of a long bone usually contains **yellow marrow,** which is a fat-storage tissue.

Bones are not inert. They contain cells and perform various functions, including the production of blood cells.

wider and thicker, making him weigh about one thousand times as much. This amount of weight would break even giant-size femurs.

In the lower leg the larger of the two bones, the **tibia** (fig. 11.7), has a ridge we call the shin. Both of the bones of the lower leg have a prominence that contributes to the ankle—the tibia on the inside of the ankle and the **fibula** on the outside of the ankle. Although there are seven **tarsal** bones in the ankle, only one receives the weight and passes it on to the heel and the ball of the foot. If you wear high-heeled shoes, the weight is thrown even further toward the front of your foot. The **metatarsal** bones form the arches of the foot. There is a longitudinal arch from the heel to the toes and a transverse arch across the foot. These provide a stable, springy base for the body. If the tissues that bind the metatarsals together become weakened, flatfeet are apt to result. The bones of the toes are called *phalanges,* just like those of the fingers, but in the foot the phalanges are stout and extremely sturdy.

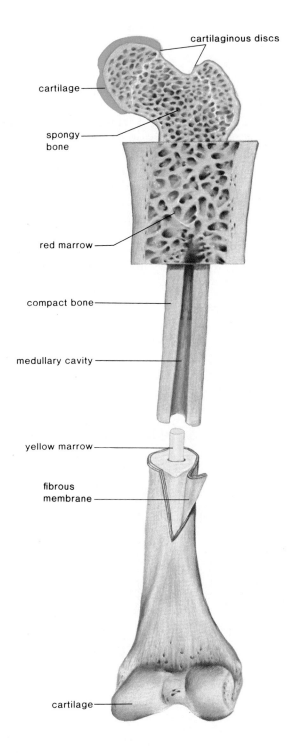

FIGURE 11.8 *Anatomy of a long bone. A long bone is encased by fibrous membrane except at the ends where it is covered by cartilage. The central shaft is composed of compact bone, but the ends are spongy bone, which can contain red marrow. A central medullary cavity contains yellow marrow.*

cartilaginous discs

cartilage

spongy bone

red marrow

compact bone

medullary cavity

yellow marrow

fibrous membrane

cartilage

Growth and Development

Most of the bones of the skeleton are cartilaginous during prenatal development. Later the cartilage is replaced by bone due to the action of bone-forming cells known as **osteoblasts.** At first, there is only a primary ossification center at the middle of a long bone, but later secondary centers form at the ends of the bones. There remains a *cartilaginous disc* between the primary ossification center and each secondary center. The length of a bone is dependent on how long the cartilage cells within the disc continue to divide. Eventually, though, the discs disappear, and the bone stops growing as the individual attains adult height.

In the adult, bone is continually being broken down and built up again. Bone-absorbing cells, called **osteoclasts,** are derived from cells carried in the bloodstream. As they break down bone, they remove worn cells and deposit calcium in the blood. Apparently after a period of about three weeks they disappear. The destruction caused by the work of osteoclasts is repaired by osteoblasts. As they form new bone, they take calcium from the blood. Eventually some of these cells get caught in the matrix they secrete and are converted to **osteocytes,** the cells found within Haversian systems (p. 57).

Because of continual renewal, the thickness of bones can change according to the amount of physical use or due to a change in certain hormone balances (see chapter 13). In most adults, the bones become weaker due to a loss of mineral content. Strange as it may seem, adults seem to require more calcium in the diet than do children in order to promote the work of osteoblasts. Many older women, due to a lack of estrogen, suffer from *osteoporosis,* a condition in which weak and thin bones cause aches and pains and tend to fracture easily. A tendency toward osteoporosis may also be augmented by lack of exercise and too little calcium in the diet.

Bone is a living tissue, and it is always being rejuvenated.

Joints

Bones are joined together at the joints, which are often classified according to the amount of movement they allow. Some bones, such as those that make up the cranium, are sutured together and are *immovable.* Other joints are *slightly movable,* such as the joints between the vertebrae. The vertebrae are separated by discs, described earlier, that increase their flexibility. Similarly, the two hipbones are slightly movable where they are ventrally joined by cartilage. Owing to hormonal changes, this joint becomes more flexible during late pregnancy, which allows the pelvis to expand during childbirth. Most joints are *freely movable* **synovial joints,** in which the two bones are separated by a cavity. **Ligaments** composed

FIGURE 11.9 *The knee joint, an example of a freely movable synovial joint. Notice that there is a cavity between the bones that is encased by ligaments and lined by synovial membrane. The kneecap protects the joint.*

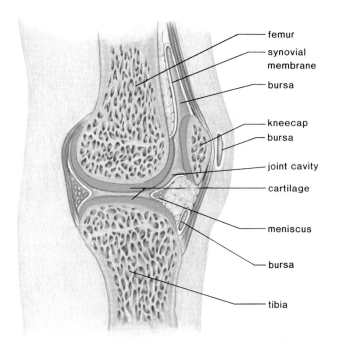

- femur
- synovial membrane
- bursa
- kneecap
- bursa
- joint cavity
- cartilage
- meniscus
- bursa
- tibia

of fibrous connective tissue bind the two bones to one another, holding them in place as they form a capsule. In a "double-jointed" individual, the ligaments are unusually loose. The joint capsule is lined by synovial membrane, which produces *synovial fluid,* a lubricant for the joint. The knee is an example of a synovial joint (fig. 11.9). In the knee, as in other freely movable joints, the bones are capped by cartilage, although there are also crescent-shaped pieces of cartilage between the bones called **menisci.** These give added stability, helping to support the weight placed on the knee joint. Unfortunately, athletes often suffer injury of the menisci, known as torn cartilage. The knee joint also contains thirteen fluid-filled sacs called bursae, which ease friction between tendons and ligaments and between tendons and bones. Inflammation of the bursae is called bursitis. Tennis elbow is a form of bursitis.

There are different types of movable joints. The knee and elbow joints are *hinge joints* because, like a hinged door, they largely permit movement in one direction only. More movable are the ball-and-socket joints; for example, the ball of the femur fits into a socket on the hipbone. *Ball-and-socket joints* allow movement in all planes and even a rotational movement. The various movements of body parts at joints are depicted in figure 11.10.

FIGURE 11.10 *Movement of body parts in relation to joints. Muscles are attached to bones across a joint. One of these bones remains still and the other moves.*

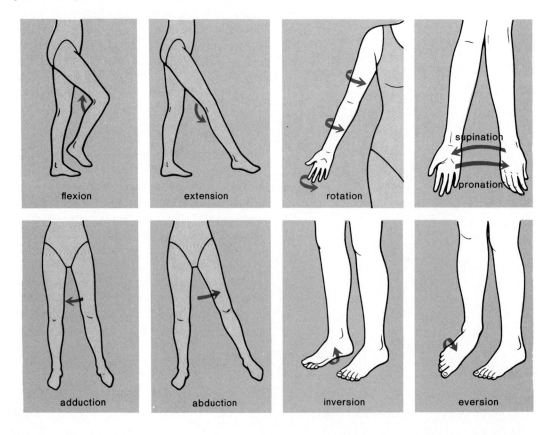

flexion

extension

rotation

supination
pronation

adduction

abduction

inversion

eversion

Synovial joints are subject to *arthritis*. In rheumatoid arthritis, the synovial membrane becomes inflamed and grows thicker. Degenerative changes take place that make the joint almost immovable and painful to use. There is evidence that these effects are brought on by an autoimmune reaction. In old-age arthritis, or osteoarthritis, the cartilage at the ends of the bones disintegrates so that the two bones become rough and irregular. This type of arthritis is apt to affect the joints that have received the greatest use over the years.

Joints are classified according to the degree of movement. Some joints are immovable, some are slightly movable, and some are freely movable.

SKELETAL MUSCLES

Muscles are effectors; they enable the organism to respond to a stimulus (p. 206). Skeletal muscles are attached to the skeleton, and their contraction accounts for voluntary movements. Involuntary muscles, both smooth and cardiac, were discussed previously on page 58. Here we will divide our discussion of skeletal muscle into macroscopic anatomy and physiology and microscopic anatomy and physiology.

Macroscopic Anatomy and Physiology

Muscles are typically attached to bone by **tendons** made of fibrous connective tissue. Tendons extend from a muscle to the far side of joints (fig. 11.11). When the central portion of the muscle, called the belly, contracts, one bone remains fairly stationary and the other one moves. The **origin** of the muscle is on the stationary bone, and the **insertion** of the muscle is on the bone that moves.

When a muscle contracts, it shortens. Therefore, muscles can only pull; they cannot push. Because opposite motions are needed at a joint, muscles generally work in *antagonistic pairs*. For example, the biceps and triceps are a pair of muscles that move the lower arm up and down (fig. 11.11).

FIGURE 11.11 *Attachment of skeletal muscles as exemplified by the biceps and triceps. The origin of a muscle remains stationary, while the insertion moves. These muscles are antagonistic. When the biceps contracts, the lower arm is raised, and when the triceps contracts, the lower arm is lowered.*

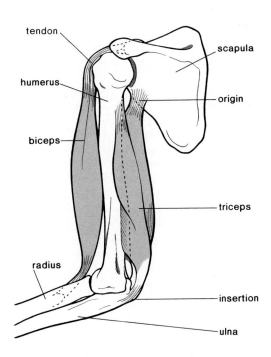

EXERCISE: CHANGING PERCEPTIONS

In the 1970s, reams of scientific studies showing that vigorous activity helped stave off heart disease were condensed to the battle cry "Exercise!" . . . Today, exercise physiologists are busy dismantling some of the myths that arose from earlier interpretations of their research. Exercise, they now stress, need not be jarring or exhausting to produce health benefits. Walking a few flights of stairs up to your office each day, mowing the lawn with a push mower, and parking your car at the far end of the lot can represent real strides toward a healthier heart.

Although research has mostly focused on how exercise helps prevent heart disease, the benefits don't stop there. Even moderately increasing your activity level can yield important, almost immediate benefits to your life now, at whatever age. It can help teenage girls keep their weight down at a time when fat cells may be multiplying; it can ease backache in office-bound workers; and it can help restore bone mass to older people with osteoporosis.

Ironically, the fitness craze that began in the 1970s (and in which only 15% of Americans participated in the first place) "turned off the very people who needed it most," says Dr. J. W. Yates, an exercise physiologist at the University of Louisville, Kentucky. "They figured if they couldn't meet the requirements, they might as well not bother."

That, it turns out, was a very unhealthy assumption. A study published in the *Journal of the American Medical Association* in 1987 showed that men who spent an hour a day doing moderate activities—things like walking, yard work, and home repairs, as well as weekend bicycling, fishing, and dancing—cut their risk of fatal heart attack by one-third.

Another study by Dr. Henry Miller of the Bowman Gray School of Medicine in Winston-Salem, N.C., showed that heart patients benefited just as much from an exercise program that elevated the heart rate to only 45% of its capacity as from a program that elevated the heart rate to 70% of its capacity. Another of Miller's studies showed that when sedentary people began a vigorous exercise program—jogging, calisthenics, and the like—50% of them dropped out within two months because of injuries. But when a similar group started out with a low-impact program—like stationary bicycling—the whole group stuck with the program for three months.

This new data is causing numerous exercise experts to question or revise old fitness guidelines, re-evaluate "thresholds" for gaining benefits, and focus more on physical activity—just moving around—than regimented exercise alone.

"It's fair to say epidemiology tends now to suggest the dose [of exercise] required for health benefits is less than the standard we've been prescribing," says Pate of his study for the American College of Sports Medicine. "Our whole field is confronted with the task of developing guidelines that are more appropriate." . . .

If a single researcher could be considered responsible for the fitness craze, it would be Dr. Ralph Paffenbarger, a specialist in heart disease and exercise with posts at both Stanford and Harvard Universities. His landmark research, involving 17,000 Harvard alumni and 6,000 San Francisco longshoremen, showed that the men who exercised vigorously and burned at least 2,000 calories a week doing so cut their risk of dying from heart disease by half. The reduction was dramatic even if the men smoked or if their parents had both died of heart disease.

The problem was the public—and many health professionals—interpreted these findings as meaning that anything less didn't do any good. If you weren't up for running 20 miles a week, or an hour of tennis five times a week, or cross-country skiing for a half hour a day—well, you'd be just as well off sitting in front of the tube with a beer. The 2,000 calorie threshold was transformed into a "magic number"— above it, you were fit. Below it, you were a basket case.

Paffenbarger says he never believed that the 2,000-calorie-figure was carved in stone. And neither did Dr. Arthur Leon, an epidemiologist at University of

Minnesota's School of Public Health. But both were surprised when Leon's research showed that a far more modest level of activity could exert a very powerful protective effect on the heart.

"People who don't want to do formal, sweaty exercise can be told that less can be beneficial," says Leon. "Just moving around more is a big help." He backs up his assertion with the results of his study, published in the *Journal of the American Medical Association* [in the fall of 1987]. That study analyzed the realtionship between heart attack and physical activity off the job in 12,138 men who participated in the Multiple Risk Factor Intervention Trial, a nationwide study conducted at twenty-two medical centers.

It found that men who were only moderately active—spending an average of 48 minutes a day on leisure-time physical activity—had one-third fewer heart attacks than their peers who moved around during leisure time an average of 16 minutes each day. And the moderately active group didn't spend all, or even most, of their exercise time huffing and puffing. Mostly, the report

found, their activities were in the light-to-moderate range: lawn and garden work, bowling, ballroom dancing—activities we don't often even think of as exercise.

Indeed, some of the most dramatic gains are made by the sedentary folks whose initial efforts fall short of the magic 2,000 calories and whose activities never include much bouncing around. For example, after only four months of attending a twice-weekly, low-impact aerobics class at Northwestern University, men and women with rheumatoid arthritis—a chronic, severe form of joint disease—reported much less pain, swelling, fatigue, and depression than before they began exercising.

"The lower end of the spectrum gets the most benefit from its effort," says cardiologist Miller. "If you've just had a heart transplant, it takes very little—a walk around the gym—to get improvement. If you take a sedentary office worker it doesn't take much more than adding a mile of walking a day to get improvement."

continued

The enclosed shopping mall, intended as a boon for buyers, has now also become a haven for fitness walkers. Climate controlled, its passages unimpeded by curbs or stoplights, the mall provides a perfect environment for those determined to put in their daily mileage, rain or shine, while eliminating many of the outdoor hazards that may deter older pedestrians. Some mall walkers, like the Galleria Mall GoGetters in Glendale, California have formed their own clubs. A few malls now issue special walking maps, while others will open on holidays, even when the stores are closed, just to accommodate the local ramblers.

EXERCISE: CHANGING PERCEPTIONS
continued

Leon and Paffenbarger point out that it's relatively easy to program greater amounts of activity into your normal day. Walk down to the accounting department at work instead of calling the accountant. Take stairs instead of elevators. Stroll during your lunchtime. A brisk walk—especially if you swing your arms—can get your heart rate up without the jouncing of jogging that many people dislike.

Leon's study showed that the benefits of exercise start to level off at a certain point—at least as far as fatal heart attack is concerned. The rate of fatal heart attack among the moderately active men was the same as the most active men in the study, who devoted an average of 134 minutes a day to leisure-time physical activity.

Does this mean the most active group was no more fit than the moderately active group? Of course not. Fitness, stresses Bud Getchell, executive director of the nonprofit National Institute for Fitness and Sport in Indianapolis, includes flexibility, muscular strength, endurance, and body composition (lean mass vs. fat) as well as cardiovascular health. To achieve the type of fitness that would allow you to take on most vigorous activities with ease—from lifting a heavy box at work, to shoveling snow without hurting your back, to enjoying a game of pickup basketball—most fitness experts say you should spend about three hours (see table A),

spaced out during the week, doing activities that strengthen muscles and enhance flexibility, as well as challenge the heart.

Your body will reward your efforts. Although most people think the main benefit of fitness is reducing the risk of heart disease, there's mounting evidence that other tissues and organs benefit as well: Several studies at the Institute for Aerobics Research in Dallas, Harvard, and other institutions have found that more active people have lower rates of colon, brain, kidney, and reproductive cancers, as well as leukemia, than their more sedentary counterparts. And studies that looked at exercise along with other variables found this still held true when factors like age, diet, and socioeconomic background were taken into account. Researchers speculate this may be because activity promotes the delivery of more nutrients and oxygen to these organs and tissues.

Indeed, according to Dr. Everett L. Smith, director of the Biogerontology Laboratory at the University of Wisconsin–Madison, half of the functional decline between ages 30 and 70 could be prevented if we simply used our bodies more. Bone density, nerve function, and kidney efficiency, as well as overall strength and flexibility, can be largely preserved into our later years simply by keeping up an active life.

First appeared in *The Boston Globe* May 1, 1988. Copyright © 1988 Sy Montgomery. Reprinted by permission.

TABLE A *A CHECKLIST FOR STAYING FIT*

CHILDREN, 7–12	TEENAGERS, 13–18	ADULTS, 19–55	SENIORS, 55 AND UP
Vigorous activity 1 to 2 hours daily	Vigorous activity 3 to 5 times a week	Vigorous activity for one-half hour, 3 times a week	Moderate exercise 3 times a week
Free play	Build muscle with calisthenics	Exercise to prevent lower back pain: aerobics, stretching, yoga	Plan a daily walk
Build motor skills through team sports, dance, swimming	Plan aerobic exercise to control buildup of fat cells	Take active vacations: hike, bicycle, cross-country ski	Daily stretching exercises
Encourage more exercise outside of physical education classes	Pursue tennis, swimming, riding—sports that can be enjoyed for a lifetime	Find exercise partners: join running club, bicycle club, outing group	Learn a new sport: golf, fishing, ballroom dancing
Initiate family outings: bowling, boating, camping, hiking	Continue team sports, dancing, hiking, swimming		Try low-impact aerobics
			Before undertaking new exercises, consult your doctor

FIGURE 11.12 *Ventral view of superficial skeletal muscles.*

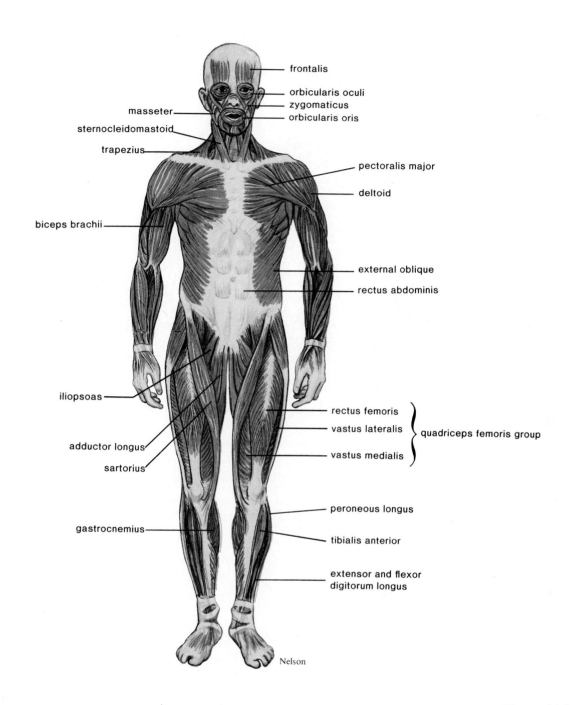

When the *biceps* contracts, the lower arm flexes, and when the *triceps* contracts, the lower arm extends.

Figures 11.12 and 11.13 depict and tables 11.2 and 11.3 list various superficial muscles of the body. Note these antagonistic pairs:

Sternocleiodomastoid	Flexes head
Trapezius	Extends head
Gluteus medius	Abducts thigh
Adductor longus	Adducts thigh

Iliopsoas	Flexes thigh
Gluteus maximus	Extends thigh
Quadriceps femoris group	Extends lower leg
Hamstring group	Flexes lower leg
Tibialis anterior	Inverts foot
Peroneous longus	Everts foot

FIGURE 11.13 *Dorsal view of superficial skeletal muscles.*

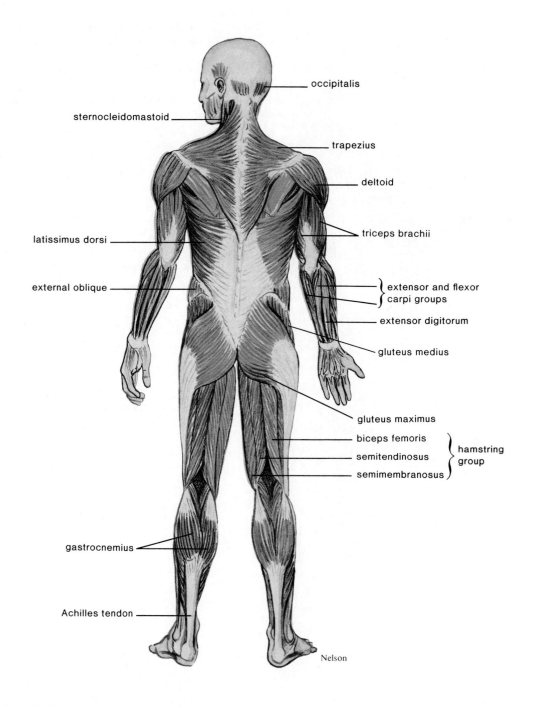

occipitalis

sternocleidomastoid

trapezius

deltoid

latissimus dorsi

triceps brachii

external oblique

extensor and flexor carpi groups

extensor digitorum

gluteus medius

gluteus maximus

biceps femoris

semitendinosus

semimembranosus

hamstring group

gastrocnemius

Achilles tendon

Nelson

Typically muscles are attached across a joint. When a muscle contracts, the movable end (insertion) is pulled toward the fixed end (origin), and movement occurs at the joint. This system requires that muscles work in antagonistic pairs.

In the laboratory, it is possible to study the contraction of individual whole muscles. Customarily, muscle tissue is removed from a frog, and its activity is recorded by an apparatus called a **physiograph** (fig. 11.14). The muscle is stimulated electrically, and when it contracts, it pulls on a lever. The lever's movement is recorded, and the resulting pattern is called a **myogram.**

TABLE 11.2 MUSCLES (VENTRAL VIEW)

NAME	FUNCTION
Head and Neck	
Frontalis	Wrinkles forehead and lifts eyebrows
Orbicularis oculi	Closes eye (winking)
Zygomaticus	Raises corner of mouth (smiling)
Masseter	Closes jaw
Orbicularis oris	Closes and protrudes lips (kissing)
Arm and Trunk	
External oblique	Compresses abdomen; rotates trunk
Rectus abdominis	Flexes spine
Pectoralis major	Flexes and adducts shoulder and arm ventrally (pulls arm across chest)
Deltoid	Abducts and raises arm at shoulder joint
Biceps brachii	Flexes forearm and supinates hand
Legs	
Adductor longus	Adducts thigh
Iliopsoas	Flexes thigh
Sartorius	Rotates thigh (sitting cross-legged)
Quadriceps femoris group	Extends lower leg
Peroneous longus	Everts foot
Tibialis anterior	Dorsiflexes and inverts foot
Flexor and extensor digitorum longus groups	Flex and extend toes

TABLE 11.3 MUSCLES (DORSAL VIEW)

NAME	FUNCTION
Head and Neck	
Occipitalis	Moves scalp backward
Sternocleidomastoid	Turns head to side; flexes neck and head
Trapezius	Extends head; raises and adducts shoulders dorsally (shrugging shoulders)
Arm and Trunk	
Latissimus dorsi	Extends and adducts shoulder and arm dorsally (pulls arm across back)
Deltoid	Abducts and raises arm at shoulder joint
External oblique	Rotates trunk
Triceps brachii	Extends forearm
Flexor and extensor carpi groups	Flex and extend hand
Flexor and extensor digitorum groups	Flex and extend fingers
Buttocks and Legs	
Gluteus medius	Abducts thigh
Gluteus maximus	Extends thigh (forms buttock)
Hamstring group	Flexes lower leg
Gastrocnemius	Extends foot (tiptoeing)

All-or-None Response

Either a *single* muscle fiber responds to a stimulus and contracts, or it does not. At first, the stimulus may be so weak that no contraction occurs, but as soon as the strength of the stimulus reaches the *threshold stimulus,* the muscle fiber will contract completely. Therefore, a muscle fiber obeys the **all-or-none law.**

Contrary to that of an individual fiber, the strength of contraction of a whole muscle can increase according to the degree of stimulus beyond the *threshold stimulus.* A whole muscle contains many fibers, and the degree of contraction is dependent on the total number of fibers contracting. The maximum stimulus is the one beyond which the degree of contraction does not increase.

Muscle fibers obey the all-or-none law, but whole muscles do not obey this law.

FIGURE 11.14 *Physiograph. This apparatus can be used to record a myogram, a visual representation of the contraction of a muscle that has been dissected from an animal.*

Muscle Twitch

If a muscle is placed in a physiograph and given a maximum stimulus, it will contract and then relax (fig. 11.15). This action—a single contraction that lasts only a fraction of a second—is called a muscle **twitch.** Figure 11.15a is a myogram of a twitch, which is customarily divided into a *latent period,* or the period of time between stimulation and initiation of contraction; the period of *contraction;* and the period of *relaxation.*

If a muscle is exposed to two maximum stimuli in quick succession, it will respond to the first but not to the second stimulus. This is because it takes an instant following a contraction for the muscle fibers to recover in order to respond to the next stimulus. The very brief moment following stimulation during which a muscle remains unresponsive is called the refractory period.

Summation and Tetanus

If a muscle is given a rapid series of threshold stimuli, it may respond to the next stimulus without relaxing completely. In this way, muscle tension **summates** until maximal sustained **tetanus** is achieved (fig. 11.15b). The myograms no longer show individual contractions; rather, they are completely fused and blended into a straight line. Tetanus continues until

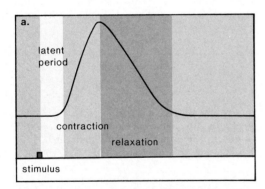

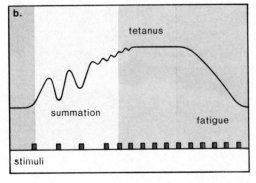

FIGURE 11.15 Physiology of muscle contraction. a. Simple muscle twitch is composed of three periods: latent, contraction, and relaxation. b. Summation and tetanic contraction. When a muscle is not allowed to relax completely between stimuli, the contractions increase in size and then the muscle remains contracted until it fatigues.

the muscle fatigues, due to depletion of its energy reserves. **Fatigue** is apparent when a muscle relaxes even though stimulation is continued.

Tetanus occurs whenever skeletal muscles are being actively used. Ordinarily, however, only a portion of any particular muscle is involved—while some fibers are contracting, others are relaxing. Because of this, intact muscles rarely fatigue completely.

Muscle twitch, summation, and tetanus are related to the frequency with which a muscle is stimulated.

Muscle Tone

Intact skeletal muscles also have **tone,** a condition in which there are always some fibers contracted. Muscle tone is particularly important in maintaining posture. If the muscles of the neck, trunk, and legs suddenly become relaxed, the body collapses.

The maintenance of the right amount of tone requires the use of special sense receptors called **muscle spindles** (fig. 11.16). A muscle spindle consists of a bundle of modified muscle fibers with sensory nerve fibers wrapped around a short, specialized region somewhere near the middle of their length. A spindle contracts along with muscle fibers, but thereafter it sends stimuli to the CNS that enable the CNS to regulate muscle contraction so that tone is maintained.

Effect of Contraction on Size of Muscle

As mentioned previously, forceful muscular activity over a prolonged period of time causes muscles to increase in size. This increase, called *hypertrophy,* occurs only if the muscle contracts to at least 75% of its maximum tension. However, only a few minutes of forceful exercise a day are required for hypertrophy to occur. The number of muscle fibers do not increase; instead, the size of each fiber increases (table 11.4). The fibers show a gain in metabolic potential as well as in the number of myofibrils. This means that the muscle can work longer before it gets tired. Some athletes take steroids, either testosterone or related chemicals, to promote muscle growth.

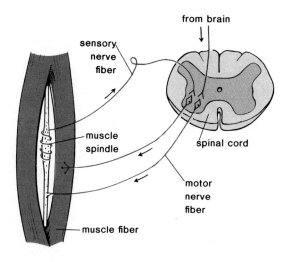

TABLE 11.4 *CHARACTERISTICS OF SKELETAL MUSCLE*

Weight: 40% of total body weight

Total number of muscles in body: Over 600

Number of fibers (cells) in a muscle: A few hundred to over a million

Size of muscle cell: 10–100 μm diameter; up to 20 cm long

Size of filaments: Thick = 15 nm diameter; thin = 5 nm diameter

Time of contraction: Fast fiber = 10 msec; slow fiber = 100 msec

Time for contraction-relaxation cycle: 0.15 second

David W. Deamer: *Being Human.* Copyright © 1981 Holt Rinehart and Winston, College Division.

When muscles are not used or are used for only very weak contractions, they decrease in size, or atrophy. Atrophy can occur when a limb is placed in a cast or when the nerve serving a muscle is damaged. If nerve stimulation is not restored, the muscle fibers will gradually be replaced by fat and fibrous tissue. Unfortunately, atrophy causes the fibers to shorten progressively, leaving body parts in contorted positions.

Microscopic Anatomy and Physiology

A whole skeletal muscle is composed of muscle fibers (fig. 11.17a and b). Each fiber is a cell and contains the usual cellular components, but special terminology has been assigned to certain ones, as indicated in table 11.5. Also, a muscle fiber has some unique anatomical characteristics. For one thing, the **sarcolemma,** or cell membrane, forms tubules that penetrate or dip down into the cell so that they come into contact, but do not fuse, with expanded portions of modified ER, which is termed the **sarcoplasmic reticulum** in muscle cells. The tubules comprise the *T* (for transverse) *system* (fig. 11.18). The expanded portions of the sarcoplasmic reticulum, called *calcium-storage sacs,* contain calcium, Ca^{++}, an element that is essential to muscle contraction. The sarcoplasmic reticulum encases hundreds, and sometimes even thousands, of **myofibrils,** which are the contractile portions of the fibers. Myofibrils are cylindrical in shape and run the length of the fiber. The light microscope shows that the myofibrils have light and dark bands called striations (figs. 11.17c and 11.17d). It is the banding pattern of myofibrils that cause skeletal muscle to be striated. Electron micrographs (figs. 11.18 and 11.19a) have revealed that there are even areas of light and dark within the bands themselves. These areas can be studied in relation to a unit of a myofibril called a **sarcomere.**

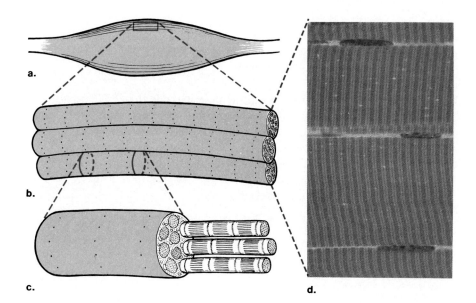

FIGURE 11.17 *Anatomy of a muscle as revealed by the light microscope.*
a. *Whole muscle.* b. *Muscle fibers.*
c. *Single muscle fiber with a few exposed myofibrils.* d. *Light micrograph shows the striations characteristic of skeletal muscle. The striations are due to the light and dark banding pattern of myofibrils.*

TABLE 11.5 MUSCLE CELLS	
COMPONENT	**TERM**
Cell membrane	Sarcolemma
Cytoplasm	Sarcoplasm
Endoplasmic reticulum	Sarcoplasmic reticulum

FIGURE 11.18 Anatomy of a muscle fiber as revealed by the electron microscope. A muscle fiber contains numerous myofibrils, each of which is enclosed by sarcoplasmic reticulum. The sarcolemma forms tubules that dip down and come in contact with the sarcoplasmic reticulum.

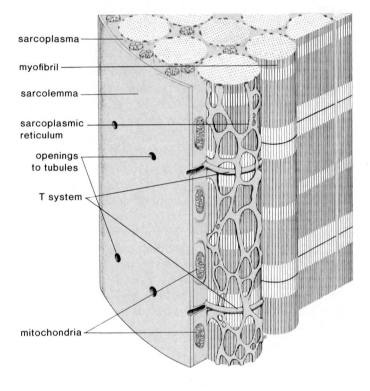

- sarcoplasma
- myofibril
- sarcolemma
- sarcoplasmic reticulum
- openings to tubules
- T system
- mitochondria

Myofibrils are the contractile portions of muscle fibers.

Sarcomere Structure and Function

A sarcomere extends between two dark lines called the *Z lines* (fig. 11.19b). The *I band* is a light region that takes in the sides of two sarcomeres; therefore, an I band includes a Z line. The center dark region of a sarcomere is called the *A band*. The A band is interrupted by a light center, the *H zone*, and a fine, dark stripe called the *M line* cuts through the H zone.

These bands and zones relate to the placement of filaments within each sarcomere. A sarcomere contains two types of filaments, *thin filaments* and *thick filaments* (fig. 11.19b). The thin filaments are attached to a Z line, and the thick filaments are anchored by an M line. The I band is light because it contains only thin filaments; the A band is dark because it contains both thin and thick filaments, except at the center where, in the lighter H zone, only thick filaments are found.

Sarcomere contraction is dependent on two proteins, **actin** and **myosin,** that make up the thin and thick filaments respectively (table 11.6). When a sarcomere contracts (fig. 11.19c), the actin filaments slide past the myosin filaments and approach one another. This causes the I band to become shorter and the H zone to almost or completely disappear. The movement of actin in relation to myosin is called the **sliding filament theory** of muscle contraction. During the sliding process, the sarcomere shortens even though the filaments themselves remain the same length (fig. 11.20).

The overall formula for muscle contraction can be represented as follows:

$$ATP \longrightarrow ADP + \textcircled{P}$$
$$actin + myosin \xrightarrow{} actomyosin$$
$$Ca^{++}$$

The participants in this reaction have the functions listed in table 11.7. Even though the actin filaments slide past the myosin filaments, it is the latter that does the work. In the presence of Ca^{++}, portions of the myosin filaments, called *cross bridges* (fig. 11.20b), reach out and attach to the actin filaments, pulling them along. Following attachment, ATP is broken down as detachment occurs. Myosin brings about ATP breakdown, and therefore it is not only a structural protein, it is also an ATPase enzyme. The cross bridges attach and detach some 50 to 100 times as the thin filaments are pulled to the center of a sarcomere. If, by chance, more ATP molecules are not available, detachment cannot occur. This explains *rigor mortis,* permanent muscle contraction after death.

The sliding filament theory states that actin filaments slide past myosin filaments because myosin has cross bridges that pull the actin filaments inward.

It is obvious from our discussion that ATP provides the energy for muscle contraction. In order to assure a ready supply, muscle fibers contain **creatine phosphate** (phosphocreatine), which stores a ready supply of high-energy phosphate. Creatine phosphate does not participate directly in muscle contraction. Instead, it is used to regenerate ATP by the following reaction:

$$creatine \sim P + ADP \rightarrow ATP + creatine$$

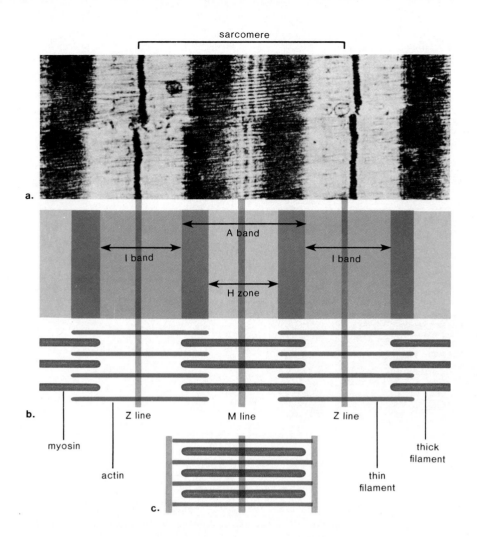

sarcomere

a.

A band

I band

I band

H zone

b.

myosin

actin

Z line

M line

c.

Z line

thick filament

thin filament

FIGURE 11.19 a. *Electron micrograph of a sarcomere showing the typical striations of skeletal muscle.* b. *The striations contain various bands and dark lines. The I band contains the Z line and thin filaments. The A band contains both thin and thick filaments except at the center where the H zone has only thick filaments anchored by the M line.* c. *Notice that the I band has decreased in size and the H zone has disappeared in the contracted sarcomere because the thin filaments have moved to the center.*

TABLE 11.6   *CONTRACTILE ELEMENTS*

COMPONENT	*DEFINITION*
Myofibril	Muscle cell contractile subunit
Sarcomere	Functional unit of myofibril
Myosin	Thick filament
Actin	Thin filament

Oxygen Debt

When all of the creatine phosphate has been depleted and there is no oxygen available for aerobic respiration, a muscle fiber can generate ATP by using anaerobic respiration (p. 44). Anaerobic respiration, which is apt to occur during strenuous exercise, can supply ATP for only a short time because lactic acid buildup produces muscular aching and fatigue.

We all have had the experience of having to continue deep breathing following strenuous exercise. This continued intake of oxygen is required to complete the metabolism of lactic acid, which has accumulated during exercise and represents an **oxygen debt** that the body must pay to rid itself of lactic acid. The lactic acid is transported to the liver, where 1/5 of it is completely broken down to carbon dioxide and water by means of the Krebs cycle and respiratory chain. The ATP gained by this respiration is then used to convert 4/5 of the lactic acid back to glucose.

Muscle contraction requires a ready supply of ATP. Creatine phosphate is used to generate ATP rapidly. If oxygen is in limited supply, anaerobic respiration produces ATP and results in oxygen debt.

Muscle Fiber Contraction and Relaxation

Nerves innervate muscles, and nerve impulses cause muscles to contract. A motor axon within a nerve sends branches to several muscle fibers, which collectively are termed a *motor unit*. Each branch has terminal knobs that contain synaptic vesicles filled with the neuromuscular transmitter acetylcholine (ACh). A terminal knob lies in close proximity to the

FIGURE 11.20 *Sliding filament theory. a. Relaxed sarcomere. b. Contracted sarcomere. Note that during contraction, the I band and H zone decrease in size. This indicates that the thin filaments slide past the thick filaments. Even so, the thick filaments do the work by pulling the thin filaments by means of cross bridges.*

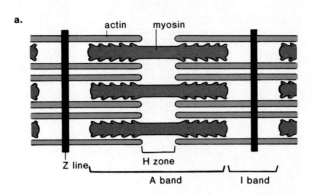

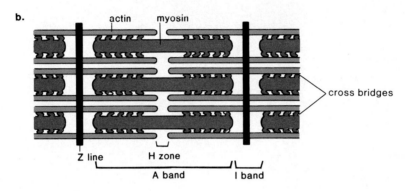

TABLE 11.7 MUSCLE CONTRACTION	
NAME	**FUNCTION**
Actin filaments	Slide past myosin causing contraction
Ca^{++}	Needed for actin to bind to myosin
Myosin filaments	a. Enzyme that splits ATP b. Pulls actin by means of cross bridges
ATP	Supplies energy for bonding between actin and myosin = actomyosin

sarcolemma of the muscle fiber. This region, called a **neuromuscular junction** (fig. 11.21), has the same components as a synapse: a presynaptic membrane, a synaptic cleft, and a postsynaptic membrane. Only in this case, the postsynaptic membrane is a portion of the sarcolemma of a muscle fiber. The sarcolemma, just like a neural membrane, is polarized; the inside is negatively charged, and the outside is positively charged.

Nerve impulses cause synaptic vesicles to merge with the presynaptic membrane and release ACh into the synaptic cleft. When ACh reaches the sarcolemma, it is depolarized. The result is a **muscle action potential** that spreads over the

sarcolemma and down the T system to where calcium ions are stored in calcium-storage sacs of the sarcoplasmic reticulum. When the action potential reaches a sac, Ca^{++} ions are released and diffuse into the sarcoplasm, after which they attach to the thin filaments. A thin filament is a twisted double strand of globular actin molecules. Associated with the actin filaments are two other proteins: *tropomyosin,* which forms threads that twist about the actin filaments, and *troponin,* which is located at intervals along the tropomyosin. The calcium combines with the troponin, and this causes the tropomyosin threads to shift in position so that myosin binding sites are exposed (fig. 11.22).

The thick myosin filament is a bundle of myosin molecules, each having a globular head. These heads form the cross bridges that attach to the binding sites on actin. Also, as they break down ATP, detachment and reattachment to a site farther along occurs. In this way, actin filaments are pulled along the myosin filament.

Contraction discontinues when nerve impulses no longer stimulate the muscle fiber. With the cessation of a muscle action potential, Ca^{++} ions are pumped back into their storage sacs by active transport. Relaxation now occurs.

A neuromuscular junction can be compared to a synapse. A muscle action potential occurs that causes calcium to be released from calcium-storage sacs, and thereafter contraction occurs.

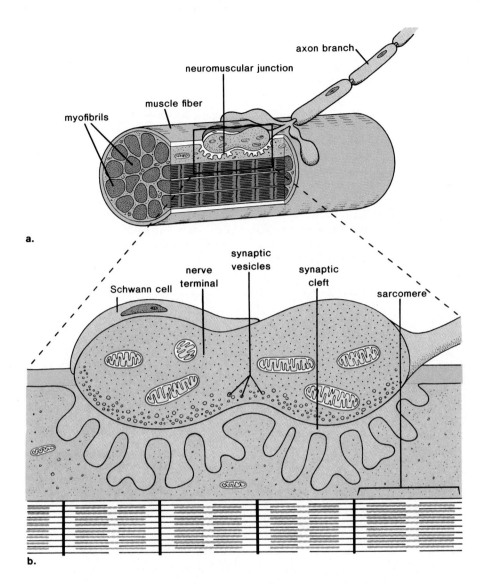

a.

axon branch

neuromuscular junction

muscle fiber

myofibrils

b.

Schwann cell

nerve terminal

synaptic vesicles

synaptic cleft

sarcomere

FIGURE 11.21 *Anatomy of a neuromuscular junction.* a. *A neuromuscular junction occurs where a synaptic ending of an axon branch comes in close proximity to a muscle fiber.* b. *The synaptic ending contains synaptic vesicles filled with ACh. When these vesicles fuse with the presynaptic membrane, ACh diffuses across the synaptic cleft to initiate a muscle action potential.*

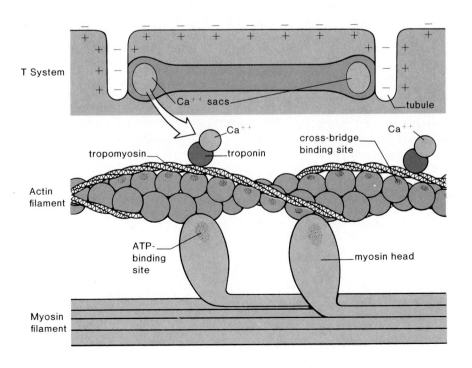

T System

Ca^{++} sacs

tubule

Ca^{++}

tropomyosin

troponin

cross-bridge binding site

Ca^{++}

Actin filament

ATP-binding site

myosin head

Myosin filament

FIGURE 11.22 *Detailed structure and function of sarcomere contraction. After calcium (Ca^{++}) is released from its storage sacs, it combines with troponin, a protein that occurs periodically along tropomyosin threads. This causes the tropomyosin threads to shift their position so that cross-bridge binding sites are revealed along the actin filaments. The myosin filament extends its globular heads, forming cross bridges that bind to these sites. The breakdown of ATP by myosin causes the cross bridges to detach and reattach farther along the actin. In this way, the actin filaments are pulled along past the myosin filaments.*

SUMMARY

The skeleton aids movement of the body while it also supports and protects the body. Bones serve as deposits for inorganic salts, and some bones are sites for blood-cell production. The skeleton is divided into two parts: (1) the axial skeleton, which is made up of the skull, ribs, and vertebrae; and (2) the appendicular skeleton, which is composed of the appendages and their girdles. Joints are regions where bones are linked together.

Whole skeletal muscles work in antagonistic pairs and have degrees of contraction. Muscle fibers obey the all-or-none law; it is possible to study a single contraction (muscle twitch) and sustained contraction (summation and tetanus) by using a physiograph.

Muscle fibers are cells that contain myofibrils in addition to the usual components of cells. Longitudinally, myofibrils are divided up into sarcomeres, where it is possible to note the arrangement of actin and myosin filaments. When a sarcomere contracts, the actin filaments slide past the myosin filaments. Myosin has cross bridges that attach to and pull the actin filaments along. ATP breakdown by myosin is necessary for detachment to occur.

Innervation of a muscle fiber begins at a neuromuscular junction. Here synaptic vesicles release ACh into the synaptic cleft. When the sarcolemma receives the ACh, a muscle action potential moves down the T system to calcium-storage sacs. When calcium is released, contraction occurs. When calcium is actively transported back into the storage sacs, muscle relaxation occurs.

OBJECTIVE QUESTIONS

1. The skull, ribs, and sternum are all in the _____ skeleton.
2. The foramen magnum is an opening in the _____ bone of the cranium.
3. The _____ bones are the cheekbones.
4. The vertebral column protects the _____ cord.
5. The two bones of the lower arm are the _____ and _____ .
6. In the leg, the _____ group extends the lower leg, and the _____ group flexes the lower leg.
7. As question 6 illustrates, muscles work in _____ pairs.
8. Actin and myosin filaments are found within cell inclusions called _____ , which are divided into units called _____ .
9. The molecule _____ serves as an immediate source of high-energy phosphate for ATP production in muscle cells.
10. The region between an axon ending and muscle cell sarcolemma is called a _____ junction.

Answers

1. axial 2. occipital 3. zygomatic 4. nerve 5. radius, ulna 6. quadriceps, hamstring 7. antagonistic 8. myofibrils, sarcomeres 9. creatine phosphate 10. neuromuscular

STUDY QUESTIONS

1. Distinguish between the axial and appendicular skeletons. (p. 224)
2. List the bones of the skull, including those of the cranium and the face. (p. 226)
3. List the bones that form the pectoral and the pelvic girdles. (p. 228)
4. Describe the anatomy of a long bone, including its growth and development. (p. 228)
5. Describe the anatomy of a freely movable joint, and list the different types of joint movements. (p. 230)
6. Describe how muscles are attached to bones. (p. 231) Give several examples of antagonistic pairs of muscles. (p. 235)
7. The study of whole muscle physiology often includes observing both threshold and maximum stimulus, muscle twitch, summation, and tetanus. Describe the significance of each of these. (p. 238)
8. How is the tone of a muscle maintained, and how do muscle spindles contribute to the maintenance of tone? (p. 238)
9. Discuss the microscopic anatomy of a muscle fiber and the structure of a sarcomere. What is the sliding filament theory? (pp. 239–240)
10. Give the function of each participant in the following reaction: (p. 240)

$$\text{actin} + \text{myosin} \xrightarrow[\text{Ca}^{++}]{\overset{\text{ATP} \quad \text{ADP} + \textcircled{P}}{}} \text{actomyosin}$$

11. What is oxygen debt, and how is it repaid? (p. 241)
12. What causes a muscle action potential? How does the muscle action potential bring about sarcomere and muscle fiber contraction? (p. 242)

THOUGHT QUESTIONS

1. Is there any similarity between the way that nerves use ATP and the way muscles use ATP? If not, what does that say about the use of ATP?

2. What is the value of having a jointed skeleton, and how might this be demonstrated by discussing the different types of joints?

3. Muscles vary in bulk. Why do the muscles that have the most bulk move the appendages?

KEY TERMS

actin (ak'tin)   one of the two major proteins of muscle; makes up thin filaments in myofibrils of muscle fibers. *See* myosin.   *240*

appendicular skeleton (ap''en-dik'u-lar skel'ě-ton)   portion of the skeleton forming the upper extremities, pectoral girdle, lower extremities, and pelvic girdle.   *224*

axial skeleton (ak'se-al skel'ě-ton)   portion of the skeleton that supports and protects the organs of the head, neck, and trunk.   *224*

compact bone (kom-pakt' bōn)   bone composed of osteocytes located in lacunae that are arranged concentrically around Haversian canals.   *228*

creatine phosphate (kre'ah-tin fos'fāt)   a compound unique to muscles that contains a high-energy phosphate bond.   *240*

insertion (in-ser'shun)   the end of a muscle that is attached to a movable bone.   *231*

muscle action potential (mus'el ak'shun po-ten'shal)   an electrochemical change, due to increased sarcolemma permeability, that is propagated down the T system and results in muscle contraction.   *242*

myofibrils (mi''o-fi'brilz)   the contractile portions of muscle fibers.   *239*

myosin (mi'o-sin)   one of two major proteins of muscle; makes up thick filaments in myofibrils and is capable of breaking down ATP. *See* actin.   *240*

neuromuscular junction (nu''ro-mus'ku-lar jungk'shun)   the point of contact between a nerve cell and a muscle fiber.   *242*

origin (or'i-jin)   end of a muscle that is attached to a relatively immovable bone.   *231*

osteocyte (os'te-o-sīt)   a mature bone cell.   *229*

oxygen debt (ok'sĭ-jen det)   oxygen that is needed to metabolize lactic acid, a compound that accumulates during vigorous exercise.   *241*

red marrow (red mar'o)   blood-cell-forming tissue located in spaces within bones.   *228*

synovial joint (si-no've-al joint)   a freely movable joint.   *229*

tetanus (tet'ah-nus)   sustained muscle contraction without relaxation.   *238*

tone (tōn)   the continuous partial contraction of muscle; due to contraction of a small number of muscle fibers at all times.   *238*

CHAPTER TWELVE

SENSES

CHAPTER CONCEPTS

1 Sense organs are sensitive to environmental stimuli and are therefore termed receptors. Each receptor responds to one type of stimulus, but they all initiate nerve impulses.

2 The sensation realized is the prerogative of the region of the cerebrum receiving the nerve impulses.

3 There are sense receptors that respond to mechanical stimuli, chemical stimuli, and light. Our knowledge of the outside world is dependent on our response to these stimuli.

Scanning electron micrograph of stirrup (stapes), a small bone in the middle ear attached to the membrane of the oval window.

INTRODUCTION

We view the world through our sense organs, and we can be fooled. The sense organs send information to be processed by the brain, and its interpretation is not always accurate. The boy and the man in figure 12.1 seem to be equally distant; therefore, our brain tells us that the boy is larger than the man. Actually the man's corner is farther away and has a higher ceiling than the boy's corner in this room[1] that is specially prepared to fool the mind.

S ince sense organs receive external and internal stimuli, they are called **receptors.** Each type of receptor is sensitive to only one type of stimulus. Table 12.1 lists the receptors discussed in this chapter and the stimulus to which each reacts.

[1]The room is a trapezoid with this shape:

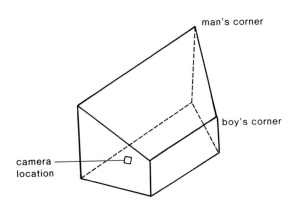

TABLE 12.1 RECEPTORS

RECEPTORS	SENSE	STIMULUS
General		
Temperature	Hot-cold	Heat flow[a]
Touch	Touch	Mechanical displacement of tissue[b]
Pressure	Pressure	Mechanical displacement of tissue[b]
Pain	Pain	Tissue damage[c]
Proprioceptors	Limb placement	Mechanical displacement[b]
Special		
Eye	Sight	Light[d]
Ear	Hearing	Sound waves[b]
	Balance	Mechanical displacement[b]
Taste buds	Taste	Chemicals[c]
Olfactory cells	Smell	Chemicals[c]

[a]Radioreceptor
[b]Mechanoreceptor
[c]Chemoreceptor
[d]Photoreceptor

Receptors are the first component of the reflex arc, which was described in chapter 10. When a receptor is stimulated, it generates nerve impulses that are transmitted to the spinal cord and/or brain, but only if the impulses reach the cerebrum are we conscious of a sensation. The sensory portion of the cerebrum can be mapped according to the parts of the body and the type of sensation realized at different loci (fig. 10.21).

GENERAL RECEPTORS

Microscopic receptors (table 12.1) are present in the skin, in the visceral organs, and in the muscles and joints. They are all specialized nerve endings for the detection of touch, pressure, pain, temperature (hot and cold), and proprioception. Proprioception refers to the sense of knowing the position of the limbs; for example, if you close your eyes and move your arm about slowly, you still have a sense of where your arm is located.

Skin

The skin (fig. 12.2) contains receptors for touch, pressure, pain, and temperature. It is a mosaic of these tiny receptors, as you can determine by passing a metal probe slowly over your skin. At certain points there will be a feeling of pressure; at others, a feeling of hot or cold (depending on the temperature of the probe). Certain parts of the skin contain more receptors for a particular sensation; for example, the fingertips have an abundance of touch receptors.

A simple experiment suggests that temperature receptors are sensitive to the flow of heat. Fill three bowls with water—one cold, one warm, and one hot. Put your left hand in the cold water and your right in the hot water for a few moments. Your hands will adjust, or adapt, to these temperatures, so that when you put both hands in warm water, each hand will indicate a different temperature of the water. Therefore, it seems that when the outside temperature is higher than the temperature to which we have adjusted, we detect a sensation of warmth or hotness as heat flows into the skin. When the outside temperature is low enough that heat flows out from the skin, we detect coolness or cold.

Other skin receptors besides those for temperature also demonstrate adaptation. **Adaptation** occurs when the receptor becomes so accustomed to stimuli that it stops generating impulses even though the stimulus is still present. The touch receptors are of this type. They can quickly adapt to the clothing we put on so that we are not constantly aware of the feel of the clothes against our skin.

FIGURE 12.2 *Receptors in human skin.* a. *Free nerve endings are pain receptors; Pacinian corpuscles are pressure receptors; Merkel's disks and Meissner's corpuscles are touch receptors, as are the nerve endings surrounding the hair follicle.* b. *Enlargements of three of these.*

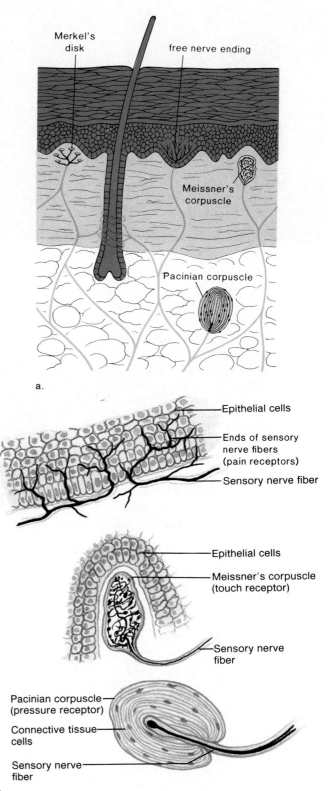

The receptors of the skin can be used to illustrate that sensation actually occurs in the brain and not in the sense organ itself. If the nerve fiber from the sense organ is cut, there is no sensation. Also, since a nerve impulse is always the same electrochemical change, the particular sensation realized does not have to do with the nerve impulse. It is the brain that is responsible for the type of sensation felt and the localization of the sensation. For example, if we connected a pain receptor in the foot to a nerve normally receiving impulses from a heat receptor in the hand, and then proceeded to stick the pain receptor in the foot, the subject would report the feeling of warmth in the hand. The brain indicates the sensation and the localization. This realization is mildly disturbing because it makes us aware of how dependent we are on the anatomical wholeness of the body in order to be properly aware of our surroundings.

Specialized receptors in the human skin respond to temperature, touch, pressure, and pain.

Muscles and Joints

The sense of position and movement of limbs (i.e., proprioception) is dependent upon receptors termed **proprioceptors.** Muscle spindles, discussed earlier in chapter 11, are sometimes considered to be proprioceptors. Stretching of associated muscle fibers causes muscle spindles to increase the rate at which they fire, and for this reason, they are sometimes called **stretch receptors.** The *knee jerk* is a common example of the manner in which muscle spindles act as stretch receptors (fig. 12.3). When the legs are crossed at the knee and the tendon at the knee is tapped, the tendon and muscles in the thigh are stretched. Stimulated by the stretching, muscle spindles transmit impulses to the spinal cord and thereafter the thigh muscles contract. This causes the lower leg to jerk upward in a kicking motion.

There are proprioceptors located in the joints and associated ligaments and tendons that respond to stretching, pressure, and pain. Nerve endings from these receptors are integrated with those received from other types of receptors so that the person knows the position of body parts.

Receptors in the muscles and joints, called proprioceptors, give us a sense of how our body parts are positioned.

SPECIAL SENSES

The special senses include the chemoreceptors for taste and smell, the light receptors for sight, and the mechanoreceptors for hearing and balance.

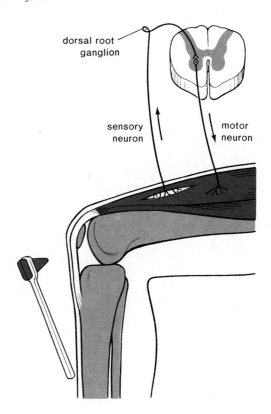

dorsal root ganglion

sensory neuron

motor neuron

Chemoreceptors

Taste and smell are called the *chemical senses* because these receptors are sensitive to certain chemical substances in the food we eat and the air we breathe.

Taste buds are located on the tongue. Many lie along the walls of the papillae (fig. 12.4), the small elevations visible to the naked eye. Isolated ones are also present on the palate, pharynx, and epiglottis.

Taste buds (fig. 12.4b) are pockets of cells that extend through the tongue epithelium and open at a taste pore. Within the oval pocket are supporting cells and a number of elongated cells that end in microvilli. These cells, which have associated nerve fibers, are sensitive to chemicals dissolved in the pore. Nerve impulses are most probably generated when the chemicals bind to receptor sites found on the microvilli.

The sense of taste has been shown to be genetically inherited, and foods taste different to various people. This might very well account for the fact that some persons dislike a food that is preferred by others.

It is believed that there are four types of tastes (bitter, sour, salty, sweet) and that taste buds for each are concentrated on the tongue in particular regions. Sweet receptors are most plentiful near the tip of the tongue. Sour receptors occur primarily along the margins of the tongue. Salt receptors are most common on the tip and the upper front portion of the tongue, and bitter receptors are located toward the back of the tongue.

FIGURE 12.4 *Taste buds.* a. *Elevations called papillae indicate the presence of taste buds. The location of those containing taste buds responsive to sweet, sour, salt, and bitter is indicated.* b. *Drawing of a taste bud shows the various cells that make up a taste bud. Sensory cells in the bud end in microvilli that have receptors for the chemicals that exhibit the tastes noted in (a). When the chemicals combine with the receptors, nerve impulses are generated.*

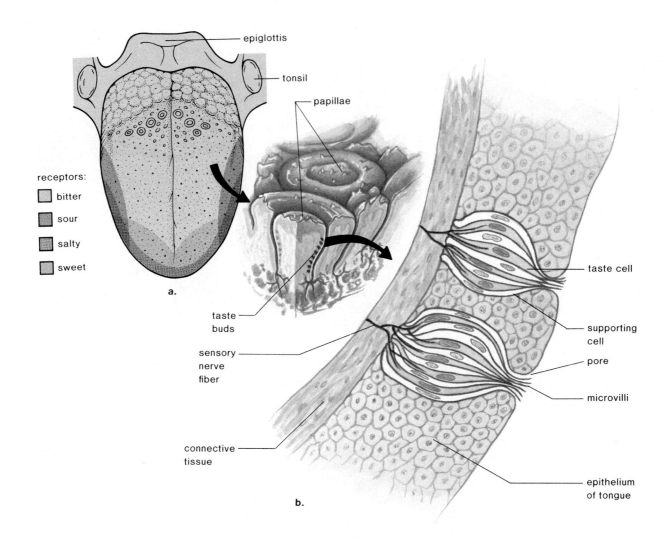

receptors:
bitter
sour
salty
sweet

a.

epiglottis
tonsil
papillae
taste buds
sensory nerve fiber
connective tissue

taste cell
supporting cell
pore
microvilli
epithelium of tongue

b.

The **olfactory cells** (fig. 12.5) are located high in the roof of the nasal cavity. These cells, which are specialized endings of the fibers that make up the olfactory nerve, lie among supporting epithelial cells. Each cell ends in a tuft of six to eight cilia, which probably bear receptor sites for various chemicals. Research, resulting in the stereochemical theory of smell, suggests that different types of smell are related to the various shapes of molecules rather than to the atoms that make up the molecules. When chemicals combine with the receptor sites, nerve impulses are generated. The olfactory receptors, like the touch and temperature receptors, also adapt to outside stimuli. In other words, we can become accustomed to a smell and no longer take notice of it.

The sense of taste and the sense of smell supplement each other, creating a combined effect when interpreted by the cerebral cortex. For example, when we have a cold, we think that our food has lost its taste, but actually we have lost the ability to sense its smell. This may work in the reverse also. When we smell something, some of the molecules move from the nose down into the mouth region and stimulate the taste buds there. Thus, part of what we refer to as smell may actually be taste (fig. 12.6).

The receptors for taste (taste buds) and the receptors for smell (olfactory microvilli) work together to give us our sense of taste and our sense of smell.

FIGURE 12.5 a. *Position of olfactory epithelium in a nasal passageway.* b. *The olfactory receptor cells, which have cilia projecting into the nasal cavity, are supported by columnar epithelial cells. When these cells are stimulated by chemicals in the air, nerve impulses are conducted to the brain by olfactory nerve fibers.*

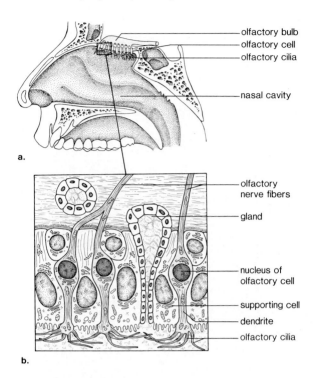

FIGURE 12.6 *If you're on a diet, it is safer to eat the same monotonous food each day instead of tempting the palate. A variety of interesting tastes and smells can cause you to eat more than you desire to.*

FIGURE 12.7 *Anatomy of the human eye. Notice that the sclera becomes the cornea; the choroid becomes the ciliary body and the iris. The ciliary body is thrown into seventy to eighty radiating folds that contain the ciliary muscle and ligaments that hold and adjust the shape of the lens. The retina contains the receptors for sight, and vision is most acute in the fovea centralis, where there are only cones. A blind spot occurs where the optic nerve leaves the retina, and there are no receptors for sight.*

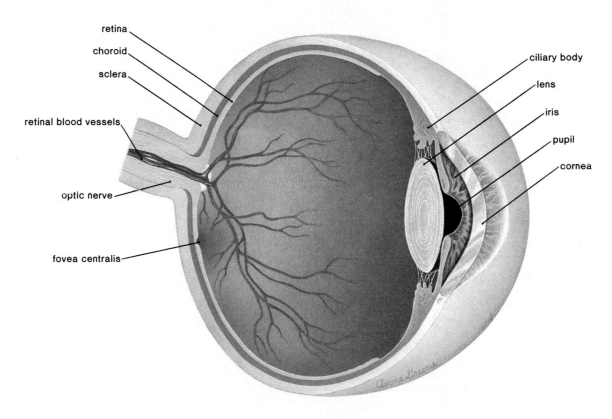

Photoreceptor—the Eye

The eye (fig. 12.7 and table 12.2), an elongated sphere about 1 inch in diameter, has three layers or coats. The outer **sclera** is a white fibrous layer except for the transparent *cornea,* the window of the eye. The middle thin, dark-brown layer, the **choroid,** contains many blood vessels and absorbs stray light rays. Toward the front, the choroid thickens and forms a ring-shaped structure, the *ciliary body,* containing the **ciliary muscle,** which controls the shape of the lens for near and far vision. Finally the choroid becomes a thin, circular, muscular diaphragm, the iris, which regulates the size of the pupil. The **lens,** attached to the ciliary body by ligaments, divides the eye into two cavities. A viscous and gelatinous material, the **vitreous humor,** fills the large cavity behind the lens. The chamber between the cornea and the lens is filled with an alkaline, watery solution secreted by the ciliary body and called the **aqueous humor.**

TABLE 12.2 *NAME AND FUNCTION OF PARTS OF THE EYE*

PART	FUNCTION
Lens	Refraction and focusing
Iris	Regulates light entrance
Pupil	Admits light
Choroid	Absorbs stray light
Sclera	Protection
Cornea	Refraction of light
Humors	Refraction of light
Ciliary body	Holds lens in place
Retina	Contains receptors
Rods	Black-and-white vision
Cones	Color vision
Optic nerve	Transmits impulse
Fovea	Region of cones in retina
Ciliary muscle	Accommodation

FIGURE 12.8 a. *Structure of retina. Rods and cones are located toward the back of the retina, followed by the bipolar cells, then the ganglionic cells whose fibers become the optic nerve. Notice that the rods share bipolar cells but the cones do not. Cones, therefore, distinguish more detail.* b. *The blind spot where the optic nerve pierces the eyeball is clearly visible in eye examinations.*

Retina

The inner layer of the eye, the **retina,** has three layers of cells (fig. 12.8). The layer closest to the choroid contains the sense receptors for sight, the **rods** and **cones** (fig. 12.9); the middle layer contains bipolar cells; and the innermost layer contains ganglionic cells whose fibers become the **optic nerve.** Only the rods and cones contain light-sensitive pigments, and therefore light must penetrate to the back of the retina before nerve impulses are generated. Nerve impulses initiated by the rods and cones are passed to the bipolar cells, which in turn pass them to the ganglion cells, whose fibers pass in front of the retina forming the optic nerve. The optic nerve turns to pierce the layers of the eye. You'll notice in figure 12.8 that there are many more rods and cones than there are nerve fibers leaving ganglionic cells. In fact, the retina has as many as 150 million rods but only one million optic nerve fibers. This means that there will be considerable mixing of messages and a certain amount of integration before nerve impulses are sent to the brain. There are no rods and cones where the optic nerve passes through the retina; therefore, this is a **blind spot** (fig. 12.8) where vision is impossible.

The retina contains a very special region called the **fovea centralis,** an oval yellowish area with a depression where there are only cone cells. Vision is most acute in the fovea centralis.

The eye has three layers: the outer sclera, the middle choroid, and the inner retina. Only the retina contains receptors for sight.

Glaucoma A small amount of aqueous humor is continually produced each day. Normally it leaves the anterior chamber by way of tiny ducts that are located where the iris meets the cornea. If these drainage ducts are blocked, pressure rises and compresses the retinal arteries whose capillaries feed nerve fibers located in the retina. With the passage of time, some of these fibers die, and the result is partial or total blindness.

Physiology

Focusing When we look at an object, light rays are **focused** on the retina (fig. 12.10). In this way, an image of the object appears on the retina. The image on the retina occurs when the rods and cones in a particular region are excited. Obviously, the image is much smaller than the object. In order to produce this small image, light rays must be bent (refracted) and brought to a focus. They are bent as they pass through the cornea. Further bending occurs as the rays pass through the lens and humors.

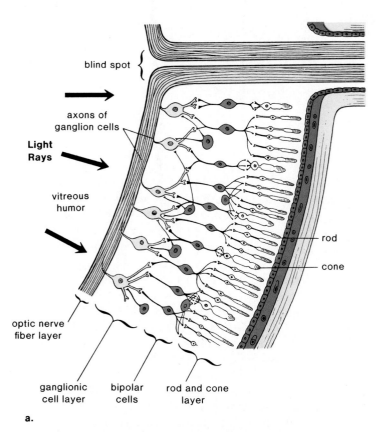

blind spot

axons of ganglion cells

Light Rays

vitreous humor

rod

cone

optic nerve fiber layer

ganglionic cell layer   bipolar cells   rod and cone layer

a.

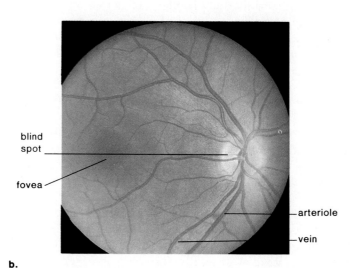

blind spot

fovea

arteriole

vein

b.

FIGURE 12.9 *Receptors for sight. a. Drawing of rod and cone. The photosensitive pigment is located in the discs of the outer segment, which apparently is a modified cilium. A cilium-like stalk with nine sets of microtubules connects the outer segment to the inner segment. b. Scanning electron micrograph of rods and cones. The cones in the foreground are responsible for color vision, and the rods in the background are responsible for night vision.*

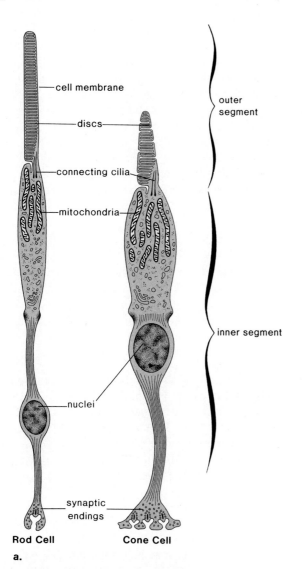

b.

FIGURE 12.10 Focusing. Light rays from each point on an object are bent by the cornea in such a way that they are directed to a single point after emerging from the lens. By this process an inverted image of the object forms on the retina.

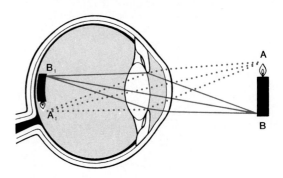

Accommodation Light rays are reflected from an object in all directions. If the eye is distant from an object, only nearly parallel rays enter the eye and the cornea alone is needed for focusing. But if the eye is close to the object, many of the rays are at sharp angles to one another and additional focusing is required. The lens provides this additional focusing power. Although the lens remains flat when we view distant objects, it rounds up when we view close objects. This **accommodation** of the lens provides the additional refraction required to bring the diverging light rays to a sharp focus on the retina (fig. 12.11). The shape of the lens is controlled by the ciliary muscles within the ciliary body. When we view a distant object, the ciliary muscles are relaxed, causing the ligaments attached to the ciliary body to be under tension; therefore, the lens remains relatively flat. When we view a close object, the ciliary muscles contract, releasing the tension on the ligaments and, therefore, the lens rounds up due to its natural elasticity (table 12.3). Since close work requires contraction of the ciliary muscles, it very often causes "eye strain."

With aging, the lens loses some of its elasticity and is unable to accommodate in order to bring close objects into focus. This usually necessitates the wearing of glasses, as is discussed on page 258.

The lens is also subject to *cataracts;* it can become cloudy and unable to transmit rays of light. Special cells within the interior of the lens contain proteins called crystallin. Recent research suggests that cataracts are brought on when these proteins become oxidized, causing their three-dimensional shape to change. If so, researchers believe that they may eventually be able to find ways to restore the normal configuration of crystallin so that cataracts can be treated medically instead of surgically. Currently, however, surgery is used to correct cataracts. First, a surgeon opens the eye near the rim of the cornea, then a tiny needle about the size of the point of a ballpoint pen is placed into the eye. An ultrasonic beam is then used to break up the lens into tiny particles. Next a hollow needle is used to suck out the fragments

FIGURE 12.11 *Accommodation.* a. *When the eye focuses on a far object, the lens is flat because the ciliary muscles are relaxed and the suspensory ligaments are taut.* b. *When the eye focuses on a near object, the lens rounds up because the ciliary muscles contract, causing the suspensory ligaments to relax.*

FIGURE 12.12 *Eye surgery for a cataract.* a. *After sound waves are used to fragment the lens, the fragments are sucked up by a hollow needle.* b. *Often a plastic intraocular lens is implanted to restore the eye's focusing power.*

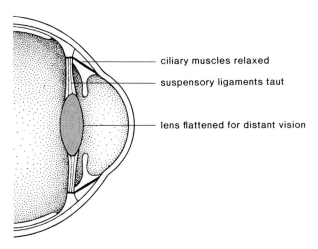

ciliary muscles relaxed

suspensory ligaments taut

lens flattened for distant vision

**a. Normal
Distant Focus**

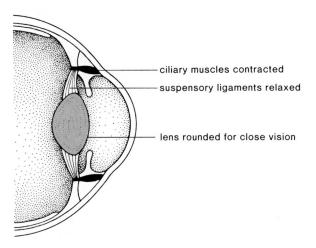

ciliary muscles contracted
suspensory ligaments relaxed

lens rounded for close vision

b. Near Focus

a.

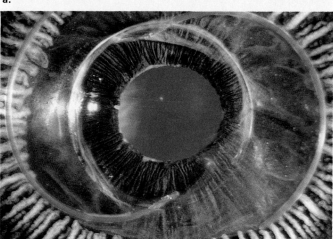

b.

TABLE 12.3 ACCOMMODATION		
OBJECT	**CILIARY MUSCLES**	**LENS**
Near object	Ciliary muscles contract, ligaments relax	Lens becomes round
Far object	Ciliary muscles relax, ligaments under tension	Lens is flattened

(fig. 12.12). Most often at the time of removal of the cataract, a plastic intraocular lens is implanted in the eye and attached to the iris. Usually the power of the lens, based on measurements taken by the ophthalmologist before surgery, corrects for distant vision only, but recently bifocal intraocular lenses that correct for both distant and close vision have become available.

The lens, assisted by the cornea and humors, focuses images on the retina.

Inverted Image The image on the retina is upside down (fig. 12.10), and it is thought that perhaps by experience this image is righted in the brain. In one experiment, scientists wore glasses that inverted the field. At first, they had difficulty adjusting to the placement of the objects, but then they soon became accustomed to their inverted world. Experiments such as these suggest that if we see the world upside down, the brain learns to see it right side up.

FIGURE 12.13   *Both eyes "see" the entire object, but information from the right half of each retina goes to the right visual cortex and information from the left half of the retina goes to the left visual cortex. When the information is pooled, the brain "sees" the entire object and "sees" it in depth.*

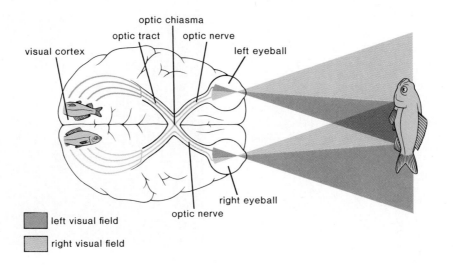

Stereoscopic Vision   We can see well with either eye alone, but the two eyes functioning together provide us with stereoscopic vision (3-D vision). Normally, the two eyes are directed by the eye muscles toward the same object, and therefore the object is focused on corresponding points of the two retinas. But each eye sends to the brain its own information about the placement of the object because each forms an image from a slightly different angle. These data are pooled to produce depth perception by a two-step process. First, because the optic nerves cross at the optic chiasma (fig. 12.13), half of the brain receives information from both eyes about the same part of an object. Later, the two halves of the brain communicate to arrive at a complete three-dimensional interpretation of the whole object.

The workings of the brain allow us to see the world right side up and in three dimensions.

Biochemistry   In *dim light,* the pupils enlarge so that more rays of light can enter the eyes. As the rays of light enter, they strike the rods and cones, but only the 150 million rods located in the periphery, or sides, of the eyes are sensitive enough to be stimulated by this faint light. The rods do not detect fine detail or color, so at night, for example, all objects appear to be blurred and have a shade of gray. Rods do detect even the slightest motion, however, because of their abundance and position in the eyes.

Although it has been known for some time that vision generated by the rods is dependent on the presence of rhodopsin, the actual events have only recently been worked out

in detail. Many molecules of rhodopsin, also called visual purple, are located within the membrane of the discs (lamellae) found in the outer segment (fig. 12.9) of the rods. **Rhodopsin** is a complex molecule that contains a protein (*opsin*) and a pigment molecule called *retinal,* which is a derivative of vitamin A. When retinal absorbs light energy, it changes shape and opsin is activated (fig. 12.14). The reactions that follow eventually end when many molecules of GMP (guanosine monophosphate) are converted to cyclic GMP.[2] Cyclic GMP, in turn, initiates an action potential in the cell membrane of the rod that travels along the membrane until it reaches the synaptic endings, where a transmitter substance is released to excite a bipolar cell (fig. 12.8). Now rhodopsin returns to its former configuration so that the series of events can reoccur and reoccur. Each stimulus generated lasts about 1/10 of a second. This is why we continue to see an image if we close our eyes immediately after looking at an object. It also allows us to see motion if still frames are presented at a rapid rate, as in "movies."

In *bright light* the pupils get smaller, so that less light enters the eyes. The cones, located primarily in the fovea, are active and detect the fine detail and color of an object. In order to perceive depth, as well as to see color, we turn our eyes so that reflected light from the object strikes the fovea. Color vision has been shown to depend on three kinds of cones that contain pigments sensitive to either blue, green, or red light. The nerve impulses generated from one type of cone not only stimulate certain cells in the visual cortex of the brain, they also inhibit the reception of impulses from other cone types. For example, when we see red, certain cells in

---

[2]GMP is a nucleotide that contains the base guanine, the sugar ribose, and one phosphate group. In cyclic GMP, the single phosphate is attached to ribose in two places.

FIGURE 12.14 *Biochemistry of vision. Rhodopsin, a complex molecule containing the protein opsin and the pigment retinal, is present in disc membrane within rod cells (fig. 12.9). When retinal absorbs light energy, it changes shape, and this activates opsin to begin a series of reactions that end when GMP is converted to cyclic GMP, a molecule that can trigger an action potential in rod cell membrane.*

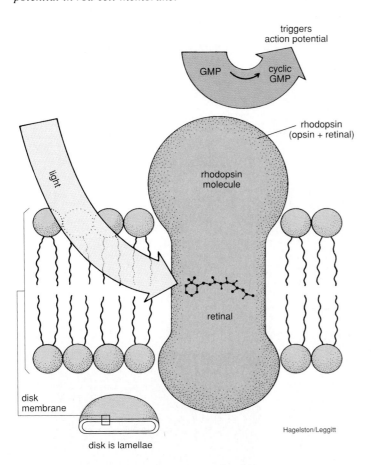

triggers action potential

GMP → cyclic GMP

light

rhodopsin (opsin + retinal)

rhodopsin molecule

retinal

disk membrane

disk is lamellae

Hagelston/Leggitt

FIGURE 12.15 *Color vision. Light is radiant energy having different wavelengths. Visible white light can be divided into wavelengths of different sizes and colors. When all the wavelengths of the visible spectrum enter the eye at the same time in nearly equal quantities, we see the light as colorless or white. This is why sunlight or the light from an electric lamp appears colorless. An object has color when it absorbs some wavelengths but not others. A black object absorbs all wavelengths, a white object absorbs none, and a red object absorbs all colors but red. Red wavelengths, for example, are reflected from red apples to our eyes, and therefore, if we have cones that are sensitive to red light, we will see the red dots in this chart. If we have cones sensitive to green light, we will be able to make out the number 96. However, red-green color blindness is the most common type.*
The above has been reproduced from Ishihara's Tests for Colour Blindness *published by Kanehara and Co. Ltd., Tokyo, Japan, but tests for color blindness cannot be conducted with this material. For accurate testing, the original plates should be used.*

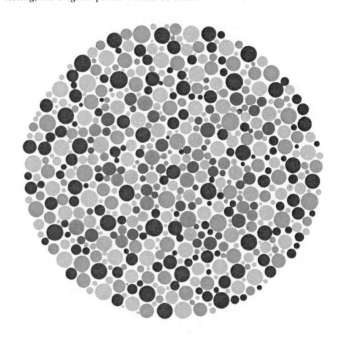

the brain are prohibited from receiving impulses from green cones. Similarly, impulses sent through blue cones tend to oppose the combination of signals sent by red and green cones—which together produce yellow. This process assists integration and enables the brain to tell the location of various colors in the environment (fig. 12.15). Complete color blindness is extremely rare. In most instances, a particular type of cone is lacking or deficient in number. The lack of red or green cones is the most common, affecting about 5% of the American population. If the eye lacks red cones, the green colors become accentuated, and vice versa.

The receptors for sight are the rods and cones. The rods are responsible for vision in dim light, and the cones are responsible for vision in bright light and for color vision. When either is stimulated, nerve impulses begin and are transmitted in the optic nerve to the brain.

Corrective Lenses
The majority of people can see what is designated as a size "20" letter 20 feet away and therefore are said to have 20/20 vision. Persons who can see close objects but cannot see the letters from this distance are said to be nearsighted. Nearsighted persons can see near better than they can see far. These individuals often have an elongated eyeball, and when they attempt to look at a far object, the image is brought to focus in front of the retina (fig. 12.16). They can see near because they can adjust the lens to allow the image to focus on the retina; but to see far, these people must wear concave lenses that diverge the light rays so that the image can be focused on the retina.

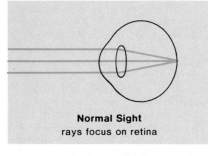

Normal Sight
rays focus on retina

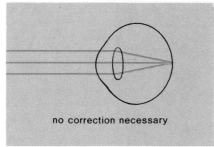

no correction necessary

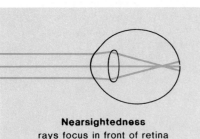

Nearsightedness
rays focus in front of retina

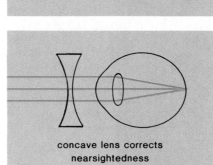

concave lens corrects
nearsightedness

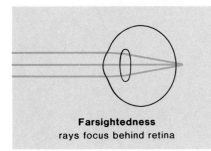

Farsightedness
rays focus behind retina

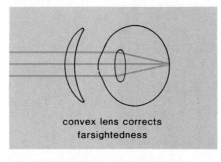

convex lens corrects
farsightedness

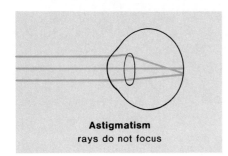

Astigmatism
rays do not focus

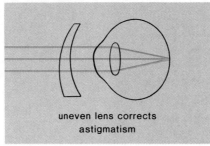

uneven lens corrects
astigmatism

a.   b.   c.

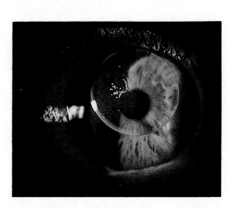

FIGURE 12.16 a. *Diagram illustrating common abnormalities of the eye. Both the cornea and the lens function in bringing light rays (lines) to focus, but sometimes they are unable to compensate for the shape of the eyeball or for an uneven cornea. b. Corrective lenses can be prescribed to allow the individual to see normally. c. Ophthalmologists can examine the fit of a contact lens by using a narrow beam of light from a "slit lamp" while looking through a biomicroscope.*

HUMAN ISSUE

A Russian ophthalmologist, Svyatoslav Fyodorov, has developed the eye operation known as radial keratotomy, or radial K, to correct nearsightedness. Eight to sixteen cuts are made in the cornea so that they radiate out from the center like spokes in a wheel. When the cuts heal, the cornea is flattened. Although some patients are satisfied with the result, others complain of glare and varying visual acuity. If you were an ophthalmologist, would you recommend this operation to patients who simply want to improve their appearance by not wearing glasses? Under what circumstances would you make the recommendation?

Persons who can easily see the optometrist's chart but cannot see close objects well are farsighted; these individuals can see far away better than they can see near. They often have a shortened eyeball, and when they try to see near, the image is focused behind the retina. When the object is far away, the lens can compensate for the short eyeball, but when the object is close, these persons must wear a convex lens to increase the bending of light rays so that the image will be focused on the retina.

When the cornea or lens is uneven, the image is fuzzy because the light rays cannot be focused evenly on the retina. This fault, called astigmatism, can be corrected by an unevenly ground lens to compensate for the uneven cornea. Table 12.4 summarizes these conditions and their corrections.

TABLE 12.4 COMMON ABNORMALITIES OF THE EYE

NAME	EFFECT	FAULT	RESULT	CORRECTION
Nearsighted	Can't see far, can see near	Long eyeball	Image focused in front of retina	Concave lens
Farsighted	Can't see near, can see far	Short eyeball	Image focused behind retina	Convex lens
Astigmatism	Can't focus	Irregular eyeball	Image not focused	Irregular lens

TABLE 12.5 THE EAR

	OUTER EAR	MIDDLE EAR	INNER EAR	
			Cochlea	*Sacs plus semicircular canals*
FUNCTION	Directs sound waves to tympanic membrane	Picks up and amplifies sound waves	Hearing	Maintains equilibrium
ANATOMY	Pinna Auditory canal	Tympanic membrane Ossicles	Contains organ of Corti Auditory nerve starts here	Saccule and utricle Semicircular canals
MEDIA	Air	Air (eustachian tube)	Fluid	Fluid

Note: Path of vibration: Sound waves—vibration of tympanic membrane—vibration of hammer, anvil, and stirrup—vibration of oval window—fluid pressure waves in canals of inner ear lead to stimulation of hair cells—bulging of round window.

Bifocals As mentioned earlier, with normal aging, the lens loses some of its ability to change shape in order to focus on close objects. Since nearsighted individuals still have difficulty seeing objects clearly in the distance, they must wear bifocals, which means that the upper part of the lens is for distant vision and the remainder is for near vision.

Shape of the eyeball determines the need for corrective lenses; the inability of the lens to accommodate as we age also requires a corrective lens for close vision.

Mechanoreceptor—the Ear

The ear accomplishes two sensory functions: balance and hearing. The sense cells for both of these are located in the inner ear and consist of hair cells with cilia that respond to mechanical stimulation. Each hair cell has from 30–150 extensions that are called cilia despite the fact that they contain tightly packed filaments rather than microtubules. When the cilia of any particular hair cell are displaced in a certain direction, the cell generates nerve impulses, which are sent along a cranial nerve to the brain.

Anatomy
Table 12.5 lists the parts of the ear, and figure 12.17 is a drawing of the ear. The ear has three divisions: outer, middle, and inner. The **outer ear** consists of the **pinna** (external flap) and **auditory canal.** The opening of the auditory canal is lined with fine hairs and sweat glands. In the upper wall are modified sweat glands that secrete earwax, a substance that helps guard the ear against the entrance of foreign materials such as air pollutants.

The **middle ear** begins at the **tympanic membrane** (eardrum) and ends at a bony wall in which are found two small openings covered by membranes. These openings are called the **oval** and **round windows.** Three small bones are found between the tympanic membrane and the oval window. Collectively called the **ossicles,** individually they are the **hammer** (malleus), **anvil** (incus), and **stirrup** (stapes) (fig. 12.17) because their shapes resemble these objects. The hammer adheres to the tympanic membrane, and the stirrup touches the oval window. The posterior wall of the middle ear also has an opening that leads to many air spaces within the mastoid process.

FIGURE 12.17 *Anatomy of the human ear. a. In the middle ear, the hammer, anvil, and stirrup amplify sound waves. Otosclerosis is a condition in which the stirrup becomes attached to the inner ear so is unable to carry out its normal function. It can be replaced by a plastic piston, and thereafter the individual hears normally because sound waves are transmitted as usual to the cochlea, which contains the receptors for hearing. b. Inner ear. The sense organs for balance are in the inner ear: the vestibule contains the utricle and saccule, and the ampullae are at the bases of the semicircular canals. The receptors for hearing are also in the inner ear: the cochlea has been cut to show the location of the organ of Corti.*

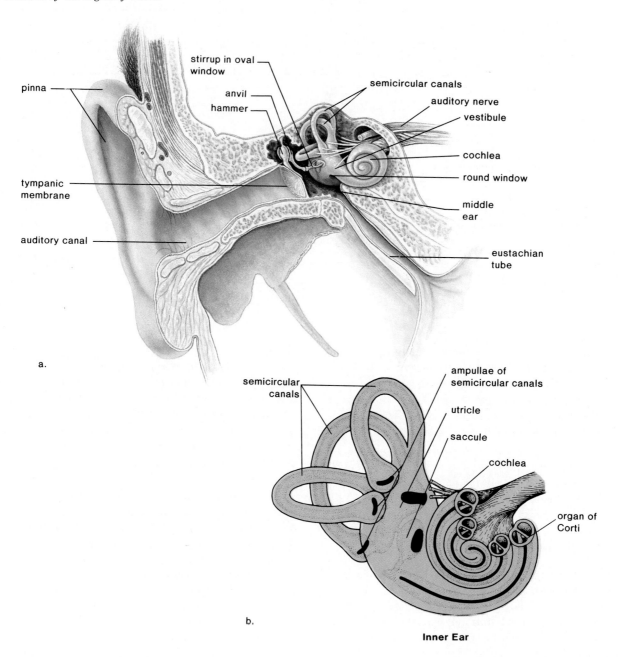

Inner Ear

The **eustachian tubes** (fig. 12.17) extend from the middle ear to the nasopharynx and permit equalization of air pressure. Chewing gum, yawning, and swallowing in elevators and airplanes help move air through the eustachian tubes upon ascent and descent.

Whereas the outer ear and middle ear contain air, the inner ear is filled with fluid. The **inner ear** (fig. 12.17b), anatomically speaking, has three areas: the first two, called the vestibule and semicircular canals, are concerned with balance; and the third, the cochlea, is concerned with hearing.

The **semicircular canals** are arranged so that there is one in each dimension of space. The base of each canal, called an **ampulla,** is slightly enlarged. Within the ampullae (fig. 12.17b) are little hair cells having cilia that are inserted into a gelatinous medium.

The **vestibule** is a chamber that lies between the semicircular canals and the cochlea. It contains two small sacs called the **utricle** and **saccule.** Within both of these are little hair cells having cilia that protrude into a gelatinous substance. Resting on this substance are calcium carbonate granules, or **otoliths.**

The **cochlea** resembles the shell of a snail because it spirals. Within the tubular cochlea are three canals: the *vestibular canal,* the **cochlear canal,** and the *tympanic canal.* Along the length of the basilar membrane, which forms the lower wall of the cochlear canal, are little hair cells having cilia that come into contact with another membrane called the tectorial membrane. The hair cells plus the **tectorial membrane** are called the **organ of Corti.** When this organ sends nerve impulses to the cerebral cortex, it is interpreted as sound.

The outer ear, middle ear, and cochlea are necessary for hearing. The vestibule and semicircular canals are concerned with the sense of balance.

Physiology

Balance (Equilibrium) The sense of balance has been divided into two senses: *static equilibrium,* referring to knowledge of movement in one plane, either vertical or horizontal, and *dynamic equilibrium,* referring to knowledge of angular and/or rotational movement.

When the body is still, the otoliths in the utricle (fig. 12.18) and saccule rest on the hair cells. When the head and/or body moves horizontally or vertically, the granules in the utricle and saccule are displaced. Displacement causes the cilia to bend slightly so that the cell generates nerve impulses that travel by way of a cranial nerve to the brain.

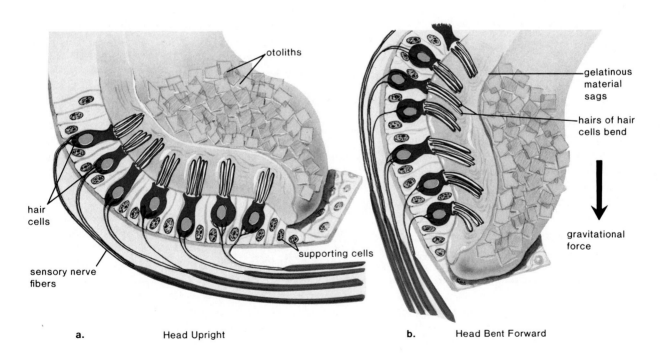

a.   Head Upright   b.   Head Bent Forward

FIGURE 12.18 *Receptor hair cells in the utricle and saccule are involved in our sense of static equilibrium: responsiveness to movement sideways or up and down. a. When the head is upright otoliths are balanced directly on the cilia of hair cells. b. When the head is bent forward the otoliths shift, and the cilia are bent, causing nerve impulses to be generated.*

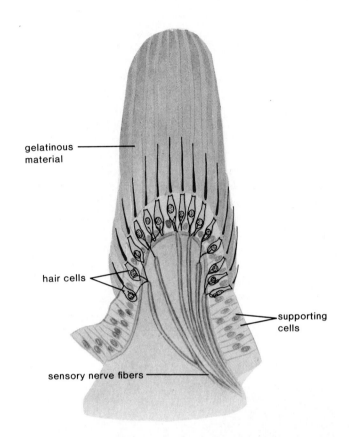

gelatinous material

hair cells

supporting cells

sensory nerve fibers

When the body is moving about, the fluid within the semicircular canals moves back and forth. This causes bending of the cilia attached to hair cells within the ampullae (fig. 12.19), and they initiate nerve impulses that travel to the brain. Continuous movement of the fluid in the semicircular canals causes one form of motion sickness.

Movement of the otoliths within the utricle and saccule are important for static equilibrium. Movement of fluid within the semicircular canals contributes to our sense of dynamic equilibrium.

Hearing The process of hearing begins when sound waves enter the auditory canal. Just as ripples travel across the surface of a pond, sound travels by the successive vibrations of molecules. Ordinarily, sound waves do not carry much energy, but when a large number of waves strike the eardrum, it moves back and forth (vibrates) ever so slightly. The hammer then takes the pressure from the inner surface of the eardrum and passes it by way of the anvil to the stirrup so that the pressure is multiplied about twenty times as it moves from the eardrum to the stirrup. The stirrup strikes the oval window, causing it to vibrate, and in this way the pressure is passed to the fluid within the inner ear.

If the cochlea is unwound, as shown in figure 12.20, the vestibular canal is seen to connect with the tympanic canal; therefore, as the figure indicates, pressure waves move from one canal to the other toward the round window, a membrane that can bulge to absorb the pressure. As a result of the movement of the fluid within the cochlea, the basilar membrane moves up and down, and the cilia of the hair cells rub against the tectorial membrane. This bending of the cilia initiates nerve impulses that pass by way of the auditory nerve to the brain, where the impulses are interpreted as a sound.

The organ of Corti is narrow at its base but widens as it approaches the tip of the cochlear canal. Each part of the organ is sensitive to different wave frequencies, or pitch. Near the tip, the organ of Corti responds to low pitches such as a tuba, and near the base, it responds to higher pitches such as a bell or whistle. The neurons from each region along the length of the cochlea lead to slightly different areas in the brain. The pitch sensation we experience depends upon which of these areas of the brain is stimulated. Volume is a function of the amplitude of sound waves. Loud noises cause the fluid of the cochlea to move back and forth to a greater degree, and this, in turn, causes the basilar membrane to move up and down to a greater extent. The resulting increased stimulation is interpreted by the brain as loudness. It is believed that tone is an interpretation of the brain based on the distribution of hair cells stimulated.

The sense receptors for sound are hair cells on the basilar membrane (the organ of Corti). When the basilar membrane vibrates, the delicate hairs touch the tectorial membrane and nerve impulses begin and are transmitted in the auditory nerve to the brain.

Deafness There are two major types of deafness: *conduction* and *nerve*. Conduction deafness can be due to a congenital defect, such as those that occur when a pregnant woman contracts German measles during the first trimester of pregnancy. (For this reason every female should be sure to be immunized against rubella before the childbearing years.) Conduction deafness can also be due to infections that have caused the ossicles to fuse together, restricting their ability to magnify sound waves. As was mentioned in chapter 8, respiratory infections can spread to the ear by way of the eustachian tubes; therefore, every cold and ear infection should be taken seriously.

FIGURE 12.20 *Organ of Corti. a. Enlarged cross section through the organ of Corti, showing the receptor hair cells from the side. b. Cochlea unwound, showing the placement of the organ of Corti along its length. The arrows represent the pressure waves that move from the oval window to the round window. These cause the basilar membrane to vibrate and the cilia of at least a portion of the 15,000 hair cells to bend against the tectorial membrane. The generated nerve impulses result in hearing.*

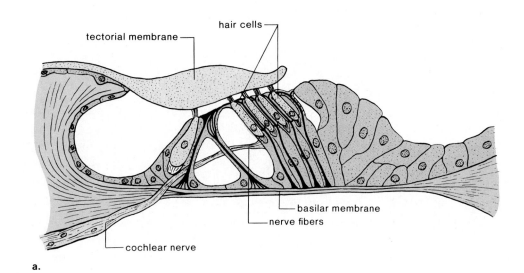

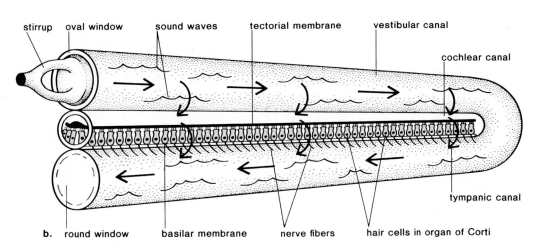

HUMAN ISSUE

Loss of sight is probably more critical than a hearing loss. Could this be why many young people carelessly listen to rock music that is many times louder than is safe for retention of their hearing? If not, what are the pressures that cause people to endanger their sense of hearing? The government has regulations requiring that employers protect the ears of their employees from loud noises. Should the government also make it illegal to play music at concerts and in other public places so loudly that it endangers the hearing of listeners?

Nerve deafness most often occurs when cilia on the sense receptors within the cochlea have worn away. Since this may happen with normal aging, old people are more likely to have trouble hearing; however, nerve deafness also occurs when young people listen to loud music amplified to 130 decibels. Because the usual types of hearing aids are not helpful for nerve deafness, it is wise to avoid subjecting the ears to any type of continuous loud noise. Costly cochlear implants that directly stimulate the auditory nerve are available, although those who have these electronic devices report that the speech they hear is like that of a robot.

PICTURING THE EFFECTS OF NOISE

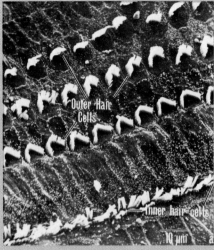
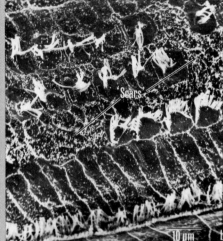

a.

b.

Damage to the organ of Corti due to loud noise. a. Normal organ of Corti. b. Organ of Corti after twenty-four- hour exposure to a noise level typical of rock music. Note scars where cilia have worn away.

"We have an idea of what noise does to the ear," David Lipscomb [of the University of Tennessee Noise Laboratory] says. "There's a pretty clear cause-effect relationship." And these photomicrographs of the cochlea's tiny structures graphically document noise trauma to the inner ear.

Hair cells transmit the mechanical energy of sound waves into those neural impulses that the brain interprets as sound. Loud noise can damage or destroy hair cells as these scanning electron micrographs illustrate.

Hair cells come in two varieties: a single row of inner cells and a triple row of outer ones. "Outer cells degenerate before inner cells," notes Clifton Springs, N.Y., otolaryngologist Stephen Falk. The most subtle change wrought by noise is a development of vesicles, or blister-like protrusions, along the walls of the hair cells' stereocilia. Continued assault by noise will lead to a rupturing of the vesicles and damage. In addition, the "cuticular plate"—base tissue supporting the stereocilia—may soften followed by a swelling and ultimate degeneration of hair cells.

But sensory hair cells are not the only structures at risk. Adjacent inner-ear cells . . . may undergo vacuolation—development of degenerative empty spaces in cells. Even nerve fibers synapsing at the hair cells' roots may die. In the final phase of noise-induced cochlear damage, the organ of Corti—of which hair cells and supporting cells are a part—is completely denuded and covered by a layer of scar tissue.

SUMMARY

All receptors are the first part of a reflex arc, and they initiate nerve impulses that eventually reach the cerebrum where sensation occurs. Among the general receptors are those located in the skin and the chemoreceptors for taste and smell.

Vision is a specialized sense dependent on the eye, optic nerve, and the brain. The rods, receptors for vision in dim light, and the cones, receptors that depend on bright light and provide color and detailed vision, are both located in the retina, the inner layer of the eyeball. The

cornea, humors, and especially the lens bring the light rays to focus on the retina. To see a close object, accommodation occurs as the lens rounds up. Due to the optic chiasma, both sides of the brain must function together to give us three-dimensional vision.

Hearing is a specialized sense dependent on the ear, auditory nerve, and the brain. The outer and middle portions of the ear simply convey and magnify the sound waves that strike the oval window. Its vibrations set up pressure waves within

the cochlea, which contains the organ of Corti, consisting of hair cells with the tectorial membrane above. When the hair cells strike this membrane, nerve impulses are initiated that finally result in hearing.

The ear also contains receptors for our sense of balance. Static equilibrium relies on the stimulation of hair cells by otoliths within the utricle and saccule. Dynamic equilibrium is dependent on the stimulation of hair cells within the ampullae of the semicircular canals.

OBJECTIVE QUESTIONS

1. The sense organs for position and movement are called _____ .
2. Taste buds and olfactory receptors are termed _____ because they are sensitive to chemicals in the air and food.
3. The receptors for sight, the _____ and _____ , are located in the _____ , the inner layer of the eye.
4. The cones give us _____ vision and work best in _____ light.
5. The lens _____ for viewing close objects.
6. People who are nearsighted cannot see objects that are _____ . A _____ lens will restore this ability.
7. The ossicles are the _____ , _____ , and _____ .
8. The semicircular canals are involved in our sense of _____ .
9. The organ of Corti is located in the _____ canal of the _____ .
10. Vision, hearing, taste, and smell do not occur unless nerve impulses reach the proper portion of the _____ .

Answers

1. proprioceptors 2. chemoreceptors 3. rods, cones, retina 4. color, bright (day) 5. rounds up 6. distant, concave 7. hammer, anvil, stirrup 8. balance 9. cochlear, cochlea 10. brain.

STUDY QUESTIONS

1. Name three factors that all receptors have in common. (pp. 247–248)
2. What type receptors are categorized as general and what type are categorized as special receptors? (p. 247)
3. Discuss the receptors of the skin, viscera, and joints. (pp. 248–249)
4. Discuss the chemoreceptors. (pp. 249–251)
5. Describe the anatomy of the eye (p. 252), and explain focusing and accommodation. (p. 253)
6. Describe sight in dim light. What biochemical events are responsible for vision in dim light? (p. 256) Discuss color vision. (p. 256)
7. Relate the need for corrective lenses to three possible shapes of the eye. (p. 258) Discuss bifocals. (p. 259)
8. Describe the anatomy of the ear and how we hear. (pp. 259–260)
9. Describe the role of the utricle, saccule, and semicircular canals in balance. (p. 261)
10. Discuss the two types of deafness, including why young people frequently suffer loss of hearing. (p. 262)

THOUGHT QUESTIONS

1. In general, list the steps, from sense organ to brain, by which sensation occurs. Discuss the weaknesses and strengths of depending on such a method for knowledge of the outside world.
2. How would you change the steps listed in question 1 so that it is more applicable to vision? Do your changes lend strength or weakness to the method for visual sensation?
3. Contrast the type of detailed data needed from vision and from hearing. Show that this difference in need may be related to the visual pathway versus the auditory pathway.

accommodation (ah-kom''o-da'shun) lens adjustment in order to see close objects. *254*

choroid (ko'roid) the vascular, pigmented middle layer of the wall of the eye. *252*

ciliary muscle (sil'e-er''e mus'el) a muscle that controls the curvature of the lens of the eye. *252*

cochlea (kok'le-ah) that portion of the inner ear that resembles a snail's shell and contains the organ of Corti, the sense organ for hearing. *261*

cones (kōnz) bright-light receptors in the retina of the eye that detect color and provide visual acuity. *253*

fovea centralis (fo've-ah sen-tral'is) region of the retina, consisting of densely packed cones, that is responsible for the greatest visual acuity. *253*

lens (lenz) a clear membranelike structure found in the eye behind the iris. The lens brings objects into focus. *252*

organ of Corti (or'gan uv kor'ti) a portion of the inner ear that contains the receptors for hearing. *261*

otoliths (o'to-liths) granules associated with ciliated cells in the utricle and saccule. *261*

proprioceptor (pro''pre-o-sep'tor) receptor that assists the brain in knowing the position of the limbs. *249*

retina (ret'i-nah) the innermost layer of the eyeball that contains the rods and cones. *253*

rhodopsin (ro-dop'sin) visual purple, a pigment found in the rods. *256*

rods (rodz) dim-light receptors in the retina of the eye that detect motion but no color. *253*

saccule (sak'ūl) a saclike cavity that makes up part of the membranous labyrinth of the inner ear; receptors for static equilibrium. *261*

sclera (skle'rah) white, fibrous outer layer of the eyeball. *252*

semicircular canals (sem''e-ser'ku-lar kah-nal') tubular structures within the inner ear that contain the receptors responsible for the sense of dynamic equilibrium. *261*

tympanic membrane (tim-pan'ik mem'brān) membrane located between outer and middle ear that receives sound waves; the eardrum. *259*

utricle (u'tre-k'l) saclike cavity that makes up part of the membranous labyrinth of the inner ear; receptors for static equilibrium. *261*

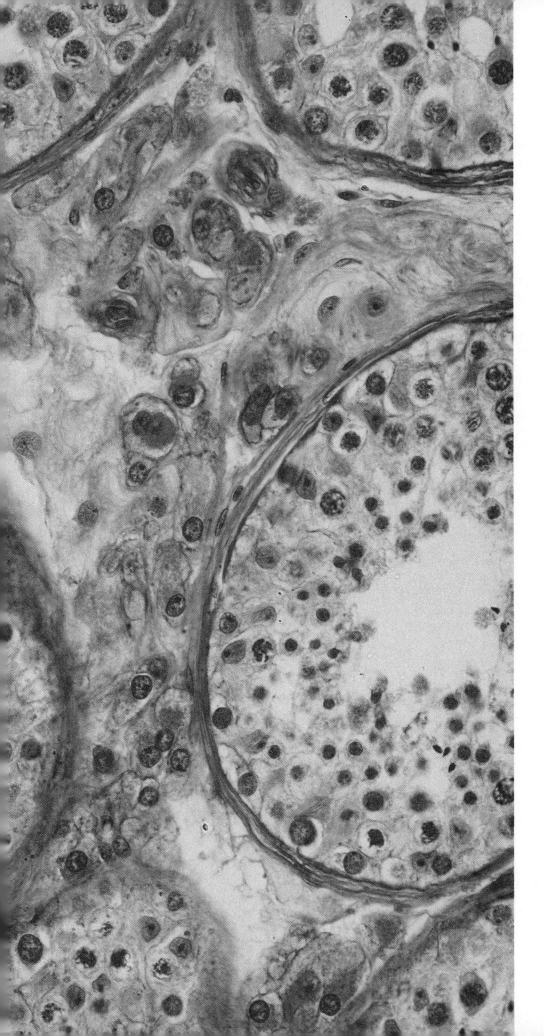

CHAPTER THIRTEEN

HORMONES

CHAPTER CONCEPTS

1 The hormonal glands are usually ductless glands that secrete directly into the bloodstream.

2 Hormonal secretions coordinate the biochemical functioning of the body by acting on target organs.

3 In general, hormonal glands are controlled by a negative feedback mechanism.

4 Malfunctioning of hormonal glands can bring about a dramatic change in appearance and cause early death.

Cross section of testis where seminiferous tubules produce sperm and interstitial cells produce testosterone, the main male sex hormone.

INTRODUCTION

The first pregnancy test was developed in 1928 after it was discovered that the urine of a pregnant woman contains a gonadotropic hormone that is secreted in such excess it spills over into the urine. When the urine of a pregnant woman was injected into a frog, it caused the frog to ovulate and produce eggs. The basis of a pregnancy test is the same today (fig. 13.1) except that antibodies to the hormone are added to a blood or urine sample. An observable antibody-antigen reaction (called agglutination) indicates that the woman is pregnant.

Along with the nervous system, hormones coordinate the functioning of body parts. Their presence or absence affects our metabolism, our appearance, and our behavior. It is now known that hormones are chemical regulators of cellular activity.

Hormones are organic substances that fall into two basic categories: (1) peptides (used here to include amino acid, polypeptide, and protein hormones) and (2) steroid hormones. Steroids are complex rings of carbon and hydrogen atoms (fig. 1.22). The difference between steroids is due to the atoms attached to these rings.

FIGURE 13.2 *Cellular activity of hormones.* a. *Peptide hormones combine with receptors located in the cell membrane. This promotes the production of cAMP, which in turn leads to activation of a particular enzyme.* b. *Steroid hormones pass through the membrane to combine with receptors, and the complex is believed to activate certain genes, leading to protein synthesis.*

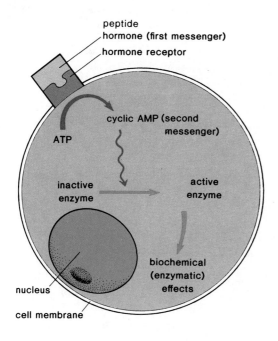

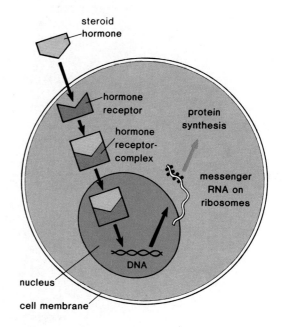

When hormones of the first type are received by a cell, they bind to specific receptor sites (fig. 13.2a) in the membrane. This hormone-receptor complex activates an enzyme that produces, for example, cyclic adenosine monophosphate (cAMP). Cyclic AMP is a compound made from ATP, but it contains only one phosphate group attached to ribose at two locations. The cAMP now activates the enzymes of the cell to carry out their normal functions.

Unlike peptide hormones, steroid hormones pass through the cell membrane with no difficulty because they are relatively small and lipid soluble. There are receptor molecules present in the cytoplasm (fig. 13.2b). After the hormone has combined with the receptor, the hormone-receptor complex moves into the nucleus, where it binds with chromatin at a location that promotes activation of a particular gene. Protein synthesis follows. In this manner, steroid hormones lead to protein synthesis.

Hormones are chemical messengers that influence the metabolism of the cell either directly or indirectly, depending on the hormone type.

Hormones are produced by glands (fig. 13.3) called **endocrine glands** that secrete their products internally, placing them directly in the blood. Since these glands do not have ducts for the transport of their secretions, they are sometimes called *ductless glands.* All hormones are carried throughout the body by the blood, but each one affects only a specific part or parts, appropriately termed *target organs.*

Table 13.1 lists the major endocrine glands in humans, the hormones produced by each, and the associated disorders that occur when there is an abnormal level of the hormones—either too much or too little. The adrenal cortex and sex glands produce steroid hormones; the other glands produce hormones that are either amino acids, polypeptides, or proteins.

The endocrine glands secrete hormones into the bloodstream for transport to target organs.

FIGURE 13.3 *Anatomical location of major endocrine glands in the body. The hypothalamus controls the pituitary, which in turn controls the hormonal secretions of the thyroid, adrenal cortex, and sex organs. Both sets of sex organs are shown; ordinarily an individual has only one set of these.*

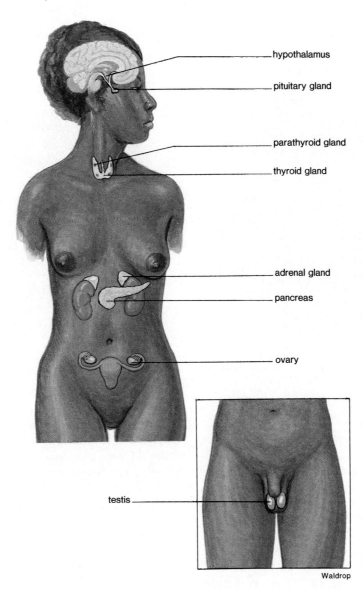

TABLE 13.1   THE PRINCIPAL ENDOCRINE GLANDS AND THEIR MAJOR HORMONES

GLAND	HORMONES	CHIEF FUNCTIONS	DISORDERS TOO MUCH/TOO LITTLE
Hypothalamus	Releasing hormones	Stimulates anterior pituitary	*See* anterior pituitary
Anterior pituitary	Thyroid-stimulating (TSH, thyrotropic)	Stimulates thyroid	*See* thyroid
	Adrenocorticotropic (ACTH)	Stimulates adrenal cortex	*See* adrenal cortex
	Gonadotropic	Stimulates gonads	*See* testes and ovary
	Follicle-stimulating (FSH)	Egg and sperm	
	Leuteinizing (LH)	Sex hormones	
	Lactogenic (LTH, prolactin)	Milk production	
	Growth (GH, somatotropic)	Growth	Giant, acromegaly/midget
Posterior pituitary	Antidiuretic (ADH, vasopressin)	Water retention by kidneys	Diverse[a]/diabetes insipidus
	Oxytocin	Uterine contraction	
Thyroid	Thyroxin	Increases metabolic rate (cellular respiration)	Exophthalmic goiter/simple goiter, myxedema, cretinism
	Calcitonin	Plasma level of calcium	Tetany/weak bones
Parathyroid	Parathormone (PTH)	Plasma levels of calcium and phosphorus	Weak bones/tetany
Adrenal cortex	Glucocorticoids (cortisol)	Gluconeogenesis	
	Mineralocorticoids (aldosterone)	Sodium retention; potassium excretion by kidneys	Cushing syndrome/Addison disease
Adrenal medulla	Adrenalin (epinephrine)	Fight or flight	
Pancreas	Insulin	Lowers blood sugar	Shock/diabetes mellitus
	Glucagon	Raises blood sugar	
Testes	Androgens (testosterone)	Secondary male characteristics	Diverse/eunuch
Ovary	Estrogen (by follicle)	Secondary female characteristics	Diverse/masculinization
	Progesterone (by corpus luteum)		

[a]The word *diverse* in this chart means that the symptoms have not been described as a syndrome in the medical literature.

PITUITARY GLAND

The pituitary gland, which has two portions called the **anterior pituitary** and the **posterior pituitary,** is a small gland, about 1 cm in diameter, that lies at the base of the brain.

Posterior Pituitary

The posterior pituitary is connected by means of a stalk to the hypothalamus, the portion of the brain that is concerned with homeostasis. The hormones released by the posterior pituitary are made in nerve cell bodies in the hypothalamus. These hormones then migrate through axons that terminate in the posterior pituitary (fig. 13.4). They are stored in these terminals until their release.

The posterior pituitary releases *antidiuretic hormone (ADH),* sometimes called **vasopressin. ADH,** as discussed in chapter 9, promotes the reabsorption of water from the collecting duct, a portion of the kidney tubules. It is believed that the hypothalamus contains osmoreceptors, or cells that are sensitive to the amount of water in the blood. When these cells detect that the blood lacks sufficient water, ADH is released from the posterior pituitary (fig. 13.4). As the blood becomes more dilute, the hormone ceases to be released.

FIGURE 13.4   *The posterior pituitary is connected to the hypothalamus by a stalk. The hypothalamus produces the hormones (ADH and oxytocin) that are secreted by the posterior pituitary. These are sent to the posterior pituitary by way of nerve fibers.*

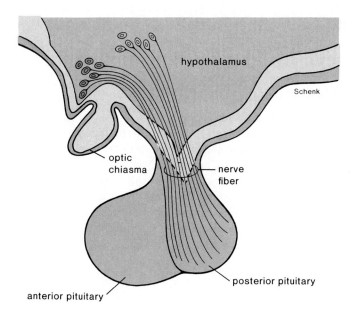

FIGURE 13.5 *Regulation of ADH secretion. When the blood is concentrated, the hypothalamus produces ADH, which is released by the posterior pituitary. This acts on the kidneys to retain more water so that the blood is diluted. Thereafter the hypothalamus does not produce ADH.*

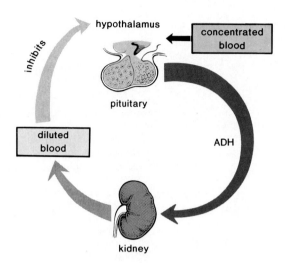

Figure 13.5 illustrates how the level of this hormone is controlled by a circular pattern in which the effect of the hormone (dilute blood) acts to shut down the release of the hormone. This is an example of control by negative feedback. Negative feedback mechanisms regulate the activities of most hormonal glands.

Inability to produce ADH causes **diabetes insipidus** (watery urine) in which a person produces copious amounts of urine with a resultant loss of salts from the blood. The condition can be corrected by the administration of ADH.

Oxytocin is another hormone released by the posterior pituitary that is made in the hypothalamus. Oxytocin causes the uterus to contract and may be used to artificially induce labor. It also stimulates the release of milk from the breast when a baby is nursing.

The hormones of the posterior pituitary, ADH and oxytocin, are produced in the hypothalamus.

Anterior Pituitary

There is a portal system composed of tiny blood vessels that connects the anterior pituitary to the hypothalamus. The hypothalamus controls the anterior pituitary by producing hypothalamic-releasing hormones, which are transported to the anterior pituitary by the blood within the portal system that connects the two organs. Each type of releasing hormone causes the anterior pituitary either to secrete or to stop secreting a specific hormone. The anterior pituitary produces at least six different hormones (fig. 13.6).

Three of the hormones produced by the anterior pituitary have a direct effect on the body. **Growth hormone (GH),** or somatotropin, affects the physical appearance dramatically since it determines the height of the individual (fig. 13.7). If little or no growth hormone is secreted by the anterior pituitary during childhood, the person could become a midget—perfect proportions but quite small in stature. If too much growth hormone is secreted, the person could become a giant. Giants usually have poor health, primarily because growth hormone has a secondary effect on blood sugar level, promoting an illness called diabetes mellitus (sugar diabetes), which is discussed in a following section on the pancreas.

Growth hormone promotes cell division, protein synthesis, and bone growth. It stimulates the transport of amino acids into cells and increases the activity of ribosomes, both of which are essential to protein synthesis. In bones, it promotes growth of the cartilaginous plates and causes osteoblasts to form bone (p. 229). Evidence suggests that the effects on cartilage and bone may actually be due to hormones called somatomedins, released by the liver. Growth hormone causes the liver to release somatomedins.

If the production of GH increases in an adult after full height has been obtained, only certain bones respond. These are the bones of the jaw, eyebrow ridges, nose, fingers, and toes. When these begin to grow, the person takes on a slightly grotesque look with huge fingers and toes, a condition called **acromegaly** (fig. 13.8).

Lactogenic hormone (LTH), also called **prolactin,** is produced in quantity only after childbirth. It causes the mammary glands in the breasts to develop and produce milk.

FIGURE 13.6 *The anterior pituitary is connected to the hypothalamus only by a portal system. The hypothalamus sends releasing hormones to the anterior pituitary by this circulatory route. The releasing hormones specifically promote or inhibit the secretion of anterior pituitary hormones.*

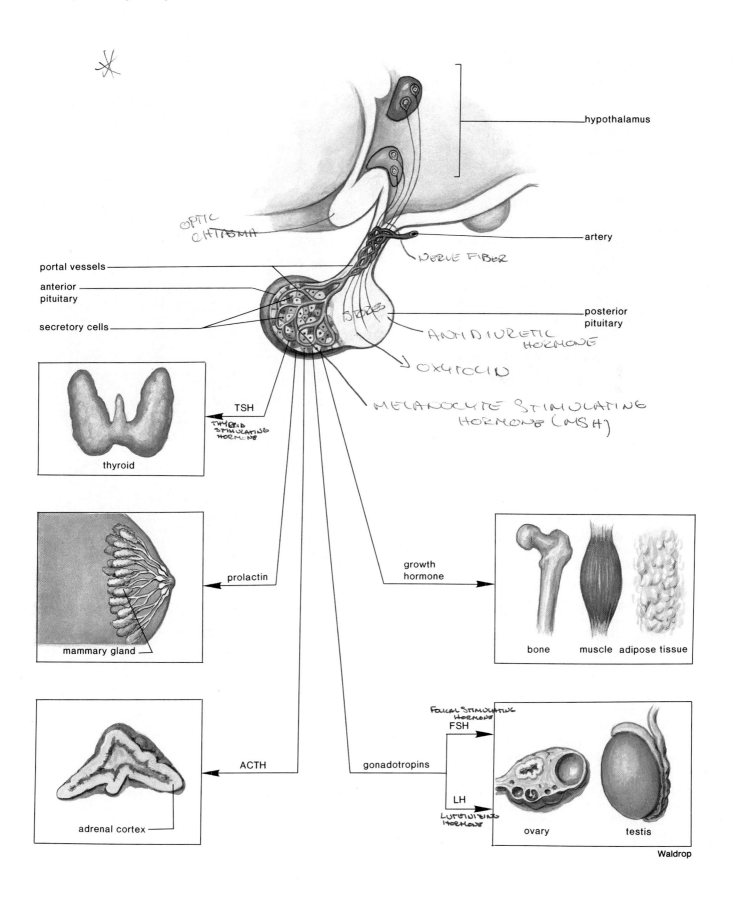

hypothalamus

OPTIC CHIASMA

artery

NERVE FIBER

portal vessels

anterior pituitary

posterior pituitary

secretory cells

STORES

ANTIDIURETIC HORMONE

OXYTOCIN

MELANOCYTE STIMULATING HORMONE (MSH)

TSH
THYROID STIMULATING HORMONE

thyroid

prolactin

mammary gland

growth hormone

bone   muscle   adipose tissue

ACTH

adrenal cortex

gonadotropins

FOLICAL STIMULATING HORMONE
FSH

LH
LUTEINIZING HORMONE

ovary   testis

Waldrop

FIGURE 13.7 *Sandy Allen is one of the world's tallest women, due to a higher than usual amount of growth hormone produced by the anterior pituitary.*

FIGURE 13.8 *Acromegaly is caused by the production of growth hormone in the adult. It is characterized by an enlargement of the bones in the face and fingers of an adult. a. At age 20, this individual was normal. b. At age 24, there is some enlargement of nose and jaw.*

a.

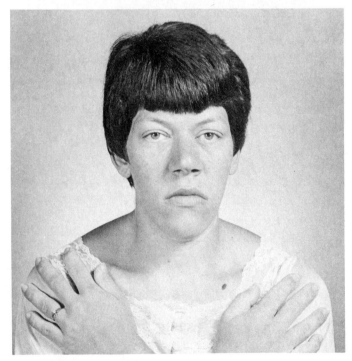

b.

GH and LTH are two hormones produced by the anterior pituitary. GH influences the height of children and brings about a condition called acromegaly in adults. LTH promotes milk production after childbirth.

Master Gland

The anterior pituitary is called the master gland because it controls the secretion of other endocrine glands (fig. 13.6). As indicated in table 13.1, the anterior pituitary secretes the following hormones, which have an effect on other glands:

1. **TSH,** thyroid-stimulating hormone
2. **ACTH,** a hormone that stimulates the adrenal cortex
3. **Gonadotropic hormones** (FSH and LH) that stimulate the gonads, the testes in males and the ovaries in females

TSH causes the thyroid to produce thyroxin; ACTH causes the adrenal cortex to produce cortisol; and gonadotropic hormones cause the gonads to secrete sex hormones. Notice that it is now possible to indicate a three-tiered relationship between the hypothalamus, pituitary, and other endocrine glands. The hypothalamus produces releasing hormones that control the anterior pituitary, and the anterior pituitary produces hormones that control the thyroid, adrenal cortex, and gonads. Figure 13.9 illustrates the feedback mechanism that controls the activity of these glands.

FIGURE 13.9 Control of hormone secretion. The level of thyroxin in the body is controlled in three ways, as shown: (a) the level of TSH exerts feedback control over the hypothalamus; (b) the level of thyroxin exerts feedback control over the anterior pituitary; and (c) over the hypothalamus, which secretes TRH (thyroid releasing hormone). In this way thyroxin controls its own secretion. Substitution of the appropriate terms would also allow this diagram to illustrate control of cortisol and sex hormone levels.

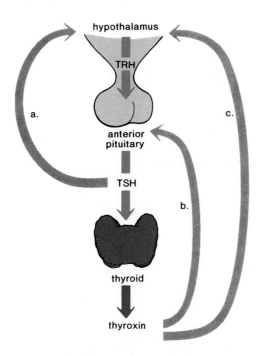

FIGURE 13.10 Thyroid gland. The thyroid is composed of many follicles, lined by epithelial cells, that secrete a precursor to thyroxin. (ES = epithelial cell surface; Lu = lumen of follicle; CT = connective tissue.)
From: Tissues and Organs: A Text-Atlas of Scanning Electron Microscopy by R. G. Kessel and R. Kardon. W. H. Freeman and Company, © 1979.

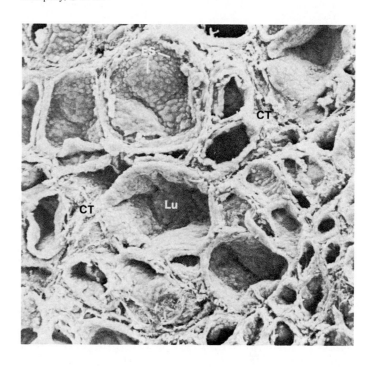

The hypothalamus, anterior pituitary, and the other endocrine glands controlled by the anterior pituitary are all involved in a self-regulating feedback loop.

THYROID GLAND

The thyroid gland (fig. 13.3) is located in the neck and is attached to the trachea just below the larynx. Internally (fig. 13.10), the gland is composed of a large number of follicles filled with thyroglobulin, the storage form of thyroxin. The production of both of these requires iodine. Iodine is actively transported into the thyroid gland, where the concentration may become as much as twenty-five times that of the blood. If iodine is lacking in the diet, the thyroid gland enlarges, producing a goiter (fig. 13.11). The cause of this enlargement becomes clear if we refer to figure 13.9. When there is a low level of thyroxin in the blood, a condition called hypothyroidism, the anterior pituitary is stimulated to produce **TSH**. TSH causes the thyroid to increase in size so that enough **thyroxin** usually is produced. In this case, enlargement continues because enough thyroxin is never produced. An enlarged thyroid that produces some thyroxin is called a **simple goiter.**

FIGURE 13.11 Simple goiter. An enlarged thyroid gland is often caused by a lack of iodine in the diet. Without iodine the thyroid is unable to produce thyroxin, and continued anterior pituitary stimulation causes the gland to enlarge.

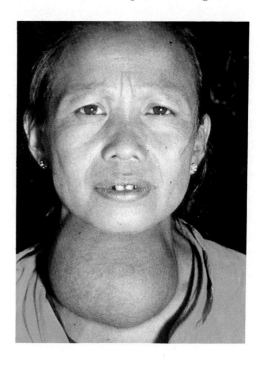

Activity and Disorders

Thyroxin increases the metabolic rate. It does not have a target organ; instead, it stimulates most of the cells of the body to metabolize at a faster rate. The number of respiratory enzymes in the cell increases, as does oxygen uptake.

If the thyroid fails to develop properly, a condition called **cretinism** results. Cretins (fig. 13.12) are short, stocky persons who have had extreme hypothyroidism since childhood and/or infancy. Thyroid therapy can initiate growth, but unless treatment is begun within the first two months of the condition's onset, mental retardation results. The occurrence of hypothyroidism in adults produces the condition known as **myxedema** (fig. 13.13), which is characterized by lethargy, weight gain, loss of hair, slower pulse rate, decreased body temperature, and thickness and puffiness of the skin. The administration of adequate doses of the thyroid hormone restores normal function and appearance.

In the case of hyperthyroidism (too much thyroxin), the thyroid gland is enlarged and overactive, causing a goiter to form and the eyes to protrude for some unknown reason. The type of goiter formed is called **exophthalmic goiter** (fig. 13.14). The patient usually becomes hyperactive, nervous, irritable, and suffers from insomnia. Removal or destruction of a portion of the thyroid by means of radioactive iodine is sometimes effective in curing the condition.

Calcitonin

In addition to thyroxin, the thyroid gland also produces the hormone **calcitonin.** This hormone helps regulate the calcium level in the blood and opposes the action of parathyroid hormone. The interaction of these two hormones is discussed on page 184.

The anterior pituitary produces TSH, a hormone that promotes the production of thyroxin by the thyroid. Thyroxin, which speeds up metabolism, can affect the body as a whole as exemplified by cretinism and myxedema.

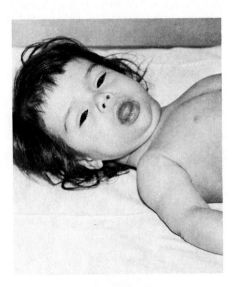

FIGURE 13.12 *Cretinism. Cretins are individuals who have suffered from thyroxin insufficiency since birth or early childhood. Skeletal growth is usually inhibited to a greater extent than soft tissue growth; therefore, the child appears short and stocky. Sometimes the tongue becomes so large that it obstructs swallowing and breathing.*

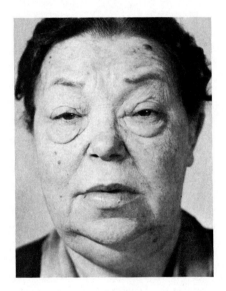

FIGURE 13.13 *Myxedema is caused by thyroid insufficiency in the older adult. An unusual type of edema leads to swelling of the face and bagginess under the eyes.*

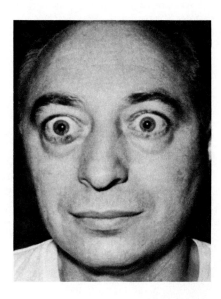

ADRENAL GLANDS

The adrenal glands, as their name implies (*ad* = near; *renal* = kidneys), lie atop the kidneys (fig. 13.3). Each consists of an outer portion, called the *cortex,* and an inner portion, called the *medulla.* These portions, like the anterior and posterior pituitary, have no connection to one another.

The hypothalamus exerts control over the activity of both portions of the adrenal glands. The hypothalamus can initiate nerve impulses that travel by way of the brain stem, nerve cord, and sympathetic nerve fibers (fig. 10.17) to the adrenal medulla, which then secretes its hormones. The hypothalamus, by means of ACTH-releasing hormone, controls the anterior pituitary's secretion of **ACTH,** which in turn stimulates the adrenal cortex. Stress of all types, including both emotional and physical trauma, prompts the hypothalamus to stimulate the adrenal glands.

Scientists were very interested of late to learn that ACTH and the opioids (endorphins and enkephalins) are all chemically related. Although enkephalins are found in other parts of the nervous system, they are not predominant in the pituitary as are ACTH and beta-endorphin. It's possible that both ACTH and beta-endorphin are released as a response to stress. The runner's high (fig. 13.15) has been attributed to the release of endorphins from the anterior pituitary.

The adrenal glands have two parts, an outer cortex and an inner medulla. The adrenal medulla is under nervous control and the cortex is under hormonal control—ACTH is an anterior pituitary hormone.

FIGURE 13.15 *It is now known that the anterior pituitary produces beta-endorphin, an internal opioid that, like morphine, can produce a feeling of euphoria and a higher threshold of pain. It is possible that physical activity causes the release of endorphin and can elevate the mood. This would account for what is called "the runner's high."*

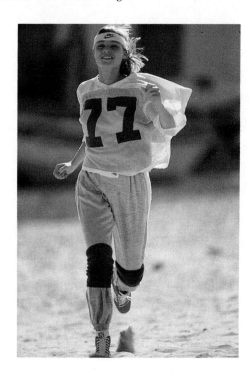

FIGURE 13.16 *Adrenal glands.* a. *The adrenal glands lie atop the kidneys and consist of the cortex and the medulla.* b. *Reaction to stress. Stress causes the hypothalamus to produce a releasing hormone that stimulates the anterior pituitary to produce ACTH. This hormone causes the adrenal cortex to produce cortisol, a hormone that brings about gluconeogenesis, which is thought to relieve stress.*

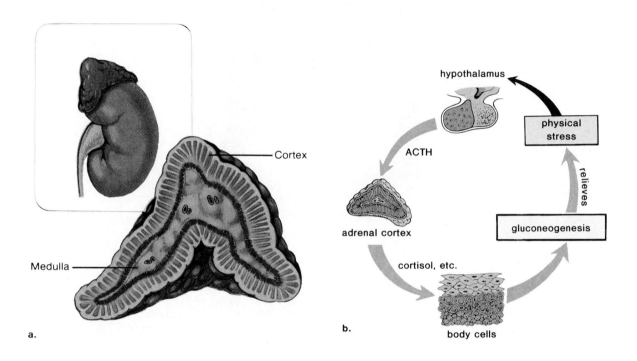

Adrenal Medulla

The adrenal medulla (fig. 13.16a) secretes **adrenalin** and **noradrenalin.** The postganglionic fibers of the sympathetic nervous system also secrete noradrenalin. In fact, the adrenal medulla is often considered to be an adjunct to the sympathetic nervous system.

Adrenalin and noradrenalin are involved in the body's immediate response to stress. They bring about all those effects that occur when an individual reacts to an emergency. Blood glucose level rises and the metabolic rate increases, as do breathing and the heart rate. The blood vessels in the intestine constrict, but those in the muscles dilate. Increased circulation to the muscles causes them to have more strength than usual. The individual has a wide-eyed look and is extremely alert. Adrenalin has such a profound effect on the heart that it is often injected directly into a heart that has stopped beating in an attempt to stimulate its contraction.

The adrenal medulla releases adrenalin and noradrenalin into the bloodstream, helping us cope with situations that seem to threaten our survival.

Adrenal Cortex

Although the adrenal medulla may be removed with no ill effects, the adrenal cortex is absolutely necessary to life. The two major types of hormones made by the adrenal cortex are the glucocorticoids and the mineralocorticoids. It also secretes a small amount of male and an even smaller amount of female sex hormones. All of these hormones are steroids.

Glucocorticoids

Of the various glucocorticoids, the hormone responsible for the greatest amount of activity is cortisol. The secretion of cortisol helps an individual recover from stress (fig. 13.16b). **Cortisol** raises the level of amino acids in the blood, which, in turn, leads to an increased level of glucose when the liver converts these amino acids into glucose. It is said that the adrenal cortex brings about gluconeogenesis, or the production of glucose from nonglucose substances. Gluconeogenesis aids recovery because it allows an individual to maintain cellular respiration, especially in the brain, even when the body is not being supplied with dietary glucose. The amino acids not converted to glucose can be used for tissue repair should injury occur.

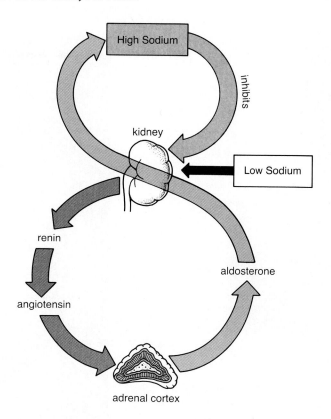

adrenal cortex

Cortisol also counteracts the inflammatory response (p. 134). During the inflammatory response, capillaries become more permeable and fluid leaks out, causing swelling in surrounding tissues. This causes the pain and swelling of joints that accompany arthritis and bursitis. The administration of cortisol aids these conditions because it reduces inflammation.

Mineralocorticoids

The secretion of mineralocorticoids, the most important of which is **aldosterone,** is not under the control of the anterior pituitary. These hormones maintain the level of sodium (Na^+) and potassium (K^+) in the blood, their primary target organ being the kidney, where they promote renal reabsorption of sodium and renal excretion of potassium. The levels of Na^+ and K^+ in the blood are critical for nerve conduction and muscle contraction; in fact, cardiac failure may result from too low a level of potassium.

The level of Na^+ is particularly important to the maintenance of blood pressure, and the concentration of this ion indirectly regulates the secretion of aldosterone. When the sodium level is low the kidneys secrete renin. Renin is an enzyme that leads to the conversion of the plasma protein angiotensinogen to angiotensin, and this molecule stimulates the adrenal cortex to release aldosterone (fig. 13.17). This is

called the *renin-angiotensin-aldosterone system*. The effect of this system is to raise the blood pressure. First angiotensin constricts the arteries directly, and secondly, when aldosterone causes the kidneys to reabsorb sodium, blood volume is raised as water is reabsorbed.

Cortisol contributing to gluconeogenesis and aldosterone contributing to sodium retention are two hormones secreted by the adrenal cortex.

Disorders

Addison Disease When there is a low level of adrenal cortex hormones in the body, the person begins to suffer from Addison disease. Because of the lack of cortisol, the patient is unable to maintain the glucose level of the blood, tissue repair is suppressed, and there is high susceptibility to any kind of stress. Even a mild infection can cause death. Due to the lack of aldosterone, the blood sodium level is low, and the person experiences low blood pressure along with acidosis and low pH. In addition, the patient has a peculiar bronzing of the skin (fig. 13.18).

FIGURE 13.18 *Addison disease. This condition is characterized by a peculiar bronzing of the skin, as seen in the face and the thin skin of the nipples of this patient.*

FIGURE 13.19 *Cushing syndrome. Persons with this condition tend to have an enlarged trunk and moonlike face. Masculinization may occur in women due to the excessive androgens in the body.*

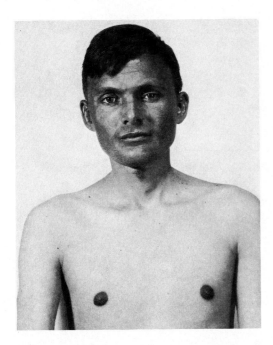

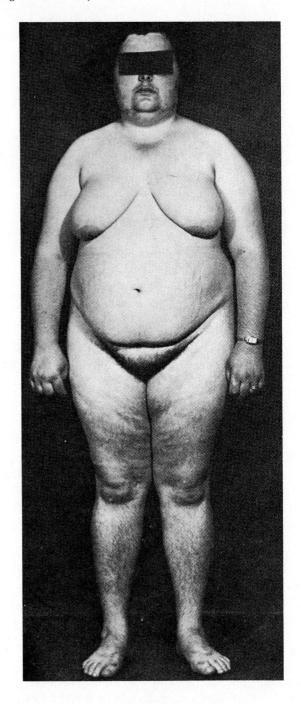

Cushing Syndrome   When there is a high level of adrenal cortex hormones in the body, the person suffers from Cushing syndrome (fig. 13.19). Cortisol causes a tendency toward diabetes mellitus, a decrease in muscular protein, and an increase in subcutaneous fat. Because of these effects, the person usually develops thin arms and legs and an enlarged trunk. Due to the high level of sodium in the blood, the patient has alkaline blood, hypertension, and edema of the face, which gives the face a moonlike shape. Masculinization may occur in women due to oversecretion of adrenal male sex hormone.

Addison disease is due to adrenal cortex hyposecretion, and Cushing syndrome is due to adrenal cortex hypersecretion.

SEX ORGANS

The sex organs are the testes in the male and the ovaries in the female. As will be discussed in detail in the following chapter, the testes produce testosterone, the primary male sex hormone, and the ovaries produce estrogen and progesterone, the female sex hormones. The hypothalamus and pituitary gland control the hormonal secretions of these organs in the same manner as described for the thyroid gland in figure 13.9.

The sex hormones control the secondary sex characteristics of the male and female (p. 297 and p. 307). Among other traits, males have greater muscle strength than do females. Generally, athletes believe that the intake of anabolic steroids—that is, the male sex hormone testosterone or synthetically related steroids—will cause greater muscle strength. The reading on the next page discusses the pros and cons of taking these steroids, which are considered illegal by the International Olympic Committee. Any Olympic athlete whose urine tests positive for steroids at the time of an event is immediately disqualified from winning a medal.

Inside a glitzy health club in an affluent North Shore suburb of Boston, 19-year-old Matthew Creighton stands in front of a mirror that covers an entire wall and inspects his bulging biceps.

Creighton has been pumping iron for six years, going through the same routine five mornings a week. He began entering weightlifting tournaments last year and quickly realized that good training alone was not going to make him very competitive. On a September morning last year, Creighton met a man outside a health club and for $50 bought his first bottle of steroids. He has been a regular user since. . . .

While much of the attention surrounding the nation's drug epidemic has focused on cocaine and heroin, muscle-enhancing drugs such as steroids are being used in steadily increasing numbers, and especially alarming is their growing popularity among youngsters.

Because steroids by and large have not been the target of federal law-enforcement authorities, there are no accurate figures on the number of steroid users. But the best estimate, according to federal officials, is that 1 million–3 million Americans use steroids, a figure that has increased steadily since the early 1970s.

The uproar surrounding the use of steroids by Canadian sprinter Ben Johnson in the Seoul Olympics has underscored an issue that has received little public attention in the past. For many years, steroids were used primarily by bodybuilders, weightlifters and other athletes, such as professional football players, to help them in competition. But in recent years, steroid use has expanded to fitness buffs, high school and college athletes, and skinny youngsters wanting to gain bulk quickly.

Anabolic steroids are synthetic hormones taken either orally or by injection. They were developed in the 1930s to prevent the atrophying of muscles in patients with debilitating illnesses. In some cases, they also were given to burn victims and surgery patients to speed recovery. . . .

Steroids, which can be dispensed legally only through a prescription, also have been used to treat certain forms of rare anemia and breast cancer.

But those using steroids illegally to enhance their appearance take them in large quantities, sometimes as much as 40 or 50 times the recommended dosage, according to FDA officials. Prolonged use of such a large quantity can lead to stunted growth in youngsters and high blood cholesterol. In men, it can result in baldness, acne, shrunken testes, feminized breasts and infertility.

In women, who normally produce very low levels of testosterone and therefore gain much more from steroids, such overuse can promote facial hair, deepening of the voice and an enlarged clitoris.

While scientific research has not been exhaustive, some studies have linked steroid abuse to cancer of the liver, prostate and testes, as well as kidney diseases and atherosclerosis.

"No one has really conclusively determined the longterm effects of steroid use because most doctors don't want to appear to condone a practice many consider unhealthy." said Dr. Gloria Troendle, senior medical officer at the FDA (Federal Drug Administration).

But another FDA official in Washington, who asked not to be identified said, "It seems the evidence is becoming clear that youngsters who use anabolic steroids in large doses for two or three months face the possibility of dying in their 30s or 40s."

Growing evidence also points to another conclusion that steroids, which some in the medical community say appear to be addictive, can damage the mind. In psychotic side effects sometimes referred to as " 'roid mania," users of large quantities of steroids have experienced mania, wild aggression and delusions, said Dr. Harrison Pope, a psychiatrist at McLean Hospital in Belmont who has studied the effects of steroids on the mind.

One patient Pope examined had a friend videotape him while he deliberately drove a car into a tree at 35 m.p.h. "My hunch is that we are seeing

continued

THE GROWING THREAT OF STEROIDS

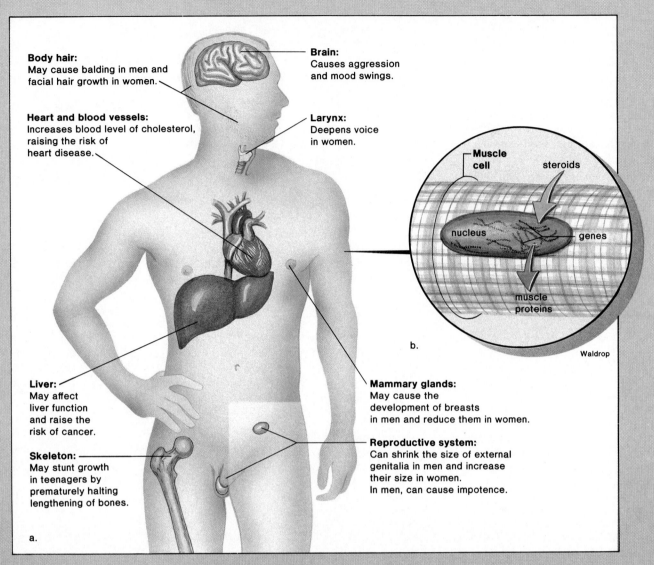

Body hair:
May cause balding in men and facial hair growth in women.

Brain:
Causes aggression and mood swings.

Heart and blood vessels:
Increases blood level of cholesterol, raising the risk of heart disease.

Larynx:
Deepens voice in women.

Muscle cell   **steroids**

nucleus   genes

muscle proteins

b.

Waldrop

Liver:
May affect liver function and raise the risk of cancer.

Mammary glands:
May cause the development of breasts in men and reduce them in women.

Skeleton:
May stunt growth in teenagers by prematurely halting lengthening of bones.

Reproductive system:
Can shrink the size of external genitalia in men and increase their size in women. In men, can cause impotence.

a.

How steroid drugs affect the body.
a. *Suspected harmful effects.* b. *How steroids build muscle.*

only a small part of . . . the magnitude of the psychiatric effects of steroids," Pope said in a recent interview in *Currents,* a medical trade journal.

Steroids had their first nonmedical use in World War II when Nazi doctors gave them to soldiers in an attempt to make them more aggressive. Following the war, the Soviet Union and other Eastern European nations began dispensing steroids to their athletes. In the 1950s, an American doctor working with the U.S. weightlifting team learned that Soviet athletes were using steroids and markedly improving their performance. The doctor introduced steroids to lifters at the York (Pa.) Barbell Club, where they were an immediate hit.

In the 1960s, as their dangerous side effects started being documented, steroids fell into disfavor in the medical community. Since then, the FDA has banned most steroids. Currently, only about a dozen are approved for limited medical use.

The limited availability of steroids by prescription has helped trigger a flourishing black market. Smuggled into the United States from Mexico and, to a lesser degree, Eastern Europe, the illicit trafficking has become a $100-million-a-year business, according to federal authorities.

Steriods are sold through the mail or in gyms and health clubs, federal authorities said. "You've got kids in the club making maybe $5 or $6 an hour, but if they're selling steroids, they can make as much as $900–$1,000 a week," said the operator of several health clubs in the Los Angeles area who did not want to give his name. . . .

Until a mandate from Congress changes the status of the drug, the DEA (Drug Enforcement Administration) does not have the authority to mount a large-scale assault on steroid dealers. A bill in the House would make steroids a controlled substance, but the DEA is opposed to the measure because it says it does not have the resources to enforce it.

There have been some successes, however. According to the U.S. Justice Department, the federal government has obtained guilty pleas or convictions from 60 persons in the past 30 months, and 120 more individuals face charges related to trafficking in steroids. And the anti-drug bill ratified recently by Congress includes a provision raising the penalty for the distribution of steroids from a misdemeanor to a felony, with jail terms of three years for dispensing to adults and six years to children.

Still, most medical, sports and law enforcement officials agree that the nation's craving for steroids will not subside soon without more stringent law enforcement and better education about the dangers of the drugs. . . .

Despite the health risks, steroids will continue to be popular among younger users who take them to "feel better about themselves." said Dr. Jack Freinhar, a psychiatrist at Del Amo Hospital in Torrance, Calif., who treats adolescent steroid abusers.

"Physical risk doesn't matter when you're talking about saving the self," said Freinhar, who believes steroids can become an addiction he calls "reverse anorexia." "With anorexia, one is never thin enough. In our culture, there's this push to be muscular. Most of the youngsters I treat have had a deprived childhood and they have holes in the self. They need acceptance. One way is to look good, and steroids provide a fast method. But they can never get enough because each time they look in the mirror they still see themselves as too thin."

"The Growing Threat of Steroids," a three-part series by John Powers and Diego Rabadeneira, *The Boston Globe,* October 1988. Reprinted courtesy of *The Boston Globe.*

Androgens (specifically testosterone) are the male sex hormones. Estrogen and progesterone are the female sex hormones.

PARATHYROID GLANDS

The parathyroid glands are embedded in the posterior surface of the thyroid gland, as shown in figure 13.20. Many years ago, these four small glands were sometimes removed by mistake during thyroid surgery. Under the influence of **parathyroid hormone (PTH),** also called parathormone, the calcium (Ca^{++}) level in the blood increases and the phosphate (PO^-) level decreases. The hormone stimulates the absorption of calcium from the gut, the retention of calcium by the kidneys, and the demineralization of bone. In other words, PTH promotes the activity of osteoclasts, the bone-resorbing cells. Although this also raises the level of phosphate in the blood, PTH acts on the kidneys to excrete phosphate in the urine. When the ovaries produce decreasing amounts of the female sex hormone estrogen following menopause, a woman is more likely to suffer from osteoporosis, characterized by a thinning of the bones. It is therefore reasoned that estrogen makes bones less sensitive to PTH. As discussed on page 91, a diet sufficient in calcium may help prevent osteoporosis to a small degree.

If insufficient parathyroid hormone is produced, the level of calcium in the blood drops, resulting in **tetany.** In tetany, the body shakes from continuous muscle contraction. The effect is really brought about by increased excitability of the nerves, which fire spontaneously and without rest. Calcium plays an important role in both nervous conduction and muscle contraction.

The level of PTH secretion is controlled by a feedback mechanism involving calcium (fig. 13.20). When the calcium level rises, PTH secretion is inhibited, and when the calcium level lowers, PTH secretion is stimulated.

As mentioned previously, the thyroid secretes calcitonin, which also influences blood calcium level. Although calcitonin has the opposite effect of PTH, particularly on the bones, its action is not believed to be as significant. Still, the two hormones function together to regulate the level of calcium in the blood.

Parathyroid hormone maintains a high blood level of calcium by promoting its absorption in the gut, its reabsorption by the kidneys, and demineralization of bone. These actions are opposed by calcitonin produced by the thyroid.

FIGURE 13.20 *Parathyroid glands. a. These small glands are embedded in the posterior surface of the thyroid gland. Yet the parathyroids and thyroid glands have no anatomical or physiological connection with one another. b. Regulation of parathyroid hormone secretion. A low blood level of calcium causes the parathyroids to secrete parathyroid hormone, which causes the kidneys and gut to retain calcium and osteoclasts to break down bone. The end result is an increased level of calcium in the blood. A high blood level of calcium inhibits hormonal secretion of parathyroid hormone.*

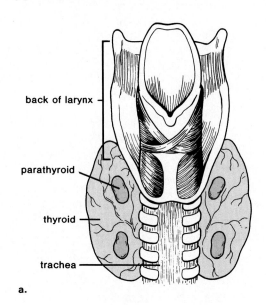

back of larynx

parathyroid

thyroid

trachea

a.

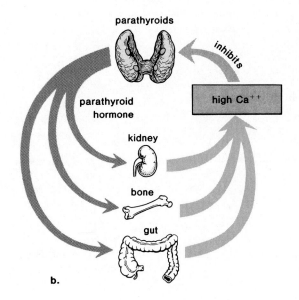

parathyroids

inhibits

parathyroid hormone

high Ca^{++}

kidney

bone

gut

b.

TABLE 13.2 *SYMPTOMS OF INSULIN SHOCK AND DIABETIC COMA*

INSULIN SHOCK	DIABETIC COMA
Sudden onset	Slow, gradual onset
Perspiration, pale skin	Dry, hot skin
Dizziness	No dizziness
Palpitation	No palpitation
Hunger	No hunger
Normal urination	Excessive urination
Normal thirst	Excessive thirst
Shallow breathing	Deep, labored breathing
Normal breath odor	Fruity breath odor
Confusion, disorientation, strange behavior	Drowsiness and great lethargy leading to stupor
Urinary sugar absent or slight	Large amounts of urinary sugar
No acetone in urine	Acetone present in urine

Reprinted from *How To Live With Diabetes* by Henry Dolger, M.D., and Bernard Seeman. By permission of W. W. Norton and Company, Inc. Copyright © 1972, 1965, 1958 by Henry Dolger and Bernard Seeman.

PANCREAS

The pancreas is a long, soft organ that lies transversely in the abdomen (fig. 13.21) between the kidneys and near the duodenum of the small intestine. It is composed of two types of tissues; one of these produces and secretes the digestive juices that go by way of the pancreatic duct to the small intestine, and the other type, called the **islets of Langerhans,** produces and secretes the hormones insulin and glucagon directly into the blood. Insulin and glucagon are hormones that affect the blood glucose level in opposite directions—**insulin** decreases the level and **glucagon** increases the level of glucose.

Insulin is secreted when there is a high level of glucose in the blood, which usually occurs just after eating. Once the glucose level returns to normal, insulin is not secreted, as illustrated in figure 13.22. Insulin is believed to cause almost all of the cells of the body to take up glucose. When the liver and muscles take up glucose, they convert to glycogen any glucose not needed immediately. Therefore, insulin promotes the storage of glucose as glycogen.

Diabetes Mellitus

The symptoms of **diabetes mellitus** (sugar diabetes) include the following: sugar in the urine; frequent, copious urination; abnormal thirst; rapid loss of weight; general weakness; drowsiness and fatigue; itching of the genitals and skin; visual disturbances, blurring; and skin disorders, such as boils, carbuncles, and infection.

Many of these symptoms develop because sugar is not being metabolized by the cells. The liver fails to store glucose as glycogen, and all of the cells fail to utilize glucose as an energy source. This means that the blood glucose level rises very high after eating, causing glucose to be excreted in the urine. More water than usual is therefore excreted so that the diabetic is extremely thirsty.

Since carbohydrates are not being metabolized, the body turns to the breakdown of proteins and fat for energy. Unfortunately, the breakdown of these molecules leads to the buildup of acids in the blood (acidosis) and respiratory distress. It is the latter that can eventually cause coma and death of the diabetic. The symptoms that lead to coma (table 13.2) develop slowly.

There are two types of diabetes. In *Type I* diabetes, formerly called juvenile-onset diabetes, the pancreas is not producing insulin. Therefore, the patient must have daily insulin injections. These injections control the diabetic symptoms but may still cause inconveniences since either an overdose of insulin or the absence of regular eating can bring on the symptoms of insulin shock (table 13.2). These symptoms appear because the blood sugar level has decreased below normal levels. Since the brain requires a constant supply of sugar, unconsciousness results. The cure is quite simple: an immediate source of sugar, such as a sugar cube or fruit juice, can counteract insulin shock immediately.

Obviously, insulin injections are not the same as a fully functioning pancreas that responds on demand to a high glucose level by supplying insulin. For this reason, some doctors advocate a pancreas transplant, which does however require that the recipient take immunosuppressive drugs.

Of the over six million people who now have diabetes in the United States, at least five million have *Type II,* formerly called maturity-onset diabetes. In this type of diabetes, now known to occur in obese people of any age, the pancreas is producing insulin, but the cells do not respond to it. At first the cells lack the receptors necessary to detect the presence of insulin, and later the cells are even incapable of taking up glucose. If Type II diabetes is left untreated, the results can be as serious as Type I diabetes. Diabetics are prone to blindness, kidney disease, and circulatory disorders, including strokes. Pregnancy carries an increased risk of diabetic coma, and the child of a diabetic is somewhat more likely to be stillborn or to die shortly after birth. It is important, therefore, that Type II diabetes be prevented or at least controlled. The best defense is a nonfattening diet and regular exercise. If that fails, there are oral drugs that make the cells more sensitive to the effects of insulin or that stimulate the pancreas to make more of it.

The most common illness due to hormonal imbalance is diabetes mellitus, caused by a lack of insulin or insensitivity of cells to insulin. Insulin lowers blood glucose levels by causing the cells to take it up and the liver to convert it to glycogen.

FIGURE 13.21 *Gross and microscopic anatomy of the pancreas. The pancreas lies in the abdomen between the kidneys near the duodenum. As an exocrine gland it secretes digestive enzymes that enter the duodenum by the pancreatic duct. As an endocrine gland it secretes insulin and glucagon into the bloodstream. (Top right) The "alpha" cells of the islets of Langerhans produce glucagon, and the "beta" cells produce insulin.*

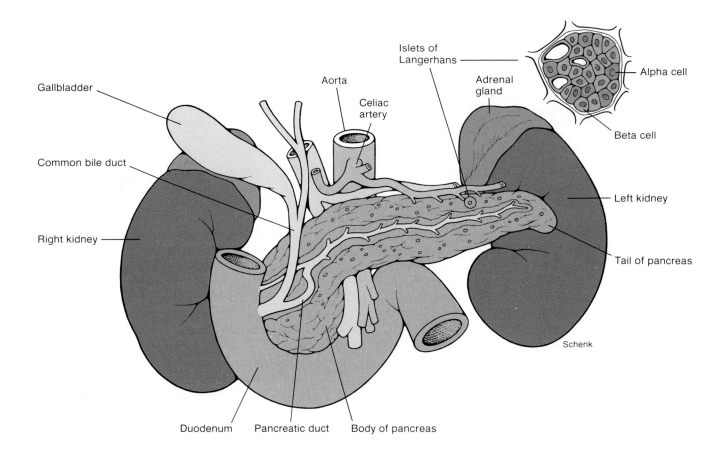

FIGURE 13.22 *Regulation of insulin secretion. In response to a high blood sugar level, the pancreas secretes insulin, which promotes the uptake of glucose in body cells, muscles, and the liver. As a result of a low blood glucose level, the pancreas stops secreting insulin.*

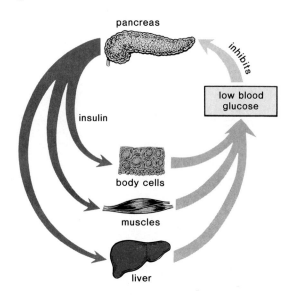

HUMAN ISSUE

Two physicians that do pancreas transplants differ in their approach. Dr. Sutherland of the University of Minnesota Hospital uses pancreatic tissue from live donors if possible, but Dr. Rohrer of New England Deaconess Hospital in Boston prefers to use only tissue from cadavers. He feels that pancreas donors are unnecessarily endangering their own health because they too might one day develop diabetes—the speed of the disease's onset is partially determined by the total amount of functioning pancreas. The pancreas donor should be a relative whose antigens best match those of the recipient, and diabetes is known to run in families. In addition, sometimes a pancreatic transplant does not arrest the devastating effects of diabetes in the recipient. Given these facts, what policies should be developed regarding pancreas transplants? Should relatives be obligated to be pancreas donors? Who should make the final decision when the potential donor is a child under the age of 18?

FIGURE 13.23 *Local hormones versus nonlocal hormones.* a. *Some hormones, called autocoids, are produced and effective in the immediate environment and are therefore called local hormones. It would be possible to include neurotransmitters in this category.* b. *Some hormones are secreted into the bloodstream and transported by the blood to target organs. It would be possible to include hypothalamic-releasing hormones in this category.*

OTHER HORMONES

It is appropriate to discuss certain hormones in other chapters of the text. For example, the gonadotropic and sex hormones will be discussed at greater length in chapter 14. Previously it was mentioned that the stomach lining produces the hormone gastrin, which helps regulate the secretion of pepsinogen. The intestinal lining produces hormones that cause the pancreas and gallbladder to send secretions to the small intestine (p. 77). It should also be mentioned that the thymus gland secretes thymosins, a family of hormones that assure the ability of the immune system to combat disease.

LOCAL HORMONES

Aside from hormones produced by endocrine glands and carried in the bloodstream, there are hormones, called *autocoids,* that act on tissues or cells in their immediate vicinity. Such hormones could even be said to include the well-known neurotransmitters (fig. 13.23) and the opioids, the endorphins and enkephalins.

Other autocoids we have mentioned previously are bradykinin, histamine, and prostaglandins. All of these are involved in the inflammatory reaction (p. 134) and the origination of stimuli that result in the sensation of pain.

Prostaglandins

Prostaglandins (PG) are derived from cell membrane phospholipids. First discovered in semen, they were named after the prostate gland from which they were thought to originate. The prostaglandins in semen cause the uterus to contract during sexual intercourse, which is believed to assist the movement of sperm as they traverse this organ. Because of their effect on the uterus, they are sometimes used to abort a fetus. Prostaglandins are also being considered for possible use as a safe, self-administered, once-a-month means of preventing pregnancy. As a contraceptive, prostaglandins cause menstruation to begin soon after administration. Prostaglandins are involved in many other aspects of reproduction and development.

Prostaglandins also function in nonreproductive organs. They are being considered as possible treatment for ulcers because they reduce gastric secretion; as treatment of hypertension because they lower blood pressure; and in the prevention of thrombosis because they inhibit platelet aggregation.

Sometimes prostaglandins have contrary effects. For example, one type helps prevent blood clots, but another helps blood clots to form. Also a large dose of PG may have an effect that is opposite to that of a small dose. Therefore, it has been very difficult to standardize PG therapy, and in most instances, prostaglandin drug therapy is still considered experimental.

SUMMARY

Hormones are chemical messengers having a metabolic effect on cells. The hypothalamus produces the hormones ADH and oxytocin, released by the posterior pituitary, and also controls the anterior pituitary by means of releasing hormones. In addition to LTH and GH, which affect the body directly, the anterior pituitary secretes hormones that control other endocrine glands: TSH stimulates the thyroid to release thyroxin; ACTH stimulates the adrenal cortex to release glucocorticoids and mineralocorticoids; gonadotropins stimulate the gonads to release the sex hormones. The secretion of hormones is controlled by negative feedback. In the case of the hormones just mentioned, this mechanism involves the hypothalamus and the anterior pituitary in addition to the hormonal gland in question.

The most common illness due to hormonal imbalance is diabetes mellitus (sugar diabetes). This condition occurs when the islets of Langerhans within the pancreas fail to produce insulin. Insulin promotes the uptake of glucose by the cells and the conversion of glucose to glycogen, thereby lowering the blood glucose levels. Without the production of insulin, the blood sugar level rises and some of it spills over into the urine. The real problem in diabetes mellitus, however, is acidosis, which may cause the death of the diabetic if therapy is not begun.

In addition to hormones produced by endocrine glands, it has become evident that there are many hormones that work locally.

OBJECTIVE QUESTIONS

1. The hypothalamus _____ the hormones _____ and _____ released by the posterior pituitary.
2. The _____ secreted by the hypothalamus control the anterior pituitary.
3. Generally hormone production is self-regulated by a _____ _____ mechanism.
4. Growth hormone is produced by the _____ pituitary.
5. Simple goiter occurs when the thyroid is producing inactive _____ .
6. ACTH, produced by the anterior pituitary, stimulates the _____ of the adrenal glands.
7. An overproductive adrenal cortex results in the condition called

_____ .
8. Parathyroid hormone increases the level of _____ in the blood.
9. Type I diabetes mellitus is due to a malfunctioning _____ , but Type II diabetes is due to malfunctioning _____ .
10. Prostaglandins are not carried in the _____ as are hormones that are secreted by the endocrine glands.

Answers

1. produces; ADH, oxytocin 2. releasing hormones 3. feedback 4. anterior 5. thyroxin 6. cortex 7. Cushing syndrome 8. calcium 9. pancreas, cells 10. blood

STUDY QUESTIONS

1. Hormones fall into which two groups from a chemical standpoint? (p. 268)
2. What mechanisms of action have been suggested to explain how hormones work? (p. 269)
3. Define endocrine gland and target organ. (p. 269)
4. How does the hypothalamus control the posterior pituitary? the anterior pituitary? (pp. 271, 272)
5. Discuss two hormones secreted by the anterior pituitary that have an effect on the body proper rather than on other glands. (p. 272) Why is the anterior pituitary called the master gland? (p. 274)
6. For each of the following endocrine glands name the hormone(s) secreted, the effect of the hormone(s), and the medical illnesses, if any, that result from too much or too little of each hormone: posterior pituitary, thyroid, parathyroids, adrenal cortex, adrenal medulla, pancreas. (p. 271)
7. Give the anatomical location of each of the endocrine glands listed in question 6. (p. 270)
8. Draw a diagram to describe the action and control of ADH, thyroxin, glucocorticoids (e.g., cortisol), aldosterone, parathyroid hormone, and insulin. (pp. 272, 275, 278, 279, 284, 286)

THOUGHT QUESTIONS

1. Use figure 13.9 to explain, with examples, that control by negative feedback will result in a fluctuation of hormonal blood levels about a mean.

2. Name several hormones that control the concentration of blood glucose, and show that this is not really an unnecessary duplication of effects.

3. Argue that it is incorrect to speak of the nervous system versus the endocrine system and that instead the two systems should be considered one system.

KEY TERMS

acromegaly (ak″ro-meg′ah-le) a condition resulting from an increase in growth hormone production after adult height has been achieved. *274*

anterior pituitary (an-te′re-or pi-tu′i-tār″e) a portion of the pituitary gland that produces at least six hormones and is controlled by hypothalamic-releasing hormones. *271*

cretinism (kre′tin-izm) a condition resulting from a lack of thyroid hormone in an infant. *276*

diabetes insipidus (di″ah-be′tēz in-sip′i-dus) condition characterized by an abnormally large production of urine due to a deficiency of antidiuretic hormone. *272*

diabetes mellitus (di″ah-be′tēz mě-li′-tus) condition characterized by a high blood glucose level and the appearance of glucose in the urine due to a deficiency of insulin production or uptake by cells. *285*

endocrine gland (en′do-krin gland) a gland that secretes hormones directly into the blood or body fluids. *269*

exophthalmic goiter (ek″sof-thal′mik goi′ter) an enlargement of the thyroid gland accompanied by an abnormal protrusion of the eyes. *276*

islets of Langerhans (i′lets uv lahng′er-hanz) distinctive groups of cells within the pancreas that secrete insulin and glucagon. *285*

myxedema (mik″sě-de′mah) a condition resulting from a deficiency of thyroid hormone in an adult. *276*

posterior pituitary (pos-tēr′e-or pi-tu′i-tār″e) portion of the pituitary gland connected by a stalk to the hypothalamus. *271*

simple goiter (sim′p'l goi′ter) condition in which an enlarged thyroid produces low levels of thyroxin. *275*

FURTHER READINGS FOR PART III

Aoki, C. and Siekevitz, P. December 1988. Plasticity in brain development. *Scientific American.*

Barr, M. L. 1979. *The human nervous system, an anatomical viewpoint,* 3rd. ed. New York: Harper and Row Publishers.

Bloom, F. E. October 1981. Neuropeptides. *Scientific American.*

Fincher, J. 1981. *The brain: mystery of matter and mind.* Washington, D.C.: U.S. News Books.

Hudspeth, A. J. January 1983. The hair cells of the inner ear. *Scientific American.*

Keynes, R. D. March 1979. Ion channels in the nerve cell membrane. *Scientific American.*

Koretz, J. F. and Handelman, G. H. July 1988. How the human eye focuses. *Scientific American.*

Llinas, R.R. October 1982. Calcium in synaptic transmission. *Scientific American.*

Loeb, G. E. February 1985. The functional replacement of the ear. *Scientific American.*

Mishkin, M. and Appenzeller, T. June 1987. The anatomy of memory. *Scientific American.*

Nauta, W. J. H., and M. Feirtag. September 1979. The organization of the brain. *Scientific American.*

Norman, D. A. 1982. *Learning and memory.* San Francisco: W. H. Freeman.

Norton, W. T., and P. Morell. May 1980. Myelin. *Scientific American.*

Orci, L. et al. September 1988. The insulin factory. *Scientific American.*

Rubenstein, E. March 1980. Diseases caused by impaired communication among cells. *Scientific American.*

Schnapt, J. L. and D. A. Baylor. April 1987. How photoreceptor cells respond to light. *Scientific American.*

Schwartz, J. H. April 1980. The transport of substances in nerve cells. *Scientific American.*

Selim, R. D. 1982. *Muscles: the magic of motion.* Washington, D.C.: U.S. News Books.

Shashoua, V. E. July–August 1985. The role of extracellular proteins in learning and memory. *American Scientist.*

Snyder, S. H. October 1985. The molecular basis of communication between cells. *Scientific American.*

Stevens, C. F. September 1979. The neuron. *Scientific American.*

Stryer, L. July 1987. The molecules of visual excitation. *Scientific American.*

Wertenbaker, L. 1981. *The eye: Window to the world.* Washington, D.C.: U.S. News Books.

Wurtman, R. J. April 1982. Nutrients that modify brain function. *Scientific American.*

Zwislocki, J. J. 1981. Sound analysis in the ear: A history of discoveries. *American Scientist* 69:184.

PART FOUR

HUMAN REPRODUCTION

We are in the midst of a sexual revolution; we have the freedom to engage in varied sexual practices and to reproduce by alternative methods. With freedom comes a responsibility to be familiar with the biology of reproduction and its health consequences, not only to ourselves but to potential offspring.

Reproductive freedom is exemplified by the variety of birth control measures available today and by the willingness of the infertile to use alternative methods of conception such as in vitro fertilization. Unfortunately, our newfound freedom has also increased the occurrence of sexually transmitted diseases. The viral diseases AIDS and herpes are difficult to treat, but presently the bacterial diseases gonorrhea, chlamydia, and syphilis are curable by antibiotic therapy, although certain strains are becoming resistant.

The way humans reproduce is an adaptation to life on land. The male introduces semen, sperm within a fluid medium, into the body of the female, where a resulting zygote will be protected from the drying effects of the environment. Humans develop within a watery environment provided by an extraembryonic membrane in the uterus of the female. Nourishment is supplied by the placenta, an organ composed of maternal and fetal tissues. The embryo develops into a fetus within the body of the female, and birth occurs when there is a reasonable chance for independent existence. Induction, the ability of one tissue to influence the development of another, can help account for the orderliness of development. The steps in human development can be outlined from the fertilized egg to the birth of a child.

CHAPTER FOURTEEN

REPRODUCTIVE SYSTEM

CHAPTER CONCEPTS

1 The male reproductive system is designed for the continuous production of a large number of sperm within a fluid medium.

2 The female reproductive system is designed for the monthly production of an egg and preparation of the uterus for possible implantation of the fertilized egg.

3 Hormones control the reproductive process and the sex characteristics of the individual.

4 Birth control measures vary in effectiveness from those that are very effective to those that are minimally effective.

5 There are alternative methods of reproduction today, including in vitro fertilization followed by artificial implantation.

Female reproductive organs superimposed on X ray and enhanced silhouette.

FIGURE 14.1 *When frogs mate they shed their eggs and sperm right in the water where fertilization takes place. The watery environment protects the gametes and zygote from drying out. When humans mate, the male deposits his sperm inside the female. It's her body that protects the sperm, egg, and zygote from drying out.*

INTRODUCTION

Organisms that reproduce in the water deposit their eggs and sperm in the water (fig. 14.1) because the aquatic environment protects them from drying out. But organisms that reproduce on the land need a mechanism to protect the gametes and developing zygote from the drying effects of the air. In humans the egg stays within the body of the female, where it is fertilized and the zygote undergoes development. The sperm pass from the male within seminal fluid, which if exposed would indeed dry out and be useless. Sexual intercourse prevents this eventuality. During sexual intercourse sperm are deposited into the female's vagina, which is lubricated by secretions. The human sex act is an adaptation to the land environment.

HUMAN ISSUE

Most public schools now have some sort of sex education program, and these have wide acceptance because of our concerns over child molestation, teenage pregnancy, AIDS, and other sexually transmitted diseases. Sex education, however, raises a number of controversial issues, such as in which grade sex education should begin and how explicit the course should be. For example, should the course include a description of diseases like herpes and genital warts, and should all types of sex acts be discussed? If so, should the discussion confine itself to the medical consequences of certain acts or should it also include opinions on the morality of these behaviors? Where do you stand on these issues?

FIGURE 14.2 *Side view of male reproductive system. Trace the path of the genital tract from a testis to the exterior. The seminal vesicles, Cowper's glands, and prostate gland produce seminal fluid and do not contain sperm. Notice that the penis in this drawing is not circumcised since the foreskin is present.*

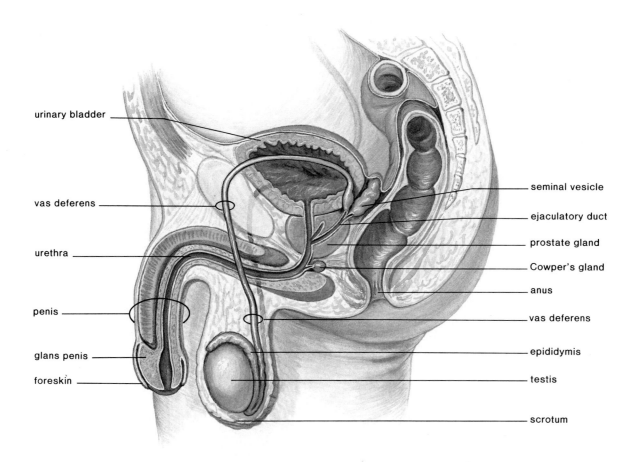

I n advanced forms of sexual reproduction there are two types of gametes (sex cells), both of which contribute the same number of chromosomes to the new individual (fig. 17.4). The sperm are small and swim to the stationary egg, a much larger cell that contains food for the developing embryo. It seems reasonable that there should be a large number of sperm to ensure that a few will find the egg. In humans, the male continually produces sperm, which are temporarily stored before being released.

MALE REPRODUCTIVE SYSTEM

Figure 14.2 shows the reproductive system of the male, and table 14.1 lists the anatomical parts of this system.

Testes

The **testes** lie outside the abdominal cavity of the male within the **scrotum.** The testes begin their development inside the abdominal cavity but descend into the scrotal sacs during the last two months of fetal development. If by chance the testes do not descend and the male is not treated or operated on to

TABLE 14.1 *MALE REPRODUCTIVE SYSTEM*

ORGAN	FUNCTION
Testes	Produce sperm and sex hormones
Epididymis	Maturation and some storage of sperm
Vas deferens	Conducts and stores sperm
Seminal vesicles	Contribute to seminal fluid
Prostate gland	Contributes to seminal fluid
Urethra	Conducts sperm
Cowper's glands	Contribute to seminal fluid
Penis	Organ of copulation

place the testes in the scrotum, sterility—the inability to produce offspring—usually follows. This is because the internal temperature of the body is too high to produce viable sperm.

Seminiferous Tubules

Connective tissue forms the wall of each testis and divides it into lobules (fig. 14.3). Each lobule contains one to three tightly coiled **seminiferous tubules,** which have a combined

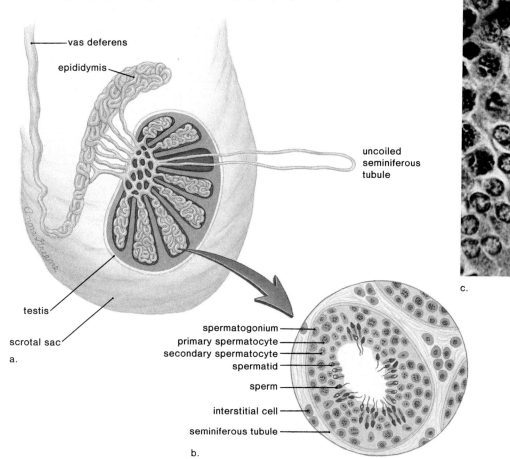

FIGURE 14.3 Sections through a testis. a. Longitudinal section showing lobules containing seminiferous tubules. b. Cross section of a tubule showing germ cells in various stages of spermatogenesis. c. Micrograph of a portion of (b).

length of approximately 250 m. A microscopic cross section through a tubule shows it is packed with cells undergoing spermatogenesis. These cells are derived from undifferentiated germ cells, called spermatogonia (singular spermatogonium), that lie just inside the outer wall and divide mitotically, always producing new spermatogonia. Some spermatogonia move away from the outer wall to increase in size and become primary spermatocytes that undergo meiosis, a type of cell division described in chapter 17. Although these cells have forty-six chromosomes, they divide to give secondary spermatocytes, each with twenty-three chromosomes. Secondary spermatocytes divide to give spermatids that also have twenty-three chromosomes, but are single stranded. Spermatids then differentiate into spermatozoa, or mature sperm.

Sperm The mature sperm, or spermatozoan (fig. 14.4), has three distinct parts: a head, a midpiece, and a tail. The *tail* contains the 9 + 2 pattern of microtubules typical of cilia and flagella (fig. 2.15), and the *midpiece* contains energy-producing mitochondria. The *head* contains the twenty-three chromosomes within a nucleus. The tip of the nucleus is covered by a cap called the **acrosome,** which is believed to con-

FIGURE 14.4 *Microscopic anatomy of sperm.*

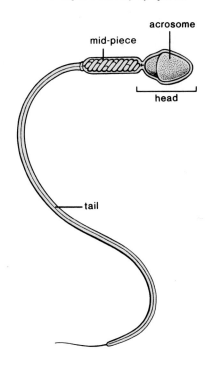

tain enzymes needed for fertilization. The human egg is surrounded by several layers of cells and a mucoprotein substance. The acrosome enzymes digest a portion of these cells, allowing a single sperm to penetrate the egg. It is hypothesized that each acrosome contains such a minute amount of enzyme that it requires the action of many sperm to allow just one to actually penetrate the egg. This may explain why so many sperm are required for the process of fertilization. A normal human male usually produces several hundred million sperm per day, an adequate number for fertilization. Sperm are continually produced throughout a male's reproductive life.

In males, spermatogenesis occurs within the seminiferous tubules of the testes. Sperm have a head, capped by an acrosome, where twenty-three chromosomes reside in the nucleus; a mitochondria-containing midpiece; and a tail with a 9 + 2 pattern of microtubules.

Interstitial Cells

The male sex hormones, the androgens, are secreted by cells that lie between the seminiferous tubules and are therefore called **interstitial cells.** The most important of the androgens is testosterone. Its functions are discussed on page 298.

Genital Tract

Sperm are produced in the testes, but they mature in the **epididymis** (fig. 14.2), a tightly coiled tubule about 6 m in length that lies just outside each testis. During the two-to-four-day maturation period, the sperm develop their characteristic swimming ability. Also, it is possible that during this time defective sperm are removed from the epididymis. Each epididymis joins with a **vas deferens,** which ascends through a canal called the *inguinal canal* and enters the abdomen where it curves around the bladder and empties into the urethra. Sperm are stored in the last part of the epididymis and the first part of the vas deferens. They pass from each vas deferens into the urethra only when ejaculation is imminent.

Seminal Fluid

At the time of ejaculation, sperm leave the penis in a fluid called **seminal fluid.** This fluid is produced by three types of glands—the seminal vesicles, the prostate gland, and the Cowper's glands. The **seminal vesicles** lie at the base of the bladder, and each has a duct that joins with a vas deferens. The **prostate gland** is a single doughnut-shaped gland that surrounds the upper portion of the urethra just below the bladder. In older men, the prostate may enlarge and cut off the urethra, making urination painful and difficult. This condition may be treated medically or surgically. **Cowper's glands** are pea-sized organs that lie posterior to the prostate on either side of the urethra.

Each component of seminal fluid seems to have a particular function. Sperm are more viable in a basic solution; seminal fluid, which is white and milky in appearance, has a slightly basic pH (about 7.5). Swimming sperm require energy, and seminal fluid contains the sugar fructose, which presumably serves as an energy source. Seminal fluid also contains prostaglandins, chemicals that cause the uterus to contract. Some investigators now believe that uterine contraction is necessary to propel the sperm and that the sperm swim only when they are in the vicinity of the egg.

Penis

The **penis** has a long shaft and an enlarged tip called the glans penis. At birth the penis is covered by a layer of skin called the **foreskin** or prepuce. Gradually over a period of five to ten years, the foreskin becomes separated from the penis and may be retracted. During this time there is a natural shedding of cells between the foreskin and penis. These cells, along with an oil secretion that begins at puberty, is called smegma. Therefore, in the child no special cleansing method is needed to wash away smegma, but in the adult the foreskin can be retracted to do so. **Circumcision** is the surgical removal of the foreskin soon after birth.

The penis is the copulatory organ of males. When the male is sexually aroused, the penis becomes erect and ready for intercourse (fig. 14.5). **Erection** is achieved because blood sinuses within the erectile tissue of the penis become filled with blood. Parasympathetic impulses dilate the arteries of the penis, while the veins are passively compressed so that blood flows into the erectile tissue under pressure. If the penis fails to become erect, the condition is called **impotency.** Although it was formerly believed that almost all cases of impotency were due to psychological reasons, it has recently been reported that some cases may be due to hormonal imbalances. Treatment consists of finding the precise imbalance and restoring the proper level of testosterone.

Ejaculation

As sexual stimulation becomes intense, sperm enter the urethra from each vas deferens and the accessory glands (seminal vesicles, prostate gland, and Cowper's glands) secrete seminal fluid. Sperm and seminal fluid together are called **semen.** Once semen is in the urethra, rhythmical muscle contractions cause it to be expelled from the penis in spurts. During ejaculation, a sphincter closes off the bladder so that no urine enters the urethra. (Notice that the urethra carries either urine or semen at different times.)

The contractions that expel semen from the penis are a part of male **orgasm,** the physiological and psychological sensations that occur at the climax of sexual stimulation. The psychological sensation of pleasure is centered in the brain, but the physiological reactions involve the genital (reproductive) organs and associated muscles as well as the entire body. Marked muscle tension is followed by contraction and relaxation.

FIGURE 14.5 *Erection contrasted to a flaccid penis and relaxed scrotum. The erectile tissue fills with blood when the penis becomes erect. Note that the penis lacks a foreskin, due to circumcision.*

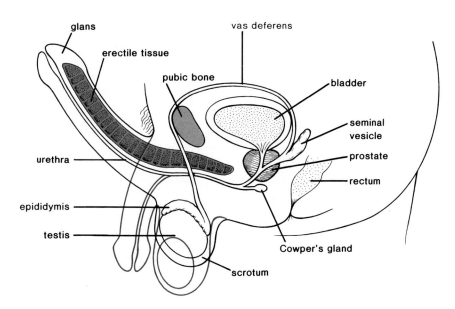

Following ejaculation and/or loss of sexual arousal, the penis returns to its normal flaccid state. After ejaculation, a male typically experiences a period of time, called the refractory period, during which stimulation does not bring about an erection.

There may be in excess of 400 million sperm in 3.5 ml of semen expelled during ejaculation. The sperm count can be much lower than this, however, and fertilization (fig. 14.6) will still take place.

Sperm mature in the epididymis and are also stored in the vas deferens before entering the urethra just prior to ejaculation. The accessory glands (seminal vesicles, prostate gland, and Cowper's glands) produce seminal fluid. Semen, which contains sperm and seminal fluid, leaves the penis during ejaculation.

Hormonal Regulation in the Male

The hypothalamus has ultimate control of the testes' sexual functions because it secretes a releasing hormone that stimulates the anterior pituitary to produce the gonadotropic hormones. Two gonadotropic hormones, **FSH** (follicle-stimulating hormone) and **LH** (luteinizing hormone), are named for their function in females but exist in both sexes, stimulating the appropriate gonads in each. It is believed that FSH promotes spermatogenesis in the seminiferous tubules and that LH promotes the production of testosterone in the interstitial cells. Sometimes LH in males is given the name interstitial cell-stimulating hormone (ICSH).

The hormones mentioned are involved in a feedback process (fig. 14.7) that maintains the production of testosterone at a fairly constant rate. For example, when the amount of testosterone in the blood rises to a certain level, it causes the hypothalamus to decrease its secretion of releasing hormone, which causes the anterior pituitary to decrease its secretion of LH. As the level of testosterone begins to fall, the hypothalamus increases secretion of the releasing hormone and the anterior pituitary increases its secretion of LH, and stimulation of the interstitial cells reoccurs. It should be emphasized that only minor fluctuations of testosterone level occur in the male and that the feedback mechanism in this case acts to maintain testosterone at a normal level. It has long been suspected that the seminiferous tubules produce a hormone that blocks FSH secretion. This substance, termed inhibin, has recently been isolated.

Testosterone

The male sex hormone, **testosterone,** has many functions. It is essential for the normal development and functioning of the primary sex organs, those structures we have just been discussing. It is also necessary for the maturation of sperm, probably after diffusion from the interstitial cells into the seminiferous tubules.

Greatly increased testosterone secretion at the time of puberty stimulates the growth of the penis and testes. Testosterone also brings about and maintains the secondary sex characteristics in males that develop at the time of puberty. Testosterone causes growth of a beard, axillary (underarm)

FIGURE 14.6 *Fertilization.* a. *Scanning electron micrographs of an echinoderm egg surrounded by a large number of sperm.* b. *Only one sperm penetrates the egg to achieve fertilization.*

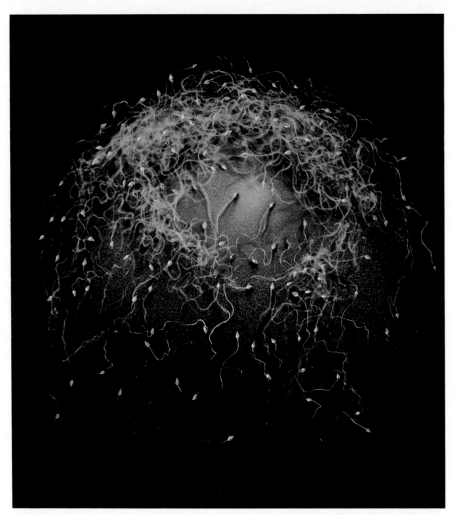

a.

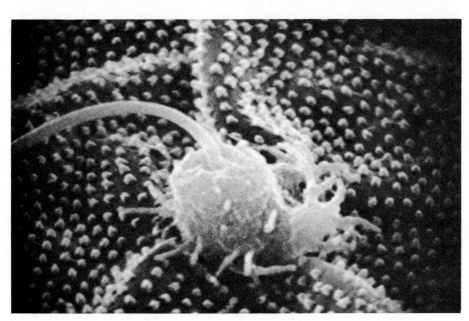

b.

FIGURE 14.7 *Hypothalamic-pituitary-gonad system as it functions in the male. GnRH is a hypothalamic-releasing hormone that stimulates the anterior pituitary to secrete LH and FSH. These gonadotropic hormones act on the testes. LH promotes the production of testosterone, and FSH promotes spermatogenesis. Negative feedback controls the level of all hormones involved.*

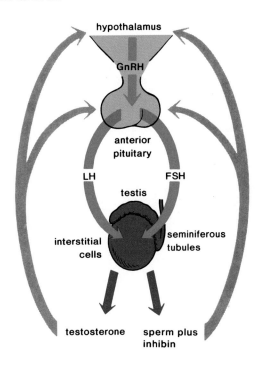

hair, and pubic hair. It prompts the larynx and vocal cords to enlarge, causing the voice to change. It is responsible for the greater muscle strength of males, the reason some athletes take supplemental amounts of *anabolic steroids,* which are either testosterone or related chemicals. (The pros and cons of taking anabolic steroids are discussed in the reading on page 281.) Testosterone also causes oil and some sweat glands in the skin to secrete; therefore, it is largely responsible for acne and body odor. Another side effect of testosterone activity is baldness. Genes for baldness are probably inherited by both sexes, but baldness is seen more often in males because of the presence of testosterone.

Testosterone is believed to be largely responsible for the sex drive and may even contribute to the supposed aggressiveness of males.

In males, FSH promotes spermatogenesis and LH promotes testosterone production. Testosterone stimulates growth of the male genitals during puberty and is necessary for the maturation of sperm and the development of the secondary sex characteristics.

TABLE 14.2 *FEMALE REPRODUCTIVE SYSTEM*

ORGAN	FUNCTION
Ovaries	Produce egg and sex hormones
Fallopian tubes (oviducts)	Conduct egg
Uterus (womb)	Location of developing fetus
Cervix	Contains opening to uterus
Vagina	Organ of copulation and birth canal

FEMALE REPRODUCTIVE SYSTEM

Table 14.2 lists the anatomical parts of the female reproductive system, and figure 14.8 shows this system.

Ovaries

The **ovaries** lie in shallow depressions, one on each side of the upper pelvic cavity. A longitudinal section through an ovary shows that it is made up of an outer cortex and an inner medulla. The cortex contains ovarian **follicles** at various stages of maturation. A female is born with a large number of follicles (400,000) in both ovaries, each containing a primary oocyte. There are no new follicles after a female is born. Since primary oocytes are present at birth, they age as the woman ages. This is one possible reason older women are more likely to produce children with genetic defects. Only a small number of follicles (about 400) ever mature because a female produces only one egg per month during her reproductive years.

As the follicle undergoes maturation, it develops from a primary to a secondary to a **Graafian follicle.** In a primary follicle an oocyte divides meiotically into two cells, each having twenty-three chromosomes (fig. 17.15). One of these cells, termed the secondary oocyte, receives almost all the cytoplasm, nutrients, and enzymes. The other is a polar body that disintegrates. A secondary follicle contains the secondary oocyte pushed to one side of a fluid-filled cavity. In a Graafian follicle, the fluid-filled cavity increases to the point that the follicle wall balloons out on the surface of the ovary and bursts, releasing the secondary oocyte surrounded by the zona pellucida, a gel-like layer, and a few cells; this is called **ovulation.** Once a follicle has lost its oocyte, which we will now term an egg, it develops into a **corpus luteum,** a glandlike structure. If pregnancy does not occur, the corpus luteum begins to degenerate after about ten days. If pregnancy does occur, the corpus luteum persists for three to six months.

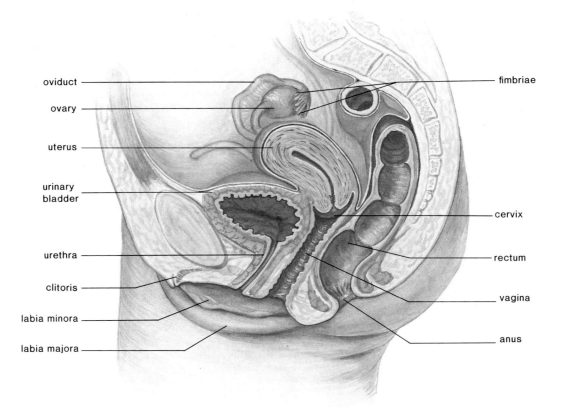

The follicle and corpus luteum secrete the female sex hormones estrogen and progesterone, as discussed on page 304.

In females, oogenesis occurs within the ovaries, where one follicle reaches maturity each month. This follicle balloons out of the ovary and bursts to release the egg. The ruptured follicle develops into a corpus luteum. The follicle and corpus luteum produce the female sex hormones estrogen and progesterone.

Genital Tract

The female genital tract includes the oviducts, uterus, and vagina.

Oviducts

The oviducts, also called uterine or Fallopian tubes, extend from the uterus to the ovaries. The oviducts are not attached to the ovaries but instead have fingerlike projections, called **fimbriae,** that sweep over the ovary at the time of ovulation. When the egg bursts (fig. 14.9) from the ovary during ovulation, it is usually swept up into an oviduct by the combined action of the fimbriae and the beating of cilia that line the tubes.

Since the egg must traverse a small space before entering an oviduct, it is possible for the egg to get lost and instead enter the abdominal cavity. Such eggs usually disintegrate but in some rare cases have been fertilized in the abdominal cavity and have implanted themselves in the wall of an abdominal organ. Very rarely, such embryos have come to term, the child being delivered by surgery.

Once in the oviduct, the egg is propelled slowly by cilia movement and tubular muscular contraction toward the uterus. Fertilization usually occurs in an oviduct because the egg only lives approximately six to twenty-four hours. The developing zygote normally arrives at the uterus after several days and then embeds, or implants, itself in the uterine lining, which has been prepared to receive it. Occasionally, the zygote becomes embedded in the wall of an oviduct, where it begins to develop. These tubular pregnancies cannot succeed because the tubes are not anatomically capable of allowing full development to occur.

Uterus

The **uterus** is a thick-walled, muscular organ about the size and shape of an inverted pear. Normally it lies above and is tipped over the urinary bladder. The oviducts join the uterus anteriorly, and posteriorly the cervix enters into the vagina nearly at a right angle. A small opening in the cervix leads to the vaginal canal. Development of the embryo takes place

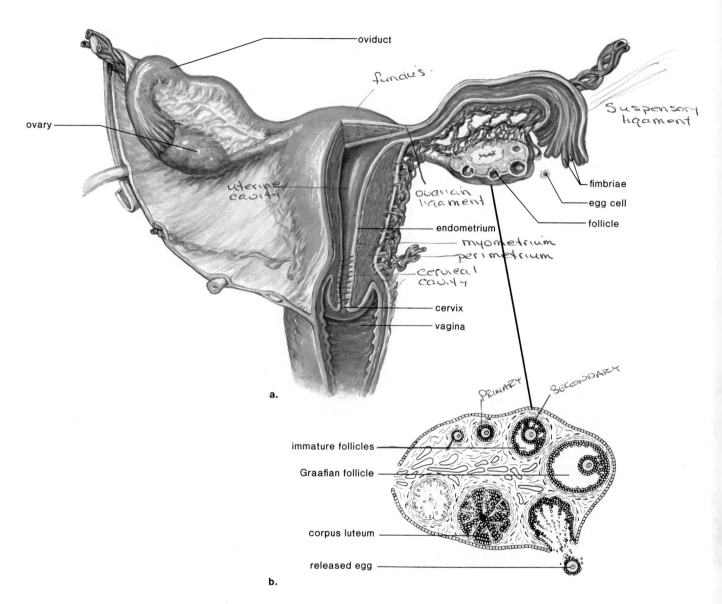

oviduct

fundus

ovary

Suspensory ligament

uterine cavity

ovarian ligament

fimbriae

egg cell

endometrium

follicle

myometrium
perimetrium

cervical cavity

cervix

vagina

a.

PRIMARY

SECONDARY

immature follicles

Graafian follicle

corpus luteum

released egg

b.

in the uterus. This organ, sometimes called the womb, is approximately 5 cm wide in its usual state but is capable of stretching to over 30 cm to accommodate the growing baby. The lining of the uterus, called the **endometrium,** participates in the formation of the placenta (p. 306), which supplies nutrients needed for embryonic and fetal development. The endometrium has two layers: a basal layer and an inner functional layer. In the nonpregnant female, the functional layer of the endometrium varies in its thickness according to a monthly reproductive cycle, called the uterine cycle (p. 304).

Cancer of the cervix is a common form of cancer in women. Early detection is possible by means of a **Pap test,** which requires that the doctor remove a few cells from the region of the cervix for microscopic examination. If the cells

are cancerous, a hysterectomy may be recommended. A hysterectomy is the removal of the uterus. Removal of the ovaries in addition to the uterus is termed an ovariohysterectomy. Since the vagina remains, the woman may still engage in sexual intercourse.

Vagina

The **vagina** is a tube that makes a 45° angle with the small of the back. The mucosal lining of the vagina lies in folds that extend as the fibromuscular wall stretches. This capacity to extend is especially important when the vagina serves as the birth canal. It may also facilitate intercourse when the vagina receives the penis during copulation.

THE REASONS FOR ORGASM

Why does orgasm exist? A glance at the reproduction of primitive forms of animal life shows that such a fancy thing far exceeds what's necessary to pass genes from one generation to another, which even in mammals can be attained by a no-frills male ejaculation and mere passive receptivity on the part of the female. Yet it's unlikely that nature, as economical as it is, would create something so baroque without good reason.

The simpler the creature, the more difficult it is to determine whether its sexual relations are pleasurable, much less orgasmic. However, all mammals show a marked interest in sex, and will even work for it—a good indication that it's rewarding. Scientists prefer low-key terms like "ejaculatory reflex" and "estrus behavior" when discussing animal sexuality; they're reluctant to apply the term "orgasm" because it can't be verified in animals. The best gauge of orgasm is uniquely human: Did you or didn't you?

Male sexual responses, particularly erection and ejaculation, have been better studied than female processes. Even the humble rat, that scrupulously observed mammal, exhibits the basic criteria associated with male orgasm: ejaculation followed by refraction, accompanied by characteristic movements. "Males of most mammalian species have what at least looks like a precursor to human orgasm, demonstrated by skeletal and facial patterns," says Benjamin Sachs, a reproductive behavior researcher at the University of Connecticut. "You may as well call it orgasm. I'd be more cautious about females." Moreover, in addition to appearances, males have an excellent motive for ejaculation: progeny.

Alas for researchers, human females don't always show characteristic muscular movements during orgasm. Even more vexing, they need not have orgasms to conceive—or even to seek and enjoy mating. The best evidence for female orgasm is subjective and anecdotal—the yea or nay of the woman in question. Until there's a way to acquire such information from animals, subhuman female orgasm can only be inferred. Does this mean that for female animals mating is simply a selfless matter of lie still and think of England?[1] Ronald Nadler, a primatologist at the Yerkes Regional Primate Research Center in Atlanta, believes it's unlikely that women are alone among female primates in having orgasms, but the real controversy is not whether, but why, females have orgasms at all.

There are two schools of thought on this. The first maintains that the female has orgasms because the male does, though hers serve no essential reproductive function (an analogy is the male nipple). The female genitals are poorly designed for easy stimulation to orgasm, particularly during intercourse, and if orgasm were really important to procreation, anatomy would have evolved differently to facilitate it. This idea of female orgasm as vestigial is supported by the fact that the inability to have orgasms is common in women yet rare in men, and that orgasm sometimes requires considerable education and practice for women to achieve.

The second school says that female orgasm is designed for a reason or reasons. Sarah Blaffer Hardy, a primatologist at the University of California at Davis, notes that a female could accumulate sufficient stimulation for orgasm as well as increase her chances of conception via the repeated matings typical of primates during estrus. Furthermore, if the motions of the vagina and uterus during orgasm can be proved to enhance the mobility of sperm, as some researchers speculate, orgasm might be a way to enable a female to be somewhat selective about when and even by whom she becomes pregnant. Although neither camp can prove its thesis, a majority holds that female orgasm has a purpose—without knowing what the purpose is.

While scholars debate, couples unwittingly contend with evolutionary traits that have more practical application. As sex therapists are wont to put it, "Men heat up like light bulbs and women like irons." Trying as this difference in timing can be, it makes sociobiological sense: a female motivated to seek repeated sexual encounters increases her opportunities for pregnancy and assures male benignity to a greater extent than if she were quickly aroused and satisfied. On the other hand, a male, always on the defensive, can't afford lengthy dalliances with his back to potential enemies.

Though many theories about the evolution of orgasm are specific to males or females, the response also makes sense in social, as well as reproductive, terms for both. While the sex lives of other mammals are regulated by hormones—totally so in nonprimates like rats, partially so in nonhuman primates—humans appear to be freed from such controls, and even castrated men and women can enjoy sex and have orgasms. There are reliable reports of children seven years old having orgasms, and there's no limit at the other end of the age spectrum—further indications that there's more to sex than reproduction. Roger Short, a reproductive biologist at Monash University in Melbourne, Australia, thinks that because human females, unlike those of other species, will accept intercourse at any time, whether or not they are in a fertile period, most couplings are performed for social purposes such as strengthening pair-bonding and releasing sexual tension, so that the more mundane business of life can be attended to.

Winifred Gallagher © 1986 Discover Publications, Inc.

[1]Traditional advice to Victorian brides on their wedding nights.

FIGURE 14.10 *External genitalia of female. Vascular erectile tissue found on either side of the vestibule is shown in red. At birth, the opening of the vagina is partially occluded by a membrane called the hymen. Physical activities and sexual intercourse disrupt the hymen.*

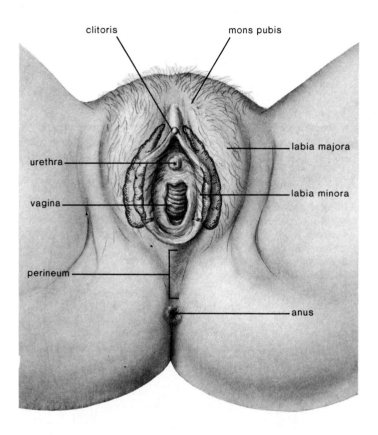

Orgasm

Sexual response in the female may be more subtle than in the male, as discussed in the reading on the opposite page, but there are certain corollaries. The clitoris is believed to be an especially sensitive organ for initiating sexual sensations. It is possible for the clitoris to become ever so slightly erect as its erectile tissues become engorged with blood, but vasocongestion is more obvious in the labia minora, which expand and deepen in color. Erectile tissue within the vaginal wall also expands with blood, and the added pressure in these blood vessels causes small droplets of fluid to squeeze through the vessel walls and lubricate the vagina.

Release from muscle tension occurs in females, especially in the region of the vulva and vagina, but also throughout the entire body. Increased uterine motility may assist the transport of sperm toward the oviducts. Since female orgasm is not signaled by ejaculation, there is a wide range in normalcy regarding sexual response.

The egg must cross a small space to enter the oviducts, which conduct it toward the uterus. The vagina, the copulatory organ in females, opens into the vestibule where the urethra also opens. The vestibule is bounded by the labia minora, which come together at the clitoris, a highly sensitive organ. Outside the labia minora are the labia majora. There is no ejaculation in the female, and therefore orgasm is harder to detect and varies widely in normalcy.

External Genitalia

The external genital organs of the female (fig. 14.10) are known collectively as the **vulva.** The vulva includes two large, hair-covered folds of skin called the **labia majora.** They extend backward from the *mons pubis,* a fatty prominence underlying the pubic hair. The **labia minora** are two small folds lying just inside the labia majora. They extend forward from the vaginal opening to encircle and form a foreskin for the *clitoris,* an organ that is homologous to the penis. Although quite small, the clitoris has a shaft of erectile tissue and is capped by a pea-shaped glans. The clitoris also has sense receptors that allow it to function as a sexually sensitive organ.

The *vestibule,* a cleft between the labia minora, contains the openings of the urethra and the vagina. The vagina may be partially closed by a ring of tissue called the hymen. The hymen is ordinarily ruptured by initial sexual intercourse; however, it can also be disrupted by other types of physical activities. If the hymen persists after sexual intercourse, it can be surgically ruptured.

Notice that the urinary and reproductive systems in the female are entirely separate. For example, the urethra carries only urine and the vagina serves only as the birth canal and the organ for sexual intercourse.

HORMONAL REGULATION IN THE FEMALE

Hormonal regulation in the female is quite complex, so we will begin with a simplified presentation and follow with a more in-depth presentation for those who wish to study the matter in more detail. The following glands and hormones are involved in hormonal regulation.

Hypothalamus: secretes GnRH (gonadotropic-releasing hormone)
Anterior pituitary: secretes *FSH* (follicle-stimulating hormone) and *LH* (luteinizing hormone), the gonadotropic hormones
Ovaries: secrete estrogen and progesterone, the female sex hormones

Hormonal Regulation (Simplified)

Ovarian Cycle

The gonadotropic and sex hormones are not present in constant amounts in the female and instead are secreted at different rates during a monthly **ovarian cycle,** which lasts an average of twenty-eight days but may vary widely in specific

TABLE 14.3 OVARIAN AND UTERINE CYCLES (SIMPLIFIED)

OVARIAN CYCLE PHASES	EVENTS	UTERINE CYCLE PHASES	EVENTS
Follicular, Days 1–13	FSH secretion by pituitary	Menstruation, Days 1–5	Endometrium breaks down
	Follicle maturation and secretion of estrogen	Proliferation, Days 6–13	Endometrium rebuilds
Ovulation, Day 14[a]			
Luteal, Days 15–28	LH secretion by pituitary	Secretory, Days 15–28	Endometrium thickens and glands are secretory
	Corpus luteum formation and secretion of progesterone		

[a]Assuming a 28–day cycle

individuals. For simplicity's sake it is convenient to emphasize that during the first half of a twenty-eight-day cycle (days one to thirteen, table 14.3), FSH from the anterior pituitary is promoting the development of a follicle in the ovary and this follicle is secreting estrogen. As the blood estrogen level rises, it exerts feedback control over the anterior pituitary secretion of FSH so that this follicular phase comes to an end (fig. 14.11). The end of the follicular phase is marked by ovulation on the fourteenth day of the twenty-eight-day cycle. Similarly it may be emphasized that during the last half of the ovarian cycle (days fifteen to twenty-eight, table 14.3) anterior pituitary production of LH is promoting the development of a corpus luteum, which is secreting progesterone. As the blood progesterone level rises, it exerts feedback control over anterior pituitary secretion of LH so that the corpus luteum begins to degenerate. As this luteal phase comes to an end, menstruation occurs.

Uterine Cycle

The female sex hormones estrogen and progesterone have numerous functions, one of which is discussed here. The effect these hormones have on the endometrium of the uterus causes the uterus to undergo a cyclical series of events known as the **uterine cycle** (table 14.3). Cycles that last twenty-eight days are divided as described in the following paragraphs.

During *days one to five* there is a low level of female sex hormones in the body, causing the uterine lining to disintegrate and its blood vessels to rupture. A flow of blood, known as the *menses,* passes out of the vagina during a period of **menstruation,** also known as the menstrual period.

During *days six to thirteen* increased production of estrogen by an ovarian follicle causes the endometrium to thicken and become vascular and glandular. This is called the proliferation phase of the uterine cycle.

Ovulation usually occurs on the fourteenth day of the twenty-eight-day cycle.

During *days fifteen to twenty-eight* increased production of progesterone by the corpus luteum causes the endometrium to double in thickness and the uterine glands to become mature, producing a thick mucoid secretion. This is

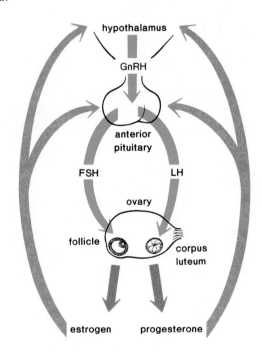

FIGURE 14.11 Hypothalamic-pituitary-gonad system (simplified) as it functions in the female. GnRH is a hypothalamic-releasing hormone that stimulates the anterior pituitary to secrete LH and FSH. These gonadotropic hormones act on the ovaries. FSH promotes the development of the follicle that later, under the influence of LH, becomes the corpus luteum. Negative feedback controls the level of all hormones involved.

called the secretory phase of the uterine cycle. The endometrium is now prepared to receive the developing zygote, but if pregnancy does not occur, the corpus luteum degenerates and the low level of sex hormones in the female body causes the uterine lining to break down. This is evident due to the menstrual discharge that begins at this time. Even while menstruation is occurring, the anterior pituitary begins to increase its production of FSH and a new follicle begins maturation. Table 14.3 indicates how the ovarian cycle controls the menstrual cycle.

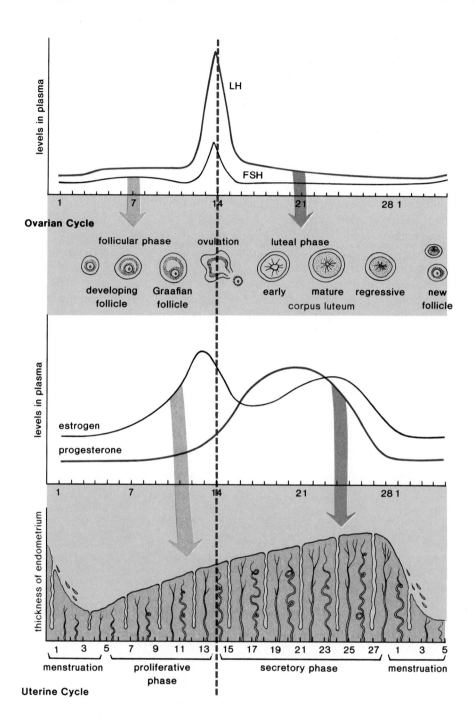

FIGURE 14.12 *Ovarian and uterine cycles. Above shows that the plasma levels of FSH and LH control the ovarian cycle, which consists of a follicular and luteal phase. Notice how ovulation is associated with a sudden spurt of these hormones. Below shows that the plasma levels of estrogen and progesterone control the uterine cycle, which consists of menstruation, proliferative phase, and secretory phase.*

Hormonal Regulation (Detailed)

Figure 14.12 shows the changes in blood concentration of all four hormones participating in the ovarian and uterine cycles. Notice that all four of these hormones (FSH, LH, estrogen, and progesterone) are present during the entire twenty-eight days of the cycle. Thus, in actuality, both FSH and LH *are* present during the follicular phase, and both are needed for follicle development and maturation of the egg. The follicle secretes primarily estrogen and a very minimal amount of progesterone. Similarly, both LH and FSH are present in decreased amounts during the luteal phase. LH may be primarily responsible for corpus luteum formation, but the corpus luteum secretes both progesterone and estrogen. The effect that these hormones have on the endometrium has already been stated. Estrogen stimulates growth of the endometrium and readies it for the reception of progesterone, which causes it to thicken and become secretory.

Feedback Control

It has been frequently mentioned that a hormone can exert feedback inhibition. Therefore it comes as no surprise to find that as the estrogen level increases during the first part of the follicular phase, FSH secretion begins to decrease. However, toward the end of the follicular phase there is a sharp increase in FSH and LH secretion at the point when the estrogen level is the highest. Regarding this phenomenon, it is believed that the high level of estrogen exerts *positive feedback on the hypothalamus,* causing it to secrete gonadotropic-releasing hormone, after which the pituitary momentarily produces an unusually large amount of FSH and LH. It is the surge of LH that is believed to promote ovulation. During the luteal phase, estrogen and progesterone bring about feedback inhibition as expected and the levels of both LH and FSH decline steadily. In this way all four hormones eventually reach their lowest levels, causing menstruation to occur. It still is not known what causes the corpus luteum to degenerate if pregnancy does not occur. In some mammals there is evidence to suggest that prostaglandins (p. 287) cause degeneration, but this is not believed to be the case in humans.

Hormonal regulation in the female results in an ovarian cycle. During the first half of the cycle, FSH causes maturation of the follicle, which secretes estrogen. After ovulation, LH converts the follicle into the corpus luteum, which produces progesterone. Estrogen and progesterone regulate the uterine cycle. Estrogen causes the endometrium to rebuild. Ovulation occurs on the fourteenth day of a twenty-eight-day cycle. As progesterone is produced by the corpus luteum, the endometrium thickens and becomes secretory. Then a low level of hormones causes the endometrium to break down as menstruation occurs.

Pregnancy

If pregnancy occurs, menstruation does not occur. Instead, the developing zygote embeds itself in the endometrial lining several days following fertilization. This process, called **implantation,** is what causes the female to become *pregnant.* During implantation, an outer layer of cells surrounding the zygote produces a gonadotropic hormone (**HCG,** or **h**uman **c**horionic **g**onadotropic hormone) that prevents degeneration of the corpus luteum and instead causes it to secrete even larger quantities of progesterone. The corpus luteum may be maintained for as much as six months, even after the placenta is fully developed.

The **placenta** (fig. 16.13) originates from both maternal and fetal tissue and is the region of exchange of molecules between fetal and maternal blood although there is no mixing of the two types of blood. After its formation, the placenta continues production of HCG and begins production of progesterone and estrogen. The latter hormones have two effects: they shut down the anterior pituitary so that no new follicles mature, and they maintain the lining of the uterus so that the corpus luteum is not needed. There is no menstruation during the period of pregnancy.

Pregnancy Tests

Pregnancy tests, which are readily available in hospitals, clinics, and now even drug and grocery stores, are based on the fact that HCG is present in the blood and urine of a pregnant woman.

Before the advent of monoclonal antibodies, only a hospital blood test requiring the use of radioactive material was available to detect pregnancy before the first missed menstrual period. Now there is a monoclonal antibody (p. 155) test for the detection of pregnancy as early as ten days after conception. This test can be done on a urine sample, and the results are available within the hour.

The physical signs that might prompt a woman to have a pregnancy test are cessation of menstruation, increased frequency of urination, morning sickness, and increase in the size and fullness of the breasts, as well as darkening of the areolae (fig. 14.13).

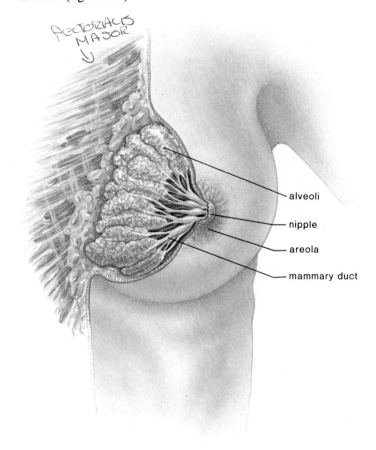

FIGURE 14.13 Anatomy of breast. The female breast contains lobules consisting of ducts and alveoli. The alveoli are lined by milk-producing cells in the lactating (milk-producing) breast.

Female Sex Hormones

The female sex hormones, estrogen and progesterone, have many effects on the body. In particular, estrogen secreted at the time of puberty stimulates the growth of the uterus and vagina. Estrogen is necessary for egg maturation and is largely responsible for the secondary sex characteristics in females. For example, it is responsible for the onset of the uterine cycle, as well as female body hair and fat distribution. In general, females have a more rounded appearance than males because of a greater accumulation of fat beneath the skin. Also, the pelvic girdle enlarges in females so that the pelvic cavity has a larger relative size compared to males; this means that females have wider hips. Both estrogen and progesterone are also required for breast development.

Breasts

A female breast contains fifteen to twenty-five lobules (fig. 14.13), each with its own milk duct. A milk duct begins at the nipple and divides into numerous other ducts that end in blind sacs called *alveoli.* In a nonlactating (nonmilk-producing) breast, the ducts far outnumber the alveoli because alveoli are made up of cells that can produce milk.

Milk is not produced during pregnancy. *Lactogenic hormone* (prolactin) is needed for lactation (milk production) to begin, and the production of this hormone is suppressed because of the feedback inhibition estrogen and progesterone have on the pituitary during pregnancy. It takes a couple of days after delivery for milk production to begin. In the meantime, the breasts produce a watery, yellowish-white fluid called *colostrum,* which differs from milk in that it contains more protein and less fat. The continued production of milk requires continued breast feeding, because sucking causes the continued production of lactogenic hormone. Also, when a breast is suckled, the nerve endings in the areola are stimulated and nerve impulses travel to the hypothalamus, which causes oxytocin to be released by the posterior pituitary. When this hormone arrives at the breasts, it causes contraction of the lobules so that milk flows into the ducts (called milk letdown).

Menopause

Menopause, the period in a woman's life during which the menstrual cycle ceases, is likely to occur between ages 45 and 55. Due to changes in the woman's body that involve the hypothalamus-pituitary-ovary axis, the ovaries produce decreasing amounts of estrogen or progesterone. At the onset of menopause, the menstrual cycle becomes irregular, but as long as menstruation occurs it is still possible for a woman to conceive and become pregnant. Therefore, a woman is usually not considered to have completed menopause until there has been no menstruation for a year. The hormonal changes during menopause often produce physical symptoms, such as "hot flashes," that are caused by circulatory

irregularities, dizziness, headaches, insomnia, sleepiness, and depression. Again, there is great variation among women, and any of these symptoms may be absent altogether.

Women sometimes report an increased sex drive following menopause, and it has been suggested that this may be due to androgen production by the adrenal cortex.

Estrogen and progesterone affect the female genitals, promote development of the egg, and maintain the secondary sex characteristics. Lactogenic hormone causes the breasts to begin to secrete milk after delivery, but another hormone, oxytocin, is responsible for milk letdown. When menopause occurs, FSH and LH are still produced but the ovaries are no longer able to respond.

HUMAN ISSUE

Teenage pregnancy is a matter of utmost current concern. Girls who are not of an age to care for themselves are getting pregnant and either seeking abortions or having their babies. Since both of these present hardships to all concerned, solutions to the problem have been sought. Do you feel that prevention of teenage pregnancy is a private matter between parents and children, or do you feel that society should be actively involved, perhaps by making contraceptives available to young people? For example, some high schools now have birth control clinics. Do you approve of this action, or is there a better way to prevent teenage pregnancy?

BIRTH CONTROL

The use of birth control (contraceptive) methods decreases the probability of pregnancy but does not, except where noted, offer any protection against contracting a sexually transmitted disease such as AIDS. A common way to discuss pregnancy rate is to indicate the number of pregnancies expected per 100 women per year. For example, it is expected that 80 out of 100 young women, or 80%, who are regularly engaging in unprotected intercourse will be pregnant within a year. Another way to discuss birth control methods is to indicate their effectiveness, in which case the emphasis is placed on the number of women who will not get pregnant. For example, with the least effective method given in figure 14.14, we expect that 70 out of 100, or 70%, sexually active women will not get pregnant, and 30 women will get pregnant within a year. The very best and surest method of birth control is total abstinence.

FIGURE 14.14 *Effectiveness (the percentage of women who are not expected to be pregnant within one year) of various birth control measures. Sterilization and the pill offer the best protection, while creams and so forth offer the least protection from the occurrence of pregnancy. This graph assumes that users are properly and faithfully using the various means of birth control.*
Data based on Guttmacher, Alan F., Pregnancy, Birth, and Family Planning. *New York: New American Library 1973.*

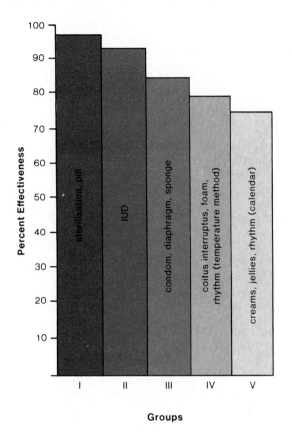

Group I

Sterilization is a surgical procedure that renders the individual incapable of reproduction. Sterilization operations do not affect the secondary sex characteristics nor sexual performance.

In the male, a **vasectomy** consists of cutting and tying the vas deferens on each side so that the sperm are unable to reach the seminal fluid that is ejected at the time of orgasm. The sperm are then largely reabsorbed. Following this operation, which can be done in a doctor's office, the amount of ejaculate remains normal because sperm account for only about 1% of the volume of semen. Also, there is no effect on the secondary sex characteristics since testosterone continues to be produced by the testes.

In the female, **tubal ligation** consists of cutting or otherwise sealing the oviducts. Pregnancy rarely occurs because the passage of the egg through the oviducts has been blocked. Whereas major abdominal surgery was formerly required for a tubal ligation, today there are simpler procedures. Using a method called *laparoscopy,* which requires only two small incisions, the surgeon inserts a small, lighted telescope to view the oviducts and a small surgical blade to sever them. An even newer method called hysteroscopy uses a telescope within the uterus to seal the tubes by means of an electric current.

Although recently developed microsurgical methods allow either a vas deferens or oviduct to be rejoined, it is still wise to view a vasectomy or tubal ligation as permanent. Even following successful resectioning, fertility is usually reduced by about 50%.

The *birth control pill* (fig. 14.15d) is usually a combination of estrogen and progesterone that is taken for twenty-one days of a twenty-eight-day cycle (beginning at the end of menstruation). The estrogen and progesterone in the pill effectively shut down the pituitary production of both FSH and LH so that no follicle begins to develop in the ovary; and since ovulation does not occur, pregnancy cannot take place. Both beneficial and adverse side effects have been linked to the pill. Women report relief of menstrual discomforts and acne. They also report several minor adverse side effects such as nausea and vomiting. Less common complaints are weight gain, headaches, and chloasma, areas of darkened skin on the face. One serious side effect of the pill is increased incidence of thromboembolism—almost exclusively in women who are over 35 and who smoke. Since there are possible side effects, those taking the pill should always be seen regularly by a physician.

Group II

An *IUD* (intrauterine device) (fig. 14.15a) is a small piece of molded plastic that is inserted into the uterus by a physician. It is generally used by women who have given birth to at least one child. Most likely, an IUD prevents implantation by the embryo since there is often an inflammatory reaction where the device presses against the endometrium. The minor side effects of the IUD are expulsion, pain, irregular bleeding, or profuse menses. The major side effects of the IUD are rare and include pelvic infection and perforation of the uterus. The infection can usually be treated by antibiotics, but deaths have been known to occur. Therefore they are in limited use in the United States at present.

Group III

The *diaphragm* (fig. 14.15c) is a soft rubber or plastic cup with a flexible rim that lodges behind the pubic bone and fits over the cervix. Each woman must be properly fitted by a physician, and the diaphragm can be inserted into the vagina two hours at most before sexual relations. It must also be used with a spermicidal jelly or cream and should be left in place for at least six hours after sexual relations. If intercourse is repeated during this time, more jelly or cream should be inserted by means of a plastic insertion tube.

The *cervical cap,* a widely used contraceptive device popular in Europe, is currently being introduced in this country. The cervical cap is thicker and smaller than the diaphragm. The thimble-shaped rubber or plastic cup fits snugly

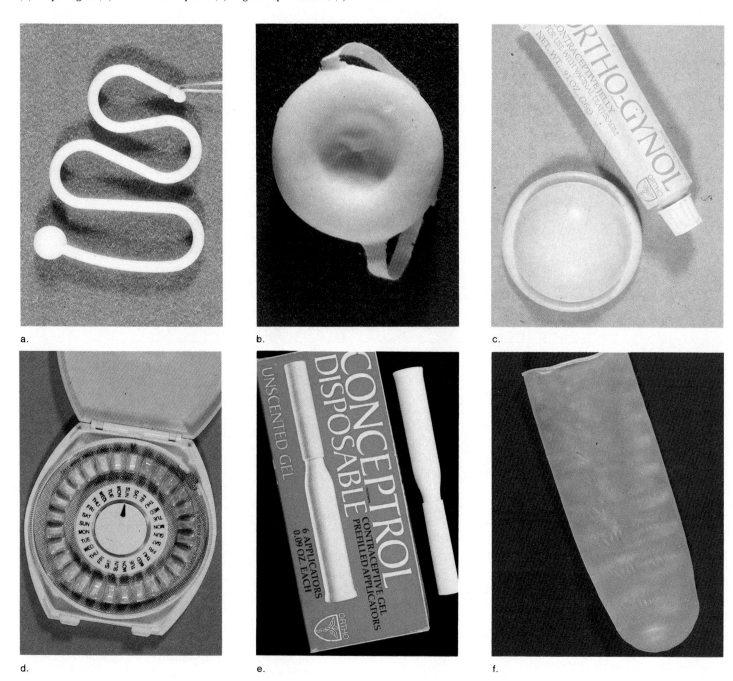

a.

b.

c.

d.

e.

f.

around the cervix. Unlike the diaphragm, the cervical cap is effective even if left in place for several days.

A *condom* (fig. 14.15f) is a thin latex rubber sheath that fits over the erect penis. The ejaculate is trapped inside the sheath and thus does not enter the vagina. When used in conjunction with a spermicidal foam, cream, or jelly, the protection is better than with the condom alone. Today it is possible to purchase condoms that are already lubricated with a spermicide. The condom is generally recognized as giving protection against sexually transmitted diseases such as those discussed in the following chapter.

A *vaginal sponge* (fig. 14.15b) permeated with spermicide and shaped to fit the cervix is a new contraceptive recently made available to the general public after seven years of testing. Unlike the diaphragm and cervical cap, the sponge need not be fitted by a physician since one size fits everyone. It is effective immediately after placement in the vagina and remains effective for twenty-four hours. Like the other means of birth control in this category, the sponge is about 85% effective in preventing pregnancy.

1 menstruation begins	2	3	4	5	6	7
8	9	10 intercourse leaves sperm to fertilize egg	11	12 egg may be released	13	14
15 egg may also be released	16	17 egg may still be present	18	19	20	21
22	23	24	25	26	27	28
1 menstruation begins						

Group IV

It is possible for the male to withdraw the penis just before ejaculation so that the semen is deposited away from the vaginal area. This method of birth control, called *coitus interruptus,* has a relatively high failure rate because a few drops of seminal fluid may escape from the penis before ejaculation takes place. Even a small amount of semen can contain numerous sperm.

Spermicidal jellies, creams, and foams (fig. 14.15e) contain sperm-killing ingredients and may be inserted into the vagina with an applicator up to thirty minutes before each occurrence of intercourse. Foams are considered the most effective of this group of contraceptives. When used alone, these are not highly effective means of birth control for those who have frequent intercourse. They do offer some protection against sexually transmitted disease; nonoxynol 9 is a viral inhibitor giving some protection against AIDS.

Group V

Natural family planning, formerly called the *rhythm method* of birth control, is based on the realization that a woman ovulates only once a month and that the egg and sperm are viable for a limited number of days. If the woman has a consistent twenty-eight-day cycle, then the period of "safe" days can be determined, as in figure 14.16. This method of birth control is not very effective because the days of ovulation can vary from month to month, and the viability of the egg and sperm varies perhaps monthly but certainly from person to person.

A more reliable way to practice natural family planning is to await the day of ovulation each month and then wait three more days before engaging in intercourse. The day of ovulation can be more accurately determined by noting the body temperature early each morning (body temperature rises at ovulation) or by taking the pH of the vagina each day (near the day of ovulation the vagina becomes more alkaline) or by noting the consistency of the mucus at the cervix (at ovulation the mucus is thicker and heavier). Physicians can instruct women how to do these procedures.

Only a medically recognized method of birth control such as those discussed here should be used. Douching is of little value and position of intercourse will not prevent pregnancy at all. In fact, the proximate location of the penis (at the time of ejaculation) near but not in the vagina has been known to result in pregnancy.

Numerous birth control methods and devices are available for those who wish to prevent pregnancy. The more effective methods are sterilization, the pill, the IUD, the sponge, and the diaphragm. A condom used with a spermicidal jelly or foam is also effective. The less effective methods are spermicidal foam and jelly alone, coitus interruptus, and natural family planning.

Future Means of Birth Control

There are three areas in which birth control investigations have been directed. There is need for a morning-after medication, a long-lasting method, and a medication that is specifically for males.

In this country, DES, a synthetic estrogen that affects the uterine lining making implantation difficult, is sometimes given following intercourse. Since large doses are required, causing nausea and vomiting, DES is usually given only for incest or rape. There is a new birth control pill on the market in France consisting of a synthetic steroid that prevents progesterone from acting on the uterine lining because it has a high affinity for progesterone receptors. In clinical tests the uterine lining sloughed off within four days in 85% of women who were less than a month pregnant. To improve the success rate, investigators have considered adding a small quantity of prostaglandins that cause contraction of the uterus and disintegration of the corpus luteum. The promoters of this treatment are using the term "contragestation" to describe its effects; however, it should be recognized that this medication, rather than preventing implantation, brings on an *abortion,* the loss of an implanted fetus. It is expected that the medication will be used by many women who are experiencing delayed menstruation without knowing whether they are actually pregnant.

Depo-Provera is an injectable contraceptive that is commercially available in many countries outside the United States. The injection contains crystals that gradually dissolve over a period of three months. The crystals contain a chemical related to progesterone and this chemical suppresses ovulation. The drug has not been approved for use in the United States because cancer developed in some test animals receiving the injections. More animal studies are now underway. An even more potent progesterone-like molecule is now being tested for implantation under the skin. The *implant* consists of narrow tubes that slowly release the drug over a period of five years.

Various possibilities exist for a *"male pill."* Scientists have made analogs of gonadotropic-releasing hormones that interfere with the action of this hormone and prevent it from stimulating the pituitary. Experiments in both animal and human subjects suggest that one of these might possibly inhibit spermatogenesis in males (and ovulation in females) without affecting the secondary sex characteristics. The seminiferous tubules produce a hormone termed inhibin that inhibits FSH production by the pituitary (p. 299). It is hoped that this chemical may someday be produced commercially and made available in pill form for males. Testosterone and/or related chemicals can be used to inhibit spermatogenesis in males, but there are usually feminizing side effects because the excess is changed to estrogen by the body.

There are numerous well-known birth control methods and devices available to those who wish to prevent pregnancy. These differ as to their effectiveness. In addition, new methods are expected to be developed.

Infertility

Sometimes couples do not need to prevent pregnancy; conception or fertilization does not occur despite frequent intercourse. The American Medical Association estimates that 15% of all couples in this country are unable to have any children and are therefore properly termed sterile; another 10% have fewer children than they wish and are therefore termed *infertile.*

Infertility can be due to a number of factors. It is possible that fertilization takes place, but the embryo dies before implantation takes place. One area of concern is that radiation, chemical mutagens, and the use of psychoactive drugs can contribute to sterility, possibly by causing chromosomal mutations that prevent development from proceeding normally. The lack of progesterone can also prevent implantation, and therefore the proper administration of this hormone is sometimes helpful.

It is also possible that fertilization never takes place. There may be a congenital malformation of the reproductive tract, or there may be an obstruction of the oviduct or vas deferens. Endometriosis, the spread of uterine tissue beyond the uterus, is also a cause of infertility, as discussed in the reading on the next page. Sometimes physical defects can be corrected surgically. If no obstruction is apparent, it is possible to give females a substance rich in FSH and LH that is extracted from the urine of postmenopausal women. This treatment causes multiple ovulations and sometimes multiple pregnancies, however.

ENDOMETRIOSIS

Endometriosis, an often unrecognized disease that afflicts anywhere from four million to ten million American women, is a major cause of infertility. The condition is caused by the spread and growth of tissue from the lining of the uterus (or endometrium) beyond the uterine walls (see figure). These endometrial cells form bandlike patches and scars throughout the pelvis and around the ovaries and Fallopian tubes, resulting in a variety of symptoms and degrees of discomfort. Because endometriosis has been associated with delayed childbearing, it is sometimes called the "career woman's disease." But recent studies have shown that the disorder strikes women of all socioeconomic groups and even teenagers, though those with heavier, longer, or more frequent periods may be especially susceptible. Says Dr. Donald Chatman of Chicago's Michael Reese Hospital, "Endometriosis is an equal-opportunity disease."

How the disease begins is something of a mystery. One theory ascribes it to "retrograde menstruation." Instead of flowing down through the cervix and vagina, some menstrual blood and tissue back up through the Fallopian tubes and spill out into the pelvic cavity (see figure). Normally this errant flow is harmlessly absorbed, but in some cases the stray tissue may implant itself outside the uterus and continue to grow. A second theory suggests that the disease arises from misplaced embryonic cells that have lain scattered around the abdominal cavity since birth. When the monthly hormonal cycles begin at puberty, says Dr. Howard Judd, director of gynecological endocrinology at UCLA Medical Center, "some of these cells get stirred up and could be a major cause of endometriosis."

If anything about endometriosis is clear, it is that once the disease has begun, it will probably get worse. Stimulated by the release of estrogen, the implanted tissue grows and spreads. Cells from the growths break away and are ferried by lymphatic fluid throughout the body, sometimes, although rarely, forming islands in the lungs, kidneys, bowel, or even the nasal passages. There they respond to the menstrual cycle, causing monthly bleeding from the rectum or wherever else they have settled.

The most common symptom of endometriosis is pain, which can occur during menstruation, urination, and sexual intercourse. Unfortunately, these warnings are often overlooked by women and their doctors. . . . To confirm that a patient has endometriosis, doctors look for the telltale tissue by peering into the pelvic cavity with a fiber-optic instrument called a laparoscope. After diagnosis, a number of treatments can be prescribed. One is pregnancy—if it is still feasible; the nine-month interruption of menstruation can help shrink misplaced endometrial tissue. Taking birth-control pills may also help, but more effective is a drug called danazol, a synthetic male hormone that stops ovulation and causes endometrial tissue to shrivel. But it can also produce acne, facial-hair growth, weight gain, and other side effects.

A new experimental treatment with perhaps fewer ill effects involves a synthetic substance called nafarelin, similar to gonadotropic-releasing hormone. Normally GnRH is released in bursts by the hypothalamus gland, eventually triggering the process of ovulation. But "if the GnRH stimulation is given continuously instead of in pulses," explains Dr. Robert Jaffe of the

---

When reproduction does not occur in the usual manner, couples today are seeking alternative reproductive methods that may include the following techniques:

Artificial Insemination by Donor (AID)   Since the 1960s, there have been hundreds of thousands of births following artificial insemination in which sperm are placed in the vagina by a physician. Sometimes a woman is artificially inseminated with her husband's sperm. This is especially helpful if the husband has a low sperm count—the sperm can be collected over a period of time and concentrated so that the sperm count is sufficient to result in fertilization. Often, however, a woman is inseminated by sperm acquired from a donor who is a complete stranger to her.

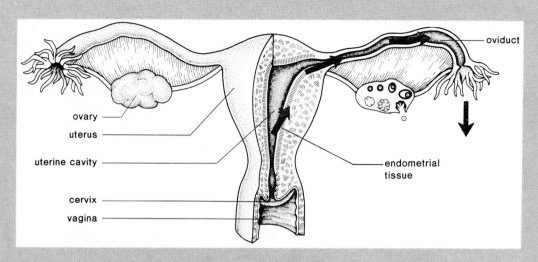

ovary

uterus

uterine cavity

cervix

vagina

oviduct

endometrial tissue

During endometriosis (caused by the spread and growth of tissue from the lining of the uterus, or endometrium, beyond the uterine walls), menstrual blood flows up through the fallopian tubes and into the pelvic cavity, as shown by arrows.

University of California, San Francisco, "the whole [ovulatory] system shuts off," and the endometrial implants "virtually melt away."

For severe cases of endometriosis, surgical removal of the ovaries and uterus may be the only solution. But less extreme surgery can often help. At Atlanta's Northside Hospital, Dr. Camran Nezhat has had success with a high-tech procedure called videolaseroscopy, which employs a laparoscope rigged with a tiny video camera and a laser. The camera images, enlarged on a video screen, enable Nezhat to zero in on endometrial tissue and vaporize it with the laser. In a study of 102 previously infertile patients,

Nezhat found that 60.7% were able to conceive within two years of videolaseroscopy treatment.

Like many other doctors who see the unfortunate consequences of endometriosis, Nezhat is concerned that a "lot of women do not seek help for this problem." Any serious pain, he notes, needs investigating. Agrees Cheri Bates (a victim), "If a doctor tells you that suffering is a woman's lot in life, get another doctor."

In Vitro Fertilization (IVF)   Over a hundred babies have been conceived using this method. First a woman is given appropriate hormonal treatment. Then laparoscopy may be done. The laparoscope is a metal tube about the size of a pencil that is equipped with a tiny light and telescopic lens. In this instance, it is also fitted with a tube for retrieving eggs. After insertion through a small incision near the woman's naval, the physician guides the laparoscope to the ova-ries where the eggs are sucked up into the tube. Alternately, it is possible to place a needle through the vaginal wall and guide it, by the use of ultrasound, to the ovaries where the needle is used to retrieve the eggs. This method is called transvaginal retrieval.

Concentrated sperm from the male is placed in a so-lution that approximates the conditions of the female genital tract. When the eggs are introduced, fertilization occurs. The resultant zygotes begin development, and after about two to four days they are inserted into the uterus of the woman, who is now in the secretory phase of her menstrual cycle. If implantation is successful, development is normal and continues to term.

Gamete Intrafallopian Transfer (GIFT) This method was devised as a means to overcome the low success rate (15 to 20%) of in vitro fertilization. The method is exactly the same as in vitro fertilization except the eggs and sperm are immediately placed in the oviducts after they have been brought together. This procedure would be helpful to couples whose eggs and sperm never make it to the oviducts; sometimes the egg gets lost between the ovary and the oviducts, and sometimes the sperm never reach the oviducts. GIFT has an advantage in that it is a one-step procedure for the woman—the eggs are removed and reintroduced all in the same time period. For this reason it is less expensive—$1,500 compared with $3,000 and up for in vitro fertilization.

Surrogate Mothers Over a hundred babies have been born to women paid to have them by other individuals who have contributed sperm (or egg) to the fertilization process. If all the alternative methods discussed here are considered, it is possible to imagine that a baby could have five parents: (1) sperm donor, (2) egg donor, (3) surrogate mother, and (4) and (5) adoptive mother and father.

Some couples are infertile. There may be a hormonal imbalance or a blockage of the oviducts. When corrective medical procedures fail, it is possible today to consider an alternative method of reproduction.

FIGURE 14.17 *Sometimes couples utilize alternative methods of reproduction in order to experience the joys of parenthood.*

SUMMARY

In males, spermatogenesis occurs within the seminiferous tubules of the testes, which also produce testosterone within the interstitial cells. Sperm mature in the epididymis and are stored here and in the first part of the vas deferens before entering the urethra, along with seminal fluid, prior to ejaculation. Hormonal regulation involving secretions from the hypothalamus, anterior pituitary, and the testes in the male maintains testosterone at a fairly constant level.

In females, egg production occurs within the ovaries where one follicle produces an egg each month. Fertilization, if it occurs, takes place in the oviducts, and the resulting embryo travels to the uterus where it implants itself in the uterine lining. In the nonpregnant female, hormonal regulation in the female involves the ovarian and uterine cycle, dependent upon the hypothalamus, anterior pituitary, and the female sex hormones, estrogen and progesterone.

Numerous birth control methods and devices are available for those who wish to prevent pregnancy. Infertile couples are increasingly resorting to alternative methods of reproduction.

OBJECTIVE QUESTIONS

1. If you are tracing the path of sperm, the structure that follows the epididymis is the _____ .

2. The prostate gland, Cowper's glands, and the _____ all contribute to seminal fluid.

3. The primary male sex hormone is _____ .

4. An erection is caused by the entrance of _____ into the penis.

5. In the female reproductive system, the uterus lies between the oviducts and the _____ .

6. In the ovarian cycle, once each month a _____ produces an egg. In the uterine cycle, the _____ lining of the uterus is prepared to receive the zygote.
7. The female sex hormones are _____ and _____ .

8. Pregnancy in the female is detected by the presence of _____ in the blood or urine.
9. Aside from abstinence, the most effective means of birth control are _____ in males and _____ in females.

10. In vitro fertilization occurs in _____ .

STUDY QUESTIONS

1. Discuss the anatomy and physiology of the testes. (p. 294) Describe the structure of sperm. (p. 295)
2. Give the path of sperm. (p. 296)
3. What glands produce seminal fluid? (p. 296)
4. Discuss the anatomy and physiology of the penis. (p. 296) Describe ejaculation. (p. 296)
5. Discuss hormonal regulation in the male. Name three functions of testosterone. (p. 297)
6. Discuss the anatomy and physiology of the ovaries. (p. 299) Describe ovulation. (p. 299)
7. Give the path of the egg. Where do fertilization and implantation occur? (p. 300) Name two functions of the vagina. (p. 301)
8. Describe the external genitalia in females. (p. 303)
9. Compare male and female orgasm. (pp. 296, 303)
10. Discuss hormonal regulation in the female, either simplified and/or detailed. (p. 303) Give the events of the uterine cycle, and relate them to the ovarian cycle. (p. 304) In what way is menstruation prevented if pregnancy occurs? (p. 306)
11. Name four functions of the female sex hormones. (p. 307) Describe the anatomy and physiology of the breast. (p. 307)
12. Aside from abstinence, discuss the various means of birth control and their relative effectiveness. (p. 308)

THOUGHT QUESTIONS

1. As a continuation of the introduction to this chapter, state specifically the anatomical and physiological means by which humans are adapted to reproduce on land.
2. Men retain their reproductive potential much longer than women do. How is this suited to a difference in gamete production and to their different contributions to reproduction of offspring?
3. All organisms have a life that includes a reproductive strategy. For example, some insects spend much of their lives as larvae, undergo metamorphosis into winged forms, reproduce, and die all within a single season. They do not spend any time at all caring for their offspring and instead produce a large number of which a few may survive. Outline and discuss the human life strategy. Include a reference to culture, mentioned on page 000 of the text.

KEY TERMS

endometrium (en''do-me'tre-um) the lining of the uterus that becomes thickened and vascular during the uterine cycle. *301*

erection (ĕ-rek'shun) referring to a structure, such as the penis, that is turgid and erect as opposed to being flaccid or lacking turgidity. *296*

Graafian follicle (graf'e-an fol'li-k'l) mature follicle within the ovaries that houses a developing egg. *299*

implantation (im''plan-ta'shun) the attachment of the embryo to the lining (endometrium) of the uterus. *306*

interstitial cells (in''ter-stish'al selz) hormone-secreting cells located between the seminiferous tubules of the testes. *296*

menopause (men'o-pawz) termination of the ovarian and uterine cycles in older women. *307*

menstruation (men''stroo a'shun) loss of blood and tissue from the uterus at the end of a uterine cycle. *304*

ovarian cycle (o-va're-an si'k'l) monthly occurring changes in the ovary that affect the level of sex hormones in the blood. *303*

ovaries (o'var-ez) the female gonads, the organs that produce eggs and estrogen and progesterone. *299*

semen (se'men) the sperm-containing secretion of males; seminal fluid plus sperm. *296*

seminiferous tubules (sem''i-nif'er-us tu'bŭlz) highly coiled ducts within the male testes that produce and transport sperm. *294*

testes (tes'tēz) the male gonads, the organs that produce sperm and testosterone. *294*

uterine cycle (u'ter-in si'k'l) monthly occurring changes in the characteristics of the uterine lining. *304*

CHAPTER FIFTEEN

SEXUALLY TRANSMITTED DISEASES

CHAPTER CONCEPTS

1 Viruses have to reproduce inside a living cell and are, for this reason, obligate parasites.

2 AIDS and herpes are both caused by viruses; therefore, it is difficult to find a cure.

3 Bacteria are independent cells that lack the organelles found in human cells.

4 Gonorrhea, chlamydia, and syphilis are caused by bacteria; therefore, they are curable by antibiotic therapy.

5 There are also other fairly common sexually transmitted diseases caused by a protozoan, fungus, and even a louse.

Coccidiodomycosis, a fungal disease. Sporangia have formed in the patient's tissue, releasing spores capable of spreading infection.

INTRODUCTION

Can you imagine that a sexually transmitted disease could have altered the course of history? At the end of the fifteenth century, syphilis spread throughout Europe. Some believe that Columbus's sailors were infected by the Indians and carried the bacterium to Europe on their return from the New World. In any case, Henry VIII (fig. 15.1) contracted syphilis just before he married Catherine of Aragon. She bore him four sons but all had congenital syphilis and were stillborn or fatally malformed. He blamed her for this tragedy and sought an annulment from the Catholic Church. When it was denied, he broke with the church so that he could divorce Catherine and take another wife. England has been a Protestant country since that time.

HUMAN ISSUE

Sexually transmitted diseases are especially prevalent among those who are promiscuous. Unfortunately, many who are afflicted with these diseases are reluctant to seek treatment, perhaps because they feel there is a stigma associated with having such a disease. Should society "de-stigmatize" these diseases so that they are viewed in the same light as other types of diseases? Or would you say that these individuals get what they deserve and, therefore, significant sums of money should not be spent on research, public education programs, or medical treatment? In thinking about this, consider situations in which a promiscuous spouse may pass on the condition to a faithful husband or wife and in which children born to an afflicted woman are also infected.

S exually transmitted diseases (venereal diseases) are contagious diseases caused by microorganisms. Microorganisms are generally too small to be seen by the naked eye, but they are present in large numbers in the environment. Although most microorganisms are harmless and indeed are even used by humans to make various products, a few cause diseases in humans. Those that cause sexually transmitted diseases can be passed from one human to another by sexual contact.

VIRAL IN ORIGIN

Viruses cause numerous diseases in humans (table 15.1) including AIDS and herpes, two sexually transmitted diseases of great concern today. Viruses are tiny particles that always have at least two parts: an outer *coat of protein* and an inner *core of nucleic acid*. Viruses have a definite shape because the coat consists of a geometric arrangement of protein molecules. The viruses that cause AIDS and the one that causes herpes are polyhedral in shape (fig. 15.3). Many viruses also have an outer envelope that contains lipid as well as protein molecules. The lipid molecules are derived from the host's membrane, but the protein is unique to the virus.

Since viruses are not cellular, they are incapable of independent reproduction and reproduce only inside a living cell. For this reason, they are called *obligate parasites*. A **parasite** requires a **host** organism in order to function properly, complete its life cycle, and reproduce. In the laboratory, viruses are maintained by injecting them into live chick embryos (fig. 15.2). Outside living cells, viruses are nonliving and can be stored just as chemicals are stored.

Most viruses are extremely specific. Not only do they prefer a particular type of organism, such as humans, they also prefer a particular tissue type. This specificity is due to the ability of the virus to combine with a particular molecular configuration, such as a receptor on the cell surface. Within a half hour, the virus, or simply the nucleic acid core depending on the virus, has entered the cell. Most often the viral genes, whether they are DNA or RNA viruses, immediately take over the machinery of the cell, and the virus undergoes reproduction. These are the steps required for a DNA virus to reproduce. 1. Viral DNA replicates repeatedly, utilizing the nucleotides within the host. Multiple copies of viral DNA result. 2. Viral DNA is transcribed into mRNA, which undergoes translation. Multiple copies of coat protein result. 3. Assemblage occurs. Viral DNA is packaged inside a coat protein. If the virus has an envelope, it is formed at the cell membrane just before the virus leaves the cell (fig. 15.3a).

There are a few types of viruses that do not immediately undergo reproduction; instead, viral DNA becomes integrated into the host DNA. RNA viruses that do this are called **retroviruses** because they have a special enzyme called *reverse transcriptase* through which their RNA is transcribed into cDNA (a DNA copy of the RNA gene) that

becomes incorporated into host DNA. Sometimes the virus remains *latent,* and during this time the viral DNA is replicated along with host DNA. Eventually, viral reproduction may occur (fig. 15.3b). Certain environmental factors, such as ultraviolet radiation, can cause a latent virus to undergo reproduction.

Viruses reproduce only inside host cells. Some viruses have the ability to incorporate their DNA within host DNA. Retroviruses are RNA viruses that are able to perform reverse transcription.

TABLE 15.1	*INFECTIOUS DISEASES CAUSED BY VIRUSES*
RESPIRATORY TRACT	**NERVOUS SYSTEM**
Common colds	Encephalitis
Flu[a]	Polio[a]
Viral pneumonia	Rabies[a]
SKIN REACTIONS	**LIVER**
Measles[a]	Yellow fever[a]
German measles[a]	Hepatitis A & B
Chickenpox[a]	
Smallpox[a]	**OTHER**
Warts	Mumps[a]
	Cancer
SEXUALLY TRANSMITTED	
AIDS	
Herpes	

[a]Vaccines available. Yellow fever, rabies, and flu vaccines are only given if the situation requires them. Smallpox vaccinations are no longer required.

FIGURE 15.2 Inoculation of live chick eggs with virus particles. A virus only reproduces inside a living cell, not because it uses the cell as nutrients but rather because it takes over the machinery of the cell.

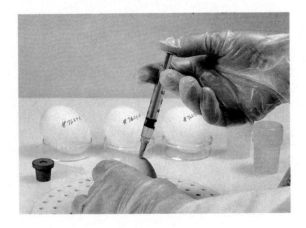

FIGURE 15.3 *Viral life cycles.* a. *Some DNA animal viruses, like the one depicted here, immediately reproduce within cells. Others integrate their DNA into host DNA for a time.* b. *Retroviruses must first reverse transcribe RNA into DNA (called cDNA because it contains a copy of the viral RNA) before integration is possible.*

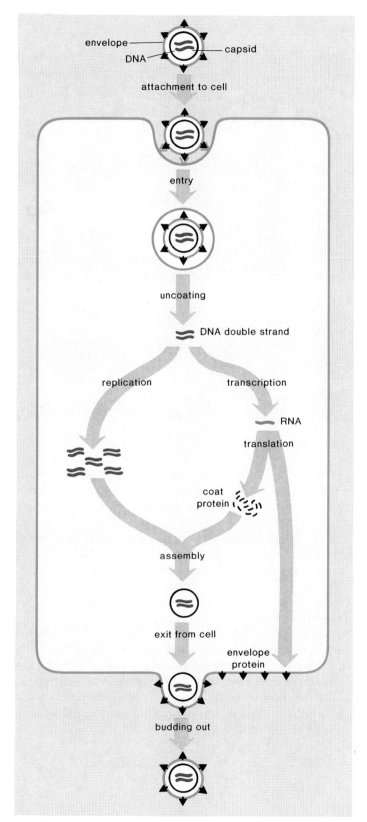

a. DNA virus

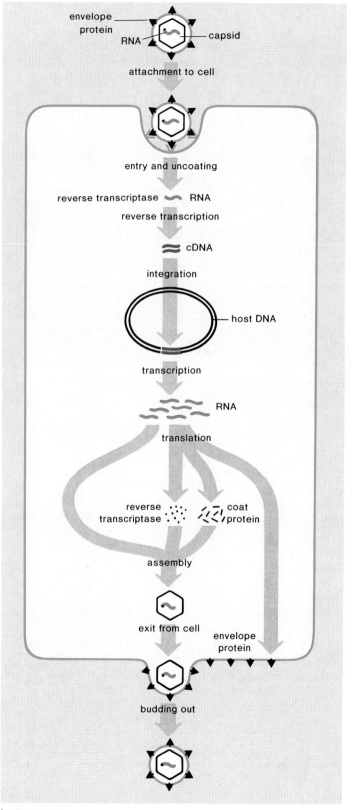

b. RNA virus

FIGURE 15.4   a. *AIDS viruses erupting from T cell. Several virus particles are budding from the periphery of a T cell. Other virus particles cluster outside the cell membrane (×30). b. An enlargement of several virus particles (×90). c. Virus particle in the process of budding from the cell membrane (×50). d. Mature AIDS virus following the budding process.*

AIDS virus

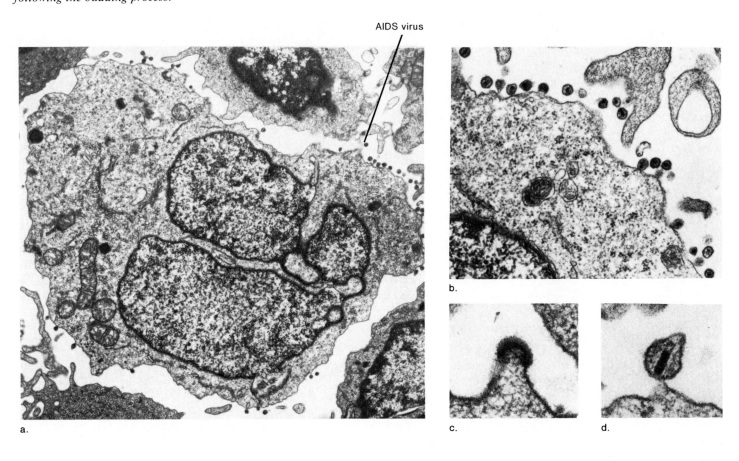

a.

b.

c.

d.

AIDS

The organisms that cause **AIDS** (acquired immune deficiency syndrome) are retroviruses that attack helper T cells and macrophages, the very cells that protect us from disease. The viruses, now called **HIV** (human immunodeficiency viruses), usually stay hidden inside macrophages, but on occasion do bud from T cells (fig. 15.4).

Symptoms

AIDS has three stages of infection. During the first stage, the individual usually has detectable antibodies in the bloodstream and may exhibit swollen lymph nodes. During the second stage, called AIDS-related complex (**ARC**), symptoms also include weight loss, night sweats, fatigue, fever, and diarrhea. Finally, the person may develop full-blown AIDS, especially characterized by the development of pneumonia, skin cancer, and also neuromuscular and psychological disturbances.

It takes several years before an individual passes through the first two stages and develops AIDS itself. Although a few individuals have had AIDS as long as six years, most die within this time period.

Transmission

As stated in the accompanying reading (p. 322), AIDS is transmitted by blood, semen, and vaginal fluid. In the United States the two main afflicted groups are homosexual men and intravenous drug users (and their sexual partners). In Africa and some parts of South America, though, AIDS is transmitted chiefly through heterosexual intercourse and an equal number of men and women are infected.

Certain portions of the country have been harder hit than other sectors (fig. 15.5). Even in New York City, which reports the highest level of infectivity, there are regions that have the most cases. Here, too, the number of AIDS deaths among needle-using addicts is higher than among homosexual men. The AIDS viruses can cross the placenta, and the Bronx currently has one in forty-three babies born with HIV antibodies in the blood. Some of these newborns may have the antibodies without having the virus, but 30–50% are most likely infected.

Although intravenous drug users are liable to spread the disease to the general heterosexual population, to date infection among the general population is still less than 1%. Health officials emphasize that unprotected intercourse with multiple partners or a single infected partner increases the chance of transmission. The use of a condom reduces the risk, but the very best preventive measure at this time is monogamy with a sexual partner who is free of the disease. Casual contact with someone who is infected, such as shaking hands, eating at the same table, or swimming in the same pool, is not a mode of transmission.

HUMAN ISSUE

Despite assurances to the contrary, there are those who fear casual contact with an AIDS victim. Studies of households have consistently shown the risk of catching AIDS due to close but non-sexual contact with an infected person—even if that person is a family member—is virtually nonexistent. Do you trust these findings, or do you think that, regardless of such studies, it behooves all of us to be as cautious as possible? Do you think AIDS patients should be quarantined or prevented from holding jobs or going to school?

Treatment

The drug AZT has been found to be helpful in prolonging the lives of AIDS patients. When RNA from the viruses is transcribed to DNA, the cell tends to use the drug instead of thymidine, and this results in nonfunctioning DNA. Other drugs that interfere with the life cycle of the viruses are also being developed.

As discussed on page 146, investigators are trying to develop a vaccine for AIDS. The AIDS viruses mutate frequently but researchers have identified portions of the coat protein that they believe are relatively stable. When these are injected into the bloodstream, antibodies do develop but it is not yet known whether such antibodies will offer protection against infection. After all, persons with AIDS do have antibodies but for some reason or other these antibodies do not stem the course of the disease. The viruses rarely enter the bloodstream; they cause T cells to fuse, and they may spread from cell to cell in this manner.

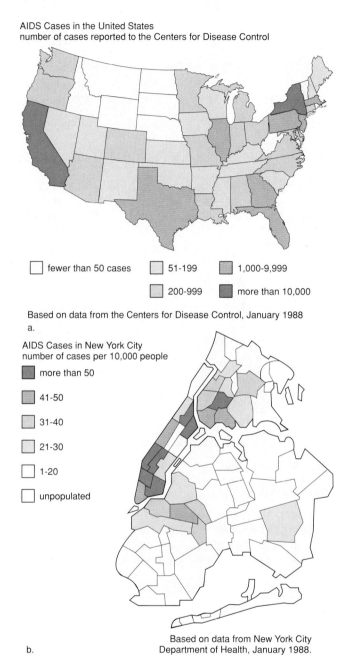

FIGURE 15.5 *Geography of AIDS.* a. *States are shaded according to the number of cases reported (see key).* b. *New York City reports the greatest number of cases but some regions of the city are harder hit than others (see key).*
Maps by Nina Wallace © 1988 Discover Publications, Inc.

AIDS Cases in the United States
number of cases reported to the Centers for Disease Control

☐ fewer than 50 cases   ☐ 51-199   ☐ 1,000-9,999

☐ 200-999   ■ more than 10,000

Based on data from the Centers for Disease Control, January 1988
a.

AIDS Cases in New York City
number of cases per 10,000 people

■ more than 50

▨ 41-50

▨ 31-40

☐ 21-30

☐ 1-20

☐ unpopulated

b.   Based on data from New York City
Department of Health, January 1988.

AIDS PREVENTION

1. The AIDS viruses are transmitted by blood, semen, and vaginal fluid. Physical contact does not transmit the viruses unless one of these fluids is exchanged. Thus, casual, occupational, or household contact is safe.
2. There is no known reason to fear spread by insects, by objects that an infected person has touched, or by coughs and sneezes.
3. Transmission through IV drug use can be prevented by not sharing needles or by sterilizing needles and syringes with Clorox solution or boiling water.
4. Intercourse, anal or vaginal, is the sexual practice most likely to transmit the viruses.
5. Sexual activities that do not involve exposing a mucous membrane (rectum, vagina, mouth) to body fluids are presumed to be safe.
6. Completely monogamous relationships between uninfected people are safe.
7. "Knowing your partner" is not as easy as it sounds. Establishing whether a partner has been, or is likely to have been, exposed to AIDS is very difficult. Screening tests are improving but still fallible.
8. Use of condoms is strongly encouraged. Laboratory evidence indicates that an intact condom prevents the viruses from passing. How much safety condoms provide during actual use is not known. However, it seems very probable that they work well when properly employed. Condoms should not be used with oil-based lubricants, which dissolve the latex. If the condom is not snugly fitted to the penis, it may be more likely to break.
9. A single behavior may be regarded as "low, but not zero, risk." If such a behavior is frequently repeated, the cumulative risk of transmitting the viruses increases.
10. Much remains to be learned about AIDS prevention.

Excerpted with permission from the April 1987 issue of *The Harvard Medical School Health Letter,* © 1986 by the President and Fellows of Harvard College.

Genital Herpes

The different herpes viruses are large, and their DNA code for about 100 specific proteins. Chickenpox and mononucleosis, among other ailments, are due to herpes viruses. **Genital herpes** is caused by **herpes simplex virus** (fig. 15.6), of which there are two types: *type 1,* which usually causes cold sores and fever blisters, and *type 2,* which more often causes genital herpes, but the delineation is not complete and crossovers do occur.

Transmission and Symptoms
Genital herpes is one of the more prevalent sexually transmitted diseases today (fig. 15.7); an estimated 20.5 million persons in the United States have it, with an estimated 500,000 new cases appearing each year. Some have even estimated that one out of every six sexually active individuals may be capable of spreading the disease to another. Immediately after infection there are no symptoms, but once inside, the virus begins to multiply rapidly. The individual may experience a tingling or itching sensation before blisters appear at the infected site within two to twenty days. Once the blisters rupture, they leave painful ulcers that may take as long as three weeks or as little as five days to heal. These symptoms may be accompanied by fever, pain upon urination, and swollen lymph nodes.

After the ulcers heal, the disease is only dormant, and blisters can reoccur repeatedly at variable intervals. Sunlight, sex, menstruation, and stress seem to cause the symptoms of genital herpes to reoccur. While the virus is latent, it resides in nerve cells. Type 1 resides in a group of nerves located near the brain, and type 2 resides in nerve cells that lie near the spinal cord. Type 1 occasionally travels via a nerve fiber to the eye and causes an eye infection that can lead to blindness. Type 2 has been known to cause a form of meningitis and was formerly thought to cause a form of cervical cancer. This is no longer believed to be the case.

Infection of the newborn can occur if the child comes in contact with a lesion in the birth canal. In one to three weeks the infant is gravely ill and may become blind or have neurological disorders, including brain damage, or may die. The incidence of these occurrences is steadily increasing. In

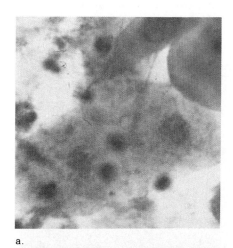

FIGURE 15.6 a. *Cell infected with herpes virus. b. Enlarged model of herpes virus.*

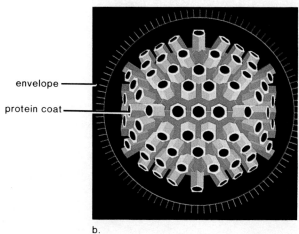

envelope
protein coat

a.

b.

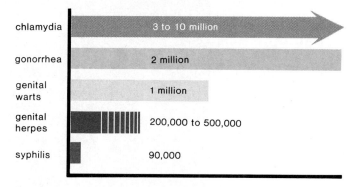

FIGURE 15.7 *Statistics (one year) for the most common sexually transmitted diseases show that chlamydia, gonorrhea, and genital warts all are much more common than herpes and syphilis. Chlamydia, gonorrhea, and syphilis are all curable with antibiotic therapy.*

Sexually Transmitted Disease

chlamydia	3 to 10 million
gonorrhea	2 million
genital warts	1 million
genital herpes	200,000 to 500,000
syphilis	90,000

the Seattle area, there were 2.6 cases per 100,000 births between 1966 and 1969. In 1982 the figure was 17.2 cases per 100,000 births. Birth by cesarean section may have prevented these instances. In two cases husbands had not told their wives about their infection. To prevent a herpes infection of the newborn, it is best to be honest with your spouse and your doctor.

Treatment

There is no present cure for herpes. The ointment form of acyclovir relieves initial symptoms, but the oral form, only recently developed, prevents the occurrence of outbreaks. Work is also being done to develop a vaccine.

Genital Warts

Genital warts are caused by the human *papillomavirus (HPV)*, which is a cubical DNA virus that reproduces in the nuclei of cells.

Transmission and Symptoms

HPV is sexually transmitted. Sometimes carriers do not have any sign of warts, although flat lesions may be present. The warts, when present, are commonly seen on the penis and foreskin of males and the vaginal opening in females. If the warts are removed, they may reoccur.

HPV, rather than genital herpes, is now associated with cancer of the cervix as well as tumors of the vulva, vagina, anus, and penis. Some researchers believe that the virus is involved in 90–95% of all cases of cancer of the cervix. If cancer develops, other environmental influences are believed also to be involved. Smoking, birth control pills, and herpes or other venereal infections are possible cofactors.

Physicians are disheartened that teenagers with multiple sex partners seem to be particularly susceptible to HPV infections (and warts). More and more cases of cancer of the cervix are being seen among this age group.

Treatment

Presently there is no cure for an HPV infection. A suitable medication to treat HPV before cancer occurs is being sought, and efforts are also underway to develop a vaccine.

HUMAN ISSUE

How conscientious should people be about telling potential sex partners that they have a sexually transmitted disease, especially a noncurable one like AIDS or herpes? Is it like a "buyers-beware market"—it's up to the individual to inquire about the health and/or sexual practices of a potential partner? At the other extreme, should people with AIDS or herpes become celibates and refrain from having sex? How would you feel if you fell in love with someone and then were told that this person had AIDS or herpes?

FIGURE 15.8 *Scanning electron micrographs of bacteria. a. Spherical-shaped bacteria. b. Rod-shaped bacteria. c. Spiral-shaped bacteria with flagella used for locomotion. See figure 15.9 for a generalized drawing of a bacteria.*
c. *Dr. R. G. Kessel and Dr. C. Y. Shih, From* Scanning Electron Microscopy, *Springer-Verlag, Berlin, Heidelberg, New York, 1976.*

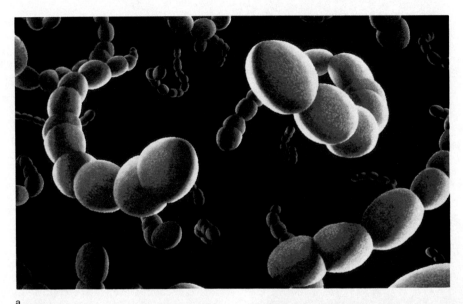

a.

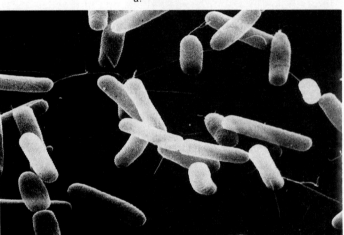

b.

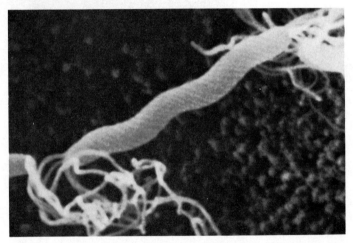

c.

BACTERIAL IN ORIGIN

Although **bacteria** are generally larger than viruses, they are still quite small. Since they are microscopic, it is not always obvious that they are abundant in the air, water, soil, and on most objects. It has even been suggested that the combined weight of all bacteria would exceed that of any other type of organism on earth. Bacteria occur in three basic shapes (fig. 15.8): rod (bacillus); round, or spherical (coccus); and spiral (a curved shape called a spirillum). Some bacteria can locomote by means of flagella.

The bacterial cell is termed a *prokaryotic* (meaning, before the nucleus) cell to distinguish it from a *eukaryotic* (meaning, true nucleus) cell. Human cells are eukaryotic and contain numerous organelles in addition to a nucleus. Prokaryotic cells lack these organelles, except for ribosomes, but still they are functioning cells. They do have DNA; it is just not contained within a nuclear envelope, and although they have no mitochondria, they do have respiratory enzymes located in the cytoplasm. Notice in figure 15.9 that the bacterial cell is surrounded by a cell wall in addition to a cell membrane. Some bacteria are also surrounded by a polysaccharide or polypeptide capsule that enhances their **virulence** (that is, ability to cause disease).

FIGURE 15.9 *Bacteria are prokaryotic cells and lack the organelles found in eukaryotic (e.g., human) cells.*

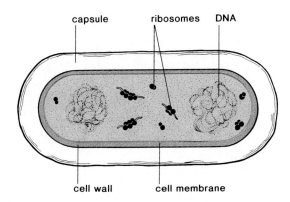

capsule   ribosomes   DNA

cell wall   cell membrane

FIGURE 15.10 *Reproduction in bacteria. Above, the single chromosome is seen to be attached to the cell membrane where it is replicating. As the cell membrane and cell wall lengthen, the two chromosomes separate. Once fission has taken place, each bacterium has its own chromosome.*

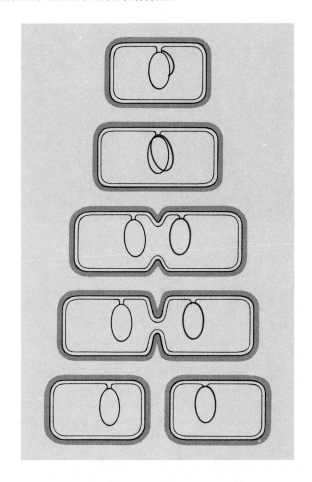

Bacteria reproduce asexually by **binary fission.** First, the single chromosome duplicates, and then the two chromosomes move apart into separate areas. Next the cell membrane and cell wall grow inward and partition the cell into two daughter cells, each of which now has its own chromosome (fig. 15.10). Under favorable conditions, growth may be very rapid with cell division occurring as often as every twelve to fifteen minutes. When faced with unfavorable environmental conditions, some bacteria can form **endospores.** During spore formation, the cell shrinks, rounds up within the former cell membrane, and secretes a new, thicker wall inside the old one. Endospores are amazingly resistant to extreme temperatures, drying out, and harsh chemicals, including acids and bases. When conditions are again suitable for growth, the spore absorbs water, breaks out of the inner shell, and becomes a typical bacterial cell.

Most bacteria are free living **saprophytes** that perform many useful services in the environment. Saprophytes send out digestive enzymes into the environment to break down large molecules into small molecules that can be absorbed across the cell membrane. Most bacteria are aerobic and require a constant supply of oxygen as we do, but a few are anaerobic, even being killed by the presence of oxygen. Table 15.2 lists the human diseases caused by bacteria; only a few serious illnesses are caused by anaerobic bacteria, such as botulism, gas gangrene, and tetanus. These bacteria and others produce toxins, chemicals that seriously interfere with the normal functioning of the body. Sometimes just a bacterial toxin is used to make a vaccine.

Bacteria have long been used by humans to produce various products commercially. Chemicals, such as ethyl alcohol, acetic acid, butyl alcohol, and acetone are produced by bacteria. Bacterial action is involved in the production of butter, cheese, sauerkraut, rubber, cotton, silk, coffee, and cocoa. By means of gene splicing, bacteria are now used to produce human insulin and interferon, as well as other types of proteins (p. 428). Even the antibiotics mentioned previously and in the reading on page 327 are produced by bacteria.

TABLE 15.2 *INFECTIOUS DISEASES CAUSED BY BACTERIA*

RESPIRATORY TRACT	NERVOUS SYSTEM
Strep throat	Tetanus[a]
Pneumonia	Botulism
Whooping cough[a]	Meningitis
Diphtheria[a]	
Tuberculosis[a]	DIGESTIVE TRACT
	Food poisoning
SKIN REACTIONS	Typhoid fever[a]
Staph (pimples and boils)	Cholera[a]
Gas gangrene[a] (wound	
infections)	SEXUALLY TRANSMITTED
	Gonorrhea
	Syphilis
	Chlamydia

[a]Vaccines are available. Tuberculosis vaccine is not used in this country. Typhoid fever, cholera, and gas gangrene vaccines are given if the situation requires it. Others are routinely given.

FIGURE 15.12 Gonorrhea bacteria in the bloodstream. If you look carefully, you will notice that these round bacteria occur in pairs; for this reason they are called diplococci.

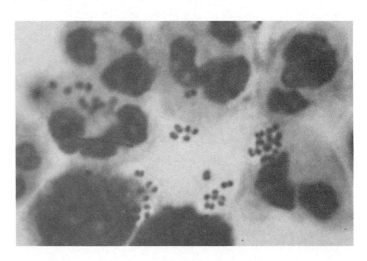

General cleanliness is the first step toward preventing the spread of infectious bacteria. Disinfectants and antiseptics also help reduce the number of infectious bacteria. **Sterilization,** a process that kills all living things, even endospores, is used whenever all bacteria must be killed. Sterilization can be achieved by use of an autoclave (fig. 15.11), a container that admits steam under pressure. If bacteria do invade our bodies and cause an infection, antibiotic therapy is often helpful.

Bacteria are prokaryotic cells capable of independent existence. Most are free living, but a few cause human diseases that can be cured by antibiotic therapy.

FIGURE 15.13 Gonorrhea infection of the eye is possible whenever the bacteria comes in contact with the eyes. This can happen when the newborn passes down the birth canal. Manual transfer from the genitals to the eyes is also possible.

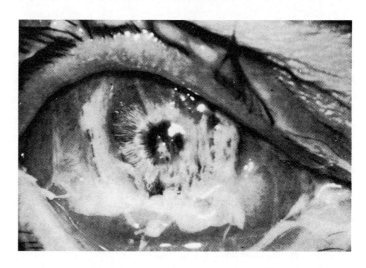

Gonorrhea

Gonorrhea is caused by the bacterium *Neisseria gonorrhoeae,* which is a nonmotile (lacks flagella), nonencapsulated, and nonspore-forming diplococcus, meaning that two cells generally stay together (fig. 15.12). Figure 15.7 indicates that there are usually two million new cases of this sexually transmitted disease, commonly called clap, each year. This is equivalent to about one new case every twenty seconds.

The diagnosis of gonorrhea in the male is not difficult as long as he displays typical symptoms (as many as 40% of males may be asymptomatic). The patient complains of pain on urination and has a thick, greenish yellow urethral discharge three to five days after contact. In the female, the bacteria are apt to first settle within the urethra and about the cervix from which they may spread to the oviducts, causing an inflammation known as **pelvic inflammatory disease (PID).** As the inflamed tubes heal, they may become completely or partially blocked by scar tissue. Now the female is sterile or at best subject to ectopic pregnancy (a pregnancy

that begins at a location other than the uterus). Unfortunately, 60–80% of females are asymptomatic until they develop severe pains in the abdominal region due to PID. PID affects about one million women a year in the United States. Similar to females, in untreated males there may be inflammation followed by scarring of the vas deferens.

Homosexual males develop gonorrhea proctitis, or infection of the anus, for which the symptoms are pain in the anus and blood or pus in the feces. Oral sex can cause infection of the throat and tonsils. Gonorrhea may also spread to other parts of the body, causing heart damage or arthritis. If by chance the person touches infected genitals and then the eyes, a severe infection of the eyes can result (fig. 15.13).

An antibiotic is a chemical that selectively kills bacteria when it is taken into the body as a medicine. There has been a dramatic reduction in the number of deaths due to pneumonia, tuberculosis, and other infections since 1900, and this can in part be attributed to the increasing use of antibiotic therapy.

Most antibiotics are produced naturally by soil microorganisms. Penicillin is made by the fungus *Penicillium;* streptomycin, tetracycline, and erythromycin are all produced by a bacterium, *Streptomyces.* Sulfa, an analog of a bacterial growth factor, can be produced in the laboratory.

Antibiotics are metabolic inhibitors specific for bacterial enzymes. This means that they poison bacterial enzymes without harming host enzymes. Penicillin blocks the synthesis of the bacterial cell wall; streptomycin, tetracycline, and erythromycin block protein synthesis; and sulfa prevents the production of a coenzyme.

There are problems associated with antibiotic therapy. Some patients are allergic to antibiotics and the reaction may even be fatal. Antibiotics not only kill off disease-causing bacteria, they also reduce the number of beneficial bacteria in the intestinal tract. The latter may have held in check a pathogen that now is free to multiply and invade the body. The use of antibiotics sometimes prevents natural immunity from occurring, leading to the necessity for reoccurring antibiotic therapy. Most important, perhaps, is the growing resistance of certain strains of bacteria. While penicillin used to be 100% effective against hospital strains of *Staphylococci aureus,* today it is far less effective. Tetracycline and penicillin, long used to cure gonorrhea, now have a failure rate of more than 20% against certain strains of *Gonococcus.* Most physicians believe that antibiotics should only be administered when absolutely necessary, and some believe that if this

Penicillium chrysogenum, *the mold that produces penicillin.*

ANTIBIOTICS AND ANTIVIRAL DRUGS

is not done then resistant strains of bacteria will completely replace present strains and antibiotic therapy will no longer be effective at all. They are very much opposed to the current practice of adding antibiotics to livestock feed in order to make animals grow fatter because resistant bacteria are easily transferred from animals to humans.

The development of antiviral drugs has lagged far behind the development of antibiotics. Viruses lack most enzymes and instead utilize the metabolic machinery of the host cell. Rarely has it been possible to find a drug that successfully interferes with viral reproduction without also interfering with host metabolism. One such drug, however, called vidarabine, was approved in 1978 for treatment of viral encephalitis, an infection of the nervous system. Acyclovir (ACV) seems to be helpful in treating genital herpes, and the drug AZT (azidothymidine) is being used in AIDS patients.

If a pregnant woman has gonorrhea bacteria in the vagina, her newborn child will be exposed to them during the process of birth. A resultant bacterial infection of the eyes leading to blindness can occur. Because of this, all newborn infants receive eye drops containing antibacterial agents, such as silver nitrate, tetracycline, or penicillin, as a protective measure.

Transmission and Treatment

Gonococci live for a very short time outside the body; therefore, most infections are spread by intimate contact, usually sexual intercourse. A woman has a 50–60% risk of contracting the disease after a single exposure to an infected man, whereas a man has a 20% risk after exposure to an infected woman. To test for gonorrhea, a physician simply takes a sample of the discharge, examines it microscopically for the presence of the gonococcus bacteria, and/or cultures the discharge and identifies the bacteria biochemically (fig. 15.14). Gonorrhea can be cured by antibiotic treatment as described in the reading for this chapter. There is no vaccine for gonorrhea, and immunity does not seem possible. Therefore, it is possible to contract the disease many times over.

Gonorrhea is one of the oldest known and most common of the sexually transmitted diseases. Untreated, an infection can cause sterility in either sex. If sought, appropriate antibiotic therapy will cure the condition.

Chlamydia

This sexually transmitted disease is named for the tiny bacterium that causes it, *Chlamydia trachomatis.* For years chlamydiae were considered to be more closely related to viruses than to bacteria, but today it is known that they are prokaryotic cells. Even so, they are obligate parasites due to their inability to produce ATP molecules. After a cell phagocytizes them, they develop inside the phagocytic vacuole, which eventually bursts and liberates many new infective chlamydiae.

New infections of **chlamydia** occur at even a faster rate than gonorrheal infections (fig. 15.7). They are the most common cause of **nongonococcal urethritis (NGU).** About eight to twenty-one days after exposure, men experience a mild burning sensation upon urinating and a mucoid discharge. Women may have a vaginal discharge along with the symptoms of a urinary tract infection. Unfortunately, a physician may mistakenly diagnose a gonorrheal or urinary tract infection and prescribe the wrong type antibiotic, or the person may never seek medical help. In either case the infection continues, leading to PID and sterility or ectopic pregnancy in females.

If a newborn comes in contact with chlamydia during delivery, inflammation of the eyes or pneumonia may result. There are also those who believe that chlamydia infections increase the possibility of premature and stillborn births.

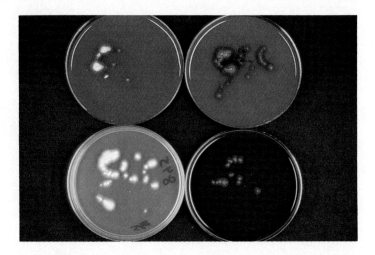

FIGURE 15.14 Culture plate with bacterial colonies. Visual examination and biochemical tests allow medical personnel to determine the type of bacteria growing on culture plates.

Detection and Treatment

New and faster laboratory tests are now available for chlamydia detection. Their expense sometimes prevents public clinics from using them, however. It's been suggested that these criteria could help physicians decide which women should be tested: no more than 24 years old; having had a new sex partner within the preceding two months; having a cervical discharge; bleeding during parts of the vaginal exam; and using a nonbarrier method of contraception. Some doctors, however, are routinely prescribing additional antibiotic appropriate to treating chlamydia for anyone who has gonorrhea because 40% of women and 20% of men with gonorrhea also have chlamydia.

The use of condoms serves as a protection for both gonorrheal and chlamydial infections. Also, it has just recently been reported that nonoxynol 9, the active ingredient in most spermicidal gels and foams and also in the contraceptive sponge, can help prevent chlamydial and AIDS infections.

Chlamydia is now believed to be the most common of the sexually transmitted diseases. PID and sterility are common effects of the infection in the female. This condition often accompanies gonorrheal infection.

Syphilis

Syphilis is caused by the bacterium *Treponema pallidum,* an actively motile, corkscrewlike organism that is classified as a spirochete. Because this bacterium is difficult to stain, it shows up only when viewed with a dark-field microscope (fig. 15.15). Syphilis is less common than gonorrhea (fig. 15.7), but it is the more serious of the two infections.

Syphilis has three stages, which may be separated by latent stages in which the bacteria are resting before multiplying again. During the primary stage, a hard chancre (ulcerated sore with hard edges) indicates the site of penetration some ten to ninety days after infection. The chancre

FIGURE 15.15 Treponema pallidum, *the cause of syphilis. Notice the characteristic spiral shape and the presence of flagella.*

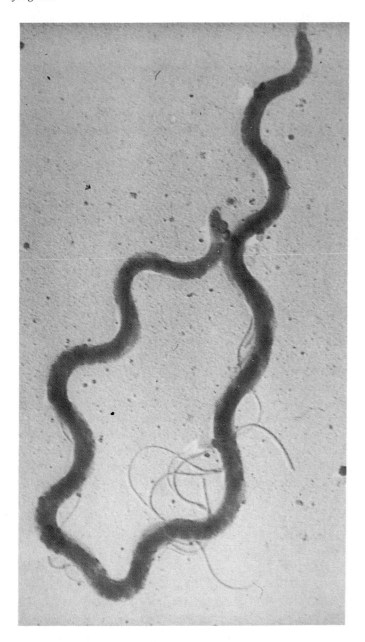

FIGURE 15.16 *Secondary stage of syphilis is a body rash that occurs even on the palms of the hands and soles of the feet.*

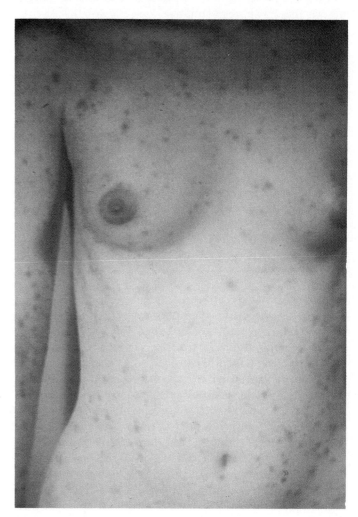

FIGURE 15.17 *A symptom of the tertiary stage of syphilis is gummas, large ulcerating sores, shown here on the hand.*

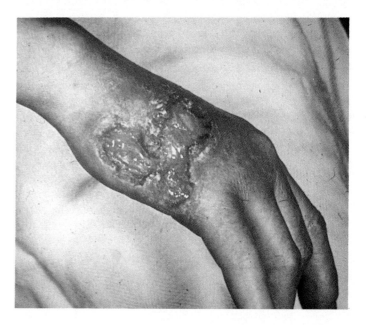

may go unnoticed, especially since it usually heals spontaneously within four to six weeks, leaving little scarring. During the secondary stage, which occurs in about a month, it is evident that bacteria have invaded and spread throughout the body because the victim breaks out in a rash (fig. 15.16). Curiously, the rash does not itch and is seen even on the palms of the hands and soles of the feet. There may be hair loss and gray patches on the mucous membranes, including the mouth. These symptoms may reoccur over a period of five years. Finally, in the tertiary stage, **gummas** (large destructive ulcers) appear on the skin (fig. 15.17) and within the internal organs, especially the small arteries (cardiovascular syphilis) and brain (neurosyphilis).

Congenital syphilis is caused by syphilis bacteria crossing the placenta. The newborn is blind and/or has numerous anatomical malformations.

HUMAN ISSUE

There are a number of controversial issues pertaining to children born to women having a sexually transmitted disease. In the case of AIDS and syphilis, the fetus can become infected while yet in the womb. Should women who have been treated for these conditions be discouraged from having children? If so, how should this be effectively handled?

Herpes, chlamydia, and gonorrhea can infect the newborn as the child passes through the birth canal. Should women who have been treated for these conditions insist on a cesarean birth so that their babies are sure to be protected? What would your reaction be if your spouse did not tell you that he or she had herpes and then a child was infected during birth?

Transmission and Treatment

The syphilis spirochete is not present in the environment; it is only present within human beings. Only close intimate contact, such as sexual intercourse, transmits the condition from one person to another.

Since a special microscope is needed to view the spirochete, an infection is usually detected by a blood test. The blood tests, of which there are several, all rely on detecting the presence of antibodies to syphilis, rather than the presence of the organism itself. You might think that since antibodies are present, the person could become immune. However, immunity is not achieved since it is possible for the person to contract syphilis again after being cured.

Because the first two stages of syphilis disappear of their own accord, the individual may believe that he or she is well, but this is far from true. The disease is simply latent, and each successive stage is more damaging than the previous stage. However, syphilis can be cured by antibiotic therapy even if it has reached the tertiary stage, so a person should never feel that it is too late for treatment. If organ damage has occurred, this cannot be reversed.

Syphilis, which has three stages (chancre, rash, gummas), can end in cardiovascular and brain impairment. However, this is unnecessary because syphilis is curable with antibiotic therapy. Syphilis can be acquired prenatally and causes birth defects.

OTHER DISEASES

Vaginitis

Two other types of organisms are of interest: protozoans and fungi. **Protozoans** are eukaryotes and usually exist as single cells. Each type protozoan has its own mode of locomotion. There are some that move by extensions of the cytoplasm, called pseudopodia, and others that move by cilia, and still others that have flagella. Protozoans are most often found in an aquatic environment, such as freshwater ponds, and the ocean simply teems with them. All protozoans require an outside source of nutrients, and the parasitic ones take this nourishment from their host.

Similarly, although most **fungi** are saprophytic, as are most bacteria, there are parasitic forms. Usually the body of a fungus is made up of filaments called hyphae, but yeasts are an exception since they are single cells. Almost everyone is familiar with yeast; because of its ability to ferment, it is used to make bread rise and to produce wine, beer, and whiskey.

Females very often have vaginitis, or infection of the vagina, that is caused by the flagellated protozoan *Trichomonas vaginalis* (fig. 15.18) or the yeast *Candida albicans*. The protozoan infection causes a frothy white or yellow foul-smelling discharge, accompanied by itching, and the yeast infection causes a thick, white, curdy discharge, also accompanied by itching. *Trichomonas* is most often acquired by sexual intercourse, and the asymptomatic male is usually the reservoir of infection. *Candida albicans,* however, is a normal organism found in the vagina; its growth simply increases beyond normal under certain circumstances. Women taking the birth control pill are sometimes prone to yeast infections, for example.

Vaginitis is a common occurrence often caused by *Trichomonas vaginalis,* a protozoan, or by the yeast *Candida albicans.*

Crabs

Small lice, animals that look like crabs under the microscope (fig. 15.19), infect the pubic and underarm hair of humans. The females lay their eggs around the base of the hair, and these eggs hatch within a few days to produce a large number of animals that suck blood from their host and cause itching.

Crabs can be contracted by direct contact with an infected person or by contact with his/her clothing or bedding. Self-treatment is possible as proper, effective medication may be purchased without a doctor's prescription.

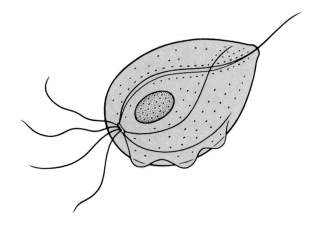

FIGURE 15.18 Trichomonas vaginalis. *This protozoan is pear-shaped and uses flagella to move about.*

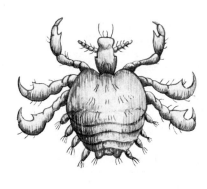

FIGURE 15.19 Phthirus pubis, *the parasitic louse that infects the pubic hair of humans.*

SUMMARY

Human beings are subject to sexually transmitted diseases (STDs). Some of these are viral in origin (AIDS, genital herpes, genital warts). Viruses are tiny particles that always have an outer coat of protein and an inner core of nucleic acid that may be DNA or RNA. DNA viruses are apt to undergo reproduction inside a cell immediately, while RNA viruses may be retroviruses that incorporate cDNA into host DNA for some time before reproduction. Some STDs are bacterial in origin (gonorrhea, chlamydia, syphilis). Bacteria are prokaryotic cells that reproduce by binary fission and may form endospores. Most bacteria are aerobic saprophytes that perform useful services for humans, but a few cause disease. Other STDs are caused by a protozoan (vaginitis), a fungus (yeast infection), or animals (crabs).

OBJECTIVE QUESTIONS

1. All viruses have an inner core of _____ and a coat of _____ .
2. AIDS is caused by a type of RNA viruses known as a _____ .
3. Although a sexually transmitted disease, AIDS is actually spread by _____ contact.
4. Herpes simplex virus type 1 causes _____ , and type 2 causes _____ .

5. Bacterial cells have one of three shapes: _____ , _____ , and _____ .
6. Females are often asymptomatic for gonorrhea until they develop _____ .
7. The most prevalent sexually transmitted disease today is _____ .
8. The tertiary stage of syphilis is characterized by large sores called _____ .

9. These three sexually transmitted diseases are curable by antibiotic therapy: _____ , _____ , and _____ .
10. Women who take the birth control pill are subject to a _____ infection.

Answers

1. nucleic acid, protein 2. retrovirus 3. blood 4. cold sores, genital herpes 5. rod, round, spiral 6. PID 7. chlamydia 8. gummas 9. gonorrhea, chlamydia, syphilis 10. yeast

STUDY QUESTIONS

1. Describe the structure and life cycle of viruses, including those that reproduce immediately and those that undergo a latent period. (p. 318)
2. Give the cause and discuss the geographic distribution of AIDS in the U.S. (pp. 320–321)
3. What are the symptoms for the three stages leading up to and including AIDS? (p. 320)
4. What are the problems associated with and how might investigators devise a vaccine for AIDS? (p. 321)

5. Give the cause and the expected yearly increase in the number of cases of herpes, gonorrhea, chlamydia, and syphilis. (p. 323)
6. Describe the progressive symptoms of a herpes infection. (p. 322)
7. List the three shapes of bacteria, and describe the structure of a prokaryotic cell. (p. 324)
8. Describe the symptoms of a gonorrheal infection in the male and in the female. What is PID, and how does it affect reproduction? (p. 326)

9. Describe the symptoms of chlamydia in the male and in the female, and compare this disease to a gonorrheal infection. (p. 328)
10. Describe the three stages of syphilis. (pp. 328–329)
11. How does the newborn acquire an infection of AIDS, herpes, gonorrhea, chlamydia, or syphilis? What effects do these infections have on infants? (pp. 321, 322, 326, 328, 328)
12. List other common sexually transmitted diseases and describe the associated symptoms. (p. 330)

THOUGHT QUESTIONS

1. Parasites reproduce inside living hosts, and they all must have a means of transmission from host to host. Explain the necessity for this stage in the organism's life cycle.

2. Reexamine figure 15.3 and list all the steps in the life cycle of a retrovirus where a drug (medication) might be designed to interfere with the cycle. Discuss in specific terms.

3. From your knowledge of the human body, discuss the mode of entry by parasites in general. Give specific examples that may have been mentioned in previous chapters of the text.

KEY TERMS

AIDS (ādz)   acquired immune deficiency syndrome, a disease caused by retroviruses and transmitted via body fluids; characterized by failure of the immune system. *318*

bacteria (bak-te′re ah)   prokaryotes that lack the organelles of eukaryotic cells. *324*

binary fission (bi′na-re fish′un)   reproduction by division into two equal parts by a process that does not involve a mitotic spindle. *325*

chlamydia (klah-mid′e-ah)   an organism that causes a sexually transmitted disease particularly characterized by urethritis. *328*

endospore (en′do-spor)   a resistant body formed by bacteria when environmental conditions worsen. *325*

fungus (fung′gus)   an organism, usually composed of strands called hyphae, that lives chiefly on decaying matter; e.g., mold and yeasts. *330*

genital (jen′i-tal)   pertaining to the external sexual organs as in genital herpes and genital warts. *322*

gonorrhea (gon″o-re′ah)   contagious sexually transmitted disease caused by bacteria and leading to inflammation of the urogenital tract. *326*

gummas (gum′ahz)   large unpleasant sores that occur during the tertiary stage of syphilis. *329*

herpes simplex virus (her′pēz sim′plex vi′rus)   a virus of which type I causes cold sores and type II causes genital herpes. *322*

HIV   viruses responsible for AIDS; human immunodeficiency virus. *320*

host (hōst)   an organism on or in which another organism lives. *318*

parasite (par′ah-sīt)   an organism that resides externally on or internally within another organism and does harm to this organism. *318*

pelvic inflammatory disease (pel′vik in-flam′ah-to″re di-zēz) PID;   a disease state of the reproductive organs in females caused by an organism that is sexually transmitted. *326*

retrovirus (ret″ro-vi′rus)   virus capable of integrating its genes into those of the host by utilizing RNA to DNA transcription. *318*

saprophyte (sap′ro-fīt)   a heterotrophic organism such as bacteria and fungi that externally breaks down dead organic matter before absorbing the products. *325*

sterilization (ster″i-li-za′shun)   the inability to reproduce; a surgical procedure eliminating reproductive capability; the absence of living organisms due to exposure to environmental conditions that are unfavorable to sustain life. *326*

syphilis (sif′i-lis)   chronic, contagious sexually transmitted disease caused by a spirochete bacterium. *328*

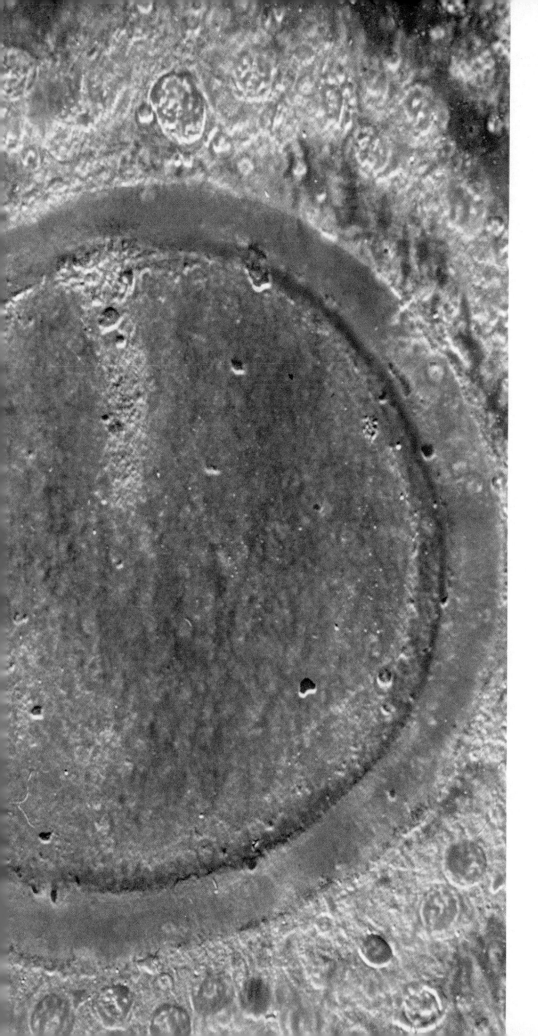

CHAPTER SIXTEEN

DEVELOPMENT

CHAPTER CONCEPTS

1 The first stages of human development lead to the establishment of the embryonic germ layers.

2 Induction, or the ability of one tissue to influence the development of another, can explain differentiation, or specialization of parts, and can account for the orderliness of development.

3 Human embryos are dependent on the presence of the extraembryonic membranes, the chorion and amnion, to nourish and protect them during development.

4 Human development is divided into the embryonic period when structures first arise and fetal development when there is a refinement of these structures.

Transmission electron micrograph of a single human egg surrounded by many sperm.

INTRODUCTION

How do you define environment? Figure 16.1 shows a human embryo surrounded by a watery sac as it develops within the uterus of its mother. No doubt this is the embryo's environment, but is the mother's environment also the embryo's environment? In many ways it is. If the mother is x-rayed, has AIDS, or takes a harmful drug, the embryo can be harmed irrevocably. On the other hand, if the mother takes care of herself and receives adequate nutrition, the child is more likely to be both physically and mentally healthy.

Well, what about psychological assaults on the mother? Can exposure to stressful situations, for example, be harmful to the embryo and later to the fetus? Perhaps so. One study shows that during and after World War II, the incidence of birth defects in Germany was markedly increased. The possibility exists that this was due to mental anguish as well as to inadequate prenatal care. It has even been suggested that the personality of the child and adult is also affected by prenatal environmental factors including those within the womb itself. Perhaps we are psychologically affected by whether the womb is large or small, or whether or not the womb contains another developing embryo, for example.

F ertilization occurs in the upper third of an oviduct (fig. 16.2), and development begins even as the embryo passes down this tube to the uterus. For two months the developing human being is called an **embryo** because the organ systems are forming. Only when we can recognize that this will be a human being is the term **fetus** used. The fetal period is largely a time of maturation and growth of structures that have already formed.

FIGURE 16.1 An embryo floating within the watery sac, which will be its home until birth occurs.

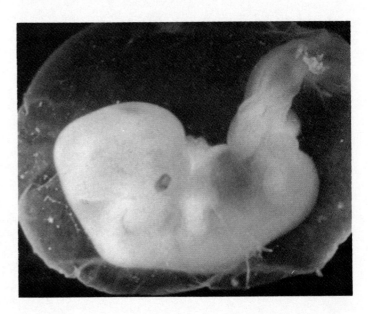

FIGURE 16.2 Fertilization and implantation of human embryo. After the egg is fertilized, it begins cleavage as it moves toward the uterus. At the time of implantation, the embryo is developing the germ layers.

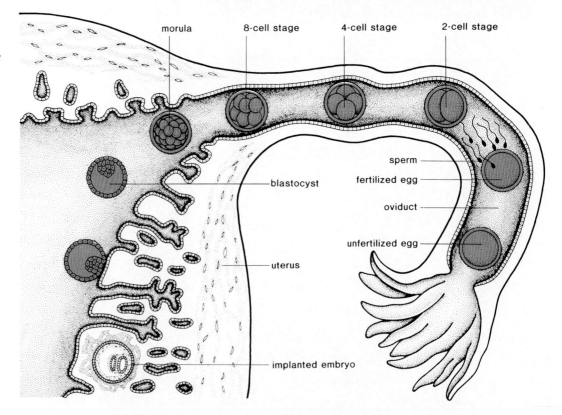

DEVELOPMENTAL PROCESSES

While development is occurring, these processes take place.

1. **Cleavage:** Within thirty hours after fertilization, the zygote begins to divide so that at first there are two, then four, eight, sixteen, and thirty-two cells, and so forth. Since increase in size does not accompany these divisions, the embryo is at first no larger than the zygote (the fertilized egg) was. Cell division during cleavage is mitotic, and each cell receives a full complement of chromosomes and genes.
2. **Growth:** Later cell division is accompanied by an increase in size of the daughter cells, and growth in the true sense of the term takes place.
3. **Morphogenesis:** Morphogenesis refers to the shaping of the embryo and is first evident when certain cells are seen to move, or migrate, in relation to other cells. By these movements, the embryo begins to assume various shapes.
4. **Differentiation:** When cells take on a specific structure and function, differentiation occurs. The first system to become visibly differentiated is the nervous system.

EMBRYONIC DEVELOPMENT

First Week

By the time the embryo has reached the uterus on the third day it is a **morula,** a ball of many cells. The morula is not much larger than the zygote because there has been no growth. By about the fifth day, the morula has been transformed into the **blastocyst,** a hollow ball of about one hundred cells (fig. 16.3a). The blastocyst has a single layer of outer cells called the trophoblast and an inner cell mass. The **trophoblast,** which will give rise to the **chorion,** is not to be part of the embryo, and for that reason it is called an **extraembryonic membrane** (fig. 16.3). The **inner cell mass** will eventually become the fetus. Each cell within the inner cell mass has the genetic capability of becoming a complete individual. Sometimes during human development, the inner cell mass splits, and two embryos start developing rather than one.

FIGURE 16.3 *Early development of fetus and umbilical cord.* a. *The blastocyst with its inner cell mass and surrounding trophoblast.* b. *Amniotic cavity and yolk sac appear.* c. *Chorionic villi first appear.* d. *Embryo is connected to chorion by a body stalk.* e.–g. *Embryo becomes more differentiated as the umbilical cord forms.*

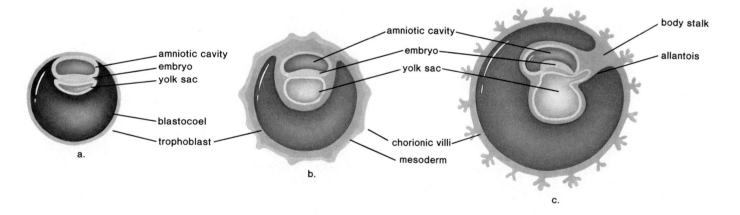

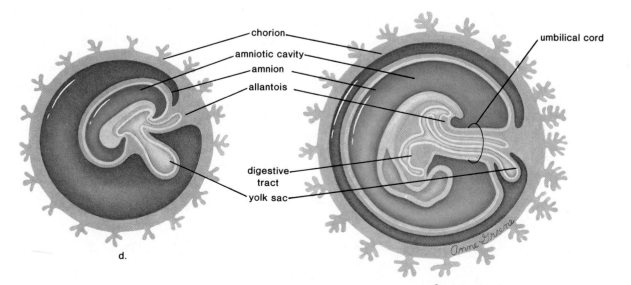

FIGURE 16.4 *Conception of fraternal versus identical twins. a. Fraternal twins are formed when two eggs are released and fertilized. Fraternal twins receive a different genetic inheritance from both the mother and father. They can even have different fathers. b. Identical twins occur when the embryo breaks in two during an early stage of development. Identical twins have the exact same genetic inheritance from both the mother and father.*

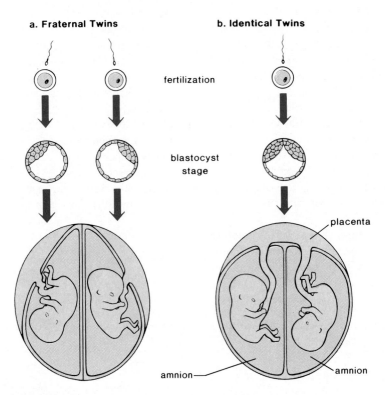

These two embryos will be *identical twins* (fig. 16.4) because they have inherited exactly the same chromosomes. *Fraternal twins,* who arise when two different eggs are fertilized by two different sperm, do not have identical chromosomes. It has even been known to happen that these "twins" have different fathers.

During the first week, the embryo undergoes cleavage, and then the morula becomes the blastocyst having two main parts, the outer trophoblast (to become the chorion) and the inner cell mass (to become the fetus).

Second Week

At the end of the first week, the embryo begins the process of **implanting** itself in the wall of the uterus as the trophoblast secretes enzymes to digest away some of the tissue and blood vessels of the uterine wall (fig. 16.2). The embryo is now about the size of the period at the end of this sentence.

The trophoblast begins to secrete human chorionic gonadotropin (HCG), the hormone that is the basis for the pregnancy test and serves to maintain the corpus luteum past the time it would normally disintegrate. Because of this, the endometrium is maintained and menstruation does not occur.

Ectopic Pregnancy

An **ectopic** (out of place) **pregnancy** occurs whenever the embryo implants itself in a location other than the uterus. In most of these instances, the embryo implants itself in the oviduct, but implantation in an ovary or even the abdominal cavity has been known to occur. The oviduct is unable to expand sufficiently to permit the pregnancy to continue, and it will rupture if the embryo is not removed.

PID (p. 326) due to a sexually transmitted disease is the most common cause of ectopic pregnancy because the resultant scarring entraps the embryo and prevents it from traveling to the uterus. Other conditions that predispose a woman to ectopic pregnancies are an infection following childbirth or abortion, a history of ectopic pregnancy, endometriosis, pelvic surgery, failed tubal ligation, and the use of fertility drugs. Fertility drugs cause multiple ovulations, and not all of the fertilized eggs may make it to the uterus.

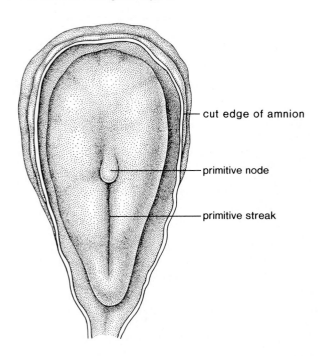

- cut edge of amnion
- primitive node
- primitive streak

TABLE 16.1 *ORGANS DEVELOPED FROM THE THREE GERM LAYERS*

ECTODERM	MESODERM	ENDODERM
Skin epidermis including hair, nails, and sweat glands	All muscles	Lining of digestive tract, trachea, bronchi, lungs, gallbladder, and urethra
Nervous system including brain, spinal cord, ganglia, and nerves	Dermis of skin	Liver
Retina, lens, and cornea of eye	All connective tissue including bone, cartilage, and blood	Pancreas
Inner ear	Blood vessels	Thyroid, parathyroid, and thymus glands
Lining of nose, mouth, and anus	Kidneys	Urinary bladder
Tooth enamel	Reproductive organs	

The classical signs of ectopic pregnancy are lower abdominal pain, often more noticeable on one side of the abdomen, a missed menstrual period accompanied by irregular vaginal bleeding, and the presence of a tender mass on one side of the pelvis. A positive pregnancy test along with an ultrasound scan that does not indicate a uterine pregnancy is also highly suggestive of an ectopic pregnancy.

It is very important for women to be aware of and consider the possibility of an ectopic pregnancy because undiagnosed cases can lead to a tubal rupture, internal bleeding, shock, and possible death.

During the second week, the embryo implants itself in the uterine lining, and the trophoblast begins to produce HCG.

Third Week

During the third week of development, the inner cell mass detaches itself from the trophoblast as two more extraembryonic membranes (fig. 16.3) form. The **yolk sac,** which forms below the embryo, has no nutritive function but it is the first site of blood-cell formation in the embryo. The **amnion** and its cavity, which form above the embryo, are also very important because it is here that the embryo and later the fetus will develop in a fluid that acts as an insulator against cold and heat. It also absorbs any shock such as a blow to the mother's abdomen.

The inner cell mass has now flattened into the embryonic disc composed of two layers of cells: the ectoderm above and the endoderm below. Once the embryonic disc elongates to become the *primitive streak* (fig. 16.5), another layer, the mesoderm, forms by invagination of cells along the streak. The mesoderm lies between the ectoderm and the endoderm, where it forms blocks of tissue called *somites* along the midline.

It's possible to relate the development of future organs to these layers, called **germ layers** (table 16.1). In general, the ectoderm will become the nervous system, skin, hair, and nails; the endoderm will produce the inner linings of the digestive, respiratory, and urinary tracts; and the mesoderm will produce the muscles, skeleton, and circulatory systems (fig. 16.6).

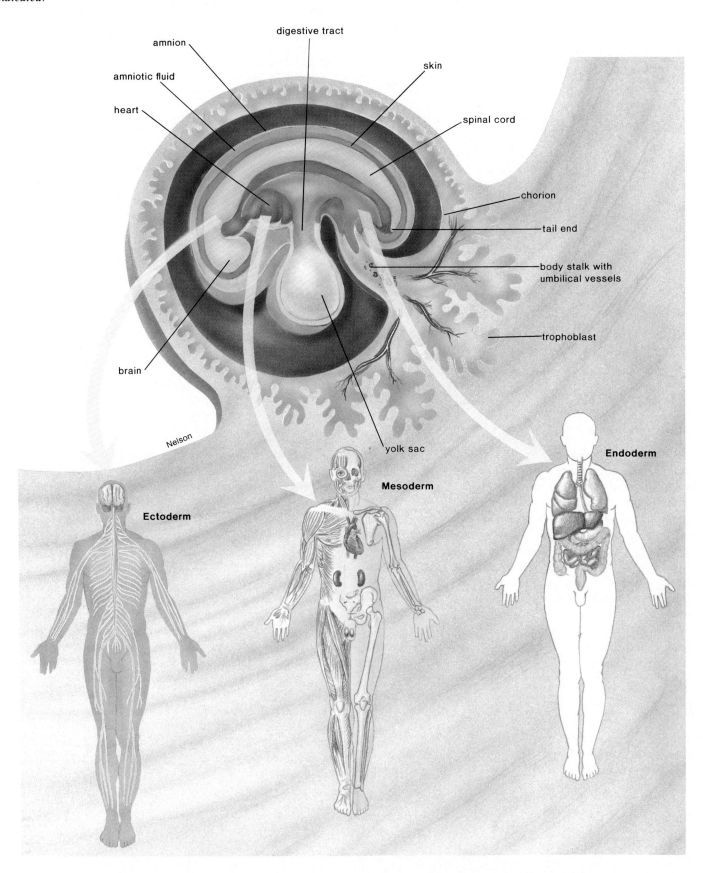

FIGURE 16.7 *Extraembryonic membranes are not part of the embryo. These membranes are also found during the development of chicks and humans, where each has a specific function. a. In the chick, the chorion lies just beneath the shell and performs gas exchange. The allantois collects nitrogenous wastes. The yolk sac provides nourishment, and the amnion provides a watery environment. b. In humans, only the chorion and amnion have comparable functions to those of the chick. The chorion forms the fetal half of the placenta, where exchange occurs with mother's blood, and the amnion provides a watery environment. The yolk sac has no nutritive function but is the first site of blood cell production; the allantoic membrane gives rise to the umbilical vessels.*

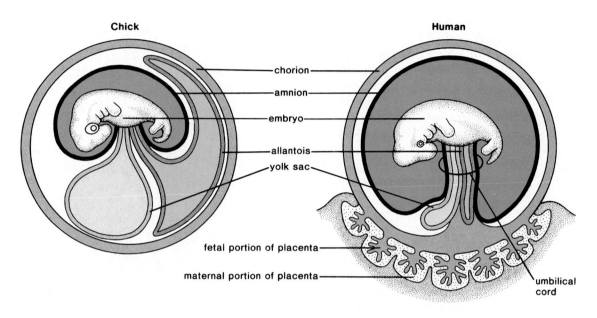

Chick **Human**

chorion
amnion
embryo
allantois
yolk sac
fetal portion of placenta
maternal portion of placenta
umbilical cord

FIGURE 16.8 a. *Neural plate is an ectoderm layer that will give rise to the neural tube.* b. *As microfilaments contract, cells begin to invaginate.* c. *Continued constriction causes the neural tube to pinch off from the outer ectoderm layer.*

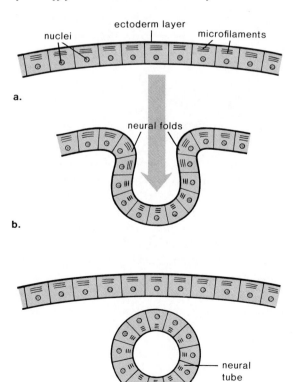

nuclei   ectoderm layer   microfilaments

a.

neural folds

b.

neural tube

c.

The germ layers (ectoderm, endoderm, and mesoderm) are laid down during the third week of development. Development of each of the organs can be related to certain of the germ layers.

Fourth Week

At the fourth week, there is a bridge of mesoderm, called the *body stalk,* connecting the caudal end of the embryo with the chorion. The fourth extraembryonic membrane, the **allantois** (fig. 16.7), is contained within this stalk, and its blood vessels become the umbilical blood vessels. The umbilical cord that connects the developing embryo to the placenta is fully formed after the head and tail lift up and the body stalk is moved toward the ventral side by constriction (fig. 16.3c–d).

Neurulation

The nervous system is the first organ system to be visually evident, and its mode and cause of differentiation have been studied. At first a thickening appears along the entire dorsal length of the embryo, and then invagination occurs as neural folds appear. When the neural folds meet (fig. 16.8) at the midline, the **neural tube,** which later develops into the brain and nerve cord, is formed. Figure 16.9 shows a human embryo during the fourth week when neurulation is in progress. A cross section of this embryo (fig. 16.9b) indicates that the **notochord** derived from mesoderm lies beneath the developing neural tube. The notochord is a supporting rod that will later be replaced by the vertebral column.

Experiments, particularly in the frog, have shown that if the presumptive (potential) nervous system lying just above the notochord is cut out and transplanted to another region of the embryo, it will not form a neural tube. On the other hand, if the presumptive notochord is cut out and transplanted beneath what would be belly ectoderm, this ectoderm now differentiates into neural tissue. These experiments indicate that notochord mesoderm causes the overlying ectoderm to form the nervous system, and it is said that the dorsal mesoderm induces the formation of the neural tube. The process of **induction** can help explain the orderly development of the embryo. One tissue is induced by another, and this in turn induces another tissue, and so on, until development is complete. Most likely the inducing tissue produces chemical messengers that are received by the tissue being induced.

Development of the Heart

Development of the heart also begins in the third week and continues into the fourth week. At first there are right and left heart tubes; when these fuse, the heart begins pumping blood even though the chambers of the heart are not yet fully formed. The veins enter posteriorly and the arteries exit anteriorly from this largely tubular heart, but later the heart will twist so that all major blood vessels enter or exit anteriorly.

During the fourth week major organs like the central nervous system and the heart make their appearance.

Fifth Week

At five weeks the embryo (fig. 16.10) is barely larger than the height of this print. Little flippers called limb buds have appeared and are developing into arms and legs, from which hands and feet and fingers and toes will soon form. Notice in figure 16.11 that within the limb buds the cartilaginous skeleton is also growing and differentiating. A cartilaginous skull is evident as the head enlarges, and facial features are becoming obvious. The mouth breaks through when ectoderm invaginates to meet the newly closed digestive tube, which begins to have various parts, including the accessory digestive organs.

Figure 16.10 shows that not only does the 5-week-old embryo have a tail, it also has gill arches. Obviously only fishes and amphibian larvae have actual use for gill slits as functioning structures. The fact that humans go through this

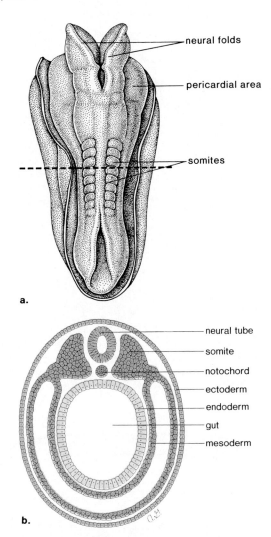

FIGURE 16.9   a. *Human embryo at twenty-one days. The neural folds still need to close at the anterior and posterior of the embryo. The pericardial area contains the primitive heart, and the somites are the precursors of the muscles. b. Cross section of (a) where the dotted line appears. Notice that the notochord lies just beneath the neural tube. In this colored drawing, ectoderm-derived structures are in blue, mesoderm-derived structures are in red, and endoderm-derived structures are in yellow.*

embryonic stage of the lower forms indicates a relationship between the two. However, in humans actual gill slits never form; instead, the first gill arch becomes the cavity of the middle ear and eustachian tube. The second arch becomes the tonsils, and the third and fourth arches become the thymus and parathyroids, respectively. The fifth arch disappears. Thus, gill arches develop because they are necessary to later development.

During the fifth week, human features like the head, arms, and legs begin to make their appearance.

FIGURE 16.10 *Human embryo at beginning of fifth week.* a. *Scanning electron micrograph.* b. *Drawing. The embryo is curled so that the head touches the heart, two organs whose development is further along than the rest of the body. The organs of the gastrointestinal tract are forming. The presence of the tail is an evolutionary remnant; its bones will regress and become those of the coccyx. The arms and legs will develop from the bulges that are called limb buds.*

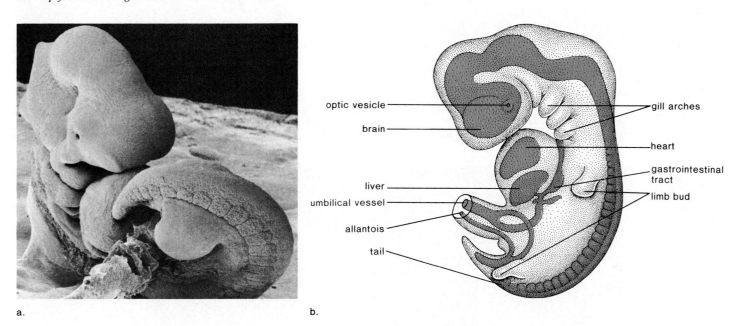

a.

b.

optic vesicle
brain
liver
umbilical vessel
allantois
tail
gill arches
heart
gastrointestinal tract
limb bud

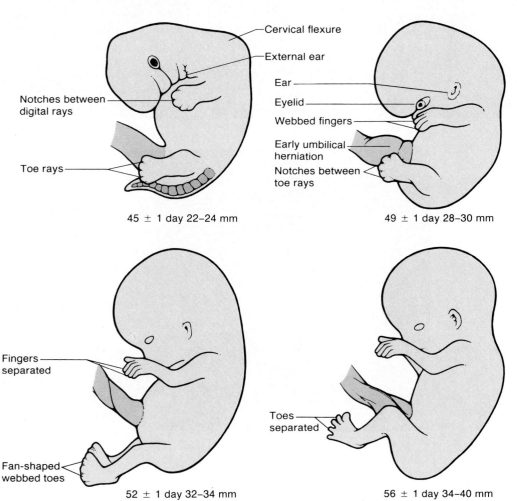

FIGURE 16.11 *Human embryo at the days indicated and at the size noted. Ten millimeters (mm) equals 0.4 inches.*

Cervical flexure
External ear
Notches between digital rays
Toe rays

45 ± 1 day 22–24 mm

Ear
Eyelid
Webbed fingers
Early umbilical herniation
Notches between toe rays

49 ± 1 day 28–30 mm

Fingers separated
Fan-shaped webbed toes

52 ± 1 day 32–34 mm

Toes separated

56 ± 1 day 34–40 mm

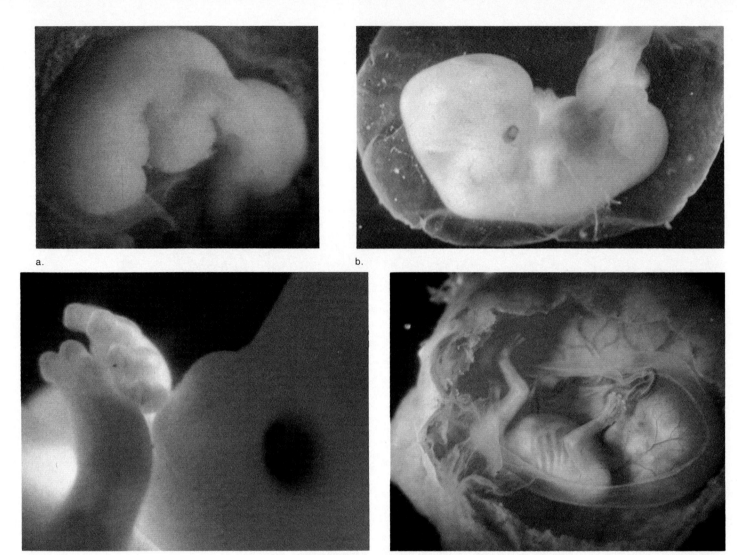

a.

b.

c.

d.

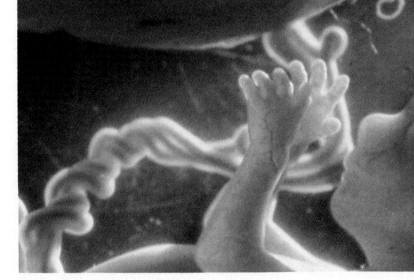

FIGURE 16.12 *Human development from the fourth to the sixteenth week.* a. *In the four-week-old embryo, the body is flexed and C shaped.* b. *At the end of six weeks, the head becomes disproportionately large.* c. *In the eight-week-old embryo, the nose is flat, the eyes are far apart, and the eyelids are fused.* d. *Surrounded by the extraembryonic membranes, this twelve-week-old fetus appears to be sucking its thumb.* e. *At sixteen weeks, the blood vessels are easily visible through the transparent skin.*

e.

FIGURE 16.13 *FIGURE 16.13 Anatomy of the placenta. The placenta is composed of both fetal and maternal tissues. Chorionic villi penetrate the uterine lining and are surrounded by maternal blood. Exchange of molecules between fetal and maternal blood takes place across the walls of the villi.*

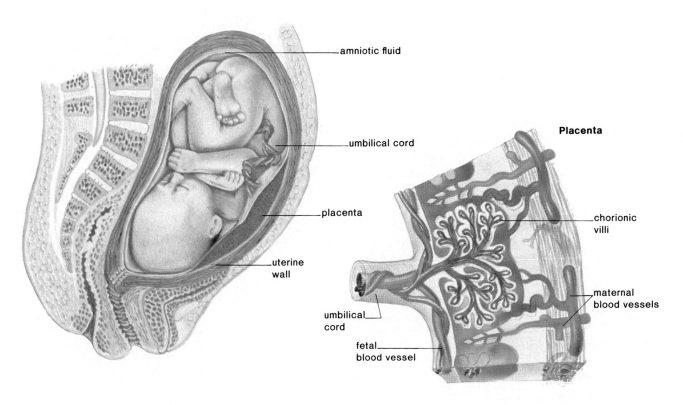

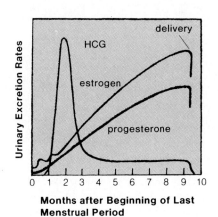

Sixth to Eighth Week

There is a remarkable change in external appearance during these weeks (figs. 16.11 and 16.12) from a form that would be difficult to recognize as human to one that is easily recognizable as human. Concurrent with brain development the head achieves its normal relationship with the body as a neck region develops. The nervous system is well enough developed to permit reflex actions such as a startle response to being touched. At the end of the second month, the embryo is about an inch (2.5 cm) long and weighs no more than an aspirin tablet (2.25 gm).

Placenta

The *placenta* begins formation once the embryo is fully implanted. Treelike extensions of the chorion called **chorionic villi** project into the maternal tissues. Later these disappear in all areas except the one where the placenta develops. By the tenth week, the placenta (fig. 16.13) is fully formed and begins to produce progesterone and estrogen (fig. 16.14). These hormones have two effects; due to their negative feedback effect on the hypothalamus and anterior pituitary, they prevent any new follicles from maturing, and they maintain the lining of the uterus so that the corpus luteum is not needed. There is no menstruation during the period of pregnancy.

FIGURE 16.14 Urinary excretion of HCG (human chorionic gonadotropin), estrogen, and progesterone during pregnancy. Urinary excretion rates are an indication of blood concentrations of these hormones.

BIRTH DEFECTS

It is believed that at least one in sixteen newborns has a birth defect, either minor or serious, and the actual percentage may be even higher. Most likely only 20% of all birth defects are due to heredity. Those that are can sometimes be detected before birth by subjecting embryonic and/or fetal cells to various tests following chorionic villi sampling or amniocentesis (p. 378).

Treatment of the fetus in the womb is a rapidly developing area of medical expertise. Biochemical defects can sometimes be treated by giving the mother appropriate medicines. For example, if a baby is unable to make use of vitamin B and/or is unable to use biotin efficiently, the mother can take these substances in doses large enough to prevent any untoward effects. Structural defects can sometimes be corrected by surgery. For example, if the fetus has water on the brain or is unable to pass urine, tubes that temporarily allow the fluid to pass out into the amniotic fluid can be inserted even while the fetus is still in the womb. Physicians are hopeful that eventually all sorts of structural defects can be corrected by lifting the fetus from the womb just long enough for corrective surgery to be done.

Birth defects not due to heredity, called congenital defects, cannot be passed on to the next generation. These are often caused by microbes or substances that have crossed the placenta and altered normal development. Hundreds of babies have contracted AIDS while in their mother's womb, and the number is expected to increase to more than 3,000 by 1991.

Medications can also sometimes cause problems. When the synthetic hormone DES was given to pregnant women to prevent miscarriage, their daughters showed various abnormalities of the reproductive organs and an increased tendency toward cervical cancer. Other sex hormones, including birth control pills, can possibly cause abnormal fetal development, including abnormalities of the sex organs.

Drugs of all types should be avoided. Most people are aware that women taking the tranquilizer thalidomide produced children with deformed arms and legs. Also, aspirin, caffeine (present in coffee, tea, and cola), and alcohol should be severely limited. Mood-altering drugs (p. 216) most likely should not be taken at all. It is not unusual for babies of drug addicts and alcoholics to display withdrawal symptoms and to have various abnormalities. Babies born to women who have about forty-five drinks a month and as many as five drinks on one occasion are apt to have fetal alcohol syndrome (FAS) with decreased weight, height, and head size; malformation of the head and face; and mental retardation (b). A recent new threat is the use of cocaine during pregnancy; "cocaine babies" now make up 60% of drug-affected babies. Severe fluctuations in blood pressure accompany the use of cocaine, and these temporarily deprive a fetal brain of oxygen. Cocaine babies have visual problems, lack coordination, and are mentally retarded. Other fetotoxic chemicals should also be avoided. These include pesticides and many organic industrial chemicals.

The placenta has a fetal side contributed by the chorion and a maternal side consisting of uterine tissues. Notice how the chorionic villi are surrounded by maternal blood sinuses, yet the blood of the mother and fetus never mix since exchange always takes place across cell membranes. Carbon dioxide and other wastes move from the fetal to the maternal side, and nutrients and oxygen move from the maternal to the fetal side of the placenta. The umbilical cord stretches between the placenta and the fetus. Although it may seem that the **umbilical cord** travels from the placenta to the intestine, actually the umbilical cord is simply taking fetal blood to and from the placenta. The umbilical cord is the lifeline of the fetus because it contains the umbilical arteries and vein, which transport waste molecules (carbon dioxide and urea) to the placenta for disposal and take oxygen and nutrient molecules from the placenta to the rest of the fetal circulatory system.

Harmful chemicals can also cross the placenta, and this is of particular concern during the embryonic period when various structures are first forming. Each organ or part seems to have a sensitive period, during which a substance can alter its normal formation. The reading on these pages concerns the origination of birth defects.

Cigarette smoke includes some of these very same chemicals so that babies born to smokers are often underweight and subject to convulsions.

It is recommended that all women who are capable of reproduction and who are sexually active take precautions to protect developing embryos. An Rh negative woman who has an Rh positive child should receive a RhoGam injection to prevent the production of Rh antibodies (p. 136), which can cause birth defects, including nervous system and heart defects, during a subsequent pregnancy. Also, women should be immunized for German measles (rubella) before the childbearing years. German measle viruses can cause such defects as deafness.

X-ray diagnostic therapy should be avoided during pregnancy because X rays are mutagenic to a developing embryo or fetus. Children born to women who have received X-ray treatment are apt to have birth defects and/or develop leukemia later on. As discussed in chapter 15, children born to women with herpes, gonorrhea, and chlamydia are subject to blindness and other mental and physical defects when they become infected after passing through the birth canal. Birth by cesarean section could prevent these occurrences.

Now that physicians and laypeople are aware of the various ways in which birth defects can be prevented, it is hoped that all types of birth defects, both genetic and congenital, will decrease dramatically.

a.

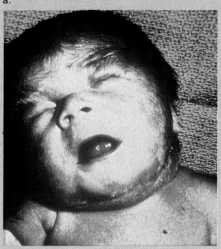

b.

a. *New York bars are now required to post a notice warning that the consumption of alcohol by a pregnant woman can affect the child she carries.* b. *Photo of a baby with fetal alcohol syndrome.*

At the end of the embryonic period all organ systems have been established, and there is a mature and fully functioning placenta. The embryo is only about 1 inch (2.5 cm) long.

FETAL PERIOD

Third and Fourth Months

At the beginning of the third month, the head is still very large, the nose is flat, the eyes are far apart, and the ears are distinctively present. Head growth will now begin to slow down as the rest of the body increases in length. Epidermal refinements such as eyelashes, eyebrows, hair on head, fingernails, and nipples appear.

FIGURE 16.15 *Development of the external genitalia. At first it is difficult to tell male (a) from female (d), but later the penis enlarges while the clitoris regresses.*

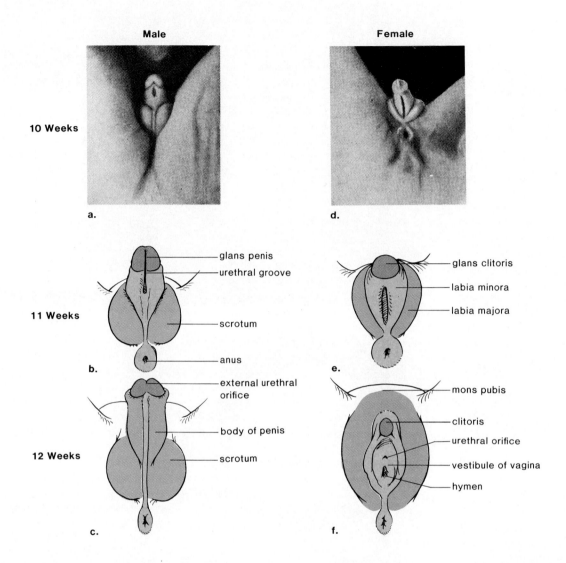

Cartilage is replaced by bone as ossification centers appear in most of the bones. Cartilage remains at the ends of the long bones, and ossification will not be complete until age 18 or 20. The fetal skull has six large membranous areas called **fontanels,** which permit a certain amount of flexibility as the head passes through the birth canal and allow rapid growth of the brain during infancy. The fontanels will disappear by 2 years of age.

At the third month, it is possible to distinguish males from females. Apparently the Y chromosome brings about the production of a protein called the H-Y antigen (because females form antibodies to it) that triggers the differentiation of gonads into testes. Once the testes differentiate they produce androgens, the male sex hormones. It is the androgens, especially testosterone, that stimulate the growth of the male external genitalia (fig. 16.15). In the absence of androgens, female genitalia form. The ovaries have no need to produce estrogen because there is plenty of it circulating in the mother's bloodstream.

FIGURE 16.16 *Newborn intensive care unit staff at Mount Sinai Hospital in New York. There are six nurses, a newborn (neonatologist) specialist, a resident, and two psychologists who care for each baby and the parents. Only babies who have a low birth weight (below 5.5 lbs.) are placed in intensive care, which if prolonged, can be very expensive.*

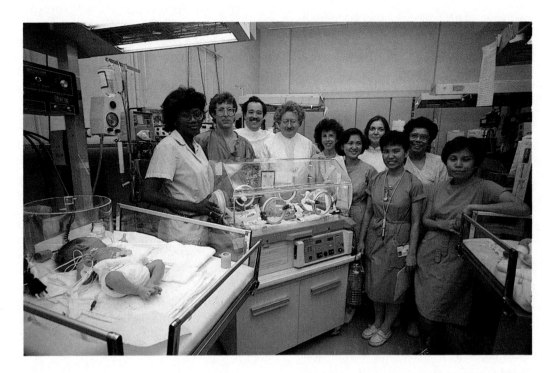

At this time, both testes and ovaries are located within the abdominal cavity but later, in the last trimester of fetal development, the testes will descend into the scrotal sacs (scrotum). Sometimes the testes fail to descend and in that case an operation may later be done to place them in their proper location.

During the fourth month, the fetal heartbeat is loud enough to be heard when a physician applies a stethoscope to the mother's abdomen. By the end of the fourth month, the fetus is less than ½ foot (140 mm) in length and weighs a little more than ½ pound (200 g).

During the third and fourth months, it is obvious that the skeleton is changing from cartilage to bone. The sex of the individual is now distinguishable.

Fifth to Seventh Months

During this period of time, the mother begins to feel movement. At first there is only a fluttering sensation, but as the fetal legs grow and develop, kicks and jabs will be felt. The fetus, though, is in the **fetal position** with the head bent down and in contact with the flexed knees.

The wrinkled, translucent, pinkish-colored skin is covered by a fine down called **lanugo.** This in turn is coated with a white, greasy, cheeselike substance called **vernix caseosa,** which probably protects the delicate skin from the amniotic fluid. The eyelids are now fully open, however.

At the end of this period, the weight has increased to almost 3 pounds (1,350 gm) and the length to almost a foot (300 mm). It's possible that if born now, the baby will be able to survive.

Premature Babies

About 7% of all newborns in the United States weigh less than 5½ pounds. Most of these are premature babies (fig. 16.16) who face the following difficulties.

Respiratory Distress Syndrome (Hyaline Membrane Disease) The lungs don't produce enough of a chemical surfactant that helps the alveoli stay open. Therefore, the lungs tend to collapse instead of expanding to be filled with air.

Retinopathy of Prematurity The high level of oxygen needed to insure adequate gas exchange by the immature lungs can lead to proliferation of blood vessels within the eyes with ensuing blindness.

Intracranial Hemorrhage The delicate blood vessels in the brain are apt to break, causing swelling and inflammation of the brain. If not fatal, this can lead to brain damage.

Jaundice The immature liver fails to excrete the waste product bilirubin; instead, it builds up in the blood, possibly causing brain damage.

Infections There is a low level of antibodies in the body, and the various medical procedures performed could possibly introduce germs. Also, an infection of the bowels is common, along with perforation, bleeding, and shock.

Circulatory Disorders Fetal circulation, discussed in the following section, has two features—the oval opening between the atria, and the arterial duct that allows blood to bypass the lungs. If these features persist in the newborn, a mixing of oxygenated blood with deoxygenated blood will result and circulation of the blood will be impaired, perhaps leading to blue baby syndrome or heart failure.

The reasons for premature birth have been investigated, and it's been concluded that prenatal care, including good nutrition and the willingness to refrain from excessive drinking of alcohol and smoking cigarettes, could reduce the incidence of premature birth and/or low birth weight.

Fetal Circulation

As figure 16.17 shows, the fetus has four circulatory features that are not present in adult circulation.

1. **Oval opening,** or foramen ovale: an opening between the two atria. This opening is covered by a flap of tissue that acts as a valve.
2. **Arterial duct,** or ductus arteriosus: a connection between the pulmonary artery and the aorta.
3. **Umbilical arteries and vein:** vessels that travel to and from the placenta, leaving waste and receiving nutrients.
4. **Venous duct,** or ductus venosus: a connection between the umbilical vein and the vena cava.

All of these features may be related to the fact that the fetus does not use the lungs for gas exchange since oxygen and nutrients are received from the mother's blood by way of the placenta (fig. 16.13).

If we trace the path of blood in the fetus, we may begin with the right atrium (fig. 16.17). From the right atrium, the blood may pass directly into the left atrium by way of the oval opening, or it may pass through the atrioventricular valve into the right ventricle. From the right ventricle the blood goes into the pulmonary artery, but because of the arterial duct, most of the blood then passes into the aorta. Thus, by whatever route the blood takes, most of the blood will reach the aorta instead of the lungs.

Blood within the aorta travels to the various branches, including the iliac arteries that connect to the umbilical arteries leading to the placenta, where exchange between maternal blood and fetal blood takes place. It is interesting to note that the blood in the umbilical arteries, which travels to the placenta, is low in oxygen, but the blood in the umbilical vein, which travels from the placenta, is high in oxygen. The umbilical vein enters the venous duct, which passes directly through the liver. The venous duct then joins with the posterior vena cava, which goes to the right atrium again.

At birth, the oval opening usually closes. With the tying of the cord and the expansion of the lungs, blood enters the lungs in quantity. Return of this blood to the left side of the heart usually causes a flap to cover over the opening. Even if it doesn't, passage of the blood from the right to the left atrium rarely occurs because either the opening is small or it closes when the atria contract. In a small number of cases, the passage of impure blood from the right to the left side of the heart is sufficient to cause a "blue baby." Such a condition may now be corrected by open heart surgery.

The arterial duct closes because endothelial cells divide and block off the duct. Remains of the arterial duct and parts of the umbilical arteries and vein later are transformed into connective tissue.

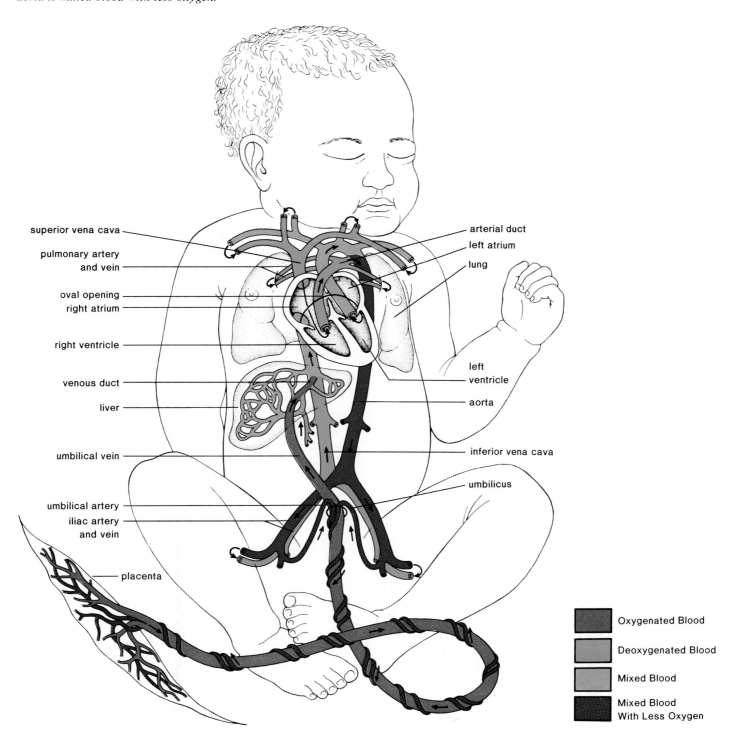

FIGURE 16.17 *Fetal circulation. Oxygenated blood becomes mixed with deoxygenated blood when the umbilical vein joins with the posterior vena cava via the venous duct. This mixed blood is routed to the left ventricle by way of the oval opening and then passes to the aorta and brain. Deoxygenated blood from the anterior vena cava is routed to the aorta via the arterial duct and therefore blood in the dorsal aorta is mixed blood with less oxygen.*

superior vena cava

pulmonary artery
and vein

oval opening
right atrium

right ventricle

venous duct

liver

umbilical vein

umbilical artery
iliac artery
and vein

placenta

arterial duct
left atrium
lung

left
ventricle

aorta

inferior vena cava

umbilicus

Oxygenated Blood

Deoxygenated Blood

Mixed Blood

Mixed Blood
With Less Oxygen

FIGURE 16.18 *Three stages of partuition. a. Position of fetus just before birth begins.* b. *Dilation of cervix.* c. *Birth of baby.* d. *Expulsion of afterbirth.*

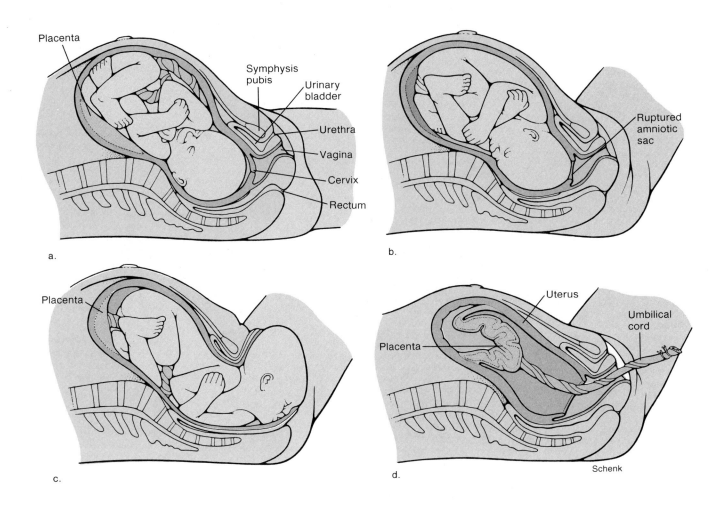

Eighth and Ninth Months

As the time of birth approaches, the fetus rotates so that the head is pointed toward the cervix (fig. 16.18a). If the fetus does not turn, then the likelihood of a breech birth (rump first) may call for a cesarean section because it is very difficult for the cervix to expand enough to accommodate this form of birth and asphyxiation of the baby is more likely. At the end of this time period, the fetus is about 21 inches (530 mm) long and weighs about 7½ pounds (3,400 gm). Weight gain is largely due to an accumulation of fat beneath the skin.

From the fifth to the ninth month, the fetus continues to grow and gain weight. Babies born after six to seven months may survive but are subject to various illnesses that may have lasting effects or cause an early death.

BIRTH

The uterus characteristically contracts throughout pregnancy. At first, light, often indiscernible contractions last about twenty to thirty seconds and occur every fifteen to twenty minutes, but near the end of pregnancy they become stronger and more frequent so that the woman may falsely think that she is in labor. The onset of true labor is marked by uterine contractions that occur regularly every fifteen to twenty minutes and last for forty seconds or more. **Parturition,** which includes labor and expulsion of the fetus, is usually considered to have three stages.

Stages

During the *first stage,* the cervix dilates; during the *second,* the baby is born; and during the *third,* the afterbirth is expelled.

The events that cause parturition are still not entirely known but there is now evidence suggesting the involvement of prostaglandins. It may be, too, that the prostaglandins cause the release of oxytocin from the maternal posterior pituitary. Both prostaglandins and oxytocin do cause the uterus to contract, and either can be given to induce parturition.

Stage 1

Prior to the first stage of parturition or concomitant with it, there may be a "bloody show" caused by the expulsion of a mucous plug from the cervical canal. This plug prevents bacteria and sperm from entering the uterus during pregnancy.

Uterine contractions during the first stage of labor occur in such a way that the cervical canal slowly disappears (fig. 16.18b) as the lower part of the uterus is pulled upward toward the baby's head. This process is called *effacement,* or "taking up the cervix." With further contractions, the baby's head acts as a wedge to assist cervical dilation. The baby's head usually has a diameter of about 4 inches, and therefore the cervix has to dilate to this diameter in order to allow the head to pass through. If it has not occurred already, the amniotic membrane is apt to rupture now, releasing the amniotic fluid, which escapes out the vagina. The first stage of labor ends once the cervix is completely dilated.

Stage 2

During the second stage, the uterine contractions occur every one to two minutes and last about one minute each. They are accompanied by a desire to push or bear down. As the baby's head gradually descends into the vagina, the desire to push becomes greater. When the baby's head reaches the exterior, it turns so that the back of the head is uppermost (fig. 16.18c). Since the vagina may not expand enough to allow passage of the head without tearing, an *episiotomy* is often performed. This incision, which enlarges the opening, is stitched later and will heal more perfectly than a tear would. As soon as the head is delivered, the baby's shoulders rotate so that the baby faces either to the right or left. The physician may at this time hold the head and guide it downward while one shoulder and then the other emerges. The rest of the baby follows easily.

Once the baby is breathing normally, the umbilical cord is cut and tied, severing the child from the placenta. The stump of the cord shrivels and leaves a scar, which is the navel.

Stage 3

The placenta, or **afterbirth,** is delivered during the third stage of labor (fig. 16.18d). About fifteen minutes after delivery of the baby, uterine muscular contractions shrink the uterus and dislodge the placenta. The placenta is then expelled into the vagina. As soon as the placenta and its membranes are delivered, the third stage of labor is complete.

During the first stage of birth, the cervix dilates; during the second, the child is born; and during the third, the afterbirth is expelled.

DEVELOPMENT AFTER BIRTH

Development does not cease once birth has occurred; it continues throughout the stages of life: infancy, childhood, adolescence, and adulthood.

Infancy

Infancy lasts until about 2 years of age. The newborn has certain innate reflexes that help it establish a relationship with its caretakers. For example, there is the *rooting reflex.* A baby will turn its head in the direction of a touch on its cheek and open its mouth. This helps the baby find the nipple. In manifesting the *sucking reflex,* a baby will suck on any object that touches the mouth. This reflex assures that the infant will obtain nourishment from the breast. Babies cry (*crying reflex*) when they are uncomfortable or are hungry. This helps to assure that their needs will be met by caretakers. The *smile reflex* is obvious at 2 to 3 months of age, when an infant develops the "social smile," a means of interacting with caretakers (fig. 16.19). Before this time, what appears to be a smile may simply be due to gas pains.

FIGURE 16.19 *Babies and parents interact by smiling at one another.*

The importance of the primary caretaker in the emotional development of the infant has been investigated. A series of experiments with rhesus monkeys has shown that the psychological well-being of primates is enhanced by close bodily contact between mother and offspring and between the offspring themselves. Monkeys raised in isolation with no mother show marked signs of abnormality. They stare fixedly into space, rock back and forth aimlessly, and even gnaw and bite themselves.

Monkeys raised with a soft terry-cloth surrogate mother do better. If given the choice between two surrogate mothers, the monkeys cling to the soft terry-cloth one in preference to a wire-frame mother that can nurse. But these monkeys are still not completely normal, showing signs of fear and aggression when placed with other monkeys, and they usually reject their own young. However, monkeys who lack a normal mother can still develop normally if they are allowed to play with peers for as little as fifteen minutes per day. The peer-to-peer affection system is extremely important to normal development in primates.

Infants are not only responsive to touch; they also are capable of seeing, hearing, smelling, and tasting. Their sensorimotor development is outlined in table 16.2. Their ability to perform ever more difficult operations can be correlated to the development of the brain during this time period.

TABLE 16.2 MILESTONES IN SENSORIMOTOR DEVELOPMENT

AGE	SENSORIMOTOR ACHIEVEMENTS	VOCALIZATION AND LANGUAGE
3 months	Grasps objects, smiles spontaneously, holds head steady when sitting, lifts up head when on stomach, follows object readily with eyes.	Squeals, coos especially in response to social interaction, laughs.
4½ months	Reaches awkwardly for some objects that are seen, frequently looks at hands, readily brings objects in hand to mouth, holds head steady in most positions, sits with props, bears some weight on legs.	Eyes seem to search for speaker, some consonants are mixed in with cooing sounds, babbling begins.
6 months	Usually reaches for near objects that are seen and looks at objects that are grasped, sits without support leaning forward on hands, bears weight on legs but must be balanced by adult, brings feet to mouth when lying on back.	Simple babbling of sounds like "mamama" and "bababa," turns to voice, laughs easily.
9 months	Grasps small objects with thumb and fingertips, shows first preference for one hand (usually the right), sits upright with good control, stands holding on, crawls, often imitates.	Repetition of sounds in babbling becomes common, some production of intonation patterns of parents' language, some imitation of sounds, understands "no."
1 year	Neat grasp of small objects, stands alone, walks with one hand held by adult, seats self on floor, mouths objects much less, drinks from cup but messily, often imitates simple behaviors, cooperates in dressing.	Produces a few words such as "mama" and "dada," understands a few simple words and commands, produces sentencelike intonation patterns called expressive jargon.
1½ years	Puts cubes in bucket, dumps contents from bottle, walks alone and falls only rarely, uses spoon with little spilling, undresses self, scribbles spontaneously.	Produces between 5 and 50 single words, produces complex intonation patterns, understands many words and simple sentences.
2 years	Turns single pages in a book, builds tower of blocks, shifts easily between sitting and standing, runs, throws and kicks balls, washes hands, puts on clothing.	Produces more than 50 words, produces a few short "sentences," understands much in concrete situations, shows much interest in language and communication.

Note: These behaviors represent a few of the sensorimotor milestones of infancy. The ages shown here are averages for middle-class American children; normal infants show wide variations in the ages at which they develop these behaviors.
From *Human Development* by Kurt W. Fisher and Arlyne Lazerson. Copyright © 1984. Reprinted with the permission of W. H. Freeman and Company.

Childhood and Adolescence

Childhood lasts until puberty (around 10 years in girls and 12 years in boys), and then adolescence continues until adulthood, a term that has a social as well as a biological definition. The growth of girls and boys during this time is shown in figure 16.20.

At the time of puberty, the sex organs mature, and the secondary sex characteristics begin to appear. The cause of puberty is related to the level of sex hormones in the body. It is now recognized that the hypothalamic-pituitary-gonad system functions long before puberty, but the level of hormones is low because the hypothalamus is supersensitive to feedback control. At the start of puberty, the hypothalamus becomes less sensitive to feedback control and begins to increase its production of releasing hormones, causing the pituitary and the gonads to mature and increase their production of hormones. The sensitivity of the hypothalamus continues to decrease until the gonadotropic and sex hormones reach the adult level.

The sex hormones, in conjunction with other hormones, have a profound effect on the body during puberty. There is an acceleration of growth leading to changes in height, weight, fat distribution, and body proportions. Males commonly experience a growth spurt later than females; thus, they grow for a longer period of time. This means that males are generally taller than females and have broader shoulders and longer legs relative to trunk length. Androgens are responsible for the greater muscular development in males, and estrogens are responsible for a greater accumulation of fat beneath the skin in females. The latter causes females to have a more rounded appearance than males. Early growth of the breasts, or mammary glands, is referred to as "budding" of the breasts. Budding is followed by the development of lobes, the functional portions of the breast, and the deposition of fat tissue that gives the breasts their adult shape.

FIGURE 16.20 *The average height of females* (above) *and males* (below) *from infancy through adolescence. Each has an adolescent growth spurt, but the male's growth spurt occurs about two years after the female's.*
From J. M. Tanner, "Growing Up" in Scientific American, *September 1973. Copyright © 1973 by Scientific American, Inc. All rights reserved.*

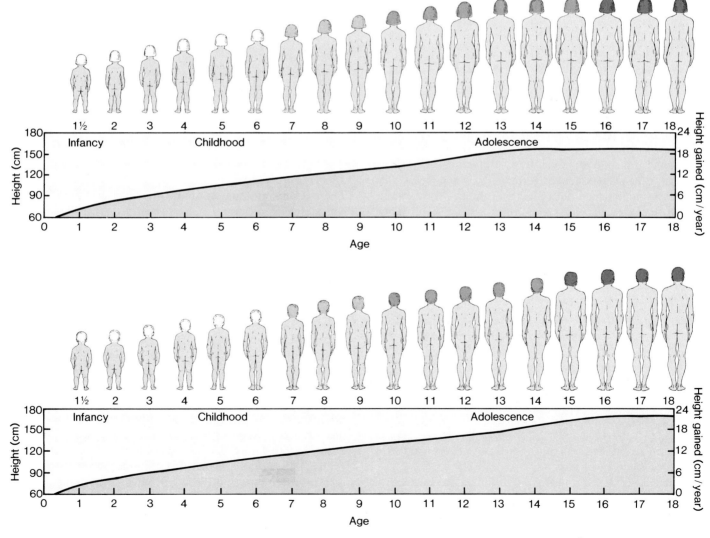

In females the pelvic cavity usually has a larger relative size compared to males. This means that females have wider hips than males and that the thighs converge at a greater angle toward the knees. Because the female pelvis tilts forward, females tend to have protruding buttocks, a more pronounced lower back curve, and an abdominal bulge. Both males and females develop axillary (underarm) and pubic hair. In females the upper border of pubic hair is horizontal, and in males it tapers toward the naval. Males develop noticeable hair on the face, chest, and even occasionally on other regions of the body such as the back, but females do not. Oil- and sweat-producing glands become active in both sexes, and acne may accompany puberty due to clogged oil-producing glands.

Developmental changes keep occurring throughout infancy, childhood, adolescence, and adulthood.

Adulthood and Aging

Young adults are at their physical peaks in muscle strength, reaction time, and sensory perception. The organ systems at this time are best able to respond to altered circumstances in a homeostatic manner. From now on there will be an almost imperceptible gradual loss in certain of the body's abilities. **Aging** encompasses these progressive changes that contribute to an increased risk of infirmity, disease, and death (fig. 16.21).

Today, there is great interest in **gerontology,** the study of aging, because there are now more older individuals in our society than ever before, and the number is expected to rise dramatically. In the next half-century, those over age 75 will rise from the present 13 million to 35–45 million, and those over age 80 will rise from 3 million to 6 million individuals. The human life span is judged to be a maximum of 110 to 115 years. The present goal of gerontology is not necessarily to increase the life span but to increase the health span, the number of years that an individual will enjoy the full function of all body parts and processes.

HUMAN ISSUE

Almost 30% of today's federal health-care budget is spent on medical care for the elderly. With the oncoming demographic shift toward elderly people, can we afford to take care of ourselves when we become old? A former Colorado governor bluntly suggested that "we all have the duty to die" when we grow old. This may be an extreme view, but it points to several age-related issues that are becoming increasingly common. Should the government pay most of the health-care costs of the elderly? Should physicians not prescribe expensive treatments if a person is past a certain age limit, say 75, and cannot afford to pay? Should the elderly not be considered for organ transplants if the organs are needed by younger individuals?

FIGURE 16.21 *Aging is a slow process during which the body undergoes changes that eventually will bring about death even if no marked disease or disorder is present. Although the human life span probably cannot be expanded, it most likely is possible to expand the health span, the length of time the body functions normally.*

Theories of Aging

There are many theories about what causes aging. We will consider four of these.

Genetic in Origin Several lines of evidence indicate that aging has a genetic basis. For example, 1. the children of long-lived parents tend to live longer than those of short-lived parents. Perhaps the genes are programmed to control aging and the time of death. The maximum life span of animals is species-specific; for humans it is about 110 years. 2. The number of times a cell will divide is also species-specific. The maximum number of times human cells divide is around fifty times. Perhaps as we grow older more and more cells are unable to divide any longer, and instead they tend to undergo degenerative changes and die. 3. Some cell lines may become nonfunctional long before the maximum number of divisions has occurred. Whenever DNA replicates, mutations can occur, and this may lead to the production of nonfunctional proteins. Eventually the number of inadequately functioning cells may build up, and this contributes to the aging process.

FIGURE 16.22 *Collagen is a protein that makes up the white fibers within connective tissue, a tissue found in or around most organs of the body. It has been well substantiated that collagen undergoes the cross-linking depicted here, but other proteins and even DNA may undergo a similar process. The number of affected molecules most likely builds up over the years and may contribute to the decline noticed in body parts.*

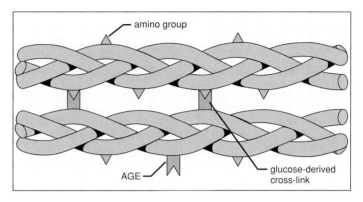

Whole Body Processes   A decline in the hormonal system may affect many different organs of the body. For example, diabetes Type II is a common development in older individuals. The pancreas makes insulin but the cells lack the receptors that enable them to respond. Menopause in women occurs for a similar reason. There is plenty of FSH in the bloodstream but the ovaries do not respond. Perhaps aging results from the loss of hormonal activities and a decline in the functions they control.

The immune system, too, no longer performs as it once did, and this may affect the body as a whole. The thymus gland gradually decreases in size, and eventually most of it is replaced by fat and connective tissue. The incidence of cancer increases among the elderly, which may signify that the immune system is no longer functioning as it should. This idea is substantiated, too, by the increased incidence of autoimmune diseases in older individuals.

It's possible, though, that aging is not due to the failure of a particular system that can affect the body as a whole, but to a specific type of tissue change that affects all organs and even the genes. It has been noticed for some time that proteins, such as collagen which makes up the white fibers (p. 55) and is present in many support tissues, become increasingly cross-linked as people age (fig. 16.22). Undoubtedly this cross-linking contributes to the stiffening and loss of elasticity characteristic of aging tendons and ligaments. It may also account for the inability of organs such as the blood vessels, heart, and lungs to function as they once did. Some researchers have now found that glucose has the tendency to attach to any type of protein, which is the first step in a cross-linking process that ends with the formation of "advanced glycosylation end products (AGEs)." The presence of AGE-derived cross-links not only explains why cataracts develop, it may also contribute to the development of atherosclerosis and to the inefficiency of the kidneys in diabetics and older individuals. Even DNA-associated proteins

seems capable of forming AGE-derived cross-links, and perhaps this increases the rate of mutations as we age. These researchers are presently experimenting with the drug aminoguanidine, which can prevent the development of AGEs.

Extrinsic Factors   The present data about the effects of aging are often based on comparisons of the characteristics of the elderly to younger age groups. But today's elderly were not as aware when they were younger, perhaps, of the importance of, for example, diet and exercise to general health. It's possible, then, that much of what we attribute to aging is instead due to years of poor health habits.

For example, osteoporosis is associated with a progressive decline in bone density in both males and females so that fractures are more likely to occur after only minimal trauma. Osteoporosis is common in the elderly—by age 65, ⅓ of women will have vertebral fractures, and by age 81, ⅓ of women and ⅙ of men will have suffered a hip fracture. While there is no denying that there is a decline in bone mass as a result of aging, certain extrinsic factors are also important. The occurrence of osteoporosis itself is associated with cigarette smoking, heavy alcohol intake, and inadequate calcium intake. Not only is it possible to eliminate these negative factors by personal choice, it is also possible to add a positive factor. A moderate exercise program has been found to slow down the progressive loss of bone mass.

Rather than collecting data on the average changes observed between different age groups, it might be more useful to note the differences within any particular age group. If this type of comparison is done, extrinsic factors that contribute to a decline and extrinsic factors that promote the health of an organ can be identified.

Effect of Aging on Body Systems

Keeping in mind that we want to accept such data with reservations, we will still discuss in general the effects of aging on the various systems of the body. Figure 16.23 compares the percentage of function of various organs in a 75- to 80-year-old man to that of a 20 year old whose organs are assumed to function at 100% capacity. When making this comparison we may note that the body has a vast functional reserve so that it can still perform well even when not at 100% capacity.

Skin As aging occurs the skin becomes thinner and less elastic because the number of elastic fibers decreases and the collagen fibers undergo cross-linking as discussed previously. Also, there is less adipose tissue in the subcutaneous layer so that older people are more likely to feel cold. The loss of thickness accounts for sagging and wrinkling of the skin.

Homeostatic adjustment to heat is also limited because there are fewer sweat glands for sweating to occur. There are fewer hair follicles so that the hair on the scalp and extremities thins out. The number of sebaceous glands is reduced and the skin tends to crack.

There is a decrease in the number of melanocytes, making the hair turn gray and the skin paler. In contrast, some of the remaining pigment cells are larger, and pigmented blotches appear in the skin.

Processing and Transporting Cardiovascular disorders are the leading cause of death. The heart shrinks because there is a reduction in cardiac muscle cell size. This leads to loss of cardiac muscle strength and reduced cardiac output. Still, it is observed that the heart, in the absence of disease, is able to meet the demands of increased activity. It can increase its rate to double or triple the amount of blood pumped each minute even though the maximum possible output declines.

Because the middle coat of arteries contains elastic fibers that are most likely subject to cross-linking, the arteries become more rigid with time, and their size is also further reduced by the presence of plaque (p. 114) so that blood pressure readings gradually rise. Such changes are common in individuals living in western industrialized countries but not in agricultural societies. As mentioned earlier, diet has been suggested as a way to control degenerative changes in the cardiovascular system (p. 115).

There is reduced blood flow to the liver, and this organ does not metabolize drugs as efficiently as before. This means that as a person gets older, less medication is needed to maintain the same level in the bloodstream.

Circulatory problems are often accompanied by respiratory disorders and vice versa. Growing inelasticity of lung tissue means that ventilation is reduced. But since we rarely use the entire vital capacity, these effects will not be noticed unless there is increased demand for oxygen.

There is also reduced blood supply to the kidneys. The kidneys become smaller and less efficient at filtering off wastes. Salt and water balance are difficult to maintain, and the elderly dehydrate faster than younger people. Difficulties involving urination include incontinence and inability to urinate. In men, the prostate gland may enlarge and reduce the diameter of the urethra making urination so difficult that surgery is often needed.

The loss of teeth, which is often seen in elderly people, is more apt to be the result of long-term neglect than a result of aging. The digestive tract loses tone, and secretion of saliva and gastric juice is reduced but there is no indication of reduced absorption. Therefore, an adequate diet, rather than vitamin and mineral supplements, is recommended. There are common complaints of constipation, increased amount of gas, and heartburn, but gastritis, ulcers, and cancer may also occur.

Integration and Coordination It is often mentioned that while most tissues of the body regularly replace their cells, some at a faster rate than others, the brain and muscles do not. No new nerve or skeletal muscle cells are formed in the adult. However, contrary to previous opinion, recent studies show that few neural cells of the cortex are lost during the normal aging process. This means that cognitive skills remain unchanged even though there is characteristically a loss in short-term memory. Although the elderly learn more slowly than the young, they can acquire new material and remember it as well as the young. It's noted that when more time is given for the subject to respond, age differences in learning decrease.

Neurons are extremely sensitive to oxygen deficiency, and if neuron death does occur, it may not be due to aging itself but to reduced blood flow in narrowed blood vessels. Specific disorders, such as depression, Parkinson disease, and Alzheimer disease (p. 214), are sometimes seen but they are not common. Reaction time, however, does slow and more stimulation is needed for hearing, taste, and smell receptors to function as before. After age 50, there is a gradual reduction in the ability to hear tones at higher frequencies, and this may make it difficult to identify individual voices and to understand conversation in a group. The lens of the eye does not accommodate as well and also may develop a cataract. Glaucoma is more likely to develop because of a reduction in the size of the anterior chamber of the eye.

Loss of skeletal muscle mass is not uncommon but it can be controlled by a regular exercise program. There is a reduced capacity to do heavy labor but routine physical work should be no problem. A decrease in the strength of the respiratory muscles and inflexibility of the rib cage contributes to the inability of the lungs to expand as before and reduced muscularity of the urinary bladder contributes to difficulties in urination.

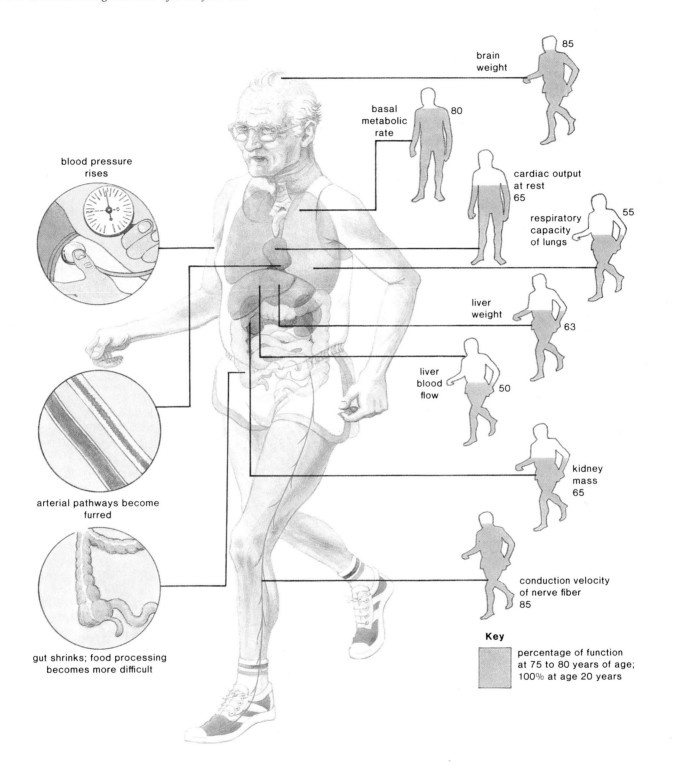

FIGURE 16.23 *Percentage of function remaining of various organs in a 75- to 80-year-old as measured against that of a 20-year-old.*

brain
weight
85

basal
metabolic
rate
80

cardiac output
at rest
65

respiratory
capacity
of lungs
55

blood pressure
rises

liver
weight
63

liver
blood
flow
50

arterial pathways become
furred

kidney
mass
65

conduction velocity
of nerve fiber
85

gut shrinks; food processing
becomes more difficult

Key

percentage of function
at 75 to 80 years of age;
100% at age 20 years

As noted before, aging is accompanied by a decline in bone density. Osteoporosis, characterized by a loss of calcium and minerals from bone, is not uncommon, but there is evidence that proper health habits may prevent its occurrence. Arthritis, which restricts the motility of joints, is also seen. In arthritis, as the articular cartilage deteriorates, ossified spurs develop. These cause pain upon movement of the joint.

Weight gain occurs because the metabolic rate (p. 92) decreases and inactivity increases. Muscle mass is replaced by stored fat and retained water.

Reproductive System   Females undergo menopause, and thereafter the level of female sex hormones in the blood falls markedly. The uterus and cervix are reduced in size, and there is a thinning of the walls of the oviducts and vagina. The external genitals become less pronounced. In males, the level of androgens falls gradually over the age span 50–90, but sperm production continues until death.

It is of interest that, as a group, females live longer than males. Often females have more physical complaints but these do not tend to be fatal. Males, on the other hand, lead in cases of emphysema, heart disease, atherosclerosis, and strokes. Although their health habits may be poorer, it is also possible that the female sex hormone, estrogen, offers to women some protection against circulatory disorders. Males suffer a marked increase in heart disease in their forties but an increase is not noted in females until after menopause.

Conclusion

We have listed many adverse effects due to aging but it is important to emphasize that while such effects are seen they are not a necessary occurrence (fig. 16.24). We must discover

FIGURE 16.24   *The aim of gerontology is to allow the elderly to enjoy living. This requires studying the possible debilities that can occur as one ages and then making recommendations as to how best to forestall or prevent their occurrences.*

any extrinsic factors that might precipitate these adverse effects so that they can be guarded against. Just as it is wise to make the proper preparations to remain financially independent when older, it is also wise to realize that biologically successful old age begins with the health habits developed when you are younger.

SUMMARY

During development, cleavage, growth, morphogenesis, and differentiation occur. Embryonic development occurs during the first eight weeks, and fetal development occurs during the third to the ninth month. The first three weeks are marked by the development of the extraembryonic membranes and the germ layers from which the various organs are derived. The CNS and the heart are the first organs to appear, but by the end of the eighth week all organ systems are present, and externally the fetus has a human appearance.

The fetus is dependent upon the placenta for gas exchange and as a source of nutrient molecules. During this fetal period there is a refinement of all organs, and a bony skeleton replaces the cartilaginous skeleton. A suitable weight increase is important to survival after birth. Birth has three stages: during the first stage, the cervix dilates; during the second, the child is born; and during the third, the afterbirth is expelled.

Development continues after birth. Infancy lasts until 2 years of age, and by this time the child has developed all sorts of motor skills and is usually able to form sentences. Puberty marks the division between childhood and adolescence. Females have a growth spurt at about age 12 while boys have theirs at age 15. Aging encompasses progressive changes, from about age 20 on, that contribute to an increased risk of infirmity, disease, and death.

OBJECTIVE QUESTIONS

1. When cells take on a specific structure and function, _____ occurs.
2. The morula becomes the _____ , a structure that contains the inner cell mass.
3. The _____ membranes include the chorion, _____ , yolk sac, and allantois.
4. The blastocyst _____ itself in the uterine lining.
5. The notochord _____ the formation of the nervous system.

6. During embryonic and fetal development, gas exchange occurs at the _____ .
7. During development, there is a connection between the pulmonary artery and the aorta called the _____ .
8. Fetal development begins with the _____ month.
9. The fontanels are commonly called _____ .
10. In most births, the _____ appears before the rest of the body.

11. As we age, the proteins in the body undergo _____ , a process that causes body parts to become stiff and rigid.
12. Most deaths are due to failure of the _____ system.

Answers

1. differentiation 2. blastocyst
3. extraembryonic, amnion 4. implants
5. induces 6. placenta 7. arterial duct
8. third 9. soft spots 10. head
11. cross-linking 12. cardiovascular

STUDY QUESTIONS

1. List the processes of development. (p. 335)
2. List the events of embryonic development and fetal development in a sequential manner. (pp. 336–344)
3. What are the three germ layers? What structures are associated with each germ layer? (p. 337)
4. What is induction? Discuss an experiment that was done to show that induction takes place. (p. 340)

5. Draw a generalized cross section of a four-week embryo, and label the parts. (p. 340)
6. What are the extraembryonic membranes of the chick and embryo? What are their respective functions? (p. 339)
7. Give several ways in which birth defects can be prevented. (p. 344)

8. Trace the path of blood in the fetus from the umbilical vein to the aorta using two different routes. (p. 349)
9. Describe the three stages of parturition. (pp. 350–351)
10. List and discuss the stages of development after birth. (pp. 351–354)
11. Explain the current theories in regard to the causes of aging. (pp. 354–356)
12. What are the major changes in body systems that have been observed as people age? (pp. 356–358)

THOUGHT QUESTIONS

1. Differentiation is still a process that is not well understood. What are some of the genetic and metabolic factors that need to be understood to explain induction, for example?
2. Very often evolution makes use of previously evolved structures for a different purpose. Why does this seem reasonable, and what example of this methodology has been given in this chapter?
3. The theories on aging have been divided into genetic and whole body. Develop an overall theory that would draw from both of these types of theories.

KEY TERMS

allantois (ah-lan'to-is) one of the extraembryonic membranes; in reptiles and birds a pouch serving as a repository for nitrogenous waste; in mammals a source of blood vessels to and from the placenta. *339*

amnion (am'ne-on) an extraembryonic membrane; a sac around the embryo containing fluid. *337*

chorion (ko're-on) an extraembryonic membrane that forms an outer covering around the embryo; in reptiles and birds it functions in gas exchange; in mammals it contributes to the formation of the placenta. *335*

differentiation (dif''er-en''she-a'shun) the process and developmental stages by which a cell becomes specialized for a particular function. *335*

extraembryonic membranes (eks''trah-em''bre-on'ik mem'brānz) membranes that are not a part of the embryo but are necessary to the continued existence and health of the embryo. 335

gerontology (jer''on-tol'o-je) the study of aging; those progressive changes that contribute to an increased risk of infirmity, disease, and death. *354*

induction (in-duk'shun)   a process by which one tissue controls the development of another, as when the embryonic notochord induces the formation of the neural tube.   *340*

lanugo (lah-nu'go)   downy hair with which a fetus is born; fetal hair.   *347*

morphogenesis (mor''fo-jen'ĭ-sis)   the movement of cells and tissues to establish the shape and structure of an organism.   *335*

parturition (par''tu-rish'un)   the processes that lead to and include the birth of a human and the expulsion of the extraembryonic membranes through the terminal portion of the female reproductive tract.   *350*

trophoblast (trof'o-blast)   the outer membrane that surrounds the human embryo and, when thickened by a layer of mesoderm, becomes the chorion, an extraembryonic membrane.   *335*

umbilical cord (um-bil'ĭ-kal kord)   cord connecting the fetus to the placenta through which blood vessels pass.   *344*

vernix caseosa (ver'niks ka''se-o'sah)   cheeselike substance covering the skin of the fetus.   *347*

yolk sac (yōk sak)   one of the extraembryonic membranes within which yolk is found; in mammals, the first site of blood-cell formation in the embryo.   *337*

FURTHER READINGS FOR PART IV

Baconsfield, P., G. Birdwood, and R. Baconsfield. August 1980. The placenta. *Scientific American.*

Cerami, A. et al. May 1987. Glucose and aging. *Scientific American.*

DeRobertis, E. M., and J. B. Gurdon. December 1979. Gene transplantation and the analysis of development. *Scientific American.*

Frisch, R. E. March 1988. Fatness and fertility. *Scientific American.*

Goldberg, S., and B. DeVitto. 1983. *Born too soon: Preterm birth and early development.* San Francisco: W. H. Freeman and Company.

Guttmacher, A. F. 1973. *Pregnancy, birth, and family planning.* New York: New American Library.

Leach, P. 1980. *Your baby and child from birth to age five.* New York: Alfred A. Knopf.

Lein, A. 1979. *The cycling female: Her menstrual rhythm.* San Francisco: W. H. Freeman.

Mader, S. S. 1980. *Human reproductive biology.* Dubuque, IA: Wm. C. Brown Publishers.

Nilsson, L. 1977. *A child is born,* rev. ed. New York: Delacorte Press.

Rugh, R., et al. 1971. *From conception to birth: The drama of life's beginnings.* New York: Harper & Row, Publishers.

Scientific American. October 1988. Entire issue devoted to AIDS.

Vander, A. J., et al. 1980. *Human physiology,* 3d ed. New York: McGraw-Hill.

Volpe, E. P. 1983. *Biology and human concerns,* 3d ed. Dubuque, IA: Wm. C. Brown Publishers.

Wantz, M. S., and J. E. Gay. 1981. *The aging process: A health perspective.* Cambridge, MA: Winthrop.

Wassarman, P. M. December 1988. Fertilization in mammals. *Scientific American.*

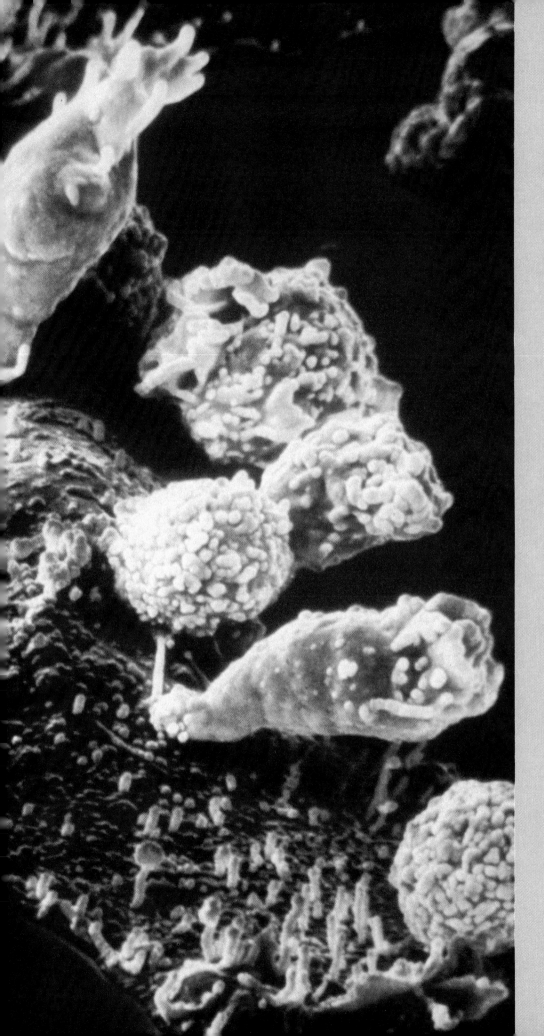

CANCER

Normally round T cells become elongated active fighters as they chemically attack a cancer cell.

FIGURE C.1 a. *Normal fibroblasts are flat and extended.* b. *After being infected with a cancer-causing virus, the cells become round and cluster together in piles.*

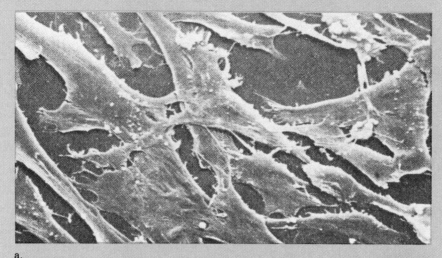

a.

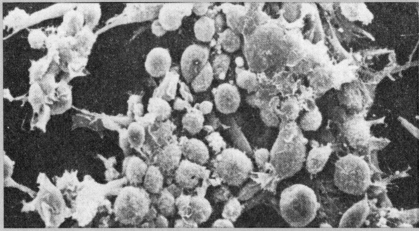

b.

CANCER

Since one out of every five deaths in the United States is due to cancer, it is wise to be familiar with its causes, treatments, and ways to protect yourself from its development.

Characteristics of Cancer

Despite the fact there are several hundred different types of cancer, it is still possible to list certain general characteristics. Cancer is a cellular disorder, and so these general characteristics pertain to cancer cells.

First, cancer cells experience uncontrolled and disorganized growth. You'll recall that normal cells only divide about fifty times, but cancer cells enter the cell cycle (fig. 17.10) over and over again and never differentiate. In tissue culture, cancer cells (fig. C.1) lack the contact inhibition exhibited by normal cells. Normal cells in tissue culture grow in only one layer because they adhere to the glass and stop dividing once they make contact with their neighbors. Cancer cells have lost all restraint and grow in multiple layers, most likely because of cell surface changes. In the body, a cancer cell divides to form a growth, or *tumor,* that invades and destroys neighboring tissue (fig. C.2). This new growth, termed *neoplasia,* is made up of cells that are disorganized, a condition termed *anaplasia.* The cells are disorganized because they don't differentiate into the tissues of the organ and they never help fulfill the function of the organ. To support their growth, cancer cells release a growth factor that causes neighboring blood vessels to branch into the cancer tissue. This phenomenon has been termed vascularization, and some modes of treatment are aimed at preventing vascularization from occurring.

FIGURE C.2 *In the body, cancer cells form a tumor, a disorganized mass of cells undergoing uncontrolled growth. As a tumor grows, it invades underlying tissues, and some of the cells leave this primary tumor and move through layers of tissue into blood or lymphatic vessels. After traveling through these vessels, the cells start new tumors elsewhere in the body. A carcinoma is a cancer that begins in epithelial tissue; a sarcoma is one that begins in connective tissue.*

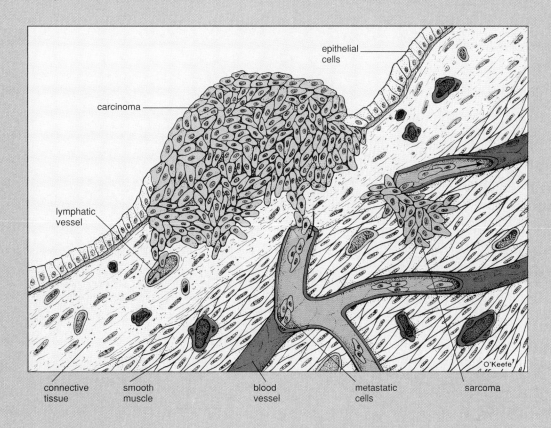

Second, cancer cells detach from the tumor and spread around the body. To accomplish this, the cells often must make their way across a basement membrane (p. 55) and into a blood vessel or lymphatic vessel (fig. C.2). It's been discovered that cancer cells have receptors that allow them to adhere to basement membranes; they also produce proteolytic enzymes that degrade the membrane and allow them to invade the underlying tissues. Cancer cells also tend to be motile. This may be associated with a disorganized internal cytoskeleton and their lack of intact microfilament bundles. After traveling through the blood or lymph, cancer cells start new tumors elsewhere in the body. This process is called *metastasis.*

If the original tumor is found before metastasis has occurred, the chances of cure are greatly increased. This is the rationale for the early detection of a cancer. In fact, sometimes a person isn't even considered to have cancer unless metastasis has occurred. *Benign tumors* are those that remain in one place; *malignant tumors* are those that metastasize.

Causes of Cancer

Carcinogenesis (the development of cancer) is often described as a two-step process involving *(1) initiation* and *(2) promotion.* During initiation, a DNA mutation occurs such that a cell is "transformed," making a cancerous growth possible at a later date. Three major groups of initiators have been identified: (1) viruses, (2) radiations, and (3) certain chemicals. You'll recall, for example, that the papillomavirus that causes genital warts has been associated with the development of cervical cancer. Radiations, including both UV radiation from the sun and X rays (fig. C.3a), damage the bonding between the components of a DNA molecule (p. 27). It has been firmly established that skin cancers are associated with UV radiation. Chemical carcinogens, which include pesticides (fig. C.3b) and pollutants, are known to bring about base-sequence changes in DNA. Cigarette smoke contains organic pollutants that play a role in the development of lung cancer.

FIGURE C.3 a. *Patient undergoing X-ray diagnosis. While many people are concerned about radiation from the nuclear power industry, medical diagnostic radiation actually accounts for at least 90% of human exposure to radiation. Everyone should be aware of this potential danger and have X rays only when necessary. All nonvisible short wavelengths of radiation are believed to be mutagenic, but the greater the amount, the greater the risk. b. Chemical mutagens include pesticides, manufacturing chemicals, and hallucinogenic drugs. Suspected mutagens are often tested in bacteria, fruit flies, and finally mice. If cancer develops in the mice, then the government either allows the product to be sold with a warning label or else it bans the sale of the product entirely.*

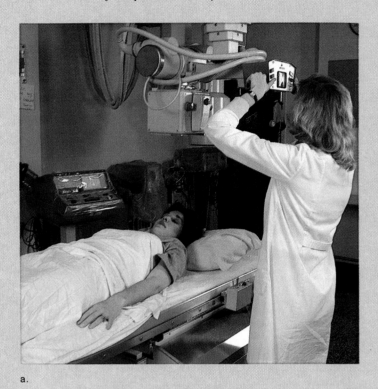

a.

b.

A promoter is any influence that causes a cancer cell to grow in an uncontrolled manner. It's possible that promotion can simply involve a second change in the DNA brought about by the same factors just discussed. In other words, cumulative DNA changes can finally result in the growth of cancerous cells. A promoter might also provide the environment that causes transformed cells to form a tumor. For example, there is some evidence to suggest that a diet rich in saturated fats and cholesterol is a possible promoter for cancer (fig. 4.18). Considerable time may elapse between initiation and promotion, which is why cancer is more apt to be seen in older rather than younger individuals.

Oncogenes

It is now clear that cells contain proto-oncogenes, genes that can be transformed into oncogenes, or cancer-causing genes (fig. C.4). These genes are not alien to the cell; they are normal, essential genes that have undergone a mutation. Investigators have shown that an oncogene that causes both lung cancer and bladder cancer differs from a normal gene by a change in only one nucleotide. It is now believed that almost any type of mutation can convert a proto-oncogene into an oncogene. For instance, in addition to genetic mutations, a chromosomal rearrangement (p. 378) may place a normally dormant structural gene next to an active regulatory gene. If this structural gene is a proto-oncogene, it may now become an oncogene. Alternately, an oncogene can be introduced into a cell by a virus. All cancer-causing viruses discovered so far are retroviruses (p. 318). The fact that retroviruses insert their DNA into host DNA allows them to plant a foreign oncogene into host DNA. Viruses that cause cancer contain oncogenes that they picked up from a previous host.

Having one oncogene may not cause a cell to become cancerous. Rather, the introduction of another oncogene or an enhancer, a gene that regulates the activity of an oncogene, will activate the cell to form a tumor (fig. C.5). As mentioned previously, this may be correlated with the two-step process involving an initiator and a promoter.

FIGURE C.4 *Summary of the development of cancer. A virus can pass an oncogene to a cell. A normal gene, called a proto-oncogene, can become an oncogene due to a mutation caused by a chemical or radiation. The oncogene either expresses itself to a greater degree than normal or else expresses itself inappropriately. Thereafter the cell becomes cancerous. Cancer cells are usually destroyed by the immune system, and the individual only develops cancer when the immune system fails to perform this function.*

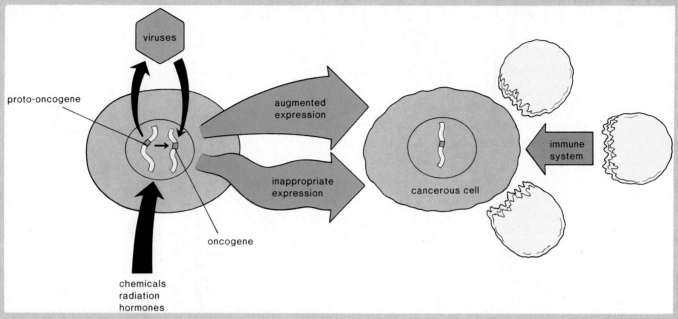

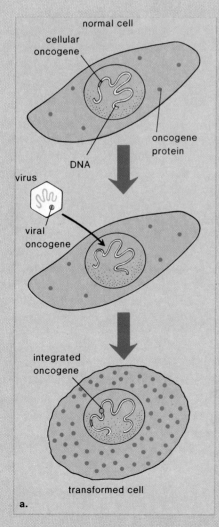

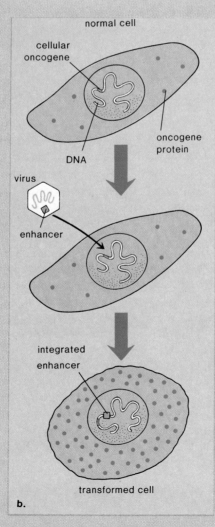

FIGURE C.5 *Transformation of normal cells into cancerous cells by means of a viral infection.* a. *The normal cell contains one oncogene but is not cancerous. The virus also contains an oncogene, which it passes to the cell. Now the cell is transformed and becomes cancerous because it produces a protein that causes the cell to become abnormal.* b. *The normal cell contains an oncogene but is not cancerous. The virus contains an enhancer that it passes to the cell. Now the cell is transformed and becomes cancerous because it produces a protein that causes the cell to become abnormal.*

FIGURE C.6 *Growth of cells involves many elements that begin at the cell membrane. Receptor at left receives growth factor (GF), and this initiates chemical reaction in cytoplasm. Receptor at right is malfunctioning since reaction takes place without GF being in place. Note oncogene in nucleus.*

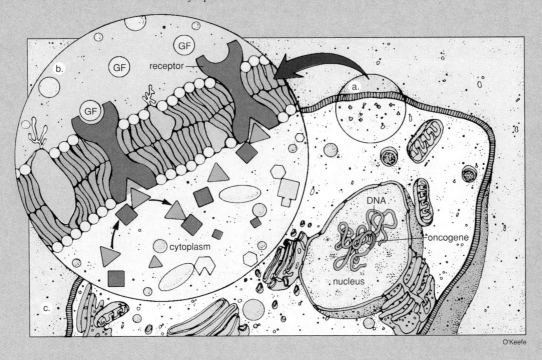

O'Keefe

Function of Oncogenes   Several oncogenes have been found and studied. All oncogenes found thus far are mutated forms of normal genes that regulate growth and development, but they don't all have the same type of function. Figure C.6 shows that growth of cells is governed by a series of elements that stretch from the surface of the cell to the nucleus. These elements include growth factor hormones and growth factor inhibitors that act on the surface of the cell, receptors for these hormones, proteins that carry signals from the receptors, and nuclear functions that regulate the response to these influences. Oncogenes are genes that cause one of these elements to malfunction in such a way that the cell divides repeatedly to produce a tumor. Exactly how this comes about is not now known.

Types of Cancer

Cancers are classified according to the type of tissue from which they arise. *Carcinomas* are cancers of the epithelial tissues (p. 51), and adenocarcinomas are cancers of glandular epithelial cells. Since epithelial tissue covers the surface of the body and lines internal cavities, carcinomas include cancer of the skin, breast, liver, pancreas, intestines, lung, prostate, and thyroid. These cancers may be spread to other body parts by the lymphatic system. *Sarcomas* are cancers that arise in connective tissue (p. 55), such as muscle, bone, and fibrous connective tissue. These cancers may be spread to other body parts by the bloodstream. *Leukemias* are cancers of the blood, and *lymphomas* are tumors of lymphoid tissue.

Of the Skin

Skin cancer (p. 60) is the most common form of cancer. It is most often caused by the ultraviolet rays in sunlight. These rays, like X rays, bring about genetic mutations that lead to cancer. (Note that in reference to the terminology mentioned previously, ultraviolet rays and X rays would be considered initiators.)

There are three types of skin cancer. The most common and least dangerous is basal-cell cancer, which develops in the deepest layer of the epidermis (fig. 3.14), usually where oil gland distribution is greatest. Often, the cancer may look like a pimple, but then it forms a gray border and ulcerates. Squamous-cell cancer occurs in the upper layers of the epidermis. It is generally characterized by extremely dry, rough, scaly patches but can vary greatly in appearance. The third type of skin cancer is more rare but much more virulent. Malignant melanomas usually appear as dark brown or black patches like moles. They may also develop from moles.

Skin cancers are usually surgically removed, sometimes by electrosurgery, the use of an electric current to destroy the cancerous cells. Cryosurgery destroys the cells by freezing them. Following their removal, the person is free of cancer but will most likely have a scar.

The best protection against skin cancer is to avoid the sun, especially in the middle of the day. Also the use of sunscreens that contain para-aminobenzoic acid, a chemical that absorbs UV rays, is recommended.

Of the Female Reproductive Organs

Cancer of the breast (fig. 14.13) is the most common cancer in females. Most cases of breast cancer occur beyond the age of 40 in women who have at least two close relatives who have also had breast cancer. It's possible that diet is a factor because breast cancer is rare among the Japanese, who eat less meat and fat. Early detection is extremely advisable, and all women should personally examine their breasts once a month in a manner recommended by the medical profession. Secretions from and changes in the appearance of the nipple should be noted, and all lumps, especially those that are hard and rough, should be reported to a physician immediately. Some doctors recommend that mammography, during which the breast is x-rayed, be routinely done in older women. The X-ray image is called a mammogram.

For years the surgical treatment for breast cancer has been a radical mastectomy, a procedure that involves total removal of the affected breast together with the axillary lymph nodes and muscles of the chest wall. There are those who believe that a simple mastectomy, which is much less disfiguring, is just as effective in certain cases.

The second most common cancer of the female reproductive organs is cancer of the endometrium. It is more common during or after menopause. Cancer of the cervix (p. 300) is the third most common cancer of the female reproductive system. It is most frequent in women from 30 to 50 years of age. Sexually active women having multiple partners are at greater risk and so are those who have a genital warts infection (p. 323). Cancer of the cervix can be detected by a Pap smear, a microscopic examination of the cervical cells. All women are encouraged to have a Pap test every year. Cancer of the uterus and cervix often requires a hysterectomy in which these organs are removed.

Of the Respiratory Tract

Cancerous growths in the mouth make up one in twenty of all cases of human cancers. Malignant growths on the lips are a frequent finding in areas of the world where strong sunlight is usual. The tongue is another site of cancer, often starting as an innocuous-looking area of whitening of the normally pink surface. A frequent cause of tongue and mouth cancer is tobacco smoking, especially in pipes and cigars.

Lung cancer (p. 173) has been on the increase since the turn of the century. Recently, however, there has been a decline in the number of men with lung cancer but a large increase in the number of women with lung cancer. Coal miners and radioactive ore miners inhale fumes that can lead to lung cancer, and persons who manufacture products containing asbestos are subject to a tumor of the pleura (fig. 8.17b). Only about 1% of lung cancer patients have alveolar cell carcinoma. The rest have cancer of the bronchus (fig. C.7). Other types of bronchial cancer are oat (small) cell cancer, large-cell cancer, and adenocarcinoma that arises in the glandular

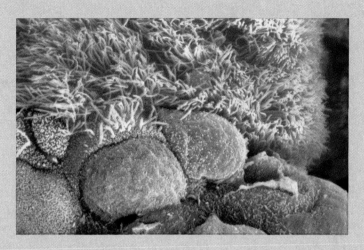

FIGURE C.7 *Cancer cells in the bronchi, respiratory tubes that conduct air to the lungs. The normal cells are pink and have cilia that sweep impurities up into the throat. The cancerous cells are green and lack cilia. If an individual stops smoking, there is a good chance the tissue will become normal again.*

cells. For the ordinary individual, smoking is the most frequent cause of lung cancer. Coughing, spitting blood, and loss of weight are symptoms of lung cancer. In some cases, there are no symptoms until a shadow is noticed on a routine chest X ray. Usually, lung cancer is initially treated by an attempted removal of the cancerous tissue.

Of the Digestive Tract and Accessory Organs

Cancer of the esophagus, stomach, colon, liver, and pancreas are common enough to be listed among the top ten most frequently diagnosed types of cancer. We mentioned before that diet has been linked to cancer of the breast, and it is certainly not surprising that the same would be true for the digestive organs. Since 1900 the incidence of colon cancer has risen, while that of stomach cancer has fallen in the United States. The average diet has also changed during this time period. The intake of fats has risen by 40% while that of pickled and smoked food has decreased dramatically.

Cancers of the esophagus and stomach are usually carcinomas. Cancers of the esophagus metastasize early, and surgery may not be helpful. Surgery is helpful in about 10% of patients with cancer of the stomach. Persistent indigestion is sometimes a warning sign for cancer of the stomach. Cancers of the colon and rectum often arise from precancerous polyps (protruding growths from the mucous membrane). These are usually adenocarcinomas that can be detected by a sigmoidoscope and subsequently surgically removed. Cancers of the liver may occasionally be successfully treated if identified in time to remove all of the tumor surgically. On the other hand, the average survival for persons with cancer of the pancreas is relatively short even if treatment is attempted.

Of the Blood and Lymphatic Organs

Leukemias are a group of cancers of the blood-forming tissues, which usually result in an overproduction of abnormal white blood cells. Two common forms of leukemia are acute lymphocytic leukemia with malignant lymphocytes, which affects mostly young children, and AML (acute myelocytic leukemia) with malignant granular leukocytes (p. 132), which strikes middle-age adults. In chronic leukemia, the cells are fairly normal and the disease progresses slowly. In acute leukemia, the cells may be immature and poorly differentiated, and the disease progresses rapidly. As the cells accumulate in the blood and lymph, the patient becomes anemic, bleeds easily, and is susceptible to infections. In the end, the leukemic cells invade the body's vital organs, impeding their functions and leading to death. A combination of chemotherapy and radiation treatments has proven helpful to patients with leukemia.

Lymphomas are cancers of the lymphoid organs such as lymph nodes, spleen, and the thymus gland. Hodgkin disease, which is characterized by the presence of two kinds of abnormal lymph cells, is also classified as a lymphoma. Lymphomas can spread from one lymph area to another by way of the lymphatic system.

Detection and Treatment

Individuals should be aware of the seven danger signals for cancer (table C.1) and inform their doctors when they notice any one of these. Thereafter, cancer can be detected by physical examination assisted by various means to view the internal organs. For example, ultrasound (high frequency sound waves) and CAT (computer-assisted axial tomography) scanner, which uses X rays along with a computer to produce cross-sectional pictures of the body, are newer methods to detect tumors.

Surgery, Radiation, and/or Chemotherapy

Tumors can be surgically removed, but there is always the danger that malignant tumors have already metastasized. Therefore, when a growth is malignant, surgery is often preceded by or followed by radiation therapy. Radiation destroys the more rapidly dividing cancer cells but causes less damage to the more slowly dividing normal cells.

Chemotherapy is another means of trying to catch any stray cells left behind when a growth is malignant. A combination of drugs is given, each one of which has its own specific effect. Certain types of leukemia and various cancers of the lymphatic system are often effectively treated by this

---

TABLE C.1 DANGER SIGNALS FOR CANCER

1. Unusual bleeding or discharge
2. A lump or thickening in the breast or elsewhere
3. A sore that does not heal
4. Change in bowel or bladder habits
5. Persistent hoarseness or cough
6. Persistent indigestion or difficulty in swallowing
7. Change in a wart or mole

---

means. A combination of drugs is required because cancer cells can become resistant to therapy when only one drug is used for a time. In some instances, it appears that the cell membrane of resistant cancer cells contains a carrier that pumps the drug out of the cell.

Recently, monoclonal antibodies produced by the technique described in figure 7.14 have been used to carry a drug directly to individual cancer cells. When the drug is attached to a monoclonal antibody specifically developed to attack the patient's tumor cells, it will bring the drug right to any tumor cell present in the patient's body.

Immunotherapy

Although it's possible to produce monoclonal antibodies that will seek out a specific type cancer cell, actually, the development of cancer signifies a failure in cell-mediated immunity (p. 151). Most likely cancer cells do have altered cell membrane antigens that normally subject them to attack by T cells. Whenever cancer develops, it would seem that T cells have not been activated. If this is correct, then, the use of lymphokines might awaken the cell-mediated immunity and lead to destruction of the cancer.

Interferon was the first type of lymphokine therapy to be investigated. Unfortunately, the results in general have been disappointing. However, interferon has been found to be effective in up to 90% of patients with a type of leukemia known as hairy-cell leukemia (because of the hairy appearance of the malignant cells).

The interleukins are also being clinically tested. In one technique being used, T cells are withdrawn from the patient and activated by culturing them in the presence of an interleukin. The cells are then reinjected into the patient, who is also given doses of interleukin to maintain the killer activity of the T cells. This type therapy shows some promise but is still regarded as experimental.

Survival Rates

When cancer patients live five years beyond the time of diagnosis and treatment, it is generally considered that they are cured. Presently, less than 50% of all cancer patients survive the five-year period, but there has been an increase in the survival rate for certain cancers. For localized breast cancer, the survival rate has risen from 78% in the 1940s to 96% today. The survival rate for prostate cancer has steadily improved over the past twenty years. The five-year survival rate for prostate cancer has risen from 50% to 68% for whites and from 35% to 58% for blacks. There has been a recent dramatic increase in survival rate for children with acute lymphocytic leukemia. In some medical centers, the overall children's survival rate has risen to 75%.

Avoidance of Cancer

There is clear evidence that the risk of certain types of cancer can be reduced by adopting the proper behaviors. For example, avoiding excessive sunlight certainly reduces the risk of skin cancer (p. 60); increasing dietary roughage (p. 87) and reducing dietary fat is believed to reduce the risk of colon cancer; not smoking cigarettes reduces the risk not only of lung cancer, but of other types of cancer as well (p. 174).

While pickled and smoked foods, saturated fats, and cholesterol may possibly be promoters of cancer (fig. 4.18), other dietary items are believed to be antipromoters (fig. C.8). For example, fresh fruits rich in vitamin A and C and vegetables like broccoli, cabbage, and cauliflower may offer some protection against cancer. Beans, rice, and potatoes also are believed to contain inhibitors of the proteolytic enzymes that allow cancer cells to metastasize.

FIGURE C.8 *There are some data to suggest that the diet can influence the development of cancer. Fresh fruits, especially those high in vitamin A and C, and vegetables, especially those in the cabbage family, are believed to reduce the risk of cancer.*

FURTHER READINGS FOR CANCER SUPPLEMENT

Bishop, J. M. March 1982. Oncogenes. *Scientific American.*

Cohen, L. A. November 1987. Diet and Cancer. *Scientific American.*

Feldman, M. and Eisenback, L. November 1988. What makes a tumor Cell Metastatic? *Scientific American.*

Golde, D. W. and Gasson, J. C. July 1988. Hormones that stimulate the growth of blood cells. *Scientific American.*

Hunter, T. August 1984. The proteins of oncogenes. *Scientific American.*

Nicholon, G. L. March 1979. Cancer metastasis, *Scientific American.*

Old, L. J. May 1988. Tumor Necrosis factor. *Scientific American.*

Pardee, A. B. and G. Prem veer Reddy. *Cancer: Fundamental Ideas.* Burlington, N.C. Carolina Biological Supply Co.

Varmus, H. September 1987. Reverse transcription. *Scientific American.*

Weinberg, R. A. September 1988. Finding the anti-oncogene. *Scientific American.*

———. November 1983. A molecular basis of cancer. *Scientific American.*

PART FIVE

HUMAN INHERITANCE

Human beings practice sexual reproduction, which requires meiosis as a part of gamete production, fertilization, and zygote development. Gametes carry half of the total number and different combinations of chromosomes. Therefore, except for identical twins, no child receives exactly the same combination of chromosomes as another. It is sometimes possible to determine the chances of an offspring receiving a particular chromosome or gene. If the genetic makeup of the parent is known, for example, it is sometimes possible to determine the chances of a child inheriting a genetic disease.

Genes, now known to be constructed of DNA, control not only the metabolism of the cell but also, ultimately, the characteristics of the individual. DNA contains a code for the sequence of amino acids in proteins, which are synthesized at the ribosomes. The step-by-step procedure by which the DNA code is transcribed and translated to assure the formation of a particular protein has been discovered.

CHAPTER SEVENTEEN

CHROMOSOMES AND CHROMOSOMAL INHERITANCE

CHAPTER CONCEPTS

1 The nucleus contains the gene-bearing chromosomes, which occur as homologous pairs.

2 Ordinary cell division assures that each new cell will receive a full complement of chromosomes and genes.

3 Reduction division is required to produce the sex cells, which contain half of the full number of chromosomes. When the sperm (male sex cell) fertilizes the egg (female sex cell), the full number of chromosomes is restored.

4 Chromosomal inheritance includes the autosomal and sex chromosomes. Females are XX and males are XY.

5 The genes are on the chromosomes, and all possible combinations of chromosomes and genes occur among the gametes.

Sperm determine the sex of the new individual because each sperm carries either an X or a Y chromosome.

Most mothers are devoted to their children. Is this because children inherit half of their chromosomes from their mothers?

INTRODUCTION

Mother's love (fig. 17.1) has been extolled by poets and religious leaders for centuries as selfless devotion. However, behavioral geneticists have a possible alternative explanation. First of all, a woman's chances of having children are not as great as a man's— she produces only one egg a month for a limited number of years, but a man produces millions of sperm each day even into old age. Children have some of the same chromosomes as their mother, and therefore a child represents a way for her to be immortal in a biological sense. Then, too, you have to consider that a woman is certain in a way that no man can be that a child is hers. Although this explanation of motherly love is based on a certain amount of selfishness, the result is the same. Most mothers will do all in their power to protect and teach their children so that they can become productive members of society.

CHROMOSOMES

Genes carried on chromosomes determine what the cell is like and what the individual is like. An examination of the body cells of a multicellular organism shows that all the nuclei have the same number of chromosomes. This number is characteristic of the organism—corn plants have twenty chromosomes, houseflies have twelve, and humans have forty-six.

Karyotype

In a nondividing cell the chromosomes are the indistinct and diffuse *chromatin* (fig. 2.2), but in a dividing cell the short and thick *chromosomes* appear. A cell may be photographed just prior to division so that a picture of the chromosomes is obtained. The chromosomes may be cut out of the picture

FIGURE 17.2 *Karyotype preparation. a. Already this human embryo is made up of millions of cells. Due to the process of division, each cell contains the same number and kinds of chromosomes, copies of the very ones that were inherited from its parents. It is possible to retrieve some embryonic cells and view these chromosomes. b. To view the chromosomes, cell division is halted by chemical means, and then the cells are microscopically magnified and photographed. An enlargement of the photograph permits the chromosomes to be cut out and arranged by pairs. The resulting display is called a karyotype.*

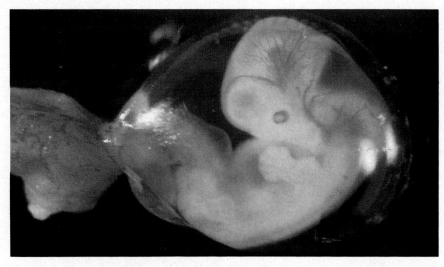

a.

cells   colchicine stops cell division   remove sample   prepare slide   observe cells

prepare karyotype   cut out individual chromosomes   photograph and enlarge chromosomes

b.

FIGURE 17.3 a. *Karyotype of a male. Note the pairs of autosomes numbered from 1 to 22 and the one pair of sex chromosomes, X and Y.* b. *Drawing of an enlarged chromosome.*
From John W. Hole, Jr., Human Anatomy and Physiology, *4th ed. Copyright © 1987 Wm. C. Brown Publishers, Dubuque, Iowa. All Rights Reserved. Reprinted by permission.*

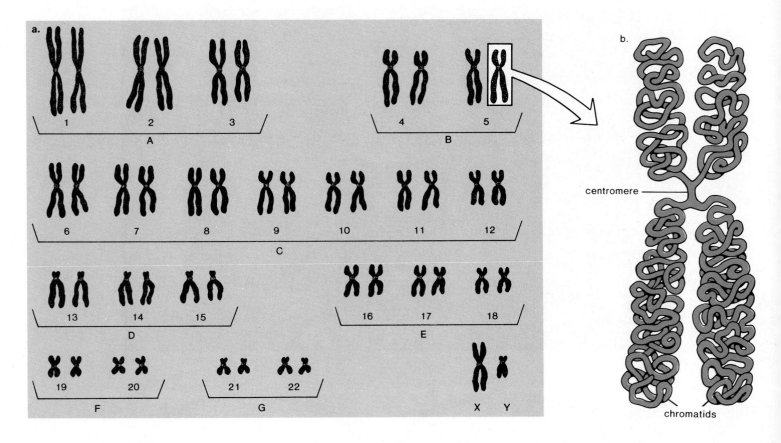

and arranged by pairs (fig. 17.2) of homologous chromosomes. **Homologous chromosomes** are recognized by the fact that each member of the pair is of the same size and has the same general appearance. The resulting display of pairs of chromosomes is called a **karyotype** (fig. 17.3). Although both males and females have twenty-three pairs of chromosomes, one of these pairs is of unequal length in males. The larger chromosome of this pair is called the X and the smaller is called the Y. Females have two X chromosomes in their karyotype. The X and Y chromosomes are called the **sex chromosomes** because they contain the genes that determine sex. The other chromosomes, known as **autosomes,** include all of the pairs of chromosomes except the X and Y chromosomes. Each pair of autosomes in the human karyotype is numbered.

Notice, as further illustrated in figure 17.3b, that each chromosome prior to division is composed of two identical parts, called **chromatids.** These two sister (twin) chromatids are genetically identical and contain the same *genes,* the units of heredity that control the cell. The chromatids are held together at a region called the **centromere.**

A karyotype shows the individual's total number of chromosomes arranged by homologous pairs.

Life Cycle

The human life cycle requires sexual reproduction in which the sperm of the male fertilizes the egg in the female. The resulting zygote develops into the newborn infant, who grows to be an adult (fig. 17.4). Two types of cell division occur during the human life cycle: **mitosis** and **meiosis.**

FIGURE *17.4* *Human life cycle. Gamete production requires meiotic cell division.*
After the sperm fertilizes the egg, the resulting zygote grows and matures. Growth
requires mitotic cell division.
Source: Antenatal Diagnosis, *HEW, 1979, pp. 1–48.*

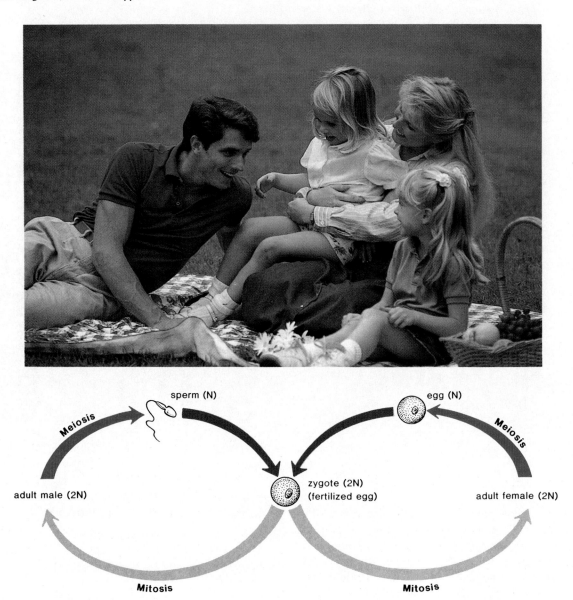

Meiosis (reduction division) occurs as a part of gametogenesis, the production of **gametes,** which is a term used collectively to mean the sperm and egg. Because of meiosis, the sperm contains twenty-three chromosomes and the egg contains twenty-three chromosomes. A new individual comes into existence when the sperm of the male fertilizes the egg of the female. The resulting **zygote** contains twenty-three pairs of chromosomes; one of each pair was contributed by the father, and one of each pair was contributed by the mother. As the zygote grows to become the adult, mitosis occurs so that each and every cell has forty-six chromosomes. In this way each body cell contains the full complement of chromosomes and genes. The full complement of chromosomes is called the **diploid,** or **2N,** number of chromosomes.

TABLE 17.1 *MITOSIS VERSUS MEIOSIS*

CELL TYPE	CELL DIVISION	DESCRIPTION	RESULT
Somatic or body cells	Mitosis	2N (diploid)→2N(diploid)	More body cells = growth
Germ cells in gonads of animals	Meiosis	2N(diploid)→N(haploid)	Gamete or sex cell production

Meiosis occurs only in the sex organs, or **gonads**—the testes in males and the ovaries in females. Here diploid germ cells develop into the gametes that have half the total number, called the **haploid**, or **N**, number of chromosomes. The haploid number always has one of each kind of chromosome. For example, in humans the germ cells have forty-six chromosomes, or twenty-three pairs, but gametes contain only twenty-three chromosomes—one of each of the pairs.

When a haploid sperm fertilizes a haploid egg, the new individual has the diploid number of chromosomes, half of which came from the father and half of which came from the mother. Thus, each parent contributes one of each of the pairs of chromosomes possessed by the new individual. Table 17.1 summarizes the major differences between mitosis and meiosis.

The life cycle of humans requires two types of cell divisions: mitosis and meiosis. Mitosis is responsible for growth and repair, while meiosis is required for gamete production.

MITOSIS

Overview

Mitosis[1] is *cell division in which the daughter cells retain the same number and kinds of chromosomes as the mother cell.* Therefore, the newly formed cells are genetically identical to the original cell. The *mother cell* is the cell that divides, and the *daughter cells* are the resulting cells. Although *humans have forty-six chromosomes,* each cell in figure 17.5 contains only four chromosomes for simplicity's sake. (In determining the number of chromosomes it is necessary to count only the number of independent centromeres.) A cell prepares for mitosis by replication of the genetic material contained within each chromosome (fig. 17.5). **Replication** is the process by which DNA makes a copy of itself, as is described in detail in chapter 19. Because of replication, each chromosome contains two sister chromatids, sometimes called a

[1]The term *mitosis* technically refers only to nuclear division but for convenience is used here to refer to division of the entire cell.

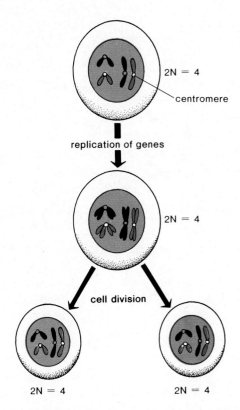

FIGURE 17.5 *Mitosis overview. Following replication of genes, each chromosome in the mother cell contains two sister chromatids. During mitotic division, the sister chromatids separate so that daughter cells have the same number and kinds of chromosomes as the mother cell.*

2N = 4
centromere
replication of genes
2N = 4
cell division
2N = 4   2N = 4

chromatid pair, or *dyad*. During mitosis, the chromatid pairs separate, ensuring that each new cell will receive a copy of each chromosome rather than two copies of one chromosome and none of another. Different genes are on different chromosomes, and it is necessary for every cell to have a copy of each chromosome in order to have a full complement of genes. As an aid in describing the events of mitosis, the process has been divided into four phases: prophase, metaphase, anaphase, and telophase (fig. 17.6).

FIGURE 17.6 *Mitosis has four stages, excluding interphase and daughter cells.*
Notice that in these drawings of mitosis for animal cells that the centrioles have
doubled. There are two pairs of centrioles in the late interphase cell at the start of
mitosis.

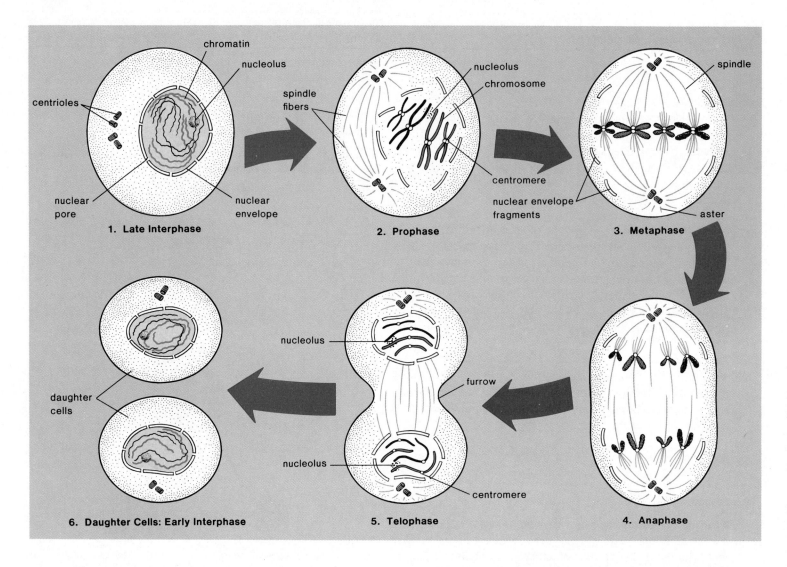

Mitosis ensures that each cell in the body is genetically identical. At the time of division, a chromosome cell consists of sister chromatids. When these separate, each newly forming cell receives the same number and kinds of chromosomes as the original cell.

In between cell divisions, the cell is said to be in interphase. During **interphase** an animal cell resembles the representation in figure 2.2. The nuclear envelope and the nucleoli are intact. The chromosomes, however, are not visible because the chromosomal material is dispersed in the form of chromatin. In animal cells there is a pair of centrioles just outside the nucleus.

It used to be said that interphase was a resting stage, but we now know that this is not the case. During interphase the organelles are metabolically active and are carrying on their normal functions. If cell division is about to occur, each centriole replicates so that there are two pairs of centrioles outside the nucleus. DNA replicates so that each chromosome consists of sister chromatids.

FIGURE 17.7 *The spindle apparatus from an animal cell consists of the structures shown. Each spindle fiber is a bundle of microtubules. Some believe that the centrioles are involved in the production of microtubules and thus of the spindle.*

FIGURE 17.8 *Microtubules can assemble at one end and disassemble at the other. During assembly, protein dimers join together, and during disassembly the protein dimers separate from one another.*

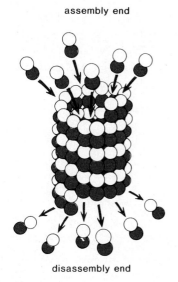

assembly end

disassembly end

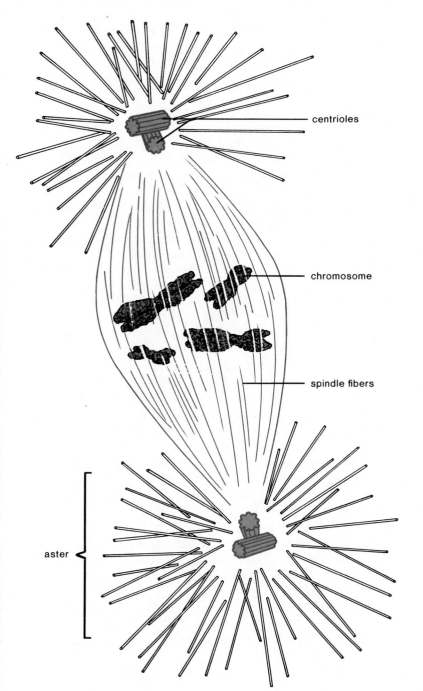

centrioles

chromosome

spindle fibers

aster

Stages

Prophase

It is apparent during **prophase** that cell division is about to occur. The chromatin material shortens and thickens so that the chromosomes are readily visible. The pairs of centrioles begin separating and moving toward opposite ends of the nucleus. **Spindle fibers** appear between the separating pairs of centrioles. As the spindle appears, the nuclear envelope begins to fragment and the *nucleolus* begins to disappear.

Toward the end of prophase the chromosomes are randomly placed even though the spindle appears to be fully formed.

Function of Centrioles

The entire spindle apparatus is shown in figure 17.7. It consists of asters, spindle fibers, and centrioles. Both the short asters radiating from the centrioles and the long spindle fibers are composed of microtubules. It is known that microtubules are capable of assembling and disassembling (fig. 17.8), which would account for the appearance and disappearance of asters and spindle fibers. It is possible that the centrioles are involved in spindle formation but it could also be that their location at the poles simply ensures that each daughter cell will have a pair of centrioles.

Metaphase

As **metaphase** begins, the nuclear envelope has disappeared, and the spindle now occupies the region formerly occupied by the nucleus. Each chromosome attaches to a fiber and moves to the *equator* (center) of the spindle. Metaphase is characterized by a fully formed spindle, with the chromosomes, each composed of two chromatids, arranged across the equator (fig. 17.9a). Notice the **asters,** short lengths of microtubules that radiate out from the centrioles, located at the *poles* (ends) of the spindle.

FIGURE 17.9 *Photomicrographs of cells undergoing mitosis.* a. *During metaphase, the chromosomes are lined up along the equator of the spindle.* b. *During anaphase, the separation of sister chromatids results in chromosomes that are pulled by spindle fibers to opposite poles of the spindle.*

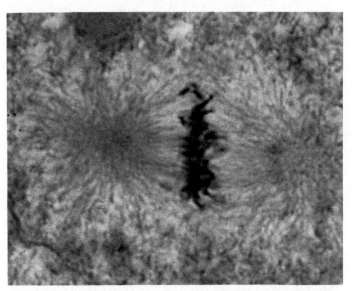

a.

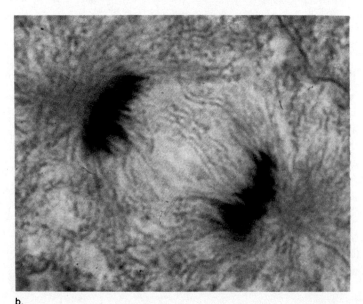

b.

Anaphase

During **anaphase** each centromere splits. Now the sister chromatids separate, and each moves toward an opposite pole of the spindle (fig. 17.9b). *Once separated, the chromatids are called chromosomes.* Separation of the sister chromatids ensures that each daughter cell will receive a copy of each type of chromosome and thus have a full complement of genes. As the newly formed chromosomes move to opposite poles, the entire cell elongates (fig. 17.6).

Telophase

During **telophase** the spindle disappears, possibly due to disassembly of the microtubules making up the spindle fibers. As the nuclear envelopes re-form and the nucleoli reappear in each daughter cell, the chromosomes become indistinct chromatin again. Following nuclear division, cytoplasmic division, sometimes called **cytokinesis,** usually occurs. In animal cells, *furrowing,* or an indentation of the membrane between the two daughter cells, divides the cytoplasm. Furrowing is complete when each daughter cell has a complete membrane enclosing it. Microfilaments are believed to take part in the furrowing process since they are always in the vicinity.

Table 17.2 is a summary of the stages of mitosis.

Cell Cycle

The cell cycle (fig. 17.10) includes interphase and the four stages of division. At the completion of a cycle, during interphase, some cells mature and become specialized. Each type of specialized cell has a characteristic life span. For example, red blood cells live about 120 days, but many nerve cells live as long as the individual does.

During interphase, some cells prepare to divide again; that is, they prepare to complete the cell cycle. There is a limit to the number of times an animal cell will divide before death follows degenerative changes. Most will divide about fifty times, and only cancer cells retain the ability to divide repeatedly. Cancer cells have abnormal chromosomes and other abnormalities of cell structure. Aging of cells in the

TABLE 17.2 *STAGES OF MITOSIS*

STAGE	EVENTS
Prophase	Replication has occurred, and each chromosome is composed of a pair of sister chromatids
Metaphase	Chromatid pairs (dyads) are at the equator of the cell
Anaphase	Chromatids separate, and each one is now termed a chromosome
Telophase	Each pole has the same number and kinds of chromosomes as the mother cell

individual appears to be a normal process most likely controlled by the nucleus. In the laboratory, aging of cells can be delayed by environmental circumstances such as reduced temperature, but it cannot be postponed indefinitely.

The cell cycle includes interphase and the four stages of division: prophase, metaphase, anaphase, and telophase. Some cells differentiate during interphase, and some prepare to enter the cycle again.

Importance of Mitosis

Mitosis assures that each body cell has the same number and kinds of chromosomes. It is important to the growth and repair of multicellular organisms. When a baby develops in its mother's womb, mitosis occurs as a component of growth. As a wound heals, mitosis occurs to repair the damage.

Mitosis also occurs during the process of asexual reproduction. In lower animals, a group of cells called a bud can give rise to an entire individual. This new individual has the same genes and is identical to the parent individual. The term *cloning* is sometimes used to refer to the asexual production of individuals from mature cells of donor animals such as mice or even humans. Thus far, only embryonic animal cells seem to retain the capability of beginning development again. An adult cell has an entire set of genes but is too specialized to begin development. Research is still going on to overcome this difficulty so that a mammal can be cloned from an adult cell.

FIGURE 17.10 *The cell cycle consists of mitosis and interphase. During interphase, there is growth before and after DNA synthesis. DNA synthesis is required for the process of replication, by which DNA makes a copy of itself. Some daughter cells "break out" of the cell cycle and become specialized cells performing a specific function.*

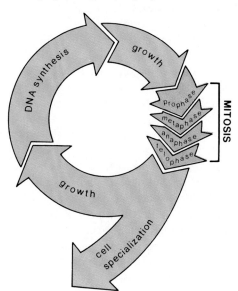

MEIOSIS

Overview

Meiosis, which requires two cell divisions, results in *four daughter cells, each having one of each kind of chromosome and therefore half the number of chromosomes as the mother cell.*[2] Figure 17.11 presents an overview of meiosis, indicating the two cell divisions, **meiosis I** and **meiosis II.** Prior to meiosis I, replication has occurred and each chromosome

[2]The term *meiosis* technically refers only to nuclear division but for convenience is used here to refer to division of the entire cell.

FIGURE 17.11 *Overview of meiosis. Following replication of genes, the mother cell undergoes two divisions, meiosis I and meiosis II. During meiosis I, homologous chromosomes separate, and during meiosis II, chromatids separate. The final daughter cells have the haploid number of chromosomes in single copy.*

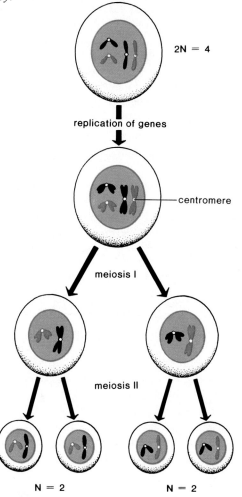

FIGURE 17.12 *During crossing over, pieces of chromosomes are exchanged between chromatid pairs.* a. *Chromatid pairs before crossing over has occurred.* b. *Chromatid pairs after* crossing over. *Notice the change in chromosome structure.* c. *Chromatid pairs have separated.*

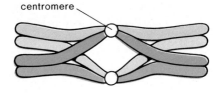

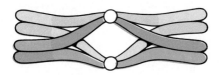

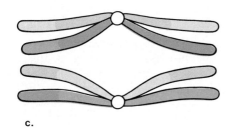

a.

b.

c.

consists of sister chromatids held together at a centromere. During meiosis I, the homologous chromosomes come together and line up side by side due to a means of attraction still unknown. This is called **synapsis** and results in **tetrads,** an association of four chromatids that stay in close proximity during the first two phases of meiosis I. Now an important process called **crossing over** may occur (fig. 17.12). During crossing over, chromatids of the homologous chromosome pairs exchange segments. This produces new combinations of genes on the chromatids.

Following synapsis, the homologous chromosomes separate during meiosis I. This separation means that one chromosome of every homologous pair will reach each gamete.[3] There are no restrictions to the separation process; either chromosome of a homologous pair may occur in a gamete with either chromosome of any other pair.[4]

Notice that at the completion of meiosis I (fig. 17.11), the chromosomes still consist of sister chromatids. During meiosis II, the sister chromatids separate, resulting in four daughter cells, each of which has the haploid number of chromosomes. Although two cell divisions have taken place, replication has occurred only once. By counting only the number of independent centromeres, you will find that the mother cell has the diploid number of chromosomes and each of the four daughter cells has the haploid number.

Meiosis is cell division that halves the chromosome number. The process requires two divisions and results in four daughter cells, each with the haploid number of chromosomes.

Stages

The stages of meiosis I are diagrammed in figure 17.13. During **prophase I,** the homologous chromosomes of each pair undergo synapsis, forming tetrads. The nuclear envelope and nucleolus do not disappear until the end of prophase I. At **metaphase I,** tetrads line up at the equator of the spindle. During **anaphase I,** the homologous chromosomes of each pair separate, and the chromosomes (each still composed of sister chromatids) move to the poles of the spindle. Each pole receives half of the total number of chromosomes. In **telophase I,** the nuclear envelope reforms and the nucleoli reappear. There is no replication of DNA between meiosis I and meiosis II.

During meiosis I, the chromatids making up a tetrad exchange chromosome pieces. When the homologous chromosomes separate, each daughter cell receives one from each pair of chromosomes.

In the second division of meiosis (fig. 17.13), the phases are referred to as prophase II, metaphase II, anaphase II, and telophase II. During **anaphase II,** centromere separation occurs, and the chromatids part to become independent chromosomes. At the end of **telophase II,** there are four cells. Each of these four cells is haploid; that is, the nucleus has half the number of chromosomes and half the DNA content of the mother cell nucleus.

Separation of sister chromatids during meiosis II produces a total of four daughter cells, each with the haploid number of chromosomes. Figure 17.14 contrasts mitosis to meiosis.

Spermatogenesis and Oogenesis

Spermatogenesis, sperm production, takes place in the testes of males. **Oogenesis,** egg production, takes place in the ovaries of females. Gamete production is different in the two sexes. Figure 17.15 shows that for each meiosis, there result four viable sperm in males. Also, spermatogenesis occurs continuously, and at the time of ejaculation, males emit as many as 400 million or more sperm.

In contrast to spermatogenesis in males, egg production occurs only once a month in females. The first meiotic division produces two cells, but one is much larger than the other. The smaller nonfunctional cell is called a **polar body** and remains attached to the large cell. At this point, ovulation occurs and the immature egg enters the oviduct (p. 334) in females. The second meiotic division does not occur, and unless fertilization takes place oogenesis is halted. If oogenesis continues, the second meiotic division is also unequal, so that in the end there is only one mature egg and at least two nonfunctional polar bodies that disintegrate.

Figure 17.16 shows how the sperm and egg are adapted to their function. The sperm is a tiny, flagellated cell that is adapted for swimming to the maturing egg, a large cell that contributes most of the cytoplasm and nutrients to the new individual.

---

[3]See Mendel's law of segregation, p. 388
[4]See Mendel's law of independent assortment, p. 391

FIGURE 17.13 *Meiosis. During meiosis I, chromosome pairs separate so that each daughter cell has only one chromosome from every original homologous pair of chromosomes. During meiosis II, chromatids separate so that each daughter cell has the haploid number of chromosomes in single copy.*

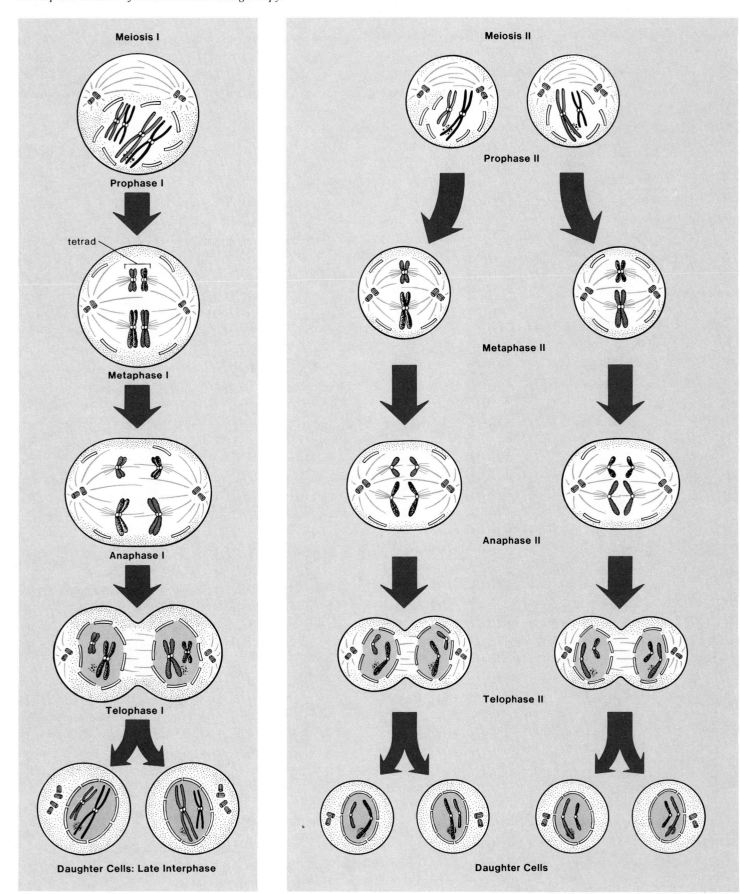

Meiosis I

Prophase I

tetrad

Metaphase I

Anaphase I

Telophase I

Daughter Cells: Late Interphase

Meiosis II

Prophase II

Metaphase II

Anaphase II

Telophase II

Daughter Cells

FIGURE 17.14 *Comparison of mitosis with meiosis.* a. *Mitosis.* b. *Meiosis.*

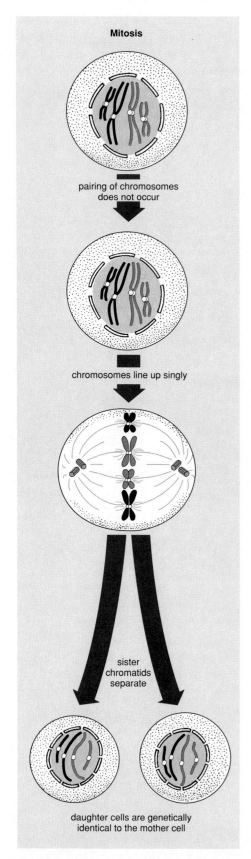

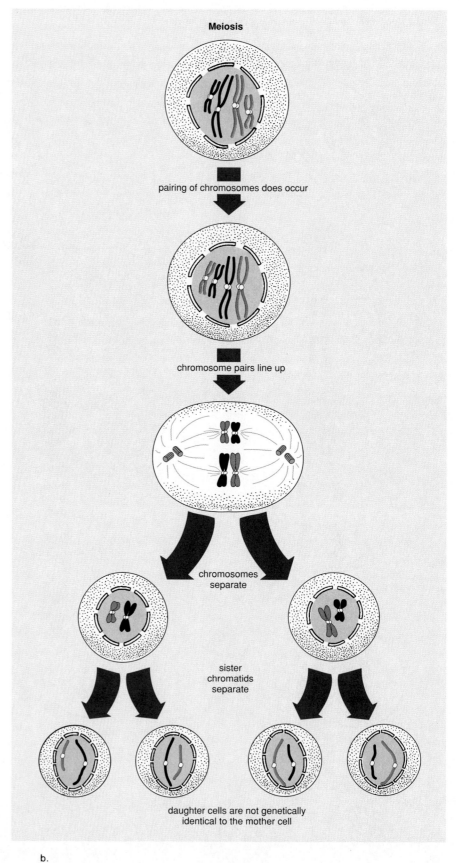

a.

b.

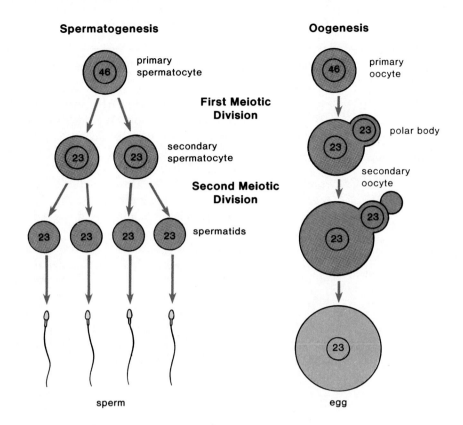

Spermatogenesis

primary spermatocyte

First Meiotic Division

secondary spermatocyte

Second Meiotic Division

spermatids

sperm

Oogenesis

primary oocyte

polar body

secondary oocyte

egg

FIGURE 17.15 *Spermatogenesis produces four viable sperm, but oogenesis produces one egg and at least two polar bodies. In humans, both sperm and egg have twenty-three chromosomes each; therefore, following fertilization the zygote has forty-six chromosomes.*

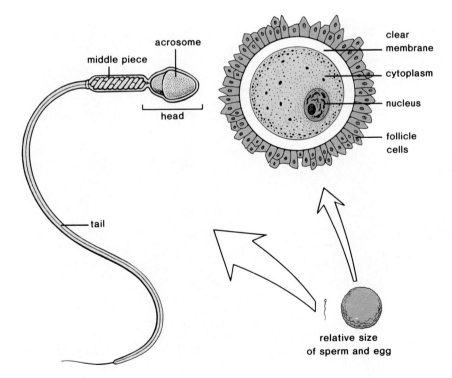

middle piece

acrosome

head

tail

clear membrane

cytoplasm

nucleus

follicle cells

relative size of sperm and egg

FIGURE 17.16 *Microscopic anatomy of sperm and egg. Note their relative sizes. The head of sperm contains the chromosomes and little else; the larger egg (surrounded by follicle cells from the ovary) contributes most of the cytoplasm to the zygote.*

Importance of Meiosis

Meiosis is nature's way of keeping the chromosome number constant from generation to generation. It assures that the next generation will have a different genetic makeup than that of the previous generation. As a result of crossing over, the chromosomes carry a new combination of genes. The egg carries half of the genes from the female parent and the sperm carries half of the genes from the male parent. When the sperm fertilizes the egg, the zygote has a different combination of genes than either parent. In this way, meiosis assures genetic variation generation after generation.

Spermatogenesis in males produces four viable sperm, but oogenesis in females produces one egg and at least two polar bodies. Each gamete is specialized for the job it does; the sperm is a tiny, flagellated cell that is propelled to the cytoplasm-laden egg.

CHROMOSOMAL INHERITANCE

Normal Inheritance

The individual normally receives twenty-two autosomal chromosomes and one sex chromosome from each parent. The sex of the newborn child is determined by the father. If a Y-bearing sperm fertilizes the egg, then the XY combination will result in the development of a male. On the other hand, if an X-bearing sperm fertilizes the egg, the XX combination results in the development of a female. All factors being equal, there is a 50% chance of having a girl or boy (fig. 17.17). It is possible to illustrate this probability by doing a **Punnett square** (fig. 17.18). In the square, all possible sperm are lined up on one side; all possible eggs are lined up on the other side (or vice versa), and every possible combination is determined. When this is done with regard to sex chromosomes, the results show one female to each male. However, for reasons that are not clear, more males than females are conceived. But from then on the death rate among males is higher; more males than females are spontaneously aborted, and this trend continues after birth until there is a dramatic reversal of the ratio of males to females (table 17.3).

The sex of a child is dependent on whether a Y-bearing or an X-bearing sperm fertilizes the X-bearing egg.

FIGURE 17.17 It's a girl. According to a somewhat controversial method advocated by Dr. Shettles of choosing the sex of your child, the X-bearing sperm is favored over the Y-bearing sperm if these requirements are met: (1) the vagina is acidic (a douche consisting of two tablespoons of white vinegar to a quart of water promotes this); (2) intercourse should be frequent, and penetration is shallow but ceases two or three days before ovulation. On the other hand, the Y-bearing sperm is favored over the X-bearing sperm if the following requirements are met: (1) the vagina is alkaline (a douche consisting of two tablespoons of baking soda to a quart of water promotes this; let it stand 15 minutes before using); (2) abstinence is practiced until the day of ovulation, when penetration is deep.

Male Gene

For some years, it has been proposed that there is a gene located on the Y chromosome that brings about maleness. Embryos begin life with no evidence of a sex, but then along about the third month of development males can be distinguished from females (p. 346). Investigators have recently reported the finding of a gene they call the testis determining factor gene (TDF) on the Y chromosome. When this gene is lacking from the Y chromosome, the individual is a female even though the chromosomal inheritance is XY. On the other hand, if the gene is present in an XX individual, this person is a male.

Abnormal Autosomal Chromosome Inheritance

Sometimes individuals are born with either too many or too few autosomal chromosomes due most likely to nondisjunction of chromosomes or sister chromatids during meiosis (fig. 17.19). It is possible also that even though there is the correct number of chromosomes, one chromosome may be defective in some way because of a chromosomal mutation (fig.

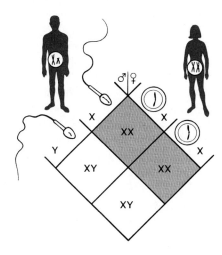

FIGURE 17.18 Inheritance of sex. An offspring is either male or female depending on whether an X or a Y chromosome is received from the male parent. In this Punnett square, the sperm and eggs are shown as carrying only a sex chromosome. Actually, of course, they also carry twenty-two autosomes.

TABLE 17.3 RATIO OF MALES TO FEMALES IN THE UNITED STATES

AGE	SEX RATIO
Conception	130:100[a]
Birth	106:100
18 years	100:100
50 years	85:100
85 years	50:100
100 years	20:100

[a]Based on spontaneous abortion data.
From Hole, John W., Jr., *Human Anatomy and Physiology*, 4th ed. © 1978, 1981, 1984, 1987 Wm. C. Brown Publishers, Dubuque, Iowa. All Rights Reserved. Reprinted by permission.

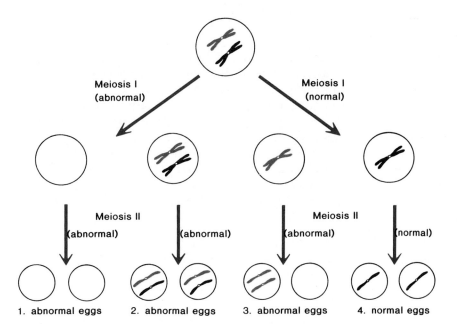

1. abnormal eggs   2. abnormal eggs   3. abnormal eggs   4. normal eggs

FIGURE 17.19 Nondisjunction during oogenesis. Nondisjunction can occur during meiosis I if the chromosome pairs fail to separate and during meiosis II if the chromatids fail to separate completely. In either case, the abnormal eggs carry an extra chromosome. Nondisjunction of the number-21 chromosome leads to Downs syndrome.

17.20). *Chromosomal mutations* are known to occur after chromosomes are broken due to exposure to radiation, addictive drugs, or pesticides, for example. When the chromosomes re-form, the pieces may be rearranged. An **inversion** results when a piece of a chromosome has turned in a direction opposite to the usual. A **deletion** occurs when a piece of a chromosome is lost and the chromosome is shorter than it was formerly. In contrast, a **duplication** is the presence of a chromosome piece more than once in the same chromosome. Deletion and duplication are apt to occur between homologous chromosomes. When nonhomologous chromosomes exchange pieces, a **translocation** has taken place. The presence of a mutated chromosome can cause the individual to have reproductive problems because the abnormal chromosome can result in nonviable zygotes or a child with birth defects.

Sometimes an individual inherits too many or too few chromosomes or a defective chromosome.

Syndromes

A **syndrome** is a group or pattern of symptoms that occur together in the same individual due to the presence of an abnormal condition. We will be discussing various syndromes that result from the inheritance of chromosomal abnormalities.

Down Syndrome   The most common autosomal abnormality is seen in individuals with **Down syndrome** (fig. 17.21). This syndrome is easily recognized. Its characteristics include a short stature; an oriental-like fold of the eyelids; stubby fingers; a wide gap between the first and second toes; a large, fissured tongue; a round head; a palm crease, the so-called simian line; and, unfortunately, mental retardation that can sometimes be severe.

Persons with Down syndrome usually have three number-21 chromosomes because the egg had two number-21 chromosomes instead of one (fig. 17.19). (In 23% of the cases studied, however, the sperm had the extra number-21 chromosome.) It would appear that **nondisjunction,** failure of the chromosome pairs or chromatids to separate completely, is most apt to occur in the older female since children with Down syndrome are usually born to women over age 40 (table 17.4). If a woman wishes to know whether or not her unborn child is affected by Down syndrome, she may elect to undergo chorionic villi testing or amniocentesis, two procedures discussed in the reading on page 381. Following this procedure, a karyotype can reveal whether the child has Down syndrome. If so, she may elect to continue or to abort the pregnancy.

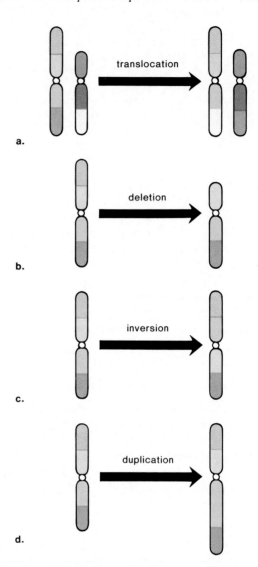

FIGURE 17.20   *Types of chromosome mutation.*
a. *Translocation is the exchange of chromosome pieces between nonhomologous parts.* b. *Deletion is the loss of a chromosome piece.* c. *Inversion occurs when a piece of a chromosome breaks loose and then rejoins in reversed direction.* d. *Duplication occurs when the same piece is repeated within the chromosome.*

It is known that the genes that cause Down syndrome are located on the bottom third of the number-21 chromosome (fig. 17.21b), and there has been a lot of investigative work to discover the specific genes responsible for the characteristics of the syndrome. Thus far, investigators have discovered several genes that may account for various conditions seen in persons with Down syndrome. For example, they have located genes most likely responsible for the increased tendency toward leukemia, cataracts, accelerated rate of aging, and mental retardation. The latter gene, dubbed the *Gart*

FIGURE 17.21 *Down syndrome.* a. *Common characteristics include a wide, rounded face and a fold of the upper eyelids. Mental retardation, along with an enlarged tongue, makes it difficult for persons with Down syndrome to learn to speak coherently.* b. *Karyotype of an individual with Down syndrome has an extra number-21 chromosome in the G set. More sophisticated technologies allow investigators to pinpoint the location of specific genes associated with the syndrome. The* Gart *gene, which leads to a high level of blood purines, may account for the mental retardation seen in persons with Down syndrome.*

a.

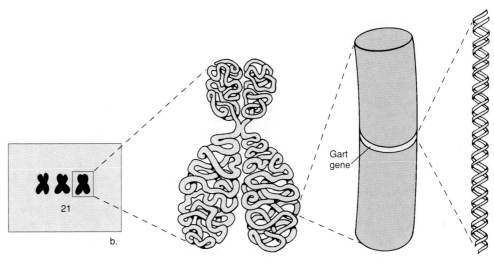

Gart gene

b.

Hagelston/Leggitt

gene, causes an increased level of purines in the blood, a finding that is associated with mental retardation. It is hoped that it will someday be possible to find a way to control the expression of the *Gart* gene even before birth so that at least this symptom of Down syndrome will not appear.

Cri du Chat Syndrome A chromosomal deletion is responsible for **cri du chat** (cat's cry) **syndrome.** Affected individuals meow like a kitten when they cry, but more important perhaps is the fact that they tend to have a small head with malformations of the face and body and that mental defectiveness usually causes retarded development. Chromosomal analysis shows that a portion of one number-5 chromosome is missing (deleted) while the other number-5 chromosome is normal.

Down syndrome is most often due to the inheritance of an extra number-21 chromosome, and cri du chat syndrome is due to the inheritance of a defective number-5 chromosome.

TABLE 17.4 *INCIDENCE OF SELECTED CHROMOSOMAL ABNORMALITIES*

NAME	FREQUENCY/100,000 LIVE BIRTHS
Down syndrome (general)	140
Down syndrome (mothers over age 40)	1,000
Turner syndrome	8
Metafemale	50
Klinefelter syndrome	80
XYY	100

From *Antenatal Diagnosis*, HEW, 1979, pp. 1–48.

MALE/FEMALE CELL DIFFERENCES

On occasion, such as at athletic competitions, it is important to be able to certify that an individual is a male or a female. Since physical examinations sometimes fail—as when a male has had a sex-change operation—officials often resort to examining the cells themselves.

You could, of course, do a karyotype, but there are shorter methods. It so happens that XX females have small, darkly staining masses of condensed chromatin, called Barr bodies (named after the person who first identified them), present in their nuclei (see figure). XY males have no comparable spots of chromatin in their nuclei. It turns out that a Barr body[1] is a condensed and at least to some degree inactive X chromosome as was proposed by Mary Lyon. The validity of the *Lyon hypothesis* means that female cells function with a single X chromosome just as males do. Still, in some cells one X is condensed and in some cells the other X is condensed so that the female body is a mosaic of genetically different cells.

No doubt, you believe that observation of Barr bodies is not a guarantee of femaleness because we have already indicated that there are XX males, albeit rarely. We need another test, and there is one. There is an antigen, called an H-Y antigen, present in the cell membrane of males but not in females. It is called an antigen because females produce antibodies against it. To test for maleness it is possible to suspend a sample of white blood cells in a solution that contains some of these antibodies. If the cells carry the H-Y antigen, indicating that the person is a male, the antibodies bind with them. Now, we can be certain who is a male and who is a female.

[1] How many Barr bodies would a person with Klinefelter syndrome have? a metafemale have?

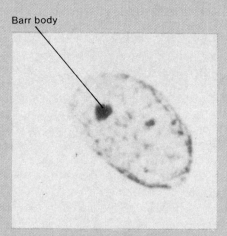

Barr body

Each female cell contains a Barr body. A Barr body is a condensed X chromosome.

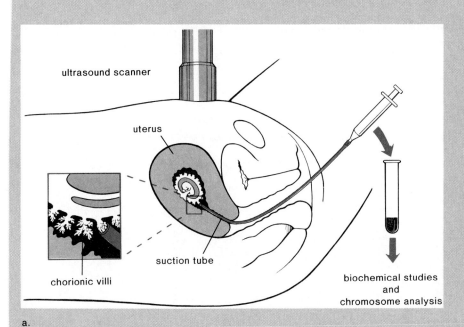

ultrasound scanner

uterus

chorionic villi

suction tube

biochemical studies
and
chromosome analysis

a.

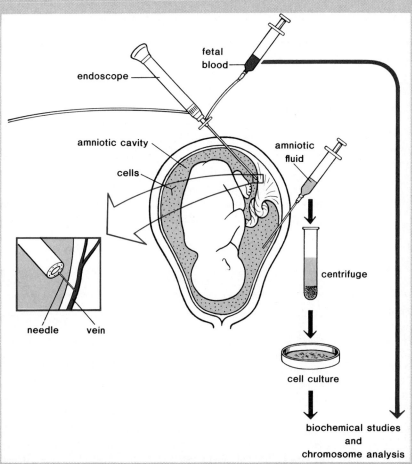

endoscope

fetal
blood

amniotic cavity

amniotic
fluid

cells

needle   vein

centrifuge

cell culture

biochemical studies
and
chromosome analysis

b.

DETECTING BIRTH DEFECTS

A new method called chorionic villi sampling (fig. a) allows physicians to collect embryonic cells as early as the fifth week of development. The doctor inserts a long thin tube through the vagina into the uterus. With the help of ultrasound, which gives a picture of the uterine contents, the tube is placed between the lining of the uterus and the chorion. Then suction is used to remove a sampling of the chorionic villi cells. Chromosomal analysis and biochemical tests to detect some genetic defects can be done immediately on these cells.

Before the development of chorionic villi sampling, physicians had to wait until about the sixteenth week of pregnancy to perform amniocentesis. In amniocentesis (fig. b), a long needle is passed through the abdominal wall to withdraw a small amount of amniotic fluid along with fetal cells. Since there are only a few cells in the amniotic fluid, testing must be delayed until these cells have grown and multiplied in cell culture. Therefore it may be another two to four weeks until the prospective parents are told whether their child has a genetic defect.

In fetoscopy, another possible procedure, the physician uses an endoscope to view the fetus. Blood can be withdrawn for prenatal diagnosis.

FIGURE 17.22 *Nondisjunction of sex chromosomes during oogenesis followed by fertilization with normal sperm results in the conditions noted. Nondisjunction of sex chromosomes during spermatogenesis followed by fertilization of normal eggs results in the conditions noted.*

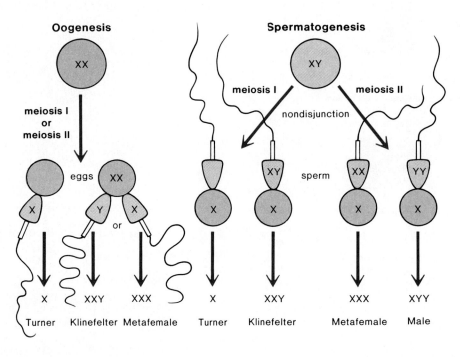

Abnormal Sexual Chromosome Inheritance

Abnormal sexual chromosome constituencies (table 17.4) are also due to the occurrence of nondisjunction. Nondisjunction of the sex chromosomes during oogenesis can lead to an egg with either two X chromosomes or no X chromosomes. Nondisjunction of the sex chromosomes during spermatogenesis can result in a sperm that has no sex chromosome, both an X and a Y chromosome, two X chromosomes, or two Y chromosomes. Assuming that the other gamete is normal, the zygote could develop into an individual with one of the conditions noted in figure 17.22.

Sometimes a person inherits an abnormal combination of sex chromosomes due to nondisjunction of these chromosomes during meiosis.

Abnormalities

An XO individual with **Turner syndrome** has only one sex chromosome, an X; the O signifies the absence of the second sex chromosome. Because the ovaries never become functional these females do not undergo puberty or menstruate, and there is a lack of breast development (fig. 17.23b). Generally, these individuals have a stocky build and a webbed neck. They also have difficulty recognizing various spatial patterns.

HUMAN ISSUE

Technological advancements associated with prenatal screening pose a number of ethical dilemmas for parents and their physicians. For example, recently a woman sued her obstetrician for not telling her that her child would be born with Down syndrome. She won her case, and the obstetrician has to pay for the support of the child. This illustrates one dilemma confronting physicians—are they obligated to inform a pregnant woman of any birth defects, or is it up to the woman to express her desire to know of any? Parents face potential dilemmas as well. For example, is it better for couples to remain uninformed about the results of the prenatal screening so that they don't put themselves through the agony of having to decide about an abortion, or is it better to know so that they can be mentally prepared for a child with an abnormality?

When an egg having two X chromosomes is fertilized by an X-bearing sperm, a **metafemale** having three X chromosomes results. It might be supposed that the XXX female with forty-seven chromosomes would be especially feminine, but this is not the case. Although there is a tendency toward learning disabilities, most metafemales have no apparent physical abnormalities and many are fertile and have children with a normal chromosome count.

When an egg having two X chromosomes is fertilized by a Y-bearing sperm, a male with **Klinefelter syndrome** results. This individual is male in general appearance, but the testes are underdeveloped and the breasts may be enlarged (fig. 17.23a). The limbs of these XXY males tend to be longer than average, body hair is sparse, and many have learning disabilities.

XYY males also occur possibly due to nondisjunction during spermatogenesis. These males are usually taller than average, suffer from persistent acne, and tend to have barely normal intelligence. At one time it was suggested that these men were likely to be criminally aggressive, but it has been shown that the incidence of such behavior is no greater than that among normal XY males.

Individuals are sometimes born with the sex chromosomes XO (Turner syndrome), XXX (metafemale), XXY (Klinefelter syndrome), and XYY. Individuals with a Y chromosome are always male no matter how many X chromosomes there may be; however, at least one X chromosome is needed for survival.

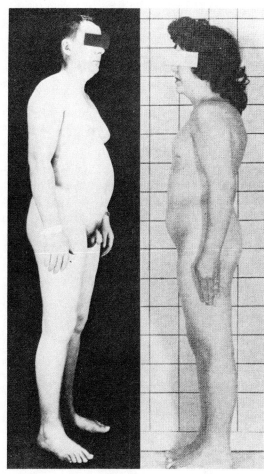

FIGURE 17.23 *Abnormal sex chromosome inheritance. a. A male with Klinefelter (XXY) syndrome, which is marked by immature sex organs and development of the breasts. b. Female with Turner (XO) syndrome, which includes a bull neck, short stature, and immature sexual features.*

a.   b.

SUMMARY

The life cycle of higher organisms requires two types of cell divisions, mitosis and meiosis. Mitosis assures that all cells in the body have the diploid number and same kinds of chromosomes. It is made up of four stages: prophase, metaphase, anaphase, and telophase. The cell cycle includes an additional stage termed interphase. During interphase, DNA replication causes each chromosome to have sister chromatids. When the chromatids separate, each newly forming cell receives the same number and kinds of chromosomes as the original cell. The cytoplasm is partitioned by furrowing in human cells.

Meiosis involves two cell divisions. During meiosis I, the homologous chromosomes (following crossing over between chromatids) separate, and during meiosis II the sister chromatids separate. The result is four cells with the haploid number of chromosomes in single copy. Meiosis is a part of gamete formation in humans. Mitosis is contrasted to meiosis in figure 17.14

An individual ordinarily receives twenty-two autosomal and one sex chromosome from each parent, but

abnormalities do occur. Also, a mutated chromosome can be inherited; inversions, deletions, translocations, and duplications of chromosomal parts are known. The major inherited autosomal abnormality is Down syndrome in which the individual inherits three number-21 chromosomes due to nondisjunction during gamete formation. Examples of abnormal sex chromosome inheritance due to nondisjunction are Turner syndrome (XO), Klinefelter syndrome (XXY), XYY males, and metafemales (XXX).

OBJECTIVE QUESTIONS

1. The arrangement of an individual's chromosomes according to homologous pairs is called a _____ .
2. The karyotype of males includes the sex chromosomes _____ , and the karyotype of females includes the sex chromosomes _____ .
3. The diploid number of chromosomes is designated the _____ number, and the haploid number is designated the _____ number.
4. If the mother cell has twenty-four chromosomes, the daughter cells following mitosis will have _____ chromosomes.
5. As the organelles called _____ separate and move to the poles, the spindle fibers appear.

6. During meiosis I, the _____ _____ separate, and during meiosis II the _____ _____ separate.
7. Meiosis in males is a part of _____ , and meiosis in females is a part of _____ .
8. There is a _____ chance of a newborn being a male or a female.

For answering questions 9–15, use this key:
   a. Down syndrome
   b. Turner syndrome
   c. Klinefelter syndrome
   d. cri du chat syndrome
   e. metafemale

9. XXY _____
10. extra number-21 chromosome _____
11. XXX _____
12. deletion in number-5 chromosome _____
13. XO _____
14. due to an autosomal nondisjunction _____
15. due to a chromosomal mutation _____

Answers

1. karyotype 2. XY, XX 3. 2N, N 4. twenty-four 5. centrioles 6. homologous chromosomes, sister chromatids 7. spermatogenesis, oogenesis 8. 50% 9. c 10. a 11. e 12. d 13. b 14. a 15. d

STUDY QUESTIONS

1. Describe the normal karyotype of a human being. What is the difference between a male and a female karyotype? (p. 365)
2. What is the makeup of a chromosome prior to the start of cell division? (p. 365)
3. How do the terms diploid (2N) and haploid (N) pertain to meiosis? (pp. 366–367)
4. Describe the stages of mitosis, including in your description the terms *centrioles, nucleolus, spindle,* and *furrowing.* (pp. 367–370)

5. Describe the stages of meiosis I, including the term *tetrad* in your description. (p. 372)
6. Draw and explain a diagram illustrating crossing over. (p. 372)
7. Compare the second series of stages of meiosis to a mitotic division. (p. 372)
8. How does spermatogenesis in males compare to oogenesis in females? (p. 372)
9. What is the importance of mitosis and meiosis in the life cycle of humans? (pp. 371, 376)

10. Which parent determines the sex of the baby? What are the chances of having a boy or a girl? (p. 377)
11. What is nondisjunction, and how does it occur? What is the most common autosomal chromosome abnormality? (p. 378)
12. What are four sex chromosome abnormalities? Describe each one. (pp. 381–383)

THOUGHT QUESTIONS

1. The spindle apparatus assures the inheritance of the diploid number of chromosomes in each daughter cell. Why is there no need to assure equal distribution of the organelles to the daughter cells?

2. Both egg and sperm contribute one member of each pair of chromosomes to the new individual, but the egg contributes most of the cytoplasm. Why does this seem appropriate, considering the manner in which humans procreate, and what implications does it have for inheritance?

3. Extremely few newborns have just one autosomal chromosome of a particular pair. It's speculated that each chromosome ordinarily carries "lethal genes," genes that are so defective that if they are not counteracted by a competent gene in the other member of the homologous pair, the individual dies. Why, then, do you suppose that a person with Down syndrome can survive?

autosomes (aw'to-sōmz)   chromosomes other than sex chromosomes.   *365*

centromere (sen'tro-mēr)   a region of attachment for a chromosome to spindle fibers that is generally seen as a constricted area.   *365*

chromatids (kro'mah-tidz)   the two identical parts of a chromosome following replication of DNA.   *365*

crossing over (kros'ing o'ver)   the exchange of corresponding segments of genetic material between chromatids of homologous chromosomes during synapsis of meiosis I.   *372*

cytokinesis (si''to-ki-ne'sis)   division of the cytoplasm of a cell.   *370*

diploid (dip'loid)   the 2N number of chromosomes; twice the number of chromosomes found in gametes.   *366*

Down syndrome (down sin'drōm)   human congenital disorder associated with an extra number-21 chromosome.   *378*

gametes (gam'ets)   reproductive cells that join in fertilization to form a zygote; most often an egg or sperm.   *366*

haploid (hap'loid)   the N number of chromosomes; half the diploid number; the number characteristic of gametes that contain only one set of chromosomes.   *367*

homologous chromosomes (ho-mol'o-gus kro'mo-sōmz)   similarly constructed; homologous chromosomes have the same shape and contain genes for the same traits.   *365*

karyotype (kar'e-o-tīp)   the arrangement of all the chromosomes within a cell by pairs in a fixed order.   *365*

Klinefelter syndrome (klīn'fel-ter sin'drōm)   a condition caused by the inheritance of a chromosomal abnormality in number; an XXY individual.   *383*

meiosis (mi-o'sis)   type of cell division that occurs during the production of gametes or spores by means of which the daughter cells receive half the number of chromosomes as the mother cell.   *365*

metafemale (met''ah-fe'māl)   a female who has three X chromosomes.   *383*

mitosis (mi-to'sis)   type of cell division in which daughter cells receive the exact chromosomal and genetic makeup of the mother cell; occurs during growth and repair.   *365*

nondisjunction (non''dis-jungk'shun)   the failure of homologous chromosomes or sister chromatids to separate during the formation of gametes.   *378*

oogenesis (o''o-jen'ĕ-sis)   production of an egg in females by the process of meiosis and maturation.   *372*

sex chromosomes (seks kro'mo-sōmz)   chromosomes responsible for the development of characteristics associated with maleness or femaleness; an X or Y chromosome.   *365*

spermatogenesis (sper''mah-to-jen'ĕ-sis)   production of sperm in males by the process of meiosis and maturation.   *372*

spindle fibers (spin'd'l fi'berz)   microtubule bundles involved in the movement of chromosomes during mitosis and meiosis.   *369*

synapsis (si-nap'sis)   the attracting and pairing of homologous chromosomes during prophase I of meiosis.   *372*

tetrads (tet'radz)   a set of four chromatids resulting from the pairing of homologous chromosomes during prophase I of meiosis.   *372*

Turner syndrome (tur'ner sin'drōm)   a condition caused by the inheritance of an abnormality in chromosome number; an X chromosome lacks a homologous counterpart—XO.   *382*

CHAPTER EIGHTEEN

GENES AND MEDICAL GENETICS

CHAPTER CONCEPTS

1 Genes, located on chromosomes, are passed from one generation to the next.

2 The Mendelian laws of genetics relate the genotype (inherited genes) to the phenotype (physical characteristics).

3 Exceptions to Mendel's laws apply to traits controlled by more than one gene and to genes located on the same homologous pair of chromosomes.

4 There are genes on the X chromosome that control traits having nothing to do with the sexual characteristics of the individual.

5 Humans are subject to many disorders due to the inheritance of faulty genes.

Close-up of a petri dish test used to check resistance toward antibodies.

INTRODUCTION

In 1896 a French professor of pediatrics, Antoin-B.-J. Marfan, published a brief case report of a 5-year-old girl who had unusually long fingers and toes, limited joint motion, and curvature of the spine. Today, many more symptoms are also associated with the condition now known as Marfan syndrome. All traits are believed to be due to the inheritance of a faulty gene that controls an aspect of connective tissue construction. More life threatening than the skeletal deformities are the circulatory difficulties. The mitral valve in the heart is leaky and allows a backward flow of blood, and the aorta is either enlarged at birth or becomes enlarged during childhood. Rupturing of the aorta can cause instant death, or else the aortic valve eventually stretches, blood flows back into the heart, and heart failure results.

Abe Lincoln, sixteenth president of the United States, was assassinated in 1865, years before Marfan syndrome was recognized. Still, there are those who have deduced that he did indeed have this condition. Certainly it would account for his lanky frame and other physical features that have been described, even including why the left foot is blurred in a photo taken in 1863 (fig. 18.1). The aortic deformities symptomatic of Marfan syndrome result in pulses of blood so strong as to shake the lower leg. If Lincoln was in such bad health, then it is reasoned that had he not been shot, he would still have died soon after this photo was taken.

Today, not only is Marfan syndrome recognized, there is even a National Marfan Foundation, a voluntary organization of people with Marfan syndrome and health professionals who treat the now controllable condition.[1]

---

[1]Further information can be obtained by writing the National Marfan Foundation at 382 Main Street, Port Washington, NY 11050.

FIGURE 18.1 *Did Abe Lincoln have Marfan syndrome? There are those who suggest that the left foot is blurred in this photograph because Lincoln's left leg was throbbing due to circulatory difficulties. Others maintain it is blurred simply because the camera didn't have it in focus.*

T oday we say that the chromosomes located within the nuclei of cells contain the genes. By this we mean that it is possible to imagine that the chromosomes can be divided up into sections and that each section controls a particular trait of the individual. We will use the word **trait** to mean some aspect of the individual, such as height. In figure 18.2b, the rectangles stand for a pair of homologous chromosomes, and the letters stand for genes that control particular traits. Genes, like the letters in the rectangles, are in a particular sequence and are at particular spots, or loci, on the chromosomes. Alternate forms of a gene having the same position on a pair of chromosomes and affecting the same trait are called **alleles.** In our example, *R* is an allele of *r,* and vice versa. Also, *S* is an allele of *s,* and vice versa. *S* could never be an allele for *R* because *S* and *R* are different genes at different loci. Each allelic pair controls some particular trait of the individual, such as color of hair, type of fingers, or length of nose.

SIMPLE MENDELIAN INHERITANCE

The first person to conduct a successful study of genetic or particulate inheritance was Gregor Mendel, a Catholic priest who grew peas in a small garden plot in 1860. Mendel knew little about cell structure, but his studies led him to conclude that inheritance is governed by factors that exist within the individual and are passed on to offspring. Mendel said *that in an individual every trait,* for example, height, *is controlled by two factors or a pair of factors.* We now call these factors alleles. He also observed that one of the factors, controlling the same trait, can be **dominant** over the other, which is **recessive.** The individual may show the dominant characteristic, for example, tallness,[2] while the recessive factor for shortness, although present, is not expressed.

---

[2]Tallness is dominant in peas, but not in humans.

FIGURE 18.2 *The genes are on the chromosomes.* a. *Giant chromosomes from the salivary gland of a fruit fly are large because they contain many chromatids that have not separated. Sometimes researchers are able to associate particular genes with the bands (light/dark regions) you see here.* b. *Diagrammatic representation of homologous pairs of chromosomes within the nucleus. Allelic pairs that control the same trait are found on a homologous pair. The letters Aa, Bb, and so forth stand for allelic pairs.*

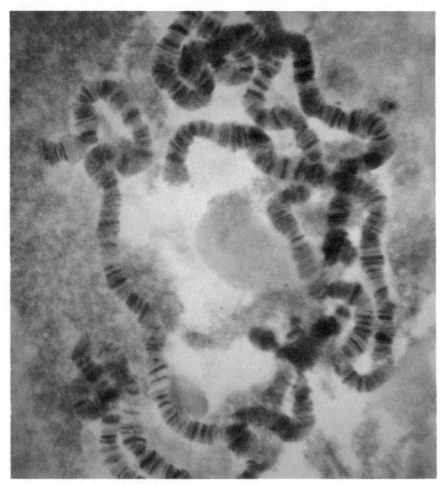

a.

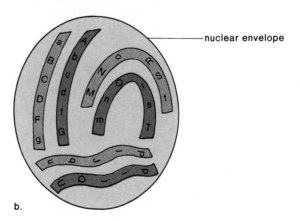

— nuclear envelope

b.

Mendel's experimental crosses made him realize that it was possible for a tall individual to pass on a factor for shortness. Therefore he concluded that while the individual has two factors for each trait, the gametes contained only one factor for each trait. This is often called Mendel's law of segregation.

Law of segregation: The factors separate when the gametes are formed, and only one factor of each pair is present in each gamete.

Inheritance of a Single Trait

Mendel suggested that letters be used to indicate factors so that *crosses* (gamete union resulting in offspring) might be more easily described. A capital letter indicates a dominant factor, and a lowercase letter indicates a recessive factor. The same procedure is used today, only the letters are now said to represent alleles. Also, Mendel's procedure and laws are applicable not only to peas but to all diploid individuals. Therefore, we will take as our example not peas but human beings. Figure 18.3 illustrates some differences between human beings who are known to be dominant or recessive.

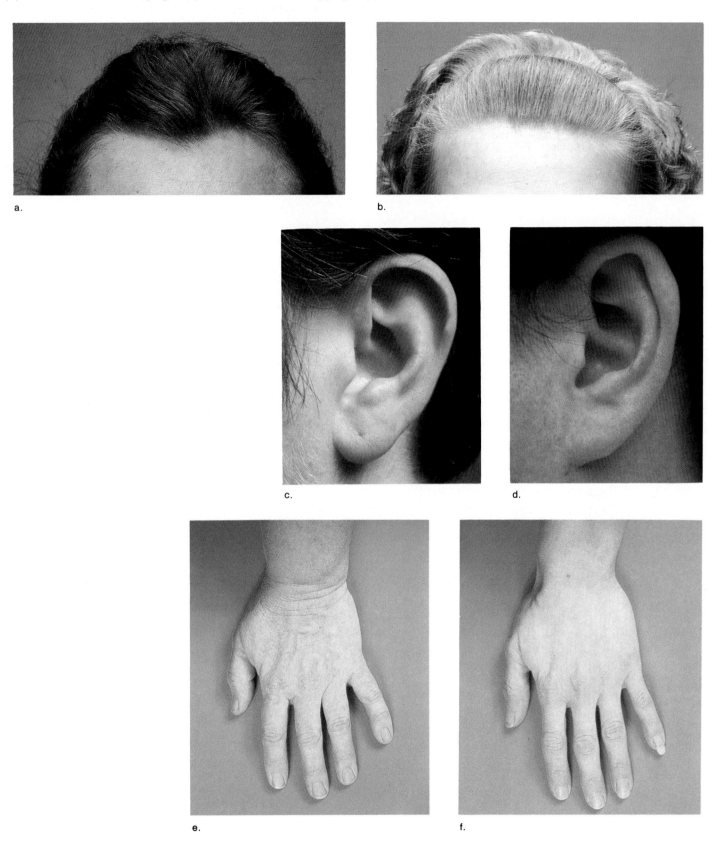

FIGURE 18.3 *Common inherited characteristics in human beings. Widow's peak (a) is dominant over (b) continuous hairline. Unattached earlobe (c) is dominant over (d) attached earlobe. Short fingers (e) are dominant over long fingers (f).*

a.

b.

c.

d.

e.

f.

TABLE 18.1 *GENOTYPE VERSUS PHENOTYPE*

GENOTYPE	GENOTYPE	PHENOTYPE
WW	Homozygous dominant	Widow's peak
Ww	Heterozygous (hybrid)	Widow's peak
ww	Homozygous recessive	Continuous hairline

In doing a problem concerning hairline, the *key* would be represented as

> *W* = Widow's peak (dominant allele)
>
> *w* = Continuous hairline (recessive allele)

The key simply tells us which letter of the alphabet to use for the gene in a particular problem and which allele is dominant, a capital letter signifying dominance.

Genotype and Phenotype

When we indicate the genes of a particular individual, two letters must be used for each trait mentioned. This is called the **genotype** of the individual. The genotype may be expressed not only by using letters but also by a short descriptive phrase, as table 18.1 shows. Thus the word **homozygous** means that the two members of the allelic pair in the zygote (*zygo*) are the same (*homo*); genotype *WW* is called *homozygous dominant* and *ww* is called *homozygous recessive*. The word **heterozygous** means that the members of the allelic pair are different (*hetero*); only *Ww* is heterozygous. Another term, **hybrid,** is sometimes used to mean heterozygous.

As table 18.1 also indicates, the word **phenotype** refers to the physical characteristics of the individual. What the individual actually looks like is the phenotype. (Also included in the phenotype are the microscopic and metabolic characteristics of the individual.) Notice that both homozygous dominant and heterozygous show the dominant phenotype.

Gamete Formation

Whereas the genotype has two alleles for each trait, the gametes have only one allele for each trait. This, of course, is related to the process of meiosis. The alleles are present on a homologous pair of chromosomes, and these chromosomes separate during meiosis (fig. 18.6). Therefore, the members of each allelic pair separate during meiosis, and there is only one allele for each trait in the gametes. When doing genetic problems it should be kept in mind that no two letters in a gamete may be the same. For this reason, *Ww* would represent a possible genotype, and the gametes for this individual could contain either a *W* or *w*. Therefore, the possible gametes for this individual are *W, w*—the comma indicating two possible gametes.

When doing genetics problems, the same alphabetic letter is used for both the dominant and recessive alleles; a capital letter indicates the dominant and a lowercase letter indicates the recessive. A homozygous dominant individual is indicated by two capital letters, and a homozygous recessive individual is indicated by two lowercase letters. The genotype of a heterozygous individual is indicated by a capital and a lowercase letter. Contrary to the individual, gametes have one letter of each type, either capital or lowercase as appropriate. All possible combinations of letters indicate all possible gametes.

Do Practice Problems 1 located on page 410.

Crosses

It is now possible for us to consider a particular cross. If a homozygous man with a widow's peak (fig. 18.3) marries a woman with a continuous hairline (fig. 18.3), what kind of hairline will their children have?

In solving the problem, we must indicate the genotype of each parent by using letters, determine what the gametes are, and what the genotypes of the children are after reproduction. In the format that follows, P_1 stands for the parental generation, and the letters in this row are the genotypes of the parents. The second row shows that each parent has only one type of gamete in regard to hairline, and therefore all the children (F_1 = first filial generation) will have a similar genotype, that is, heterozygous. Heterozygotes show the dominant characteristic, and so all the children will have a widow's peak.

P_1:	Widow's peak	×	Continuous hairline
	WW		*ww*
Gametes:	*W*		*w*
F_1:		Widow's peak	
		Ww	

These individuals are **monohybrids** because they are heterozygous for only one pair of alleles. If they marry someone else with the same genotype, what type of hairline will their children have?

P_1:	Widow's peak	×	Widow's peak
	Ww		*Ww*
Gametes:	*W, w*		*W, w*

In this problem, each parent has two possible types of gametes. In calculating F_1, it is assumed that either type of sperm has an equal chance to fertilize either type of egg. One way to assure that we have accounted for this is to use a Punnett square (p. 376).

When this is done (fig. 18.4), the results show a 3:1 phenotypic ratio; that is, three with widow's peak to one without. Such a ratio will actually be observed only if a large number of crosses of the same type take place and a large number of offspring result. Only then will all possible sperm

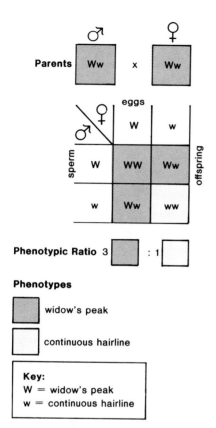

FIGURE 18.4 *Monohybrid cross. In this cross, the parents are heterozygous for widow's peak. The chances of any child having a continuous hairline are one out of four, or 25%.*

Parents | Ww | x | Ww

eggs

sperm / offspring

	W	w
W	WW	Ww
w	Ww	ww

Phenotypic Ratio 3 : 1

Phenotypes

□ widow's peak

□ continuous hairline

Key:
W = widow's peak
w = continuous hairline

have an equal chance to fertilize all possible eggs. It is obvious that we do not routinely observe hundreds of offspring from a single type cross in humans, and so it is customary to merely state that each child has three chances out of four to have a widow's peak, or one chance out of four to have a continuous hairline. It is important to realize that *chance has no memory;* for example, if two heterozygous parents have already had three children with a widow's peak and are expecting a fourth child, this child still has three chances out of four to have a widow's peak and only one chance out of four of not having one. Each individual child has the same chances.

Probability

Another way to calculate the possible results of a cross is to realize that the chance, *or probability of receiving a particular combination of alleles, is simply the product of the individual probabilities.* In the cross just considered,

$Ww \times Ww$

the offspring have an equal chance of receiving W or w from each parent. Therefore:

Probability of W = ½
Probability of w = ½

and

Probability of WW = ½ × ½ = ¼
Probability of Ww = ½ × ½ = ¼
Probability of wW = ½ × ½ = ¼
⅜ = Widow's peak
Probability of ww = ½ × ½ = ¼
¼ = Continuous hairline

Testcross

If an individual has the dominant phenotype, it is not possible to tell by inspection if the genotype is homozygous dominant or heterozygous. However, if the individual is crossed with a homozygous recessive, the results may indicate what the original genotype was. For example, figure 18.5 shows the different results if a man with a widow's peak is homozygous dominant, or if he is heterozygous, and married to a woman with a continuous hairline. (She must be homozygous recessive or she would not have a continuous hairline.) In the first case, the man can only sire children with widow's peaks, and in the second the chances are 2:2 or 1:1 that a child will or will not have a widow's peak. Thus, the cross of a possible heterozygote with an individual having the recessive phenotype gives the best chance of producing the recessive phenotype among the offspring. Therefore, this type of cross is called the **testcross.**

In doing an actual cross, it is assumed that all possible types of sperm fertilize all possible types of eggs. The results may be expressed as a probable phenotypic ratio; it is also possible to state the chances of an offspring showing a particular phenotype.

Do Practice Problems 2 located on page 410.

Inheritance of Multitraits

Two Traits (Unlinked)

Although it is possible to consider the inheritance of just one trait, actually each individual passes on to his or her offspring many genes for many traits. In order to arrive at a general understanding of multitrait inheritance, we will consider the inheritance of two traits. The same principles will apply to as many traits as we might wish to consider.

When Mendel performed two-trait crosses, he formulated his second law, the law of independent assortment.

Law of independent assortment: Pairs of factors separate independently of one another to form gametes, and therefore all possible combinations of factors may occur in the gametes.

Figure 18.6 illustrates that the law of segregation and the law of independent assortment hold because of the manner in which meiosis occurs. The law of segregation is dependent on the separation of members of homologous pairs of chromosomes; and the law of independent assortment is dependent on the random arrangement of homologous pairs with respect to one another during metaphase I prior to the separation process.

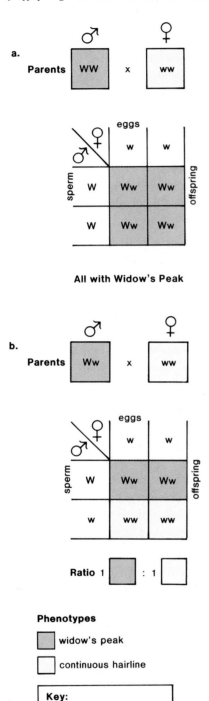

All with Widow's Peak

Ratio 1 : 1

Phenotypes

widow's peak

continuous hairline

Key:
W = widow's peak
w = continuous hairline

Crosses

When doing a two-trait cross, we realize that the genotypes of the parents require four letters because there is an allelic pair for each trait. Second, the gametes of the parents contain one letter of each kind in every possible combination, as predicted by Mendel's law of independent assortment. Finally, in order to produce the probable ratio of phenotypes among the offspring, all possible matings are presumed to occur.

To give an example, let us cross a person homozygous for widow's peak and short fingers with a person who has a continuous hairline and long fingers. The key for such a cross is

W = Widow's peak	S = Short fingers
w = Continuous hairline	s = Long fingers

P_1:   Widow's peak,   × Continuous hairline,
            Short fingers   Long fingers
                 WWSS   *wwss*

Gametes: *WS*   *ws*

F_1:   Widow's peak,
                         Short fingers
                            WwSs

In this particular cross, only one type of gamete is possible for each parent; therefore, all of the F_1 generation will have the same genotype (*WwSs*) and the same phenotype (widow's peak with short fingers). This genotype is called a **dihybrid** because the individual is heterozygous in two regards: hairline and fingers.

When a dihybrid reproduces with a dihybrid, each parent has four possible types of gametes:

F_1:	*WwSs*	×	*WwSs*
Gametes:	*WS*		*WS*
	Ws		*Ws*
	wS		*wS*
	ws		*ws*

The Punnett square (fig. 18.7) for such a cross shows the expected genotypes among sixteen offspring if all possible sperm fertilize all possible eggs. An inspection of the various genotypes in the square shows that among the offspring, *nine* will have a widow's peak and short fingers, *three* will have a widow's peak and long fingers, *three* will have a continuous hairline and short fingers, and *one* will have a continuous hairline and long fingers. This is called a 9:3:3:1 phenotype ratio, and this ratio always results when a dihybrid is mated with a dihybrid and simple dominance is present.

Probability

We can use the previous ratio to predict the chances of each child receiving a certain phenotype. For example, the possibility of getting the two dominant phenotypes together is nine out of sixteen (9 + 3 + 3 + 1 = 16) and that of getting the two recessive phenotypes together is one out of sixteen.

Key:

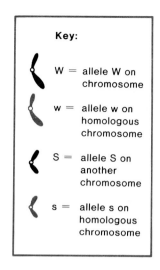

W = allele W on chromosome

w = allele w on homologous chromosome

S = allele S on another chromosome

s = allele s on homologous chromosome

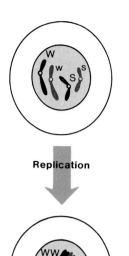

Replication

FIGURE 18.6 *Gametogenesis includes meiosis. Notice that just as chromosomes occur in pairs, so do genes; just as chromosome pairs separate during meiosis, so do pairs of genes; there are no restrictions as to the manner of separation; just as each gamete at the completion of meiosis carries one-half of the total number of chromosomes, so each gamete carries one-half of the total number of genes.*

Tetrads are present in mother cell.

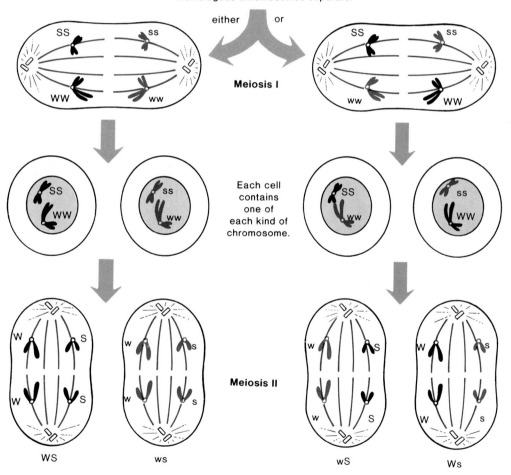

Homologous chromosomes separate.

either   or

Meiosis I

Each cell contains one of each kind of chromosome.

Meiosis II

Four different gametes are possible.

FIGURE 18.7 *Dihybrid cross. A dihybrid results when an individual homozygous dominant in two regards reproduces with an individual homozygous recessive in two regards. When a dihybrid reproduces with a dihybrid, there are four possible phenotypes among the offspring and the phenotypic ratio is 9:3:3:1 as indicated.*

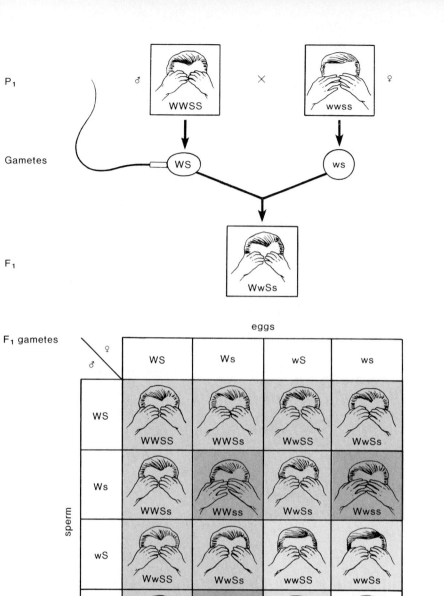

P₁

Gametes

F₁

F₁ gametes

eggs

	WS	Ws	wS	ws
WS	WWSS	WWSs	WwSS	WwSs
Ws	WWSs	WWss	WwSs	Wwss
wS	WwSS	WwSs	wwSS	wwSs
ws	WwSs	Wwss	wwSs	wwss

sperm

offspring

Phenotypic Ratio   9   :3   :3   :1

Phenotypes

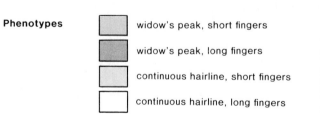

widow's peak, short fingers

widow's peak, long fingers

continuous hairline, short fingers

continuous hairline, long fingers

Key:
W = widow's peak
w = continuous hairline
S = short fingers
s = long fingers

We can also calculate the chance, or probability, of these various phenotypes occurring by knowing that the *probability of combinations of independent events is the product of the probabilities of each of the events:*

Probability of widow's peak = ¾
Probability of short fingers = ¾
Probability of continuous hairline = ¼
Probability of long fingers = ¼

Therefore:

Probability of widow's peak and short fingers = ¾ × ¾ = ⁹⁄₁₆

Probability of widow's peak and long fingers = ¾ × ¼ = ³⁄₁₆

Probability of continuous hairline and short fingers = ¼ × ¾ = ³⁄₁₆

Probability of continuous hairline and long fingers = ¼ × ¼ = ¹⁄₁₆

Testcross

An individual who shows the dominant traits can be tested for the dihybrid genotype by a mating with the recessive in both traits.

P₁: *WwSs*   ×   *wwss*

Gametes: *WS*   *ws*
                  Ws
                  wS
                  ws

The Punnett square (fig. 18.8) shows that the resulting ratio is 1 widow's peak with short fingers : 1 widow's peak with long fingers : 1 continuous hairline with short fingers : 1 continuous hairline with long fingers, or 1:1:1:1.

Table 18.2 lists all of the crosses we have studied thus far, which show a frequently observed ratio. When these types of crosses are done, these ratios are observed.

Do Practice Problems 3 located on page 410.

Genetic Disorders

Birth defects can be environmentally induced but many are genetic in origin. Some of these disorders are controlled by a single set of alleles, and therefore they are inherited in a simple Mendelian manner.

When studying human genetic disorders, biologists often construct **pedigree charts** that show the pattern of inheritance of a characteristic within a family. Let's contrast two possible patterns of inheritance in order to show how it is possible to determine whether the characteristic is an autosomal dominant or an autosomal recessive characteristic.

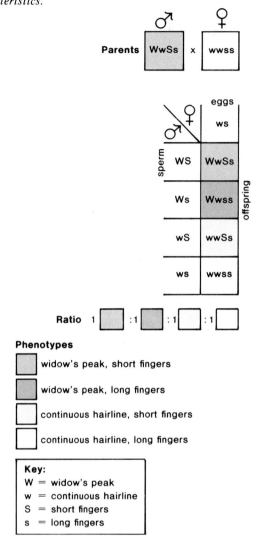

FIGURE 18.8 *Testcross. In this example it is impossible to tell by inspection if the male parent is homozygous dominant or if he is heterozygous for both traits. However, reproduction with a female who is recessive for both traits is likely to show which he is. If he is heterozygous, there is a 25% chance that the offspring will show both recessive characteristics and a 50% chance that they will show one or the other of the recessive characteristics.*

Phenotypes

widow's peak, short fingers

widow's peak, long fingers

continuous hairline, short fingers

continuous hairline, long fingers

Key:
W = widow's peak
w = continuous hairline
S = short fingers
s = long fingers

TABLE 18.2 PHENOTYPIC RATIOS OF COMMON CROSSES

Cross	Ratio
Monohybrid × monohybrid	3:1 (dominant to recessive)
Monohybrid × recessive[a]	1:1 (dominant to recessive)
Dihybrid × dihybrid	9:3:3:1 (9 both dominant, 3 other dominant, 3 other dominant, 1 both recessive)
Dihybrid × recessive[a]	1:1:1:1 (all possible combinations in equal number)

[a]Called a testcross because it can be used to test if the individual showing the dominant gene is homozygous or heterozygous. For a definition of all terms, see the glossary.

pattern I   pattern II

In both patterns, males are designated by squares and females by circles. Shaded circles and squares are afflicted individuals. A horizontal line between a square and a circle represents a marriage. A vertical line going downward leads, in these patterns, to a single child. (If there is more than one child, the entries are placed off a horizontal line.) Which pattern of inheritance would be expected for an autosomal dominant and which for an autosomal recessive characteristic?

In pattern I, the child is afflicted but neither parent is afflicted; this can only happen if the characteristic is recessively inherited. (See figure 18.9 for other ways to recognize a recessive pattern of inheritance.) In this figure heterozygotes are lightly shaded because they are carriers. Carriers are individuals who do not display the symptoms of the disease but who are capable of passing on an allele for the gene. What are the chances that any offspring from this union will be afflicted? Since the parents are monohybrids, the chances are one in four, or 25% (table 18.2).

In pattern II, the child is afflicted but one of the parents is also afflicted. When a characteristic is dominant, an afflicted child must have at least one afflicted parent. So of the two patterns, this one would have to be the dominant pattern of inheritance. (See figure 18.10 for other ways to recognize a dominant pattern of inheritance.) What are the chances that any offspring from this union will be afflicted? Since this is a monohybrid × recessive cross, the chances are 50% (table 18.2).

Pattern I is the usual one seen for a recessive genetic disorder, and pattern II is the usual one seen for a dominant genetic disorder.

Recessive Inherited Disorders
Recessive inherited genetic disorders are sometimes more prevalent among members of a particular ethnic group. The homozygous recessive genotype is more likely to occur among members of a group who tend to marry one another.

Cystic Fibrosis   Cystic fibrosis is the most common lethal recessive inherited genetic disease among Caucasians in the United States. About one in twenty Caucasians is a carrier, and about one in 2,000 children born to this group has inherited this disorder. In these children the mucus in the lungs and digestive tract is particularly thick and viscous. In the digestive tract, the thick mucus impedes the secretion of pancreatic juices and food cannot be properly digested; large, frequent, and foul-smelling stools occur. A few individuals have been known to survive childhood, but most die from recurrent lung infections.

In the past few years much progress has been made in understanding cystic fibrosis. First of all it was discovered that chloride ions fail to pass through cell membrane channels in these patients. Ordinarily after chloride has passed through the membrane, water follows; it is believed to be the lack of water in the lungs that causes the mucus to be so thick. Secondly, it is now known that the gene for cystic fibrosis is located on the number-7 chromosome. Reseachers are hopeful that they will soon have the gene isolated and know how it functions.

Lysosomal Storage Diseases   Lysosomal storage diseases are most often seen in Jewish people of Central and Eastern Europe. About 90% of U.S. Jews are of this ancestry. The two genetic disorders most often seen are *Tay-Sachs disease* and *Gaucher disease*. For both, it is estimated that one out of twenty-five Jews is a carrier and one out of every 2,500 infants born to them has the disorder. The reading on page 398 discusses both of these diseases.

Phenylketonuria (PKU)   *Phenylketonuria (PKU)* occurs in one out of 20,000 births and so is not as frequent as the disorders so far discussed. When it does occur, the parents are very often close relatives. Affected individuals lack an enzyme that is needed for the normal metabolism of the amino acid phenylalanine, and the abnormal breakdown product, a phenylketone, accumulates in the urine. All newborns are routinely tested for PKU, and if they lack the necessary enzyme, they are placed on a diet low in phenylalanine. This diet must be continued until the brain is fully developed or severe mental retardation develops. Current research suggests that a moderate restriction of phenylalanine in the diet should be continued indefinitely. If a woman who is homozygous recessive for PKU wishes to have children, she should resume the diet prior to conception because the high level of phenylalanine in her system may adversely affect development of the fetus.

HUMAN ISSUE

People with serious genetic disorders increase our tax burden and in general cost society a great deal of money. Do you think society should try to prevent the birth of children with such disorders? If you think prevention is justified, should we go to the extreme of sterilizing carriers, or should we simply rely on self-imposed control? How and who would determine which genetic disorders are serious enough to warrant preventive measures?

Dominant Inherited Disorders
At the beginning of this chapter, we discussed *Marfan syndrome,* which is recognized by skeletal, eye, and cardiovascular defects. All of these are due to an inability to produce normal connective tissue. Two other dominant disorders are discussed following.

FIGURE 18.9 *Pedigree chart for recessive characteristic. Affected individuals are shaded a darker color than carriers.*

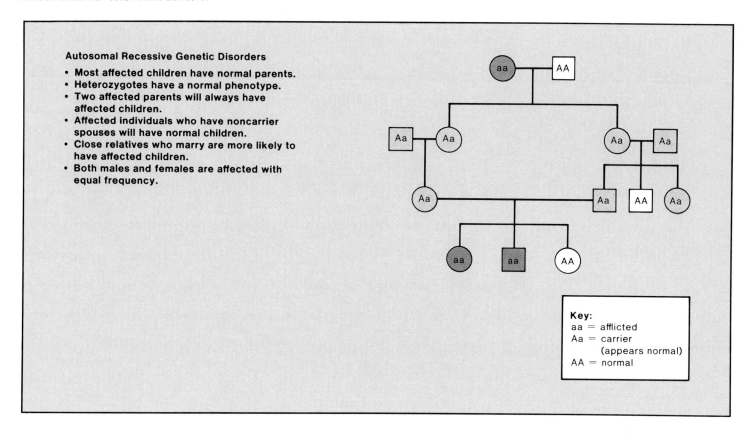

Autosomal Recessive Genetic Disorders

- Most affected children have normal parents.
- Heterozygotes have a normal phenotype.
- Two affected parents will always have affected children.
- Affected individuals who have noncarrier spouses will have normal children.
- Close relatives who marry are more likely to have affected children.
- Both males and females are affected with equal frequency.

Key:
aa = afflicted
Aa = carrier
(appears normal)
AA = normal

FIGURE 18.10 *Pedigree chart for an autosomal dominant characteristic. Affected individuals are shaded.*

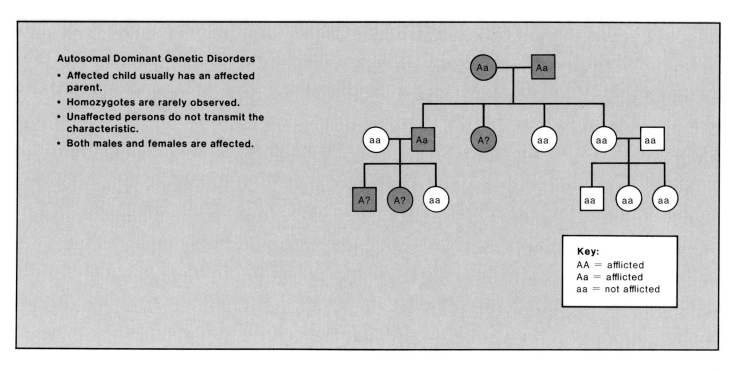

Autosomal Dominant Genetic Disorders

- Affected child usually has an affected parent.
- Homozygotes are rarely observed.
- Unaffected persons do not transmit the characteristic.
- Both males and females are affected.

Key:
AA = afflicted
Aa = afflicted
aa = not afflicted

LYSOSOMAL STORAGE GENETIC DISORDERS

Glycosphingolipids are complex lipids found in the cell membrane. Because a cell is constantly renewing its parts, on occasion these molecules must be broken down inside lysosomes. In most individuals lysosomes contain all the necessary enzymes to break down the different types of lipids, including glycosphingolipids, but in individuals with Tay-Sachs disease or Gaucher disease the lysosomes are missing an enzyme. Therefore, as one lysosome becomes filled with the nondigestible lipid, other lysosomes form that also become filled. Eventually the cell becomes crowded with ineffective lysosomes and cannot carry on its normal functions (see figure). In individuals with Tay-Sachs disease the result is always early death, but there are varying degrees of severity in Gaucher disease, and some victims are affected limitedly.

Tay-Sachs Disease

Tay-Sachs is the best-known genetic disease of the newborn among U.S. Jewish people, most of whom are of Central and Eastern European descent. At first it is not apparent that a baby has Tay-Sachs. However, the child's development begins to slow down between the fourth and eighth months as neurological impairment and psychomotor difficulties become apparent. Ophthalmologic examination reveals a characteristic red spot and a yellowish accumulation in the region of the fovea of the retina. This is due to an accumulation of yellowish glycosphingolipids in the cells surrounding the fovea of the eye. The child gradually becomes blind and helpless, develops uncontrollable seizures, and eventually becomes paralyzed. There is no treatment or cure for Tay-Sachs disease and most victims die by the age of 3 or 4 years.

Tay-Sachs disease results because of a lack of the enzyme β-hexosaminidase A (Hex A) and the subsequent storage of its substrate, a glycosphingolipid, in lysosomes. Although more and more lysosomes build up in many body cells, the primary sites of storage are the cells of the nervous system, which accounts for the onset and progressive deterioration of psychomotor functions.

There is a test to detect carriers of Tay-Sachs. The test uses a sample of serum, white blood cells, or tears to determine whether Hex A activity is present. Affected individuals have no detectable Hex A activity. Carriers have about half the level of Hex A activity found in normal individuals. Prenatal diagnosis is also possible following either amniocentesis or chorionic villi sampling (p. 378).

Gaucher Disease

Gaucher disease is just as prevalent as Tay-Sachs among the U.S. Jewish population. Gaucher disease is due to the deficiency of another lysosomal enzyme, β-glucosidase, and the buildup of its substrate, another type of glycosphingolipid, in lysosomes, particularly within macrophages, the body's scavengers (p. 134).

There are three major subtypes of Gaucher disease. In type 1, the most common type, the onset and severity of the disorder is quite varied but symptoms usually appear during childhood or adolescence. The spleen (and sometimes the liver), where macrophages are prevalent and stuffed full of glycosphingolipids, enlarges greatly. A normal spleen weighs about ½ pound, but in Gaucher patients it may weigh as much as 30 pounds. This leads

---

Neurofibromatosis (NF) Neurofibromatosis (NF), sometimes called von Recklinghausen disease, is one of the most common genetic disorders, afflicting roughly one in 3,000 people, including an estimated 100,000 in the United States. It is seen equally in every racial and ethnic group throughout the world.

At birth, or perhaps later, the individual may have six or more large tan spots on the skin. Such spots may increase in size and number and get darker. Small benign tumors (lumps) called neurofibromas may occur under the skin or deeper. Neurofibromas are made up of nerve cells and other cell types.

In most cases, symptoms are mild and patients live a normal life; however, in some cases the effects are severe. Skeletal deformaties including a large head are seen; eye and ear tumors can lead to blindness and hearing loss. Many children with NF have learning disabilities and may be overactive.

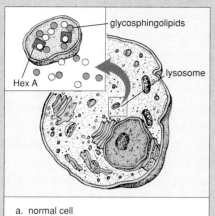

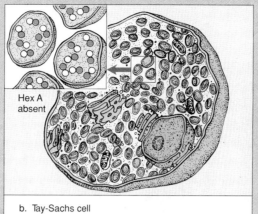

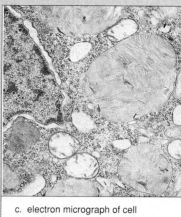

a. normal cell

b. Tay-Sachs cell

c. electron micrograph of cell

to cardiovascular complications like anemia and blood-clotting problems. Due to deterioration of the bones and joints, patients suffer from bone pain, fractures, and other orthopedic difficulties that may confine them to a wheelchair. There are no neurological abnormalities in type 1 Gaucher disease, and victims usually die from pneumonia or blood-related complications.

Type 2 Gaucher disease, a rarer form, makes its appearance during infancy, and there is neurological involvement. Most victims lose any motor skills that have developed, like rolling over, sitting up, and holding up the head. They die of breathing difficulties by about 2 years of age.

Type 3 Gaucher disease is like type 2 except that it is usually diagnosed later in childhood and the neurological impairment is not as great. Still, there may be seizures and difficulty in coordinating body movements. These individuals usually die by 20 to 30 years of age.

At present it is possible to detect carriers of Gaucher disease by determining the β-glucosidase enzyme level. Carriers have an enzyme level that

is midway between those of affected individuals and noncarrier individuals. Prenatal testing, amniocentesis or chorionic villi sampling, is also a possible means of detection.

Presently, researchers are developing a means to distinguish between the three types of Gaucher disorders, particularly in infants, before the symptoms appear. Thus far, they have found a genetic marker (see p. 408 and fig. 18.22) that accompanies Gaucher type 2 and type 3 about 80–85% of the time but does not accompany Gaucher type 1. One of the investigators, Edward Ginns of the National Institute of Mental Health in Los Angeles, says, "We can now offer carrier screening and prenatal diagnosis for a neurological disorder." The researchers have also sequenced the order of the nucleotides in a normal gene and know that it is located on the number-1 chromosome. They have begun to sequence the order of the nucleotides in Gaucher patients. Soon, then, they should know the specific metabolic differences in the cells of patients with the three different types of Gaucher disorders.

Tay-Sachs disease. a. *When Hex A is present, glycosphingolipids are broken down.* b. *When Hex A is absent, these lipids accumulate in lysosomes and lysosomes accumulate in the cell.* c. *Electron micrograph of cell crowded with lysosomes.*

Only recently, researchers have been able to determine that the NF gene is located on the number-17 chromosome. They believe the gene to be rather large because of its varying effects and because about half of all NF cases are the result of new mutations in one of the parents.

Huntington Disease   As many as one in 10,000 persons in the United States have *Huntington disease (HD)*, a neurological disorder that affects specific regions of the brain.

Minor disturbances in balance and coordination don't appear until middle age, but then the symptoms become progressively worse until the victim goes mad before death occurs.

Much has been learned about Huntington disease. The gene is located on the number-4 chromosome and there is a test (of the type described in figure 18.22, p. 408) to determine if the dominant gene has been inherited. Since treatment is not available, however, few may want to have this information.

a.

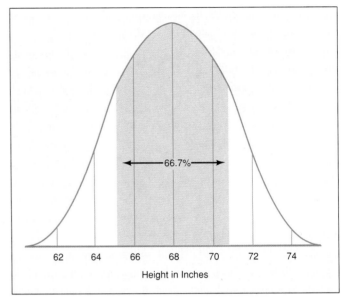

b.

Research is being conducted, though, to determine the underlying cause of the disorder. It is known that the brain of a Huntington disease victim produces more than the usual amount of quinolinic acid, an excitotoxin that can overstimulate neurons that have glutamic receptors. This is believed to lead to the death of these neurons and the subsequent symptoms of Huntington disease. Researchers are looking for a chemical that can block reception of quinolinic acid at the postsynaptic membrane or for a drug that can inhibit quinolinic acid synthesis.

Some of the best-known genetic disorders in humans are inherited in a simple Mendelian manner. Pedigree charts show the pattern of inheritance in a particular family.

BEYOND MENDEL'S LAWS

Certain traits, such as those just studied, follow the rules of simple Mendelian inheritance. There are, however, others that do not follow these rules. Examples are given following.

Polygenic Inheritance

Two or more sets of alleles may affect the same trait, sometimes in an additive fashion. Polygenic inheritance can cause the distribution of human characteristics according to a bell-shaped curve, with most individuals exhibiting the average phenotype (fig. 18.11). The more genes that control the trait, the more continuous the distribution will be.

Skin Color

Just how many pairs of alleles control skin color is not known, but a range in colors can be explained on the basis of two pairs. When a black person has children by a white person, the children are mulatto, but two mulattoes can produce children who range in skin color from black to white. If we assume that two pairs of alleles control skin color, then:

Black = *AABB*
Dark = *AABb* or *aABB*
Mulatto = *AaBb* or *AAbb* or *aaBB*
Light = *Aabb* or *aabB*
White = *aabb*

If a mulatto reproduces with a white person, the very darkest individual possible is a mulatto, but a white child is also possible (fig. 18.12).

Genetic Disorders

A number of serious genetic disorders, such as cleft lip or palate, clubfoot, congenital dislocation of the hip, and certain spinal conditions, are traditionally believed to be controlled by a combination of genes. This belief is being challenged by researchers who studied the inheritance of cleft palate in a large family in Iceland. These researchers reported the finding of a cleft palate gene on the X chromosome.

If a couple is concerned about the birth of a child with a neural tube defect, an analysis of the amniotic fluid following amniocentesis (p. 382) can reveal if there has been a leakage of neural tube substance into the fluid. If such a leakage has taken place, it is possible that the unborn child is not developing normally.

FIGURE 18.12 Inheritance of skin color. This white husband (aabb) and his mulatto wife (AaBb) had fraternal twins, one of whom was white and one of whom was mulatto.

Multiple Alleles

ABO Blood Type

Three alleles for the same gene control the inheritance of ABO blood types. These alleles determine the presence or absence of antigens on the red blood cells. Therefore, I, standing for immunogen (antigen), is used to signify the gene, and a superscript letter is used to signify the particular allele:

I^A = type A antigen on red blood cells
I^B = type B antigen on red blood cells
i^o = no antigens on the red blood cells

Each person has only two of the three possible alleles, and both I^A and I^B are dominant over i^o. Therefore, as table 18.3 shows, there are two possible genotypes for type A blood and two possible genotypes for type B blood. On the other hand, I^A and I^B are fully expressed in the presence of the other. Therefore, if a person inherits one of each of these alleles, that person will have type AB blood. Type O blood can only result from the inheritance of two i^o alleles.

An examination of possible matings between different blood types sometimes produces surprising results; for example,

P_1: $I^A i^o \times I^B i^o$
F_1: $I^A I^B$, $i^o i^o$, $I^A i^o$, $I^B i^o$

Thus, from this particular mating every possible phenotype (AB, O, A, B blood type) is possible.

Blood typing can sometimes aid in paternity suits. However, a blood test of a supposed father can only suggest that he *might* be the father, not that he definitely *is* the father. For example, it is possible, but not definite, that a man with blood type A (having genotype $I^A i^o$) is the father of a child

with blood type O. On the other hand, a blood test can sometimes definitely prove that a man is not the father. For example, a man with blood type AB could not possibly be the father of a child with blood type O. Therefore, blood tests may legally be used only to exclude a man from possible paternity.

Rh Blood Factor

It might be noted here that the blood factor called Rh is inherited separately from A, B, AB, or O type blood. In each instance it is possible to be Rh⁺ or Rh⁻, meaning in the first case that an Rh factor is present on the red blood cells and in the second that an Rh factor is not present. It may be assumed that Rh is controlled by a single allelic pair in which simple dominance prevails: Rh⁺ is dominant over Rh⁻. Complications arise when an Rh⁻ woman marries an Rh⁺ man and the child in the womb is Rh⁺. With the birth of the first child of this phenotype, the mother may begin to build up antibodies to the factor and in later pregnancies these antibodies may cross the placenta to destroy the baby's blood cells (p. 136).

TABLE 18.3 BLOOD GROUPS

PHENOTYPE	GENOTYPE
A	$I^A I^A$, $I^A i^o$
B	$I^B I^B$, $I^B i^o$
AB	$I^A I^B$
O	$i^o i^o$

I = immunogen gene

Incomplete Dominance and Codominance

Incomplete Dominance

In incomplete dominance, neither member of an allelic pair is dominant over the other and the phenotype is intermediate between the two. For example, when a curly-haired Caucasian person reproduces with a straight-haired Caucasian person, their children will have wavy hair. And two wavy-haired individuals can produce all possible phenotypes: straight hair, curly hair, and wavy hair. To signify that neither allele is dominant, one allele is designated as *H* and the other as *H'*, as is done in figure 18.13.

Codominance

In codominance, each member of an allelic pair is dominant, and the phenotype exhibits both characteristics. For example, an individual with the genotype $I^A I^B$ has the blood type AB.

Sickle Cell Anemia Most cases of *sickle cell anemia* occur among blacks and Hispanics of Caribbean ancestry. About one in every 400–600 blacks and one in every 1,000–1,500 Hispanics inherits two affected alleles and has the disorder. About one out of every ten U.S. blacks has inherited one affected allele, and is a carrier. There is a test that can be done to detect carriers. Also, prenatal testing is possible.

Sickle cell anemia occurs when the individual has the genotype $Hb^S Hb^S$. In these individuals, the red blood cells are sickle shaped (fig. 18.14b) because the abnormal hemoglobin molecule is less soluble than the normal hemoglobin, Hb^A. Sickle-shaped cells have a limited ability to transport oxygen. Inheritance involves codominance in the following manner: Individuals with the genotype $Hb^A Hb^A$ are normal; those with $Hb^S Hb^S$ have sickle cell anemia; and those with $Hb^A Hb^S$ have *sickle cell trait,* a condition in which the cells are sometimes sickle shaped, as described in the paragraphs that follow. Two individuals with sickle cell trait can produce children with all three phenotypes, as indicated in figure 18.14a.

Sickle cell anemia is prevalent among members of the black race because the structure and function of the cells seems to give protection against the malaria parasite, which utilizes red blood cells during its life cycle. Although infants with sickle cell anemia often die, those with the trait are protected from malaria, especially during ages 2 to 4. This means that in Africa these children survive and grow up to reproduce and pass on the allele to their offspring. In malaria-infected regions of Africa, as many as 40% of individuals have the allele. In the United States, about 10% of the black population carry it.

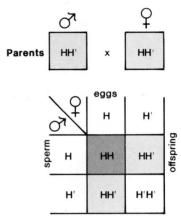

FIGURE 18.13 *Incomplete dominance. Among Caucasians, neither straight nor curly hair is dominant. When two wavy-haired individuals reproduce, the offspring has a 25% chance of having either straight or curly hair and a 50% chance of having wavy hair, the intermediate phenotype.*

The red blood cells in persons with sickle cell anemia cannot easily pass through small blood vessels. The sickle-shaped cells either break down or they clog blood vessels. Thus the individual suffers from poor circulation, anemia, and sometimes internal hemorrhaging. Jaundice, episodic pain of the abdomen and joints, poor resistance to infection, and damage to internal organs are all symptoms of sickle cell anemia. Few patients live beyond age 40.

Persons with sickle cell trait do not usually have any difficulties unless they undergo dehydration or mild oxygen deprivation. At such time, the cells become sickle shaped, clogging their blood vessels and leading to pain and even death. A study of the occurrence of sudden deaths during army basic training showed that a black with sickle cell trait was forty times more likely to die compared to normal recruits. This has caused the Army Medical Corps to advise drill instructors to train recruits more gradually, to give them enough to drink, and to make allowances for heat and humidity when planning their workouts.

There are many exceptions to Mendel's laws. These include polygenic inheritance, multiple alleles, incomplete dominance, and codominance.

Do Practice Problems 4 located on page 410.

FIGURE 18.14 a. *Inheritance of sickle cell anemia. In this example, both parents have the sickle cell trait. Therefore, each child has a 25% chance of having sickle cell anemia or of being perfectly normal and a 50% chance of having the sickle cell trait. b. Sickled cells. Individuals with sickle cell anemia have sickled red blood cells that tend to clump as illustrated here.*

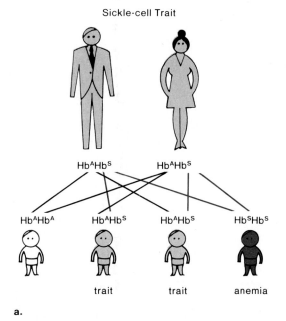

Sickle-cell Trait

$Hb^A Hb^S$ $Hb^A Hb^S$

$Hb^A Hb^A$ $Hb^A Hb^S$ $Hb^A Hb^S$ $Hb^S Hb^S$

trait   trait   anemia

a.

b.

SEX-LINKED INHERITANCE

The genes that determine the development of the sexual organs are on the sex chromosomes (p. 365). Even so, most of the genes on the sex chromosomes have nothing to do with sexual development and instead are concerned with other body traits. These genes are said to be **sex linked** because they are on the sex chromosomes. A few sex-linked genes are on the Y chromosome, but the most important ones discovered so far are only on the much larger X chromosome.

X-Linked Genes

X-linked genes have alleles on the X chromosome but determine body traits unrelated to sex. Since there are no alleles for these genes on the Y chromosome, any recessive allele present on the X chromosome in males will be expressed. As our first example of this type of inheritance, we will consider color blindness.

Color Blindness

In human beings there are three genes involved in distinguishing color because there are three different types of cones, the receptors for color vision (p. 256). Two of these genes are X-linked; one affects the green-sensitive cones, whereas the other affects the red-sensitive cones. About 6% of men in the United States are color blind due to a mutation involving green perception, and about 2% are color blind due to a mutation involving red perception. In our example, we will consider only the most prevalent type of *color blindness.*

When doing an X-linked genetics problem, the allele on the X chromosome appears as a letter attached to the X chromosome. Therefore, the key for color blindness is

X^B = normal vision
X^b = color blindness

The possible genotypes in both males and females are

$X^B X^B$ = a female with normal color vision
$X^B X^b$ = a carrier female with normal color vision
$X^b X^b$ = a female who is color blind
$X^B Y$ = a male with normal vision
$X^b Y$ = a male who is color blind

Note that the second genotype is a carrier female because although a female with this genotype appears normal, she is capable of passing on an allele for color blindness. Color-blind females are rare because they must receive the allele from both parents, but color-blind males are more common since they need only one recessive allele in order to be color blind. The allele for color blindness had to have been inherited from their mother because it is on the X chromosome; males only inherit the Y chromosome from their father.

Example Cross   If a heterozygous woman is married to a man with normal vision, what are their chances of having a color-blind daughter? a color-blind son?

Parents: $X^B X^b \times X^B Y$

Inspection indicates that all daughters will have normal color vision because they will all receive an X^B combination from their father. The sons, however, have a 50:50 chance of being color blind, depending on whether they receive an X^B or X^b from their mother. The inheritance of a Y from their father cannot offset the inheritance of an X^b from their mother. Figure 18.15 illustrates the use of the Punnett square in doing sex-linked problems.

FIGURE 18.15 *Cross involving X-linked genes. The male parent is normal, but the female parent is a carrier; an allele for color blindness is located on one of her chromosomes. Therefore, each son stands a fifty-fifty chance of being color blind.*

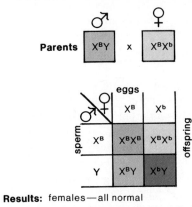

Parents: X^BY x X^BX^b

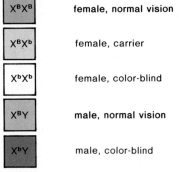

Results: females—all normal
males—1 normal : 1 color-blind

Genotypes	Phenotypes
X^BX^B	female, normal vision
X^BX^b	female, carrier
X^bX^b	female, color-blind
X^BY	male, normal vision
X^bY	male, color-blind

Key:
X^B = normal vision
X^b = color-blind

Other Recessive X-Linked Genetic Disorders

Some of the ways it is possible to recognize X-linked traits are given in figure 18.16.

Hemophilia   There are about 100,000 hemophiliacs in the United States. The most common type of hemophilia is hemophilia A, due to the absence, or minimal presence, of a particular clotting factor called Factor VIII. *Hemophilia* is called the bleeder's disease because the afflicted person's blood is unable to clot. Although hemophiliacs do bleed externally after an injury, they also suffer from internal bleeding, particularly around joints. Hemorrhages can be checked with transfusions of fresh blood (or plasma) or concentrates of the clotting protein. Unfortunately, some hemophiliacs have contracted AIDS after using whole blood, but this cannot occur if they use a purified form of the concentrate.

At the turn of the century, hemophilia was prevalent among the royal families of Europe. All of the afflicted males could trace their ancestry to Queen Victoria of England (fig. 18.17). Since none of Queen Victoria's forebearers or relatives was afflicted, it seems that the gene she carried arose by mutation either in Victoria or in one of her parents. Her carrier daughters, Alice and Beatrice, introduced the gene into the ruling houses of Russia and Spain. Alexis, the last heir to the Russian throne before the Russian Revolution, was a hemophiliac. The present British royal family has no hemophiliacs because Victoria's eldest son, King Edward VII, did not receive the gene and therefore could not pass it on to any of his descendents.

Muscular Dystrophy   Muscular dystrophy, as the name implies, is characterized by a wasting away of the muscles. The most common form, *Duchenne type*, is X-linked and occurs in about one out of every 25,000 male births. Symptoms such as waddling gait, toe walking, frequent falls, and difficulty in rising may appear as soon as the child starts to walk. Muscle weakness intensifies until the individual is confined to a wheelchair. Death usually occurs during the teenage years.

Recently the gene for muscular dystrophy was isolated, and it was discovered that the absence of a protein, now called dystrophin, is the cause of the disorder. Much investigative work determined that dystrophin is involved in the release of calcium from the calcium storage sacs (p. 239) in muscle fibers. The lack of dystrophin causes a leakage of calcium into the cell, which promotes the action of an enzyme that dissolves muscle fibers. When the body attempts to repair the tissue, the formation of fibrous tissue occurs, and this cuts off the blood supply so that more and more cells die.

A test is now available to detect carriers for Duchenne muscular dystrophy.

Most sex-linked genes are carried on the X chromosome and the Y is blank. There are four ways, noted in figure 18.16, to recognize X-linked inheritance, which also can be used to help solve X-linked genetics problems.

Do Practice Problems 5 located on page 410.

Sex-Influenced Traits

Not all traits we associate with the sex of the individual are due to sex-linked genes. Some are simply sex-influenced traits. Sex-influenced traits are characteristics that often appear in one sex but only rarely appear in the other. It's believed that these traits are governed by genes that are turned on or off by hormones. For example, the secondary sex characteristics, such as the beard of a male and the breasts of a female, probably are controlled by hormone balance.

FIGURE 18.16 *Pedigree chart for an X-linked recessive characteristic in which afflicted individuals are shaded more heavily than carriers.*

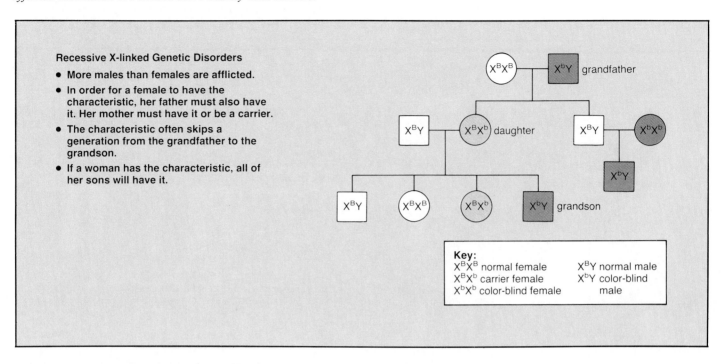

Recessive X-linked Genetic Disorders
- More males than females are afflicted.
- In order for a female to have the characteristic, her father must also have it. Her mother must have it or be a carrier.
- The characteristic often skips a generation from the grandfather to the grandson.
- If a woman has the characteristic, all of her sons will have it.

Key:
$X^B X^B$ normal female $X^B Y$ normal male
$X^B X^b$ carrier female $X^b Y$ color-blind
$X^b X^b$ color-blind female   male

FIGURE 18.17 *A simplified pedigree showing the X-linked inheritance of hemophilia in European royal families. Because Queen Victoria was a carrier, each of her sons had a 50% chance of having the disease and each of her daughters had a 50% chance of being a carrier. This pedigree shows only the affected individuals. Many others are unaffected, such as the members of the present British royal family.*

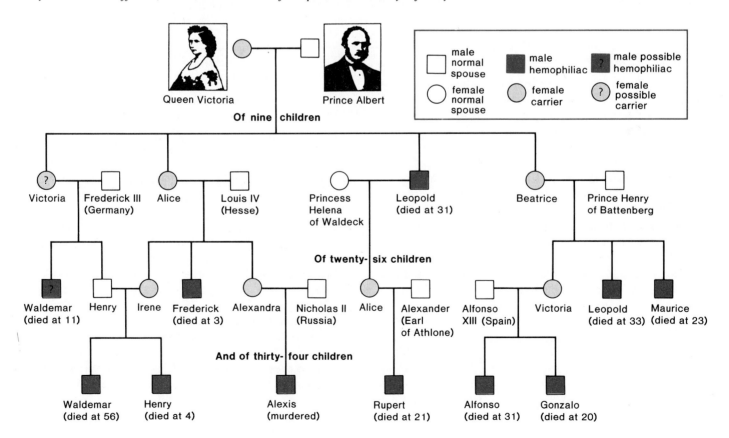

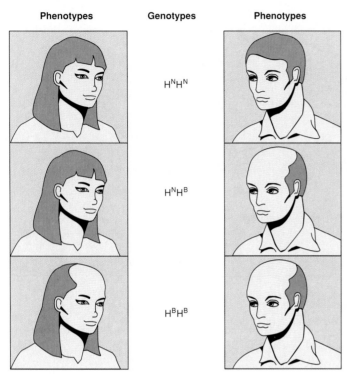

Phenotypes	Genotypes	Phenotypes
	$H^N H^N$	
	$H^N H^B$	
	$H^B H^B$	

H^N- normal hair growth

H^B- pattern baldness

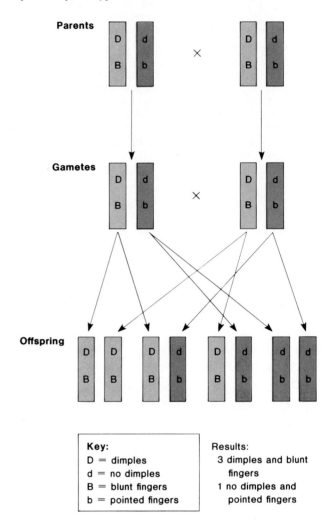

Key:
D = dimples
d = no dimples
B = blunt fingers
b = pointed fingers

Results:
3 dimples and blunt fingers
1 no dimples and pointed fingers

Baldness (fig. 18.18), is believed to be caused by the male sex hormone testosterone because males who take the hormone to increase masculinity begin to lose their hair. A more detailed explanation has been suggested by some investigators. It has been reasoned that due to the effect of hormones, males require only one gene for the trait to appear, whereas females require two genes. In other words, the gene acts as a dominant in males but as a recessive in females. This means that males born to a bald father and a mother with hair *at best* would have a 50% chance of going bald. Females born to a bald father and a mother with hair at *worst* would have a 25% chance of going bald.

Another sex-influenced trait of interest is the length of the index finger. In women the index finger is at least equal to if not longer than the fourth finger. In males the index finger is shorter than the fourth finger.

MAPPING THE HUMAN CHROMOSOMES

A chromosome map indicates the various gene loci on a particular chromosome. There are several techniques that can be used to determine the loci of genes, and we will be exploring several. Some of these techniques have been in existence for some time, and others have only just now been perfected.

Once we know the location of a gene it can be excised, the sequence of its nucleotides can be determined, and most important, its effects on the phenotype can be studied.

Linkage Data

As illustrated in figure 18.19, a chromosome pair has a series of genes. Alleles on the same chromosome are said to form a **linkage group.** Mendel's law of independent assortment cannot hold for linked genes since they tend to appear together in the same gamete. Therefore, traits controlled by linked genes tend to be inherited together.

To take a hypothetical example, let us remember that dimples are dominant over no dimples and blunt fingers are dominant over pointed fingers. If a dihybrid were married to a dihybrid, you would expect four possible phenotypes among the offspring. But as figure 18.19 shows, only two phenotypes would appear if the genes were absolutely linked in the manner illustrated. When doing linkage problems, it is better

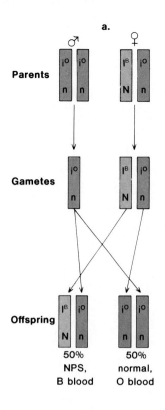

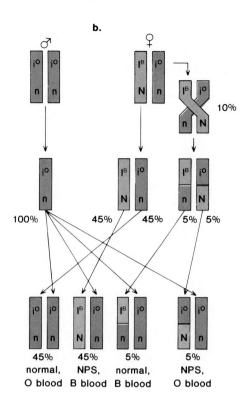

a.

b.

FIGURE 18.20 *Linkage (in practice).* a. *If linkage is complete, then only two phenotypes would appear among the offspring resulting from this cross.* b. *In practice, linkage is not complete because of crossing over. In this actual example, 10% of the offspring had recombinant characteristics because crossing over had occurred.*

Key:
I^B i^o = type B blood
i^o i^o = type O blood
N = nail-patella syndrome (NPS)
n = normal

to use the method illustrated in figure 18.19 rather than a Punnett square so that the genes can be shown on a single chromosome.

When a dihybrid is crossed with a recessive, you normally would expect all possible phenotypes among the offspring. If linkage is present, however, the number of possible phenotypes could possibly be reduced to two types. To take an actual example, it has been reported that the ABO blood type gene and the gene for a very unusual dominant condition called nail-patella syndrome (NPS) are on the same chromosome. A person with NPS has fingernails and toenails that are reduced or absent and a kneecap (patella) that is small. In one family, the female parent had the genotype $I^B i^o$ for blood type and the genotype Nn for NPS; furthermore, it could be established that the allele I^B was on the same chromosome as N and that the allele i^o was on the same chromosome as n. Notice in figure 18.20a that if linkage holds, this individual would form only two possible gametes.

The male parent in this example had the recessive genotype for both traits and therefore could form only one type of gamete carrying the recessive alleles of each gene as illustrated in figure 18.20a. Therefore, assuming linkage, the children of this couple should have only two possible phenotypes: blood type B with NPS and blood type O without NPS.

However, at least 10% of all offspring showed a phenotype in which blood type B was found without NPS and blood type O was found with NPS. This indicates that crossing over occurred (fig. 18.20b).

Crossing Over

When tetrads form during meiosis, the chromatids may exchange portions by a process of breaking and then reassociating (fig 17.12). The gametes that receive recombined chromosomes are called recombinant gametes. Recombinant gametes indicate that the linkage between two genes has been broken by crossing over. Figure 18.20 shows how recombinant gametes produced the unexpected phenotypes in our example.

Sometimes, it has been possible to begin to map chromosomes by studying the crossover frequency of linked genes. Alleles distant from one another are more likely to be separated by crossing over than alleles that are close together. Thus, the crossover frequency indicates the distance between two alleles on a chromosome. Each percentage of crossing over is taken to mean a distance of one map unit. Using these frequencies, then, it is possible to indicate the order of the alleles on the chromosome.

FIGURE 18.21 *In the presence of a fusing agent, human fibroblast cells sometimes join with mouse tumor cells to give hybrid cells having nuclei that contain both types of chromosomes. Subsequent cell division of the hybrid cell produces clones that have lost most of their human chromosomes, allowing the investigator to study these chromosomes separate from all other human chromosomes.*

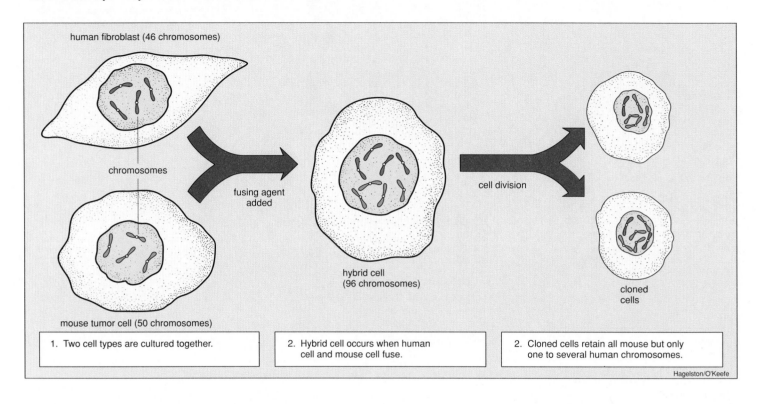

human fibroblast (46 chromosomes)

chromosomes

fusing agent added

hybrid cell (96 chromosomes)

cell division

cloned cells

mouse tumor cell (50 chromosomes)

1. Two cell types are cultured together.

2. Hybrid cell occurs when human cell and mouse cell fuse.

2. Cloned cells retain all mouse but only one to several human chromosomes.

Hagelston/O'Keefe

The presence of linkage groups changes the expected results of genetic crosses. The frequency of recombinant gametes that occur due to the process of crossing over has been used for some time to map chromosomes.

Human-Mouse Cell Data

Human and mouse cells are mixed together in a laboratory dish and, in the presence of inactivated virus of a special type, they fuse (fig. 18.21). As the cells grow and divide, some of the human chromosomes are lost, and eventually the daughter cells contain only a few human chromosomes, each of which can be recognized by their distinctive banding pattern (fig. 18.2a). Analysis of the proteins made by the various human-mouse cells enables scientists to determine which genes are to be associated with which human chromosome.

Sometimes it is possible to obtain a human-mouse cell that contains only one human chromosome or even a portion of a chromosome. This technique has been very helpful to those researchers who have been studying the genes that are located on the number-21 chromosome (p. 378).

Genetic Marker Data

A new mapping technique works directly with DNA and uses genetic markers to tell if an individual has a defective allele. The exact location of the defective allele is not known but the close proximity of the allele and marker can be assumed because they are almost always inherited together. For a marker to be dependable it should be inherited with the defective allele at least 98% of the time.

Detection of genetic markers makes use of special bacterial enzymes, called **restriction enzymes,**[3] each of which cuts the DNA strand at a specific nucleotide sequence producing a particular pattern of DNA fragments. A marker which is really a DNA mutation alters the normal pattern of DNA fragments resulting from restriction enzyme use. Notice in figure 18.22 the different sizes of the fragments—the polymorphism that exists—between the normal and the affected individual. Scientists refer to the differences in the observed

[3]These enzymes are called restriction enzymes because bacteria use them to restrict the growth of viruses whenever viral DNA enters bacteria. Scientists extract the enzymes from bacterial cells and use them in order to cleave DNA in the laboratory.

FIGURE 18.22 *Use of a genetic marker to detect the approximate location of a gene or to test for a genetic disorder when the exact location of the gene is unknown. a. DNA from the normal individual has certain restriction enzyme cleavage sites near the gene in question. b. DNA from another individual lacks one of the cleavage sites, and this loss indicates that they almost certainly have the genetic disorder because experience has shown that this genetic marker is almost always present when an individual has the disorder. In other cases, a gain in a cleavage site is the genetic marker.*

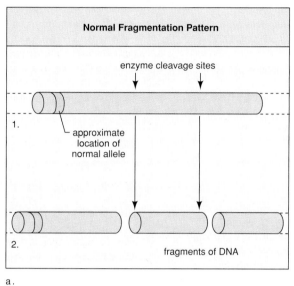

a.

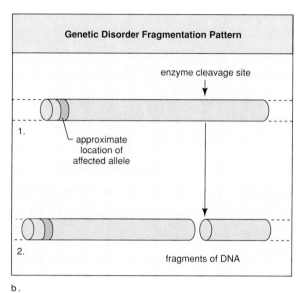

b.

fragment lengths as "restriction fragment length polymorphisms (RFLPs)." They usually discover these polymorphisms by comparing the DNA fragment pattern in a large number of closely related normal and abnormal individuals. The current tests for sickle cell anemia, Huntington disease, Gaucher Type 2, and Duchenne muscular dystrophy are all based on the presence of a marker.

HUMAN ISSUE

Suppose some of your relatives have Huntington disease so that the possibility exists that you have inherited the allele. Would you want to be tested to see if you did indeed inherit the allele, or would you rather just wait to see if the condition develops? Now, assume that you have tested positive and yet you would like to have a child. Would you go ahead and have one even though the child has a 50% chance of also inheriting the disorder? These questions are difficult to answer and they raise a fundamental issue: Is the screening test for Huntington disease an asset to society?

DNA Probe Data

On occasion it has been possible to determine the unique sequence of nucleotides in a gene. Then an exact but radioactive sequence of nucleotides can be prepared in the laboratory and used as a probe. A DNA probe seeks out the chromosome, among any others, that shares the same sequence of nucleotides and binds to it. The chromosomes can then be exposed to a photographic film that shows this pair bound together.

DNA probes, when available, more efficiently map chromosomes and even test whether an individual is a carrier for or is afflicted by a genetic disorder. This type of test is expected one day to replace those based on genetic markers discussed above.

Several new techniques—human-mouse cell preprations, genetic markers and probes—now make it possible to map the human chromosomes at a faster rate than was formerly possible.

HUMAN ISSUE

Assume that a genetic screening test shows you are a carrier for one of the more serious and common genetic diseases, then answer these questions. Should whomever you marry be legally required to take a screening test to find out if he or she also is a carrier, or would it be more appropriate to make discreet inquiries as to disorders prevalent among family members? If you were to fall in love with another carrier, would you go ahead and marry this person, but not have any children? If you did decide to have children, would you rely on prenatal testing and abortion, or would you simply take your chances and be prepared to deal with caring for a child who has a serious genetic defect?

SUMMARY

In keeping with Mendel's laws of inheritance, it is customary to use letters to indicate the genotype and gametes of individuals. Homozygous dominant is indicated by two capital letters, and homozygous recessive is indicated by two lowercase letters. Heterozygous is indicated by a capital letter and a lowercase letter. Contrary to the individual, gametes have one letter of each type, either capital or lowercase as appropriate. All possible combinations of letters can occur in the gametes (except if the genes are linked). In doing an actual cross, it is assumed that all possible types

of sperm fertilize all possible types of eggs. The results of some crosses may be determined by simple inspection, but certain others that commonly reoccur are given in table 18.2.

There are many exceptions to Mendel's laws, and these include polygenic inheritance (skin color), multiple alleles (ABO blood type), and incomplete dominance (curly hair).

Studies of human genetics have shown that there are many autosomal genetic disorders that can be explained on the basis of simple Mendelian inheritance. There are other disorders that are

controlled by several genes (polygenic) or controlled by codominant alleles.

Some genes are sex-linked, most of which occur on the X chromosome while the Y is blank. More males than females have X-linked genetic disorders. Such disorders often skip a generation, passing from grandfather to grandson.

There are several techniques that have contributed to the mapping of the human chromosomes. These include the occurrence of linkage and crossing over between linked genes; human-mouse cell fusions; genetic markers; and DNA probes.

PRACTICE PROBLEMS

Practice Problems I
1. For each of the following genotypes, give all possible gametes.
   a. *WW*
   b. *WWSs*
   c. *Tt*
   d. *Ttgg*
   e. *AaBb*
2. For each of the following, state whether a genotype or a gamete is represented.
   a. *D*
   b. *Ll*
   c. *Pw*
   d. *LlGg*

Practice Problems II
Using the information provided in figure 18.3, solve the following problems.

1. Both a man and a woman are heterozygous for freckles. What are the chances that their children will have freckles?
2. A woman is homozygous dominant for short fingers. Will any of her children have long fingers?
3. Both you and your sister or brother have attached earlobes, yet your parents have unattached ones. What are the genotypes of your parents?
4. A father has dimples, the mother does not have dimples; all the children have dimples. Dimples are dominant over no dimples. Give the probable genotype of all persons concerned.

Practice Problems III
Using the information in figure 18.3, solve these problems.

1. What is the genotype of the offspring if a man homozygous recessive for type of earlobes and homozygous dominant for type of hairline is married to a woman who is homozygous dominant for earlobes and homozygous recessive for hairline?
2. If the offspring of this cross marries someone of the same genotype, then what are the chances that this couple will have a child with a continuous hairline and attached earlobes?
3. A person who has dimples and freckles marries someone who does not. This couple produces a child who does not have dimples or freckles. What is the genotype of all persons concerned?

Practice Problems IV
1. What is the genotype of a person with straight hair? Could this individual ever have a child with curly hair?
2. What is the darkest child that could result from a mating between a light individual and a white individual?
3. What is the lightest child that could result from a mating between two mulatto individuals?
4. From the following blood types, determine which baby belongs to which parents:

Mrs. Doe	Type A
Mr. Doe	Type A
Mrs. Jones	Type A
Mr. Jones	Type AB
Baby 1	Type O
Baby 2	Type B

5. Prove that a child does not have to have the blood type of either parent by indicating what blood types *might* be possible when a person with type A blood reproduces with a person with type B blood.

Practice Problems V
1. Both the mother and father of a hemophilic son appear to be normal. From whom did the son inherit the gene for hemophilia? What is the genotype of the mother, the father, and the son?
2. A woman is color blind. What are the chances that her sons will be color blind? If she is married to a man with normal vision, what are the chances that her daughters will be color blind? will be carriers?
3. Both parents are right handed (R = right handed, r = left handed) and have normal vision. Their son is left handed and color blind. Give the genotype of all persons involved.
4. Both the husband and wife have normal vision. A woman has a color-blind daughter. What can you deduce about the girl's father?

Practice Problems 1
1. a. Wb, WS, Ws c. T, t d. Tg, tg e. AB, Ab, aB, ab
2. a. gamete b. genotype c. gamete
   d. genotype
Practice Problems 2
1. 75%
2. No
3. Heterozygous
4. $DD \times dd$; Dd

Practice Problems 3
1. Dihybrid
2. $\frac{1}{16}$
3. $DdFf \times ddff$; $ddff$
Practice Problems 4
1. $H^H H^h$, No
2. Light
3. White

Practice Problems 5
1. his mother, $X^H X^h$, $X^H Y$, $X^h Y$
2. 100%, None, 100%
3. $RrX^c X^c \times RrX^c Y$; $rrX^c Y$
4. The husband is not the father.
5. AB, O, A, B
6. 3:1
4. Baby 1 = Doe;
   Baby 2 = Jones

OBJECTIVE QUESTIONS

1. Whereas an individual has two genes for every trait, the gametes have _____ gene for every trait.
2. The recessive allele for the dominant gene W is _____ .
3. Mary has a widow's peak and John has a continuous hairline. This would be a description of their _____ .
4. W = widow's peak and w = continuous hairline; therefore, only the phenotype _____ could be heterozygous.
5. Two heterozygotes, each having a widow's peak, already have a child with a continuous hairline. The next child has what chance of having a continuous hairline? _____

6. In a testcross, an individual having the dominant phenotype is crossed with an individual having the _____ phenotype.
7. How many letters are required to designate the genotype of a dihybrid individual? _____
8. If a dihybrid is crossed with a dihybrid, how many offspring out of sixteen are expected to have the dominant phenotype for both traits? _____
9. How many different phenotypes among the offspring are possible when a dihybrid is crossed with a dihybrid? _____
10. According to Mendel's law of independent assortment, a dihybrid can produce how many types of gametes having different combinations of genes? _____
11. Do sex-linked genes determine the sex of the individual? _____
12. If a male is color blind, he inherited the allele for color blindness from his _____ .
13. What is the genotype of a female who has a color-blind father but a homozygous normal mother? _____
14. In a pedigree chart, it is observed that although the children have a characteristic, neither parent has it. The characteristic must be inherited as a _____ gene.

Answers

1. one 2. w 3. phenotype 4. widow's peak 5. 25% 6. recessive 7. four 8. nine 9. four 10. four 11. no 12. mother 13. $X^B X^b$ 14. recessive

STUDY QUESTIONS

1. Explain why there is a pair of alleles for every trait except for sex-linked traits in males. (pp. 387, 393)
2. Relate Mendel's laws of inheritance to one-trait and two-trait problems. (pp. 388, 391)
3. What is the difference between genotype and phenotype? (p. 390)
4. What are the expected results from these crosses: heterozygous × heterozygous; heterozygous × recessive; dihybrid × dihybrid; dihybrid × double recessive? (pp. 391, 392, 394, 395)
5. What does the phrase "chance has no memory" mean? (p. 391)
6. Give four examples of exceptions to Mendel's laws. (pp. 400–403)
7. What is sex linkage? (p. 403) Give all possible genotypes for an X-linked trait, and discuss each. (p. 403)
8. How could you determine if a pedigree chart is depicting the inheritance of a dominant, recessive, or X-linked characteristic? (pp. 397, 405)
9. How would linkage (assuming dominant alleles on one chromosome and recessive on the other) affect the results of the last two crosses mentioned in question 4? (p. 406)

THOUGHT QUESTIONS

1. Mendel worked with garden peas, not humans. What is the significance of being able to apply Mendel's laws to humans?
2. Another investigator, Thomas H. Morgan, worked with fruit flies in the early 1900s. He performed this cross:

	females	males
$P_1 \times P_1$:	red-eyed ×	white-eyed
F_1:	red-eyed	
F_2:	3:1	

   What data did Morgan need to show that the white eye in fruit flies is X-linked?
3. When carriers of genetic disorders for enzyme defects are tested, they have about half the level of enzyme activity as noncarriers. How does this information affect the tradition to state that certain disorders are either recessive or dominant? How might it change our perspective of the homozygous dominant individual?

ADDITIONAL GENETICS PROBLEMS

1. A woman heterozygous for polydactyly (dominant) is married to a normal man. What are the chances that their children will have six fingers and toes? (p. 392)
2. John cannot curl his tongue (recessive), but both his parents can curl their tongues. Give the genotypes of all persons involved. (p. 396)
3. Parents who do not have Tay-Sachs disease (recessive) produce a child who has Tay-Sachs. What are the chances that each child born to this couple will have Tay-Sachs? (p. 396)
4. A man with a widow's peak (dominant) who cannot curl his tongue (recessive) is married to a woman with a continuous hairline who can curl her tongue. They have a child who has a continuous hairline and cannot curl the tongue. Give the genotype of all persons involved. (p. 391)
5. Both Mr. and Mrs. Smith have freckles (dominant) and attached earlobes (recessive). Some of the children do not have freckles. What are the chances that the next child will have freckles and attached earlobes? (p. 391)
6. Mary has wavy hair (incomplete dominance) and marries a man with wavy hair. They have a child with straight hair. Give the genotype of all persons involved. (p. 402)

7. A man has type AB blood. What is his genotype? Could this man be the father of a child with type B blood? If so, what blood types could the child's mother have? (p. 401)
8. A woman with white skin has mulatto parents. If this woman married a light man, what is the darkest skin color possible for their children? the lightest? (p. 400)
9. What is the genotype of a man who is color blind (X-linked recessive) and has a continuous hairline? If this man has children by a woman who is homozygous dominant for normal color vision and widow's peak, what will be the genotype and phenotype of the children? (pp. 388, 403)
10. Is the characteristic represented by the darkened individuals inherited as a dominant, recessive, or X-linked recessive? (pp. 397, 405)

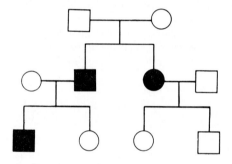

11. Fill in this pedigree chart to give the probable genotypes of the twins pictured in figure 18.12.

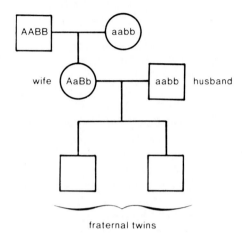

fraternal twins

Answers to Additional Genetics Problems

1. 50%
2. John = *tt*; parents = *Tt*
3. 25%
4. man = *Wwtt*; woman = *wwTt*; child = *wwTt*; child = *Hrhʹ*
5. 75%
6. Mary and husband = *HH*; child = *Hʹhʹ*
7. *Iᴬ Iᴮ*; yes; A, B, O, AB
8. light; white
9. girls = *Xᶜ Xᶜ Ww*; boys = *Xᶜ Y Ww*; both = normal vision and widow's peak
10. Autosomal recessive
11. *AaBb* and *aabb*

KEY TERMS

allele (ah-lēl′) an alternative form of a gene that occurs at a given chromosomal site (locus). *387*

dihybrid (di-hi′brid) the offspring of parents who differ in two ways: shows the phenotype governed by the dominant alleles but carries the recessive alleles. *392*

dominant (dom′i-nant) hereditary factor that expresses itself or a characteristic that is present even when the genotype is heterozygous. *387*

genotype (jen′o-tīp) the genetic makeup of any individual. *390*

heterozygous (het″er-o-zi′gus) having two different alleles (as *Aa*) for a given trait. *390*

homozygous (ho″mo-zi′gus) having identical alleles (as *AA* or *aa*) for a given trait; pure breeding. *390*

linkage (lingk′ij) alleles on the same chromosome are linked in the sense that they tend to move together to the same gamete; crossing over interferes with linkage. *406*

monohybrid (mon″o-hi′brid) the offspring of parents who differ in one way only; shows the phenotype of the dominant allele but carries the recessive allele. *390*

pedigree chart ( ) a representation showing the pattern of inheritance of a condition among family members. *395*

phenotype (fe′no-tīp) the outward appearance of an organism caused by the genotype and environmental influences. *390*

recessive (re-ses′iv) hereditary factor that expresses itself or a characteristic that is present only when the genotype is homozygous. *387*

sex linked (seks lingkt) alleles located on sex chromosomes that determine traits unrelated to sex. *403*

testcross (test kros) the crossing of a heterozygote with an organism homozygous recessive for the characteristic(s) in question in order to determine the genotype. *391*

trait (trāt) specific term for a distinguishing feature studied in heredity. *387*

X-linked (eks lingkt) an allele located on X chromosome that determines a characteristic unrelated to sex. *403*

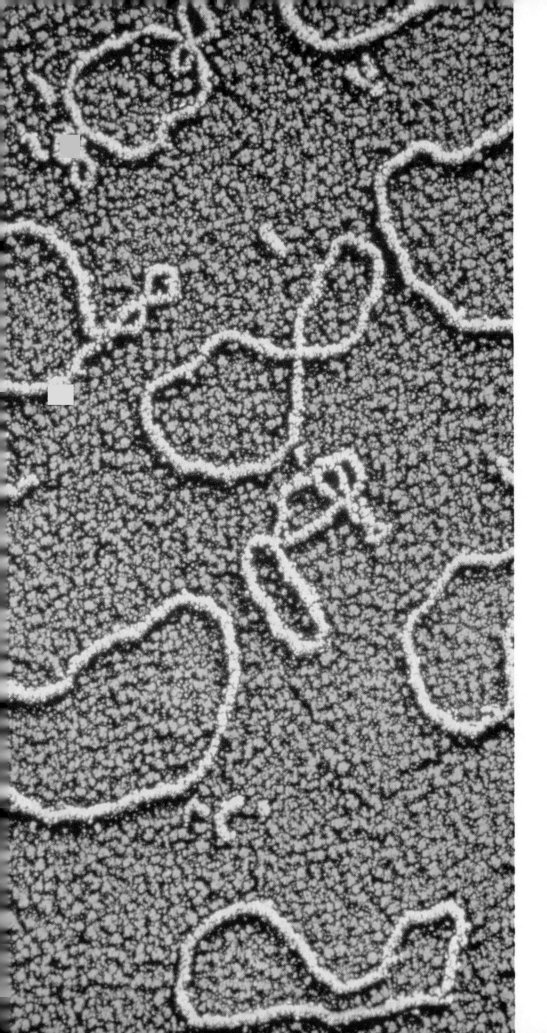

CHAPTER NINETEEN

DNA AND *BIOTECHNOLOGY*

CHAPTER CONCEPTS

1 DNA is the genetic material, and therefore its structure and function constitute the molecular basis of inheritance.

2 DNA is able to replicate, mutate, and control the phenotype.

3 DNA controls the phenotype by controlling protein synthesis, a process that also requires the participation of RNA.

4 Genetic engineering, the manipulation of DNA, is of utmost importance today.

5 Molecular genetics, including genetic engineering, has expanded the capabilities of an industry now called biotechnology.

6 Modern-day biotechnology is expected to revolutionize medical care, agriculture, and the chemical industry. It also has many environmental applications.

Plasmids of bacterial DNA from the Escherichia coli *commonly used in genetic engineering.*

Molecular genetics provides evidence for evolution. The same four types of nucleotides found in human DNA make up the DNA of the bacterium *Escherichia coli* (fig. 19.1). Further, you can put a human gene into this bacterium, and the gene will perform its normal cellular function. It's not surprising, then, that you can also transfer genes between plants and animals. For certain, the very first cell or cells must have evolved into all the life forms we see about us. The evolutionary process must be dependent upon slight changes in the DNA. Only in recent times have we come to realize that in order to truly understand the evolution of plants and animals we have to look inside the workings of a cell.

Our new knowledge of the workings of DNA have pushed us into a new era—one in which it may even be possible to cure genetic diseases. For example, we may develop the capability to splice genes that correct specific genetic defects into human zygotes. Exploring and developing this capability would require experimentation with human zygotes and artificial implantation into a woman's uterus. Does the prospect of children who are free of significant genetic diseases justify extensive experimentation with human zygotes and the risk of creating children who are the result of failed genetic experiments? Who should answer this question?

We have previously examined chromosomal and genetic inheritance. Now it is time to examine the genetic material itself to see how it functions at the cellular level. Only then is it possible to understand the development of modern-day biotechnology.

NUCLEIC ACIDS WITHIN THE CELL

The chromosomes located within the nucleus of the cell (fig. 19.2) contain DNA, a nucleic acid having a structure we examined in chapter 1. Another nucleic acid, called RNA, is located within the nucleus and the cytoplasm.

You will recall from the discussion on page 26 that the **DNA** structure is a double helix in which the two nucleotide strands are held together by hydrogen bonds between purine and pyrimidine bases. Thymine (T) is always paired with adenine (A), and guanine (G) is always paired with cytosine (C). This is called **complementary base pairing.** Besides one of these bases, each nucleotide in DNA contains a phosphate molecule and the sugar deoxyribose.

RNA is a single-stranded nucleic acid, and each polymer has a sequence of nucleotides that will also pair with a portion of DNA. However, the sugar within the nucleotides that make up an RNA strand is ribose—not deoxyribose. Also, the pyrimidine thymine does not appear in RNA; it is replaced by the pyrimidine uracil. Therefore, RNA contains these four bases: adenine (A), guanine (G), cytosine (C), and uracil (U). Figure 19.3 shows the structure of RNA, and table 19.1 summarizes the differences between DNA and RNA structure.

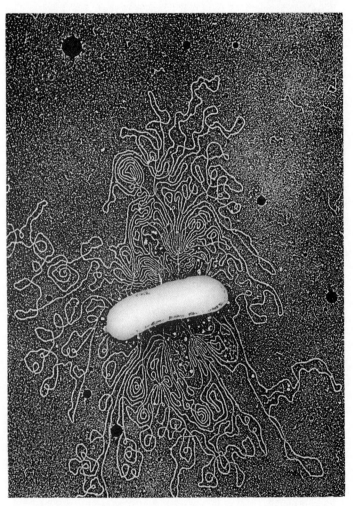

FIGURE 19.1 *A photograph of a transmission electron micrograph showing the bacterium* E. coli *has been broken open, and its DNA has escaped its normal cellular boundaries. All of this DNA is found within a single circular chromosome that lies free within the cytoplasm. Unlike human cells, bacteria don't have a membrane-bound nucleus.*

TABLE 19.1 DNA STRUCTURE COMPARED TO RNA STRUCTURE

	DNA	RNA
Sugar	Deoxyribose	Ribose
Bases	Adenine, guanine, thymine, cytosine	Adenine, guanine, uracil, cytosine
Strands	Double stranded with base pairing	Single stranded
Helix	Yes	No

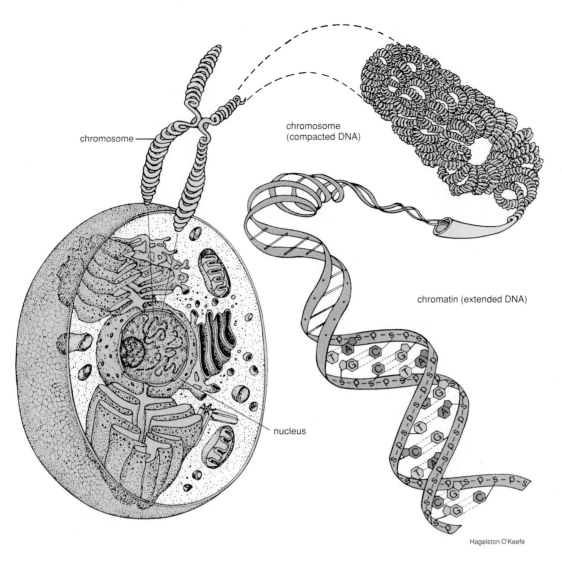

chromosome

chromosome
(compacted DNA)

nucleus

chromatin (extended DNA)

Hagelston/O'Keefe

FIGURE 19.2 DNA is located in the nucleus of a human cell. DNA is highly coiled in chromosomes, but it is extended as chromatin during interphase. It is during this time that DNA can be extracted from a cell and its structure studied.

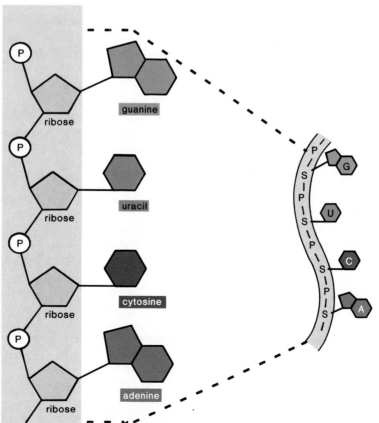

ribose

guanine

ribose

uracil

ribose

cytosine

ribose

adenine

FIGURE 19.3 Structure of RNA. RNA is single stranded. The backbone contains the sugar ribose instead of deoxyribose. The bases are guanine, uracil, cytosine, and adenine.

DNA has a structure like a twisted ladder: sugar-phosphate backbones make up the sides of the ladder; hydrogen-bonded bases make up the rungs of the ladder. The base A is always paired with the base T, and the base C is always paired with the base G. RNA differs from DNA in several respects (table 19.1).

DNA, the Hereditary Material

DNA, the hereditary material, has these four functions:

1. stores the genetic information that is needed by cells to carry on their routine activities;
2. replicates—makes copies of itself—that are passed on from cell to cell and from generation to generation;
3. controls the activities of the cell, thereby producing the phenotypic characteristics of the individual and the species;
4. undergoes mutations—permanent genetic changes passed on to the offspring—that ultimately account for the evolutionary history of life.

Replication

The double-stranded structure of DNA lends itself to replication because each strand can serve as a template for the formation of a complementary strand. A **template** is most often a mold used to produce a shape opposite to itself. In this case, the word *template* is appropriate because each new strand of DNA has a sequence of bases complementary to the bases of the old strand of DNA.

Replication requires the following steps (fig. 19.4):

1. The two strands that make up DNA become "unzipped" (i.e., the weak hydrogen bonds between the paired bases are broken).
2. New complementary nucleotides, drawn from a "pool" of free nucleotides that are synthesized from simpler materials, move into place by the process of complementary base pairing.
3. The adjacent nucleotides become joined through their sugar-phosphate components to form the new chain.
4. When the process is finished, two complete DNA molecules are present, identical to each other and to the original molecule.

This replication process is described as *semiconservative* because each double strand of DNA contains one old strand and one new strand. Although DNA replication can be easily explained, it is in actuality an extremely complicated process involving many steps and enzymes. The enzyme helicase assists the unwinding process, and the enzyme DNA polymerase joins the nucleotides together, for example. On occasion, errors are made that cause a change in the DNA and, in this way, a mutation can arise.

During replication, DNA becomes "unzipped" and a complementary strand forms opposite to each original strand. This is called semiconservative replication.

Genes and Enzymes

Many novel experiments and years of research allowed scientists to conclude that there is a relationship between genetic inheritance and the structure of enzymes and other proteins. For example, the metabolic pathway outlined in figure 19.5b was discovered in the early twentieth century. In this pathway, three genetic diseases are known. In the disorder called *phenylketonuria (PKU),* which was discussed on page 396, phenylpyruvic acid accumulates in the body and spills over into the urine because the enzyme needed to convert phenylalanine (a building block of protein) to tyrosine is missing. If the condition is not treated, the continued accumulation of phenylpyruvic acid can cause mental retardation. *Albinism* results because tyrosine cannot be converted to melanin, the natural pigment in human skin. The genetic disease *alkaptonuria* results if the enzyme needed to metabolize homogentisic acid is missing.

At first these conditions were simply called *inborn errors of metabolism,* and only later was it confirmed that the genetic fault lay in the absence of particular enzymes. This gave rise to the suggestion that genes in some way control the presence of enzymes in the individual. This was called the *one gene–one enzyme* theory.

Since enzymes are proteins, the concept was soon broadened to be the one gene–one protein theory. However, some proteins like hemoglobin have more than one type of polypeptide chain (fig. 6.6). In persons with sickle cell anemia, it is only the β polypeptide chain that has an altered sequence of amino acids (fig. 19.6) compared to the normal β chain. Therefore, it may be more appropriate to state that a gene controls the sequence of amino acids in a polypeptide. Today we define a gene as a section of a DNA molecule that determines the sequence of amino acids in a single polypeptide chain of a protein.

It was a breakthrough to discover that there is a relationship between genetic inheritance and the primary structure of proteins in an individual.

FIGURE 19.4 DNA replication. Replication is called semiconservative because each new double helix is composed of an old parental strand and a new daughter strand.

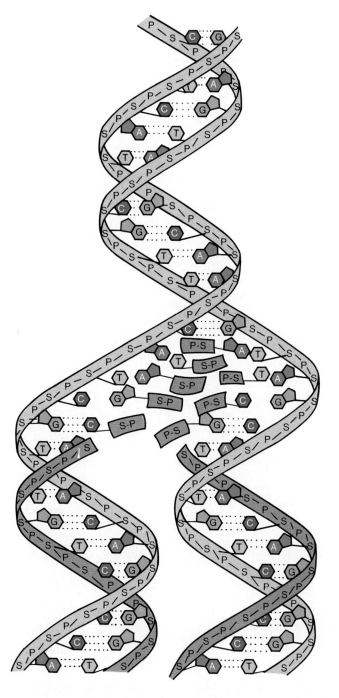

Region of parental DNA helix. (Both backbones are light.)

Region of replication. Parental DNA is unzipped and new nucleotides are pairing with those in parental strands.

Region of completed replication. Each double helix is composed of an old parental strand (light) and a new daughter strand (dark). Notice that each double helix is exactly like the other one and like the original parental strand.

FIGURE 19.5 a. *An albino is unable to produce the pigment melanin.* b. *Metabolic pathway by which phenylalanine is converted to other metabolites.* 1. *If the enzyme that converts phenylalanine to tyrosine is defective, phenylalanine is converted to phenylpyruvic acid instead, and the accumulation of this substance leads to PKU (phenylketonuria).* 2. *If the enzyme that converts tyrosine to melanin is defective, albinism results.* 3. *If homogentisic acid cannot be metabolized, alkaptonuria (meaning that the urine turns a dark color) results.*

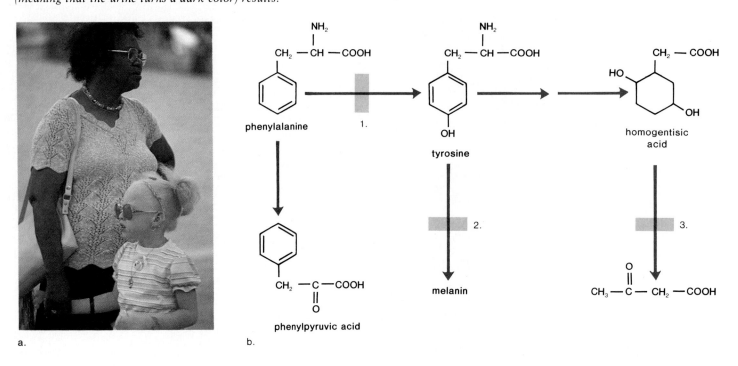

a.

b.

FIGURE 19.6 *Sickle cell anemia in humans.* a. *The first seven amino acids found in the normal and in the abnormal β chain. The substitution of a single amino acid (valine substituted for glutamic acid) at the sixth position results in sickle cell anemia.* b. *Photomicrographs of normal (left) and sickled (right) red blood cells.*

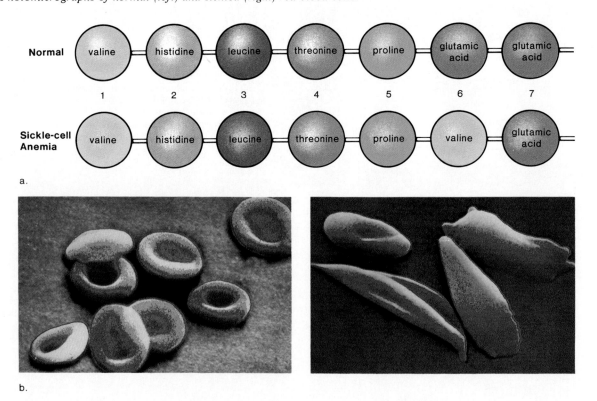

a.

b.

PROTEIN SYNTHESIS

The fact that DNA controls the production of proteins may at first seem surprising when we consider that genes are located in the nucleus of higher cells but proteins are synthesized at the ribosomes in the cytoplasm. However, you will recall that although DNA is found only in the nucleus, RNA exists in both the nucleus and the cytoplasm.

Biochemical genetic research indicates that a type of RNA called **messenger RNA (mRNA)** serves as a go-between for DNA in the nucleus and the ribosomes in the cytoplasm. This is possible because a strand of DNA can serve as a template for the production of a complementary strand of RNA, as well as a template for another strand of DNA. The RNA molecule contains a sequence of nucleotides that are complementary to those of a single gene (fig. 19.7). The mRNA moves from the nucleus to the ribosomes in the cytoplasm, where it dictates the sequence of amino acids in a polypeptide. This concept is often called the **central dogma** of modern genetics and can be diagrammed as follows:

$$\text{DNA} \xrightarrow{\text{transcription}} \text{mRNA} \xrightarrow{\text{translation}} \text{protein}$$

The diagram indicates that the control of protein synthesis requires transcription and translation. During the process of **transcription,** complementary mRNA is formed in the nucleus, and during **translation,** its message is used in the cytoplasm to produce the correct order of amino acids in a polypeptide.

Code of Heredity

DNA provides mRNA with a message that directs the order of amino acids during protein synthesis, but what is the nature of the message? The message cannot be contained in the sugar-phosphate backbone because it is constant in every DNA molecule. However, the order of the bases in DNA and mRNA can and does change. Therefore, it must be the bases that contain the message. The order of the bases in DNA must code for the order of the amino acids in a polypeptide. Can four bases provide enough combinations to code for twenty amino acids? It can if the code is a *triplet code;* each amino acid is dictated by a sequence of three bases.

Just as our alphabet forms words that can be arranged to provide information, so the bases of DNA form the triplets, whose sequences provide the information needed by an organism to develop, maintain itself, and reproduce. To understand the language of DNA, it was necessary to decipher the DNA code. To crack the code, artificial RNA was added to a medium containing bacterial ribosomes and a mixture of amino acids. Comparison of the bases in the RNA with the resulting polypeptide allowed investigators to decipher the code. Each three-letter unit of a messenger RNA is called a **codon.** All sixty-four codons have been determined (table 19.2). Each of sixty-one triplets correspond to a particular amino acid; the remaining three code for chain termination. The one codon that stands for the amino acid methionine also signals polypeptide initiation.

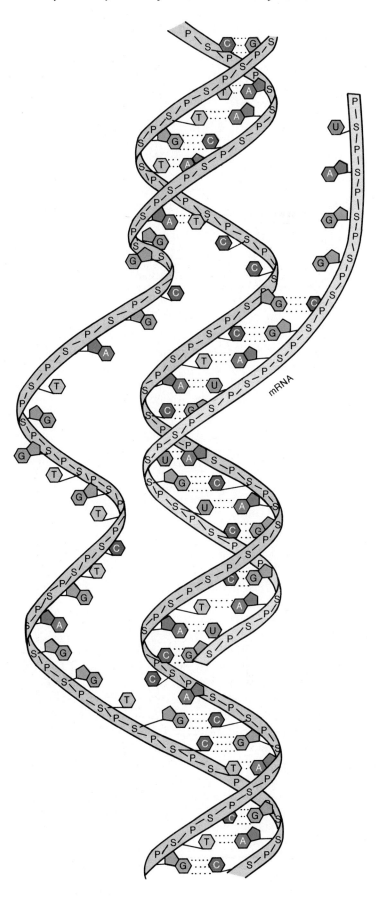

FIGURE 19.7 *Transcription. When mRNA is formed, its bases are complementary to those found in one strand of DNA.*

TABLE 19.2 *THREE-LETTER CODONS OF MESSENGER RNA, AND THE AMINO ACIDS SPECIFIED BY THE CODONS*

AAU ⎤ AAC ⎦	Asparagine	CAU ⎤ CAC ⎦	Histidine	GAU ⎤ GAC ⎦	Aspartic acid	UAU ⎤ UAC ⎦	Tyrosine
AAA ⎤ AAG ⎦	Lysine	CAA ⎤ CAG ⎦	Glutamine	GAA ⎤ GAG ⎦	Glutamic acid	UAA ⎤ UAG ⎦	(Stop)[a]
ACU ⎤ ACC ACA ACG ⎦	Threonine	CCU ⎤ CCC CCA CCG ⎦	Proline	GCU ⎤ GCC GCA GCG ⎦	Alanine	UCU ⎤ UCC UCA UCG ⎦	Serine
AGU ⎤ AGC ⎦	Serine	CGU ⎤ CGC CGA CGG ⎦	Arginine	GGU ⎤ GGC GGA GGG ⎦	Glycine	UGU ⎤ UGC ⎦	Cysteine
AGA ⎤ AGG ⎦	Arginine					UGA UGG	(Stop)[a] Tryptophan
AUU ⎤ AUC AUA ⎦	Isoleucine	CUU ⎤ CUC CUA CUG ⎦	Leucine	GUU ⎤ GUC GUA GUG ⎦	Valine	UUU ⎤ UUC ⎦	Phenylalanine
AUG	Methionine (or start)					UUA ⎤ UUG ⎦	Leucine

[a]Stop codons signal the end of the formation of a polypeptide chain.

From E. Peter Volpe, *Biology and Human Concerns*, 3d ed. Copyright © 1983 Wm. C. Brown Publishers, Dubuque, Iowa. All Rights Reserved. Reprinted by permission.

Research indicates that the code is essentially universal. The same codons stand for the same amino acids in all living things, including bacteria, plants, and animals. This illustrates the remarkable biochemical unity of living things and suggests that all living things have a common evolutionary ancestor.

DNA controls the phenotype because it controls protein synthesis. DNA and several forms of RNA participate in protein synthesis. DNA, which always stays within the nucleus, contains a triplet code, a series of three bases that codes for one particular amino acid.

Transcription

During transcription, the DNA code is passed to mRNA, and thus the code is "transcribed," or rewritten.

Messenger RNA

When mRNA forms, a segment of the DNA helix unwinds and complementary RNA nucleotides pair with DNA nucleotides of one of the strands. After the enzyme RNA polymerase joins the nucleotides by way of their sugar-phosphate components, the resulting mRNA molecule carries codons that are complementary to the DNA triplet code sequences (fig. 19.7). The mRNA strand then passes from the cell nucleus into the cytoplasm, carrying the transcribed DNA code.

During transcription, messenger RNA is made complementary to a portion of one DNA strand. It then contains codons for one polypeptide and moves to the cytoplasm to become associated with the ribosomes.

Translation

During translation (fig. 19.8) the sequence of codons in mRNA dictates the order of amino acids in a polypeptide. This is called translation because the sequence of bases in DNA is finally translated into a particular sequence of amino acids. Translation requires the involvement of several enzymes and two other types of RNA: ribosomal RNA (rRNA) and transfer RNA (tRNA).

Ribosomal RNA

Ribosomal RNA (rRNA) is found in the ribosomes (p. 36), which are composed of two subunits, each with characteristic RNA and protein molecules. The rRNA molecules are transcribed from DNA in the region of the nucleolus; the proteins are manufactured in the cytoplasm, but then migrate to the nucleolus, where the ribosomal subunits are assembled before they migrate into the cytoplasm. Ribosomes play an important role in coordinating protein synthesis.

Transfer RNA

Located in the cytoplasm are small molecules of **transfer RNA (tRNA)** that transfer the amino acids from the cytoplasm to the ribosomes. Each molecule of tRNA attaches at one end to a particular amino acid. Attachment requires ATP energy and results in a high-energy bond. Therefore, this bond is indicated by a wavy line and the entire complex is designated by *tRNA ~ amino acid*. At the other end of each tRNA there is a specific **anticodon** complementary to an mRNA codon. (Each tRNA molecule is transcribed from DNA and then, due to intramolecular binding of complementary bases, the anticodon is exposed.)

FIGURE 19.8 *During translation, a ribosome moves along an mRNA. A codon on the mRNA attracts an anticodon of a tRNA. When a codon pairs with an anticodon, the tRNA brings an amino acid to the ribosome. a. In this diagram, translation is already in progress and a single tRNA molecule is in place on a ribosome as another approaches. b. Both tRNA molecules are positioned at the ribosome. The first of these, with the anticodon UGG, bears a peptide chain, while the second, having the anticodon UUU, bears an amino acid. c. In this diagram, the first tRNA has passed the peptide chain to the second tRNA before departing. d. The ribosome has moved to the right so that the tRNA with the anticodon UUU is now in the first position on the ribosome. Another tRNA-amino acid complex with the anticodon CAU is approaching the ribosome. The same sequence of events (a–d) will now reoccur.*

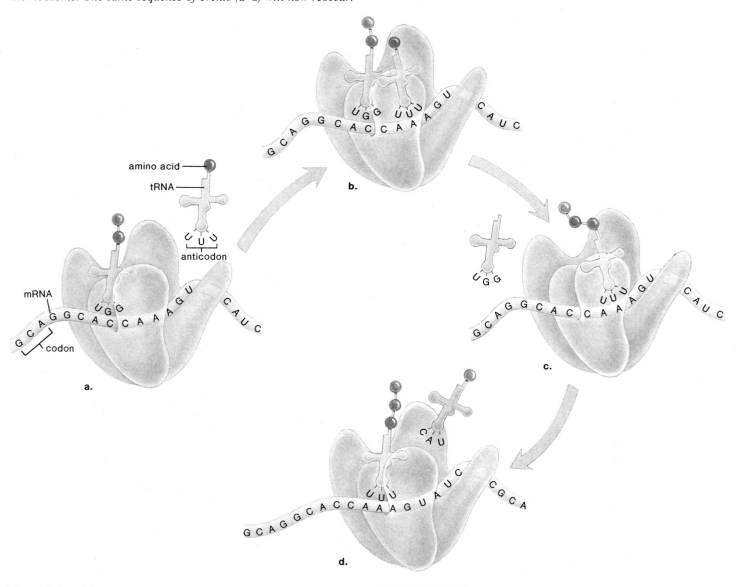

The Polypeptide

Complementary base pairing between codons and anticodons determines the order in which tRNA⊥amino acid complexes come to a ribosome, and this in turn determines the final sequence of the amino acids in a polypeptide. So, for example, if the codon sequence is ACC AAA GUA, then the tRNAs coming to the mRNA for these three codons will have the anticodons UGG, UUU, and CAU, and the eventual amino acid sequence will be threonine, lysine, valine (fig. 19.8 and table 19.2).

During translation, tRNA molecules, each carrying a particular amino acid, travel to the mRNA, and through complementary base pairing between anticodons and codons, the tRNAs and therefore the amino acids in a polypeptide chain become sequenced in a predetermined order.

TABLE 19.3 *STEPS IN PROTEIN SYNTHESIS*

NAME OF MOLECULE	SPECIAL SIGNIFICANCE	DEFINITION
DNA	Code	Sequence of bases in threes
mRNA	Codon	Complementary sequence of bases in threes
tRNA	Anticodon	Sequence of three bases complementary to codon
Amino acids	Building blocks	Transported to ribosomes by tRNAs
Protein	Enzyme	Amino acids joined in a predetermined order

Summary of Protein Synthesis

The following list, along with table 19.3, provides a brief summary of the steps involved in protein synthesis.

1. DNA, which always remains in the nucleus, contains a series of bases that serve as a *triplet code* (every three bases code for an amino acid).
2. During transcription, one strand of DNA serves as a template for the formation of messenger RNA (mRNA), which contains *triplet codons* (sequences of three bases complementary to the DNA code).
3. Messenger RNA goes into the cytoplasm and becomes associated with the *ribosomes*, which are composed of ribosomal RNA (rRNA) and proteins.
4. Transfer RNA (tRNA) molecules, each of which is bonded to a particular amino acid, have *anticodons* that pair complementarily to the codons in mRNA.
5. As the ribosome moves along mRNA, a newly arrived tRNA~amino acid complex receives the growing polypeptide chain from a tRNA molecule that thereafter leaves the ribosome to pick up another amino acid. During translation, therefore, the linear sequence of codons determines the order in which the tRNA molecules arrive at the ribosomes and thus determines the *primary structure* of a protein (i.e., the order of its amino acids).

Table 19.3 and figure 19.9 provide a brief summary of the steps involved in protein synthesis.

OTHER FEATURES OF DNA

For the sake of simplicity and clarity, we have thus far omitted certain features of DNA chemistry. In order for you to have an expanded appreciation of the workings of the cell, these features will be introduced here.

Structural versus Regulatory Genes

Certain genes, called **regulatory genes**, control or influence whether or not other genes, called **structural genes**, are transcribed. It is the structural genes that code for metabolic enzymes and various cell components. One model by which regulatory genes may function is shown in figure 19.10. The model allows us to see that, in order to understand the phenotype, it is necessary to consider the entire genotype rather than each gene separately. In other words, the presence of a structural gene does not necessarily mean that it will be expressed (i.e., transcribed and translated) since it can be turned on or off by a regulatory gene.

The presence of regulatory genes is important to the occurrence of differentiation, the existence of specialized cells in the body. All cells have the same chromosomes and genes, yet some are intestinal cells, others are muscle cells, others are blood cells, etc. There is evidence that as development proceeds, certain genes get turned on and certain other genes get turned off so that specialization occurs.

Knowledge about regulatory genes is extremely important because genes can only be manipulated when we know how to turn them on and off. Also, mutations of regulatory genes probably account for some genetic diseases and/or the development of cancer.

Structural genes code for proteins, and regulatory genes control the expression of structural genes.

Genetic Mutations

We have previously studied chromosomal mutations (fig. 17.20), but there are also genetic mutations. As you can see in table 19.4 a **genetic mutation** is any alteration in the code of a single gene or any change in its expression. Genetic mutations do not necessarily have a deleterious effect; some may have no effect at all and some may even have a beneficial effect.

Substitutions, Alterations, and Deletions of Bases

Mutations involving a change in the DNA sequence of bases are of three types (table 19.5). *Substitutions*, which involve a change in a single base, may have no effect at all if the new codon happens to stand for the same amino acid (table 19.2). Other base substitutions may lead to an amino acid substitution in the protein product. For example, a change from GAG to GUG would cause glutamic acid to be replaced by valine as it is in sickle cell hemoglobin (fig. 19.6). This particular substitution causes a drastic deleterious effect in the phenotype. It's also possible that a base substitution might result in a new "stop" codon, and this would cause transcription to be terminated before the polypeptide is fully formed. This type of base change is being investigated as a possible cause for hemophilia, the sex-linked blood clotting disorder (p. 404).

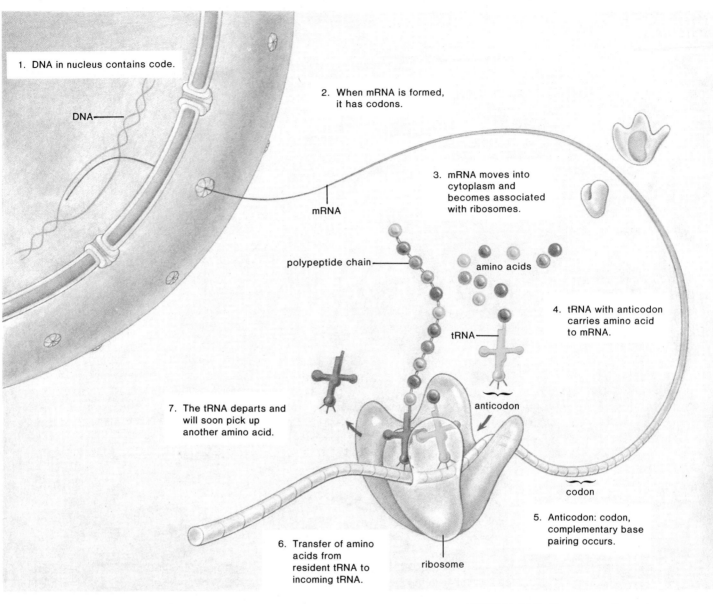

1. DNA in nucleus contains code.

DNA

2. When mRNA is formed, it has codons.

mRNA

3. mRNA moves into cytoplasm and becomes associated with ribosomes.

polypeptide chain

amino acids

4. tRNA with anticodon carries amino acid to mRNA.

tRNA

anticodon

7. The tRNA departs and will soon pick up another amino acid.

6. Transfer of amino acids from resident tRNA to incoming tRNA.

ribosome

5. Anticodon: codon, complementary base pairing occurs.

codon

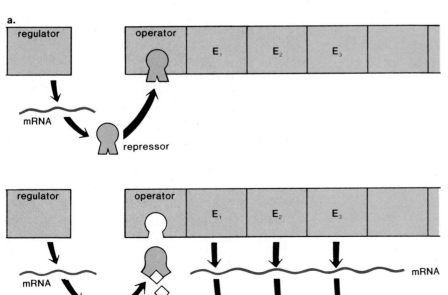

a.

regulator

operator

E₁ E₂ E₃

mRNA

repressor

regulator

operator

E₁ E₂ E₃

mRNA

mRNA

inducer

enzyme 1   2   3

FIGURE 19.9 Summary of protein synthesis. Transcription occurs in the nucleus, and translation occurs in the cytoplasm (blue). During translation, the ribosome moves along the mRNA. In the diagram, as the ribosome moves to the left, a tRNA bearing an amino acid comes to the ribosome. Thereafter the polypeptide chain will be passed to this tRNA = amino acid complex. Each time the ribosome moves a tRNA departs.

FIGURE 19.10 Regulation of protein synthesis model. a. Regulatory gene codes for a protein, called a repressor, that attaches to an operator gene, preventing the transcription of enzyme-producing genes. b. When a specific inducer is present, it combines with the repressor molecule, preventing it from attaching to the operator gene. Now transcription of the enzyme-producing genes takes place.

TABLE 19.4	MUTATIONS

TYPES OF MUTATIONS	DEFINITION
Chromosomal mutation	A rearrangement of chromosome parts, as described in figure 17.20, which may or may not result in a change of the phenotype
Genetic mutation	A change in the genetic code for a gene or in the expression of the gene. Usually results in a change of the phenotype
Germinal mutation	A mutation that manifests itself in the gametes so that it is passed on to offspring
Somatic mutation	A mutation that occurs in the body cells and that very likely is not passed on to offspring

TABLE 19.5	GENE MUTATIONS

BASE CHANGE		WORSE RESULT
Normal	TAC'GGC'ATG	
Substitution	TAG'GGC'ATG	Change in one amino acid or change to stop signal
Deletion	ACG'GCA'TG	Polypeptide can be completely altered
Addition	ATA'CGG'CAT'G	Polypeptide can be completely altered

Additions and *deletions* of bases are expected to result in profound alterations in the DNA code (table 19.5). If the altered allele codes for an enzyme, the enzyme would most likely be nonfunctional because the sequence of amino acids would be so greatly affected.

Transposons

Transposons are specific DNA sequences that have the remarkable ability to move within and out of chromosomes. Their movement to a new location sometimes alters neighboring genes, particularly by increasing or decreasing their expression. Although "movable elements" in corn were described forty years ago, their significance was only recently realized. So-called "jumping genes" have now been discovered in bacteria, fruit flies, and humans, and it is likely that all organisms have such elements.

Mutagenic Agents

Certain environmental influences like radiation, either UV light or X rays, and organic chemicals can cause genetic mutations to occur. Ultraviolet light most often causes two adjacent thymine bases to bond together. Such *thymine dimers* prevent replication of DNA and prevent mRNA transcription. Some chemicals can cause one type of base to be converted to a different type, but others are base analogues that are incorporated into DNA in place of a normal base. These, like AZT that is used as a medication for AIDS patients (p. 146), also disrupt DNA replication and mRNA transcription.

Split Genes

Human structural genes, like those of other higher organisms, are now known to be interrupted by sections of DNA that are not part of the gene (fig. 19.11). These portions are called *introns* because they are *intra*gene segments. The other

portions of the gene are called *exons* because they are ultimately *ex*pressed. When DNA is transcribed, the mRNA contains bases that are complementary to both exons and introns, but before the mRNA exits from the nucleus, it is *processed*—the nucleotides complementary to the introns are enzymatically removed.

Figure 19.11 shows that the human chromosome also contains *repetitive DNA*—the same short-to-intermediate length DNA sequence is repeated over and over. The exact function of repetitive DNA has not been determined, but perhaps these sequences are transposons that are now acting as introns. In other words, movement of transposons sometimes causes a structural gene to be interrupted by portions that are not part of the gene. Another view is that repetitive DNA has no function and simply is getting a free ride as it is duplicated and passed from generation to generation.

BIOTECHNOLOGY

Biotechnology, the use of a natural biological system to produce a product or to achieve an end desired by human beings, is not new. On the organismal level, the selective breeding of plants and animals to achieve a particular phenotype has been utilized since the dawn of civilization. Also, the biochemical capabilities of unicellular organisms have been utilized many times before. For example, even the baking of bread and the production of wine is dependent on the presence of the microorganism yeast.

Today, however, we have a new form of biotechnology (fig. 19.12) that is derived from a knowledge of molecular genetics such as we have just presented (pp. 413–422). Techniques have been developed that allow the manipulation and restructuring of an organism's genes. Bacteria can be *genetically engineered* to do our bidding. Some can be made to produce great quantities of an organic chemical of interest or a protein that is useful as a vaccine or a drug to promote human health. Other engineered bacteria may soon be released into the environment to clean up pollutants, to increase the fertility of the soil, or to kill insect pests. And there

FIGURE 19.11 *The chromosomes of higher organisms contain segments that do not code for cytoplasmic polypeptides. Repetitive DNA is noncoding and consists of sequences of base pairs repeated many times, one after another. Structural genes, themselves, are interrupted by intervening sequences or* introns, *which do not code for polypeptides. During transcription, the entire gene, including these segments, are copied into mRNA. Before the mRNA leaves the nucleus, these segments are spliced out. Structural genes are flanked at one end by regulatory genes that control whether transcription takes place or not. There is a start code at the beginning and a stop code at the end of a structural gene.*

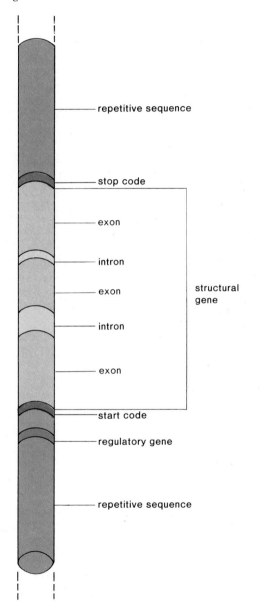

- repetitive sequence
- stop code
- exon
- intron
- exon
- intron
- exon
- start code
- regulatory gene
- repetitive sequence

structural gene

FIGURE 19.12 *Biotechnology is an industrial endeavor. a. Laboratory procedures must be adapted to mass-produce the product. b. Microbes are grown in huge tanks called fermenters because they were first used for yeast fermentation to produce wine. c. The product must be purified, and subjected to quality control. d. The product is packaged, and stored in a warehouse until shipped to a customer.*

a.

b.

c.

d.

is no need to stop with unicellular organisms. The technology is in place to alter the genotype and subsequently the phenotype of plants and animals. An agricultural revolution of unequaled magnitude is not inconceivable even in the near future. And there is no need to stop here either. Gene therapy in humans may be just over the horizon.

There is no doubt that these achievements are within our grasp and therefore it behooves us to be familiar with how such achievements will be accomplished. Also, as a society we should be aware not only of the expected benefits but also the possible risks of biotechnology.

We will begin our discussion with the basic procedures and applications of biotechnology.

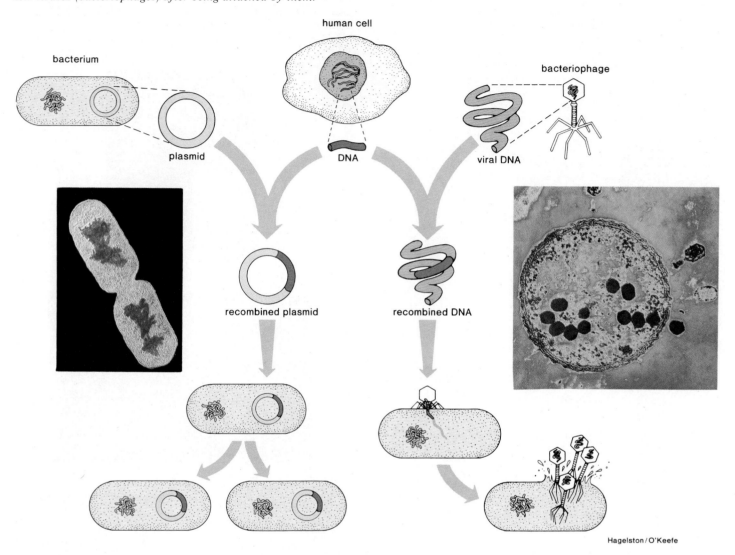

FIGURE 19.13 *Gene cloning is accomplished by using either recombined plasmid DNA or recombined viral DNA. (left) Photomicrograph of the workhorse of biotechnology, the bacterium* E. coli. *(right) Electron micrograph of* E. coli *releasing new viruses (bacteriophages) after being attacked by them.*

bacterium

human cell

bacteriophage

plasmid

DNA

viral DNA

recombined plasmid

recombined DNA

Hagelston/O'Keefe

Genetic Engineering of Cells

Genetic engineering can produce cells that are transformed. These cells contain a foreign gene and are capable of producing a new and different protein. Often, **vectors** are used to carry the foreign gene into a bacterium, a yeast cell, a plant cell, or an animal cell, including a human cell.

Vectors

One common type of vector is a plasmid. **Plasmids** are small accessory rings of DNA found in some bacteria (fig. 19.13) that carry genes not present in the bacterial chromosome. Plasmids used as vectors have been removed from bacteria and have had a foreign gene inserted into them. Treated cells will take up a plasmid, and after it enters, the host cell continues to reproduce as usual. Whenever the host reproduces, the plasmid, including the foreign gene, is copied. Eventu-

ally, there are many copies of the plasmid and therefore many copies of the foreign gene. This gene is now said to have been **cloned**.

Plasmids are not taken up by animal cells, but viruses can be used as vectors in these cells. When a virus attacks a cell, the viral DNA enters the cell. Here it may direct the reproduction of many more viruses (p. 318). Each virus derived from a viral vector will contain a copy of the foreign gene. Therefore viral vectors also allow cloning of a particular gene.

Viruses are also commonly used as vectors to introduce a foreign gene into bacteria. In fact, bacteriophages (viruses that attack bacteria) have been used to create genomic libraries. A *genomic library* is a collection of engineered bacteriophages that together carry all the genes of a species. Since each bacteriophage carries only a short sequence it takes about 10 million bacteriophages to carry all the genes of a mouse.

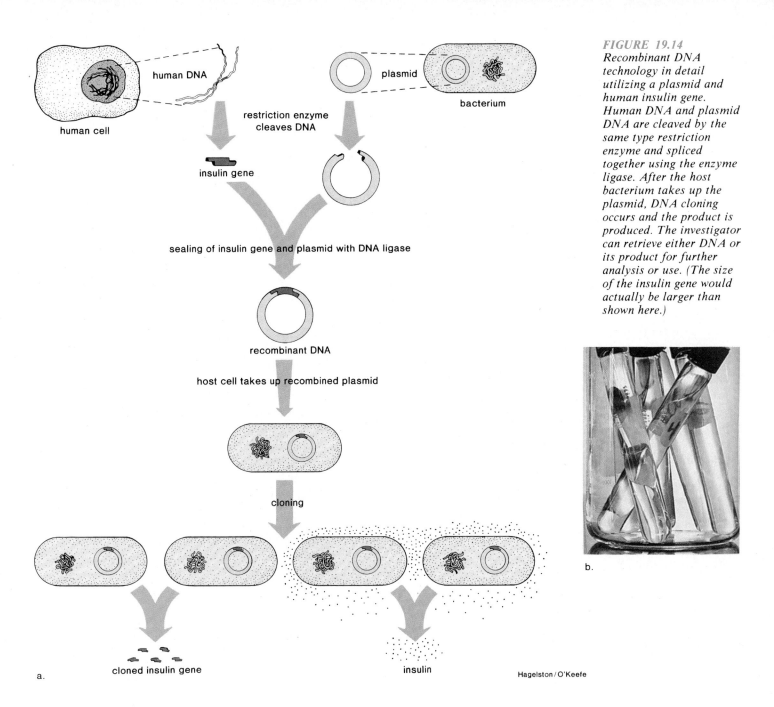

FIGURE 19.14
Recombinant DNA technology in detail utilizing a plasmid and human insulin gene. Human DNA and plasmid DNA are cleaved by the same type restriction enzyme and spliced together using the enzyme ligase. After the host bacterium takes up the plasmid, DNA cloning occurs and the product is produced. The investigator can retrieve either DNA or its product for further analysis or use. (The size of the insulin gene would actually be larger than shown here.)

human DNA

human cell

plasmid

bacterium

restriction enzyme cleaves DNA

insulin gene

sealing of insulin gene and plasmid with DNA ligase

recombinant DNA

host cell takes up recombined plasmid

cloning

cloned insulin gene

insulin

a.

b.

Hagelston/O'Keefe

Recombinant DNA

The introduction of foreign DNA into plasmid or viral DNA to give a cloning vector is a two-step process (fig. 19.14). The plasmid (or viral) DNA must be cut open, and then the foreign DNA must be inserted into this opening. Both of these steps are under the control of a specific type of enzyme.

Restriction enzymes have the capability of cutting a molecule of DNA into discrete pieces. These enzymes occur naturally in bacteria. If present, they protect bacteria from infection by cutting up viral DNA so that the viruses are unable to reproduce. They are called restriction enzymes because they *restrict* the growth of viruses. Each type of restriction enzyme, and hundreds are now known, cleaves DNA

at specific locations called the *restriction sites*. For example, there is a restriction enzyme that always cleaves double-stranded DNA when it has this sequence of bases:

```
. . . . .G A A T T C. . . . .
. . . . .C T T A A G. . . . .
```

Furthermore, this enzyme always cleaves each strand between a G and an A in this manner:

```
. . . . .G          A A T T C. . . . .
. . . . .C T T A A          G. . . . .
```

Notice that there is now a gap into which a piece of foreign DNA could be placed. It is only necessary that the foreign DNA piece end with bases complementary to those exposed by using the restriction enzyme. This is assured by cleaving the foreign DNA with the same type restriction enzyme. It must be the same type of restriction enzyme because different enzymes recognize and cleave different nucleotide sequences in DNA. The single-stranded ends that result from restriction enzyme use are called "sticky ends" because they facilitate the insertion of foreign DNA into the vector DNA.

The second enzyme needed is DNA **ligase**, an enzyme present in all cells, that ordinarily seals any breaks in a DNA molecule. Genetic engineers use this enzyme to seal the foreign piece of DNA into the vector. Gene splicing is now complete and a **recombinant DNA** molecule has been prepared. A recombinant DNA molecule is one that contains DNA from two different sources.

Once the recombined plasmid has been taken up by the host cell, it is expected to function normally—reproducing along with the host DNA and producing the protein of interest. For example, in figure 19.14 it is assumed that the gene for human insulin was spliced into the plasmid taken up by a bacterium. After such bacteria have metabolized and reproduced, the biotechnologist can recover either the cloned gene or the protein product (i.e., insulin).

Although we have not mentioned it previously, it should be noted that genes spliced into vectors very often are accompanied by regulatory genes. In summary, these procedures are common in experiments involving genetic engineering:

1. Isolation of a gene; that is, removal of a particular portion of DNA from a cell. Thereafter it is possible to determine the sequence of nucleotides in the gene.
2. Manufacture of a gene; that is, joining nucleotides together in the sequence of the normal gene, or creating a mutated gene by altering the sequence.
3. Joining the regulatory regions of a viral or bacterial gene to an isolated or machine-made structural gene so that transcription is assured.
4. Insertion of the regulator-structural gene complex into a vector.
5. Placement of the isolated or constructed gene linked to appropriate regulatory regions in another cell, where it undergoes replication and directs protein synthesis.

Cells can be genetically engineered (transformed) to produce a new protein product. Often vectors are used to carry the new gene into the cell.

Today's Biotechnological Products

Table 19.6 shows you at a glance the types of the biotechnological products that are now available. There are three categories in the table: hormones and similar types of proteins, DNA probes, and vaccines. Organic chemicals are also mentioned in the following, and monoclonal antibodies are discussed on page 155.

Hormones and Similar Types of Proteins

One impressive advantage of biotechnology is that it can make proteins available that are very difficult to come by otherwise. Consider, for example, human growth hormone. Previously, the hormone was extracted from the pituitary gland of cadavers, and it took fifty glands to achieve enough hormone for one dose. Only a few thousand adolescents had ever been treated, yet it is estimated that there are even today 10,000 to 15,000 children with growth hormone inadequacy. Similarly, insulin was previously extracted from the pancreas glands of slaughtered cattle and pigs; it was expensive and sometimes caused allergic reactions in recipients. Yet there are millions of diabetics who require daily insulin shots to control their condition. And few of us had ever heard of tPA (tissue plasminogen activator), a protein present in the body in minute quantities that activates an enzyme to dissolve blood clots. Now tPA is a product that is used to treat heart attack victims.

A study of the list of prospective protein products indicates that some of the most troublesome and serious afflictions in humans will soon be treatable: Factor VIII will be available for hemophilia; human lung surfactant will be available for premature infants with respiratory distress syndrome; atrial natriuretic factor may be helpful to many with hypertension. The list will grow because bacteria (or other cells) can be engineered to produce virtually any protein.

Hormones are also being produced for use in animals. No longer is it necessary to feed steroids to farm animals; they can be given growth hormone, which produces a leaner meat that is healthier for humans. Also, when cows are given growth hormone they produce 25% more milk than usual, which should make it possible for the dairy farmer to maintain fewer cows and cut down on his overhead expenses.

HUMAN ISSUE

Although new genetically engineered drugs are becoming available, the price for such drugs can be very expensive. One such drug is tPA, which is quite costly, but is effective in dissolving blood clots following heart attacks. Should the ability to pay, age, or general health be considered in determining who receives treatment? For example, should a 45-year-old person who has medical insurance and an 85-year-old person who does not have insurance receive equal consideration for treatment?

DNA Probes

Biotechnology is contributing greatly to diagnostics by making DNA probes (table 19.6) and monoclonal antibodies.[1] To construct a DNA probe you need only acquire or make a small piece of a gene. For example, it is possible to determine the sequence of amino acids in a polypeptide and from that deduce a short sequence of nucleotides in the gene.

[1]Even though monoclonal antibodies are not produced by recombinant DNA technology, they are still considered to be a biotechnological product. Monoclonal antibodies are discussed on page 155.

TABLE 19.6 *REPRESENTATIVE BIOTECHNOLOGICAL PRODUCTS*

HORMONES AND SIMILAR TYPES OF PROTEINS	DNA PROBES	VACCINES
FOR TREATMENT OF HUMANS	**FOR DIAGNOSTICS IN HUMANS**	**FOR HUMAN USE**
Insulin	Legionnaires' disease	Hepatitis B
Growth hormone	Walking pneumonia	Herpes[a] (oral and genital)
tPA (tissue plasminogen activator)	Sickle cell anemia	AIDS[a]
Interferon (alpha, beta[a], gamma[a])	Cystic fibrosis	Hepatitis A[a]
Erythropoietin[a]	Tendency for emphysema, thalassemia, retinoblastoma	Group B meningococci[a]
Factor VIII[a]	Hemophilia B	Hepatitis non-A and non-B[a]
Human lung surfactant[a]	Sexually transmitted diseases	Malaria[a]
Atrial natriuretic factor[a]	Polycystic kidney disease	
Superoxide dismutase[a]	Paternity testing	
Alpha-1-antitrypsin[a]	Monitoring bone marrow transplants	
	Susceptibility to atherosclerosis, diabetes, and hypertension	
	Periodontal disease	
	Huntington disease[a]	
	Duchenne muscular dystrophy[a]	
	Cancers (chronic myelogenous leukemia, T- and B-cell cancer, others[a])	
	Detection of criminals[a]	
FOR TREATMENT OF ANIMALS		**FOR ANIMAL USE**
Growth hormone (cows, pigs, chickens[a])		Hoof-and-mouth disease
Lymphokines[a]		Scours
		Brucellosis
		Rift Valley fever
		Herpes
		Influenza
		Pseudorabies
		Feline leukemia virus[a]
		Rabies[a]

[a]Expected in the near future. Any of these products that are presently available are not made by recombinant DNA technology.

Then, a laboratory machine will string the nucleotides together in the correct order so that you now have the piece of the gene you want. This piece can be cloned to give multiple copies. If the cloned DNA is denatured, the double helix will unzip, giving single strands. Each of these is a DNA probe because it will seek out and bind to any complementary DNA strand present in body fluids or removed from the body's cells. DNA probes can presently be used in paternity suits (consider that one chromosome of each pair is inherited from the father) and sometimes in forensics to put the finger on who really committed a crime. At times, it is also possible to use DNA probes to indicate whether the gene of an infectious organism is present in a discharge or in the blood. If the gene is present, the organism is present, and the cause of the infection is known. When available, probes can also tell us whether a gene coding for an hereditary defect or causing a cell to be cancerous is present.

Vaccines
Recombinant DNA technology can produce safe vaccines. In the old days, a vaccine was made by treating the infective agents (bacteria or viruses) so that they lost much of their

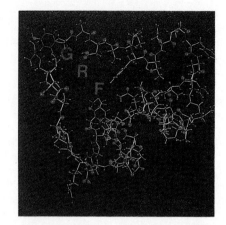

a.

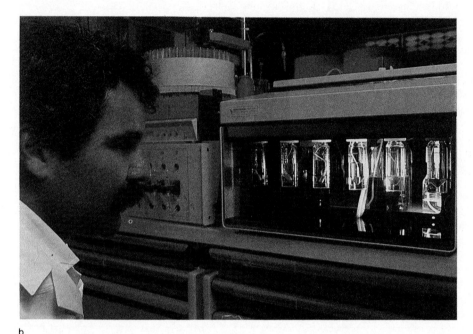

b.

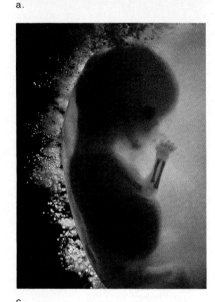

c.

virulence. The problem was that sometimes the vaccine caused the very condition it was supposed to prevent. Now we can make "subunit" vaccines, and the chance of these causing an infection is almost nil.

Biotechnology can make vaccines because bacteria and viruses have surface proteins. A gene for just one of the proteins or for a polypeptide can be extracted from an infectious organism and inserted into a plasmid. The host cells for the plasmid will produce many copies of the surface protein, which is antigenic when injected into humans. Right now only a vaccine for hepatitis B is on the market, but you can see that the ones soon expected are of great importance: a vaccine for malaria and another for AIDS. Malaria has been a scourge on humanity since the beginning of time, and AIDS is the most deadly disease known to modern humans.

Vaccines have also been produced for the inoculation of farm animals. Each of the illnesses listed in table 19.6—hoof-and-mouth disease, infant diarrhea, scours, etc.—causes an untold number of illnesses and deaths each year. These were formerly a severe drain on the time, energy, and resources of farmers.

Organic Chemicals

Organic chemicals are often synthesized in vitro (in laboratory glassware outside a living cell) by utilizing catalysts and appropriate precursor molecules or by using bacteria that carry on the synthesis. Today, there is even more interest in having bacteria produce organic chemicals because it is now possible to manipulate the genes that code for the enzymes needed to produce the organic chemical.

For example, chemists isolated a strain of bacteria that is especially good at producing phenylalanine, an organic chemical needed to make aspartame, the dipeptide sweetener better known as NutraSweet. They isolated, altered, and formed a vector for the appropriate genes so that any bacterial host can be especially geared up to produce phenylalanine.

Biotechnological products include hormones and similar types of proteins, DNA probes, and vaccines. These products are of enormous importance to the field of medicine.

FIGURE 19.16   *Cloning of entire plants from tissue cells.* a. *Sections of carrot root are cored, and thin slices are placed in a nutrient medium.* b. *After a few days, the cells form a callus, a lump of undifferentiated cells.* c. *After several weeks, the callus begins sprouting cloned carrot plants.* d. *Eventually the carrot plants can be moved from culture medium to pots.*

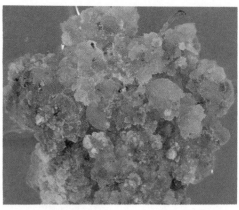

b.

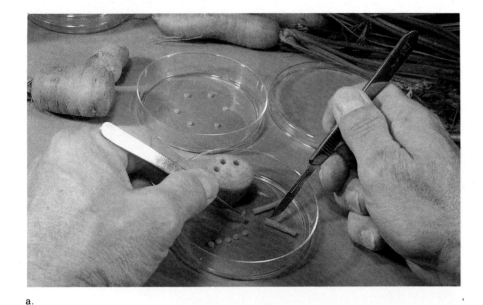

a.

c.

d.

Transgenic Organisms

Thus far we have considered only the insertion of foreign genes into bacteria. Suppose you were to put a foreign gene into an immature cell that was capable of developing into a multicellular organism. Would the gene be expressed in the cells of that organism? Indeed, this has been achieved. These organisms are called *transgenic organisms* because a gene has been transplanted into them.

Suppose a multicellular animal, such as a human, has an inborn error of metabolism. Could you in some way insert a gene into some of the body's cells so that these cells would produce the normal or lacking protein? In fact, research is underway, and we call this possibility *gene therapy.*

In the following we will consider the work in this area in plants and in animals, including humans.

Plants

Plants in particular lend themselves to biotechnological manipulation because it is not only possible to grow plant cells in tissue culture, but each one can be stimulated to produce an entire plant. For example, you can place carrot tissue from an adult plant in a nutrient-rich liquid medium and from each cell eventually get an adult plant (fig. 19.16). Previously this type of cloning has been used to provide genetic carbon copies of strawberry, asparagus, and oil palm plants for commercial use. It has also been used to isolate any variant cells to give different strains of carrots, celery, and even wheat. Plant tissue culture can be used to screen for cells that have a particular property, such as resistance to an herbicide. It is obviously less time consuming and less costly to screen millions of cells in a flask than to screen adult plants.

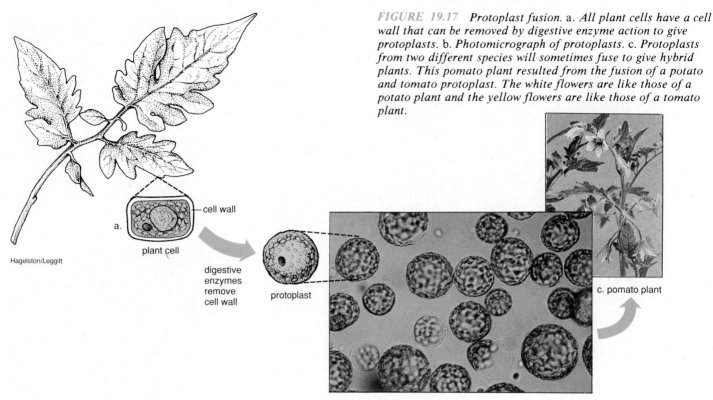

FIGURE 19.17   *Protoplast fusion.* a. *All plant cells have a cell wall that can be removed by digestive enzyme action to give protoplasts.* b. *Photomicrograph of protoplasts.* c. *Protoplasts from two different species will sometimes fuse to give hybrid plants. This pomato plant resulted from the fusion of a potato and tomato protoplast. The white flowers are like those of a potato plant and the yellow flowers are like those of a tomato plant.*

cell wall

a.

plant cell

Hagelston/Leggitt

digestive
enzymes
remove
cell wall

protoplast

c. pomato plant

b. protoplasts

Besides these procedures, it is also possible to apply recombinant DNA technology to plant cells. The presence of the plant cell wall (p. 000) hampers the possibility of plasmid uptake, and so it will be necessary to work with "naked" cells, called *protoplasts*. Protoplasts have had their walls removed (fig. 19.17). In order to have protoplasts undergo development, they need special handling. This should present no difficulty because plant breeders have had previous experience coaxing protoplasts to grow into adult plants. In some experiments, protoplasts from two different species were fused, and it was still possible to achieve mature plants. A pomato plant (potato × tomato fusion) produced both types of vegetables, but they were small and the seed quality was poor.

Genetic engineers have introduced genes that convey resistance to herbicides and insect attack, usually into tobacco protoplasts. The vector of choice is the Ti (tumor-inducing) plasmid from the bacterium *Agrobacterium tumefaciens*. Ordinarily the Ti plasmid invades plant cells and causes a cancerlike growth called crown gall disease, but when used as a vector the plasmid is engineered to lack virulence.

Plants that develop from genetically engineered protoplasts have the desired characteristics and breed true. Thus far these plants are experimental and not of commercial importance. In one instance the gene for the enzyme luciferase was transplanted from a firefly into a tobacco plant. Whenever the plant was sprayed with luciferin it glowed, proving that the inserted gene was indeed present and active (fig. 19.18).

Several problems remain before there will be transformed cereal plants for commercial use. Tobacco plants are dicots, whereas the cereals are monocots. Only recently has it been possible to get engineered corn, a monocot, to regenerate from a protoplast. Monocot protoplasts do not ordinarily take up the Ti plasmid in tissue culture. Techniques are being developed to have these cells take up genetic material directly. For example, it has just been reported that lasers make tiny self-sealing holes in cell membranes through which genetic material can enter. Individual microinjection has been used to insert foreign genes into protoplasts, but obviously it would be more efficient to simply irradiate cells with a laser beam while they are suspended in a liquid containing foreign DNA.

Since the necessary techniques are—or will be soon—perfected, it is believed that within a relatively short period of time, agriculturally significant plants will be produced that have the characteristics listed in table 19.7. In the meantime, bacteria have been engineered to promote the health of plants when they are applied to them. For example, field tests conducted to test the ability of genetically engineered bacteria, called ice-minus bacteria, to live on and protect the vegetative parts of plants from frost damage have been successful. Also, a bacterium that normally colonizes the roots of corn plants has now been endowed with genes (from another bacterium) that code for an insect-killing toxin. Field tests are scheduled.

FIGURE 19.18 *This transgenic plant glows when sprayed with luciferin, a chemical that emits light, because its cells contain the protein luciferase, a firefly enzyme that breaks down the luciferin.*

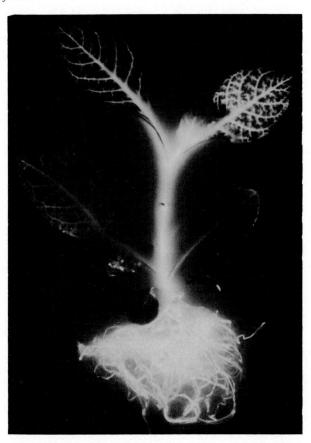

FIGURE 19.19 *From greenhouse to field testing. a. Both tobacco plants were wounded and infected with* Agrobacterium tumifaciens, *a bacterium that causes crown gall disease. The plant on the right did not develop the disease because it was engineered to have the microbial gene that provides resistance to the infection. b. Many resistant plants are grown outside to test their ability to remain healthy in an agricultural field.*

a.

b.

TABLE 19.7 *BIOTECHNOLOGY IN PLANTS*

Due to biotechnology, plants of the future may have the following characteristics:

1. a natural resistance to various diseases (fig. 19.19) and pests so that fewer pesticides will be needed
2. the ability to fix nitrogen so that less fertilizer will need to be applied to crops
3. the increased ability to grow under unfavorable environmental conditions, such as lack of water or in salty soil. Less irrigation of crops would then be necessary
4. seed proteins that contain all the essential amino acids so that a complete protein source for vegetarians will be available

Ecologists and the general public have been very much concerned about the deliberate release of these genetically engineered microbes (GEMS) into the environment. All sorts of steps, including legal action, were taken to delay field testing of the ice-minus bacteria. This stimulated a dialogue between molecular biologists and other groups that hopefully will foster more understanding on both sides in the future.

Animals, Including Humans

Genetic engineering of animals is becoming quite frequent. The reading on this page describes how the recombinant DNA technique is directed toward producing larger game fish. Genes have also been microinjected into the zygotes of laboratory mice, and these zygotes, after implantation in surrogate mothers, have developed normally. Further, they sometimes have the phenotype desired. For example, the sheep gene that codes for the milk protein betalactoglobulin (BLG) was injected into mice zygotes. (Mice normally produce a milk that has no BLG at all.) Of forty-six offspring successfully weaned, sixteen carried the BLG sequence, and the females among them later produced BLG-rich milk. One

SUPERFISH: COMING SOON TO A LAKE NEAR YOU

A Minnesota researcher injects fish eggs with growth genes.

Walleye, chinook, bluegill—to anglers, these names conjure up images of fishing trips past. Game fishing in the 1990s may call to mind a more unusual image, like that of a 100-pound chinook that has been injected with cattle growth hormone genes. Spurred by global research in aquaculture, U.S. biologists are manipulating genes and chromosomes in hopes of creating "superfish" for the table and the den wall.

"Genetic manipulation is one of the first ideas that come to mind when people want to increase the yield of some source of food," says Anne Kapuscinski, an assistant professor of fisheries at the University of Minnesota. "Some of this research is now spilling over into game-fish enhancement."

Kapuscinski, along with three Minnesota colleagues, molecular geneticists Kevin Guise, Perry Hackett, and Anthony Faras, is working under the auspices of the Minnesota Sea Grant College Program to engineer a superwalleye by inserting an extra growth hormone gene into the egg. The researchers hope to produce some of these "transgenic" walleyes by the summer of 1989. They dream of economical "cradle-to-grave" nurseries supplying the nation with walleye, lake trout, and other traditionally slow-growing game fish of the upper Midwest. The Minnesota legislature, the

project's major funder, hopes the superfish will someday spark a tourist bonanza.

Before that can happen, the products of genetic tinkering must be safe to release into the environment. Says Guise, "We are looking closely at being able to control the growth hormone gene so that we can turn it on while it's in the hatchery and turn it off when it's released into the environment. That way the fish will grow and act like a normal fish in the wild, but the DNR [Department of Natural Resources] will still see an increase in its hatchery efficiency—getting more fish for the buck." . . .

Scientists soon hope to splice into a fish's DNA code the ability to resist disease, to grow faster on less food, and perhaps even to perform special functions. "There are genes, for instance, that have been isolated from bacteria that will degrade dioxin," says Guise. "It might be possible one day to create a vacuum-cleaner type of fish. Take a bottom feeder and put in the right genes . . . I wouldn't want to eat that fish, but you may be able to engineer biological systems to clean up a damaged environment. We're looking at all sorts of things, but you have to be cautious."

The transgenic research at Minnesota parallels work being done in the People's Republic of China, where aquaculture is a 4,000-year-old tradition. Working with carp and loach, produced BLG at five times the concentration found in sheep milk. Some of the transformed mice passed on the BLG to their offspring.

The results of these types of experiments are not uniform—not all the zygotes injected survive the procedure; not all of those born express the gene to the same degree; and not all pass the gene on to their offspring. The possibility of using this genetic engineering method in humans is still believed to be quite distant because successful results cannot be guaranteed. However, there is a distinct possibility of using this genetic engineering method in farm animals. It's conceivable that genetic engineers could eventually produce farm animals with all sorts of desirable traits.

In the near future, though, we will probably see much progress in the area of gene therapy in adult humans. Even now, W. F. Anderson and collaborators at the National Institutes of Health and Memorial Sloan Kettering Cancer Center are preparing for clinical tests to correct a rare hereditary disease that results from a deficiency of the enzyme adenosine deaminase. They will be inoculating bone marrow stem cells as described in figure 19.20. Other investigators are looking at the possibility of using liver cells rather than bone marrow stem cells to correct some inborn errors of metabolism. In the opinion of Dr. Anderson, it is unethical to delay human gene therapy experiments any longer. Some

two valuable food species in Asia, Professor Zuo-Yan Zhu of the Institute of Hydrobiology in Wuhan Hubei Province has succeeded in transplanting growth genes that were then passed on naturally to successive generations. While many loach grew 1½ times faster than normal loach, some with the extra gene grew 2 to 4½ times faster. The Minnesota team hopes to achieve similar results with walleyes. That effort may get a boost when Zhu travels to Minnesota this spring on a four-month scientific exchange.

From mainland China to the labs of Purina Mills, a U.S.-based subsidiary of British Petroleum, researchers share Guise's exhilaration over ichthyological eugenics. But many of them stress his note of caution—especially when it comes to releasing genetically engineered fish into the wild. "As a scientist with a strong interest in conservation," says Kapuscinski, "I am disturbed that people think we can circumvent nature by just doing fantastic genetic manipulations. We don't really know that much yet about the performance of these kinds of animals in nature, and I think great caution has to be exercised." . . .

Researchers at fisheries typically use growth hormone genes extracted from cattle because these genes are easy to obtain. That they work in fish is an example of conservation evolution—nature's tendency to reuse effective biochemical systems up and down the

evolutionary ladder. "Cow and sheep growth hormones, as well as those from other fish, all work in a fish to produce larger fish," says Guise. Blue shark growth hormone, for example, spurs growth in carp. Frightening as this may sound to the post-*Jaws* generation, shark hormone works no more or less effectively than that from catfish.

Still, says Kapuscinski, there may be advantages to using species-specific genes. "It might be better regulated inside the fish," she says, "but we don't know at this point." To this end the team is hoping to clone walleye growth hormone genes for future experiments.

Of the many challenges facing the Minnesota team, one of the most nettlesome is the mass transfer of the genes. Walleye produce 20 to 30 thousand eggs during each yearly breeding season, and it's extremely time-consuming to microinject each one individually (see figure). Today, a technician working alone can treat only about 250 eggs per hour. This must be done after fertilization but before cell division begins; otherwise the transplanted gene will appear in a mosaic pattern in the tissues of the adult fish.

The Minnesota team is working on a promising alternative—a process known as electroporation in which an electrical pulse creates tiny, temporary pores in the egg membrane thereby creating a means of introducing the DNA into the fertilized egg. Ascertaining whether the

gene transfer was successful used to mean waiting for the fingerlings to grow up—a major slowdown to research. To expedite the process, Guise now attaches to the growth hormone gene a second gene that provides resistance to a specific poison. The hatchlings that survive the poison are presumed to have also accepted the growth gene.

The ethical ramifications of their work are not lost on either Guise or Kapuscinski. Unlike triploid and crossbred fish, which occur naturally on occasion, transgenic fish are engineered mutants. "A giant carp with an extra growth hormone gene is fairly new," says Guise. "You have to look closely at what you're doing and figure out what the rules are for the next step." . . .

In the 1800s, say Guise, carp were introduced into the Midwest with the best of intentions. States fought over the limited supply of carp eggs. Before long, though, the carp had taken over many of the waters in which they had been planted, displacing such valuable game fish as bass. "The DNR is not too happy about it right now," says Guise.

Does he ever worry that the zeal to produce game fish for the 1990s might lead to a rerun of the carp fad of the 1800s? "Sure," says Guise. "That's why we're studying as many aspects of this thing as possible."

FIGURE 19.20 The virus selected for gene therapy is a
retrovirus. Retroviruses have an RNA chromosome. When this
chromosome enters a cell, reverse transcription, during which
RNA is transcribed to DNA, precedes reproduction of the virus
(fig. 15.3). You can see this adds to those steps described here
for the virus to serve as a vector for the purpose of gene therapy.

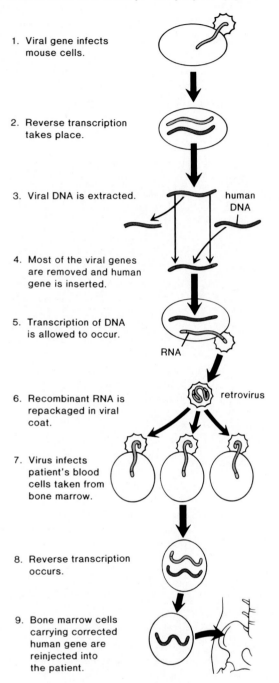

1. Viral gene infects
mouse cells.

2. Reverse transcription
takes place.

3. Viral DNA is extracted.

human
DNA

4. Most of the viral genes
are removed and human
gene is inserted.

5. Transcription of DNA
is allowed to occur.

RNA

6. Recombinant RNA is
repackaged in viral
coat.

retrovirus

7. Virus infects
patient's blood
cells taken from
bone marrow.

8. Reverse transcription
occurs.

9. Bone marrow cells
carrying corrected
human gene are
reinjected into
the patient.

FIGURE 19.21 *Gel electrophoresis is used to sequence DNA.
The DNA is cleaved into fragments that vary in length by only
one nucleotide. Each fragment ends with an A, T, C, or G,
marked by one of four fluorescent dyes. The fragments are
subjected to gel electrophoresis during which they migrate on a
gel in an electric field. Shorter fragments migrate further than
longer fragments. Under proper lighting, the investigator (or a
computer) can simply read off the sequence of the nucleotides.*

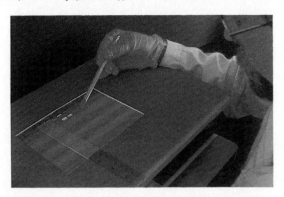

laypeople, in particular, have directly the opposite opinion.
Most believe that the issue should be publicly discussed until
there is a general consensus to guide us all.

Transformed plants and animals (transgenic organisms) have
been achieved. This genetic engineering feat is expected to
revolutionize agriculture and animal husbandry. Work has
also begun in the area of human gene therapy.

The Future

Laboratory Research

The techniques, enzymes, vectors, and hosts that are needed
for biotechnology were discovered by investigators doing basic
research in the laboratory. Basic research will continue, and
it would appear that eventually biotechnological experiments will be carried out in a cell-free system. In that case,
it will be possible to bypass vectors and hosts completely because procedures like cloning will simply take place in vitro;
i.e., within laboratory glassware. This possibility is being actively pursued.

The U.S. government has recently committed to providing the funds necessary to have the entire human genome
sequenced (fig. 19.21). The human genome is estimated to
contain up to 300,000 genes. Until this time, only about 100
genes have ever been sequenced completely because of the
time needed to accomplish the task. Now there are auto-
mated DNA sequencers that use computers to accomplish
the sequencing at a fairly rapid rate. Even so, it's estimated
that the job will take about thirty years unless even newer
and better technology is developed. In any case, it is difficult
to imagine the size of the computer data base that will be
needed to store this information. The amount of paper needed
to record the results will be staggering, requiring on the order
of 200 books the size of an encyclopedia volume.

Some researchers believe that it will be of little benefit to simply know the sequence of the bases in the human genome. Others believe that this knowledge will take us one step closer to being able to manufacture genes (and their control elements) without having to slice them out from their neighbors in a hit-or-miss fashion. Then it will be possible to study the protein, coded by the gene, to determine how it functions normally in the cell. This knowledge will contribute greatly, for example, to our understanding of how the immune system functions and of how defective genes wreak havoc on the body.

Applications

It's easy to predict that there will be at least three major revolutions, one in the area of medical diagnostics and treatment, another in agriculture, and a third in improvement of the environment.

In medicine, human gene therapy will become the order of the day. Diagnostic DNA probes will be available to indicate whether a genetic disease is present and, if possible, the defect will simply be corrected by inserting the proper gene into the body. In some instances, if the individual has inherited only a tendency toward a particular condition, proper steps may be taken to prevent its development. For example, if you know you have a tendency toward emphysema, you could be careful to avoid polluted air. If a medical condition does develop and there is no genetic cure, then new and novel medicines, such as the ones mentioned previously, will be available to keep the condition under control.

In agriculture, plants will be given the ability to grow and produce under adverse conditions. They will need a limited amount of inputs, like fertilizer, pesticides, herbicides, and water. Farm animals will be healthier and more productive, and therefore fewer animals could possibly produce the same amount of food. Altogether, then, agriculture will no longer make a major contribution to the degradation of the environment.

Bacteria, even today, have been fashioned to eat up oil and various contaminants (fig. 19.22), but thus far they have not been used. In the future, it may be possible to simply order the correct strains to clean up any contaminated area. Environmental pollution may cease to be a great threat to continuance of the human species.

The picture painted here is one-sided and very optimistic. Could it be that we would use our newfound knowledge to fashion and change the human genome to suit an idealized model and that we will be even less tolerant than today of individual differences? Could the modified plants, animals, and microbes cause environmental problems that we cannot even begin to imagine? These are the questions we have to try to answer as the new era of biotechnology descends upon us.

Biotechnological advances are expected in laboratory research, and applications of more and more significance are anticipated in all areas of human concern.

SUMMARY

During replication, DNA becomes "unzipped," and then a complementary strand forms opposite to each original strand. DNA controls protein synthesis. During transcription, mRNA is made complementary to one of the DNA strands. It then contains codons and moves to the cytoplasm and becomes associated with the ribosomes. During translation, tRNA molecules, attached to their own particular amino acid, travel to the mRNA, and through complementary base pairing, the tRNAs and therefore the amino acids in a polypeptide chain become sequenced in a predetermined way. Some genes, called structural genes, code for proteins that are enzymes or structural components of cells. Other genes, called regulatory genes, control the expression of structural genes. Genes are subject to mutations that may be substitutions, additions, or deletions of bases. Transposons are movable elements that cause mutations. Mutagenic agents also promote genetic mutations.

The DNA of higher organisms contains repetitive DNA with no known function. Other parts are introns and exons; mRNA is processed before it leaves the nucleus—the introns are removed, while the exons remain to be eventually expressed.

Biotechnology is the use of a natural biological system to produce a product or to achieve an end desired by human beings. The newest advances in biotechnology stem from the use of the recombinant DNA technique by which foreign genes are inserted into cells, very often by utilizing vectors. Transformed cells produce products of interest to humans, such as hormones and related proteins, DNA probes, and vaccines. Transgenic organisms have also been achieved. Plants lend themselves to genetic manipulation because it is possible to grow an adult plant from a cell that was originally maintained in tissue culture. Genetic engineering of animal zygotes has not as yet led to uniform results and so genetic engineering of the human zygote has not yet begun. But human gene therapy attempts are underway and may lead to success sometime within the foreseeable future. Biotechnology is expected to revolutionize almost every area of human concern.

OBJECTIVE QUESTIONS

1. Replication of DNA is semiconservative, meaning that each new double helix is composed of an _____ strand and a _____ strand.
2. The DNA code is a _____ code, meaning that every three bases stands for an _____ .
3. The three types of RNA that are necessary to protein synthesis are _____ , _____ , and _____ .

4. Which of these types of RNA carries amino acids to the ribosomes? _____
5. Genetic engineering commonly makes use of both _____ and _____ as vectors.
6. The two types of enzymes needed to accomplish gene splicing are _____ and _____ .
7. An rDNA molecule contains DNA from at least _____ different sources.
8. Biotechnology has given us _____ that can be used to diagnose illnesses and genetic diseases.

9. _____ are naked plant cells that have the capability of developing into an adult plant.
10. Plasmids and viruses that carry recombinant DNA into host cells are called _____ .

Answers

1. old, new 2. triplet, amino acid 3. mRNA, tRNA, rRNA 4. tRNA 5. plasmids, viruses 6. restriction enzymes, DNA ligase 7. two 8. DNA probes 9. Protoplasts 10. vectors

STUDY QUESTIONS

1. Compare and contrast the structure of RNA with that of DNA. (p. 414)
2. What are the four functions of the hereditary material? (p. 416)
3. Explain how DNA replicates. Why is this replication called semiconservative? (p. 416)
4. If the code is TTA'TGC'TCC'TAA, what are the codons and what is the sequence of amino acids? Show how a deletion or duplication could affect the code. (pp. 420, 424)

5. List the five steps involved in protein synthesis, mentioning the process of transcription and translation and the roles of DNA, mRNA, rRNA, and tRNA. (p. 422)
6. You are a scientist who has decided to "clone a gene." Tell precisely how you would proceed. (p. 426)
7. Name three categories of biotechnological products that are now available, and discuss their advantages. (p. 428)

8. Discuss the application of biotechnology to agriculture. Contrast the accomplishments to date to those expected in the future. (p. 431)
9. Discuss transgenic experiments in animals and tell why such experimentation is not expected in the near future in humans. (p. 434)
10. Discuss in general the many ways that biotechnology is expected to change our lives in the next few years. (p. 436)

THOUGHT QUESTIONS

1. Drawing from your knowledge of the reproductive process, support the suggestion that we are nothing but transport vehicles or hosts for DNA. What arguments can you think of to counter this suggestion?

2. Both mRNA and tRNA are transcribed from DNA in the nucleus, but they have different functions in the cytoplasm. How might they be "guided" to take up their respective functions?

3. How does the success of genetic engineering techniques lend support to the belief that organisms are chemical and physical machines?

KEY TERMS

anticodon (an″ti-ko′don) a "triplet" of three nucleotides in transfer RNA that pairs with a complementary triplet (codon) in messenger RNA. *420*

cloned (klōnd) genes from an external source that have been reproduced by bacteria. *426*

codon (ko′don) a "triplet" of three nucleotides in messenger RNA that directs the placement of a particular amino acid into a polypeptide chain. *419*

complementary base pairing (kom″plĕ-men′tă-re bās pār′ing) pairing of bases between nucleic acid strands; adenine is always paired with either thymine (DNA) or uracil (RNA) and cytosine is always paired with guanine. *414*

messenger RNA (mes′en-jer) mRNA; a nucleic acid (ribonucleic acid) complementary to genetic DNA and bearing a message to direct cell protein synthesis at the ribosome. *419*

plasmid (plaz′mid) a circular DNA segment that is present in bacterial cells but is not part of the bacterial chromosome. *426*

recombinant DNA (re-kom′bĭ-nant) DNA having genes from two different organisms often produced in the laboratory by introducing foreign genes into a bacterial plasmid. *428*

regulatory genes (reg′u-lah-tor″e jēnz) genes that code for proteins involved in regulating the activity of structural genes. *422*

replication (re″plĭ-ka′shun) the duplication of DNA; occurs when the cell is not dividing. *416*

ribosomal RNA (ri′bo-sōm″al) rRNA; RNA occurring in ribosomes, structures involved in protein synthesis. *420*

structural genes (struk′tŭr-al jēnz) genes that direct the synthesis of enzymes and also structural proteins in the cell. *422*

template (tem′plāt) a pattern that serves as a mold for the production of an oppositely shaped structure; one strand of DNA is a template for the complementary strand. *416*

transcription (trans-krip′shun) the process that results in the production of a strand of mRNA that is complementary to a segment of DNA. *419*

transfer RNA (trans′fer) tRNA; molecule of RNA that carries an amino acid to a ribosome engaged in the process of protein synthesis. *420*

translation (trans-la′shun) the process involving mRNA, ribosomes, and tRNA molecules that results in a synthesis of a polypeptide having an amino acid sequence dictated by the sequence of codons in mRNA. *419*

vector (vek′-tor) a carrier, such as a plasmid or virus, for recombinant DNA that introduces a foreign gene into a host cell. *426*

FURTHER READINGS FOR PART V

Ayala, F., and J. A. Kiger. 1984. *Modern genetics.* 2d ed. Menlo Park, CA: Benjamin/Cummings.

Baconsfield, P. G. et al. August 1980. The placenta. *Scientific American.*

Cech, T. R. November 1986. RNA as an enzyme. *Scientific American.*

Chambon, P. May 1981. Split genes. *Scientific American.*

Chilton, M. June 1983. A vector for introducing new genes into plants. *Scientific American.*

Cohen, S. N., and J. A. Shapiro. February 1980. Transposable genetic elements. *Scientific American.*

Darnell, J. E. October 1983. The processing of RNA. *Scientific American.*

———. October 1985. RNA. *Scientific American.*

Dickerson, R. E. December 1983. The DNA helix and how it is read. *Scientific American.*

Doolittle, R. F. October 1985. Proteins. *Scientific American.*

Felsenfeld, G. October 1985. DNA. *Scientific American.*

Gilbert, W., and L. Villa-Komaroff. April 1980. Useful proteins from recombinant bacteria. *Scientific American.*

Lake, J. A. August 1981. The ribosome. *Scientific American.*

Murray, A. W. and Szostak, J. W. November 1987. Artificial chromosomes. *Scientific American.*

Nomura, M. January 1984. The control of ribosome synthesis. *Scientific American.*

Novick, R. P. December 1980. Plasmids. *Scientific American.*

Patterson, D. August 1987. The causes of Down Syndrome. *Scientific American.*

Ptashne, M. January 1989. How gene activators work. *Scientific American.*

Radman, M. and Wagner, R. August 1988. The high fidelity of DNA duplication. *Scientific American.*

Rugh, R., et al. 1971. *From conception to birth: The drama of life's beginnings.* New York: Harper & Row, Publishers, Inc.

Shepard, J. F. May 1982. The regeneration of potato plants from leaf-cell protoplasts. *Scientific American.*

Stahl, F. W. February 1987. Genetic recombination. *Scientific American.*

Steitz, J. A. June 1988. "Snurps" *Scientific American.*

Volpe, E. P. 1983. *Biology and human concerns.* 3d ed. Dubuque, IA: Wm. C. Brown Publishers.

White, R. and Lalouel, J. February 1988. Chromosome mapping with DNA markers. *Scientific American.*

HUMAN EVOLUTION AND ECOLOGY

E videnices for the theory of evolution are drawn from many areas of biology. Charles Darwin was the first to recognize these evidences formally, and he suggested that those organisms best suited to the environment are the ones that survive and reproduce most successfully. Evolution causes life to have a history, and it is possible to trace the ancestry of humans even to the first cell or cells.

The diverse forms of life that are now present on earth live within ecosystems, where energy flows and chemicals cycle. Because population sizes remain constant, natural ecosystems tend to require the same amount of energy and chemicals each year.

Humans have created their own ecosystems, consisting of the country and city, which differ from natural ecosystems in that the populations constantly increase in size and ever greater amounts of energy and raw materials are needed each year. Since 1850, the human population has expanded so rapidly that some doubt there will be sufficient energy and food to permit the same degree of growth in the future. The size of the human population, however, has begun to level off, and it is suggested that energy be used efficiently and raw materials be cycled to assure a continued supply. These measures will also help curtail pollution.

The human ecosystem depends on the natural ecosystems not only because they absorb pollutants but also because they are inherently stable. Every possible step should be taken to protect natural ecosystems to help ensure the continuance of the human ecosystem.

CHAPTER TWENTY

EVOLUTION

CHAPTER CONCEPTS

1 The fossil record and comparative anatomy, embryology, and biochemistry all provide evidences of evolution.

2 Charles Darwin formulated a mechanism for an evolutionary process that results in adaptation to the environment.

3 Humans are primates, and many of their physical traits are the result of their ancestors' adaptations to living in trees.

4 Tool use, walking erect, and intelligence are all important advances made during human evolution.

5 All human races are classified as *Homo sapiens*.

An archaeologist in Chile dusting Homo sapiens remains.

INTRODUCTION

What living things are able to shape the course of evolution? Human beings, of course. When human beings alter the environment, they are influencing which organisms will survive and which will die out in a particular locale. More directly humans carry out breeding programs to select which plants or animals will reproduce more than others. The end result can be specific, discrete types (fig. 20.1); all of the different forms of domesticated dogs belong to the same species—*Canis familiaris*—and we have brought about the many varieties. It's obvious to us, today, that we are selecting out the genes we prefer because genes control the characteristics we wish to see propagated. One of the great accomplishments of twentieth-century science is to give the concept of evolution a genetic basis.

E volution is the process that explains the history and diversity of life. Data from various fields of biology give us evidence that **evolution** produced the myriad of organisms that are now present on earth.

EVIDENCES FOR EVOLUTION

Fossil Record

Our knowledge of the history of life is based primarily on the fossil record. **Fossils** (fig. 20.2) are the remains or evidence of some organism that lived long ago. Preserved fossils are most often found in sedimentary rock. Weathering produces sediments that are carried by streams and rivers into the oceans and other large bodies of water. There they slowly settle and are converted into sedimentary rock. Later sedimentary rocks are uplifted from below sea level to form new land. Now researchers are freely able to search for any fossil remains trapped in the rocks (fig. 20.3). The oldest fossils found are of prokaryotic cells dated some 3.5 billion years ago. Thereafter the fossils get more and more complex. For example, among animals simple eukaryotic forms were followed by invertebrates, and these were followed by vertebrates, until humans evolved only about two million years ago. This is clear evidence that there has been a history of

a.

b.

c.

d.

FIGURE 20. Artificial selection in dogs. All the different types of dogs like (a) Shetland sheepdog, (b) dalmation, (c) beagle, and (d) bulldog are descended from the original domesticated dog. Humans have brought about the different types by controlling how the dogs breed.

FIGURE 20.2 Fossils. a. Impression of fern leaves on a rock.
b. Saber-toothed tiger skull.

FIGURE 20.3 Field researchers examine rocks near North
Pole, Australia, for evidence of fossils. At this location,
researchers have found fossils from the earliest form of life.

a.

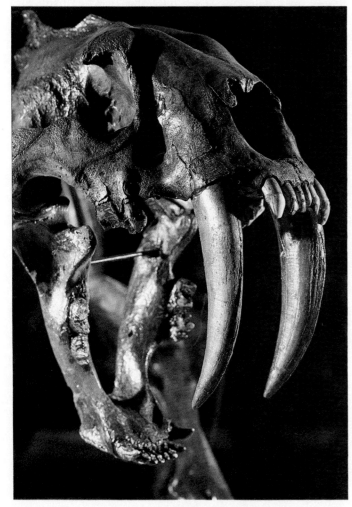

b.

life (fig. 20.4) based on evolutionary events. The later-
evolving animals are descended from the earlier ones. The
recent controversies among evolutionists are not about
whether evolution occurred but simply about the manner in
which it may have occurred.

Comparative Anatomy

A comparative study of the anatomy of groups of organisms
has shown that each has a *unity of plan.* For example, all
vertebrate animals have essentially the same type of skeleton. Unity of plans allows organisms to be classified into
various groups. Organisms most similar to one another are
placed in the same **species,** similar species are placed in a
genus, similar genera in a family; thus, we proceed from
family to **order** to **class** to **phylum** to **kingdom.** The classification of any particular organism indicates to which kingdom,
phylum, class, order, family, genus, and species the organism
belongs. According to the **binomial system** of naming organisms, each organism is given a two-part name, which consists
of the genus and species to which it belongs (table 20.1). Thus,
for example, a human is *Homo sapiens*[1] and the domesticated cat is *Felis domestica.* **Taxonomy** is the branch of biology that is concerned with classification, and biologists who
specialize in classifying organisms are called taxonomists.

[1]Varieties of the same species are also given a subspecies designation.
Neanderthals are *Homo sapiens neanderthalensis* and modern humans
are *Homo sapiens sapiens.*

There is no doubt that the fossil record supplies data to support the theory of evolution. The record shows that life began with simple organisms and advanced to the complex organisms in existence today. Also, direct ancestors for every living thing can be found in the fossil record.

A few years ago, Stephen J. Gould and Niles Eldredge, two paleontologists, wondered if the fossil record also gives us specific information regarding the tempo (rate) of evolutionary change. Based primarily on experimental genetic data in the laboratory, most biologists believe that a single species changes slowly—over thousands or even millions of years—into other species. This so-called *gradualistic model* of evolutionary change implies that the fossil record contains a plentiful supply of *intermediate forms,* which are fossil remains of life having characteristics of two different groups. Some intermediate forms have been found, but not an overwhelming number.

In contrast to the gradualistic model, Gould and Eldredge, upon close examination of the fossil record, proposed a *punctuated equilibrium model* of evolutionary change. The record shows that each species tends to remain the same for hundreds of thousands of millions of years. Then, relatively suddenly, new species appear. In other words, an *equilibrium* phase (long periods without change) is *punctuated* by a rapid burst of change. According to this model there would be few intermediate forms in the fossil record because such forms existed for only a short period of time. Gould and Eldredge also suggested that, without the presence of any other pertinent factors, such rapid change might require that natural selection select entire species for survival and not just individuals.

Thus far, evolutionists have been unable to supply additional data to support one model over the other. Very recently and unexpectedly, though, they have received some help from geologists and astronomers. For many years, evolutionists and other biologists have tried to explain the mass extinction of the dinosaurs. Dinosaurs dominated the earth for 135 million years, but suddenly they and much of the rest of life on earth vanished. In 1979, a Berkeley geologist, Walter Alvarez, and his colleagues found that Cretaceous clay contained an abnormally high level of iridium. This could have been caused by a worldwide fallout of radioactive material created by an asteroid impact at the time of the dinosaur disappearance. In 1984, paleontologists David Raup and John Sepkoski of the University of Chicago discovered that the dinosaurs are not alone; rather, the fossil record of marine animals shows that mass extinctions have occurred every 26 million years or so.

Surprisingly, astronomers, taking a hint from Alvarez, can give an explanation for these occurrences. Our solar system is in the Milky Way, a starry galaxy that is 100,000 light years[1] in diameter and 1,500 or so light years thick. Our sun moves up and down as it orbits in the Milky Way. Scientists predict that this vertical movement will cause our solar system to approach certain other members of the Milky Way every 26–33 million years, producing an unstable situation that could lead to comet bombardment of the earth. This bombardment can be likened to a worldwide atomic bomb explosion. A cloud of dust will mushroom into the atmosphere and shade out the sun, thus causing the plants to freeze and die. Mass extinction of animals will follow. Once the cosmic winter is over, plant seeds will germinate. This new growth will serve as food for the few remaining animals. Just as Darwin's finches rapidly evolved on the Galápagos Islands because of the lack of competition, these animals will also undergo rapid evolution.

This scenario suggests that the punctuated equilibrium model of evolutionary change does have merit. The equilibrium phase, during which evolutionary change is extremely slow, would occur between times of mass extinction, and the punctuated phase would occur immediately after mass extinctions. It is only necessary to return to the fossil record to see if this hypothesis holds.

[1]One light year, the distance light travels in a year, is about 6 trillion miles.

NEW QUESTIONS FROM THE FOSSIL RECORD

FIGURE 20.4 *The fossil record provides a history of life. The record indicates that primitive, single-celled organisms resembling a type of bacteria were the first life forms to appear. In time, more complex cellular forms appeared, followed by multicellular plants and animals. After that, there was an increase in complexity.*

Era	Period	Epoch	Years from Start of Period to Present	
Cenozoic	Quaternary	Recent	10,000	
		Pleistocene (Ice Age)	3 million	
	Tertiary		63 million	
Mesozoic "Age of Reptiles"	Cretaceous		135 million	
	Jurassic		181 million	
	Triassic		230 million	
Paleozoic	Permian		280 million	
	Carboniferous		345 million	
	Devonian		405 million	
	Silurian		425 million	
	Ordovician		500 million	
	Cambrian		600 million	
Proterozoic			1.5 billion	
			2.5 billion	
Archeozoic			4.5 billion	

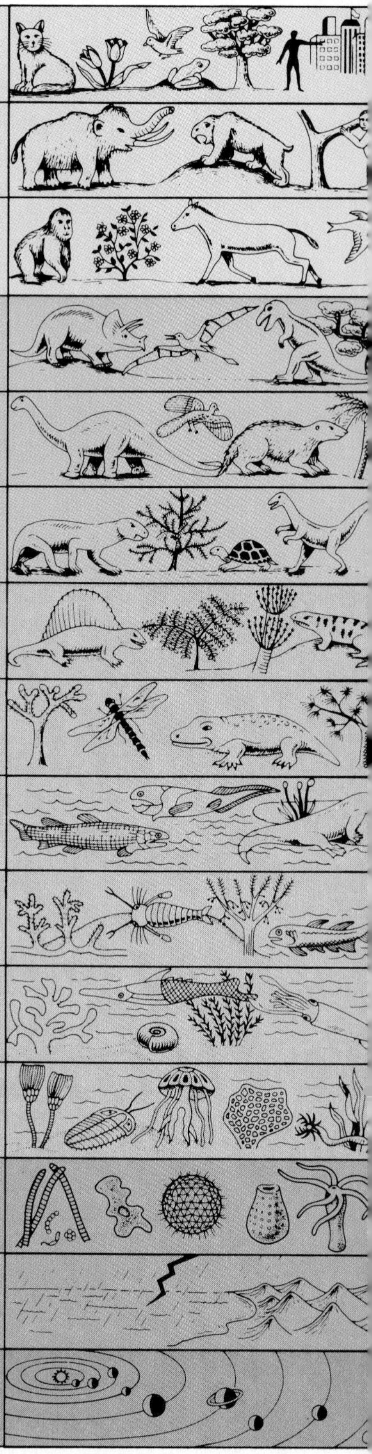

Plant Life	Animal Life
Increase in the number of herbaceous plants	Age of human civilization
Extinction of many species of plants	Great mammals such as woolly mammoth and saber-toothed tiger become extinct First humans
Dominance of land by angiosperms	Dominance of land by mammals, birds, insects Mammalian radiation First primates
Angiosperms prevalent, gymnosperms decline Trees resembling modern-day maples, oaks, and palms flourish	Dinosaurs reach peak, then become extinct Second great radiation of insects Mammals diversity
Gymnosperms such as cycads and conifers still prevalent	Dinosaurs large, specialized, more abundant First mammals appear First birds appear
Dominance of land by gymnosperms and ferns Decline of club mosses and horsetails	First dinosaurs appear Mammallike reptiles evolve
Gymnosperm and angiosperm(?) evolve	Expansion of reptiles Decline of amphibians
Age of great coal forests including club mosses, horsetails, and ferns	''Age of Amphibians'' First great radiation of insects First reptiles appear
Expansion of land plants; first forests of club mosses, horsetails, and ferns	''Age of Fishes'' First land vertebrates, the amphibians, appear
First vascular plants, modern groups of algae and fungi	First air-breathing land animals such as land scorpion appear Rise of fishes
Invasion of land by plants(?)	Diverse marine invertebrates, coral and nautaloid common First vertebrates appear as fish
Marine algae common	Diverse primitive marine invertebrates, trilobites common Animals with skeletons appear
Multicellular acoelomate and coelomate animals evolve Eukaryotic protists and fungi evolve	
Prokaryotes abundant	
Anaerobic and photosynthetic bacteria evolve Formation of earth and rest of solar system	

TABLE 20.1 THE CLASSIFICATION OF MODERN HUMANS

Kingdom	Animalia (animals)
Phylum	Chordata (chordates)
Class	Mammalia (mammals)
Order	Primates (primates)
Suborder	Anthropoidea (anthropoids)
Superfamily	Hominoidea (hominoids)
Family	Hominidae (hominids)
Genus	*Homo* (humans)
Species	*sapiens* (modern humans)

HUMAN ISSUE

That organisms have arisen by evolution is supported by investigations using accepted scientific methodology, including hypothesis testing, experimentation, observation, measurement, and predictions from scientific models. Moreover, ideas about evolution—as in all fields of science—are continually subjected to the critical scrutiny of the scientific community via scientific papers and debate. The creation story does not rely on such protocol, and is therefore considered unscientific by most scientists. If—as occasionally happens—the majority of the people in a particular school district think creationism should be taught as a bona fide scientific idea in its high school science classes, do you think science teachers should be required to teach the creation story? Do you think what is or is not science should be determined by public vote or by a local school board? To what extent do you think the public should be involved in what is taught as science in public schools?

A *unity of plan* is explainable by descent from a **common ancestor.** Species that share a recent common ancestor will share a large number of the same genes and therefore will be quite similar to each other and to this ancestor. Species that share a more distant common ancestor will have fewer genes in common and will be less similar to each other and to this ancestor because differences arise as organisms continue on their own evolutionary pathways. This principle allows biologists to construct **evolutionary trees,** diagrams that tell how various organisms are believed to be related to one another. All evolutionary trees have a branchlike pattern (fig. 20.8), indicating that evolution does not proceed in a single steplike manner; rather, evolution proceeds by way of common ancestors that often give rise to two different groups of organisms. For example, reptiles are believed to have produced both birds and mammals.

Even after related organisms have become adapted to different ways of life, they may continue to show similarities of structure. For example, the forelimb of all vertebrates contains the same fundamental bone structure (fig. 20.5) despite specific specializations. Similarities in structure that have arisen through descent from a common ancestor are called **homologous structures.** Homologous structures indicate that organisms are related. Sometimes two groups of organisms have structures that function similarly but are constructed differently. In contrast to homologous structures, **analogous structures,** such as an insect wing and bird wing, have similar functions but differ in anatomy, and therefore we know they evolved independently of one another.

Comparative Embryology

Some groups of organisms share the same type of embryonic stages. As would be expected if all vertebrates are related, their embryonic stages are similar (fig. 20.6). During development, a human embryo at one point has gill pouches—even though it will never breathe by means of gills as do fishes—and a rudimentary tail—even though it will never have a long tail as do some four-legged vertebrates. In this way, embryological observations indicate evolutionary relationships.

Vestigial Structures

An organism may have structures that are underdeveloped and seemingly useless, and yet they are fully developed and functional in related organisms. These types of structures are called **vestigial.** Figure 20.7 illustrates numerous vestigial structures in humans. The presence of these structures is understandable when we realize that related organisms have genes in common.

Comparative Biochemistry

Almost all living organisms use the same basic biochemical molecules, including DNA, ATP, and many identical or nearly identical enzymes. For example, the metabolic enzymes involved in cellular respiration and in the synthesis of cellular macromolecules are the same in all organisms. It would seem, then, that the molecules unique to living things appeared very early in the evolution of life and have been passed on ever since.

Analyses of amino acid sequences in certain proteins like hemoglobin and cytochrome C are used to determine how distantly related animals are. It is assumed that the number of differences will reflect how long ago the two species shared a common ancestor. Analyses of DNA nucleotide differences of the genome (all the genes) also have been done for the same purpose. Figure 20.8 shows the results of one such study.

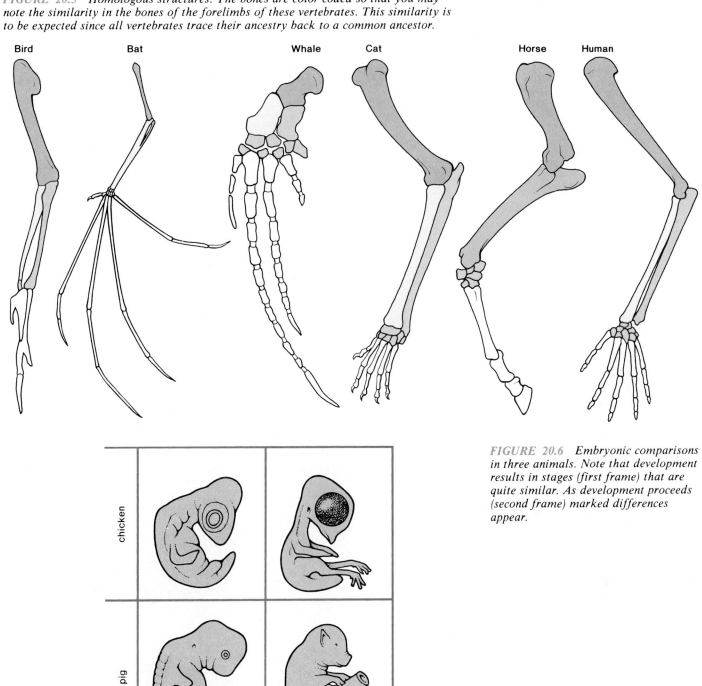

FIGURE 20.5 Homologous structures. The bones are color coded so that you may note the similarity in the bones of the forelimbs of these vertebrates. This similarity is to be expected since all vertebrates trace their ancestry back to a common ancestor.

Bird   Bat   Whale   Cat   Horse   Human

chicken

pig

human

FIGURE 20.6 Embryonic comparisons in three animals. Note that development results in stages (first frame) that are quite similar. As development proceeds (second frame) marked differences appear.

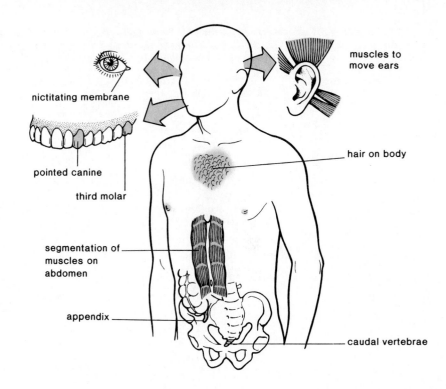

FIGURE 20.7 *Vestigial structures. Human beings have various vestigial structures such as those shown. These show our relationship to animals in which these structures are fully developed and functional.*

nictitating membrane

muscles to move ears

pointed canine

third molar

hair on body

segmentation of muscles on abdomen

appendix

caudal vertebrae

FIGURE 20.8 *Evolutionary tree of primate species based on a biochemical study of their genomes. The length of the branches indicates the approximate number of nucleotide pair differences that were found between groups.*

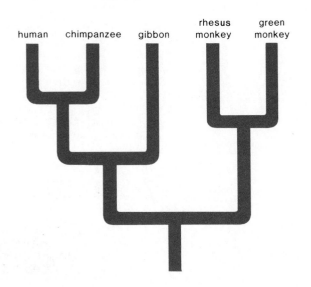

human   chimpanzee   gibbon

rhesus monkey   green monkey

FIGURE 20.9 *The quagga was striped on the front end like a zebra, but had a solid, chestnut-colored coat like a horse on the hind end. Although it became extinct a century ago, scientists have been able to clone a portion of its DNA chemically extracted from tissue removed from museum hides. Comparative biochemical studies indicate that the quagga is a zebra. Such a study emphasizes the physical and chemical aspects of life but does not in the least suggest that biochemists can bring whole organisms back from the dead.*

Investigators have been gratified to find that evolutionary trees based on biochemical data are quite similar to those based on anatomical data. Whenever the same conclusions are drawn from independent data, they substantiate scientific theory—in this case organic evolution—even more than usual.

Whereas biochemists formerly restricted their studies to living organisms, they recently have begun to study extinct organisms and even fossils. For example, investigators have extracted proteins and DNA from a scrap of muscle on the pelt of a 140-year-old museum specimen of a quagga (fig. 20.9), an animal that became extinct a century ago. Cloning provided enough DNA to establish that a quagga is a zebra, not a horse. Protein studies showed that a quagga is most closely related to a plains zebra. Such studies are only possible because all living things share the same types of chemical molecules.

FIGURE 20.10 *Parallel evolution. The grasslands in Africa and America have similarly adapted but different animals. Bison and pronghorn in North America, zebra and springbok in Africa are running animals that feed on grass; coyote and bobcat in North America, lion and cheetah in Africa are meat eaters that feed on the running animals.*

a. North American   b. African

Biogeography

It is observed that similar but geographically separate environments have different plants and animals that are similarly adapted. For example, figure 20.10 compares several North American animals with African animals that are adapted to living in a grassland environment. First, you will notice that although each group of animals could live in the other's biogeographic region, they do not. Why? They do not because geographic separation made it impossible for a common ancestor to produce descendants for both regions. On the other hand, notice that the same type of adaptations are seen in both groups of animals. This phenomenon, called **parallel evolution,** supports the belief that the evolutionary process causes organisms to be adapted to their environments.

Evolution explains the history and diversity of life. Evidences for evolution can be taken from the fossil record and comparative anatomy, embryology, and biochemistry. Also, vestigial structures and biogeography support the occurrence of evolution.

MECHANISMS OF EVOLUTION

Based on similar types of evidence, such as we have just outlined, Charles Darwin formulated his theory of natural selection of evolution around 1860. The theory of natural selection is nonteleological. A **teleological** mechanism is one in which the end result has been determined from the beginning. A nonteleological mechanism is one in which the end result cannot be determined. Do humans breathe air in order to live on land (teleological), or have circumstances brought about this ability over time (nonteleological)?

The following are critical to understanding natural selection.

Variations

Individual members of a species vary in their physical characteristics (fig. 20.11). Such variations can be passed on from generation to generation. Darwin was never able to determine the cause of variations, nor how they are passed on. Today, we realize that genes determine the appearance of an organism and that mutations (permanent genetic changes) can cause new variations to arise.

FIGURE 20.11
Variations among individuals of a population. It's easy for us to note that humans vary one from the other, but the same is true for populations of any type organism.

Struggle for Existence

In Darwin's time, a socioeconomist, Thomas Malthus, had stressed the reproductive potential of human beings. He proposed that death and famine were inevitable because the human population tended to increase faster than the supply of food. Darwin applied this concept to all organisms and saw that members of any plant or animal population must compete with one another for available resources. Competition must of necessity occur because reproduction produces more members of a population than can be sustained. Darwin calculated the reproductive potential of elephants. Assuming a life span of about 100 years and a breeding span of from thirty to ninety years, a single female will probably bear no fewer than six young. If all these young survived and continued to reproduce at the same rate, after only 750 years the descendants of a single pair of elephants would number about 19 million! Such reproductive potential necessitates a *struggle for existence,* and only certain organisms will survive and reproduce.

Survival of the Fittest

Darwin noted that in *artificial selection* humans choose which plants or animals will reproduce. This selection process brings out certain traits. For instance, there are many varieties of pigeons, each of which is derived from the wild rock dove. In a similar way, several varieties of vegetables can be traced to a single type.

In nature, it is the environment that selects those members of the population that will reproduce to a greater degree than other members. In contrast to artificial selection, the natural selection process is not teleological. Selection occurs only because certain members of a population happen to have a variation that makes them more suited to the environment. For example, any variation that increases the speed of a hoofed animal will help it escape predators and live longer; a variation that reduces water loss will help a desert plant survive; and one that increases the sense of smell will help a wild dog find its prey. Therefore, we would expect organisms with these traits to eventually reproduce to a greater extent.

Whereas Darwin emphasized that only certain organisms survived to reproduce, modern evolutionists emphasize that competition results in unequal reproduction. If certain organisms can acquire a greater share of available resources and if they have the ability to reproduce, then their chances of reproduction are greater than those of their cohorts.

Adaptation

Natural selection brings about **adaptation**—ultimately a species becomes adapted or suited to the environment. The process is slow, but each subsequent generation will include more individuals that are adapted than the previous generation. Eventually a species may become so adapted to a particular environment that it is unable to adapt to a new and changing environment. Extinction of this species may now occur (fig. 20.12).

FIGURE 20.12 *Out on an evolutionary limb? Some believe the saber-toothed tiger became extinct because its canine teeth became too long to use. But its real problem seems to have been a lack of speed. When its large, slow-moving prey became extinct, it could not adjust to catching smaller, more speedy prey.*

The following listing summarizes the theory of evolution as developed by Darwin.

1. There are inheritable variations among the members of a population.
2. Many more individuals are produced each generation than can survive and reproduce.
3. Individuals with adaptive characteristics are more likely to be selected to reproduce by the environment.
4. Gradually, over long periods of time, a population can become well adapted to a particular environment.
5. The end result of organic evolution is many different species, each adapted to specific environments.

ORGANIC EVOLUTION

Figure 20.4 outlines the evolutionary history of life on earth. When you think about it, though, organic evolution couldn't begin until the evolution of the first cell or cells. Scientists have pieced together a reasonable hypothesis (table 20.2) about the evolution of the first cell(s) that they call the origin of life.

Origin of Life

When the earth was first formed about 5 billion years ago, it was a glowing mass of free atoms (fig. 20.13a) that sorted themselves out according to weight. The heavy ones, such as

TABLE 20.2 ORIGIN OF THE FIRST CELL

Primitive Earth
> *outgassing*

Gases: No free O_2
> *energy* (e.g., lightning, ultraviolet radiation)

Small molecules: Amino acids, glucose, nucleotides, fatty acids
> *polymerization*

Macromolecules: Protein, nucleic acids
> *organization*

Protocell: Heterotrophic fermenter
> *organic evolution*

True genes: Reproduction possible

> *organic evolution*

Autotrophic cells: Including photosynthesizers
> *oxygen*

Aerobic heterotrophs: Aerobic respiration

iron and nickel, sank toward the center of the earth; the lighter atoms, such as silicon and aluminum, formed the middle shell; and the very lightest atoms, hydrogen, nitrogen, oxygen, and carbon, may have collected on the outside. The temperature was so hot that atoms could not permanently bind together; whenever bonds formed, they were quickly broken.

As the earth cooled, the heavy atoms tended to liquefy and solidify, but the intense heat at the center prevented complete solidification, and even today the earth contains a hot, thickly flowing molten core. In the middle shell, the lighter atoms congealed and formed the outer surface of the earth, the so-called crust. Cooling may also have allowed the first atmosphere to form.

Primitive Atmosphere

There are two current hypotheses about the origin of the *primitive atmosphere*. One hypothesis suggests that the gases of the primitive atmosphere came about when cooling allowed the lightest of the atoms to react with one another. Since hydrogen was the most abundant of these atoms, the hydrogen atoms combined with other hydrogen atoms and with carbon, oxygen, and nitrogen atoms to form hydrogen gas (H_2), methane gas (CH_4), water vapor (H_2O), and ammonia vapor (NH_3). The second hypothesis is also compatible with the steps listed in table 20.2. This hypothesis suggests that the gases of the primitive atmosphere were released from volcanic eruptions. Further, it is believed that the atmosphere probably also contained carbon dioxide (CO_2) and nitrogen gas (N_2).

Both hypotheses support the contention that the first atmosphere contained little or no free oxygen. Therefore, the first atmosphere is believed to have lacked O_2.

FIGURE 20.13 *A model for the origin of life.* a. *When the earth was formed, atoms sorted themselves out according to weight.* b. *The primitive atmosphere contained the gases hydrogen, methane, ammonia, and water vapor; as the latter cooled, some gases were washed into the ocean.* c. *The availability of energy, represented here by ultraviolet rays and volcanic eruption, allowed gases to form simple organic molecules that (*d*) reacted to form macromolecules in the ocean. Proteinoids formed on shore became microspheres when washed into the sea.* e. *After autotrophs arose, aerobic respiration became possible.*

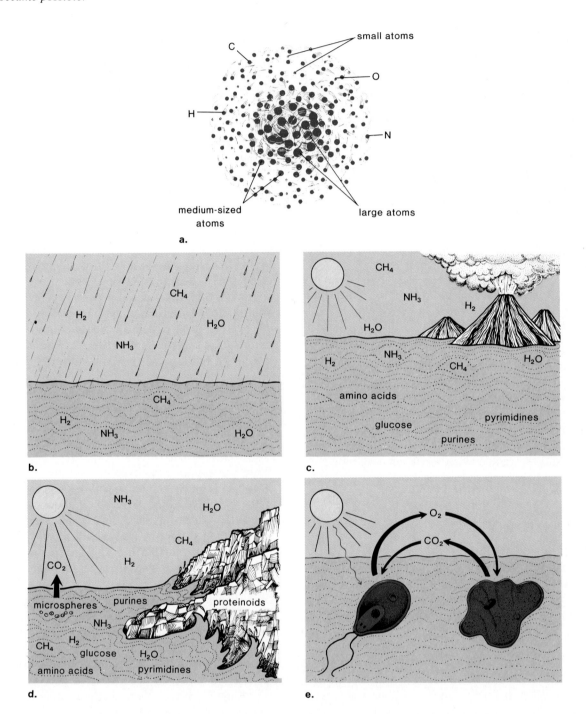

Chemical Evolution

Water, present at first as vapor in the atmosphere, formed dense, thick clouds, but cooling eventually caused the vapor to condense to liquid, and rain began to fall. This rain was in such quantity that it produced the oceans of the world (fig. 20.13b). The gases, dissolved in the rain, were carried down into the newly forming oceans.

The remaining steps shown in table 20.2 took place in the sea, where life arose. The dissolved gases, although relatively inert, are believed to have reacted together to form simple organic compounds when they were exposed to the strong outside *energy sources* present on the primitive earth (fig. 20.13c). These energy sources included heat from volcanoes and meteorites, radioactivity from the earth's crust, powerful electric discharges in lightning, and solar radiation, especially ultraviolet radiation. In a classic experiment Stanley Miller showed that an atmosphere containing methane and ammonia could have produced organic molecules. These gases were dissolved in water and circulated in a closed container past an electric spark. After a week's run, he analyzed the contents of the reaction mixture and found, among other organic compounds, amino acids and nucleotides. Other investigators have achieved the same results by utilizing carbon monoxide and nitrogen gas dissolved in water. On the basis of these experiments, it is surmised that the primitive oceans were most likely a thick organic soup.

The next step was the condensation of small organic molecules to produce the macromolecules characteristic of living things: polynucleotides, polypeptides, and polysaccharides. It is possible that macromolecules could have formed in the ocean, but it is more likely that small organic molecules were washed ashore, where they adhered to clay particles and were exposed to dry heat that encouraged polymerization. S. W. Fox has heated mixtures of amino acids at 130°–180° to form amino acid polymers that he calls proteinoids (fig. 20.13d). When proteinoids are placed in water, they form microspheres, structures that have some cell-like properties. They are, so to speak, the beginnings of cells, or **protocells.**

In contrast, Oparin showed that mixtures of macromolecules could join together in water to form **coacervate droplets.** Both microspheres and coacervates can take up molecules from the environment to form a lipid-protein film like a simple cell membrane. When they take up enzymes, they carry on some metabolic functions. Fox believes that after protocell formation, proteins and nucleic acids could have gradually evolved and increased in complexity until a true cell capable of reproduction came into being.

Some biologists disagree with this hypothesis. They believe chemical evolution may have proceeded by the initial development of nucleic acid genetic material rather than protein coacervates or microspheres. The most primitive nucleic acids might have reproduced themselves and subsequently acquired the ability to direct the synthesis of peptides. Regardless, the first true cell had to perform two functions: metabolism and reproduction.

A chemical evolution led to the protocell, which was a heterotroph that carried on anaerobic respiration. The protocell became a true cell once it could reproduce. Thereafter organic evolution began.

Organic Evolution Begins

The first primitive cell(s) possessing rudimentary enzymes, genes, and a selectively permeable membrane must have been a heterotroph living off preformed organic molecules in the primitive ocean, and it must have carried on anaerobic respiration since there was no oxygen (O_2) in the atmosphere. It was, then, a *heterotrophic fermenter.*

Once the preformed organic molecules were depleted, organic evolution would have favored any cell capable of making its own food. The first autotrophs probably lacked a light-absorbing pigment and therefore were not photosynthesizers. They could have been chemosynthetic organisms that extracted electrons from inorganic molecules (other than water) and used these electrons to generate ATP by electron transport. Or even if they did possess a pigment able to capture light energy, in order to reduce carbon dioxide, they probably did not at first use water as a hydrogen source and thus did not give off oxygen. The first autotroph to use water would have been selected because water is such a plentiful molecule. Now oxygen would have been given off as a by-product of photosynthesis (fig. 20.13e).

The presence of free oxygen (O_2) changed the character of the atmosphere; it became an oxidizing atmosphere instead of a reducing atmosphere. Abiotic synthesis of organic molecules was no longer possible because any organic molecules that happened to form would have been broken down by oxidation. As oxygen levels increased, cells capable of aerobic respiration evolved. This is the type of respiration used by the vast majority of organisms today.

The buildup of oxygen in the atmosphere caused the development of an ozone (O_3) layer. This filters out ultraviolet rays, shielding the earth from dangerous radiation. Prior to this time, organisms probably lived only deep in the oceans where they were not exposed to the intense radiation striking the earth's surface. Now life could safely spread to shallower waters and eventually move onto the land.

Plants were probably the first living organisms on land. This is reasonable because animals are dependent upon plants as a food source. The first vertebrates (animals with backbones) to dominate the land were the amphibians, whose name implies that their life cycle has two phases—one spent in the water and one spent on land. Frogs are amphibians, and during their tadpole stage they metamorphose into an adult frog. Next came the reptiles that have no need to return to water for reproductive purposes because the female lays a shelled egg. Within the egg are the extraembryonic membranes (p. 339) that service all the needs of the developing organism. Both birds and mammals evolved from the reptiles. We are mammals (table 20.3), animals characterized

TABLE 20.3 CLASSIFICATION OF HUMANS
AND RELATED ANIMALS

PHYLUM CHORDATA
SUBPHYLUM VERTEBRATA

CLASS MAMMALIA	Placental Mammals
	Insectivores (e.g., tree shrew)
	Rodents (e.g., rats)
	Marine (e.g., whales)
	Carnivores (e.g., dogs)
	Primates
ORDER PRIMATES	Prosimians
	Lemurs
	Tarsiers
	Anthropoids
	Monkeys
	Apes
	Humans
SUPERFAMILY HOMINOIDEA	*Dryopithecus*
	Modern Apes
	Humans
FAMILY HOMINIDAE	*Australopithecus*
	Homo habilis
	Homo erectus
	Homo sapiens
GENUS HOMO (**Humans**)	*Homo habilis?*
	Homo erectus
	Homo sapiens

by the presence of hair and mammary glands. Even so, the brain is most highly evolved among mammals and reaches its largest size and complexity in the primates, the order to which humans belong.

The evolution of humans didn't begin until about 5 million years ago; therefore, we have been evolving only 0.1% of the total history of the earth.

Organic evolution involves the evolution of all life forms, including vertebrates. The first land vertebrates were the amphibians, which gave rise to the reptiles, from which the birds and mammals arose. Humans are primates, a type of mammal.

Human Evolution

Humans are mammals in the order Primates (table 20.3). **Primates** were originally adapted to an arboreal life. Long and freely movable arms, legs, fingers, and toes allowed them to reach out and grasp an adjoining tree limb. The opposable

thumb and toe, meaning that the thumb and toe could touch each of the other digits, were also helpful. Nails replaced claws; this meant that primates could also easily let go of tree limbs.

The brain became well developed, especially the cerebral cortex and frontal lobes, the highest portions of the brain. Also, the centers for vision and muscle coordination were enlarged. The face became flat so that the eyes were directed forward, allowing the two fields of vision to overlap. The resulting stereoscopic (three-dimensional) vision enabled the brain to determine depth. Color vision aided the ability to find fruit or prey.

One birth at a time became the norm; it would have been difficult to care for several offspring, as large as primates, in trees. The period of postnatal maturation was prolonged, giving the immature young an adequate length of time to learn behavior patterns.

Prosimians

As diagrammed in figure 20.14, primates are believed to have evolved from mammalian *insectivores,* perhaps one resembling the modern tree shrew. The first primates were the **prosimians,** a term that means premonkeys. The prosimians are represented today by several types of animals, among them the *lemurs,* which have a squirrellike appearance, and the *tarsiers,* curious monkeylike creatures with enormous eyes.

Anthropoids

Monkeys, along with apes and humans, are anthropoids. The **anthropoids** evolved from the prosimians about 38 million years ago during the Tertiary period (fig. 20.4). There are two types of monkeys: the *new-world monkeys,* which have long prehensile (grasping) tails and flat noses, and *old-world monkeys,* which lack such tails and have protruding noses. Two of the well-known new-world monkeys are the spider monkey and the "organ grinder's monkey." Some of the better-known old-world monkeys are now ground dwellers, such as the baboon and the rhesus monkey, which has been used in medical research.

Primates evolved from shrewlike insectivores and became adapted to living in trees, as exemplified by skeletal features, good vision, and even reproduction. Primates are represented by prosimians, monkeys, apes, and humans. The latter three are anthropoids.

Hominoids

Humans are more closely related to apes (fig. 20.15) than to monkeys. **Hominoids** include apes and humans. There are four types of apes: the gibbon, orangutan, gorilla, and chimpanzee. The gibbon is the smallest of the apes with a body weight ranging from 12–25 pounds. *Gibbons* have extremely long arms that are quite specialized for swinging between tree limbs. The *orangutan* is large (165 lbs.) but nevertheless

FIGURE 20.14 *Evolution of primates took place in the Cenozoic era, which includes the Tertiary and Quaternary periods. There are at least four lines of evolution among the primates who arose from mammalian insectivores. Prosimians (before monkeys) include the lemurs and tarsiers; the monkeys include the new-world and the old-world monkeys; apes include the gibbons and great apes (orangutan, gorilla, and chimpanzee); and humans.*

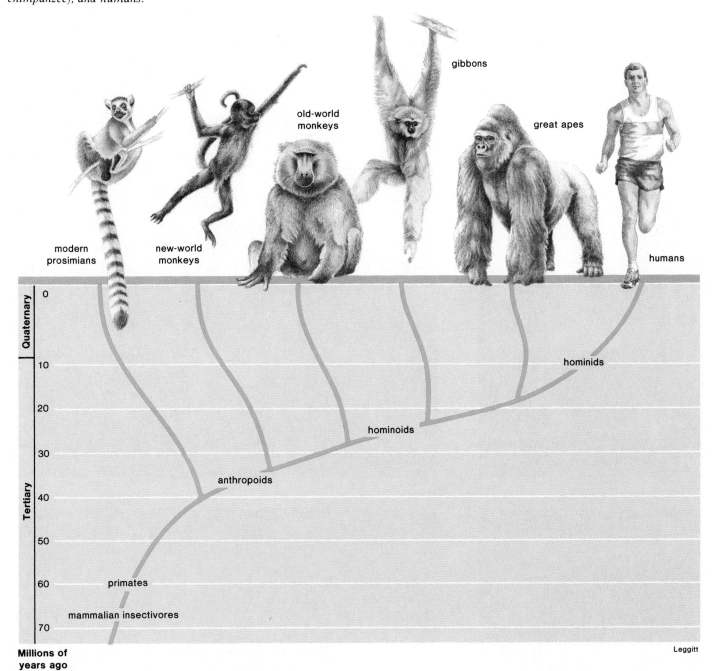

spends a great deal of time in trees, in contrast to the *gorilla,* the largest of the apes (400 lbs.), that spends most of its time on the ground. *Chimpanzees,* which are at home both in trees and on the ground, are the most humanlike of the apes in appearance, and are frequently used in psychological experiments.

Hominoid Ancestor About 25 million years ago, apes became abundant and widely distributed in Africa, Europe, and Asia. Among these, members of the genus ***Dryopithecus*** are of particular interest because they are thought to be a possible hominoid ancestor of today's apes and humans.

FIGURE 20.15 *Ape diversity.* a. *Of the apes, gibbons are the most distantly related to humans. They dislike coming down from trees, even at watering holes. They will extend a long arm into the water and then drink collected moisture from the back of the hand.* b. *Orangutans are solitary except when they come together to reproduce. Their name means "forest man"; early Malayans believed that they were intelligent and could speak but did not because they were afraid of being put to work.* c. *Gorillas are terrestrial and live in groups, in which a silver-backed male such as this one is always dominant.* d. *Chimpanzees are smaller than gorillas and have large, outstanding ears. They also live in groups.*

b.

a.

c.

d.

FIGURE 20.16   A common chimpanzee knuckle-walking. Perhaps the first hominid walked like this. Knuckle-walking does not interfere with having an opposable thumb, and thus the thumb previously used for grasping tree limbs could have eventually been used for grasping tools.

FIGURE 20.18   Comparison of the skull of Dryopithecus with that of modern apes and humans. Dryopithecus has primitive features; the common chimpanzee has apelike features, including a low brow, heavy eyebrow ridges, and a face that protrudes; humans have a high brow, reduced eyebrow ridges, and a flat face.

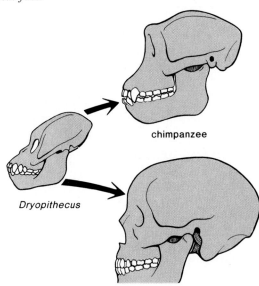

chimpanzee

Dryopithecus

FIGURE 20.17   The primate hand is capable of grasping objects because of the opposable thumb. In humans, this ability is used to grasp tools.

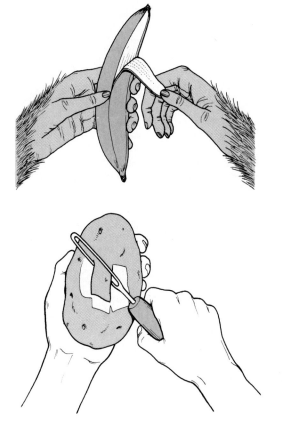

Dryopithecines were forest dwellers that probably spent most of the time in the trees. The bones of their feet, however, indicate that they also may have spent some time on the ground. When they did walk on the ground, however, they may have walked on all fours, using the knuckles of their hands to support part of their weight. "Knuckle-walking" (fig. 20.16) would have allowed retention of the opposable thumb, a primate characteristic (fig. 20.17).

The skull of *Dryopithecus* had a sloping brow, heavy eyebrow ridges, and jaws that projected forward. These are features that could have led to those of apes and humans (fig. 20.18). In apes the face projects forward forming a muzzle because the canine teeth are much larger than the adjacent teeth. In contrast to apes, humans have a high brow and lack the eyebrow ridges of apes. The face is flat and they have no muzzle because the canine teeth are comparable in size to the other human teeth.

Hominid Ancestor

It is generally agreed that **hominids,** which include humans and humanlike fossils, shared a common ancestor with apes. For example, biochemists have compared the structure of proteins and DNA in apes and humans. This evidence suggests that first gibbons began to evolve separately, followed by orangutans sometime later. This left an apelike creature that led to the evolution of other apes and humans (fig. 20.14).

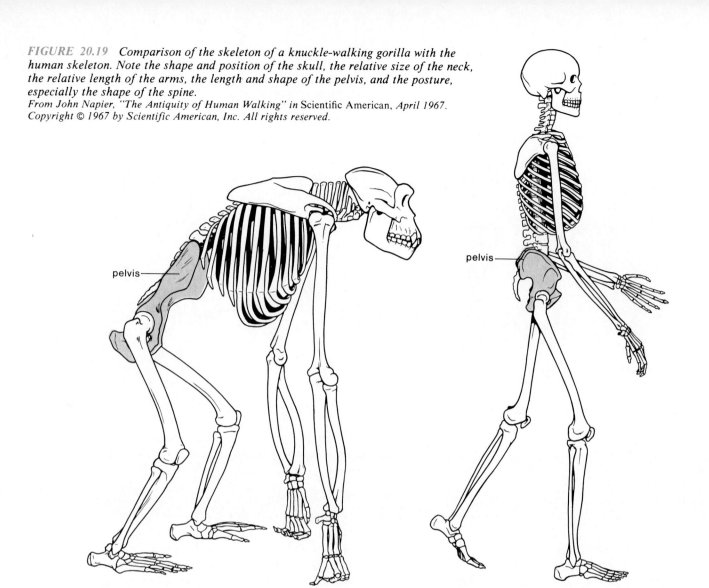

FIGURE 20.19 *Comparison of the skeleton of a knuckle-walking gorilla with the human skeleton. Note the shape and position of the skull, the relative size of the neck, the relative length of the arms, the length and shape of the pelvis, and the posture, especially the shape of the spine.*
From John Napier, "The Antiquity of Human Walking" in Scientific American, April 1967.

pelvis

pelvis

During the mid-Tertiary period, the weather was becoming cooler, and even in Africa, the tropical forests were being replaced by grasslands. At that time, the first hominids must have come down out of the trees and begun walking on two legs. It's possible this would have helped assure their survival because then they would have been able to see over tall grass to spot predators or prey and would have had their hands free to perform other functions, such as throwing rocks. The transition to erect posture required a number of adaptive changes. Figure 20.19 contrasts the skeleton of a knuckle-walking ape with a human skeleton. In the ape, the pelvic region is very long and tilts forward, but in the human, the pelvic region is short and upright. The ape shoulder girdle is more massive than that of a human, and ape arms are longer than ape legs. In apes, the head hangs forward because the foramen magnum is well to the rear of the skull. In humans, the head is erect because this opening is almost directly in the bottom center of the skull. Also in humans, the spine is flexible because of its four curvatures.

Humans and apes share a common ancestor, which may possibly have been *Dryopithecus,* a fossil commonly regarded as the first hominoid. The first hominid arose at a time when the change in weather made it advantageous to dwell on the ground.

Hominids
New ideas regarding speciation have been used to interpret the hominid fossil record. Although formerly scientists attempted to place each hominid fossil in a straight line from the most primitive to the most advanced, it is now reasoned that several hominid species could have existed at the same time. Figure 20.20 indicates a possible evolutionary tree for the known hominids.

Australopithecines   The fossil remains of those classified in the genus **Australopithecus** (Southern Apeman) were found in southern Africa and have been dated at about 4 million years ago. The australopithecines (fig. 20.21) were 4–5 feet in height, with a brain that ranged in size from 300–500

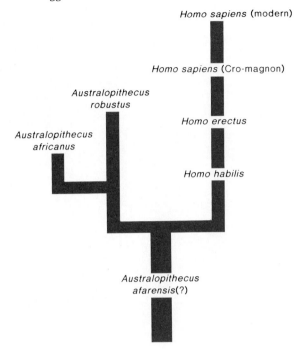

Homo sapiens (modern)

Homo sapiens (Cro-magnon)

Australopithecus robustus

Homo erectus

Australopithecus africanus

Homo habilis

Australopithecus afarensis(?)

cm³. The pelvis definitely indicates that they were capable of bipedal locomotion; supporting this contention is the non-opposability of the big toe.

Three species of *Australopithecus* have been identified—*afarensis, africanus, and robustus.* The exact relationship between the three species is in dispute. Because of its small brain size, large canine teeth, and protruding face, it has been suggested that *afarensis* is the most primitive of the three. Moreover, because of certain skeletal limb features indicating that these peoples walked erect, some believe that *afarensis,* called Lucy (from the song "Lucy in the Sky with Diamonds") by her discoverers, is also ancestral to humans, as indicated in figure 20.20. It may also be noted that the combined characteristics of this fossil indicate that an enlarged brain was *not* needed for bipedal locomotion to evolve.

Australopithecus robustus, as its name implies, was larger than *africanus.* Certain anatomical differences may be due to diet. The massive skull and face and the large cheek teeth of *robustus* indicate a vegetarian diet. The gracile facial features of *africanus* indicate a more varied diet, possibly including meat. Facial bones, including eyebrow ridges, facial muscles, and teeth, need not be as large in meat eaters. It is now believed that the australopithecines scavenged the meat that they ate, and they were not hunters.

One possible hominid evolutionary tree shows that the bipedal *Australopithecus afarensis* is a common ancestor to two other members of this genus and humans. *Australopithecus africanus* was slight of build compared to *Australopithecus robustus.*

Humans

The rest of the fossils to be mentioned are in the genus *Homo,* which is the genus for all humans, including ourselves.

Homo habilis

This newly discovered fossil is dated about 2 million years ago, making *Homo habilis* a contemporary of *Australopithecus africanus.* **Homo habilis** is significant for several reasons. 1. Formerly it was believed that humans evolved less than 2 million years ago, but this fossil and other current evidence seem to suggest that they must have begun evolving *more* than 2 million years ago. 2. Almost certainly, humans evolved in Africa. 3. Formerly only hominid fossils with a brain capacity of 1,000 cm³ were designated humans. *Homo habilis* had a maximum brain capacity of 800 cm³, and yet he has been placed in the genus *Homo* because he not only used tools, he also made them. *Homo habilis* means handyman; he was given this name because of the quality of the tools he made.

At one time it was thought that only primitive people with a brain capacity of more than 1,000 cm³ could have made tools. Since *Homo habilis* had a smaller brain and yet made tools, are we to think that the making of tools preceded the

evolution of the enlarged brain? This seems to be an unnecessary question. Increased brain capacity, no matter how slight, would have permitted better toolmaking; this combination would have been selected because toolmaking would have fostered survival in a grassland habitat. Therefore, as the brain became increasingly larger, tool use would have become more sophisticated.

Homo habilis, a fossil dated at least 2 million years ago, may not have been highly intelligent, but he did make tools. Intelligence and the making of tools probably evolved together. If *Homo habilis* is indeed considered human, then humans evolved much earlier than previously thought.

Homo erectus

Homo erectus (fig. 20.22) was prevalent throughout Eurasia and Africa during the last ice age, a time of recurrent cold weather and glaciers. *Homo erectus* had a brain size of 1,000 cm³, but the shape of the skull indicates that the areas of the brain necessary for memory, intellect, and language were not well developed.

Even so, *Homo erectus* had the grasp, posture, and striding gait of modern humans. The tools of *Homo erectus* were superior to those of *Homo habilis* because they were made of flakes chipped from a stone rather than of the stone itself. Each flake was struck off a stone "core," which was previously shaped to give just the right kind of flakes. Most likely these people were hunters who possessed a knowledge of fire and would have been able to cook their meat in order to tenderize it.

The brains of those classified as *Homo erectus* were larger by 50% than those of the australopithecines. This suggests that superior toolmaking and increased intelligence may have evolved together during the evolution of humans.

Homo sapiens

Neanderthal The **Neanderthals** (*Homo sapiens neanderthalensis*) lived over 130,000 years ago during the last ice age, and they were in existence for 70,000 years. They are called Neanderthals because they were first found in the Neanderthal Valley in Germany. Later they were also found in most of Europe and in the Middle East.

Surprisingly, the Neanderthal brain was, on the average, slightly larger than that of modern humans. The Neanderthal brain was 1,400 cm³ while that of most humans today is 1,360 cm³. The Neanderthal skeleton is robust, giving evidence that the skeletal muscles were very well developed. It's been hypothesized that a larger brain than ours was required for control of the extra musculature.

Early investigators believed that Neanderthals were apelike in appearance (fig. 20.23), but now it is believed that they had a more modernlike appearance except that they, like *Homo erectus,* had a low brow, wide nose, and receding chin. Like *Homo erectus,* too, the Neanderthal people were stone toolmakers, lived together in a kind of society, and hunted large game together.

The Neanderthals seem to have been more culturally advanced than *Homo erectus,* however. They buried their dead with flowers, as an expression of grief perhaps, and with tools, as if they thought these would be needed in a future

FIGURE 20.24
Socialization may have begun when men organized for the hunt. Since they hunted animals larger than themselves, cooperation was required.

life. Some researchers have even suggested that Neanderthals had a religion and that since great piles of bear skulls have been found in their caves, the bear must have played a role in this religion.

It's possible that the Neanderthals evolved into modern humans, called **Cro-Magnon,** because their remains appeared first in Cro-Magnon, France, but it could also be that the Neanderthals were replaced by Cro-Magnon. Another theory is that modern humans evolved in one locale, perhaps Africa, and then they interbred with other populations, including the Neanderthals, as they spread out into Europe and Asia.

Cro-Magnon Cro-Magnon (*Homo sapiens sapiens*) peoples lived about 40,000 years ago. Their brain capacity was just about like ours (1,360 cm³). They were such accomplished hunters that some researchers believe they are responsible for the extinction, during the Upper Pleistocene epoch, of many large mammalian animals, such as the giant sloth, mammoth, saber-toothed tiger, and giant ox. Since language would have facilitated their ability to hunt such large animals, it's quite possible that meaningful speech began at this time. Cooperative hunting (fig. 20.24) would also have led to socialization and the advancement of culture. Humans are believed to have lived in small groups, the men going out to hunt by day while the women remained at home with the children. It's quite possible that a hunting way of life among prehistoric men has shaped behavior even unto today.

The Cro-Magnons had an advanced form of stone technology that included the making of compound tools; the stone was fitted to a wooden handle. They were the first to throw spears (fig. 20.24), enabling them to kill animals from a distance. And they were the first to make blades, knifelike objects that do not appear to have been chipped from a stone.

HUMAN ISSUE

Which position do you support? *a.* It should be illegal to teach human evolution, and the concept should be removed from all biology textbooks. *b.* By law, biology textbooks should have to include, along with the concept of evolution, the biblical account of human origins. *c.* Since the biblical account is not a part of scientific knowledge, biology textbooks need not discuss it. *d.* It should be made illegal to include the biblical account of human origins in biology textbooks.

Cro-Magnon peoples lived during the Reindeer age, a time when great reindeer herds spread across Europe. They used every part of the reindeer, including the bones and antlers from which they sculpted many small figurines. They also painted beautiful drawings of animals on cave walls in Spain and France (fig. 20.25). Perhaps they, too, had a religion, and these artistic endeavors were an important part of their form of worship.

If the Cro-Magnon peoples did cause the extinction of many types of animals, this may help account for the transition from a hunting economy to an agricultural economy about 12,000–15,000 years ago. Warmer weather following the last ice age no doubt contributed to the occurrence of this agricultural revolution. Technological progress requiring the use of metals and energy sources led, in an amazingly short time, to the Industrial Revolution, which began about two hundred years ago. After this time, many people typically lived in cities, in large part divorced from nature and endowed with the philosophy of exploitation and control of

FIGURE 20.25 Cro-Magnon peoples are depicted painting on cave walls, and some of these paintings are still in existence for observation today. Of all the animals, only humans have developed a culture that includes technology and the arts.

nature. Only recently have we begun to realize that the human population, should work with rather than against nature. The ecology chapters that follow will stress this theme.

Human Races All human races of today are also classified as *Homo sapiens sapiens*. This is consistent with the biological definition of species because it is possible for all types of humans to interbreed and bear fertile offspring. The close relationship between the races is supported by biochemical data showing that differences in amino acid sequence between two individuals of the same race are as great as those between two individuals of different races.

It is generally accepted that racial differences developed as adaptations to climate. Although it might seem as if dark skin is a protection against the hot rays of the sun, it has been suggested that it is actually a protection against ultraviolet ray absorption. Dark-skinned persons living in southern regions and white-skinned persons living in northern regions absorb the same amount of radiation. (Some absorption is required for vitamin D production.)

Differences in body shape represent adaptations to temperature. A squat body with shortened limbs and nose retains more heat than an elongated body with longer limbs and nose. Also, the "almond" eyes, flattened nose and forehead, and broad cheeks of the Oriental are believed to be adaptations to the extremely cold weather of the Ice Age.

Although it has always seemed to some that physical differences might warrant assigning human races to different species, this contention is not borne out by the biochemical data mentioned previously.

Homo erectus had a large brain and walked with a striding gait. He also used fire. *Homo sapiens neanderthalensis* was not as primitive as formerly thought. The Neanderthals may have evolved into or may have been replaced by Cro-Magnon, the first *Homo sapiens sapiens*. Cro-Magnon was an expert hunter. Hunting promoted language and socialization. All human races belong to the same species.

SUMMARY

Evolution explains the history and diversity of life. Evidences for evolution can be taken from the fossil record and comparative anatomy, embryology, and biochemistry. Also, vestigial structures and biogeography support the occurrence of evolution.

Darwin not only presented evidences in support of organic evolution, he showed that evolution was guided by natural selection. Due to reproductive potential, there is a struggle for existence between members of the same species. Those members that possess variations more suited to the environment will most likely acquire more resources and so have more offspring than other members. Because of this natural selection process, there is a gradual change in species composition, which leads to adaptation to the environment.

Humans are primates that evolved from shrewlike insectivores and became adapted to living in trees. The first primates were prosimians, followed by the monkeys, apes, and hominids. The latter three are all anthropoids, the latter two are hominoids, and the last contains the australopithecines and humans.

Humans and apes share a common ancestor which may possibly have been *Dryopithecus,* a fossil commonly regarded as the first hominoid. The first hominid lived at a time when a change in weather made it advantageous to dwell on the ground.

One possible hominid evolutionary tree shows *Australopithecus afarensis* as a common ancestor to two other species of this genus and *Homo habilis,* the first fossil to be placed in the genus *Homo. Homo habilis* peoples may not have been highly intelligent but they did make tools.

Intelligence and making of tools probably evolved together. If *Homo habilis* is indeed considered human, then humans evolved much earlier than previously thought.

Homo erectus had a large brain and walked with a striding gait. These people used fire and probably were the first true hunters. *Homo sapiens neanderthalensis* was not as primitive as formerly thought. The Neanderthals may have evolved directly into or may have been replaced by Cro-Magnon, the first *Homo sapiens sapiens*. Cro-Magnon was an expert hunter. Hunting promoted language and socialization. In a relatively short time, humans developed an advanced culture that has tended to separate them from other organisms in the biosphere. All human races belong to the same species.

OBJECTIVE QUESTIONS

Match the phrases in questions 1–4 with those in this key:
- a. biogeography
- b. fossil record
- c. biochemistry
- d. anatomy

1. species change over time
2. forms of life are variously distributed
3. a group of related species has a unity of plan
4. same types of molecules are found in all living things

5. Darwin's theory of natural selection is _____ , meaning that the organism is unable to predetermine how it shall evolve.
6. Evolutionary success is judged by _____ success, or the number of an organism's offspring.
7. Primates are adapted to life in the _____ .
8. Anthropoids include _____ , _____ , _____ .
9. The fossil known as Lucy could probably walk _____ , but had a _____ brain.
10. The two varieties of *Homo sapiens* from the fossil record are _____ and _____ .

Answers

1. b 2. a 3. d 4. c 5. nonteleological 6. reproductive 7. trees 8. monkeys, apes, humans 9. erect, small 10. Neanderthal, Cro-Magnon

STUDY QUESTIONS

1. Show that the fossil record, comparative anatomy, vestigial structures, comparative embryology, and comparative biochemistry all give evidence that evolution occurred. (pp. 443–451)
2. What are five criteria necessary to Darwin's mechanism for evolution? (p. 453)
3. Name several primate characteristics still retained by humans. (p. 456)
4. Draw an evolutionary tree that includes all primates. (p. 457)
5. What animals mentioned in this chapter, whether living or extinct, are anthropoids? hominoids? hominids? humans? (p. 456)
6. How might adaptations to a grassland habitat have influenced the evolution of humans? (p. 460)
7. Draw a hominid evolutionary tree. (p. 461)
8. Which came first—tool use, walking erect, or intelligence? (p. 461)
9. Which humans were tool users? walked erect? had a striding gait? used fire? drew pictures? (pp. 461–464)
10. What evidence do we have that all races of humans belong to the same species? Name several races of humans. (p. 464)

THOUGHT QUESTIONS

1. The evidences for evolution are not convincing to everyone. What types of problems are involved in producing convincing evidence?
2. Do the mechanisms of evolution discussed on pages 451–453 apply to human beings today?
3. Why would you not expect the "origin of life" to occur today?

KEY TERMS

analogous structure (ah-nal'o-gus struk'tur)   similar in function but not in structure; particularly in reference to similar adaptations. *448*

anthropoids (an'thro-poidz)   higher primates, including only monkeys, apes, and humans. *456*

Australopithecus (aw''strah-lo-pith'e-cus)   the first generally recognized hominid. *460*

common ancestor (kŏ'mun an'ses-tor)   an ancestor to two or more branches of evolution. *448*

Cro-Magnon (kro-mag'non)   the common name for the first fossils to be accepted as representative of modern humans. *463*

Dryopithecus (dri''o-pith'e-cus)   a genus of extinct apes that may have included or resembled a common ancestor to both apes and humans. *457*

evolutionary trees (ev''o-lu''shun-ar-e trēz)   diagrams describing the evolutionary relationship of groups of organisms. *448*

hominid (hom'ĭ-nid)   member of a family of upright, bipedal primates that includes australopithecines and modern humans. *459*

hominoid (hom'ĭ-noid)   member of a superfamily that contains humans and the great apes and humans. *456*

Homo erectus (ho'mo ē-rek'tus)   the earliest nondisputed species of humans, named for their erect posture that allowed them to have a striding gait. *462*

Homo habilis (ho'mo hah'bĭ-lis)   an extinct species that may include the earliest humans, having a small brain but quality tools. *461*

homologous structures (ho-mol'o-gus struk'tūrz)   similar in structure but not necessarily function; homologous structures in animals share a common ancestry. *448*

Neanderthal (ne-an'der-thawl)   the common name for an extinct subspecies of humans whose remains are found in Europe, Asia, and Africa. *462*

primates (pri'māts)   animals that belong to the order Primates, the order of mammals that includes prosimians, monkeys, apes, and humans. *456*

prosimians (pro-sim'e-anz)   primitive primates such as lemurs, tarsiers, and tree shrews. *456*

taxonomy (tak-son'o-me)   the science of naming and classifying organisms. *444*

vestigial (ves-tij'e-al)   the remains of a structure that was functional in some ancestor but is no longer functional in the organism in question. *448*

CHAPTER TWENTY-ONE

The Environment

CHAPTER CONCEPTS

1 Natural ecosystems, which use solar energy efficiently and utilize chemicals that cycle, produce little pollution and waste.

2 The human ecosystem, which comprises both country and city, utilizes fossil fuel energy inefficiently and uses material resources that do not cycle. Therefore there is much pollution and waste.

3 Humans have begun to address the problem of pollution, which affects the quality of air, water, and land.

The effects of pollution are destroying the ozone of our planet Earth.

INTRODUCTION

Ecologists like to use the catchy phrase "everything is connected to everything else." By this they mean that you can't affect one part of the environment without affecting another part. For example, would you suppose that trees have anything to do with the daily environmental temperature? Well, they do and in perhaps an unexpected way. The accumulation of carbon dioxide in the air acts like the glass of a greenhouse and allows solar heat to be trapped near the surface of the earth. Environmentalists predict a gradual rise in the earth's daily temperature, especially if we continue to do away with the world's tropical rainforests. They act like a giant sponge that absorbs carbon dioxide, slowing down the greenhouse effect caused by all the carbon dioxide we pour into the environment from the burning of fossil fuels.

When the earth was first formed the outer crust was covered by ocean and barren land. Over time, plants colonized the land so that eventually it supported many complex communities of living things. Similarly, today we can also observe a series of sequential events by which bare rock becomes capable of sustaining many organisms; we call it succession (fig. 21.2). During **succession** communities replace one another in an orderly and predictable way until finally there is a **climax community,** a mix of plants and animals typical of that area that remains stable year after year. For example, in the United States not too long ago, a deciduous forest was typical of the East, a prairie was common to the Midwest, and a semidesert covered the Southwest (fig. 21.3). Humans have the habit of utilizing large communities like these, called **biomes,** for their own purposes, as described in the reading on page 469.

FIGURE 21.1   Workers preparing to plant a tree on a city street. Although trees do absorb pollutants, they will eventually succumb if overwhelmed by these same pollutants.

FIGURE 21.2   Simplified overview of succession as it may occur on abandoned farmland along the East Coast in the United States. First, lichens and mosses grow on rocks, and then annual weeds and grass colonize the area. Thereafter shrubs and trees, such as pines that can tolerate direct sunlight, invade the area. These pave the way for the germination and growth of hardwood saplings that grow and finally dominate the area.

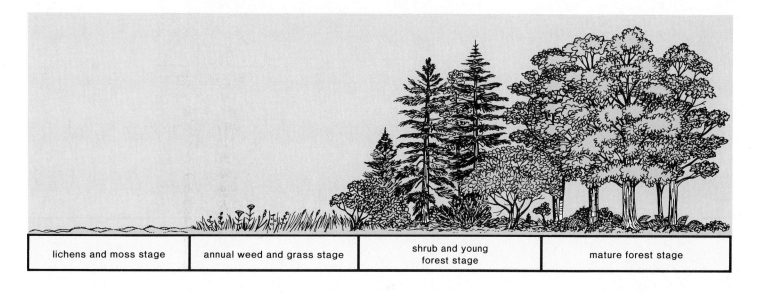

| lichens and moss stage | annual weed and grass stage | shrub and young forest stage | mature forest stage |

FIGURE 21.3 Three major biomes in the United States, each containing its own mix of plants and animals. Temperature and amount of rainfall largely determine which type biome will be found where. a. A deciduous forest is typical of eastern United States. b. A prairie biome is found in the Midwest. c. A semidesert is located in the Southwest.

a.

c.

b.

Originally, the Everglades encompassed the whole of southern Florida from Lake Okeechobee down to Florida Bay (see the accompanying map). Now largely in the Everglades National Park alone do we find the vast saw grass prairie interrupted occasionally by a cypress dome or hardwood tree island. Within these islands, both temperate and tropical evergreen trees grow amongst the dense and tangled vegetation. Mangroves are salt-tolerant trees that are found along sloughs (creeks) and at the shoreline. Only the roots of the red mangrove can tolerate the sea constantly. The prop roots of this tree protect over forty different types of juvenile fishes as they grow to maturity. During the wet season, from May to November, animals are dispersed throughout the region, but in the dry season, from December to April, they congregate wherever pools of water are found. Alligators are famous for making "gator holes" where water collects and where fishes, shrimps, crabs, birds, and a host of living things survive until the rains come again. Almost everyone is captivated by the birds that find at these holes a ready supply of the fishes they need for daily existence. The large and beautiful herons, egrets, roseate spoonbills, or anhingas fill one with awe. These birds once numbered in the millions; now they number only in the thousands. Why is this?

At the turn of the century, settlers began to drain the land just south of Lake Okeechobee in order to grow crops on the soil that had been enriched by partially decomposed saw grass. The large dike that now rings the lake prevents the water from taking its usual course: over the banks of Lake Okeechobee and slowly southward. In times of flooding, water can be shunted through the St. Lucie Canal to the Atlantic Ocean or through the canalized Caloosahatchee River to the Gulf of Mexico. In times of drought, water is contained not only in the lake, but also in three so-called conservation areas established to the south of the lake. Water must be conserved for the irrigation of the farmland and to recharge the Biscayne aquifer (underground river) that supplies drinking water for the cities on the east coast of Florida. Containing and moving the water from place to place has required the construction of over 1,400 miles of canals, 125 water control stations, and 18 large pumping stations. Now the Everglades National Park receives water only when it is discharged artificially from a conservation area. This disruption of the natural flow of water has affected the reproduction pattern of the birds, which is attuned to the natural wet-dry season turnover.

It took considerable human effort and a huge financial investment to control nature and to establish the "Everglades Agricultural Area." Has this attempt to bend nature to human will been worthwhile? The area does, in fact, produce more sugar than Hawaii and a large proportion of the vegetables

THE EVERGLADES

consumed in the United States each winter. But this has not been without a price. The rich soil, built up over thousands of years, is disappearing and most likely will be unable to sustain conventional agriculture after the year 2000. It has been suggested that *at that time* we might use the Everglades Agricultural Area for the growth of aquatic plants. Perhaps it should have been decided in the beginning to *work with nature* by growing aquatic plants instead of conventional plants. Then all the canals and pumping stations would have been unnecessary, the water would still flow from Lake Okeechobee to the glades as it had for eons, and the birds yet today would number in the millions.

a.

b.

a. *Prop roots of the red mangrove provide protective cover for fishes and other sea life during maturation.* b. *The Everglades once extended from Lake Okeechobee south to Florida Bay. Now it only encompasses three conservation areas and Everglades National Park.*

TABLE 21.1   ASPECTS OF NICHE

PLANTS	ANIMALS
Season of year for growth and reproduction	Time of day for feeding and season of year for reproduction
Sunlight, water, soil requirements	Habitat requirements
Contribution to ecosystem	Food requirements
Competition and cooperation with other organisms	Competition and cooperation with other organisms
Effect on abiotic environment	Effect on abiotic environment

All of the various climax communities of the world make up the **biosphere,** a narrow sphere or shell that encircles the earth. Although most organisms reside near the surface, others are also found a short distance in the air above and in the waters beneath the surface. Each belongs to a **population,** the members of a species that make up a community. The various populations interact with each other and the physical environment. Therefore, they form an **ecosystem** that has both a **biotic** (living) and an **abiotic** (nonliving) component. **Ecology** is defined as the study of the interactions of organisms among themselves and with the physical environment.

The process of succession leads to a climax community, which contains the plants and animals characteristic of the area. The populations within a community interact with themselves and with the physical environment, forming an ecosystem.

ECOSYSTEM COMPOSITION

Each population in an ecosystem has a habitat and a niche. The **habitat** of an organism is its place of residence; that is, the location where it may be found, such as "under a fallen log" or "at the bottom of the pond." The **niche** of an organism is its profession or total role in the community. A description of an organism's niche (table 21.1) includes its interactions with the physical environment and with the other organisms in the community. One important aspect of niche is the manner in which the organism acquires energy and chemicals. In fact, the entire ecosystem has two important aspects: *energy flow* and *chemical cycling*. These begin when photosynthesizing organisms use the energy of the sun to make their own food. Thereafter, chemicals and energy are passed from one population to another as the populations form food chains.

Within an ecosystem, energy flows and chemicals cycle.

Food Chains

Essentially, **food chains** are described by telling "who eats whom." Photosynthesizing organisms, which are at the start of a food chain, are called the **producers** because they have the ability to change what was formerly inorganic chemicals to organic food. Therefore, the producers in a food chain produce food. The other two types of populations in the biotic community are the consumers and the decomposers. **Consumers** are organisms that must take in preformed food. Herbivores are primary consumers that feed directly on producers; carnivores are secondary or tertiary consumers that feed only on consumers; omnivores are consumers that feed on both producers and consumers. **Decomposers** are organisms of decay, such as bacteria and fungi, that break down **detritus** (nonliving organic matter) to inorganic matter, which can be used again by producers. In this way the same chemicals can be used over and over again in an ecosystem. This is not true for energy. As detritus decomposes, all that remains of the solar energy taken up by the producer populations dissipates as heat (fig. 21.4). Therefore, energy does not cycle.

A typical terrestrial ecosystem is shown in figure 21.5, and a typical aquatic ecosystem is shown in figure 21.6. The arrows indicate the populations involved in various food chains; for example, here are two examples of a *grazing food chain:*

1. trees ———→ tent caterpillars ———→ red-eyed vireos ———→ hawks
2. algae ———→ water fleas ———→ catfish ———→ green herons

In some ecosystems (forests, rivers, and marshes) the detritus food chain accounts for more energy flow than does the grazing food chain because most organisms die without having been eaten. In the forest one *detritus food chain* is

detritus ———→ soil bacteria ———→ earthworms

FIGURE 21.4 *A diagram illustrating the manner in which nutrients cycle through an ecosystem. Energy does not cycle because all that is derived from the sun eventually dissipates as heat.*

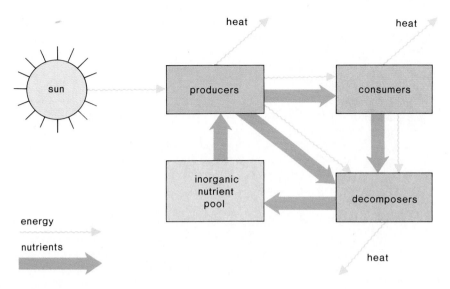

FIGURE 21.5 *This drawing depicts populations typical of a deciduous forest ecosystem. The arrows indicate food chains.*

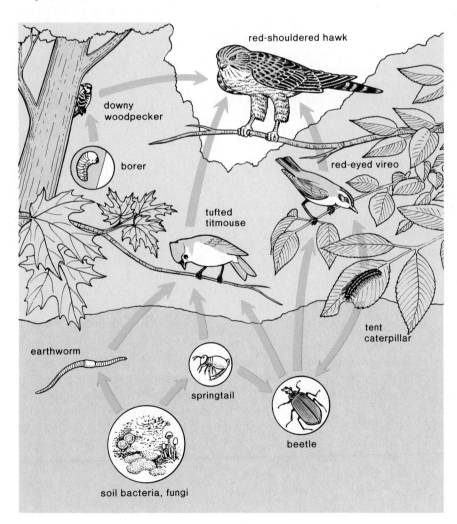

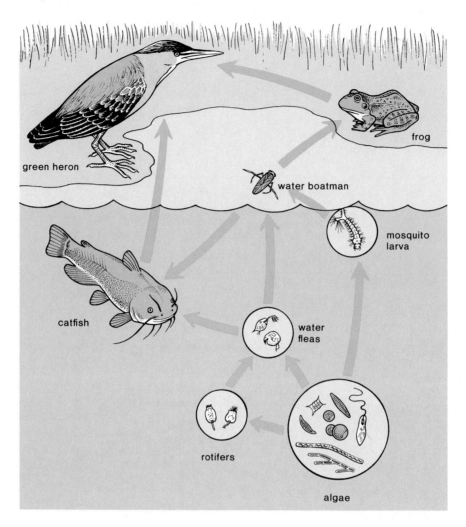

A detritus food chain often ties in with a grazing food chain, as when earthworms are eaten by a tufted titmouse. Although the grazing food chain is most important in aquatic food chains, the detritus food chain is most important in terrestrial food chains. For example, it is estimated that only about 10% of the annual leaf production is consumed by herbivores; the rest is either degraded by decomposers or eaten by scavengers, such as soil mites, earthworms, and millipedes.

Food Webs

Each food chain represents just one pathway by which chemicals and energy are passed along in an ecosystem. Natural ecosystems have numerous food chains, each linked to form a complex **food web** (figs. 21.5 and 21.6). Since organisms may belong to more than one food chain, energy flow is best described in terms of trophic (feeding) levels. The first level is the producer population, and each successive level is further removed from this population. All animals acting as primary consumers are part of the second level; all animals acting as secondary consumers are part of the third level; and so on. Each succeeding trophic level passes on less energy than was received due to a number of reasons. For example, less usable energy is available at each level because

1. not all members of the previous trophic level become food; of those that do become food, some portions, such as hair and bones, may be uneaten, or if eaten, may be undigestible;
2. some food is used for maintenance and never contributes to growth;
3. energy transformations always result in a loss of usable energy.

Eventually, as one population feeds on another and as decomposers work on detritus, all of the captured solar energy that was converted to chemical-bond energy by algae and plants is returned to the atmosphere as heat (fig. 21.7). *Therefore, energy flows through an ecosystem and does not cycle.*

FIGURE 21.7 *Dissipation of energy in an ecosystem. Producers convert inorganic matter to organic food after capturing a small amount of solar energy. When nutrient molecules are used for maintenance, their energy content eventually becomes heat. Energy, therefore, does not cycle, and ecosystems must have a continual supply of solar energy.*

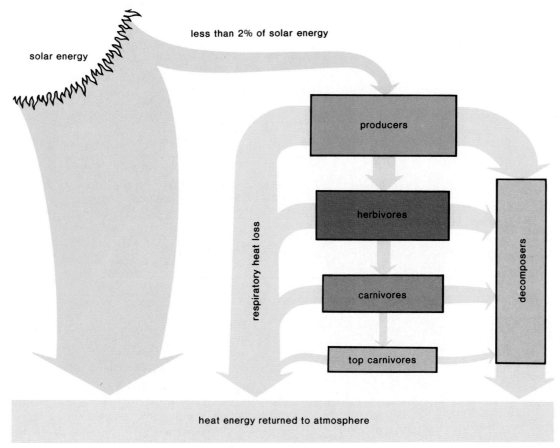

The populations in an ecosystem form food chains in which the producers produce food for the other populations by being able to capture the energy of the sun, the ultimate source of energy for our universe. Although it is convenient to study food chains, the populations in an ecosystem actually form a food web in which food chains join and overlap with one another.

Food Pyramid

Whether you consider the number of individuals, the biomass (weight of living material), or the amount of energy, each succeeding trophic level is generally smaller than the one preceding. To illustrate this, the various trophic levels are often depicted in the form of a pyramid (fig. 21.8). The producer population is at the base of the pyramid. Obviously it must be the largest because it indirectly produces food for all the other populations.

In regard to the pyramid of energy, it is generally stated that only about 10% of the energy absorbed by one trophic level can be stored by the next level. About 90% is no longer available for the reasons previously stated. In practical terms, this means that about ten times the number of people can be sustained on a diet of grain than on a diet of meat.

Stability

Mature natural ecosystems tend to be diverse and stable. *First,* the various populations stay at a constant size for the reasons listed in table 21.2. *Density-independent* effects are those forces of nature whose magnitude of influence does not depend on the size of the population affected. For example, the severity of a drought or flood has nothing to do with how many plants or animals there are in a particular area. The severity of *density-dependent* effects does depend on the size of the population affected. For example, two large populations compete to a greater degree for the same resource than do two small populations.

FIGURE 21.8 *Food pyramid. Whether you consider numbers, biomass, or energy content, each trophic level is smaller than the one on which it depends. Therefore the various trophic levels form a pyramid in which the producer population is at the base and the last consumer population is at the peak.*

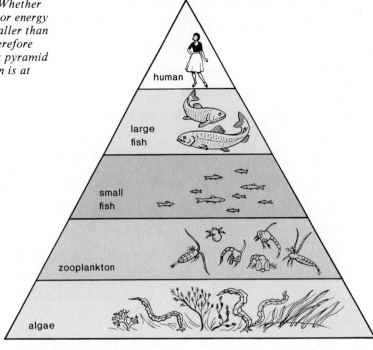

TABLE 21.2 POPULATION CONTROL

DENSITY-INDEPENDENT EFFECTS
Climate and weather changes
Natural disasters
DENSITY-DEPENDENT EFFECTS
Competition
Predation
Parasitism
Emigration

Second, diversity assists maintenance of population size. For example, if hawks eat both rabbits and mice, then you can imagine that if the rabbit population suffered an epidemic and declined in size, the mouse population would increase in size due to decreased competition for food. The increase would mean that the producer population would still be held in check, and the hawks that eat rabbits or mice would still eventually have the same amount of food. It may take a little time for the new balance to come about—a few hawks may have to migrate or starve—but essentially there would be little change in the size of the hawk population. Further, as the rabbits recover, the mouse population would decline, and eventually there would be exactly the same balance as before the epidemic. In this manner, variation of species at each level of a food pyramid gives stability to an ecosystem.

Actually, we see the same principle at work in business; a company diversifies its products so that as demand fluctuates profits will remain the same.

Constancy of population sizes means that most of the solar energy utilized by a biotic community supports stability rather than growth. The same amount of solar energy is required each year in order to maintain a highly diversified community that has little material waste due to the cycling of chemicals.

The food pyramid illustrates that each successive population has a smaller size or biomass because energy does not cycle. Mature natural ecosystems are stable because population sizes are held in check by density-independent and density-dependent factors. Also, population size is maintained by the diversity of the food web.

CHEMICAL CYCLES

In contrast to energy, matter does cycle through ecosystems. We will consider only two cycles: the carbon cycle and the nitrogen cycle. In reference to the carbon cycle, observe that carbon is taken up by producers, passes through food chains to the environment, and returns to the producers once more. The same is true for nitrogen in the nitrogen cycle. Human beings contribute to each of these cycles in a way that leads to pollution.

Carbon Cycle

The relationship between photosynthesis and respiration should be kept in mind in discussing the carbon cycle. This equation in the forward direction represents respiration, and in the other direction it represents photosynthesis:

$$C_6H_{12}O_6 \;+\; 6\,O_2 \;\rightleftharpoons\; 6\,CO_2 \;+\; 6\,H_2O$$

glucose   oxygen   carbon   water
dioxide

The equation tells us that respiration releases carbon dioxide, the molecule needed for photosynthesis. However, photosynthesis releases oxygen, the molecule needed for respiration. From figure 21.9, it is obvious that animals are dependent on green organisms not only to produce organic food and energy but also to supply the biosphere with oxygen.

In the carbon cycle on land (fig. 21.10), animals continuously release carbon dioxide into the air, but plants release carbon dioxide only when, as at night, respiration is occurring at a faster rate than is photosynthesis. After organisms die, decomposition also releases carbon dioxide. On the other hand, plants take up carbon dioxide from the air when photosynthesizing.

The carbon cycle also occurs in aquatic communities, but in this case, carbon dioxide is taken up from and returned to water. Carbon dioxide from the air combines with water to give bicarbonate ions (HCO_3^-) that serve as a source of carbon for algae, which produce food for themselves and then become food for others. Similarly, when aquatic organisms respire, the carbon dioxide they give off becomes bicarbonate ions. The amount of bicarbonate in the water is in equilibrium with the amount in the air.

Reservoirs

Living and dead organisms contain organic carbon and serve as one of the reservoirs for the carbon cycle. When we destroy forests and burn wood, we are reducing this reservoir and increasing the amount of CO_2 in the atmosphere.

In prehistoric times, certain plants and animals did not decompose, by chance, and instead were preserved in **fossil fuels** such as coal, oil, and gas. When humans burn fossil fuels, they not only add CO_2 but also other combustion products to the air. This contributes greatly to air pollution, which is discussed on page 482.

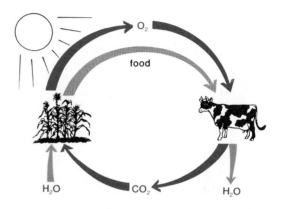

FIGURE 21.9 *Relationship between photosynthesis and respiration. Animals are dependent on plants for a supply of oxygen, and plants are dependent on animals for a supply of carbon dioxide.*

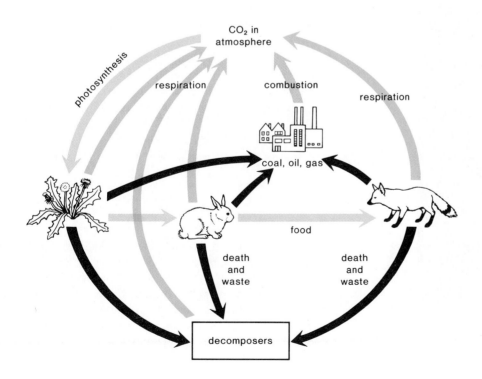

FIGURE 21.10 *Major components of the terrestrial portion of the carbon cycle. Photosynthesis utilizes CO_2 while respiration returns it to the atmosphere. The combustion of fossil fuels (coal, oil, gas) represents the human contribution to the cycle.*

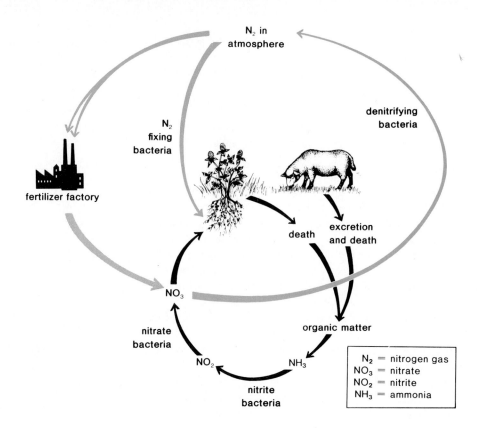

FIGURE 21.11 *Major components of the terrestrial portion of the nitrogen cycle. Three types of bacteria are at work: nitrogen-fixing bacteria convert aerial nitrogen to a form usable by plants; nitrifying bacteria, which include both nitrite and nitrate bacteria, convert ammonia to nitrite and finally to nitrate; and the denitrifying bacteria convert nitrate back to aerial nitrogen again. Humans contribute to the nitrogen cycle by converting aerial nitrogen to fertilizer.*

N_2 = nitrogen gas
NO_3 = nitrate
NO_2 = nitrite
NH_3 = ammonia

FIGURE 21.12 *Nodules on the roots of a legume (in this case, soybean). The bacteria that live in these nodules are capable of converting aerial nitrogen to a source that the plant can use.*

Another reservoir for the carbon cycle is the formation of inorganic carbonate that accumulates in limestone and in carbonaceous shells. Limestone, particularly, tends to form in the oceans. In this way, the oceans act as a sink for excess carbon dioxide. The burning of fossil fuels in the last twenty-two years has probably released 78 billion tons of carbon, and yet the atmosphere registers only an increase of 42 billion tons.

In the carbon cycle, carbon dioxide is removed from the atmosphere by photosynthesis but is returned by respiration. Living things and dead matter are reservoirs for the carbon cycle as is the ocean, particularly because it accumulates limestone. When humans cut down trees and burn wood and fossil fuels, they are changing the current balance of the carbon cycle.

Nitrogen Cycle

That portion of the nitrogen cycle that involves terrestrial organisms is represented by the diagram in figure 21.11. Aerial nitrogen, the reservoir for the nitrogen cycle, cannot be used by most organisms, but there are two types of nitrogen-fixing bacteria that do make use of it. One type is free-living in the soil, but the other type infects and lives in nodules on the roots of legumes (fig. 21.12). Here the nitrogen-fixing bacteria convert aerial nitrogen to a form that can be used by plants.

Nitrates, in particular, are taken up by the roots of plants and converted to amino acids. Thereafter, animals acquire this organic nitrogen when they eat plants. Plants and animals die, and decomposition produces ammonia that can be converted to nitrites and finally to nitrates by **nitrifying** (nitrite and nitrate) **bacteria.** To make the cycle complete, there are some bacteria, the **denitrifying bacteria,** that can convert ammonia and nitrates back to aerial nitrogen again.

Other Contributions to the Nitrogen Cycle

There are at least two other ways by which **nitrogen fixation** occurs. Nitrogen is fixed in the atmosphere when cosmic radiation, meteor trails, and lightning provide the high energy needed for nitrogen to react with oxygen. Also, humans make a most significant contribution to the nitrogen cycle when they convert aerial nitrogen to nitrates for use in fertilizers. This industrial process requires an energy input that equals that of the eventual increase in crop yield. The application of fertilizers also contributes to water pollution, as discussed on page 486. Since nitrogen-fixing bacteria do not require fossil fuel energy and do not cause pollution, research is now directed toward finding a way to make all plants capable of forming nodules (fig. 21.12) or, even better, through recombinant DNA research, to possess the biochemical ability to fix nitrogen themselves.

In the nitrogen cycle, nitrogen-fixing bacteria convert aerial nitrogen to nitrates, and denitrifying bacteria convert nitrates back to aerial nitrogen. Nitrifying bacteria convert ammonia to nitrate, the form of nitrogen most often used by plants. Humans again tap into a reservoir when they convert aerial nitrogen to nitrate for use in fertilizer.

HUMAN ECOSYSTEM

Mature natural ecosystems tend to be stable and to exhibit the characteristics listed in table 21.3. Each population is of a proper size in relation to other populations; the energy that enters and the amount of matter that cycles is appropriate to support these populations. **Pollution,** defined as any undesirable change in the environment that may be harmful to humans and other life, and excessive waste do not normally occur. Human beings have replaced natural ecosystems with one of their own making, as is depicted in figure 21.13. This ecosystem essentially has two parts: the *country,* where agriculture and animal husbandry are found, and the *city,* where most people live and industry is carried on. This representation of the human ecosystem, although simplified, allows us to see that the system requires two major inputs: *fuel energy* and *raw materials* (e.g., metals, wood, synthetic materials). The use of these two necessarily results in *pollution* and *waste* as outputs.

TABLE 21.3 ECOSYSTEMS

NATURAL	HUMAN
Independent	Dependent
Cyclical (except energy)	Noncyclical
Nonpolluting	Polluting
Renewable solar energy	Nonrenewable fossil fuel energy
Conserves resources	Uses up resources

Country

Modern U.S. agriculture produces exceptionally high yields per acre, but this bounty is dependent on a combination of five variables given here.

1. **Planting of a few genetic varieties.** The majority of farmers specialize in growing one genetic variety. Wheat farmers plant the same type of wheat, and corn farmers plant the same type of corn. This **monoculture agriculture** is subject to attack by a single type of parasite. For example, a single parasitic mold reduced the 1970 corn crop by 15%, and the results could have been much worse because 80% of the nation's corn acreage was susceptible.

2. **Heavy use of fertilizers, pesticides, and herbicides.** *Fertilizer* production requires a large energy input, and fertilizer runoff contributes to water pollution. *Pesticides* reduce soil fertility because they kill off beneficial soil organisms as well as pests, and some pesticides concentrate in food chains (p. 485), eventually producing toxic effects in predators, possibly even humans. *Herbicides,* especially those containing the contaminant dioxin, have been charged with causing negative reproductive effects and cancer.

3. **Generous irrigation.** River waters are sometimes redirected for the purpose of irrigation, in which case "used water" returns to the river carrying a heavy concentration of salt. The salt content of the Rio Grande River in the Southwest is so high that the government has built a treatment plant to remove the salt. Water is also sometimes taken from aquifers (underground rivers) whose water content can be so reduced that it becomes too expensive to pump out more water. Farmers in Texas are already facing this situation.

4. **Excessive fuel consumption.** Energy is consumed on the farm for many purposes. Irrigation pumps have already been mentioned, but large farming machines also are used to spread fertilizers, pesticides, and

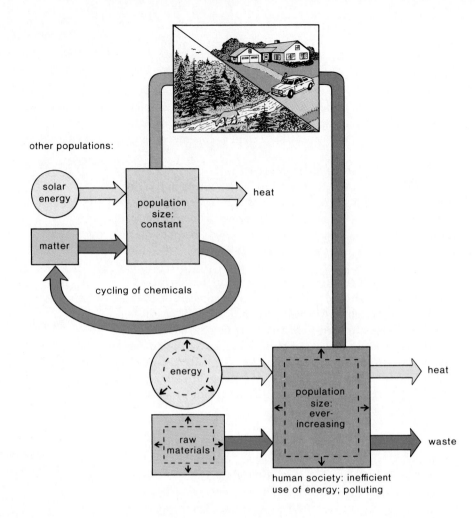

FIGURE 21.13 *In natural ecosystems, population sizes remain about the same year after year; materials cycle and energy is used efficiently. In the human ecosystem, the population size consistently increases, resulting in much pollution because of inadequate cycling of materials and inefficient use of supplemental energy.*

other populations:

solar energy → population size: constant → heat

matter → (cycling of chemicals)

energy → population size: ever-increasing → heat

raw materials → waste

human society: inefficient use of energy; polluting

herbicides and to sow and harvest the crops. It is not incorrect to suggest that modern farming methods transform fossil fuel energy into food energy.

Supplemental fossil fuel energy also contributes to animal husbandry yields. At least 50% of all cattle are kept in *feedlots* where they are fed grain. Chickens are raised in a completely artificial environment, where the climate is controlled and each has its own cage to which food is delivered on a conveyor belt. Animals raised under these conditions often have antibiotics and hormones added to their feed to increase yield.

5. **Loss of land quality.** Evaporation of excess water on irrigated lands can result in a residue of salt. This process, termed **salinization,** makes the land unsuitable for the growth of crops. Between 25% and 35% of the irrigated western croplands are thought to have excessive salinity. Soil erosion is also a serious problem. It is said that we are *mining the soil* because many farmers are not taking measures to prevent the loss of topsoil. The Department of Agriculture

estimates that erosion is causing a steady drop in the productivity of land equivalent to the loss of 1.25 million acres per year. Even more fertilizers, pesticides, and energy supplements will be required to maintain yield.

HUMAN ISSUE

Modern farming activities have had devastating effects on the environment. For example, cultivated fields do not retain the excellent soil-holding capabilities provided by natural vegetation. Soil erosion from these fields creates the murky, turbid waters characteristic of many of our rivers and streams. Water runoff from agricultural fields into streams carries with it fertilizers, leading to eutrophication of lakes and rivers (see p. 486). Therefore, in destroying the original natural vegetation of terrestrial ecosystems, modern agricultural practices have severely disrupted aquatic ecosystems. Who is responsible for preventing the profound ecological impacts exerted by modern agriculture? Is it appropriate to blame farmers?

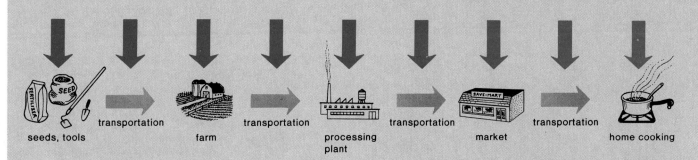

transportation   transportation   transportation   transportation

seeds, tools   farm   processing plant   market   home cooking

a.

ENERGY TO GROW FOOD

In the United States sunlight provides part of the energy for food production, but it is greatly supplemented by fossil fuel energy. Figure *a* shows that even before planting time, there is an input of fossil fuel energy for the production of seeds, tools, fertilizers, and pesticides, and for their transportation to the farm. At the farm, fuel is needed to plant the seeds, to apply fertilizers and pesticides, and to irrigate, harvest, and dry the crops. After harvesting, still more fuel is used to process the crops to make the neatly packaged products we buy in the supermarket. Most of the food we eat today has been processed in some way. Even farm families now buy at least some of their food from supermarkets in nearby towns.

Since 1940 the amount of supplemental fuel used in the American food system has increased far more sharply than the caloric content of the food we consume, as shown in figure *b*. This is partially due to the trend toward producing more food on less land. High-yielding hybrid wheat and corn plants require more care and thus about twice as much supplemental energy as the traditional varieties of wheat and corn. Cattle kept in feedlots and fed grain that has gone through the whole production process need about twenty times the amount of supplemental energy as do range-fed cattle. Our food system has been labeled energy-intensive because it requires such a large input of supplemental energy.

The intensive use of fossil fuel energy to grow and provide food in the United States is a matter for concern because the supply of fossil fuel is limited and its cost has risen overall. This, in turn, affects the cost of farming and of the food produced. What can be done? First of all, devote as much land as possible to farming and animal husbandry. Plant breeders could sacrifice some yield to develop plants that would need less supplemental energy. And we could depend more on range-fed cattle. If cattle are kept close to farmland, manure can substitute, in part, for chemical fertilizer. Biological control, the use of natural enemies to control pests, would cut down on pesticide use. Solar and wind energy could be used instead of fossil fuel energy, particularly on the farm. For example, wind-driven irrigation pumps are feasible.

Finally, of course, consumers could help matters. We could overcome our prejudice against fruits and vegetables that have slight blemishes. We could consume less processed food. We could eat less meat and buy cheaper cuts of beef, which have come from range-fed cattle. And we could avoid using electrically powered gadgets when preparing food at home.

Since the U.S. food system is so energy-intensive, it is doubtful that needy countries abroad could ever duplicate this system. Indeed, if we are concerned about feeding the hungry of the world we should cut back on our own use of supplemental energy to make more available for use by underdeveloped countries.

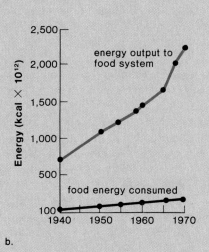

b.

a. *Energy input needed to grow food is represented by colored arrows.*
b. *Energy use in food system, 1940 through 1970, compared to the caloric content of food consumed.*

Organic Farming

Some farmers have given up these modern means of farming and instead have begun to adopt organic farming methods. This means that they do not use applications of fertilizer, pesticides, and herbicides. They use cultivation of row crops to control weeds, crop rotation to combat major pests, and the growth of legumes to supply nitrogen fertility to the soil. Some farmers use natural predators and parasites instead of pesticides to control insects (fig. 21.14). For the most part, these farmers switched farming methods because they were concerned about the health of their families and livestock and had found that the chemicals were sometimes ineffective.

A study of about forty farms showed that organic farming for the most part was just as profitable as conventional farming. Crop yields were lower, but so were operating costs. Organic farms required about ⅖ as much fossil energy to produce one dollar's worth of crop. The method of plowing and utilization of crop rotation resulted in ⅓ less soil erosion. The researchers concluded it would be well to determine how far farmers can move in the direction of reduced agricultural chemical use and still maintain the quality of the product. They noted that a modest application of fertilizer would have improved the protein content of the crop.

City

The city is dependent on the country to meet its needs. For example, each person in the city requires several acres of land for food production. Overcrowding in cities does not mean that less land is needed; each person still requires a certain amount of land to ensure survival. Unfortunately, however, as the population increases, the suburbs and cities tend to encroach on agricultural areas and rangeland (fig. 21.15).

The city houses workers for both commercial businesses and industrial plants. Solar and other renewable types of energy are rarely used; cities currently rely mainly on fossil fuel in the form of oil, gas, electricity, and gasoline. The city does not conserve resources. An office building, with constantly burning lights and windows that cannot be opened, is an example of energy waste. Another example is people who drive cars long distances instead of taking public transportation and who drive short distances instead of walking or bicycling. Materials are not recycled, and products are designed for rapid replacement.

The burning of fossil fuels for transportation, commercial needs, and industrial processes causes air and water pollution. This pollution is compounded by the chemical and solid waste pollution that results from the manufacture of many products. Consider that any product used by the average consumer (house, car, washing machine) causes pollution and waste, both during its production and when it is disposed. Humans themselves produce much sewage that is discharged into bodies of water, often after only minimal treatment.

Table 21.3 lists the characteristics of the human ecosystem as it now exists. Just as the city is not self-sufficient and requires the country to supply it with food, so the whole human ecosystem is dependent on the natural ecosystems to provide resources and absorb waste. Fuel combustion byproducts, sewage, fertilizers, pesticides, and solid wastes are all added to natural ecosystems in the hope that these systems will cleanse the biosphere of these pollutants. But we have replaced natural ecosystems with our own ecosystem and have exploited natural ecosystems for resources, adding ever more pollutants, to the extent that the remaining natural ecosystems have become overloaded.

FIGURE 21.16 *Modified human ecosystem. In order to cut down on the amount of lost heat and waste matter, heat could be used more efficiently and discarded materials could be recycled.*

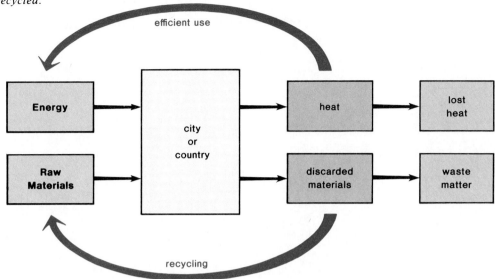

FIGURE 21.17 *Efficient use of resources. Instead of allowing cattle waste to enter a water supply, it could be sent to a conversion plant that would produce methane gas. (The residue remaining could be converted back into feed for cattle.) Excess heat, which arises from the burning of the methane gas in order to produce electricity, can be cycled back to the conversion plant.*

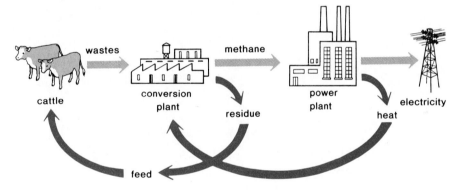

Natural ecosystems have been destroyed and overtaxed because the human ecosystem is noncyclical and because an ever-increasing number of people wish to maintain a standard of living that requires many goods and services. But we can call a halt to this spiraling process if we achieve zero population growth and if we conserve energy and raw materials. Conservation can be achieved in three ways: (1) use wisely only what is actually needed; (2) recycle nonfuel minerals such as iron, copper, lead, and aluminum; and (3) use renewable energy resources (p. 506) and find more efficient ways to utilize all forms of energy. Figure 21.16 presents a diagrammatic representation of what is needed to maintain the delicate balances of the human and natural ecosystems.

As a practical example, consider a plant that was built in Lamar, Colorado, that produces methane from feedlot animals' wastes (fig. 21.17). The methane is burned in the city's electrical plant and the heat given off is used to incubate the anaerobic digestion process that produces the methane. In addition, a protein feed supplement is produced from the residue of the digestion process. This system represents a cyclical use of material and an efficient use of energy similar to that found in nature. Many other such processes for achieving this end have been and will be devised. However, as long as the human ecosystem on the whole remains inefficient and noncyclical, it will continue to cause pollution.

FIGURE 21.18 *Air pollution.*
Pollutants enter the atmosphere from
various sources, but in most instances air
pollution is caused by the burning of
fossil fuels (coal, gas, and oil).

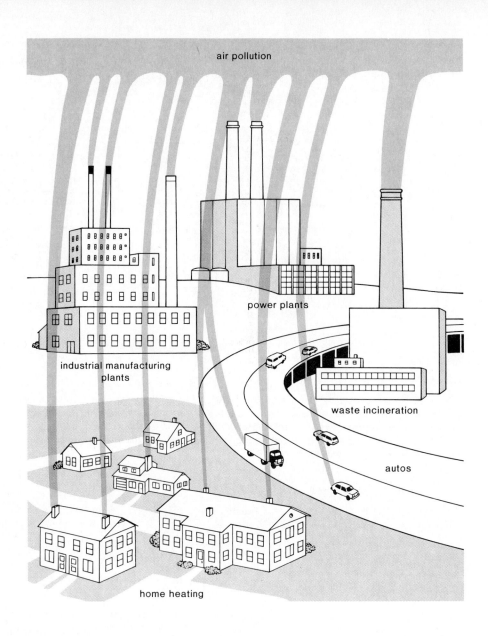

air pollution

power plants

industrial manufacturing
plants

waste incineration

autos

home heating

In contrast to mature natural ecosystems, the human ecosystem, consisting of country and city, is not stable. Nonrenewable fossil fuel energy is used inefficiently, and material resources enter the system and do not cycle. Because of these excessive inputs, the outputs to the system are much pollution and waste. Although in the past we could rely on natural ecosystems to process our wastes, such as sewage, this is no longer feasible because the size of the natural systems has been reduced and the amount of pollution has been steadily increasing. However, it is possible to change the human ecosystem so that it more nearly corresponds to a natural system by using renewable energy sources and by recycling material resources.

Air Pollution

Pollutants enter the atmosphere from various sources (fig. 21.18), but the burning of fossil fuels contributes greatest to the five categories of primary pollutants: carbon monoxide, CO; hydrocarbons, HC; nitrogen oxides, NO_x (NO, NO_2); particulates; and sulfur oxides, SO_x (SO_2, SO_3). The graph in figure 21.19 compares the sources of these pollutants. It is obvious that modes of transportation, especially the automobile, are the main cause of carbon monoxide air pollution. This chemical combines preferentially with hemoglobin to prevent circulation of oxygen within the body, causing unconsciousness. In New York City traffic, the blood concentration of carbon monoxide has been shown to reach 5.8%, a dangerous level when compared to the 1.5% that physicians consider safe. The particulates, dust and soot, can collect in

FIGURE 21.19 *Components of air pollution: CO (carbon monoxide), HC (hydrocarbons), NO$_x$ (nitrogen oxides), particulates (solid matter), SO$_x$ (sulfur oxides). Transportation contributes most to air pollution because when gasoline burns it gives off CO, a gas that interferes with the capacity of hemoglobin to carry oxygen.*

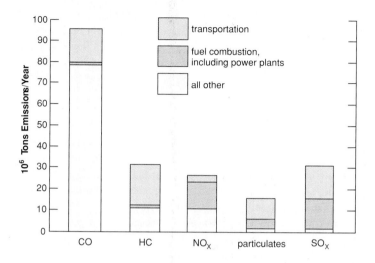

FIGURE 21.20 *The milkweed in (a) was exposed to ozone; the milkweed in (b) was grown in an enclosure with filtered air. The plant in (b) is much healthier than the plant in (a).*

a.   b.

the lungs, and nitrogen oxides and sulfur oxides irritate the respiratory tract. The hydrocarbons (various organic compounds) may be carcinogenic.

Unfortunately, these primary pollutants interact with one another, producing pollutants that are even more dangerous.

Photochemical Smog

Photochemical smog results when two pollutants from automobile exhaust—nitrogen oxide and hydrocarbons—react with one another in the presence of sunlight to produce nitrogen dioxide (NO$_2$), ozone (O$_3$), and peroxylacetyl nitrate (PAN). **Ozone** and **PAN** are commonly referred to as oxidants. Breathing ozone affects the respiratory and nervous systems, resulting in respiratory distress, headache, and exhaustion. These symptoms are particularly apt to appear in youngsters; therefore, Los Angeles schoolchildren must remain at rest inside school buildings whenever the ozone level reaches 0.35 ppm (parts per million by weight). PAN and ozone are especially damaging to plants, resulting in leaf mottling and reduced growth (fig. 21.20).

Normally, warm air near the ground rises, allowing pollutants to be dispersed and carried away by air currents. Sometimes, however, air pollutants, including smog and particulates, are trapped near the earth due to **thermal inversions.** During a thermal inversion, the cold air is at ground level and the warm air is above. (This may occur when a cold front brings in cold air and settles beneath a warm layer.) Cities surrounded by hills, such as Los Angeles and Mexico City, are particularly susceptible to the effects of a temperature inversion.

Acid Rain

A side effect of air pollution is acid rain. When sulfur oxides (SO$_x$) and nitrogen oxides (NO$_x$) are injected into the atmosphere they tend to react with water to produce sulfuric acid and nitric acid. These acids are eventually deposited on the earth's surface. Here they corrode marble, metal, and stonework and leach metals from the soil. It is the latter action that leads to the destruction of forests and causes lakes to become devoid of most forms of life.

Coal burning results in most of the sulfur oxides that lead to acid rain. There are several ways to minimize the problem. Coal could be washed prior to burning. Low-sulfur coal could be substituted for high-sulfur coal. Devices called scrubbers could be installed in the tall stacks to prevent sulfur oxides from entering the air. Under development is a new method of burning coal that uses a mixture of coal and limestone. This technique has reduced both sulfur and nitrogen oxide emissions.

The problem of NO$_x$ emissions is also solvable. Use of the catalytic converter in automobiles reduces the amount of NO$_x$ given off by about ¾. A low NO$_x$ burner is also being developed for industrial boilers.

The Weather

It is predicted that the earth's average temperature could rise as much as 8° F over the next one hundred to two hundred years because of CO$_2$ buildup due to fossil fuel combustion. CO$_2$ allows the sun's rays to pass through but absorbs and reradiates heat back toward earth. This may be compared to a greenhouse in that the glass of a greenhouse also allows

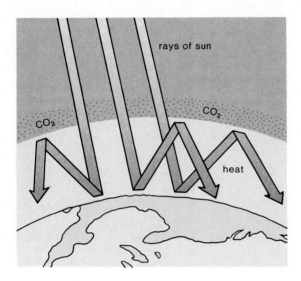

FIGURE 21.21 *Greenhouse effect. The sun's rays can penetrate CO₂ and reach the earth's surface, but the resulting heat cannot pass through a layer of CO₂ and instead is trapped near the earth's surface, just as heat is trapped by the glass of a greenhouse. The earth's temperature is forecast to substantially increase due to the accumulation of CO₂ in the atmosphere.*

sunlight to pass through, but traps the heat. This phenomenon also occurs when a car sits in the hot summer sun.

This rise in temperature from the **greenhouse effect** (fig. 21.21) could have a serious impact on agriculture and eventually on sea levels because of melting polar ice. Wind pattern changes would make the midwestern United States, the Soviet Union, and China drier. Favorable climates for the growth of crops would move north, where the soil is not so favorable for agriculture. Even though it is not certain that polar ice would melt, the sea level would rise simply because water expands when it absorbs heat. If the arctic ice should melt substantially, it is predicted that most of the world's cities would be flooded and so would some of our richest farmlands.

The concensus among researchers is that we should begin to prepare for these consequences since it is very unlikely that we will ever curtail fossil fuel combustion enough to prevent their occurrence.

Indoor Air Pollution

Although much attention has been given to the quality of outdoor air, it has been pointed out that indoor air is often more polluted. NO_2, traced to gas combustion in stoves, has been found indoors at twice the outdoor level; CO, especially in homes with gas, coal, or wood-heating or cooking stoves, often exceeds the current health standards; hydrocarbons from myriad sources appear in high concentrations; and radioactive radon gas, emitted naturally from a variety of building materials, has been detected indoors at levels that exceed outdoor levels by factors of 2 to 20. A surprising

number of household items continuously give off formaldehyde, including insulation, particle board, and plywood. Asbestos fibers are often dislodged from building materials, especially due to vibration, abrasion, cutting, grinding, sanding, or aging.

Because of these findings, some have questioned the advisability of tightening up residential buildings to save energy. Reduced ventilation unquestionably increases concentrations of pollutants. Others are calling for research studies so that the proper regulatory legislation may be prepared. For example, it would be possible to modify building codes so that safer building materials are used.

Air Quality Control

Air quality in the United States is controlled by the Clean Air Act. The Environmental Protection Agency (EPA) has established standards for the pollutants listed in figure 21.19. In addition, asbestos, beryllium, mercury, vinyl chloride, and lead are more strictly controlled because they are extremely hazardous to health. Industry is required to use special equipment, such as collectors and scrubbers, to cut down on stack emissions. Automobiles have been equipped with special devices, such as the **catalytic converter** that chemically changes HC and CO to carbon dioxide and water. The catalytic converter, which necessitates the use of nonleaded gasoline, also contributes to fuel economy. The result has been generally improved air quality throughout most of the nation.

Pollution affects all portions of the biosphere. The greatest single contribution to air pollution is the burning of fossil fuels. Acid rain and smog are caused by air pollutants. It is forecast that carbon dioxide buildup will gradually cause the weather to become warmer. Although much concern has been focused on outdoor pollution, it appears that indoor pollution may need our attention even more.

Land Pollution

Every year, the United States population discards billions of tons of solid wastes, much of it on land. **Solid wastes** include not only household trash but also sewage sludge, agricultural residues, mining refuse, and industrial wastes. Those solid wastes containing substances that cause human illness and even sometimes death are called **hazardous wastes.**

Household Trash

In 1920 the per capita production of waste was about 2.75 pounds per day; in 1970 about 5 pounds per day; and in 1980 about 8 pounds per day. The exponential growth of consumer products with "planned obsolescence" and the use of packaging materials to gain a competitive edge account for much of this increase.

Open dumping (fig. 21.22), sanitary landfills, or incineration have been the most common practices of disposing trash. These disposal methods have become increasingly expensive and also cause pollution. It would be far more satisfactory to recycle materials as much as possible and/or to use organic substances as a fuel to generate electricity. One study showed that it was possible to achieve 70–90% public participation in recycling by spending only thirty cents per household. A city the size of Washington, D.C., where 500,000 tons of waste are generated each year, could have an increase of 1,300 jobs if the community utilized solid wastes as a resource instead of throwing them away.

Hazardous Wastes

The U.S. Environmental Protection Agency, charged with the task of keeping the environment safe, estimates that in 1980 at least 57 million metric tons of the nation's total wasteload could be classified as hazardous. Hazardous wastes fall into three general categories.

1. *Heavy metals,* such as lead, mercury, cadmium, nickel, and beryllium, can accumulate in various organs, interfering with normal enzymatic actions and causing illness, including cancer.
2. *Chlorinated hydrocarbons,* also called **organochlorides,** include pesticides and numerous organic compounds, such as polychlorinated biphenyls (PCBs), in which chlorine atoms have replaced

hydrogen atoms. Research has often found organochlorides to be cancer producing in laboratory animals.
3. *Nuclear wastes* include radioactive elements that will be dangerous for thousands of years. For example, plutonium must be isolated from the biosphere for 200,000 years before it has lost its radioactivity.

Hazardous wastes are often subject to **bioaccumulation** (fig. 21.23). Bacteria in the soil do not ordinarily break down these wastes, and when other organisms take them up, they remain in the body and are not excreted. Once they enter a food chain, they become more concentrated at each trophic level. Notice in figure 21.23 that the number of dots representing DDT becomes more concentrated as the chemical is passed along from producer to tertiary consumer. Bioaccumulation is most apt to occur in aquatic food chains; there are more trophic levels in aquatic food chains than there are in terrestrial food chains. Humans are the final consumers in both types of food chains, and in some areas, mothers' milk contains a detectable amount of DDT and PCBs.

The public has become aware of hazardous dump sites that have polluted nearby water supplies. Chemical wastes buried over a quarter century ago in Love Canal, near Niagara Falls, have seriously damaged the health of some residents there. Similarly, the town of Times Beach, Missouri, is abandoned because a workman spread an organochloride (dioxin)-laced oil on the city streets, leading to a myriad of illnesses among its citizens. In other places, such as in Holbrook, Massachusetts, manufacturers have left thousands of drums containing toxic chemicals in abandoned or uncontrolled sites. These toxic chemicals are oozing out into the ground and contaminating the water supply. Illnesses, especially forms of cancer, are quite common not only in Holbrook but also in adjoining towns.

Cleanup Although the government established a superfund to clean up toxic dump sites, the enormity of the problem has not as yet led to much clean up. Specific manufacturers, however, have devised new means to clean up any accidental spills. Very often bacteria are employed. For example, bacteria have now been engineered to eat up PCBs, agent orange, and cyanide.

Legislation In an effort to control the level of toxic wastes, Congress passed the Resource Conservation and Recovery Act, which empowers the EPA to track all significant quantities of hazardous waste from wherever they are generated to their final disposal. Government regulations are designed to encourage industry to adopt new waste management strategies, including *reduction* (changing a manufacturing process so that it does not produce hazardous by-products); *recycling* (reusing waste material); and *resource recovery* (extracting valuable material from waste).

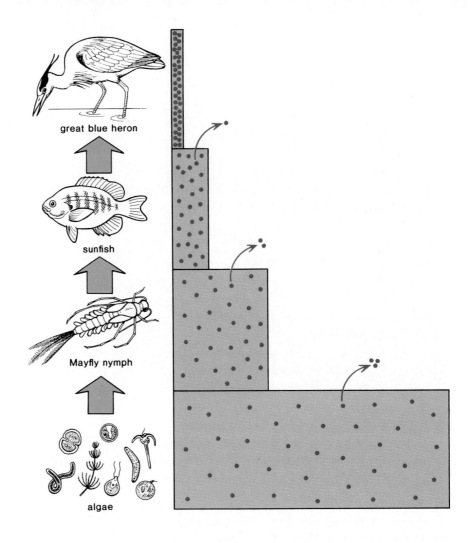

FIGURE 21.23 *Bioaccumulation. A poison (dots) that is minimally excreted (arrows) becomes maximally concentrated as it passes along a food chain due to the reduced size of the trophic levels. Because of this problem, fishermen are warned against consuming fish from the Great Lakes.*

great blue heron

sunfish

Mayfly nymph

algae

A number of companies have been able to curb their generation of wastes by employing these processes. For example, Minnesota Mining and Manufacturing Company found that it could cut the volume of its toxic wastes associated with adhesive tape production in half by using water-based glues instead of solvent-based glues. And metal processing companies in California have found that an acid they usually discard can be reused to recover zinc-iron compounds.

Water Pollution

Surface Waters

Figure 21.24 shows the many ways that humans cause surface water pollution. Some pollutants, such as fertilizers, sewage, and certain detergents, add nitrates and phosphates to freshwater lakes and ponds. This overabundance of nutrients, called **cultural eutrophication,** speeds up the tendency of bodies of water to fill in and disappear. First, the nutrients cause overgrowth of algae. The death of these algae promotes the growth of a very large decomposer population. The decomposers break down the algae, but in so doing they use up oxygen. In addition, algae also consume oxygen during the night when photosynthesis is impossible. Both of these cause a decreased amount of oxygen available to fish, ultimately causing the fish to die. The increased amount of life and death remains cause the lake to be more eutrophic, leading to a reduction in its size.

Water Quality Control The Water Pollution Control Act empowers the federal government to set minimum water quality standards for rivers and streams. The building of sewage treatment plants has helped clean up U.S. waters. In addition, large industries are no longer permitted to dump chemicals indiscriminately into water or to send their wastes to local sewage treatment plants. This two-pronged attack has helped clean up the Great Lakes, which previously were said to be dying because of pollution.

Treatment plants capable of not only digesting sewage but also of removing nutrient molecules are very expensive to build. Some communities have devised ingenious ways to

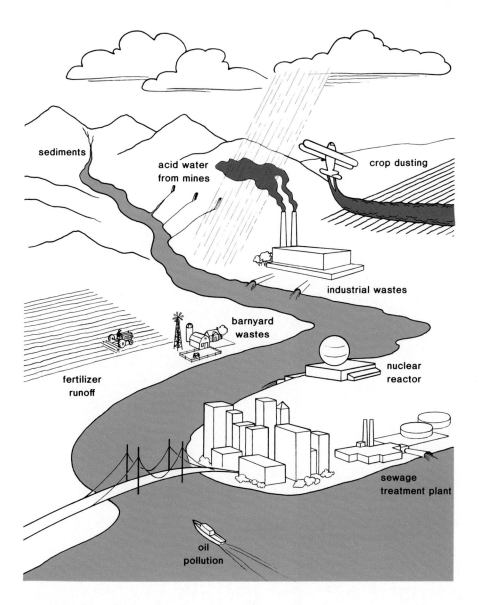

FIGURE 21.24 *Surface water pollution is caused in all the ways shown here. There is much concern of late about Chesapeake Bay because it is dying due to sediments and nutrients that enter the bay, particularly from the Susquehanna River, which begins in New York and flows through Pennsylvania.*
Adapted from "Disposal Practices and Their Effects on Ground Water" in Executive Summary, *U.S. Environmental Protection Agency, Office of Water Supply and Solid Waste Management Programs (Washington, DC, U.S. Government Printing Office).*

sediments

acid water from mines

crop dusting

industrial wastes

barnyard wastes

fertilizer runoff

nuclear reactor

sewage treatment plant

oil pollution

accomplish this end, however. Effluent from a sewage treatment plant can be passed through a swampy area or forest before it drains into a nearby waterway. Or the effluent can be used to irrigate crops and/or grow algae and aquatic plants in shallow ponds. Since these products can be used as food for animals, this represents a cyclical use of chemicals.

Groundwater

Groundwater is subject to pollution in the ways illustrated in figure 21.25. Chemicals drain from hazardous waste dumping sites, and bacteria and viruses drain from septic tanks and cesspools into the ground and may eventually reach aquifers (underground rivers). Also, previously industry was accustomed to running wastewater into a pit. The pollutants could then seep into the ground. Or pollutants were injected into deep wells from which the pollutants constantly discharged. Both of these customs have been or are in the process of being phased out. However, it is very difficult for industry to find other ways to dispose of wastes. More adequately managed and controlled waste treatment plants are needed, but because citizens do not wish to live near waste treatment plants, towns are often successful in preventing their construction. In the meantime, industries are still employing less approved methods of dealing with industrial wastes.

Hazardous wastes that are by-products of various industries have been deposited on the land but have seeped into surface water and groundwater. Unfortunately, due to bioaccumulation, these substances return to humans in a concentrated form. Although progress has been made in reducing cultural eutrophication, much has to be done in keeping our water supplies free of hazardous wastes.

FIGURE 21.25 *Groundwater pollution is caused in all the ways shown here. Discontinuance of these means of disposal for industrial wastes has been difficult to achieve because citizens do not wish to have waste disposal plants located near them. Source: Adapted from U.S. Environmental Protection Agency, Office of Water Supply and Solid Waste Management Programs, Waste Disposal Practices and Their Effects on Ground Water: Executive Summary (Washington, D.C.: Government Printing Office, 1977).*

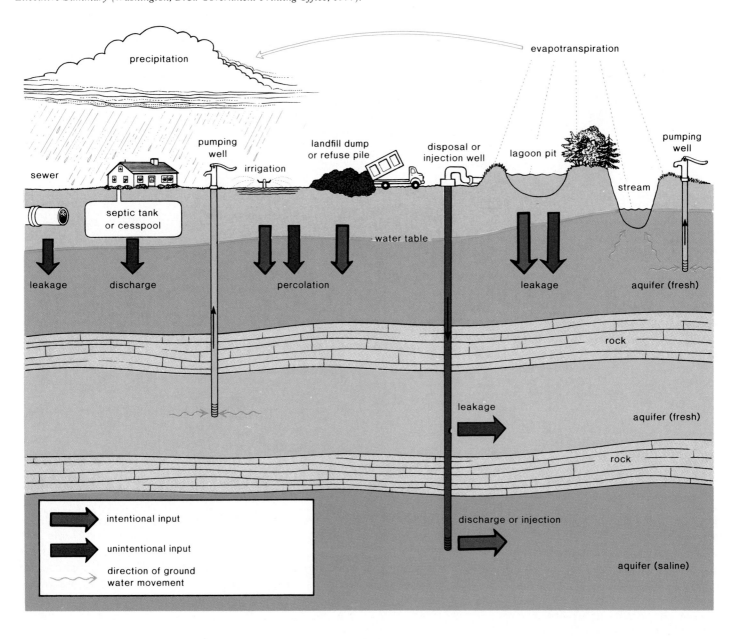

SUMMARY

In an ecosystem, the biotic component consists of populations that interact with each other and with the abiotic component, the physical environment. The populations in an ecosystem form food chains in which the producers produce food for the other populations. Chemicals cycle through the food chain and, in time, the same inorganic molecules are returned to the producer population for conversion to organic food. Energy flows through the ecosystem but does not cycle, since there is a loss of useful energy at each trophic level. The food pyramid illustrates that each successive population has a smaller size or biomass because energy does not cycle.

In contrast to mature natural ecosystems, the human ecosystem, consisting of country and city, is not stable. Nonrenewable fossil fuel energy is used inefficiently, and material resources enter the system and do not cycle. Because of these excessive inputs, the outputs to the system are much pollution and waste. Pollution affects all portions of the biosphere. The burning of fossil fuels, especially, causes air pollution, which leads to the development of smog and acid rain. Hazardous wastes, by-products of various industries, have been deposited on the land but have seeped into surface water and groundwater. Unfortunately, due to bioaccumulation, these substances return to humans in concentrated form. While progress has been made in reducing cultural eutrophication, much has to be done in keeping our water supplies free of hazardous wastes.

OBJECTIVE QUESTIONS

1. Chemicals cycle through the populations of an ecosystem, but energy is said to _____ because it is all eventually dissipated as heat.
2. The first population in a food chain is always a _____ population.
3. An ecological pyramid illustrates that each succeeding trophic level has less _____ than the previous level.
4. Natural ecosystems utilize the same amount of energy per year, but the human ecosystem utilizes an _____ _____.

5. In the carbon cycle, when organisms _____ carbon dioxide is returned to the exchange pool.
6. Humans make a significant contribution to the nitrogen cycle when they convert aerial nitrogen to _____ for use in fertilizers.
7. During the process of denitrification, nitrates are converted to _____ _____.
8. What are the three categories of hazardous wastes? _____ _____ , _____ , _____ .

9. Hazardous wastes are often subject to _____ , a process by which they become concentrated in top consumers, including humans.
10. The burning of _____ supplies most of the sulfur oxides that lead to acid rain in the United States.

Answers

1. flow 2. producer 3. energy or biomass 4. ever-increasing amount 5. respire 6. nitrates 7. aerial nitrogen 8. heavy metals, chlorinated hydrocarbons, nuclear wastes 9. bioaccumulation 10. coal

STUDY QUESTIONS

1. What is succession, and how does it result in a climax community? (p. 467)
2. Give an example of an aquatic and a terrestrial food chain. Name the producer, the consumers, and the decomposers in each chain. Explain the manner in which chemicals cycle. (pp. 470, 471)
3. A pyramid describes the size of the various trophic levels within a food web. Discuss the relationship of size and the fact that energy does not cycle. (p. 474)

4. Describe the carbon cycle. How do humans contribute to this cycle? (p. 475)
5. Describe the nitrogen cycle. How do humans contribute to this cycle? (p. 476)
6. Draw a diagram to represent the human and natural ecosystems, and discuss their inputs and outputs. (p. 478)
7. What are the primary components of air pollution? How do they contribute to smog and acid rain? (pp. 482–483)

8. How is air pollution related to the weather? (p. 483)
9. What type of chemicals are termed hazardous wastes? Why? (p. 485)
10. Why do nonbiodegradable poisons concentrate as they go from trophic level to trophic level? (p. 485)
11. What is cultural eutrophication, and how might it be prevented? (p. 486)

THOUGHT QUESTIONS

1. Argue against the suggestion that natural ecosystems are stable.
2. Sometimes we are under the impression that all bacteria are harmful. What evidence can you give from this chapter to counter this belief?
3. If all goes well, biotechnology might help with which of the environmental problems mentioned in this chapter?

KEY TERMS

abiotic (ab″e-ot′ik)   not including living organisms, especially the nonliving portions of the environment.   *470*

biosphere (bi′o-sfer)   that part of the earth's surface and atmosphere where living organisms exist.   *470*

biotic (bi-ot′ik)   pertaining to any aspect of life, especially the living portions of the environment.   *470*

climax community (kli′maks kō-mu′ni-te)   in succession, the final, stable stage.   *467*

consumers (kon-su′merz)   a population that feeds on members of other populations in an ecosystem.   *470*

cultural eutrophication (kul′tu-ral u″tro-fi-ka′shun)   enrichment of a body of water causing excessive growth of producers and then death of these and other inhabitants.   *486*

decomposers (de-kom-po′zerz)   organisms of decay (fungi and bacteria) in an ecosystem.   *470*

detritus (de-tri′tus)   nonliving organic matter.   *470*

ecology (e-kol′o-je)   the study of the relationship among organisms and their relationship with the physical environment.   *470*

ecosystem (ek″o-sis′tem)   a biological community together with the associated abiotic environment.   *470*

food web (food web)   the complete set of food links between populations in a community.   *472*

fossil fuel (fos″l fu′el)   the remains of once-living organisms that are burned to release energy, such as coal, oil, and natural gas.   *475*

greenhouse effect (grēn′hows ē-fekt)   carbon dioxide buildup in the atmosphere as a result of fossil fuel combustion; retains and reradiates heat, effecting an abnormal rise in the earth's average temperature.   *484*

habitat (hab′ĭ-tat)   the natural abode of an animal or plant species.   *470*

hazardous wastes (haz′er-dus wāsts)   wastes containing chemicals hazardous to life.   *484*

nitrogen fixation (ni′tro-jen fik-sa′shun)   a process whereby free atmospheric nitrogen is converted into compounds, such as ammonia and nitrates, usually by soil bacteria.   *477*

producers (pro-du′serz)   organisms that produce food and are capable of synthesizing organic compounds from inorganic constituents of the environment; usually the green plants and algae in an ecosystem.   *470*

succession (suk-se′shun)   a series of ecological stages by which the community in a particular area gradually changes until there is a climax community that can maintain itself.   *467*

thermal inversion (ther′mal inver′zhun)   temperature inversion such that warm air traps cold air and its pollutants near the earth.   *483*

CHAPTER TWENTY-TWO

POPULATION CONCERNS

CHAPTER CONCEPTS

1 The human population has been undergoing exponential growth since 1850.

2 The very large human population is straining the capacity of the earth to sustain it.

3 There are two points of view regarding the future supplies of nonrenewable resources. There are those who believe that we must conserve what is left and those who believe that technology will ever be able to exploit new sources.

4 Energy is required in order to exploit the environment. Most likely both coal and solar energy will be important sources of energy in the near future.

5 Thus far, the supply of food has kept up with increased population. It may be necessary to institute new agricultural methods if this is to be achieved in the future.

The population of Tokyo, Japan, is an example of a heavily crowded world.

INTRODUCTION

Imagine a watering hole that can accommodate one hundred rabbits (fig. 22.1). If at first there are only two rabbits and each pair of rabbits only produces four rabbits, how many doublings could there be without overtaxing the watering hole? You're correct if you say four and incorrect if you say five. The problem is that you can't just consider the newly arrived rabbits—you have to add in the number of rabbits already there.

2 – 4 – 8 – 16 – 32 – 64 – 128

Also, notice that when it is time to stop, there are only 62 rabbits (30 + 32). That's one of the unusual things about population growth—at one point in time it seems as if there is plenty of room and then boom! There's not enough room all of a sudden.

The human growth curve is an exponential curve (fig. 22.2). In the beginning, growth of the population was relatively slow, but as a greater number of reproducing individuals were added, growth increased until the curve began to slope steeply upward. It is apparent from the position of 1989 on the growth curve that growth is now quite rapid. The world population increases at least the equivalent of a medium-sized city every day (200,000) or the combined populations of the United Kingdom, Norway, Ireland, Iceland, Finland, and Denmark every year. These startling figures are a reflection of the fact that a very large world population is undergoing positive exponential growth.

EXPONENTIAL POPULATION GROWTH

Mathematically speaking, **exponential growth,** or geometric increase, occurs in the same manner as compound interest; that is, the percentage increase is added to the principal before the next increase is calculated. Referring specifically to populations, we wish to consider the hypothetical population sizes

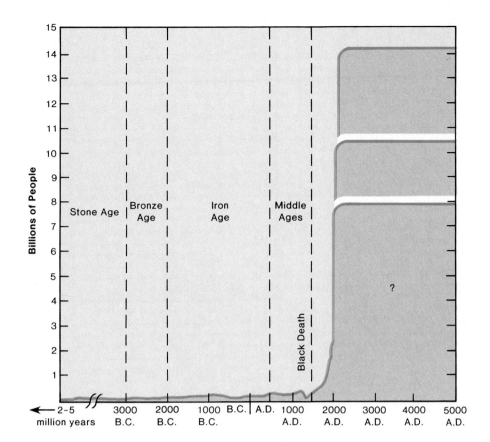

FIGURE 22.2 *Growth curve for human population. The human population has been undergoing rapid exponential growth. Now that the growth rate is declining, it is predicted that the population size will level off at 8, 10.5, or 14.2 billion, depending upon the speed with which the growth rate declines.*

TABLE 22.1 EXPONENTIAL GROWTH OF HYPOTHETICAL POPULATIONS

POPULATION SIZE	PERCENTAGE INCREASE	ACTUAL INCREASE IN NUMBERS	POPULATION SIZE	PERCENTAGE INCREASE	ACTUAL INCREASE IN NUMBERS	POPULATION SIZE
500,000,000	2.00	10,000,000	510,000,000	1.99	10,149,000	520,149,000
3,000,000,000	2.00	60,000,000	3,060,000,000	1.99	60,894,000	3,120,894,000
5,000,000,000	2.00	100,000,000	5,100,000,000	1.99	101,490,000	5,201,490,000

in table 22.1. This table illustrates the circumstances of world population growth at the moment: the percentage increase has decreased, and yet the size of the population grows by a greater amount each year. The increase in size is dramatically large because the world population is very large.

In our hypothetical examples (table 22.1), an initial increase of 2% added to the original population size, followed by a 1.99% increase, results in the third generation size listed in the last column. Notice that

1. in each instance the second generation has a larger increase than the first generation because the second generation's population was larger than the first;
2. because of exponential growth, the lower percentage increase (i.e., 1.99% compared to 2%) still brings about larger population growth;
3. the larger the population, the larger the increase for each generation.

The percentage increase is termed the **growth rate,** which is calculated per year.

Growth Rate

The growth rate of a population is determined by considering the difference between the number of persons born (birthrate, or natality) and the number of persons who die per year (death rate, or mortality). It is customary to record these rates per 1,000 persons. For example, Russia (USSR) at the present time has a birthrate of 20 per 1,000 per year, but it has a death rate of 10 per 1,000 per year. This means that Russia's population growth, or simply its growth rate, would be

$$\frac{20 - 10}{1,000} = \frac{10}{1,000} = \frac{1.0}{100} = 1.0\%$$

TABLE 22.2 RELATIONSHIP BETWEEN GROWTH RATE AND THE DOUBLING TIME OF A POPULATION

GROWTH RATE (%)	DOUBLING TIME (YEARS)
0.25	280
0.5	140
1.0	70
2.0	35
3.0	23

TABLE 22.3 ESTIMATED TIMING OF EACH BILLION OF WORLD POPULATION

	TIME TAKEN TO REACH	YEAR ATTAINED
First billion	2–5 million years	About A.D. 1800
Second billion	Approx. 130 years	1930
Third billion	30 years	1960
Fourth billion	15 years	1975
Projections:		
Fifth billion	12 years	1987
Sixth billion	11 years	1998

Source: Elaine M. Murphy. *World Population: Toward the Next Century.* (Washington, DC: Population Reference Bureau, November 1981) page 3. Reprinted by permission.

Notice that while birth and death rates are expressed in terms of 1,000 persons, the growth rate is expressed per 100 persons, or as a percentage.

After 1750 the world population growth rate steadily increased until it peaked at 2% in 1965, but it has fallen slightly since then to 1.7%. Yet there is an ever larger increase in the world population each year because of exponential growth. The explosive potential of the present world population can be appreciated by considering the doubling time.

Doubling Time

Table 22.2 shows that the **doubling time** for a population may be calculated by dividing 70 by the growth rate:

$$d = \frac{70}{gr}$$

d = doubling time
gr = growth rate
70 = demographic constant

If the present world growth rate of 1.7% should continue, the world population would double in 40 years.

$$d = \frac{70}{1.7} = 40 \text{ years}$$

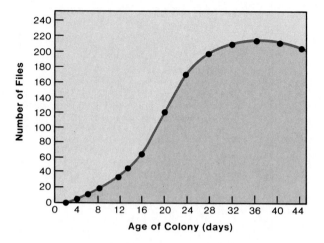

FIGURE 22.3 *Growth curve for a fruit fly colony. The fruit flies in a laboratory colony were counted every other day, and when these numbers were plotted, a sigmoidal growth curve resulted. This type of curve is expected for a population that is adjusting to the carrying capacity of the environment.*

This means that in 39 years the world would need double the amount of food, jobs, water, energy, and so on if the standard of living is to remain the same.

It is of grave concern to many individuals that the amount of time needed to add each additional billion persons to the world population has taken less and less time (table 22.3). The world reached its first billion around 1800—some 2 million years after the evolution of humans. Adding the second billion took only about 130 years, the third billion about 30 years, and the fourth billion took only about 15 years. However, if the growth rate should continue to decline, this trend would reverse itself, and eventually there would be zero population growth. Then population size would remain steady. Thus, figure 22.2 shows three possible logistic curves: the population may level off at 8, 10.5, or 14.2 billion, depending on the speed at which the growth rate declines.

Carrying Capacity

Examining the growth curves for nonhuman populations reveals that populations tend to level off at a certain size. For example, figure 22.3 gives the actual data for the growth of a fruit fly population reared in a culture bottle. At the beginning, the fruit flies were becoming adjusted to their new environment and growth was slow. Then, since food and space were plentiful, they began to multiply rapidly. Notice that the curve begins to rise dramatically just as the human population curve does now. At this time, it may be said that the population is demonstrating its **biotic potential.** Biotic potential is the maximum growth rate under ideal conditions. Biotic potential is not usually demonstrated for long because of an opposing force called **environmental resistance.** Environmental resistance includes all of the factors that cause early

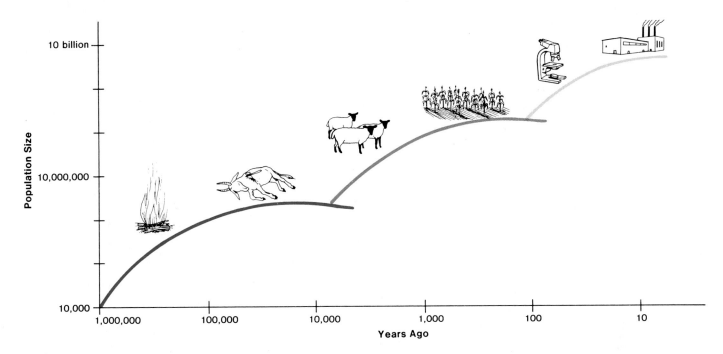

death of organisms and therefore prevent the population from producing as many offspring as it might otherwise have done. As far as the fruit flies are concerned, we can speculate that environmental resistance included the limiting factors of food and space. Also, the waste given off by the fruit flies may have begun to contribute to keeping the population size down.

The eventual size of any population represents a compromise between the biotic potential and the environmental resistance. This compromise occurs at the **carrying capacity** of the environment. The carrying capacity is the maximum population that the environment can support—for an indefinite period.

The carrying capacity of the earth for humans is not certain. Some authorities think the earth is potentially capable of supporting 50–100 billion people. Others think we already have more humans than the earth can adequately support.

The human population is expanding exponentially, and even though the growth rate has declined, there is a large increase in the population each year. Still, the doubling time has increased from thirty-five to forty years. Since a doubling of the population means a doubling of the goods and services needed to sustain the population and since the carrying capacity of the earth is unknown, a decreasing growth rate is welcomed by many.

HISTORY OF THE WORLD POPULATION

Figure 22.4 suggests that the human population has undergone three phases of exponential growth. *Toolmaking* may have been the first technological advance that allowed the human population to enter a phase of exponential growth. *Cultivation of plants* and *animal husbandry* may have allowed a second phase of growth; and the *industrial revolution,* which occurred about 1850, promoted the third phase.

Developed and Developing Countries

When discussing population trends, the countries of the world are sometimes divided according to whether they are already developed or are presently becoming developed.

Developed Countries
The industrial revolution, which was also accompanied by a medical revolution, took place in the Western world. In addition to European and North American countries, Russia and Japan also became industrialized. Collectively, these countries are often referred to as the **developed countries.** The developed countries doubled their size between 1850 and 1950 (fig. 22.5), largely due to a decline in the death rate. This decline is attributed to the influence of modern medicine and improved socioeconomic conditions. Industrialization raised personal incomes, and better housing permitted improved

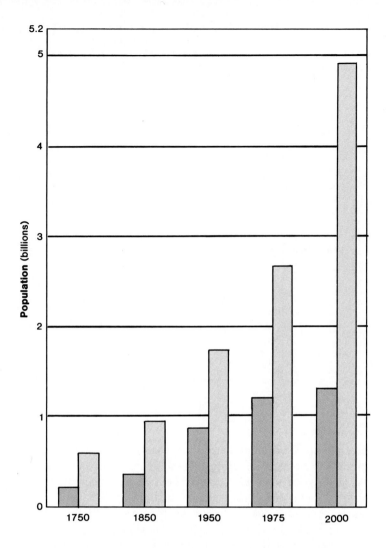

TABLE 22.4 *ANALYSIS OF ANNUAL GROWTH RATES IN DEVELOPED COUNTRIES*

PHASE	BIRTHRATE	DEATH RATE	ANNUAL RATE
I	High	High	Low
II	High	Low	High
III	Low	Low	Low

hygiene and sanitation. Numerous infectious diseases, such as cholera, typhus, and diphtheria, were brought under control.

The decline in the death rate in the developed countries was followed shortly by a decline in the birthrate. Between 1950 and 1975, populations in the developed countries showed only modest growth (fig. 22.5) because the growth rate fell from an average of 1.1% to 0.8%.

Demographic Transition Overall, the growth rate in developed countries has gone through three phases (table 22.4 and fig. 22.6). In Phase I, prior to 1850, the growth rate was

low because a high death rate canceled out the effects of a high birthrate; in Phase II, the growth rate was high because of a lowered death rate; and in Phase III, the growth rate was again low because the birthrate had declined. These phases are now known as the **demographic transition.** In seeking a reason for the transition, it has been suggested that as industrialization occurred, the population became concentrated in the cities. Urbanization may have contributed to the decline in the growth rate because in the city children were no longer the boon they were in the country. Instead of contributing to the yearly income of the family, they represented a severe drain on its resources. It could also be that urban living made people acutely aware of the problems of crowding, and for this reason the birthrate declined. Also, some investigators believe that there was a direct relationship between improvement in socioeconomic conditions and the birthrate. They point out that as the developed nations became wealthier, as infant mortality was reduced, and as educational levels increased, the birthrate declined.

Regardless of the reasons for the demographic transition, it caused the rate of growth to decline in the developed countries. The growth rate for the developed countries is now about 0.6%, and their overall population size is about ⅓ that

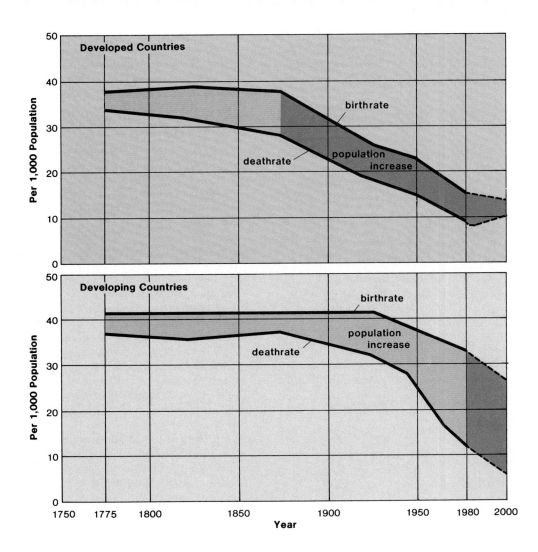

FIGURE 22.6 *Time of demographic transition. In the upper graph it is seen that in the developed countries the demographic transition occurred in the nineteenth century. In the lower graph it is seen that the demographic transition was delayed until the twentieth century.*
Source: Population Reference Bureau, based on United Nations estimates. Used by permission.

of the developing countries. A few developed countries—Austria, Denmark, East Germany, Hungary, Sweden, West Germany—are not growing or are actually losing population. As discussed in the reading on page 498, the United States has a comparatively high growth rate for a developed country.

Developing Countries

Countries such as those in Africa, Asia, and Latin America are collectively known as **developing countries** because they have not as yet become industrialized. Figure 22.6 indicates that mortality began to decline steeply in these countries following World War II. This decline was prompted not by socioeconomic development but by the importation of modern medicine from the developed countries. Various illnesses were brought under control due to the use of immunization, antibiotics, sanitation, and insecticides. Although the death rate declined, the birthrate did not decline to the same extent (fig. 22.6), and therefore the populations of the developing countries began and are today still increasing dramatically (fig. 22.5). The developing countries were unable to cope adequately with such rapid population expansion so that today

many people in these countries are underfed, ill housed, un-schooled, and living in abject poverty. Many of these poor have fled to the cities where they live in makeshift shanties on the outskirts.

HUMAN ISSUE

The world is split today between the "haves" and the "have nots." The developed countries consume most of the world's resources and have either stable or slowly growing populations. The developing countries consume far fewer resources, but their populations are growing very rapidly. In the meantime, developed and developing countries are forming new economic ties. For example, industries are transferring their places of operation to developing countries in order to make use of cheap labor, and the United States imports many raw materials from the developing countries.

Do you think that these economic ties benefit underdeveloped countries by supplying them with jobs and money, or are these countries being unfairly exploited? In your opinion, which countries—developed or developing—will be more self-sufficient and economically stable in the long run?

UNITED STATES POPULATION

The United States now has a population of 240 million. The geographic center (the point where there are just as many persons in each direction) has moved steadily westward and recently has also moved southward. Another interesting trend is the shifting emphasis from metropolitan areas to nonmetropolitan areas. In the 1970s cities increased by 9.8%, but rural small towns increased by 15.8%.

The population size of the United States is not expected to level off any time soon for two primary reasons.

1. A baby boom between 1947 and 1964 has resulted in an unusually large number of reproductive women at this time (see the figure). Thus, although each of these women is having on the average only 1.9 children, the total number of births increased from 3.1 million a year in the mid-1970s to 3.6 million in 1980–81.

2. Many people immigrate to the United States each year. In 1981 immigration accounted for 43% of the annual population growth. The number of legal immigrants was about 700,000. Even though ordinarily only 20,000 legal immigrants can come from any one country, we give special permission for large numbers of political refugees to enter the United States.

In recent years, the majority of refugees have come from Latin America (e.g., Cuba) and Asia (e.g., Indonesia).

There is also substantial illegal immigration into the United States, although the exact number is not known. Estimates range from 100,000 to 500,000 or more a year. About 50–60% of these illegal immigrants come from Mexico, according to most estimates. There has been an effort to stem the tide of illegal immigration to the United States.

Whether the United States can ever achieve a stable population size depends on the fertility rate (the average number of children each woman bears) and the net annual immigration. If the fertility rate were kept at 1.8 and immigration limited to 500,000, the population would peak at 274 million in 2050, after which it would decline. Those who favor a curtailment of our population size point out that a fertility rate of 1.8 allows couples a great deal of freedom in deciding the number of children they will have. For example, it means that 50% of all couples can have two children; 30% can have three children; 10% can have one child; 5% can have no children; and 5% can have four children.

Source: Bureau of the Census

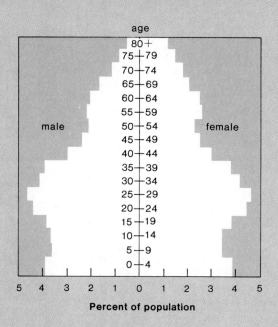

U.S. population age composition, 1985.
Source: Bureau of the Census.

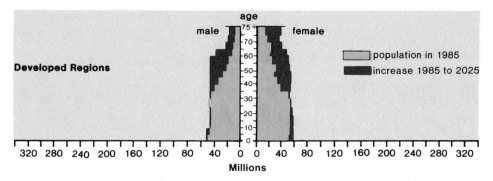

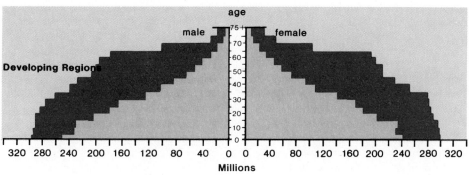

FIGURE 22.7 Age structure diagram for developed countries (above) versus developing countries (below). Because the population sizes of the developed countries are approaching stabilization, population growth will be modest, but the developing countries will expand rapidly due to their youthful profile.
Marshall Green and Robert A. Feary, "World Population: The Silent Explosion." Department of State Bulletin, *Publication 8956 (October 1978) based on U.S. Bureau of the Census Illustrative Projections of World Population to the 21st Century. Current Population Reports Special Studies.*

The growth rate of the developing countries did finally peak at 2.4% during 1960 to 1965. Since that time, the mortality decline has slowed and the birthrate has fallen. The growth rate is expected to decline to 1.8% by the end of the century. At that time, about ⅔ of the world population will be in the developing countries.

Investigators are divided as to the cause of the observed growth rate decreases in the developing countries. Previously, it was argued that this would happen only when these countries enjoyed the benefits of an industrialized society. It has now been shown, however, that countries with the greatest decline were those with the best family-planning programs. From this it may be argued that such programs can indeed help to bring about a stable population size in the developing countries. Nevertheless, it has been found that certain socioeconomic factors have also contributed to a decline in the developing countries' growth rate. Relatively high GNP (gross national product), urbanization, low infant mortality, increased life expectancy, literacy, and education all have had a dampening effect on the growth rate.

Age Structure Diagrams

Lay people are sometimes under the impression that if each couple had two children, zero population growth would immediately take place. However, **replacement reproduction,** as it is called, would still cause most countries today to continue growth due to the age structure of the population. If more young women are entering the reproductive years than there are older women leaving them behind, then replacement reproduction will give a positive growth rate.

Reproduction is at or below replacement level in some twenty developed countries, including the United States. Even so, some of these countries will continue to grow modestly, in part because there was a baby boom after World War II. Young women born in these baby boom years are now in their reproductive years, and even if each one has fewer than two children, the population will still grow. It should also be kept in mind that even the smallest of growth rates can add a considerable number of individuals to a large country. For example, a growth rate of 0.7% added over 1.6 million people to the United States population in 1984.

Whereas many developed countries have a stabilized age structure diagram (fig. 22.7), most developing countries have a youthful profile—a large proportion of the population is below the age of 15. Since there are so many young women entering the reproductive years, the population will still greatly expand even after replacement reproduction is attained. The more quickly replacement reproduction is achieved, however, the sooner zero population growth will result. In the meantime the burgeoning world population is putting such a severe strain on the biosphere that many are very much concerned about the extinction of wildlife, particularly in the developing countries that have tropical forests. The reading that follows discusses pertinent points.

WHAT WE MUST DO TO SAVE WILDLIFE

The world's jungles will continue under siege, wars over resources will increase by the year 2000, and extinctions will be greater in our lifetimes than at any equivalent period in the Earth's history since the Age of Dinosaurs.

This grim scenario for the future of our natural world was outlined by 245 scientists, government officials, and conservation experts from 65 countries—most of whom [were] involved . . . in refining a blueprint for international conservation called the World Conservation Strategy. Their views were in response to an exclusive opinion survey conducted by this magazine [*International Wildlife*]. The results . . . contain some startling predictions, but they also summarize deep and dramatic changes in the philosophical direction of those people guiding conservation in developing countries where most of the world's species of plants and animals are found.

> Biggest advance predicted:
> **"Growing awareness that nature conservation is really necessary for the benefit and survival of mankind."**
>
> *PRINCE BERNHARD SOESTLYK, THE NETHERLANDS*

The magazine's survey, which asked the experts for their views on the future of wildlife through the year 2000, comes at a time when the world's wild species and the lands that support them are under unprecedented stress. The problems stem largely from a human population expected to exceed six billion in just 14 years. As a result, the survey reveals, conservation has taken a new, more pragmatic thrust, based on the realization that saving wildlife depends on solutions to human problems.

This increasingly pro-people ethic mirrors changes in the conservationists themselves. Only a few decades ago, their movement was dominated by Europeans who often argued for saving species solely because it was the right thing to do. Now, increasingly, conservation is carried out by people in Third World countries who know declines in wildlife will accelerate unless human misery is relieved.

This does not mean that these new conservationists disagree with their colleagues in the developed world. In fact, the range of responses to the editors' questions was virtually identical from experts in both rich and poor

a.

b.

c.

Many animals, such as these—green iguana (a), ocelot (b), and blue and yellow macaw (c)—make their home in the tropical forests and will be endangered if these forests are destroyed.

nations. For both groups, the primary goal is to save large samples of wild species and their ecosystems.

Biggest setback predicted:
"Destroying the habitats which are necessary for species conservation and diverting large quantities of intellectual, monetary and natural resources for armaments."

BORIS N. VEPRINTSEV
MOSCOW, USSR

Although the issues of conservation are complex and the responses hardly unanimous, . . . world conservation is today emphasizing at least four principles designed to produce results:

(1) *Sound development is needed.* Until recently, developers and conservationists were often fierce adversaries. Developers were seen as despoilers, conservationists as obstructionists. Now, say 92 percent of the respondents to the survey, "Sound economic development and environmental protection should go hand in hand."

Development cannot be sustained unless it is environmentally sound, most experts now explain. Similarly, conservation cannot be sustained in many cases unless it pays for itself by providing a resource that benefits people on a continuing basis. A major achievement in the wildlife field, says Charles S. Luthin, conservation director of West Germany's W. W. Brehm Fund for the International Conservation of Birds, "is recognizing that conservation and development are not mutually exclusive, but rather one and the same. Without proper environmental conservation, all development initiatives are destined to failure."

(2) *Human needs must be met.* Conservationists working in poor countries have traditionally been concerned primarily with living organisms other than humans. But that emphasis is changing. On the one hand, human survival over the long term depends on a healthy Earth. On the other, conservation in crowded countries usually cannot succeed in the short run unless there are immediate payoffs for people.

In the magazine's survey, 46 percent of the experts argued that the most effective way to "sell" the wildlife message to political leaders was to stress that saving species provides direct economic benefits to people. Only 6% said it is effective to promote aesthetic or ethical reasons as a way to save wildlife.

"Unless you relate to, and accept enthusiastically, the need to satisfy human needs while protecting and improving the environment, the cause will be lost," says M. Taghi Farvar, a senior advisor to the International Union for Conservation of Nature and Natural Resources (IUCN) in Switzerland.

(3) *Entire ecosystems must be saved.* Until recently, conservationists in the developing world reacted to crises—like firemen rushing to put out a blaze. Often, this meant launching last-minute campaigns to save single species ranging from cranes to giant pandas. Increasingly today, such actions are being supplanted by long-term plans to save whole communities encompassing a variety of less-visible organisms.

The rationale for saving species, say 73% of the experts questioned, should be to preserve whole environments and as much genetic diversity—or samples of living organisms—as possible. This idea emphasizes the interrelatedness of living things and has translated into efforts to protect various types of wild communities, not just individual species. The emphasis of this holistic approach is on saving ecosystems often larger and more complex than once envisioned.

"The creation of national parks and equivalent reserves in almost all countries of the world," says Mario A. Boza, president of Costa Rica's National Parks Foundation, "is the biggest achievement in wildlife conservation during the past 15 years."

(4) *Conservation victories don't come cheap.* Increasingly, conservationists have recognized that the time for action is now, but money and other resources are in short supply. That means setting priorities and making difficult practical and ethical decisions.

Number one priority:
"To implement visible projects that are seen by all to be promoting human development needs—and at the same time conserving threatened resources."

PAUL BUTLER
ST. LUCIA, WEST INDIES

Of the experts polled, 96 percent think the rich nations of the world should spend more to help save wildlife in the Third World. But now, a large minority are also willing to grapple with more realistic trade-offs. These include the possibility of writing off the conservation problems of some countries as hopeless and spending funds elsewhere where they would do more good. . . .

Major agent for change:
"[Environmental citizen activist groups] will be the conscience of a country."

J. C. DANIEL
BOMBAY, INDIA

The experts surveyed by *International Wildlife* are particularly significant because most of them [attended] a meeting—in Ottawa, Canada 1986—to fine-tune the World Conservation Strategy, which sets priorities for immediate conservation action in many countries.

"What We Must Do To Save Wildlife," by Jon Fisher and Norman Myers, reprinted from the May/June 1986 issue of *International Wildlife,* copyright by the National Wildlife Federation.

The history of the world population shows that the developed countries underwent a demographic transition between 1950 and 1975; the developing countries are just now undergoing demographic transition. Replacement reproduction will not bring about zero population growth even in the developed countries because of a baby boom that occurred after World War II. In the developing countries, where the average age is less than 15, it will be many years before replacement reproduction will mean zero population growth.

RESOURCE CONSUMPTION

Rapid world population growth puts extreme pressure on the earth's resources, physical environment, and social organization. Although it might seem as if population increases in the developing countries are of the gravest concern, this is not necessarily the case since each person in a developed country consumes more resources and is therefore responsible for a greater amount of pollution. EI (environmental impact) is measured not only in terms of the population size but also in terms of the resources used and the pollution caused by each person in the population.

$$EI = \text{population size} \times \frac{\text{resource use}}{\text{per person}} \times \frac{\text{pollution per unit}}{\text{of resource used}}$$

Therefore, there are two types of overpopulation. The first type is due to a rapidly increased population and occurs mainly in the developing countries. The second type of overpopulation is due to increased resource consumption with its accompanying pollution; this type is most obvious in the developed countries.

Resources are either renewable or nonrenewable. The supply of **renewable resources** reoccurs. For example, rain periodically brings a renewed supply of fresh water, and harvests bring a renewed supply of food. **Nonrenewable resources** are those resources having supplies that can be used

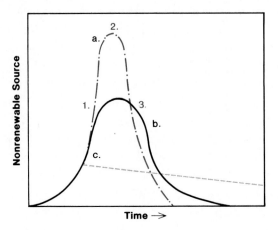

FIGURE 22.8 *Depletion patterns for a nonrenewable resource.* a. *Rapid depletion as the resource is used up quickly:* (1) *exponential consumption of a resource followed by a peak* (2) *and decline* (3) *as the resource becomes difficult to acquire.* b. *Depletion time can be extended with some recycling and less wasteful use.* c. *Efficient recycling extends the depletion curve indefinitely.*

up, or exhausted. For example, it would be possible to eventually use up all the fossil fuels that have already been deposited in the earth. There are two points of view, called the Malthusian[1] and Cornucopian views, about nonrenewable resource availability. According to the Malthusian view, each nonrenewable resource is subject to depletion, and we are rapidly approaching the time when many if not most supplies will be exhausted. Figure 22.8 shows a **depletion curve** for any nonrenewable resource that is consumed at an increasingly rapid rate. The consumption peak is followed by a consumption decline as the resource becomes more expensive to find and process. Those who uphold the Malthusian view believe that, because of exponential consumption, finding new reserves cannot sufficiently extend a depletion curve. They do not believe that technology will ever overcome inevitable shortages. They admonish us, therefore, to cut back on growth, conserve resources, and recycle whenever possible to extend the depletion curve (fig. 22.8b and c).

According to the Cornucopian view, technology will constantly be able to put off the day when no further exploitation is possible. Improved technology will enable us to (1) find new reserves, (2) exploit new reserves, and (3) substitute one mineral or energy resource for another. Sometimes we are aware of the availability of a resource but must await the development of a technology to exploit it. We now utilize offshore drilling to acquire oil; we might be able

---

[1]Malthus was an eighteenth-century economist who pointed out that since the size of a population increases geometrically and renewable resources increase only arithmetically, shortages must eventually occur. *Cornucopia* is a Latin word meaning the horn of plenty, a symbol of everlasting abundance.

FIGURE 22.9 *The cost of a product includes the cost of controlling pollution due to, for example, mining, manufacturing, and distributing the product.*

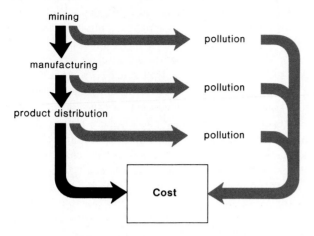

TABLE 22.5 POLLUTION CONTROL EFFICIENCY

TIME (YEARS)	POLLUTION LEVEL	POLLUTION CONTROL EFFICIENCY
—	One unit	—
15	Two units	50%
30	Four units	75%
60	Eight units	88%
120	Sixteen units	94%

to develop the means to mine the ocean floor for minerals; and we can utilize ever poorer grades of mineral ores. For example, previously only 3% copper ores were mined, but now ores with 0.3% copper are utilized.

To exploit previously inaccessible and/or less concentrated resources, a plentiful supply of energy is required. The Malthusians do not believe that increasing amounts of energy will be available, since fossil fuel reserves are being rapidly depleted. Many Cornucopians are still hoping that nuclear energy will supply the necessary energy. They point out that it is possible to design safe nuclear reactors despite nuclear accidents such as Three-Mile Island and Chernobyl, discussed in the following section. The modular high-temperature gas reactor is smaller in size than the reactors presently being used; it uses helium gas as coolant instead of water. And even if this gas should stop flowing, it will not get hot enough to melt down because it only generates 80 megawatts of power compared to the 1,000 megawatts of conventional reactors today. Also, because its components are modular, it can be mass-produced. A prototype has already been successfully tested in West Germany.

Malthusians and Cornucopians also view the problem of pollution differently. The Malthusians suggest that as resource consumption increases, the amount of pollution control needed to protect the environment becomes prohibitive. Previously, no heed was paid to the cost of pollution. Industries were allowed to pollute the environment, and the general public bore the cost of cleaning up after industry. Now, since controls are required by the government, the cost of pollution is added to the total cost of a product (fig. 22.9). The profit margin of the producer is thereby cut even as the consumer must pay more for the product. Unfortunately, also, greater and greater efficiency of control is required as output

is increased. For example, as table 22.5 shows, if output doubles every fifteen years, pollution control efficiency must increase to 94% within 120 years to stay even. Since the cost of pollution control increases manyfold per level of efficiency, it may become unprofitable to continue production.

On the other hand, Cornucopians suggest that whenever strict pollution control regulations are instigated, pollution can be reduced to acceptable levels at relatively modest costs compared to the value of total output. Given reasonable periods of time, all industries and most firms will be able to accommodate themselves to the impact of the new standards. In some instances it may be possible to change industrial processes so there is no waste or to make use of a by-product that formerly was considered a waste.

Resource consumption is dependent on population size and on consumption per individual. The developing countries are responsible for the first type of environmental impact and the developed countries are responsible for the second type. According to the Malthusian view, we are running out of nonrenewable resources, and only conservation can help extend their depletion curves. According to the Cornucopian view, we will continue to find new ways to exploit the environment.

Energy

A larger world population will require more energy in the years ahead. Developing nations are expected to be responsible for most of the increased demand for energy (fig. 22.10). The opposite trend seems to be occurring in the United States where already high demand is not increasing and might even diminish by almost 25%. Surprisingly, it has been suggested that decreased energy consumption need not mean a change

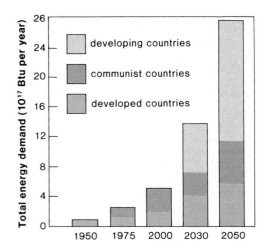

FIGURE 22.10 *History of world energy utilization. The developing countries are expected to account for much of the increase in the future.*
From Oak Ridge National Laboratory, operated by Martin Energy Systems, Inc., for the U.S. Department of Energy.

in U.S. life-style. Instead, it is possible that the GNP could rise even as less energy is used because when less money is spent on energy, more funds can be devoted to growth.

The energy sources currently available or being investigated are listed in table 22.6.

Nonrenewable Energy Sources

The depletion curve (fig. 22.8) for *petroleum* is now in the decline phase, and the depletion curve for *natural gas* has reached its peak. Supplies of these favored fossil fuels are not expected to last more than thirty years. Petroleum became the favored fuel of the present generation because it is easily transported; is versatile, serving as the raw material for gasoline and many organic compounds; and is cleaner burning than coal.

Coal is in plentiful supply, and the United States probably will depend primarily on its use to make up for decreased use of petroleum. There are environmental drawbacks to using coal in place of petroleum, however. Thousands of acres of land are strip-mined (fig. 22.11), producing huge piles of residue having the potential to leach hazardous chemicals into underground aquifers. Coal burning adds to the air many pollutants that damage human lungs, poison lakes, and erode buildings. It is even possible that the CO_2 given off will eventually cause a global warming trend (p. 484).

Three possible types of *nuclear energy* (table 22.6) can be used to generate electricity. Thus far, only light water **fission power plants,** which split [235]uranium, have been utilized in the United States. Approximately one hundred operating nuclear plants exist and twenty-five are under construction. Many proposed plants have been canceled, and there have been no new orders for the past several years. Some predict that the nuclear power industry is dead in this country. Possible reasons for this are numerous; for example, there is little need for new plants because demand for electricity is not increasing; it is cheaper to build and operate coal-firing plants; and the public is very much concerned about nuclear power safety.

In May 1986 the Chernobyl nuclear power plant in Russia suffered an accident that released so much radioactivity it triggered an alarm system in Sweden. Apparently, there was a sudden loss of coolant to the nuclear fuel cells and surrounding graphite, a material that normally functions to control a nuclear reaction. Unfortunately, in this instance the hot graphite reacted with steam to produce gases that exploded from the building. Fed by oxygen in the air, the graphite began to burn. This promoted a fuel meltdown that sent radioactivity into the atmosphere. Western nations were quick to point out that the West had no graphite-utilizing plants and that the Chernobyl plant lacked a containment tower that keeps radioactivity from entering the atmosphere should an accident occur. Even so, the drive to control the nuclear power industry was given new impetus by the Chernobyl accident, which caused several immediate deaths and may lead to hundreds more.

The public is also aware of the need to store radioactive wastes that have accumulated since the start of the nuclear age. Spent fuel rods contain low-, medium-, and high-level wastes, depending on the amount and character of the radiation they emit. High-level wastes must be stored hundreds of thousands of years before they lose their radioactivity. As yet, the government has not sanctioned any particular permanent method of disposal. Most probably, the wastes will finally be incorporated into glass or ceramic beads, packaged in metal canisters, and buried in stable salt beds, red clay deposits in the center of the ocean, or in stable rock formations. Meanwhile, some wastes have been temporarily stored above ground in tanks that on occasion have developed leaks.

HUMAN ISSUE

We are all familiar with the grim scenarios of nuclear war such as a "nuclear winter." According to this prediction, the world would turn absolutely dark and cold, making plant and animal life almost impossible. Do such predictions unfairly prejudice people against the use of nuclear power? Instead, should we be thinking about the benefits of nuclear power such as the fact it doesn't cause acid rain? All things considered, do you think that the use of nuclear power should be encouraged or discouraged in the United States?

Breeder fission reactors do not have the same environmental problems as light water nuclear power plants. They need less raw uranium, and they have little waste because they use [239]plutonium, a fuel that is actually generated from what would be reactor waste.

TABLE 22.6 ENERGY SOURCES

	ADVANTAGES	DISADVANTAGES
Nonrenewable	*Technology well established*	*Finite fuel supply*
Fossil fuels		
Coal, oil shale, and tar sands	Plentiful supply	Surface mining
		Air and water pollution
Petroleum	Cleaner burning	Limited supply
Natural gas	Cleanest burning	Limited supply
Nuclear		
Light water	Fuel availability	Thermal pollution
		Radiation pollution
Renewable	*Infinite fuel supply*	*Technology under development*
Nuclear		
Breeder	Fuel availability	Radiation pollution
		Thermal pollution
		Nuclear weapons proliferation
Fusion	Fuel availability	Radiation pollution
Geothermal	Less pollution	Availability limited
Solar and wind	Nonpolluting	Noncompetitive cost
	Large and small scale possible	
Ocean	Nonpolluting	Applicable only in certain areas
Biomass	Utilizes wastes	Air and water pollution

Plutonium, however, is a very toxic element that readily causes lung cancer; it is also the element used to make nuclear weapons. (The chances of a nuclear explosion are much greater with breeder reactors than with light water reactors.) The fear of nuclear weapons proliferation, in particular, has so far prevented the start-up of a breeder reactor in this country.

Nuclear fusion requires that two atoms, usually deuterium and tritium, be fused. Most scientists believe that a large amount of heat is needed to bring about the reaction and have been experimenting with laser-beam ignition and magnetic containment for the reaction. Since the fusion reaction gives off neutrons that can change uranium to plutonium, the best use of fusion plants might be to provide fuel for breeder reactors. One idea is to have hybrid fusion plants that would combine both fusion and breeder fission plants. But, in any case, the fusion process is still experimental and not expected to be ready for production soon.

Exploitation of the environment is dependent on a plentiful energy source. Among the nonrenewable sources, coal is in ready supply and can be utilized without developing new technologies. Each of the three types of nuclear power has a drawback that has prevented full-scale operation in this country.

FIGURE 22.11 *Aerial view of extensive surface mine spoil. Most coal today is strip-mined, a process that reduces the quality of the land unless proper careful measures are taken to restore it as it was before.*

FIGURE 22.12 *Solar energy can be utilized by various community institutions, such as Betatakin National Monument in Arizona.*

Renewable Energy Sources

Solar energy is diffuse energy that must be collected and concentrated before utilization is possible. The public, rather than the government, has provided most of the impetus for practical applications, including solar heating of homes and offices (fig. 22.12). **Solar collectors** placed on rooftops absorb radiant energy and resulting heat is later released. A fluid within the solar collector heats up and is pumped to other parts of the building for space heating, cooling, or the generation of electricity. Passive systems are also possible; specially constructed glass can be used for the south wall of a building, and building materials can be designed to collect the sun's energy during the day and to release it at night.

The U.S. government subsidized the building of Solar One, the world's largest solar-powered electrical generating facility (fig. 22.13), which began operation in 1982 at Daggett, California, in the Mojave Desert. A single boiler is located atop a large tower twenty stories high, and a large field of mirrors (called **heliostats,** which are capable of tracking the sun) reflect the sun's rays onto the tower. The water is heated to 500° C, and the steam is used to produce electricity in a conventional generator. Systems such as this require much land and cannot be placed just anywhere. For this reason, **photovoltaic (solar) cells** that produce electricity directly may be a more promising energy source. It has even been suggested that cells might be placed in orbit around the earth, where they would collect intense solar energy, generate electricity, and send it back to earth via microwaves.

Solar energy is clean energy. It does not produce air or water pollution. Nor does it add additional heat to the atmosphere since radiant energy is eventually converted to heat anyway. The problem of storage can be overcome in a number of ways, including the use of solar energy to produce hydrogen by means of hydrolysis of water. Hydrogen as either a gas or a liquid can be piped into existing pipelines and used as fuel for automobiles and airplanes. When it is burned, it forms fog, not smog (fig. 22.14).

The following types of renewable energy sources have been utilized for quite some time. *Falling water* is used to produce electricity in hydroelectric plants. *Geothermal energy* is trapped heat produced by radioactive material and from magma deep beneath the surface of the earth. Water, converted to steam by this heat, may be pumped up and used to heat buildings or to generate electricity. *Wind power* provides enough force to turn vanes, blades, or propellers attached to a shaft, which, in turn, spin the motor of a generator that produces electricity. The government has allocated a small amount of money for wind research, particularly the promotion of large windmills, such as those being constructed in the windy Columbia River valley in the Pacific Northwest. Many others believe it would be best to build numerous small windmills, such as those found in the first wind farm now operating in New Hampshire.

All of the renewable resources (falling water, geothermal, wind, etc.) still hold some promise. Solar One is now in operation, and solar cells are being perfected. Solar energy can be used to hydrolyze water to produce hydrogen gas or liquid, which can be substituted for fossil fuels.

HUMAN ISSUE

Which of these energy options—petroleum, natural gas, coal, nuclear fission, nuclear fusion, geothermal, solar, or wind—could now be used on a large scale with no significant ecological or economical disadvantages? Which energy option do you predict will be the most widely used fifty years from now? In your opinion, which energy option *currently* can be used on a large scale while having the least negative ecological impact?

FIGURE 22.13 *Solar electric power plant.* a. *Overview of Solar One, which has a field of 1,818 heliostats.* b. *Close-up of one heliostat, consisting of six sun-tracking mirrors on each side of a central support.*

a.

b.

FIGURE 22.14 *Hydrogen as a fuel. When hydrogen is burned, the product is water and water can be hydrolyzed, using solar energy, to give hydrogen again. This cycle makes hydrogen an attractive fuel.*

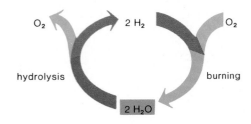

Food

There is great concern that it will not be possible to provide enough food to feed the population increases that are expected well into the twenty-first century.

Figure 22.15a and c show that the *production of food* has increased in the developed countries, and the per capita share has also increased. This means that food production has kept pace with population increase in the developed countries where the majority of people can afford the price of food and are eating better than formerly. Figure 22.15b and d show that the production of food has increased in the developing countries, but the per capita share has not substantially increased because increased food production must be divided by a large population increase. Further, in the developing countries only a segment of the population can afford the price of food and are eating well; a larger portion cannot afford the price of food and are not eating adequately. (Since the per capita share is calculated by dividing the total production by the total population size, it does not necessarily mean that all persons are eating an equal amount.) In particular, Africa, with the fastest population growth ever recorded, has had difficulty providing enough food for its populace. Of the many factors involved in producing enough food, the following are most often discussed.

Factors in Food Production

Today, most land is already being used for one purpose or another, and it would be difficult to expand the amount of farmland. Only in the tropics (sub-Saharan Africa and the Amazon Basin of Brazil) are there still sizable tracts of land not presently utilized that have enough water to grow crops. Slash-and-burn agriculture has been traditional in this area because the soil contains so few nutrients. The jungle is cut and the debris burned, providing the nutrients to allow a few years' crops. When the soil can no longer support continued cultivation, a new area is selected for cultivation. Attempts

FIGURE 22.15 *Food production in developed versus developing countries.* a. and b. *Food production has increased at similar rates in the developed and developing countries over the last three decades, but population growth has continued at a faster rate in developing than in developed countries; therefore* (c and d), *there has been a per capita gain in the developed countries but not in the developing countries.*

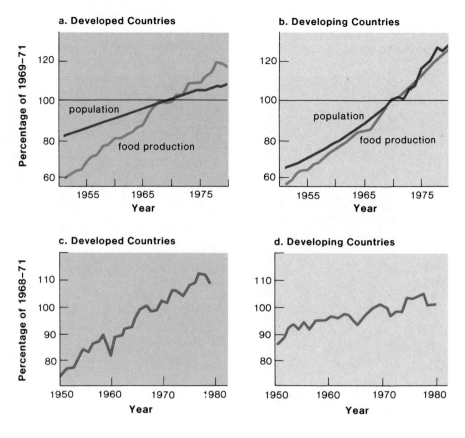

FIGURE 22.16 *Experimental rice plots. Hybridization studies are continuing in order to produce high-yield rice plants that are also resistant to pests and environmental stresses.*

to cultivate the land in a traditional manner have proven difficult because the soil tends to become hard and compact when exposed to the sun. If these problems could be solved, the food supply would be multiplied manyfold. Even so, many believe that tropical forests should remain as they are because they help control global pollution.

Since virtually all readily available, relatively fertile cropland is already in use, more effort should be directed toward safekeeping the land already being cultivated. Millions of acres of fertile lands are lost each year from encroachment by cities and suburbs, industrial development, soil erosion, and desertification. The United Nations estimates that by the end of this century twice as much land will be lost due to these factors than will be developed as farmland.

The developed countries, especially the United States, have had spectacular success in increasing yields by utilizing monoculture agriculture, and this method of agriculture, along with its many environmental problems, is now being exported to the developing countries. The term "Green Revolution" was coined to describe the introduction and rapid spread of high-yielding wheat and rice plants (fig. 22.16) that are today especially developed to grow in warmer countries even when they are not supplied with generous amounts of fertilizer and water. Research efforts are currently directed at providing plants with improved internal efficiency. The new focus is on greater photosynthetic efficiency, more efficient nutrient and water uptake, improved biological nitrogen fixation, and genetic resistance to pests and environmental stress.

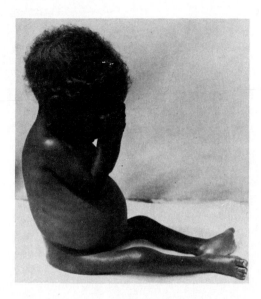

The application of fertilizer has contributed greatly to increased yield. When fertilizer is first applied to soil, there is a dramatic increase in yield, but later, as more and more fertilizer is used, the increase in yield drops off. This suggests that the developed countries could very well do with less fertilizer, but the developing countries need to apply more. Unfortunately, the rising cost of energy has caused fertilizer cost to more than double, and developing countries will most likely find it difficult to make or buy adequate amounts of fertilizer. Legumes, you will recall, increase the nitrogen content of the soil because their roots are infected with nitrogen-fixing bacteria. Research continues into the possibility of infecting more plants with these bacteria and even on the possibility of transferring the nitrogen-fixing genes to plant cells themselves by means of recombinant DNA methods.

High agricultural yields are extremely dependent on both indirect and direct uses of energy. In the United States, fossil fuels have supplied the supplemental energy, as is apparent when you consider that nearly all of the large farm machinery utilizes fossil fuel energy. In Japan, however, much of the supplemental energy required for high yields is provided by human labor. Since we are now entering a time when fossil fuels are in short supply, it would be wise for the developing countries to rely on human energy to increase yield. Keeping people in rural areas close to the source of food would also help solve the problem of transportation and packaging.

Dietary Protein Providing sufficient calories will not assure an adequate diet; there must also be enough protein in the food eaten. Due to a lack of dietary protein, children in the developing countries often have the symptoms of kwashiorkor, in which the entire body is bloated, the skin is discolored, a rash is present, and the hair has a reddish orange tinge (fig. 22.17).

Plants are low-quality, nutritionally incomplete protein sources because, although they contain some amino acids, they lack others. It is possible, however, to eat a combination of plants so that together they contain an acceptable level of all the essential amino acids. For example, wheat and beans complement each other to give a balance that is comparable to high-quality proteins such as meat and cheese.

In the developed countries most of the grain produced is fed to cattle, pigs, or chickens, which then serve as a source of high-quality protein. In many developing countries most grain must be consumed directly in order to provide sufficient calories. Therefore, beef is not expected to supply the necessary protein. Another possibility is fish consumption. Between 1950 and 1970, fish supplied an increasing portion of the human diet until the catch averaged some 18.5 kilograms per person annually. Since that time, although the catch has increased, the per capita share has decreased due to population growth. Also, overexploitation of the oceans' fisheries and pollution of coastal waters may eventually cause a decline in the yearly catch. Therefore, the expansion of aquaculture—the cultivation of fish—is highly desirable.

The production of food has become increasingly dependent on supplements, such as fossil fuel energy and fertilizer, because the amount of new agricultural land that can be cultivated is limited. Development of new high-yield plants may help the developing countries feed expected increases in population, but there may still be a problem regarding dietary protein. It is doubtful that oceanic fishing can be counted on to supply protein, and as yet, aquaculture is not sufficiently developed to do so.

STEADY STATE

Presently, the world's population increases in size each year, and an ever greater supply of energy and food is needed each year. Does this pattern have to continue in order to assure economic growth and our well-being? On the contrary, it seems unlikely that this exponential increase can long continue. The pollution caused by energy consumption and the production of goods has increased so much that it can no longer be ignored, and the once-hidden cost of pollution has now become apparent. The human population, like other populations of the biosphere, could exist in a steady state, with no increase in number of people or in resource consumption.

A stable population size would be a new experience for humans. Figure 22.18 shows the age structure for a hypothetical stable population. The following listing of stable population characteristics is evident in this figure.

1. Over 40% of the people would be fairly youthful, with only about 15% being in the senior citizen category. Further, a combination of good health habits and advances in medical science could well mean that people would remain more youthful and more productive for a longer time than is common today.

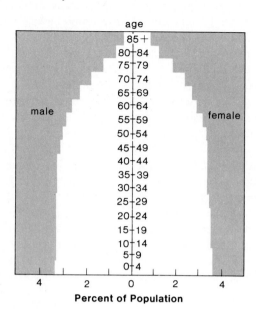

age

2. There would be proportionately fewer children and teenagers than in a rapidly expanding population. This might mean a reduction in automobile accidents and certain types of crime statistically related to the teenage years.
3. There might be increased employment opportunities for women and a generally less competitive workplace since newly qualified workers would enter the job market at a more moderate rate.
4. Creativity need not be impaired. A study of Nobel Prize winners showed that the average age at which the prize-winning work was done was over thirty.
5. The quality of life for children might increase substantially since fewer unwanted babies might be born and the opportunity would exist for each child to receive the loving attention needed.

The economy, like the population, should show no growth. For example, if the population size were to increase, the resource consumption per person should decline. In this way the amount of resource consumption would remain constant. In order to ensure that people have the goods and services they need for a comfortable but not necessarily luxurious life, goods should have a long lifetime expectancy. Frugality is envisioned as absolutely necessary to the steady-state economy. Also, materials must be recycled so that they can be used over and over again. In order to provide jobs, technology could sometimes be labor intensive instead of energy intensive. The steady state would have need of very sophisticated technology, but it is hoped that the technology would not exploit the environment; instead, it is hoped that the technology would work with the environment.

Environmental preservation would be the most important consideration in a steady-state world. Renewable energy sources, such as solar energy, would play a greater role in providing energy needs. Pollution would be minimized. Ecological diversity would be maintained, and overexploitation would cease (fig. 22.19). Ecological principles would serve as guidelines for specialists in all fields, thereby creating a unified approach to the environment. In a steady-state world all people would strive to be aware of the environmental consequences of their actions, consciously working toward achieving balance in the ecosystems of their planet.

What would our culture be like if we had steady-state manufacturing and a steady-state population? Perhaps it would be greatly improved. Certainly there are no limits to growth in knowledge, education, art, music, scientific research, human rights, justice, and cooperative human interactions. In a steady-state world the general sense of fearful competition among peoples might diminish, allowing human compassion and creativity to prosper as never before.

In a steady-state society, there would be no yearly increase in population nor resource consumption. It is forecast that under these circumstances the quality of life would improve.

SUMMARY

The human population is expanding exponentially, and even though the growth rate has declined, there is a large increase in the population each year—the doubling time is now about forty years. The developed countries underwent a demographic transition between 1950 and 1975, but the developing countries are just now undergoing demographic transition. In these countries, where the average age is less than fifteen, it will be many years before replacement reproduction will mean zero population growth.

Resource consumption is dependent on population size and on consumption per individual. The developing countries are responsible for the first type of environmental impact and the developed countries for the second type. According to the Malthusian view, we are running out of nonrenewable resources and only conservation can help extend their depletion curves. According to the Cornucopian view, we will continue to find new ways to exploit the environment.

Exploitation of the environment is dependent on a plentiful energy source. Of the nonrenewable sources, coal is in ready supply and can be utilized without developing new technologies. Each of the three types of nuclear power has a drawback that has prevented full-scale operation in this country. All of the renewable resources that hold some promise haven't been fully developed yet. Solar One is now in operation and solar cells are being perfected.

The production of food has become increasingly dependent on supplements such as fossil fuel energy and fertilizers. Development of new high-yield plants has helped developing countries feed their increasing populations, but an increased supply of protein remains a problem.

OBJECTIVE QUESTIONS

1. After a country has undergone the demographic transition, both the death rate and the birthrate are _____ (choose high or low).
2. In contrast to the developed countries, the developing countries are not as yet _____ .
3. The people of the developed countries have a higher standard of living and consume more _____ than do those of developing countries.
4. Those who hold a Cornucopian view concerning nonrenewable resources believe that _____ .
5. For the past several years, the nonrenewable energy source of choice has been _____ .
6. The steady accumulation of _____ wastes is a deterent to the use of nuclear power.
7. Solar power can be used to hydrolyze water, and the resulting _____ can be a gaseous or liquid fuel.
8. The continent least able to feed its ever-growing populace is _____ .
9. One of the chief problems in regard to a proper diet in developing countries is supplying adequate _____ .
10. Overexploitation and pollution may very well cause a _____ in the size of the yearly fish catch.

Answers

1. low 2. industrialized 3. resources 4. there is no limit to the amount of resources that can be consumed 5. petroleum 6. radioactive 7. hydrogen 8. Africa 9. protein 10. decline

STUDY QUESTIONS

1. Define exponential growth. (p. 492) Draw a growth curve to represent exponential growth, and explain why a curve representing population growth usually levels off. (p. 494)
2. Calculate the growth rate and doubling time for a population in which the birthrate is 20 per 1,000 and the death rate is 2 per 1,000. (p. 493)
3. Define demographic transition. When did the developed countries undergo demographic transition? When did the developing countries undergo demographic transition? (p. 496)
4. Give at least three differences between the developed countries and the developing countries. (p. 495)
5. Contrast the Malthusian view and the Cornucopian view concerning nonrenewable supplies. (p. 502)
6. Draw a typical depletion curve, and relate it to the consumption of fossil fuels. (p. 502)
7. Name the three types of nuclear power, and give at least one related drawback to each. (p. 504)
8. Name at least four types of renewable energy resources. (p. 506) What types of fuels might be produced from these sources? (p. 506)
9. Give at least three reasons why intensive monoculture does not seem to be a feasible solution to the food crises of developing countries. (p. 508)

THOUGHT QUESTIONS

1. Can you think of any reasons why it might be inappropriate to imply that a growth curve such as that in figure 22.3 can be applied to the human population?
2. The Mexican government at one time encouraged large families because it was believed that the greater the number of people, the greater the work force, and the greater the prosperity. What's wrong with this type of thinking?
3. Is sending food the best answer to preventing a famine? If we really wanted to help the starving people of the world, what would we be doing?

KEY TERMS

biotic potential (bi-ot′ik po-ten′shal)   the maximum population growth rate under ideal conditions.   *494*

breeder fission reactor (brēd′er fish′un re-ak′tor)   a nuclear reactor that produces more nuclear fuel than it consumes by converting nuclear wastes into ^{239}plutonium.   *504*

carrying capacity (kar′e-ing kah-pas′ĭ-te)   the largest number of organisms of a particular species that can be maintained indefinitely in an ecosystem.   *494*

demographic transition (dem-o-graf′ik tran-zi′shun)   the change from a high birthrate to a low birthrate so that the growth rate is lowered.   *496*

depletion curve (de-ple′shun kerv)   graphical depiction of a resource's dwindling amount over time.   *502*

developed countries (de-vel′opt kun′trēz)   industrialized nations that typically have a strong economy and a low rate of population growth.   *495*

developing countries (de-vel′op-ing kun′trēz)   nations that are not yet industrialized and have a weak economy and a high rate of population growth.   *497*

doubling time (dŭ′b′l-ing tīm)   the number of years it takes for a population to double in size.   *494*

environmental resistance (en-vi′′ron-men′tal re-zis′tans)   sum total of factors in the environment that limit the numerical increase of a population in a particular region.   *494*

exponential growth (eks′′po-nen′shal grōth)   growth, particularly of a population, in which the increase occurs in the same manner as compound interest.   *492*

fission power plant (fish′un re-ak′tor)   plant with a nuclear reactor that splits ^{235}uranium in order to release energy.   *504*

growth rate (grōth rāt)   percentage of increase or decrease in the size of a yearly population.   *493*

heliostat (he′le-o-stat′′)   large field of mirrors that track the sun and reflect its energy onto a mounted boiler for solar heating.   *506*

nonrenewable resource (non-re-nu′ah-b′l re′sors)   a resource that can be used up or at least depleted to such an extent that further recovery is too expensive.   *502*

nuclear fusion (nu′kle-ar fu′zhun)   a process in which the nuclei of two atoms are forced together to form the nucleus of a heavier atom with the release of substantial amounts of energy.   *505*

photovoltaic (solar) cells (fo′′to-vol-táik selz)   a manufactured mechanism that uses sunlight to produce an electromotive force.   *506*

renewable resource (re-nu′ah-b′l re′sors)   a resource that is not used up because it is continually produced in the environment.   *502*

replacement reproduction (re-plās′ment re′′pro-duk′shun)   a population in which the average number of children is one per person.   *499*

solar collector (so′lar kŏ-lek′ter)   any manufactured device that absorbs radiant energy.   *506*

FURTHER READINGS FOR PART VI

Ayala, F. J. September 1978. The mechanisms of evolution. *Scientific American.*

Batie, S. S., and R. G. Healy. February 1983. The future of American agriculture. *Scientific American.*

Borgese, E. M. March 1983. The law of the sea. *Scientific American.*

Brown, L. R., et al. 1986. *State of the world: 1986.* New York: W. W. Norton and Co.

Cloud, P. September 1983. The biosphere. *Scientific American.*

Dickerson, R. E. September 1981. Chemical evolution and the origin of life. *Scientific American 239.*

Ehrlich, P. R., R. W. Holm, and D. R. Parnell. *The process of evolution.* 2d ed. New York: McGraw-Hill.

Environment. All issues of this journal contain articles covering modern ecological problems.

Grant, P. R. 1981. Speciation and the adaptive radiation of Darwin's finches. *American Scientist* 69:653.

Gwatkin, D. R. May 1982. Life expectancy and population growth in the third world. *Scientific American.*

Himmelfarb, G. 1962. *Darwin and the darwinian revolution.* New York: W. W. Norton and Co.

Kormondy, E. J. 1976. *Concepts of ecology.* 2d ed. Englewood Cliffs, NJ: Prentice-Hall.

Lester, R. K. March 1986. Rethinking nuclear power. *Scientific American.*

Lewontin, R. C. September 1978. Adaptation. *Scientific American.*

Martin, P. I. October 1983. Labor-intensive agriculture. *Scientific American.*

Mayr, E. September 1978. Evolution. *Scientific American.*

Miller, G. T. 1981. *Living in the environment.* 3d ed. Belmont, CA: Wadsworth.

Mohnen, V. A. August, 1988. The challenge of acid rain. *Scientific American.*

Moorehead, A. 1979. *Darwin and the Beagle.* New York: Harper & Row, Publishers.

Moran, J. M., et al. 1980. *Introduction to environmental science.* San Francisco: W. H. Freeman.

Nebel, B. J. 1981. *Environmental science: The way the world works.* Englewood Cliffs, NJ: Prentice-Hall.

Odum, E. P. 1983. *Basic Ecology.* New York: CBS College Publishing.

O'Leary, P. R., Walsh, P. W., and Ham, R. H. December, 1988. Managing solid waste. *Scientific American.*

Power, J. F., and Follett, R. F. March 1987. Monoculture. *Scientific American.*

Putman, R. J., and S. D. Wratten. 1984. *Principles of Ecology.* Berkeley and Los Angeles: University of California Press.

Rasmussen, E. D. September 1982. The mechanization of agriculture. *Scientific American.*

Revelle, R. August 1982. Carbon dioxide and world climate. *Scientific American.*

Ricklefs, Robert E. 1983. *The economy of nature: A textbook in basic ecology.* New York: Chiron Press.

Scientific American. 1978. *Evolution.* San Francisco: W. H. Freeman.

Smith, R. L. 1985. *Ecology and field biology.* 3d ed. New York: Harper & Row.

Stebbins, G. L. 1977. *Processes of organic evolution.* 3d ed. Englewood Cliffs, NJ: Prentice-Hall.

Stebbins, G. L., and F. J. Ayala. July 1985. The evolution of Darwinianism. *Scientific American.*

Swaminathan, M. S. January 1984. Rice. *Scientific American.*

Turk, J., and A. Turk. 1984. *Environmental science.* New York: CBS College Publishing.

Volpe, E. P. 1981. *Understanding evolution.* 4th ed. Dubuque, IA: Wm. C. Brown Publishers.

Whittaker, R. H. 1975. *Communities and ecosystems.* 2d ed. New York: Macmillan.

APPENDIX A
Table of
Chemical
Elements

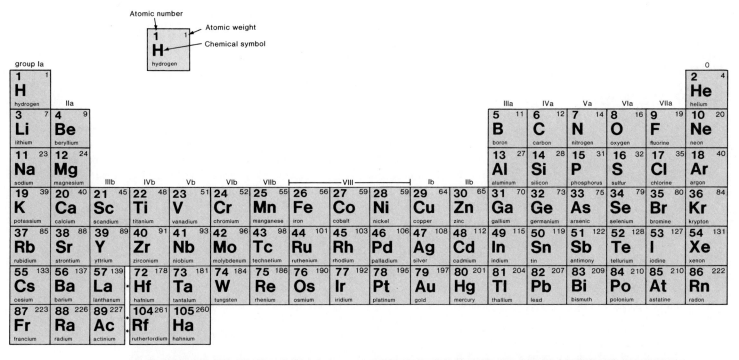

APPENDIX B
Light versus Electron Microscope

In a light microscope, a concentrated beam of light passes through the object on the stage in such a way that the emerging rays of light indicate the areas of light and dark in the object. Very often the prepared specimen has been treated with a stain to enhance these contrasts. A magnified image of the object is achieved when these rays of light are focused by the objective lens; the image is further magnified by means of the ocular lens. Since eyes are sensitive to light rays, the image may be viewed directly by the experimenter.

In an electron microscope, a concentrated beam of electrons passes through a vacuum to bombard the object. The vacuum is created when air is pumped out of the tube transmitting the electrons. The electrons are scattered by the object in such a way as to indicate areas of light and dark. The electrons are focused, and a magnified image is formed as the electrons pass the lenses of the electron microscope. The lenses of an electron microscope are not glass; they are coils of wire about which a magnetic field exists because of an electric current that travels within the wire. This field can focus electrons because they carry a negative charge that makes them sensitive to magnetic fields.

Since eyes are not sensitive to electrons, the image cannot be viewed directly; instead, it is projected onto a screen at the foot of the microscope where the electrons excite the chemical coating of the screen, producing light rays that can be seen by the viewer. A permanent record can also be made on a photographic plate; this record is an electron micrograph.

A comparison of these two microscopes shows that they both illuminate the object, magnify the object, and produce an image that eventually can be viewed by the observer. The most important difference between these two instruments is not the degree to which they magnify but instead is their resolving power. This term is used to indicate the amount of detail that can be distinguished by a microscope. The physical laws of optics tell us that resolving power is dependent on the wavelength of the light or electron beam. The theoretical limit of the resolving power of the light microscope is 200 nm (nanometers), while that of the electron microscope is 0.5 nm. This means that any two structures separated by less than 200 nm in the light microscope and 0.5 nm in the electron microscope will appear as one object. Thus, the electron microscope allows us to see much more.

One drawback to the electron microscope has been that the specimen could be viewed only after drying because a vacuum is needed to produce the electron beam. We do not know the full extent to which this drying process distorts the true appearance of the specimen. Methods are now being investigated to allow the observation of living materials in their natural state.

**Light
Microscope**

**Electron
Microscope**

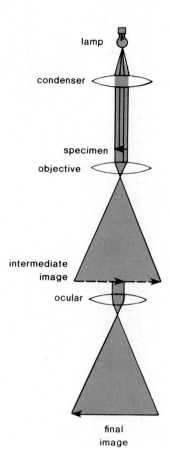

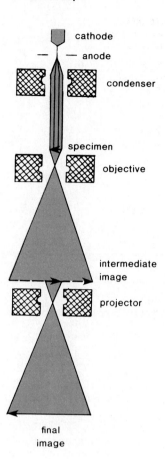

lamp

condenser

specimen

objective

intermediate
image

ocular

final
image

cathode

anode

condenser

specimen

objective

intermediate
image

projector

final
image

APPENDIX C
Metric System

THE METRIC SYSTEM

STANDARD METRIC UNITS

STANDARD METRIC UNITS		ABBREVIATIONS
Standard unit of mass	gram	g
Standard unit of length	meter	m
Standard unit of volume	liter	l

COMMON PREFIXES

COMMON PREFIXES		EXAMPLES
kilo	1,000	a kilogram is 1,000 grams
centi	0.01	a centimeter is 0.01 meter
milli	0.001	a milliliter is 0.001 liter
micro (μ)	one-millionth	a micrometer is 0.000001 (one-millionth) of a meter
nano (n)	one-billionth	a nanogram is 10^{-9} (one-billionth) of a gram
pico (p)	one-trillionth	a picogram is 10^{-12} (one-trillionth) of a gram

UNITS OF LENGTH

UNIT	ABBREVIATION	EQUIVALENT
meter	m	approximately 39 in
centimeter	cm	10^{-2} m
millimeter	mm	10^{-3} m
micrometer	μm	10^{-6} m
nanometer	nm	10^{-9} m
angstrom	Å	10^{-10} m

LENGTH CONVERSIONS

1 in	= 2.5 cm	1 mm	=	0.039 in
1 ft	= 30 cm	1 cm	=	0.39 in
1 yd	= 0.9 m	1 m	=	39 in
1 mi	= 1.6 km	1 m	=	1.094 yd
		1 km	=	0.6 mi

TO CONVERT	MULTIPLY BY	TO OBTAIN
inches	2.54	centimeters
feet	30	centimeters
centimeters	0.39	inches
millimeters	0.039	inches

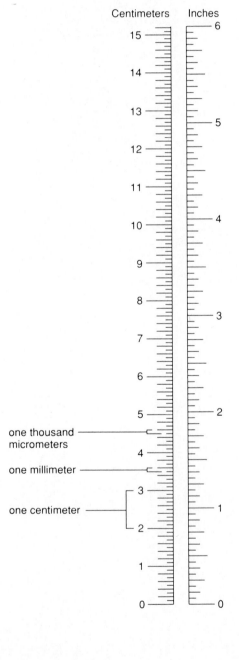

UNITS OF VOLUME

UNIT	ABBREVIATION	EQUIVALENT
liter	l	approximately 1.06 qt
milliliter	ml	10^{-3} l (1 ml = 1 cm³ = 1
microliter	µl	cc)
		10^{-6} l

VOLUME CONVERSIONS

1 tsp	= 5 ml	1 pt	= 0.47 l	1 ml	= 0.03 fl oz
1 tbsp	= 15 ml	1 qt	= 0.95 l	1 l	= 2.1 pt
1 fl oz	= 30 ml	1 gal	= 3.8 l	1 l	= 1.06 qt
1 cup	= 0.24 l			1 l	= 0.26 gal

TO CONVERT	MULTIPLY BY	TO OBTAIN
fluid ounces	30	milliliters
quarts	0.95	liters
milliliters	0.03	fluid ounces
liters	1.06	quarts

THINK METRIC

Volume

1. One can of beer (12 oz.) contains 360 ml.
2. The average human body contains between 10–12 pints of blood or between 4.7–5.6 liters.
3. One cubic foot of water (7.48 gallons) is 28.426 liters.
4. If a gallon of unleaded gasoline costs 1.00, a liter would cost 38¢.

THINK METRIC

Weight

1. One pound of hamburger is 448 grams.
2. The average human male brain weighs 1.4 kg (3 lb. 1.7 oz.).
3. A person who weighs 154 lbs weighs 70 kg.
4. Lucia Zarate weighed 5.85 kg (13 lbs.) at age 20.

UNITS OF WEIGHT

UNIT	ABBREVIATION	EQUIVALENT
kilogram	kg	10^3 g (approximately 2.2 lb)
gram	g	approximately 0.035 oz
milligram	mg	10^{-3} g
microgram	μg	10^{-6} g
nanogram	ng	10^{-9} g
picogram	pg	10^{-12} g

WEIGHT CONVERSIONS

1 oz = 28.3 g		1 g = 0.035 oz	
1 lb = 453.6 g		1 kg = 2.2 lb	
1 lb = 0.45 kg			

TO CONVERT	MULTIPLY BY	TO OBTAIN
ounces	28.3	grams
pounds	453.6	grams
pounds	0.45	kilograms
grams	0.035	ounces
kilograms	2.2	pounds

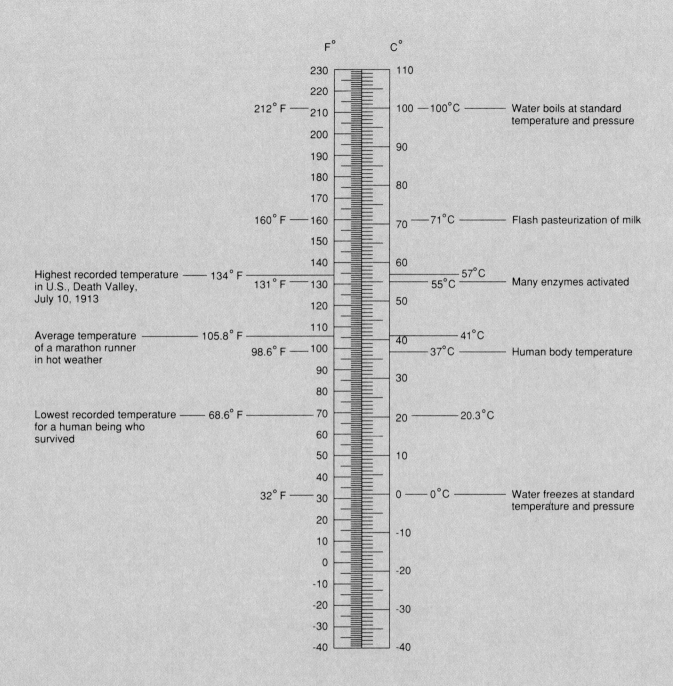

F°   C°

230   110

220

212° F — 210   100 — 100°C ——— Water boils at standard temperature and pressure

200

190   90

180   80

170

160° F — 160   70 — 71°C ——— Flash pasteurization of milk

150

140   60

Highest recorded temperature — 134° F   57°C
in U.S., Death Valley,   131° F — 130   55°C ——— Many enzymes activated
July 10, 1913

120   50

110

Average temperature — 105.8° F   41°C
of a marathon runner   40
in hot weather   98.6° F — 100   37°C ——— Human body temperature

90   30

80

Lowest recorded temperature — 68.6° F — 70   20 — 20.3°C
for a human being who
survived   60

50   10

40

32° F — 30   0 — 0°C ——— Water freezes at standard temperature and pressure

20

10   -10

0

-10   -20

-20   -30

-30

-40   -40

To convert temperature scales:

Fahrenheit to Celsius $°C = \frac{5}{9} (°F - 32)$

Celsius to Fahrenheit $°F = \frac{9}{5} (°C) + 32$

GLOSSARY

A

abiotic (ab''e-ot'ik)   not including living organisms, their effects, or products   *470*

accommodation (ah-kom''o-da'shun)   lens adjustment in order to see close objects   *254*

acetylcholine (as''e-til-ko'len)   *see* ACh

acetylcholinesterase (as''e-til-ko''lin-es'ter-as)   AChE; an enzyme that breaks down acetylcholine   *200*

ACh   a neurotransmitter substance secreted at the ends of many neurons; responsible for the transmission of a nerve impulse across a synaptic cleft   *200*

acid (as'id)   a solution in which pH is less than 7; a substance that contributes or liberates hydrogen ions in a solution   *16*

acromegaly (ak''ro-meg'ah-le)   condition resulting from an increase in growth hormone production after adult height has been achieved   *272*

acrosome (ak'ro-som)   covering on the tip of a sperm cell's nucleus believed to contain enzymes necessary for fertilization   *295*

ACTH   hormone secreted by the anterior lobe of the pituitary gland that stimulates activity in the adrenal cortex   *274, 277*

actin (ak'tin)   one of the two major proteins of muscle; makes up thin filaments in myofibrils of muscle cells. *See also* myosin   *57, 240*

action potential (ak'shun po-ten'shal)   the change in potential propagated along the membrane of a neuron; the nerve impulse   *198*

active site (ak'tiv sit)   the region on the surface of an enzyme where the substrate binds and where the reaction occurs   *40*

active transport (ak'tiv trans'port)   transfer of a substance into or out of a cell against a concentration gradient by a process that requires a carrier and the expenditure of energy   *35*

adaptation (ad''ap-ta'shun)   the fitness of an organism for its environment, including the process by which it becomes fit, in order that it may survive and reproduce; also the adjustment of sense receptors to a stimulus so that the stimulus no longer excites them   *248, 452*

Addison disease (ad'i-son di-zez')   a condition resulting from a deficiency of adrenal cortex hormones   *279*

adenosine triphosphate (ah-den'o-sen tri-fos'fat)   *see* ATP

ADH   hormone released from the posterior lobe of the pituitary gland that enhances the conservation of water by the kidneys   *188, 271*

adrenalin (ah-dren'ah-lin)   a hormone produced by the adrenal medulla that stimulates "fight or flight" reactions. Also called epinephrine   *278*

adrenocorticotropic hormone (ah-dre''no-kor''te-ko-trop'ik hor'mon)   *see* ACTH   *274, 277*

aerobic cellular respiration (a-er-o'bik sel'u-lar res''pi-ra'shun)   utilization of oxygen as the final acceptor for hydrogen atoms from a series of reactions that produce ATP   *38*

afterbirth (af'ter-berth'')   the placenta that is expelled after the birth of a child   *351*

agglutination (ah-gloo''ti-na'shun)   clumping of cells, particularly in reference to red blood cells involved in an antigen-antibody reaction   *136*

aging (aj'ing)   the degenerative biological changes that are observed as a human grows older   *354*

agranular leukocytes (ah-gran'u-lar lu'ko-sitz)   white blood cells that do not contain distinctive granules   *132*

AIDS (adz)   acquired immune deficiency syndrome, a disease caused by retroviruses and transmitted via body fluids; characterized by failure of the immune system   *320*

albumin (al-bu'min)   a plasma protein that helps to regulate the osmotic concentration of the blood   *123*

aldosterone (al''do-ster'on)   a hormone secreted by the adrenal cortex that functions in regulating sodium and potassium concentrations of the blood   *188, 279*

allantois (ah-lan'to-is)   an extraembryonic membrane that serves as a source of blood vessels for the umbilical cord   *339*

allele (ah-lel')   an alternative form of a gene that occurs at a given chromosomal site (locus)   *387*

all-or-none law (awl' or nun' law)   states that muscle fibers either contract maximally or not at all, and neurons either conduct a nerve impulse completely or not at all   *237*

alveoli (al-ve'o-li)   saclike structures that are the air sacs of a lung   *164*

amino acid (ah-me'no as'id)   a unit of protein that takes its name from the fact that it contains an amino group (NH_2) and an acid group (COOH)   *20*

ammonia (ah-mo'ne-ah) NH_3, a nitrogenous waste product resulting from deamination of amino acids   *179*

amnion (am'ne-on)   one of the extraembryonic membranes; a fluid-filled sac around the embryo   *337*

ampulla (am-pul'lah)   base of a semicircular canal in the inner ear   *261*

amylase (am'i-las)   a starch-digesting enzyme secreted by salivary glands and the pancreas   *74, 78*

analogous structure (ah-nal'o-gus struk'tur)   similar in function but not in structure, particularly in reference to similar adaptations   *448*

anaphase (an'ah-fāz) stage in mitosis during which chromatids separate, forming chromosomes 370

anaphase I (an'ah-fāz wun) phase of meiosis I during which homologous chromosomes separate and move to the poles of the spindle 372

anaphase II (an'ah-fāz too) phase of meiosis II during which the centromeres divide and sister chromatids separate and move toward the poles 372

anemia (ah-ne'me-ah) a condition characterized by a deficiency of red blood cells or of hemoglobin 128

angina pectoris (an'ji-nah pec'tor-is) condition characterized by thoracic pain resulting from occluded coronary arteries and preceding a heart attack 113

anterior pituitary (an-te're-or pi-tu'i-tar"e) the front lobe of the pituitary gland 271

anthropoids (an'thro-poidz) higher primates, including only monkeys, apes, and humans 456

antibody (an'ti-bod"e) a protein produced in response to the presence of some foreign substance in the blood or tissues 133, 143

anticodon (an"ti-ko'don) a "triplet" of three nucleotides in transfer RNA that pairs with a complementary triplet (codon) in messenger RNA 420

antidiuretic hormone (an"ti-di"u-ret'ik hor'mōn) see ADH

antigen (an'ti-jen) a foreign substance, usually a protein, that stimulates the immune system to produce antibodies 133, 142

anus (a'nus) outlet of the digestive tube 81

anvil (an'vil) also called the incus; the middle one of the auditory ossicles, which amplify the sound waves in the middle ear 259

aorta (a-or'tah) major systemic artery that receives blood from the left ventricle 105, 107

appendicular skeleton (ap"en-dik'u-lar skel'ĕ-ton) part of the skeleton forming the upper extremities, shoulder girdle, lower extremities, and hip girdle 224

appendix (ah-pen'diks) a small, tubular appendage that extends outward from the cecum of the large intestine 83

aqueous humor (a'kwe-us hu'mor) watery fluid that fills the space between the cornea and lens of the eye 252

ARAS system by which the thalamus is connected to various parts of the brain; composed of the diffuse thalamic projection system and the reticular formation 211

ARC (ark) AIDS-related complex; a prior stage to the possible development of AIDS 320

arterial duct (ar-te're-al dukt) fetal connection between the pulmonary artery and the aorta, ductus arteriosus 348

arteriole (ar-te're-ōl) a branch from an artery that leads into a capillary 100

artery (ar'ter-e) a vessel that takes blood away from the heart; characteristically possessing thick elastic walls 100

ascending reticular activating system (ah-send'ing rĕ-tik'u-lar ak'ti-vat"ing sis'tem) see ARAS

aster (as'ter) short rays of microtubules that appear at the ends of the spindle apparatus in animal cells during cell division 370

atherosclerosis (ath"er-o-sklê-ro'sis) condition in which fatty substances accumulate abnormally on the inner linings of arteries 113

atom (at'om) smallest unit of matter 13

ATP (adenosine triphosphate) a compound containing adenine, ribose, and three phosphates, two of which are high-energy phosphates. It is the "common currency" of energy for most cellular processes 27

atria (a'tre-ah) chambers; particularly the upper chambers of the heart that lie above the ventricles (sing. atrium) 102

atrioventricular (a"tre-o-ven-trik'u-lar) a structure in the heart that pertains to both the atria and ventricles; for example, an atrioventricular valve is located between an atrium and a ventricle 102

atrioventricular node (a"tre-o-ven-trik'u-lar nōd) see AV node

auditory canal (aw'di-to"re kah-nal') a tube in the outer ear that leads to the tympanic membrane 259

Australopithecus (aw"strah-lo-pith'e-kus) the first recognized hominids 460

autonomic nervous system (aw"to-nom'ik ner'vus sis'tem) portion of the nervous system that functions to control the actions of the visceral organs and skin 207

autosome (aw'to-sōm) a chromosome other than a sex chromosome 365

AV node (a-ve nōd) a small region of neuromuscular tissue located near the septum of the heart that transmits impulses from the SA node to the ventricular walls 106

axial skeleton (ak'se-al skel'ĕ-ton) portion of the skeleton that supports and protects the organs of the head, neck, and trunk 224

axon (ak'son) process of a neuron that conducts nerve impulses away from the cell body 195

B

bacteria (bak-te're ah) prokaryotes that lack the organelles of eukaryotic cells. 324

basal bodies (ba'sal bod'ēz) short cylinders having a circular arrangement of 9 microtubule triplets (9 + 0 pattern) located within the cytoplasm at the bases of cilia and flagella 40

base (bās) a solution in which pH is more than 7; a substance that contributes or liberates hydroxide ions in a solution; alkaline; opposite of acidic. Also, in genetics the chemicals adenine, guanine, cytosine, thymine, and uracil that are found in DNA and RNA 16

bile (bīl) a secretion of the liver that is temporarily stored in the gallbladder before being released into the small intestine where it emulsifies fat 76

binary fission (bi'na-re fish'un) reproduction by simple cell division that does not involve a mitotic spindle 325

binomial system (bi-no'me-al sis'tem) the assignment of two names to each organism, the first of which designates the genus and the second of which designates the species 444

bioaccumulation (bi"o-ah-ku"mu-la'shun) tendency for nonexcretable hazardous wastes to progressively concentrate through the links of a food chain 485

biome (bi'ōm) major ecological region that has the characteristics of a climax community 467

biosphere (bi'o-sfer) that part of the earth's surface and atmosphere where living organisms exist 470

biotechnology (bi"o-tek-nol'o-je) the use of biological organisms to mass-produce a product of commercial interest 424

biotic (bi-ot'ik) pertaining to any aspect of life, especially to characteristics of entire populations or ecosystems 470

biotic potential (bi-ot'ik po-ten'shal) the maximum population growth rate under ideal conditions 494

blastocyst (blas'to-sist) an early stage of embryonic development that consists of a hollow ball of cells 335

blind spot (blind spot) area where the optic nerve passes through the retina and where vision is not possible due to the lack of rods and cones 253

blood (blud) connective tissue composed of cells separated by plasma 57

blood pressure (blud presh'ur) the pressure of the blood against a blood vessel. Blood pressure is usually taken on the brachial artery where 120/80 (systoic/diastolic) mm Hg is considered normal 110

B lymphocyte (be lim'fo-sit) type of lymphocyte that is responsible for antibody-mediated immunity 142

bone (bōn) connective tissue having a hard matrix of calcium salts deposited around protein fibers 57

Bowman's capsule (bo'manz kap'sūl) a double-walled cup that surrounds the glomerulus at the beginning of the kidney tubule 183

bradykinin (brad"e-ki'nin) a substance found in damaged tissue that initiates nerve impulses resulting in the sensation of pain 134

breathing (brēth′ing) entrance and exit of air into and out of the lungs *160*

breeder fission (brēd′er fish′un) a nuclear reaction that produces plutonium, a nuclear power plant fuel *504*

bronchi (brong′ki) the two major divisions of the trachea leading to the lungs *164*

bronchiole (brong′ke-ōl) the smaller air passages in the lungs *164*

buffer (buf′er) a substance or compound that prevents large changes in the pH of a solution *17*

C

calcitonin (kal″si-to′nin) hormone secreted by the thyroid gland that helps to regulate the level of blood calcium *276*

calorie (kal′o-re) the amount of heat required to raise one kilogram of water one degree centigrade *84*

capillaries (kap′i-ler″es) microscopic vessels located in the tissues connecting arterioles to venules through the thin walls of which molecules either exit or enter the blood *100*

carbohydrate (kar″bo-hi′drāt) organic compounds with the general formula $(CH_2O)_n$ including sugars and glycogen *20*

cardiac (kar′de-ak) of or pertaining to the heart *57*

cardiac muscle (kar′de-ak mus′el) heart muscle (myocardium) consisting of striated muscle cells that interlock *58*

caries (kar′ēz) tooth decay caused by bacterial activity *73*

carrier (kar′e-er) a molecule that combines with a substance and actively transports it through the cell membrane; an individual that transmits an infectious or genetic disease *35*

carrying capacity (kar′e-ing kah-pas′i-te) the largest number of organisms of a particular species that can be maintained indefinitely in an ecosystem *495*

cartilage (kar′ti-lij) a connective tissue, usually part of the skeleton, that is composed of cells in a flexible matrix *56*

catalytic converter (kat″ah-lit′ik kon-vert′er) pollution-reducing device used on automobiles that converts gaseous hydrocarbons and carbon monoxide to carbon dioxide and water *484*

CCK hormone secreted by the small intestine that stimulates the release of pancreatic juice from the pancreas and bile from the gallbladder *77*

cell (sel) the structural and functional unit of an organism; the smallest structure capable of performing all the functions necessary for life *31*

cell body (sel bod′e) portion of a nerve cell that includes a cytoplasmic mass and a nucleus and from which the nerve fibers extend *195*

cell membrane (sel mem′brān) a membrane that surrounds the cytoplasm of cells and regulates the passage of molecules into and out of the cell *32*

cellular respiration (sel′u-lar res″pi-ra′shun) the reactions of glycolysis, Krebs cycle, and electron transport system that provide energy for ATP production *160*

central canal (sen′tral kah-nal′) tube within the spinal cord that is continuous with the ventricles of the brain and contains cerebrospinal fluid *209*

central dogma (sen′tral dog′mah) the proposal that genetic information always proceeds from DNA to mRNA to protein synthesis *419*

central nervous system (sen′tral ner′vus sis′tum) *see* CNS

centriole (sen′tre-ōl) a short, cylindrical organelle that contains microtubules in a 9 + 0 pattern and is associated with the formation of the spindle during cell division *40*

centromere (sen′tro-mēr) a region of attachment for a chromosome to a spindle fiber that is generally seen as a constricted area *365*

cerebellum (ser″e-bel′um) the part of the brain that controls muscular coordination *211*

cerebral hemisphere (ser′e-bral hem′i-sfēr) one of the large, paired structures that together constitute the cerebrum of the brain *211*

cerebrospinal fluid (ser″e-bro-spi′nal floo′id) fluid found within ventricles of the brain and surrounding the CNS in association with the meninges *209*

cerebrum (ser′e-brum) the main portion of the vertebrate brain that is responsible for consciousness *211*

chlamydia (klah-mid′e-ah) an organism that causes a sexually transmitted disease particularly characterized by urethritis *328*

cholecystokinin (ko″le-sis″to-ki′nin) *see* CCK

cholesterol (ko-les′ter-ol) a lipid produced by body cells that is used in the synthesis of steroid hormones and is excreted into the bile *24*

cholinesterase (ko″lin-es′ter-ās) an enzyme that breaks down acetylcholine, a neurotransmitter substance, (AChE) *200*

chorion (ko′re-on) an extraembryonic membrane that forms an outer covering around the embryo and contributes to the formation of the placenta *335*

chorionic villi (ko″re-on′ik vil′i) projections from the chorion that appear during implantation and that in one area contribute to the development of the placenta *343*

choroid (ko′roid) the vascular, pigmented middle layer of the wall of the eye *252*

chromatids (kro′mah-tidz) the two identical parts of a chromosome following replication of DNA *305*

chromatin (kro′mah-tin) threadlike network in the nucleus that is made up of DNA and proteins *36*

chromosomes (kro′mo-sōmz) rod-shaped bodies in the nucleus, particularly during cell division, that contain the hereditary units or genes *36*

cilia (sil′e-ah) hairlike projections that are used for locomotion by many unicellular organisms and that have various purposes in higher organisms *40*

ciliary muscle (sil′e-er″e mus′el) a muscle that controls the curvature of the lens of the eye *252*

circumcision (ser″kum-sizh′un) removal of the foreskin of the penis *296*

class (klas) in taxonomy, the category below phylum and above order *444*

cleavage (klēv′ij) cell division of the fertilized egg that is unaccompanied by growth so that numerous small cells result *335*

climax community (kli′maks kŏ-mu′ni-te) in succession, a final, stable community capable of sustaining itself indefinitely *467*

clone (klōn) asexually produced organisms having the same genetic makeup; also DNA fragments from an external source that have been reproduced by bacteria *426*

clotting (klot′ing) agglutination of red blood cells in which they clump together due to an antigen-antibody reaction *131*

CNS the central nervous system; the brain and spinal cord *195*

coacervate (ko-as′er-vāt) a mixture of polymers that may have preceded the origination of the first cell or cells *455*

cochlea (kok′le-ah) that portion of the inner ear that resembles a snail's shell and contains the organ of Corti, the sense organ for hearing *261*

cochlear canal (kok′le-ar kah-nal′) canal within the cochlea bearing small hair cells that function as hearing receptors *261*

codon (ko′don) a "triplet" of three nucleotides in messenger RNA that directs the placement of a particular amino acid into a polypeptide chain *419*

coenzyme (ko-en′zim) a nonprotein molecule that aids the action of an enzyme, to which it is loosely bound *42*

collecting duct (kŏ-lekt′ing dukt) a tube that receives urine from several distal convoluted tubules *183*

colon (ko′lon) the large intestine *81*

columnar epithelium (ko-lum′nar ep″ i-the-le-um) pillar-shaped cells usually having the nuclei near the bottom of each cell and found lining the digestive tract, for example *51*

common ancestor (kom′un an′ses-ter)   a predecessor in evolution that leads to two or more lines of descent   *448*

compact bone (kom-pakt′ bōn)   hard bone consisting of Haversian systems cemented together   *57, 228*

complement (kom′plĕ-ment)   a group of proteins in plasma that produce a variety of effects once an antigen-antibody reaction has occurred   *141*

complementary base pairing (kom″plĕ men′tă-re bās par′ing)   pairing of bases found in DNA and RNA; adenine is always paired with either thymine (DNA) or uracil (RNA) and cytosine is always paired with guanine   *414*

cones (kōnz)   bright-light receptors in the retina of the eye that detect color and provide visual acuity   *253*

connective tissue (kŏ-nek′tiv tish′u)   a type of tissue, characterized by cells separated by a matrix, that often contains fibers   *55*

consumers (kon-su′merz)   organisms of one population that feed on members of other populations in an ecosystem   *470*

coronary artery (kor′ŏ-na-re ar′ter-e)   an artery that supplies blood to the wall of the heart   *108*

corpus callosum (kor′pus kah-lo′sum)   a mass of white matter within the brain, composed of nerve fibers connecting the right and left cerebral hemispheres   *212*

corpus luteum (kor′pus lut′e-um)   a body, yellow in color, that forms in the ovary from a follicle that has discharged its egg   *299*

cortisol (kor′ti-sol)   a glucocorticoid secreted by the adrenal cortex   *278*

covalent bond (ko-va′lent bond)   chemical bond created by the sharing of electrons between atoms   *14*

Cowper's glands (kow′ perz ′glandz)   two small structures located below the prostate gland in males   *296*

cranial nerve (kra′ne-al nerv)   nerve that arises from the brain   *203*

creatine phosphate (kre′ah-tin fos′fāt)   a compound unique to muscles that contains a high-energy phosphate bond   *240*

creatinine (kre-at′ĭ-nin)   excretion product from creatine phosphate breakdown   *180*

cretinism (kre′tin-izm)   a condition resulting from a lack of thyroid hormone in an infant   *276*

cri du chat (kre-du-shah)   condition created by deletion of chromosome number-5 characterized by a child cry sounding like a cat's meow and having various physical abnormalities   *379*

cristae (kris′te)   membranous shelves of the mitochondrion's interior formed by infoldings of its inner membrane   *38*

Cro-Magnon (kro-mag′non)   the common name for the first fossils to be accepted as representative of modern humans   *463*

crossing over (kros′ing o′ver)   the exchange of corresponding segments of genetic material between chromatids of homologous chromosomes during meiosis   *372*

cuboidal epithelium (ku-boi′dal ep″ĭ-the′le-um)   cube-shaped cells found lining the kidney tubules   *51*

cultural eutrophication (kul′tu-ral u″tro-fĭ-ka′shun)   enrichment of a body of water causing excessive growth of producers and then death of these and other inhabitants   *486*

Cushing syndrome (koosh′ing sin′drōm)   a condition characterized by thin arms and legs and a "moon face" accompanied by high blood glucose and sodium levels due to hypersecretion of cortical hormones   *280*

cytokinesis (si″to-ki-ne′sis)   division of the cytoplasm of a cell   *370*

cytoplasm (si′to-plazm)   the ground substance of cells located between the nucleus and the cell membrane   *32*

cytoskeleton (si′to-skel′ĕ-ton)   filamentous protein structures found throughout the cytoplasm that help maintain the shape of the cell   *39*

D

deamination (de-am″i-na′shun)   removal of an amino group (—NH₂) from an amino acid or other organic compound   *80, 179*

decomposers (de-kom-po′zerz)   organisms of decay (fungi and bacteria) in an ecosystem   *470*

deletion (de-le′shun)   a chromosome mutation that results in loss of a portion of the chromosome   *378*

demographic transition (dem-o-graf′ik tran-zi′shun)   the change from a high birthrate to a low birthrate so that the growth rate is lowered   *496*

dendrite (den′drīt)   process of a neuron, typically branched, that conducts nerve impulses toward the cell body   *195*

denitrification (de-ni″tri-fĭ-ka′shun)   the process of converting nitrogen compounds to free nitrogen, liberated in the nitrogen cycle   *477*

denitrify (de-ni′-tri-fī)   to convert ammonia or nitrate to atmospheric nitrogen, as in the denitrifying bacteria   *477*

deoxyribonucleic acid (de-ok″se-ri″bo-nu-kle′ik as′id)   *see* DNA

depletion curve (de-ple′shun kurv)   graphical depiction of a resource's dwindling amount over time   *502*

depolarization (de-po″lar-ĭ-za′shun)   a loss in polarization, as when the nerve impulse or action potential occurs   *198*

dermis (der′mis)   the thick skin layer that lies beneath the epidermis   *61*

desmosome (des′mo-sōm)   a specialized junction between cells, which serves as a "spot weld"   *54*

detritus (de-tri′tus)   falling, settled debris at the land floor or water bed of an ecosystem that is subject to decomposer action   *470*

developed countries (de-vel′opt kun′trēz)   industrialized nations that typically have a strong economy and a low rate of population growth   *495*

developing countries (de-vel′op-ing kun′trēz)   nations that are not yet industrialized and have a weak economy and a high rate of population growth   *497*

diabetes insipidus (di″ah-be′tez in-sip′i-dus)   condition characterized by an abnormally large production of urine due to a deficiency of antidiuretic hormone   *272*

diabetes mellitus (di″ah-be′tez mĕ-li′tus)   condition characterized by a high blood glucose level and the appearance of glucose in the urine due to a deficiency of insulin   *285*

diaphragm (di′ah-fram)   a sheet of muscle that separates the chest cavity from the abdominal cavity. Also, a birth control device inserted in front of the cervix in females   *165*

diastole (di-as′to-le)   relaxation of heart chambers   *105*

diastolic pressure (di-a-stol′ik presh′ur)   arterial blood pressure during the diastolic phase of the cardiac cycle   *111*

differentiation (dif″er-en″she-a′shun)   the process and developmental stages by which a cell becomes specialized for a particular function   *335*

diffusion (di-fu′zhun)   the movement of molecules from an area of greater concentration to an area of lesser concentration   *33*

dihybrid (di-hi′brid)   the offspring of parents who differ in two ways; shows the phenotype governed by the dominant alleles but carries the recessive alleles   *392*

diploid (dip′loid)   the 2N number of chromosomes; twice the number of chromosomes found in gametes   *366*

disaccharide (di-sak′ah-rid)   a sugar such as maltose that contains two units of a monosaccharide   *22*

dissociation (dis-so″she-a′shun)   the breaking of a chemical bond when a compound or molecule is put into water, thereby releasing ions   *17*

distal convoluted tubule (dis′tal kon″vo-lūt-ed tu′bul)   highly coiled region of a nephron that is distant from Bowman's capsule   *183*

DNA   a nucleic acid, found especially in the nucleus where it contains a triplet genetic code   *24, 414*

dominant allele (dom′i-nant ah-lēl′)   hereditary factor that expresses itself even when there is only one copy in the genotype   *387*

dorsal root ganglion (dor′sal rōot gang′gle-on)   mass of sensory neuron cell bodies located in the dorsal root of a spinal nerve   *203*

double helix (du 'b'l he'liks)   a double spiral often used to describe the three-dimensional shape of DNA   26

doubling time (du'b'ling tīm)   the number of years it takes for a population to double in size   494

Down syndrome (down sin'drōm)   human congenital disorder associated with an extra number-23 chromosome   378

Dryopithecus (dri''o-pith'e-kus)   extinct apes that may have included or resembled a common ancestor to both apes and humans   457

duodenum (du''o-de'num)   the first portion of the small intestine into which ducts from the gallbladder and pancreas enter   76

duplication (du pli 'kā shən)   a chromosome mutation in which a chromosome contains two groups of identical genes; replication of DNA   378

E

ecology (e-kol'o-je)   the study of the relationship of organisms between themselves and the physical environment   470

ecosystem (ek''o-sis'tem)   a biological community together with the associated abiotic environment   470

ectopic pregnancy (ek-top'ik preg'nan-se)   implantation and development of a fertilized egg outside the uterus   336

edema (ĕ-de'mah)   swelling due to tissue fluid accumulation in the intercellular spaces   117

EEG   graphic recording of the brain's electrical activity   213

effector (ĕ-fek'tor)   a structure that allows a response to environmental stimuli such as the muscles and glands   203

elastic cartilage (e-las'tik kar'ti-lij)   cartilage composed of elastic fibers that allows for greater flexibility   56

electrocardiogram (e-lek''tro-kar'de-o-gram'')   ECG or EKG; a recording of the electrical activity associated with the heartbeat   106

electroencephalogram (e-lek''tro-en-sef'ah-lo-gram'')   *see* EEG

electron (e-lek'tron)   a subatomic particle that has almost no weight and carries a negative charge; travels in an orbital, called a shell, about the nucleus of an atom   13

element (el'ĕ-ment)   the simplest of substances consisting of only one type of atom; i.e., carbon, hydrogen, oxygen   13

embolus (em'bo-lus)   a moving blood clot that is carried through the bloodstream   121

embryo (em'bre-o)   the developing organism, particularly during the early stages   334

emulsification (e-mul''si-fi'ka'shun)   the act of dispersing one liquid in another   24

endocrine gland (en'do-krin gland)   a gland that secretes hormones directly into the blood or body fluids   54, 269

endocytosis (en''do-si-to'sis)   a process in which extracellular material is enclosed within a vesicle and taken into the cell. Phagocytosis and pinocytosis are forms of endocytosis   37

endometrium (en''do-me'tre-um)   the lining of the uterus that becomes thickened and vascular during the uterine cycle   301

endoplasmic reticulum (ER) (en-do-plaz'mic re-tik'u-lum)   a complex system of tubules, vesicles, and sacs in cells; sometimes having attached ribosomes   36

endospore (en'do-spor)   a resistant body formed by bacteria when environmental conditions worsen   325

environmental resistance (en-vi''ron-men'tal re-zis'tans)   sum total of factors in the environment that limit the numerical increase of a population in a particular region   494

enzyme (en'zim)   a protein catalyst that speeds up a specific reaction or a specific type of reaction   20

epidermis (ep''i-der'mis)   the outer layer of cells of an organism   60

epididymis (ep''i-did'i mis)   coiled tubules next to the testes where sperm mature and may be stored for a short time   296

epiglottis (ep''i-glot'is)   a structure that covers the glottis during the process of swallowing   74, 163

epithelial tissue (ep''i-the'le-al tish'u)   a type of tissue that lines cavities and covers the external surface of the body   51

erection (ĕ-rek'shun)   erect penis prepared for copulation   296

erythrocyte (ĕ-rith'ro-sit)   non-nucleated, hemoglobin containing blood cells capable of carrying oxygen; the red blood cell   122

esophagus (ĕ-sof'ah-gus)   a tube that transports food from the mouth to the stomach   74

essential amino acid (ē-sen'shal ah-me'no as'id)   amino acid required for health that cannot be synthesized in adequate amounts by body cells   84

eukaryotic (u''kar-e-ot'ik)   possessing the membranous organelles characteristic of complex cells   324

eustachian tube (u-sta'ke-an tūb)   an air tube that connects the pharynx to the middle ear   261

evolution (ev''o-lu'shun)   genetic changes that occur in populations of organisms with the passage of time, resulting in an adaptation to the environment   441

evolutionary tree (ev''o-lu'shun-ar-e tre)   a diagram describing the phylogenetic relationship of groups of organisms   448

excretion (ek-skre'shun)   removal of metabolic wastes   178

exocrine gland (ek'so-krin gland)   secreting externally; particular glands with ducts whose secretions are deposited into cavities, such as salivary glands   54

exocytosis (eks''o-si-to'sis)   a process in which an intracellular vesicle fuses with the cell membrane so that the vesicle's contents are released outside the cell   37

exophthalmic goiter (ek''sof-thal'mik goi'ter)   an enlargement of the thyroid gland accompanied by an abnormal protrusion of the eyes   276

expiration (eks''pi-ra'shun)   process of expelling air from the lungs; exhalation   161

exponential growth (eks''po-nen'shal grōth)   growth, particularly of a population, in which the total number increases in the same manner as compound interest   492

external respiration (eks-ter'nal res''pi-ra'shun)   exchange of oxygen and carbon dioxide between air and blood   160

extinction (eks-ting'shun)   in evolution, the complete disappearance of a group of organisms so that none of its kind ever exists again   451

extraembryonic membranes (eks''trah-em''bre-on'ik mem'brānz)   membranes that are not a part of the embryo but are necessary to the continued existence and health of the embryo   335

F

facilitated transport (fah-sil'i-tāt-ed trans'port)   transfer of a substance into or out of a cell along a concentration gradient by a process that requires a carrier   35

family (fam'i-le)   a rank in taxonomic classification above genus and below order   444

fatigue (fah-tēg')   muscle relaxation in the presence of stimulation due to energy reserve depletion   238

fatty acid (fat'e as'id)   an organic molecule having a long chain of carbon atoms and ending in an acidic group   22

feces (fe'sēz)   indigestible wastes expelled from the digestive tract; excrement   81

fermentation (fer''men-ta'shun)   anaerobic breakdown of carbohydrates that results in end products such as alcohol and lactic acid   44

fetal erythroblastosis (fe'tal e rith''ro-blas-to'sis)   destruction of newborn Rh positive red blood cells due to maternal antibodies against the Rh factor that have crossed the placenta   136

fetal position (fe'tal po-zish'un)   curled position of the body in which the head touches the knees; the position assumed by the fetus in the womb   347

fetus (fe'tus)   human development in its later stages following the embryonic stages   334

fibrin (fi'brin)   insoluble, fibrous protein formed from fibrinogen during blood clotting   131

fibrinogen (fi-brin'o-jen)   plasma protein that is converted into fibrin during blood coagulation   131

fibrocartilage (fi''bro-kar'ti-lij) cartilage with a matrix of strong collagenous fibers 57

fibrous connective tissue (fi'brus ko-nek'tiv tish'u) tissue composed mainly of closely packed collagenous fibers and found in tendons and ligaments 56

filtrate (fil'trat) the filtered portion of blood that is contained within Bowman's capsule 185

fimbriae (fim'bre-e'') fingerlike extensions from the oviduct near the ovary 300

fission power plant (fish'un pow'er plant) nuclear power plant that splits²³⁵ uranium to generate energy 504

flagella (flah-jel'ah) slender, long processes used for locomotion by the flagellated protozoans, bacteria, and sperm 40

focusing (fo'kus-ing) manner by which light rays are bent by the cornea and lens, creating an image on the retina 253

follicle (fol'i-kl) a structure in the ovary that produces the egg and particularly the female sex hormone, estrogen 299

follicle-stimulating hormone (fol'i-kl stim'u-la''ting hor'mon) see FSH

fontanel (fon''tah-nel') membranous region located between certain cranial bones in the skull of a fetus or infant 346

food chain (food chan) a sequence of organisms, each of which feeds on the previous one to acquire energy and organic building blocks. Includes the producer, various levels of consumers, and the decomposers 470

food web (food web) the complete set of food links between populations in a community 472

foramen magnum (fo-ra'men mag'num) opening in the occipital bone of the skull through which the spinal cord passes 224

foreskin (for'skin) skin covering the glans penis in uncircumcised males 296

formed element (form'd el'e-ment) a cellular constituent of blood 122

fossil fuel (fos'l fu'el) the remains of once-living organisms that are burned to release energy, such as coal, oil, and natural gas 475

fossils (fos'lz) any remains of an organism that have been preserved in the earth's crust 443

fovea centralis (fo've-ah sen-tral'is) region of the retina consisting of densely packed cones, that is responsible for the greatest visual acuity 253

frontal lobe (frun'tal lob) area of the cerebrum responsible for voluntary movements and higher intellectual processes 211

FSH hormone secreted by the anterior pituitary gland that stimulates the development of an ovarian follicle in a female or the production of sperm cells in a male 297

fungus (fung' gus) an organism, usually composed of strands called hyphae, that lives chiefly on decaying matter; e.g., mushroom and mold 330

G

gallbladder (gawl'blad-er) saclike organ associated with the liver that stores and concentrates bile 76

gallstones (gawl'stonz) precipitated crystals of cholesterol or calcium carbonate formed from bile within the gallbladder or bile duct 81

gamete (gam'et) a reproductive cell that joins with another in fertilization to form a zygote; most often an egg or sperm 366

ganglion (gang'gle-on) a collection of neuron cell bodies outside the central nervous system 203

gastric (gas'trik) of or pertaining to the stomach 75

gastric gland (gas'trik gland) gland within the stomach wall that secretes gastric juice 75

gastric inhibitory peptide (gas'trik in-hib'i-tor''e pep'tid) see GIP

gastrin (gas'trin) a hormone secreted by stomach cells to regulate the release of pepsin by the stomach wall 77

genetic mutation (je-net'ik mu-ta'shun) a permanent change in the DNA code such that there is an observed change in the phenotype of the individual 422

genital (jen'i-tal) an organ within the reproductive system; external genitals are found within the vulva in females and consist of the penis and testicles in males 296, 300

genital herpes (jen'i-tal her'pez) a sexually transmitted disease characterized by open sores on the external genitalia 322

genital warts (jen'i-tal worts) warts that develop on external genitalia and can be acquired by sexual contact 323

genotype (je'no-tip) the genetic makeup of any individual 390

genus (je'nus) a rank in taxonomic classification above species and below family 444

germ layers (jerm la'ers) primary tissues of an embryo (ectoderm, mesoderm, endoderm) that give rise to the major tissue systems of the adult animal 337

gerontology (jer''on-tol'o-je) the study of older individuals, including the biological degenerative changes associated with the elderly 354

GIP hormone produced by the small intestine that inhibits gastric acid secretion 77

glial cells (gli'al selz) cells that support, protect, and nourish neurons within the brain and spinal cord 59

globin (glo'bin) the protein portion of a hemoglobin molecule 124

glomerulus (glo-mer'u-lus) a cluster; for example, the cluster of capillaries surrounded by Bowman's capsule in a kidney tubule 185

glottis (glot'is) slitlike opening between the vocal cords 74, 163

glucagon (gloo'kah-gon) hormone secreted by the pancreatic islets of Langerhans that causes the release of glucose from glycogen 285

glucose (gloo'kos) the most common six-carbon sugar 22

glycerol (glis'er-ol) an organic compound that serves as a building block for fat molecules 22

glycogen (gli'ko-jen) a polysaccharide that is the principal storage compound for sugar in animals 22

glycolysis (gli-kol'i-sis) the metabolic pathway that converts sugars to simpler compounds 42

glycoprotein (gli''ko-pro'te-in) a substance composed of a carbohydrate combined with a protein 85

Golgi apparatus (gol'ge ap''ah-ra'tus) an organelle that consists of concentrically folded membranes and functions in the packaging and secretion of cellular products 37

gonad (go'nad) an organ that produces sex cells; the ovary, which produces eggs, and the testis, which produces sperm 367

gonadotropic hormone (go-nad''o-trop'ik hor'mon) a type of hormone that regulates the activity of the ovaries and testes; principally FSH and LH (ICSH) 274

gonorrhea (gon''o-re'ah) contagious venereal disease caused by bacteria and leading to inflammation of the urogenital tract 326

Graafian follicle (graf'e-an fol'li-k'l) mature follicle within the ovaries that houses a developing egg 299

granular leukocytes (gran'u-lar lu'ko-sitz) white blood cells that contain distinctive granules 132

greenhouse effect (gren'hows e-fekt') carbon dioxide buildup in the atmosphere as a result of fossil fuel combustion that retains and reradiates heat creating an abnormal rise in the earth's average temperature 484

growth (groth) increase in the number of cells and/or the size of these cells 335

growth hormone (groth hor'mon) a hormone released by the anterior lobe of the pituitary gland that promotes the growth of the organism; GH or somatotropin 272

growth rate (groth rat) percentage of increase or decrease in the size of a population 493

gummas (gum'ahz) large unpleasant sores that occur during the tertiary stage of syphilis 329

H

habitat (hab′i-tat) the natural abode of an animal or plant species 470

hammer (ham′er) also called the malleus; the first of the auditory ossicles that amplify the sound waves in the middle ear 259

haploid (hap′loid) the N number of chromosomes; half the diploid number; the number characteristic of gametes that contain only one set of chromosomes 367

hard palate (hard pal′at) anterior portion of the roof of the mouth that contains several bones 74

hazardous wastes (haz′ard-us wāsts) wastes containing chemicals dangerous to life 484

HCG hormone a hormone produced by the placenta that helps maintain pregnancy and is the basis for the pregnancy test 306

heart (hart) muscular organ located in thoracic cavity responsible for maintenance of blood circulation 102

heart attack (hart ah-tak′) condition resulting when circulation in the coronary arteries is blocked 100

heavy metals (hev′e met′′lz) a category of hazardous waste that comprises metals such as Hg and Pb 485

heliostat (he′le-o-stat) large-field mirrors that track the sun and reflect its energy onto a mounted boiler for solar heating 506

helper T cells (help′er T selz) T lymphocytes that stimulate certain other T and B lymphocytes to perform their respective functions; T_H cells 150

heme (hēm) the iron-containing portion of a hemoglobin molecule 124

hemoglobin (he′′mo-glo′bin) a red, iron-containing pigment in blood that combines with and transports oxygen 124, 169

hepatic portal system (hē-pat′ik por′tal sis′tem) portal system that begins at the villi of the small intestine and ends at the liver 109

hepatic portal vein (hē-pat′ik por′tal vān) vein leading to the liver formed by the merging of blood vessels from the villi of the small intestine 78, 109

herpes simplex virus (her′pēz sim′plex vi′rus) a virus of which type I causes cold sores and type II causes genital herpes 322

heterozygous (het′′er-o-zi′gus) having two different alleles (as *Aa*) for a given trait 390

histamine (his′tah-min) substance produced by basophil-derived mast cells in connective tissue that causes capillaries to dilate and release immune and other substances 134

HIV viruses responsible for AIDS; human immunodeficiency virus 320

homeostasis (ho′′me-o-sta′sis) the constancy of conditions, particularly the environment of the body's cells: constant temperature, blood pressure, pH, and other body conditions 61

hominids (hom′i-nidz) members of a family containing upright, bipedal primates including modern humans 459

hominoids (hom′i-noidz) members of a superfamily containing humans and the great apes 456

Homo erectus (ho′mo ē-rek′tus) the earliest nondisputed species of humans, named for their erect posture that allowed them to have a striding gait 462

Homo habilis (ho′mo hah′bī-lis) an extinct species that may include the earliest humans, having a small brain but quality tools 461

homologous (ho-mol′o-gus) similarly constructed; homologous chromosomes have the same shape and contain genes for the same traits; homologous structures in animals share a common ancestry 365, 448

homozygous (ho′′mo-zi′gus) having identical alleles (as *AA* or *aa*) for a given trait; pure breeding 390

host (hōst) an organism on or in which another organism lives 318

human chorionic gonadotropic hormone (hu′man ko′′re-on′ik go-nad′′o-trōp′ik hor′mōn) *see* HCG

hyaline cartilage (hi′ah-līn kar′ti-lij) cartilage composed of very fine collagenous fibers and a matrix of a clear, milk glass appearance 56

hybrid (hi′brid) an offspring resulting from the crossing of genetically different strains, populations, or species 390

hybridomas (hi-brid-o′mahz) fused lymphocytes and cancer cells used in the manufacture of monoclonal antibodies 155

hydrogen bond (hi′dro-jen bond) a weak attraction between a hydrogen atom carrying a partial positive charge and an atom of another molecule carrying a partial negative charge 15

hydrolysis (hi-drol′i-sis) the splitting of a bond within a larger molecule by the addition of water 19

hydrolytic enzyme (hi-dro-lit′ik en′zīm) describes an enzyme in which the substrate is broken down by the addition of water 74

hypertension (hi′′per-ten′shun) elevated blood pressure, particularly the diastolic pressure 112

hypertonic solution (hi′′per-ton′ik so-lu′shun) one that has a greater concentration of solute, a lesser concentration of water than the cell 34

hypothalamus (hi′′po-thal′ah-mus) a region of the brain; the floor of the third ventricle that helps maintain homeostasis 209

hypotonic solution (hi′′po-ton′ik so-lu′shun) one that has a lesser concentration of solute, a greater concentration of water than the cell 34

I

immune complex (i-mūn′ kom′pleks) the product of an antigen-antibody reaction 149

implantation (im′′plan-ta′shun) the attachment and penetration of the embryo to the lining (endometrium) of the uterus 306

impotency (im′po-ten′′se) failure of the penis to achieve erection 296

induction (in-duk′shun) in development, the ability of one body part to influence the development of another part 340

inflammation (in′′flah-ma′shun) a response to tissue damage that may include bacterial and viral invasion characterized by dilation of blood vessels and an accumulation of fluid in the affected region 134

inflammatory reaction (in-flam′ah-to′′re re-ak′shun) a tissue response to injury that is characterized by dilation of blood vessels and an accumulation of fluid in the affected region 134

inner cell mass (in′er sel mas) the portion of a blastocyst that will develop into the embryo and fetus 335

inner ear (in′er ēr) portion of the ear consisting of the vestibule, semicircular canals, and the cochlea, where balance is maintained and sound is transmitted 261

innervation (in′′er-va′shun) stimulation of muscle fiber to contract by a motor axon 195

insertion (in-ser′shun) the end of a muscle that is attached to a movable part 231

inspiration (in′′spi-ra′shun) the act of breathing in; inhalation 161

insulin (in′su-lin) a hormone produced by the pancreas that regulates carbohydrate storage 285

integration (in′′tē-gra′shun) the summing up of negative and positive stimuli within a dendrite 202

interferon (in′′ter-fēr′on) a protein formed by a cell infected with a virus that can increase the resistance of other cells to the virus 142

internal respiration (in-ter′nal res′′pi-ra′shun) exchange between blood and tissue fluid of oxygen and carbon dioxide 160

interneuron (in′′ter-nu′ron) a neuron that is found within the central nervous system and takes nerve impulses from one portion of the system to another 197

interphase (in′ter-faz) interval between successive cell divisions during which the chromosomes are extended and the cell is metabolically active 368

interstitial cells (in′′ter-stish′al selz) hormone-secreting cells located between the seminiferous tubules of the testes 296

inversion (in-ver′zhun) stagnant, nonmoving warm air that covers and traps pollutants beneath it 378

ion (i'on)   an atom or group of atoms carrying a positive or negative charge   *14*

ionic bond (i-on'ik bond)   a chemical attraction between a positive and negative ion   *14*

islets of Langerhans (i'lets uv lahng'er-hanz)   distinctive groups of cells within the pancreas that secrete insulin and glucagon   *285*

isotonic solution (i''so-ton'ik so-lu'shun)   one that contains the same concentration of solutes and water as does the cell   *34*

isotopes (i'so-tōps)   atoms with the same number of protons and electrons but differing in the number of neutrons and therefore in weight   *14*

K

karyotype (kar'e-o-tīp)   the arrangement of all the chromosomes within a nucleus by pairs in a fixed order   *365*

keratin (ker'ah-tin)   an insoluble protein present in the epidermis and in epidermal derivatives such as hair and nails   *60*

kidneys (kid'nēz)   organs in the urinary system that concentrate and excrete urine   *182*

kidney stones (kid'ne stōnz)   deposits of uric acid, calcium oxalate, calcium phosphate, or magnesium phosphate that sometimes form in the renal pelvis   *183*

killer T cells   T lymphocytes that attack cells bearing foreign bodies   *150*

kilocalorie (kil'o-kal''o-re)   unit of measurement equal to 1,000 calories, which are units of heat (a kilocalorie is the amount of heat required to raise the temperature of water by 1° C). In nutrition the kilocalorie is called a big calorie (Calorie)   *84*

kingdom (king'dum)   the largest taxonomic category into which organisms are placed: Monera, Protista, Fungi, Plants, and Animals   *444*

Klinefelter syndrome (klin'fel-ter sin'drōm)   a condition caused by the inheritance of XXY chromosomes   *383*

Krebs cycle (krebz si'kl)   a series of reactions found within mitochondria that give off carbon dioxide. Also called the citric acid cycle because the reactions begin and end with citric acid   *42*

L

labium (la'be-um)   a fleshy border or liplike fold of skin, as in the labia majora and labia minora of the female genitalia   *303*

lacteal (lak'te-al)   a lymph vessel in a villus of the intestinal wall of mammals   *78*

lactogenic hormone (lak''to-jen'ik hor'mōn)   LTH; a hormone secreted by the anterior pituitary that stimulates the production of milk from the mammary glands   *272*

lacuna (lah-ku'nah)   a small pit or hollow cavity, as in bone or cartilage, where a cell or cells are located   *229*

lanugo (lah-nu'go)   short, fine hair that is present during the later portion of fetal development   *347*

larynx (lar'ingks)   structure that contains the vocal cords; voice box   *163*

latent (la'tent)   hidden, as when a virus integrates its DNA into host DNA and does not undergo the lytic cycle   *318*

lens (lenz)   a clear, membranelike structure found in the eye behind the iris. The lens brings objects into focus   *252*

leukocyte (lu'ko-sīt)   refers to several types of colorless, nucleated blood cells that, among other functions, resist infection; white blood cells   *122*

LH   hormone produced by the anterior pituitary gland that stimulates the development of the corpus luteum in females and the production of testosterone in males   *297*

ligament (lig'ah-ment)   a strong connective tissue that joins bone to bone   *56, 229*

ligase (li'gās)   an enzyme that seals cuts in DNA such as those made after the use of a restriction enzyme   *428*

limbic system (lim'bik sis'tem)   an area of the forebrain implicated in visceral functioning and emotional responses; involves many different centers of the brain   *213*

linkage group (lingk'ij groop)   alleles on the same chromosome are linked in the sense that they tend to move together to the same gamete; crossing over interferes with linkage   *406*

lipase (li'pās)   an enzyme that digests or breaks down fats   *78*

lipid (lip'id)   a group of organic compounds that are insoluble in water; notably fats, oils, and steroids   *22*

loop of Henle (loop uv hen'le)   that part of a nephron that lies between the proximal convoluted tubule and the distal convoluted tubule; consists of a descending portion and ascending portion; concerned with reabsorption of water   *183*

loose connective tissue (loos kō-nek'tiv tish'u)   tissue composed mainly of fibroblasts that are separated by collagen and elastin fibers and found beneath epithelium   *55*

LTH   a hormone secreted by the anterior pituitary that stimulates the production of milk from the mammary glands   *271*

lumen (lu'men)   the cavity inside any tubular structure, such as the lumen of the gut   *61*

luteinizing hormone (lu'te-in-īz''ing hor'mōn)   *see* LH

lymph (limf)   fluid having the same composition as tissue fluid and carried in lymph vessels   *117*

lymphatic system (lim-fat'ik sis'tem)   vascular system that takes up excess tissue fluid and transports it to the bloodstream   *117*

lymph node (limf nōd)   a mass of lymphoid tissue located along the course of a lymphatic vessel   *119*

lymphokines (lim'fo-kīnz)   chemicals secreted by T cells that have the ability to affect the characteristics of monocytes   *150*

lymph vessels (limf ves'elz)   vessels that make up the lymphatic system, collecting excess tissue fluid (lymph) and returning it to systemic veins   *117*

lysosome (li'so-sōm)   an organelle in which digestion takes place due to the action of powerful hydrolytic enzymes   *37*

M

macrophage (mak'ro-fāj)   an enlarged monocyte that ingests foreign material and cellular debris   *134*

major histocompatibility protein (ma'jor his''to-kom-pat''i-bil'i-te pro'te-in)   *see* MHC protein

matrix (ma'triks)   the secreted basic material or medium of biological structures, such as the matrix of cartilage or bone   *55*

medulla (mĕ-dul'ah)   the inner portion of an organ; for example, the adrenal medulla   *183*

medulla oblongata (mĕ-dul'ah ob''long-gah'tah)   the lowest portion of the brain that is concerned with the control of internal organs   *165, 209*

meiosis (mi-o'sis)   type of cell division that occurs during the production of gametes by means of which the daughter cells receive the haploid number of chromosomes   *365, 381*

meiosis I (mi-o'sis wun)   that portion of meiosis during which homologous chromosomes come together and then later separate   *371*

meiosis II (mi-o'sis too)   that portion of meiosis during which sister chromatids separate, resulting in four haploid daughter cells   *371*

melanin (mel'ah-nin)   a pigment found in the skin and hair of humans that is responsible for their coloration   *60*

memory cells (mem'o-re selz)   cells derived from B and T lymphocytes that remain within the body for some time and account for the presence of active immunity   *144, 151*

meninges (mĕ-nin'jēz)   protective membranous coverings about the central nervous system   *209*

meniscus (mĕ-nis'kus)   a piece of fibrocartilage that separates the surfaces of bones in the knee   *230*

menopause (men'o-pawz)   termination of the uterine cycle in older women   *307*

menstruation (men''stroo a'shun)   loss of blood and tissue from the uterus at the end of a female uterine cycle   *304*

messenger RNA (mes''n-jer ar-en-a)   a nucleic acid complementary to genetic DNA and bearing a message to direct cell protein synthesis at the ribosome   *419*

metabolism (mė-tab′o-lizm)   all of the chemical changes that occur within cells considered together   *40*

metafemale (met″ah-fe′māl)   a female who has three X chromosomes   *383*

metaphase (met′ah-fāz)   stage in mitosis during which chromosomes assemble at the mitotic spindle equator   *370*

metaphase I (met″ah-fāz wun)   phase of meiosis I during which tetrads line up at the equator of the spindle   *372*

MHC protein   a surface molecule that serves as a genetic marker   *149*

microfilament (mi″kro-fil′ah-ment)   an extremely thin fiber found within the cytoplasm that is involved in the maintenance of cell shape and movement of cell contents   *39*

microtubule (mi″kro-tu′bul)   an organelle composed of thirteen rows of globular proteins; found in multiple units in several other cell organelles such as the centriole, cilia, and flagella   *39*

midbrain (mid′brān)   a small region of the brain stem located between the diencephalon and pons   *209*

middle ear (mid′l ēr)   portion of the ear consisting of the tympanic membrane, the oval and round windows, and the ossicles, where sound is amplified   *259*

mineral (min′er-al)   an inorganic, homogeneous substance   *90*

mitochondrion (mi″to-kon′dre-on)   an organelle in which aerobic respiration produces the energy molecule, ATP   *38*

mitosis (mi-to′sis)   cell division by means of which two daughter cells receive the exact chromosome and genetic makeup of the mother cell; occurs during growth and repair   *365, 367*

mixed nerves (mikst nervs)   nerves that contain both the long dendrites of sensory neurons and the long axons of motor neurons   *203*

monoclonal antibodies (mon″-o-klōn′al an′ti-bod″ez)   antibodies of one type that are produced by cells derived from a lymphocyte that has fused with a cancer cell   *155*

monoculture agriculture (mon′o-kul″tūr ag′ri-kul″tūr)   planting of a single type of crop that is subject to destruction by a single type of pest or parasite   *477*

monohybrid (mon″o-hi′brid)   the offspring of parents who differ in one way only; shows the phenotype of the dominant allele but carries the recessive allele   *390*

mononucleosis (mon″o-nu″kle-o′sis)   viral disease characterized by the presence of an increase in atypical lymphocytes in the blood   *134*

monosaccharide (mon″o-sak′ah-rīd)   a simple sugar; a carbohydrate that cannot be decomposed by hydrolysis   *20*

morphogenesis (mor″fo-jen′i-sis)   the establishment of shape and structure in an organism   *335*

morula (mor′u-lah)   an early stage in development in which the embryo consists of a mass of cells, often spherical   *335*

motor nerve (mo′tor nerv)   nerve containing only the long axons of motor neurons   *203*

motor neuron (mo′tor nu′ron)   a neuron that takes nerve impulses from the central nervous system to an effector   *195*

muscle action potential (mus′el ak′shun po-ten′shal)   an electrochemical change due to increased sarcolemma permeability that is propagated down the T system and results in muscle contraction   *241*

muscle spindle (mus′el spin′dul)   modified skeletal muscle fiber that can respond to changes in muscle length   *238*

muscular tissue (mus′ku-lar tish′u)   a type of tissue that contains cells capable of contracting; skeletal muscles are attached to the skeleton, smooth muscle is found within walls of internal organs, and cardiac muscle comprises the heart   *57*

mutation (mu-ta′shun)   a chromosomal or genetic change that is inherited either by daughter cells following mitosis or by an organism following reproduction   *378, 422*

myelin (mi′ė-lin)   the fatty cell membranes that cover long neuron fibers and give them a white, glistening appearance   *197*

myocardium (mi″o-kar′de-um)   heart (cardiac) muscle consisting of striated muscle cells that interlock   *102*

myofibrils (mi″o-fi′brilz)   the contractile portions of muscle fibers   *239*

myoglobin (mi″o-glo′-bin)   respiratory pigment in muscles

myogram (mi′o-gram)   a recording of a muscular contraction   *236*

myosin (mi′o-sin)   the thick filament in myofibrils made of protein and capable of breaking down ATP   *57, 240*

myxedema (mik″ sė-de′mah)   a condition resulting from a deficiency of thyroid hormone in an adult   *276*

N

NA   excitatory neurotransmitter active in the peripheral and central nervous systems; norepinephrine   *200*

natural selection (nat′u-ral sė-lek′shun)   the process by which better adapted organisms are favored to reproduce to a greater degree and pass on their genes to the next generation   *451–453*

Neanderthal (ne-an′der-thawl)   the common name for an extinct subspecies of humans whose remains are found in Europe, Asia, and Africa   *462*

negative feedback (neg′ah-tiv fēd′bak)   mechanism that is activated by a surplus imbalance and acts to correct it by stopping the process that brought about the surplus   *66*

nephron (nef′ron)   the anatomical and functional unit of the vertebrate kidney; kidney tubule   *183*

nerve (nerv)   a bundle of long nerve fibers that run to and/or from the central nervous system   *58*

nerve impulse (nerv im′puls)   an electrochemical change due to increased neurolemma permeability that is propagated along a neuron from the dendrite to the axon following excitation   *198*

neural tube (nu′ral tūb)   during development a tube formed from ectoderm located just above the notochord   *339*

neurilemma (nu″ri-lem′ah)   a thin, membranous covering surrounding the myelin of a nerve fiber   *197*

neuroglial cells (nu-rog′le-ahl selz)   supporting cells within the brain and spinal cord that perform functions other than transmission of nerve impulse   *205*

neuromuscular junction (nu″ro-mus′ku-lar jungk′shun)   the point of contact between a nerve cell and a muscle cell   *242*

neuron (nu′ron)   nerve cell that characteristically has three parts: dendrite, cell body, axon   *58, 195*

neurotransmitter substance (nu″ro-trans mit′er sub′stans)   a chemical made at the ends of axons that is responsible for transmission across a synapse   *200*

neutron (nu′tron)   a subatomic particle that has a weight of one atomic mass unit, carries no charge, and is found in the nucleus of an atom   *13*

NGU   an infection of the urinary tract by an organism other than *N. gonorrheae*   *328*

niche (nich)   the functional role and position of an organism in the ecosystem   *470*

nitrifying bacteria (ni′tri-fi″ing bak-te′re-ah)   bacteria active in the nitrogen cycle that oxidize ammonia (NH_4^+) to nitrite (NO_2^-) and nitrate (NO_3^-)   *477*

nitrogen fixation (ni′tro-jen fik-sa′shun)   a process whereby free atmospheric nitrogen is converted into compounds, such as ammonia and nitrates, usually by soil bacteria   *477*

nondisjunction (non″dis-jungk′shun)   the failure of homologous chromosomes or sister chromatids to separate during the formation of gametes   *378*

nongonococcal urethritis (non″gon-o-kok′al u″rē-thri′tis)   *see* NGU

nonrenewable resource (non-re-nu′ah-b′l re′sors)   a resource that can be used up or at least depleted to such an extent that further recovery is too expensive   *502*

noradrenalin (nor″ah-dren′ah-lin)   *see* NA

nose (nōz)   the part of the human face that bears the nostrils and contains the nasal cavities   *163*

notochord (no'to-kord) dorsal supporting rod that exists only during embryonic development and is replaced by the vertebral column *340*

nuclear envelope (nu'kle-ar en've-lōp) double-layered membrane enclosing the nucleus *36*

nuclear fusion (nu'kle-ar fu'zhun) a process in which the nuclei of two atoms are forced together to form the nucleus of a heavier atom with the release of substantial amounts of energy *505*

nucleic acid (nu-kle'ik as'id) a large organic molecule made up of nucleotides joined together; for example, DNA and RNA *24*

nucleolus (nu-kle'o-lus) an organelle found inside the nucleus; composed largely of RNA for ribosome formation; (*pl:* nucleoli) *36*

nucleotide (nu'kle-o-tīd) a molecule consisting of three subunits: phosphoric acid, a five-carbon sugar, and a nitrogenous base; a building block of a nucleic acid *25*

nucleus (nu'kle-us) a large organelle containing the chromosomes and acting as a control center for the cell; also the center of an atom *36*

O

obesity (o-bēs'i-te) an excessive accumulation of adipose tissue; usually the condition of exceeding the desirable weight by more that 20% *95*

obligate parasite (ob'li-gāt par'ah-sīt) an organism, such as a virus, that always causes disease *318*

occipital lobe (ok-sip'i-tal lōb) area of the cerebrum responsible for vision, visual images, and other sensory experiences *211*

olfactory cells (ol-fak'to-re selz) cells located high in the nasal cavity that bear receptor sites on cilia for various chemicals and the stimulation of which results in smell *250*

oogenesis (o''o-jen'ē-sis) production of the egg in females by the process of meiosis and maturation *372*

optic nerve (op'tik nerv) nerve composed of the ganglion cell fibers that form the innermost layer of the retina *253*

order (or'der) in taxonomy, the category below class and above family *444*

organelle (or''gah-nel') specialized structures within cells, such as the nucleus, mitochondria, endoplasmic reticulum *32*

organochloride (or''gah-no-klo'rīd) chlorinated hydrocarbons such as pesticides and PCBs *485*

organ of Corti (or'gan uv kor'ti) the organ that contains the hearing receptors in the inner ear *261*

orgasm (or'gazm) physical and emotional climax during sexual intercourse; results in male ejaculation *296*

origin (or'i-jin) end of a muscle that is attached to a relatively immovable part *231*

osmosis (oz-mo'sis) the movement of water from an area of greater concentration of water to an area of lesser concentration of water across a semipermeable membrane *34*

osmotic pressure (oz-mot'ik presh'ur) pressure generated by the osmotic flow of water *34*

ossicles (os'i-k'lz) the tiny bones found in the middle ear: hammer, anvil, and stirrup *259*

osteoblast (os'te-o-blast'') a bone-forming cell *229*

osteoclast (os'te-o-klast'') a cell that causes the erosion of bone *229*

osteocyte (os'te-o-sīt) a mature bone cell *229*

otoliths (o'to-liths) granules that lie above and whose movement stimulates ciliated cells in the utricle and saccule *261*

outer ear (out'er ēr) portion of the ear consisting of the pinna and the auditory canal *259*

oval opening (o'val o'pen-ing) an opening between the two atria in the fetal heart; also foramen ovale *348*

oval window (o'val win'do) membrane-covered opening between the stapes and the inner ear *259*

ovarian cycle (o-va're-an si'k'l) monthly occurring changes in the ovary that affect the level of sex hormones in the blood *303*

ovaries (o'var-ez) female gonads, the organs that produce eggs and estrogen and progesterone; the base of the pistil in angiosperms *299*

ovulation (o''vu-la'shun) the discharge of a mature egg from the follicle within the ovary *299*

oxidation (ok''si-da'shun) the loss of electrons (inorganic) or the removal of hydrogen atoms (organic) *42*

oxygen debt (ok'si-jen det) the amount of oxygen needed to metabolize lactic acid that accumulates during vigorous exercise *241*

oxyhemoglobin (ok''se-he''mo-glo'bin) hemoglobin bound to oxygen in a loose, reversible way *169*

oxytocin (ok''se-to'sin) hormone released by the posterior pituitary that causes contraction of the uterus and milk letdown *272*

ozone (o'zōn) variety of oxygen (O_3) forming a harmful component of photochemical smog *483*

P

pacemaker (pās'māk-er) a small region of neuromuscular tissue that initiates the heartbeat; also SA node *106*

PAN oxidant pollutant formed from automobile exhaust that is harmful to plant tissue *483*

pancreas (pan'kre-as) a vertebrate organ located near the stomach that secretes digestive enzymes into the duodenum and produces hormones, notably insulin *76*

Pap test (pap test) sampling of the cells from the cervix that is examined to determine if a woman has cancer of the cervix *301*

parallel evolution (par'ah-lel ev''o-lu'shun) occurrence of similar structures with similar functions in distantly related but geographically separated groups due to similar evolutionary pressures *451*

parasite (par'ah-sīt) an organism that resides externally on or internally within another organism and does harm to this organism *318*

parasympathetic nervous system (par''ah-sim''pah-thet'ik ner'vus sis'tem) a portion of the autonomic nervous system that usually promotes those activities associated with a normal state *207*

parathyroid hormone (par''ah-thi'roid hor'mōn) *see* PTH

parietal lobe (pah-ri'ē-tal lōb) area of the cerebrum responsible for sensations involving temperature, touch, pressure, and pain, as well as speech *211*

parturition (par''tu-rish'un) passageway and delivery of a newborn organism through the terminal portion of the female reproductive tract *350*

pectoral girdle (pek'tor-al ger'dl) portion of the skeleton that provides support and attachment for the arms *227*

pelvic girdle (pel'vik ger'dl) portion of the skeleton to which the legs are attached *227*

pelvic inflammatory disease (pel'vik in-flam'ah-to''re di-zēz') *see* PID

pelvis (pel'vis) a bony ring formed by the innominate bones. Also a hollow chamber in the kidney that lies inside the medulla and receives freshly prepared urine from the collecting ducts *227*

penis (pe'nis) male copulatory organ *296*

pepsin (pep'sin) a protein-digesting enzyme secreted by gastric glands *76*

peptide bond (pep'tīd bond) the bond that joins two amino acids *20*

periodontitis (per''e-o-don-ti'tis) inflammation of the gums *73*

peripheral nervous system (pē-rif'er-al) *see* PNS

peristalsis (per''i-stal'sis) a rhythmical contraction that serves to move the contents along in tubular organs such as the digestive tract *75*

peritubular capillary (per''i-tu'bu-lar kap'i-lar''e) capillary that surrounds a nephron and functions in reabsorption during urine formation *185*

pH (pe āch) a measure of the hydrogen ion concentration; any pH below 7 is acid and any pH above 7 is basic *17*

phagocytosis (fag''o-si-to'sis)   the taking in of bacteria and/or debris by engulfing; cell eating   *134*

pharynx (far'ingks)   a common passageway (throat) for both food intake and air movement   *74, 163*

phenotype (fe'no-tīp)   the outward appearance of an organism caused by the genotype and environmental influences   *390*

phospholipids (fos''fo-lip'idz)   lipids containing phosphorus that are particularly important in the formation of cell membranes   *24*

photovoltaic cell (fo''to-vol-ta'ik sel)   a manufactured device that uses sunlight to produce an electromotive force   *506*

phylum (fi'lum)   a taxonomic category applied to animals that follows kingdom and lies above class   *444*

physiograph (fiz'e-o-graf)   instrument used to record a myogram   *236*

PID   an inflammation that in particular causes the oviducts to become occluded by scar tissue   *326*

pinna (pin'nah)   outer, funnellike structure of the ear that picks up sound waves   *259*

placenta (plə 'sent ə)   a structure formed from the chorion and uterine tissue through which nutrient and waste exchange occur for the embryo and later the fetus   *306*

plaque (plak)   an accumulation of soft masses of fatty material, particularly cholesterol, beneath the inner linings of arteries   *113*

plasma (plaz'mah)   the liquid portion of blood   *57, 122*

plasma cell (plaz'mah sel)   a cell derived from a B cell lymphocyte that is specialized to mass-produce antibodies   *144*

plasmid (plaz'mid)   a circular DNA segment that is present in bacterial cells but is not part of the bacterial chromosome   *426*

platelet (plāt'let)   cell-like discs formed from fragmentation of megakaryocytes that initiate blood clotting in the blood   *57, 131*

pleural membranes (ploor'al mem'brānz)   serous membranes that enclose the lungs   *165*

PNS   the nerves and ganglia of the nervous system that lie outside of the brain and spinal cord   *195*

polar bodies (po'lar bod'ēz)   nonfunctioning daughter cells that have little cytoplasm and are formed during oogenesis   *372*

pollution (pō-lu'shun)   contamination of air, water, or soil with undesirable amounts of material or heat. The material can be an overabundance of a natural substance or a small amount of a toxic substance   *477*

polymer (pol'ī-mer)   a large molecule made up of many identical subunits   *19*

polypeptide (pol''e-pep'tīd)   a molecule composed of many amino acids linked together by peptide bonds   *20*

polysaccharide (pol''e-sak'ah-rīd)   a macromolecule composed of many units of sugar   *22*

polysome (pol'e-sōm)   a cluster of ribosomes all attached to the same mRNA molecule and each one bringing about the synthesis of the same polypeptide   *37*

pons (ponz)   a portion of the brain stem above the medulla oblongata and below the midbrain   *209*

population (pop''u-la'shun)   all the organisms of the same species in one place (area/ space)   *470*

posterior pituitary (pos-tēr'e-or pi-tu'i-tār-e)   portion of the pituitary gland connected by a stalk to the hypothalamus   *271*

postganglionic axon (pōst''gang-gle-on'ik ak'son)   in the autonomic nervous system, the axon that leaves, rather than goes to, a ganglion   *207*

postsynaptic membrane (pōst''si-nap'tik mem'brān)   a membrane that is part of a synapse and receives a neurotransmitter substance   *200*

preganglionic axon (pre''gang-gle-on'ik ak'son)   in the autonomic nervous system, the axon that goes to, rather than leaves, a ganglion   *207*

pressure filtration (presh'ur fil-tra'shun)   process by which small molecules leave a capillary due to blood pressure   *185*

presynaptic membrane (pre''si-nap'tik mem'brān)   a membrane that is part of a synapse and releases a neurotransmitter substance   *200*

primates (pri'māts)   animals that belong to the order Primates, the order of mammals that includes prosimians, monkeys, apes, and humans   *456*

producers (pro-du'serz)   organisms that produce food and are capable of synthesizing organic compounds from inorganic constituents of the environment; usually the green plants and algae in an ecosystem   *470*

prokaryotic (pro''kar-e-ot'ik)   lacking the organelles found in complex cells; such as a bacterium   *324*

prolactin (pro-lak'tin)   *see* lactogenic hormone

prophase (pro'fāz)   early stage in mitosis during which chromosomes appear   *369*

prophase I (pro'fāz wun)   phase of meiosis I during which the spindle appears and the nuclear envelope and nucleolus disappear   *372*

proprioceptor (pro''pre-o-sep'tor)   sensory receptor that assists the brain in knowing the position of the limbs   *249*

prosimians (pro-sim'e-anz)   primitive primates such as lemurs, tarsiers, and tree shrews   *456*

prostaglandins (pros''tah-glan'dinz)   hormones that have various and powerful effects often within the cells that produce them   *287*

prostate gland (pros'tāt gland)   a gland in males that is located about the urethra at the base of the bladder; produces most of the seminal fluid   *296*

protein (pro'te-in)   a macromolecule composed of one or several long polypeptides   *20*

prothrombin (pro-throm'bin)   plasma protein made by liver that must be present in blood before clotting can occur   *131*

protocell (pro'to-sel)   a structure that precedes the evolution of the true cell in the history of life   *455*

proton (pro'ton)   a subatomic particle found in the nucleus of an atom that has a weight of one atomic mass unit and carries a positive charge; a hydrogen ion   *13*

protozoan (pro''to-zo'an)   unicellular, generally colorless and motile microorganism   *330*

proximal convoluted tubule (prok'si-mal kon'vo-lūt-ed tu'būl)   highly coiled region of a nephron near Bowman's capsule   *183*

pseudostratified (su''do-strat'ī-fīd)   the appearance of layering in some epithelial cells when actually each cell touches a baseline and true layers do not exist   *51*

PTH   hormone secreted by the parathyroid glands that raises the blood calcium level primarily by stimulating reabsorption of bone   *284*

pulmonary (pul'mo-ner''e)   referring to the lungs   *107*

pulmonary artery (pul'mo-ner''e ar'ter-e)   a blood vessel that takes blood away from the heart to the lungs   *104, 107*

pulmonary circuit (pul'mo-ner''e ser-ket)   the blood vessels that take deoxygenated blood to and oxygenated blood away from the lungs   *107*

pulmonary vein (pul'mo-ner''e vān)   a blood vessel that takes blood away from the lungs to the heart   *104, 107*

pulse (puls)   vibration felt in arterial walls due to expansion of the aorta following ventricle contraction   *110*

Punnett square (pun'et skwar)   a gridlike device that enables one to calculate the results of simple genetic crosses by lining gametic genotypes of two parents on the outside margin and their recombination in boxes inside the grid   *376*

purines (pu'rinz)   nitrogenous bases found in DNA and RNA that have two interlocking rings   *25*

pus (pus)   thick yellowish fluid composed of dead phagocytes, dead tissue, and bacteria   *134*

pyrimidines (pi-rim'i-dinz)   nitrogenous bases found in DNA and RNA that have just one ring   *25*

R

receptor (re-sep′tor) a sense organ specialized to receive information from the environment. Also a structure found in the membrane of cells that combines with a specific chemical in a lock-and-key manner *203, 247*

recessive allele (re-ses′iv ah-lēl′) hereditary factor that only expresses itself when two copies are present in the genotype *387*

recombinant (re-kom′bi-nant) DNA having genes from two different organisms or gametes carrying recombined chromosomes after crossing over *428*

rectum (rek′tum) the terminal portion of the intestine *81*

red blood cell (red blod sel) *see* erythrocyte

red marrow (red mar′o) blood-cell-forming tissue located in spaces within certain bones *228*

reduced hemoglobin (re-dūst′ he-mo-glo′bin) hemoglobin that has released its oxygen *124*

reduction (re-duk′shun) the gain of electrons (inorganic); the addition of hydrogen atoms (organic) *42*

reflex (re′fleks) an inborn autonomic response to a stimulus that is dependent on the existence of fixed neural pathways *206*

regulatory gene (reg′u-lah-tor′′e jēn) genes that regulate the activity of structural genes *422*

renal pelvis (re-nal pel-vis) a bony ring formed by the innominate bones. Also a hollow chamber in the kidney that lies inside the medulla and receives freshly prepared urine from the collecting ducts *183*

REM sleep (rapid eye movement) a stage in sleep that is characterized by eye movements and dreaming *213*

renewable resource (re-nu′ah-b′l re′sors) a resource that is not used up because it is continually produced in the environment *502*

replacement reproduction (re-plās′ment re′′pro-duk′shun) replacement of only one's self in the population, as when a couple has a maximum of two children *499*

replication (re′′pli-ka′shun) the duplication of DNA; occurs when the cell is not dividing *367, 416*

repolarization (re-po′′lar-i-za′shun) the recovery of a neuron's polarity to the resting potential after it ceases transmitting impulses *198*

RER endoplasmic reticulum having attached ribosomes *36*

residual volume (re-zid′u-al vol′ūm) the amount of air (about 1,200 cc) that remains in the lungs after the most forceful expiration *168*

resource (re′sors) anything needed or used by a population or organism *502*

respiratory chain (re-spi′rah-to′′re chān) a series of molecules found within mitochondria that pass electrons (sometimes accompanied by hydrogen ions) from one to the other in such a way that the energy of oxidation is captured and ATP is generated *42*

resting potential (rest′ing po-ten′shal) the voltage recorded from inside a neuron when it is not conducting nerve impulses *198*

restriction enzyme (re-strik′shun en′zīm) a type of enzyme that cuts DNA at a location that contains a certain sequence of bases; hundreds of such enzymes exist *408, 427*

retina (ret′i-nah) the innermost layer of the eyeball that contains the rods and cones *253*

retinene (ret′i-nēn) substance used in the production of rhodopsin *256*

retrovirus (ret′′ro-vi′rus) virus capable of integrating its genes into those of the host by utilizing RNA to DNA transcription *318*

Rh factor (ar′āch fak′tor) a type of antigen on the red blood cells *136*

rhodopsin (ro-dop′sin) visual purple, a pigment found in the rods of one type of receptor in the retina of the eye *256*

ribonucleic acid (ri′′bo-nu-kle′ik as′id) *see* RNA

ribosomal RNA (ri′bo-sōm′′al) RNA found in ribosomes *420*

ribosomes (ri′bo-sōmz) minute particles, found attached to endoplasmic reticulum or loose in the cytoplasm, that are the site of protein synthesis *36*

ribs (ribz) bones hinged to the vertebral column and sternum that along with muscle define the top and sides of the chest cavity *165, 226*

RNA a nucleic acid important in the synthesis of proteins that contains the sugar ribose; the bases uracil, adenine, guanine, cytosine; and phosphoric acid *24, 414*

rods (rodz) dim-light receptors in the retina of the eye that detect motion but no color *253*

rough endoplasmic reticulum (ruf en′′do-plaz′mik rē-tik′′u-lum) endoplasmic reticulum having attached ribosomes *36*

round window (rownd win′do) a membrane-covered opening between the inner ear and the middle ear *259*

S

saccule (sak′ūl) a saclike cavity that makes up part of the membranous labyrinth of the inner ear *261*

salinization (sal′′i-ni-za′shun) excessive salinity in soil created by the evaporation of excess water in irrigated land *478*

salivary gland (sal′i-ver-e gland) a gland associated with the mouth that secretes saliva *74*

SA node (sinoatrial node) a small region of neuromuscular tissue that initiates the heartbeat. Also called the pacemaker *106*

saprophyte (sap′ro-fīt) a heterotrophic organism such as bacteria and fungi that externally breaks down dead organic matter before absorbing the products *325*

sarcolemma (sar′′ko-lem′ah) the membrane that surrounds striated muscle cells *239*

sarcomere (sar′ko-mēr) the structural and functional unit of a myofibril *239*

sarcoplasmic reticulum (sar′′ko-plaz′mik re-tik′u-lum) membranous network of channels and tubules within a muscle fiber corresponding to the endoplasmic reticulum of other cells *239*

sclera (skle′rah) white fibrous outer layer of the eyeball *252*

scotopsin (sko-top′sin) protein portion of rhodopsin *256*

scrotum (skrōt-em) the sacs that contain the testes *294*

secretin (se-kre′tin) hormone secreted by the small intestine that stimulates the release of pancreatic juice from the pancreas *77*

selectively permeable (sē-lek′tiv-le per′me-ah-b′l) indicates condition in the living cell membrane in which permeability is a regulated process *33*

selective reabsorption (sē-lek′tiv re′′ab-sorp′shun) one of the processes involved in the formation of urine; involves the greater reabsorption of nutrient molecules compared to waste molecules from the contents of the kidney tubule into the blood *186*

semen (se′men) the sperm-containing secretion of males; seminal fluid plus sperm *296*

semicircular canal (sem′′e-ser′ku-lar kah-nal′) tubular structures within the inner ear that contain the receptors responsible for the sense of dynamic equilibrium *261*

semilunar valves (sem′′e-lu′nar val′vz) valves resembling half-moons located between the ventricles and their attached vessels *102*

seminal fluid (sem′i-nal floo′id) fluid produced by various glands situated along the male reproductive tract *296*

seminal vesicle (sem′i-nal ves′i-k′l) a convoluted saclike structure attached to the vas deferens near the base of the bladder in males *296*

seminiferous tubules (sem′′i-nif′er-us tu′būlz) highly coiled ducts within the male testes that produce and transport sperm *294*

sensory nerve (sen′so-re nerv) nerve containing only sensory neuron dendrites *203*

sensory neuron (sen′so-re nu′ron) a neuron that takes the nerve impulse to the central nervous system; afferent neuron *195*

septum (sep′tum) partition or wall such as the septum in the heart, which divides the right half from the left half *102*

SER endoplasmic reticulum without attached ribosomes *36*

serum (se'rum)   light-yellow liquid left after clotting of the blood *132*

sex chromosome (seks kro'mo-sōm)   a chromosome responsible for the development of characteristics associated with maleness or femaleness; an *X* or *Y* chromosome *365*

sex-linked genes (seks-linkt' jēnz)   genes found on the sex chromosomes that control somatic traits *403*

simple goiter (sim'p'l goi'ter)   condition in which an enlarged thyroid produces low levels of thyroxin *275*

sinoatrial node (si''no-a'tre-al nōd)   *see* SA node

sinus (si'nus)   a cavity, as the sinuses in the human skull *224*

skeletal muscle (skel'ĕ-tal mus'el)   the contractile tissue that comprises the muscles attached to the skeleton; also called striated muscle *58*

skin (skin)   organ system covering the body that serves in sensory-, protective-, excretory-, and temperature regulating-capacities *59*

sliding filament theory (slīd'ing fil'ah-ment the'o-re)   the movement of actin in relation to myosin in explaining the mechanics of muscle contraction *240*

smooth endoplasmic reticulum (smōoth en''do-plaz'mik rĕ-tik'u-lum)   endoplasmic reticulum without attached ribosomes *36*

smooth muscle (smōoth mus'el)   the contractile tissue that comprises the muscles found in the walls of internal organs *58*

soft palate (soft pal'at)   entirely muscular posterior portion of the roof of the mouth *74*

solar collector (so'lar ko-lek'tor)   any manufactured structure that absorbs radiant energy *506*

solar energy (so'lar en'er-je)   energy derived from the sun's rays *506*

solid wastes (sol'id wāsts)   household trash, sewage sludge, agricultural residues, mining and industrial wastes *484*

somatic nervous system (so-mat'ik ner'vus sis'tem)   that portion of the PNS containing motor neurons that control skeletal muscles *203*

species (spe'shēz)   a group of similarly constructed organisms that are capable of interbreeding and producing fertile offspring; organisms that share a common gene pool *444*

spermatogenesis (sper''mah-to-jen'ĕ-sis)   production of sperm in males by the process of meiosis and maturation *372*

sphincter (sfingk'ter)   a muscle that surrounds a tube and closes or opens the tube by contracting and relaxing *76*

spinal nerve (spi'nal nerv)   nerve that arises from the spinal cord *203*

spindle (spin'd'l)   an apparatus composed of microtubules to which the chromosomes are attached during cell division *369*

spindle fibers (spin'd'l fi'berz)   microtubule bundles involved in the movement of chromosomes during mitosis and meiosis *369*

spleen (splēn)   a large, glandular organ located in the upper left region of the abdomen that stores and purifies blood *119*

spongy bone (spun'je bōn)   bone found at the ends of long bones that consists of bars and plates separated by irregular spaces *57, 228*

squamous epithelium (skwa'mus ep''i-the'le-um)   flat cells found lining the lungs and blood vessels *51*

stapes (sta'pēz)   the innermost of the ossicles of the ear, which fits against the oval window of the inner ear; also called the stirrup *259*

stereoscopic vision (ste''re-o-skop'ik vizh'un)   the product of two eyes and both cerebral hemispheres functioning together so that depth perception results *256*

sterilization (ster''i-li-za'shun)   the inability to reproduce; a surgical procedure eliminating reproductive capability; the absence of living organisms due to exposure to environmental conditions that are unfavorable to sustain life *326*

steroid (ste'roid)   lipid-soluble, biologically active molecules having four interlocking rings; examples are cholesterol, progesterone, testosterone *24*

stirrup (stir-ep)   the innermost of the ossicles of the middle ear, which fits against the oval window; also called the stapes *259*

stratified (strat'i-fīd)   layered, as in stratified epithelium, which contains several layers of cells *51*

stretch receptors (strech re-sep'torz)   muscle fibers that upon stimulation cause muscle spindles to increase the rate at which they send impulses to the CNS *249*

stroke (strōk)   condition resulting when an arteriole bursts or becomes blocked by an embolism *113*

structural genes (struk'tūr-al jēnz)   genes that determine protein structure *422*

subcutaneous (sub''ku-ta'ne-us)   a tissue layer found in vertebrate skin that lies just beneath the dermis and tends to contain fat cells *61*

succession (suk-sĕ'shun)   a series of ecological stages by which the community in a particular area gradually changes until there is a stable community that can maintain itself *467*

summation (sum-ma'shun)   ever greater contraction of a muscle due to constant stimulation that does not allow complete relaxation to occur *238*

suppressor T cell (su-pres'or T sel)   T lymphocyte that suppresses certain other T and B lymphocytes from continuing to divide and perform their respective functions, T_s cells *151*

sympathetic nervous system (sim''pah-thet'ik ner'vus sis'tem)   that part of the autonomic nervous system that generally causes effects associated with emergency situations *207*

synapse (sin'aps)   the region between two nerve cells where the nerve impulse is transmitted from one to the other; usually from axon to dendrite *200*

synapsis (si-nap'sis)   the attracting and pairing of homologous chromosomes during meiosis *372*

synaptic cleft (si-nap'tik kleft)   small gap between presynaptic and postsynaptic membranes of a synapse *200*

synaptic ending (si-nap'tik end'ing)   the knob at the end of an axon in a synapse *200*

syndrome (sin'drōm)   a group of symptoms that together characterize a disease condition *378*

synovial joint (si-no've-al joint)   a freely movable joint *229*

synthesis (sin'the-sis)   to build up, such as the combining together of two small molecules to form a larger molecule *19*

syphilis (sif'i-lis)   chronic, contagious venereal disease caused by a spirochete bacterium *328*

systemic (sis-tem'ik)   that part of the circulatory system that serves body parts other than the gas-exchanging surfaces in the lungs *107*

systole (sis'to-le)   contraction of the heart chambers, particularly the left ventricle *105*

systolic pressure (sis-tol'ik presh'ur)   arterial blood pressure during the systolic phase of the cardiac cycle *110*

T

taste bud (tāst bud)   organ containing the receptors associated with the sense of taste *249*

taxonomy (tak-son'o-me)   the science of naming and classifying organisms *444*

tectorial membrane (tek-to're-al mem'brān)   a membrane in the organ of Corti that lies above and makes contact with the receptor cells for hearing *261*

teleological (te''le-o-loj'i-kal)   assuming that a process is directed toward a final goal or purpose *451*

telophase (tel'o-fāz)   stage of mitosis during which the diploid number of chromosomes are located at each pole *370*

telophase I (tel'o-fāz wun)   phase of meiosis I during which the nuclear envelope and nucleolus reappear as the spindle disappears *372*

telophase II (tel'o-fāz too)   phase of meiosis II during which the spindle disappears as the nuclear envelope reappears and the cell membrane furrows, yielding two haploid cells   *372*

template (tem'plāt)   a pattern that serves as a mold for the production of an oppositely shaped structure; one strand of DNA is a template for the complementary strand   *416*

temporal lobe (tem'po-ral lōb)   area of the cerebrum responsible for hearing and smelling, the interpretation of sensory experience and memory   *211*

tendon (ten'don)   a tissue that connects muscle to bone   *56, 231*

testcross (test kros)   a genetic mating in which a possible heterozygote is crossed with an organism homozygous recessive for the characteristic(s) in question in order to determine its genotype   *391*

testes (tes'tez)   the male gonads, the organs that produce sperm and testosterone   *294*

testosterone (tes-tos'te-rōn)   the most potent of the androgens   *297*

tetanus (tet'ah-nus)   sustained muscle contraction without relaxation   *238*

tetany (tet'ah-ne)   severe twitching caused by involuntary contraction of the skeletal muscles due to a lack of calcium   *284*

tetrad (tet'rad)   a set of four chromatids resulting from the pairing of homologous chromosomes during meiosis   *372*

thalamus (thal'ah-mus)   a mass of gray matter located at the base of the cerebrum in the wall of the third ventricle; receives sensory information and selectively passes it to the cerebrum   *209*

thermal inversion (ther'mal in-ver'zhun)   temperature inversion that traps cold air and its pollutants near the earth with the warm air above it   *483*

thrombin (throm'bin)   the enzyme derived from prothrombin that converts fibrinogen to fibrin threads during blood clotting   *131*

thrombocyte (throm'bo-sit)   a blood platelet   *122*

thrombus (throm'bus)   a blood clot that remains in the blood vessel where it formed   *131*

thymus (thi'mus)   a lymphatic organ that lies in the neck and chest area and is absolutely necessary to the development of immunity   *119*

thyroid stimulating hormone (thi'roid stim'u-lāt"ing hor'mōn)   *see* TSH

thyroxin (thi-rok'sin)   the hormone produced by the thyroid that speeds up the metabolic rate   *275*

tidal volume (tīd'al vol'ūm)   amount of air that enters the lungs during a normal, quiet inspiration   *168*

tight junction (tīt junk'shun)   a zipperlike junction between two cells that prevents passage of molecules   *54*

tissue fluid (tish'u floo'id)   fluid found about tissue cells containing molecules that enter from or exit to the capillaries   *117*

T lymphocyte (lim'fo-sīt)   lymphocytes that interact directly with antigen-bearing cells and are responsible for cell-mediated immunity   *142*

tone (tōn)   the continuous partial contraction of muscle; also the quality of a sound   *238*

trachea (tra'ke-ah)   the windpipe that serves as a passageway for air   *163*

tract (trakt)   a bundle of neurons forming a transmission pathway through the brain and spinal cord   *209*

trait (trāt)   specific term for a distinguishing feature studied in heredity   *387*

transcription (trans-krip'shun)   the process that results in the production of a strand of mRNA that is complementary to a segment of DNA   *419*

transfer RNA (trans'fer)   molecule of RNA that carries an amino acid to a ribosome in the process of protein synthesis   *420*

translation (trans-la'shun)   the process involving mRNA, ribosomes, and tRNA that results in a synthesis of a polypeptide having an amino acid sequence dictated by the sequence of codons in mRNA   *419*

translocation (trans"lo-ka'shun)   chromosome mutation due to the attachment of a broken chromosome fragment to a nonhomologous chromosome   *378*

transposon (trans-po'zon)   motile genetic elements called jumping genes that can move between chromosomes and thereby bring about genetic mutations   *424*

trophoblast (trof'o-blast)   the outer membrane that surrounds the human embryo and, when thickened by a layer of mesoderm, becomes the chorion   *335*

trypsin (trip'sin)   a protein-digesting enzyme secreted by the pancreas   *78*

TSH   hormone that causes the thyroid to produce thyroxin   *274, 275*

tubal ligation (tu'bal li-ga'shun)   cutting of the oviducts in females to cause sterilization   *308*

tubular excretion (tu'bu-lar eks-kre'shun)   process occurring in the distal convoluted tubule during which substances are added to urine   *187*

Turner syndrome (tur'ner sin'drōm)   a condition caused by the inheritance of a single X chromosome   *382*

twitch (twich)   a brief muscular contraction followed by relaxation   *238*

tympanic canal (tim-pan'ik kah-nal')   the most inferior of the three canals that lies in the cochlea of the inner ear; fluid movement from the vestibular canal to the tympanic canal sets up pressure waves that cause the basilar membrane to vibrate, leading to sound sensation   *260*

tympanic membrane (tim-pan'ik mem'brān)   a membrane located between the external and middle ear; the eardrum   *259*

U

umbilical arteries and vein (um-bil'i-kal ar'ter-ēz and vān)   fetal blood vessels which travel to and from the placenta   *348*

umbilical cord (um-bil'i-kal kord)   cord connecting the fetus to the placenta through which blood vessels pass   *344*

urea (u-re'ah)   primary nitrogenous waste of mammals   *180*

ureter (u-re'ter)   tube between kidney and bladder   *182*

urethra (u-re'thrah)   tube that takes urine from bladder to outside   *182*

uric acid (u'rik as'id)   waste product of nucleotide breakdown   *180*

urinalysis (u"ri-nal'i-sis)   a medical procedure in which the composition of a patient's urine is determined   *190*

urinary bladder (u'ri-ner"e blad'der)   an organ where urine is stored before being discharged by way of the urethra   *182*

uterine cycle (u'ter-in si'kl)   the female reproductive cycle that is characterized by regularly occurring changes in the uterine lining   *304*

uterus (u'ter-us)   the female organ in which the fetus develops   *300*

utricle (u'tre-k'l)   an enlarged cavity that makes up part of the membranous labyrinth of the inner ear   *261*

V

vaccine (vak'sēn)   treated antigens that can promote active immunity when administered   *154*

vagina (vah-ji'nah)   copulatory organ in females   *301*

valves (valvz)   a structure that opens and closes, insuring one-way flow; common to vessels such as the systemic veins and the lymphatic veins and to the heart   *101*

varicose veins (var'i-kōs vānz)   abnormally swollen and enlarged veins, especially in the legs   *117*

vas deferens (vas def'er-ens)   tube that leads from the epididymis to the urethra of the male reproductive tract   *296*

vasectomy (vah-sek'to-me)   cutting of the vas deferens in males to bring about sterilization   *308*

vasopressin (vas''o-pres'in)   secreted by the posterior pituitary; promotes reabsorption of water by the kidneys; also called antidiuretic hormone (ADH)   *271*

vector (vek'tor)   a carrier of genetic information that is used to transform a host cell   *426*

vein (vān)   a blood vessel that takes blood to the heart   *100*

vena cava (ve'nah ka'vah)   one of two large veins that convey deoxygenated blood to the right atrium of the heart   *104, 107*

venous duct (ve'nus dukt)   fetal connection between the umbilical vein and the inferior vena cava, ductus venosus   *348*

ventilation (ven''ti-la'shun)   breathing; the process of moving air into and out of the lungs   *165*

ventricle (ven'tri-k'l)   a cavity in an organ such as the ventricles of the heart or the ventricles of the brain   *102, 209*

venule (ven'ūl)   type of blood vessel that takes blood from capillaries to veins   *100*

vernix caseosa (ver'niks ka-se-o'sah)   cheeselike substance covering the skin of the fetus   *347*

vestibule (ves'ti-būl)   a space or cavity at the entrance of a canal such as within the inner ear at the base of the semicircular canals   *261*

vestigial (ves-tij'e-al)   the remains of a structure that was functional in some ancestor but is no longer functional in the organism in question   *448*

villi (vil'i)   fingerlike projections that line the small intestine and function in absorption   *78*

virulence (vir'u-lens)   bacterial ability to overcome host resistance   *324*

virus (vi'rus)   a genetic element that contains either DNA or RNA within a coat of protein; outside a cell it is nonliving but inside it either becomes incorporated into host DNA or undergoes a reproductive cycle   *318*

vital capacity (vi'tal kah-pas'i-te)   the maximum amount of air a person can exhale after taking the deepest breath possible   *168*

vitamin (vi'tah-min)   usually coenzymes, needed in small amounts, that the body is no longer capable of synthesizing and therefore must be in the diet   *89*

vitreous humor (vit're-us hu'mor)   the substance that occupies the space between the lens and retina of the eye   *252*

vocal cords (vo'kal kordz)   folds of tissue within the larynx that create vocal sounds when they vibrate   *163*

vulva (vul'vah)   the external genitalia of the female that lie near the opening of the vagina   *303*

W

white blood cell (wit blod sel)   *see* leukocyte

X

X-linked (eks'linkt)   refers to inherited traits not related to sex but carried by the female (X) chromosome   *403*

XYY male (eks-wi-wi māl)   a male that has an extra Y chromosome   *383*

Y

yellow marrow (yel'o mar'o)   fat storage tissue found in the cavities within certain bones   *228*

yolk sac (yōk sak)   one of the extraembryonic membranes within which yolk is found   *337*

Z

zygote (zi'gōt)   diploid cell formed by the union of two gametes, the product of fertilization   *366*

ILLUSTRATIONS

Figures 3.5a, 3.15, 3.16, and 3.17 From John W. Hole, Jr., *Human Anatomy and Physiology,* 4th ed. Copyright © 1987 Wm. C. Brown Publishers, Dubuque, Iowa. All Rights Reserved. Reprinted by permission.

Figure 4.11 From John W. Hole, Jr., *Human Anatomy and Physiology,* 4th ed. Copyright © 1987 Wm. C. Brown Publishers, Dubuque, Iowa. All Rights Reserved. Reprinted by permission. **Figure 4.16** From Vincent Hegarty, *Decisions in Nutrition.* Copyright © C. V. Mosby Company, St. Louis, Missouri.

Figures 5.2 and 5.19 From Kent M. Van De Graaff and Stuart Ira Fox, *Concepts of Human Anatomy and Physiology.* Copyright © 1986 Wm. C. Brown Publishers, Dubuque, Iowa. All Rights Reserved. Reprinted by permission. **Figures 5.6a and 5.7** From Kent M. Van De Graaff, *Human Anatomy,* 2d ed. Copyright © 1988 Wm. C. Brown Publishers, Dubuque, Iowa. All Rights Reserved. Reprinted by permission. **Figure 5.9** From Crouch, James E.: *Functional Human Anatomy,* 4th ed. Lea & Febiger, Philadelphia, 1985. Reprinted by permission. **Figures 5.10b and 5.11a** From John W. Hole, Jr., *Human Anatomy and Physiology,* 4th ed. Copyright © 1987 Wm. C. Brown Publishers, Dubuque, Iowa. All Rights Reserved. Reprinted by permission. **Figure 5.15** From Stuart Ira Fox, *A Laboratory Guide to Human Physiology: Concepts and Applications,* 4th ed. Copyright © 1987 Wm. C. Brown Publishers, Dubuque, Iowa. All Rights Reserved. Reprinted by permission. **Figure 5.17** From Stuart Ira Fox, *Human Physiology,* 2d ed. Copyright © 1987 Wm. C. Brown Publishers, Dubuque, Iowa. All Rights Reserved. Reprinted by permission.

Figure 6.3a From John W. Hole, Jr., *Human Anatomy and Physiology,* 4th ed. Copyright © 1987 Wm. C. Brown Publishers, Dubuque, Iowa. All Rights Reserved. Reprinted by permission. **Illustration, p. 127** © Anthony Hunt.

Figures 7.5 and 7.10 From Stuart Ira Fox, *Human Physiology,* 2d ed. Copyright © 1987 Wm. C. Brown Publishers, Dubuque, Iowa. All Rights Reserved. Reprinted by permission.

Figures 8.5 and 8.13 From John W. Hole, Jr., *Human Anatomy and Physiology,* 4th ed. Copyright © 1987 Wm. C. Brown Publishers, Dubuque, Iowa. All Rights Reserved. Reprinted by permission. **Figure 8.6** From Kent M. Van De Graaff and Stuart Ira Fox, *Concepts of Human Anatomy and Physiology.* Copyright © 1986 Wm. C. Brown Publishers, Dubuque, Iowa. All Rights Reserved. Reprinted by permission. **Figure 8.12** Based on an illustration from *Respiration: The Breath of Life (The Human Body Series).* © Torstar Books, Inc., 1985.

Figures 9.5 and 9.7 From Kent M. Van De Graaff, *Human Anatomy,* 2d ed. Copyright © 1988 Wm. C. Brown Publishers, Dubuque, Iowa. All Rights Reserved. Reprinted by permission.

Figure 10.2 Reprinted with permission of Macmillan Publishing Company from *Kimber-Gray-Stackpole's Anatomy and Physiology,* 17th edition by Marjorie A. Miller, Anna B. Drakontides, and Lutie C. Leavell. Copyright © 1977 Macmillan Publishing Company. **Figures 10.8b & c and 10.19** From John W. Hole, Jr., *Human Anatomy and Physiology,* 4th ed. Copyright © 1987 Wm. C. Brown Publishers, Dubuque, Iowa. All Rights Reserved. Reprinted by permission. **Figure 10.18 (bottom)** From Kent M. Van De Graaff, *Human Anatomy,* 2d ed. Copyright © 1988 Wm. C. Brown Publishers, Dubuque, Iowa. All Rights Reserved. Reprinted by permission.

Figure 11.3a,b,c From Kent M. Van De Graaff, *Human Anatomy,* 2d ed. Copyright © 1988 Wm. C. Brown Publishers, Dubuque, Iowa. All Rights Reserved. Reprinted by permission. **Figure 11.4** From Kent M. Van De Graaff and Stuart Ira Fox, *Concepts of Human Anatomy and Physiology.* Copyright © 1986 Wm. C. Brown Publishers, Dubuque, Iowa. All Rights Reserved. Reprinted by permission. **Figures 11.5, 11.8, 11.9, 11.12, and 11.13** From John W. Hole, Jr., *Human Anatomy and Physiology,* 4th ed. Copyright © 1987 Wm. C. Brown Publishers, Dubuque, Iowa. All Rights Reserved. Reprinted by permission. **Fig. 11.21** From Bernard Katz, *Nerve, Muscle, and Synapse.* Copyright © 1966 McGraw Hill Book Company, New York. Reprinted by permission.

Figures 12.2b, 12.4, 12.18, and 12.19 From John W. Hole, Jr., *Human Anatomy and Physiology,* 4th ed. Copyright © 1987 Wm. C. Brown Publishers, Dubuque, Iowa. All Rights Reserved. Reprinted by permission. **Figure 12.8a** From Crouch, James E., *Functional Human Anatomy.* Lea and Febiger, Philadelphia, 1985. Reprinted by permission. **Figure 12.20a** From Kent M. Van De Graaff and Stuart Ira Fox, *Concepts of Human Anatomy and Physiology.* Copyright © 1986 Wm. C. Brown Publishers, Dubuque, Iowa. All Rights Reserved. Reprinted by permission.

Figures 13.4 and 13.6 From Kent M. Van De Graaff and Stuart Ira Fox, *Concepts of Human Anatomy and Physiology.* Copyright © 1986 Wm. C. Brown Publishers, Dubuque, Iowa. All Rights Reserved. Reprinted by permission. **Figure 13.16a** From John W. Hole, Jr., *Human Anatomy and Physiology,* 4th ed. Copyright © 1987 Wm. C. Brown Publishers, Dubuque, Iowa. All Rights Reserved. Reprinted by permission. **Figure 13.21** From Kent M. Van De Graaff, *Human Anatomy,* 2d ed. Copyright © 1988 Wm. C. Brown Publishers, Dubuque, Iowa. All Rights Reserved. Reprinted by permission.

Figures 14.2 and 14.10 From John W. Hole, Jr., *Human Anatomy,* 4th ed. Copyright © 1986 Wm. C. Brown Publishers, Dubuque, Iowa. All Rights Reserved. Reprinted by permission. **Figures 14.9 and 14.13** From Kent M. Van De Graaff and Stuart Ira Fox, *Concepts of Human Anatomy and Physiology.* Copyright © 1986 Wm. C. Brown Publishers, Dubuque, Iowa. All Rights Reserved. Reprinted by permission.

Figures 16.6 and 16.18 From Kent M. Van De Graaff and Stuart I. Fox, *Concepts of Human Anatomy and Physiology.* Copyright © 1986 Wm. C. Brown Publishers, Dubuque, Iowa. All Rights Reserved. Reprinted by permission. **Figures 16.11a-d and 16.15b,c,e,f** From Kent M. Van De Graaff, *Human Anatomy,* 2d ed. Copyright © 1988 Wm. C. Brown Publishers, Dubuque, Iowa. All Rights Reserved. Reprinted by permission. **Figure 16.23** From *The Human Body: Growth and Development* © Torstar Books. Reprinted by permission.

Figures 18.11b and 18.18 From E. Peter Volpe, *Biology and Human Concerns,* 3d ed. Copyright © 1983 Wm. C. Brown Publishers, Dubuque, Iowa. All Rights Reserved. Reprinted by permission.

Figure 20.7 From Storer, et al., *General Zoology,* 6th ed. Copyright © McGraw-Hill Book Company, New York. Reprinted by permission. **Figure 20.8** From E. Kohne, J. A. Chiscon, and B. H. Hoyer, in *Journal of Human Evolution,* Volume 1, 627, 1972. Reprinted by permission of Academic Press, Inc., London. **Figure 20.18** From *Fossil Man* by Michael H. Day. Copyright © 1970 by the Hamlyn Group. Used by permission of Grosset and Dunlap, Inc. and the Hamlyn Publishing Group Limited, England.

Figure 22.8 From adapted figure 15.1 in *Fundamentals of Ecology,* Third Edition, by Eugene P. Odum, copyright © 1971 by Saunders College Publishing, a division of Holt, Rinehart and Winston, Inc., reprinted by permission of the publisher. **Figure 22.15a-d** From Terry N. Barr, "The World Food Situation and Global Grain Prospects" in *Science,* 214, 1087–1095, 1981, Fig. 5. Copyright © 1981 American Association for the Advancement of Science. Reprinted by permission.

ILLUSTRATORS

ANNE GREENE–Fig. 12.7, fig. 14.3, fig. 16.3.

TOM WALDROP–Fig. 5.21, fig. 7.4, fig. 10.12, fig. 13.3.

KATHLEEN HAGELSTON–Provided many of the sketches for the following illustrators:

MARJ LEGGITT–Fig. I.2, TA I.1, fig. I.6, fig. 4.18, fig. 9.13, fig. 12.14, fig. 17.21b, fig. 19.17a, fig. 20.14.

LAURIE O'KEEFE–Fig. C.2, fig. C.6, fig. 2.8, fig. 3.11, fig. 6.8, fig. 7.3, fig. 7.7b, fig. 7.8, fig. 7.9c, TA 18.2, fig. 18.21, fig. 19.2, fig. 19.13, fig. 19.14.

PRECISION GRAPHICS–Fig. 1.1, fig. 4.16, fig. 4.19, fig. 4.22, fig. 7.6b, fig. 8.11b, fig. 8.14, fig. 10.26, fig. 13.17, fig. 15.5a,b, fig. 16.22, TA 18.1, fig. 18.11b, fig. 18.16, fig. 18.18, fig. 18.22, fig. 21.19.

PHOTOGRAPHS

CONTENTS PAGE–Pages vi–vii: © Photri/Marilyn Gartman Agency; **page viii:** © CNRI/Science Photo Library/Photo Researchers, Inc.; **page ix left:** © Martha McBride/Unicorn Stock Photos, **right:** © Bill Longcore/Science Source/Photo Researchers, Inc.; **page x:** © Willie Hill, Jr./The Image Works; **page xi:** © W. Fischer/FPG International.

PART OPENERS–**Part One:** © David M. Phillips/Visuals Unlimited; **Part Two:** © Karen Holsinger Mullen/Unicorn Stock Photos; **Part Three:** © Alan Carey/Image Works, Inc.; **Part Four:** © Petit Format/Nestle/Science Source/Photo Researchers, Inc.; **Part Five:** © Bob Daemmrich/The Image Works; **Part Six:** © Philip M. Zito.

A

INDEX